THE VEGETATION OF ANTARCTICA THROUGH GEOLOGICAL TIME

The fossil history of plant life in Antarctica is central to our understanding of the evolution of vegetation through geological time, and also plays a key role in reconstructing past configurations of continents and associated climatic conditions. This book provides the first detailed overview of the development of Antarctic vegetation from the Devonian period to the present day, providing Earth scientists with valuable insights into the break-up of the ancient supercontinent of Gondwana.

Details of specific floras and ecosystems are provided within the context of changing geological, geographical and environmental conditions, alongside comparisons with contemporaneous and modern ecosystems. The authors demonstrate how palaeobotany contributes to our understanding of the palaeoenvironmental changes in the Southern Hemisphere during this period of Earth history.

The book is a complete and up-to-date reference for researchers and students in Antarctic palaeobotany and terrestrial palaeoecology.

DAVID J. CANTRILL is Chief Botanist and Director of the National Herbarium of Victoria at the Royal Botanic Gardens Melbourne, Australia. Throughout his academic life he has undertaken extensive work across the Southern Hemisphere, particularly Antarctica, researching fossil floras from the Permian to early Cenozoic. His research into Antarctic biotas has concentrated on the systematic composition of floras, palaeoecology, the role of Antarctica in mediating climate and biogeographic patterns during Gondwana break-up, and developing present-day patterns of austral plant distribution.

IMOGEN POOLE is a Senior Research Fellow at Utrecht University and Honorary Research Fellow at the University of Aberdeen. She has worked extensively on palaeovegetational, palaeoenvironmental and palaeoclimatic interpretations of Cretaceous and younger floras from both the Northern and Southern Hemispheres. A particular focus of her research has been determining the floristic composition of high latitude Antarctic floras and reconstructing the palaeoenvironment and climate of these early angiospermous ecosystems.

Authorship sequence is purely alphabetical as both authors contributed equally to the compilation of this book.

INTERNATIONAL STRATIGRAPHIC CHART

International Commission on Stratigraphy

ICS

Eonothem / Eon	Erathem / Era	System / Period	Series / Epoch	Stage / Age	Age Ma	GSSP
Phanerozoic	Cenozoic	Quaternary	Holocene		0.0117	⚲
			Pleistocene	Upper	0.126	
				"Ionian"	0.781	
				Calabrian	1.806	⚲
				Gelasian	2.588	⚲
		Neogene	Pliocene	Piacenzian	3.600	⚲
				Zanclean	5.332	⚲
			Miocene	Messinian	7.246	⚲
				Tortonian	11.608	⚲
				Serravallian	13.82	⚲
				Langhian	15.97	
				Burdigalian	20.43	
				Aquitanian	23.03	⚲
		Paleogene	Oligocene	Chattian	28.4 ±0.1	
				Rupelian	33.9 ±0.1	⚲
			Eocene	Priabonian	37.2 ±0.1	
				Bartonian	40.4 ±0.2	
				Lutetian	48.6 ±0.2	
				Ypresian	55.8 ±0.2	⚲
			Paleocene	Thanetian	58.7 ±0.2	⚲
				Selandian	~61.1	⚲
				Danian	65.5 ±0.3	⚲
	Mesozoic	Cretaceous	Upper	Maastrichtian	70.6 ±0.6	⚲
				Campanian	83.5 ±0.7	
				Santonian	85.8 ±0.7	
				Coniacian	~88.6	
				Turonian	93.6 ±0.8	⚲
				Cenomanian	99.6 ±0.9	⚲
			Lower	Albian	112.0 ±1.0	
				Aptian	125.0 ±1.0	
				Barremian	130.0 ±1.5	
				Hauterivian	~133.9	
				Valanginian	140.2 ±3.0	
				Berriasian	145.5 ±4.0	

Eonothem / Eon	Erathem / Era	System / Period	Series / Epoch	Stage / Age	Age Ma	GSSP
					145.5 ±4.0	
Phanerozoic	Mesozoic	Jurassic	Upper	Tithonian	150.8 ±4.0	
				Kimmeridgian	~155.6	
				Oxfordian	161.2 ±4.0	
			Middle	Callovian	164.7 ±4.0	
				Bathonian	167.7 ±3.5	⚲
				Bajocian	171.6 ±3.0	⚲
				Aalenian	175.6 ±2.0	⚲
			Lower	Toarcian	183.0 ±1.5	
				Pliensbachian	189.6 ±1.5	⚲
				Sinemurian	196.5 ±1.0	⚲
				Hettangian	199.6 ±0.6	⚲
		Triassic	Upper	Rhaetian	203.6 ±1.5	
				Norian	216.5 ±2.0	
				Carnian	~228.7	⚲
			Middle	Ladinian	237.0 ±2.0	⚲
				Anisian	~245.9	
			Lower	Olenekian	~249.5	
				Induan	251.0 ±0.4	⚲
	Paleozoic	Permian	Lopingian	Changhsingian	253.8 ±0.7	⚲
				Wuchiapingian	260.4 ±0.7	⚲
			Guadalupian	Capitanian	265.8 ±0.7	⚲
				Wordian	268.0 ±0.7	⚲
				Roadian	270.6 ±0.7	⚲
			Cisuralian	Kungurian	275.6 ±0.7	
				Artinskian	284.4 ±0.7	
				Sakmarian	294.6 ±0.8	
				Asselian	299.0 ±0.8	⚲
		Carboniferous	Pennsylvanian	Upper	Gzhelian	303.4 ±0.9	
				Kasimovian	307.2 ±1.0	
				Middle	Moscovian	311.7 ±1.1	
				Lower	Bashkirian	318.1 ±1.3	⚲
			Mississippian	Upper	Serpukhovian	328.3 ±1.6	
				Middle	Visean	345.3 ±2.1	⚲
				Lower	Tournaisian	359.2 ±2.5	⚲

IUGS
International Union of Geological Sciences

Eonothem / Eon	Erathem / Era	System / Period	Series / Epoch	Stage / Age	Age Ma	GSSP
Phanerozoic	Paleozoic	Devonian	Upper	Famennian	359.2 ±2.5	⚷
				Frasnian	374.5 ±2.6	⚷
					385.3 ±2.6	⚷
			Middle	Givetian	391.8 ±2.7	⚷
				Eifelian	397.5 ±2.7	⚷
			Lower	Emsian	407.0 ±2.8	⚷
				Pragian	411.2 ±2.8	⚷
				Lochkovian	416.0 ±2.8	⚷
		Silurian	Pridoli		418.7 ±2.7	⚷
			Ludlow	Ludfordian	421.3 ±2.6	⚷
				Gorstian	422.9 ±2.5	⚷
			Wenlock	Homerian	426.2 ±2.4	⚷
				Sheinwoodian	428.2 ±2.3	⚷
			Llandovery	Telychian	436.0 ±1.9	⚷
				Aeronian	439.0 ±1.8	⚷
				Rhuddanian	443.7 ±1.5	⚷
		Ordovician	Upper	Hirnantian	445.6 ±1.5	⚷
				Katian	455.8 ±1.6	⚷
				Sandbian	460.9 ±1.6	⚷
			Middle	Darriwilian	468.1 ±1.6	⚷
				Dapingian	471.8 ±1.6	⚷
			Lower	Floian	478.6 ±1.7	⚷
				Tremadocian	488.3 ±1.7	⚷
		Cambrian	Furongian	Stage 10	~ 492 *	
				Stage 9	~ 496 *	
				Paibian	~ 499	⚷
			Series 3	Guzhangian	~ 503	⚷
				Drumian	~ 506.5	⚷
				Stage 5	~ 510 *	
			Series 2	Stage 4	~ 515 *	
				Stage 3	~ 521 *	
			Terreneuvian	Stage 2	~ 528 *	
				Fortunian	542.0 ±1.0	⚷

Eonothem / Eon	Erathem / Era	System / Period	Age Ma	GSSP GSSA
Precambrian	Proterozoic	Neo-proterozoic	Ediacaran	542	⚷
			Cryogenian	~635	⏲
			Tonian	850	⏲
		Meso-proterozoic	Stenian	1000	⏲
			Ectasian	1200	⏲
			Calymmian	1400	⏲
		Paleo-proterozoic	Statherian	1600	⏲
			Orosirian	1800	⏲
			Rhyacian	2050	⏲
			Siderian	2300	⏲
				2500	⏲
	Archean	Neoarchean		2800	⏲
		Mesoarchean		3200	⏲
		Paleoarchean		3600	⏲
		Eoarchean		4000	
		Hadean (informal)		~4600	

Subdivisions of the global geologic record are formally defined by their lower boundary. Each unit of the Phanerozoic (~542 Ma to present) and the base of Ediacaran are defined by a basal Global Boundary Stratotype Section and Point (GSSP ⚷), whereas Precambrian units are formally subdivided by absolute age (Global Standard Stratigraphic Age, GSSA). Details of each GSSP are posted on the ICS website (*www.stratigraphy.org*).

Numerical ages of the unit boundaries in the Phanerozoic are subject to revision. Some stages within the Cambrian will be formally named upon international agreement on their GSSP limits. Most sub-Series boundaries (e.g. Middle and Upper Aptian) are not formally defined.

The listed numerical ages are from 'A Geologic Time Scale 2004', by F.M. Gradstein, J.G. Ogg, A.G. Smith, et al. (2004; Cambridge University Press) and 'The Concise Geologic Time Scale' by J.G. Ogg, G. Ogg and F.M. Gradstein (2008).

Sept. 2010

This chart was drafted by Gabi Ogg. Intra Cambrian unit ages with * are informal, and awaiting ratified definitions.

THE VEGETATION OF ANTARCTICA
THROUGH GEOLOGICAL TIME

DAVID J. CANTRILL

Royal Botanic Gardens Melbourne

IMOGEN POOLE

Utrecht University, The Netherlands

CAMBRIDGE
UNIVERSITY PRESS

CAMBRIDGE
UNIVERSITY PRESS

University Printing House, Cambridge CB2 8BS, United Kingdom

One Liberty Plaza, 20th Floor, New York, NY 10006, USA

477 Williamstown Road, Port Melbourne, VIC 3207, Australia

4843/24, 2nd Floor, Ansari Road, Daryaganj, Delhi - 110002, India

79 Anson Road, #06-04/06, Singapore 079906

Cambridge University Press is part of the University of Cambridge.

It furthers the University's mission by disseminating knowledge in the pursuit of education, learning and research at the highest international levels of excellence.

www.cambridge.org
Information on this title: www.cambridge.org/9781108446822

First published 2012
First paperback edition 2017

A catalogue record for this publication is available from the British Library

Library of Congress Cataloging in Publication data
Cantrill, David J., 1962–
The vegetation of Antarctica through geological time / David J. Cantrill, Royal Botanic Gardens Melbourne, Imogen Poole, Utrecht University, The Netherlands.
pages cm
ISBN 978-0-521-85598-3 (hardback)
1. Plants – Evolution – Antarctica. 2. Plants, Fossil – Antarctica.
3. Palaeobotany – Devonian. 4. Palaeoecology – Devonian. 5. Palaeontology – Devonian. 6. Geological time. I. Poole, Imogen. II. Title.
QK980.C35 2012
561′.19989–dc23
2012001241

ISBN 978-0-521-85598-3 Hardback
ISBN 978-1-108-44682-2 Paperback

Contents

Acknowledgements

The authors are indebted to a number of colleagues and friends who have provided invaluable help in a number of ways throughout the course of writing this book. Such help included the answering of questions, providing copies of (elusive) publications, allowing access to unpublished material, in-depth discussions, translating foreign literature and comments on earlier drafts of the chapters. They particularly thank Professor R. Askin, Professor K. Birkenmajer, Professor A. Gaździcki, Professor L. Ivany, Professor B. Mohr, Professor M. Reguero, Professor E. Truswell, Dr R. Carpenter, Dr H. Falcon-Lang, Dr R. Hunt, Dr V. Thorn, and particularly Professor T. Dutra for her constant discussions and unyielding enthusiasm for this project. The staff at the British Antarctic Survey, Utrecht University (the Netherlands), Leeds University (England), Naturhistoriska Riksmuseet Stockholm (Sweden), Royal Botanic Gardens Melbourne (Australia) and the Royal Botanic Gardens Kew (England) are thanked for their help in acquiring relevant data and publications and/or allowing access to facilities during the course of the original research. Professor David Jolley is thanked for his encouragement and use of facilities at Aberdeen University during the final stages of this work. The authors thank the British Antarctic Survey and the British Navy for the ability to visit many of the sites mentioned in the text between the periods 1992 to 2002. This work was the product of direct and indirect funding from a number of grants from National Environmental Research Council, Nederlandse Organisatie voor Wetenschappelijk Onderzoek Aard- en Levenswetenschappen, SYNTHESYS, and Vetenskåpsrådet for which they are extremely grateful. Finally the authors would like to thank their spouses, families and friends for their unyielding support, both mentally and physically (especially on the domestic front), that enabled this work to be finally laid to rest.

1

Historical background and geological framework

Introduction

Today Antarctica is a continent locked in ice with nearly 98% of its current terrain covered by permanent ice and snow. Yet for the vast majority of its existence the Antarctic landmass was ice-free and supported a diversity of plants and animals. The history of the vegetation of Antarctica through geological time is intimately linked with that of the geography and climate, for both geography and climate have shaped, fashioned and controlled these high latitude ecosystems over geological timescales.

Today ice sheet and sea ice processes in Antarctica are key drivers of the oceanic circulation system. The oceanic circulation ensures the Antarctic continent remains inhospitable to terrestrial plant life through extreme cold. Yet on a geological timescale the current lack of extensive vegetation is a relatively recent phenomenon, possibly as recent as Pliocene. During the 300 million years covering the Paleozoic, Mesozoic and Cenozoic the fossil record reveals evidence of extensive forests across the high latitudes. Moreover occupying the central component of Gondwana, Antarctica played an important role in creating present day Southern Hemisphere biogeographic patterns by providing terrestrial connections between what are today widely separated landmasses. An obvious example of this is the distribution of Antarctic beech (*Nothofagus*) across South America and the southwest Pacific, which can be attributed to past terrestrial connections across Antarctica. The lack of present day terrestrial diversity however makes it difficult to integrate the Antarctic region into biogeographic studies that rely on determining relationship between extant groups. This further emphasises the importance of the Antarctic fossil record for understanding present day Southern Hemisphere diversity patterns. The interplay between supercontinent fragmentation, creation of new oceanic basins and oceanic circulation patterns, and evolution of the climate system all played a critical role in the history of Antarctic vegetation.

This book sets out to explore the interplay between geological processes and climate evolution on the evolving vegetation of Antarctica, but first, to set the book in its rightful context, we summarise the history of palaeobotanical interest in Antarctica and provide a brief summary of the geology of this southernmost continent.

History of palaeobotanical discovery

Today the Antarctic region is easily defined by strong oceanic and atmospheric features. The Antarctic Circumpolar Current, a feature observed by early explorers, effectively isolates cold surface waters of the Southern Ocean from warmer waters further to the north. The dramatic change in temperature over a short distance in surface waters (termed the Antarctic Convergence) is a major biogeographic feature that supports different assemblages of marine organisms either side of the convergence (Arntz, Gutt and Klages, 1997; Linse *et al.*, 2006). Although the Antarctic region today is defined by these present day oceanographic features they have not always been present. Nevertheless, the Antarctic Circumpolar Current broadly forms a convenient boundary to discuss the early scientific expeditions of the southern high latitudes including the early voyages of Halley (1699–1700) and Cook (1772–1775) – both of which penetrated the southern polar regions yet failed to discover land.

Cook made several forays into the high latitudes in an attempt to discover land but while unsuccessful, his records played an important role in eliminating large tracts of the Southern Ocean from further exploration. These exploratory expeditions paved the way for the discovery of the continent. The voyages of Cook provided dismal descriptions of the southern polar regions depicting them simply as a frozen wasteland. Despite this, whalers and sealers continued to push further south into the southern high latitudes driven on by the depletion of sealing and whaling grounds further north. Therefore it is no surprise that sealers are acknowledged as being the discoverers of several of the Antarctic islands (e.g. Macquarie Island by F. Hasselburgh aboard the *Perserverance*; Fogg, 1992), including those lying off the Antarctic Peninsula (e.g. South Shetland Island by William Smith aboard the *Williams*; Fogg, 1992; Figure 1.1). A number of the whaling and sealing trips combined economic pursuits with science and discovery and it is amongst the records of these expeditions to the Antarctic Peninsula that the first reports of fossil plant material can be found.

In 1819, Tsar Alexander I despatched two Russian expeditions, one to the Arctic and the second to the Antarctic. Thaddeus Bellingshausen commanded the Antarctic voyage and furthered the earlier discoveries of Cook, with a circumnavigation of the Antarctic in 1820. Bellingshausen discovered several islands including Alexander Island, off the Antarctic Peninsula, and Peter I Island (Figure 1.1). More controversial is the discovery of the Antarctic continent itself. While it is clear that Bellingshausen observed the ice shelves attached to the continent, and therefore can be regarded as the discoverer of the Antarctic continent, this has been contested (Jones, 1982). The voyage of Bellingshausen made significant inroads into the discovery of Antarctica but due largely to the haste in which the voyage was put together and despatched from St Petersburg, it was poorly equipped for scientific discovery (Fogg, 1992).

The United States Exploring Expedition was a privately funded sealing trip that sailed in 1829 and returned to the United States in 1831. The naturalist on the expedition was James Eights (1833) who made geological observations on the South Shetland Islands and reported carbonised wood preserved in conglomerates, and thus penned the first scientific record of fossil plants from the Antarctic regions. However, these and other discoveries, such as the

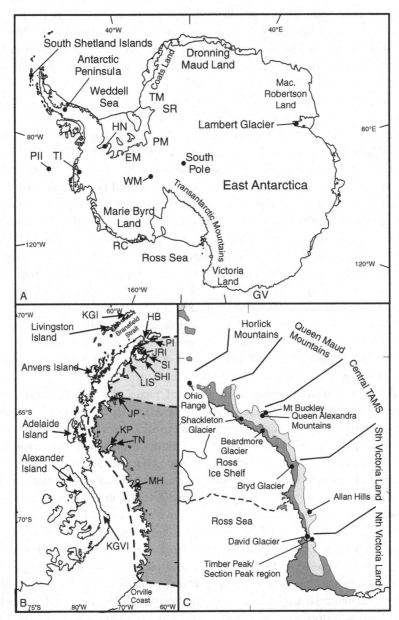

Figure 1.1 Geographic locality map showing the approximate positions of the places mentioned in the text. **A**, Antarctica. TI = Thurston Island, PII = Peter I Island, HN = Haag Nunataks, EM = Ellsworth Mountains, TM = Theron Mountains, SR = Shackleton Range, WM = Whitmore Mountains, GV = George V Coast, RC = Robertson Coast. **B**, Antarctic Peninsula region. KGI = King George Island, HB = Hope Bay, PI = Paulet Island, JRI = James Ross Island, SI = Seymour Island, SHI = Snow Hill Island, LIS = Larsen Ice Shelf, JP = Jason Peninsula, KP = Kenyon Peninsula, TN = Table Nunatak, MH = Mount Hill, KGVI = King George VI Sound. Light shading = approximate extent of the Larsen Basin, dark shading = approximate extent of the Latady Basin. **C**, Transantarctic Mountains showing main sectors and key regions mentioned in book. Dark shading shows extent of outcrop, light shading represents extent of Jurassic Ferrar Group and co-occurring Beacon Supergroup, outcrop pattern derived from Elliot and Hanson (2001).

material from the Kerguelen Islands collected by the surgeon and one of the naturalists, McCormick, on the James Clark Ross voyages of the HMS *Erebus* and HMS *Terror* (1839–1843), raised little interest scientifically and appear to have been forgotten or largely overlooked in later works. Indeed the Kerguelen material was not seriously examined for nearly another century (Edwards, 1921; Seward, 1919).

Scientific interest in polar palaeofloras began in the late 1800s with the discovery of extensive fossils in the Arctic resulting in works such as those by Oswald Heer (1868–1883), which documents Cretaceous and Cenozoic floras. This demonstrated that the northern polar regions had not always been frozen and virtually devoid of plant life, but rather during the past had supported extensive and diverse forests. The implications of a much warmer world excited scientific debate, and the diversity and composition of the Arctic Cenozoic floras led Heer to suggest that the northern temperate flora arose in the Arctic region and spread southwards, and the Arctic was thus a centre of origin for the modern flora (Heer, 1868, 1874). These ideas gained considerable momentum in the late 1800s (e.g. Nathorst, 1884) and spurred further exploration to and discovery in the Arctic. Indeed this thinking contributed greatly to the misleading interpretations of the Cenozoic fossil floras of New Zealand (von Ettingshausen, 1887, 1891), Australia (von Ettingshausen, 1888) and South America (e.g. Dusén, 1899; Engelhardt, 1891, 1895), where fossil plant material were nearly always assigned to modern northern temperate taxa with scant regard to the fact that the local flora differed markedly in floristic composition. The fossil plant discoveries in the Arctic raised the question whether floras might have been present in the south, and if so, whether these showed similar patterns to those in the Arctic.

Two expeditions were instrumental in the next phase of Antarctic palaeobotanical discovery. In 1892, Captain C. A. Larsen aboard the *Jason*, led a Norwegian whaling and sealing expedition to the Antarctic Peninsula. This expedition was primarily for commercial purposes but Larsen harboured an interest in scientific exploration. The *Jason* reached Seymour Island in November 1892 and, after first sailing south along the edge of the Larsen Ice Shelf, returned to Seymour Island in December of that year (Larsen, 1894). On December 4th a small party, including Larsen himself, landed on Seymour Island with the aim of looking for seals, but while ashore they discovered abundant fossil shells and wood, and so began the first of many collections of fossils from Antarctica (Larsen, 1894).

At the same time as Larsen's expedition was heading south another whaling party, the *Dundee Antarctic Expedition 1892–1893* was also underway. This was a much larger expedition comprising four vessels (*Active*, *Balaena*, *Diana* and *Polar Star*). Although essentially a commercial expedition, two surgeons (C. W. Donald and J. Bruce) with interests in natural history were appointed and scientific instruments were provided by the Royal Geographical Society of Edinburgh and the Metereological Office. In spite of the greater scientific weight of the *Dundee Antarctic Expedition 1892–1893* commercial interests thwarted the efforts of the naturalists. This lack of support served only to frustrate both scientists (Fogg, 1992; Zinsmeister, 1988).

From the published records of both expeditions it is unclear when the parties from the *Jason* and the *Dundee Antarctic Expedition* first met (Zinsmeister, 1988). However, the meeting on the 28th December 1892 was important since the *Jason*, which had been at sea for several months, was running low on stores particularly tobacco (Aagaard, 1930). A frustrated Bruce, joined a few days later by Donald, decided to trade tobacco for most of the fossils collected by the crew members of the *Jason*, and as a result these fossils made their way back to Scotland and were later described by Sharman and Newton (1894, 1898). However, not all the fossils collected on Seymour Island by Larsen and his crew were traded. Captain Larsen's personal collection ended up in Oslo although some specimens are also housed in the Swedish Museum of Natural History in Stockholm. In 1894 Larsen returned to the Antarctic, made other landings on Seymour Island (Larsen, 1894) and explored further inland, reporting the copious presence of fossil wood. Later Zinsmeister (1988), through fieldwork on Seymour Island and detective work of published records, determined the area described by Larsen as being the head of Cross Valley, an area now known to be rich in Paleocene wood and leaves.

When describing the fossils collected by Larsen from his two trips to the Antarctic region, Sharman and Newton (1898) followed the conventional thinking of the time regarding the origin of the Earth's biota. They noted the importance of these discoveries, and catalogued the similarities of the southern flora and fauna to those in the northern latitudes. Larsen's discoveries were also influential in the major discoveries made by the *Swedish South Polar Expedition 1901–1903* (Figure 1.2). Otto Nordenskjöld, whose uncle had been a polar explorer and the first to navigate the Northeast Passage in the Arctic Basin, led the expedition. Between 1895 and 1897 Otto Nordenskjöld had already explored southern South America before making trips to Alaska in 1898 and East Greenland in 1900. The *Swedish South Polar Expedition* aimed to further the discoveries of Larsen, and Larsen was thus selected to captain the vessel, *Antarctic*, south. In the annals of Antarctic exploration the *Swedish South Polar Expedition* was an epic, and testament to the endurance and perseverance of these early explorers and scientists.

The expedition first called at the South Shetland Islands on the west side of the Antarctic Peninsula prior to sailing to Seymour Island on the east side. The initial landing on Seymour Island at Penguin Point was disappointing as the rocks lacked fossils and this influenced Nordenskjöld's decision not to build his wintering base here (Nordenskjöld and Andersson, 1905) but rather on nearby Snow Hill Island (Figure 1.2A) – a decision strongly influenced by the discovery of ammonites made during a reconnaissance landing. However, this was a decision that Nordenskjöld later came to rue once he realised the wealth of palaeontological material on Seymour Island (Nordenskjöld and Andersson, 1905; Zinsmeister, 1988). The following summer, largely due to thick ice conditions, Larsen was unable to reach the Snow Hill base to pick up the over-wintering party. As a result Dr Gunnar Andersson, Lt Duse and Seaman Grunden were put ashore at Hope Bay with the intention of sledging to Snow Hill Island to advise Nordenskjöld of the difficulties in access, and to suggest they travel to Hope Bay (Figure 1.1). However, and somewhat ironically, the sea ice had by this time broken up, and the party was unable to reach Snow Hill Island, and so returned to Hope Bay to await the return of the *Antarctic*. Unbeknown to the Hope Bay party the *Antarctic* had become trapped in the ice, and drifted for over a month before being crushed and sinking to the east of Paulet

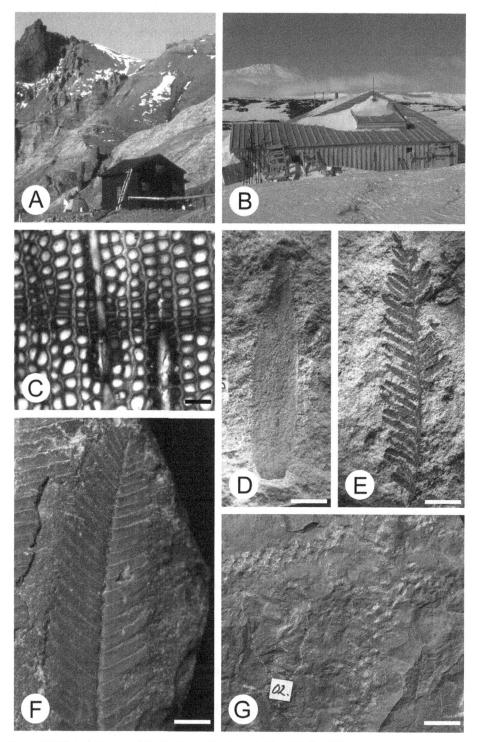

Figure 1.2 Historic bases and fossil plant collections. **A,** *Swedish South Polar Expedition 1901–1903*
Base on Snow Hill Island. Nordenskjöld's expedition discovered Jurassic, Cretaceous and Cenozoic

Island. Larsen and his crew took to the lifeboats, and managed to reach Paulet Island where they spent a miserable winter in a stone hut that they had built. All three parties spent winter in the Antarctic: Nordenskjöld at his base on Snow Hill Island (Figure 1.2A); Larsen and crew on Paulet Island; and Gunnar Andersson, together with Duse and Grunden, in a small stone hut at Hope Bay. Remarkably all parties survived the winter with the Hope Bay party finally making it to Snow Hill Island to meet Nordenskjöld the following spring. The Paulet Island party arrived on Snow Hill Island just as the relief vessel sent from Argentina was due to depart. In spite of the adversity faced by Nordenskjöld and the expedition members the *Swedish South Polar Expedition* should be regarded as one of the most successful scientific expeditions during this period of initial exploration and discovery. The scientific results of this expedition were published between 1908 and 1933 in a series of comprehensive volumes (Conrad, 1999).

Three significant palaeobotanical discoveries were the legacy of the *Swedish South Polar Expedition*. In spite of the meagre rations, and having to survive on penguins over the winter, the Hope Bay party discovered Jurassic plant fossils at Mount Flora (Halle, 1913a, 1913b; Nathorst, 1904, 1906; Figure 1.2F, G). The Hope Bay Flora was described in detail with extensive comparisons to other Jurassic floras known at the time (Halle, 1913a, 1913b). The quality of the monograph (Halle, 1913a, 1913b) ensured it became a reference flora for dating Jurassic Southern Hemisphere floras. Although in recent times the Hope Bay flora has been variously regarded as Early Jurassic (Rees, 1993; Rees and Cleal, 2004) to Early Cretaceous (Gee, 1989; Stipanicic and Bonetti, 1970) recent radiometric dating has confirmed an early Middle Jurassic age (Hunter *et al.*, 2005).

From Snow Hill Island several Late Cretaceous plants were also described (Halle, 1913b) along with a diverse leaf flora of Paleocene age from Seymour Island (Dusén, 1908) (Figure 1.2D, E) – although the treatment later received some criticism from experts on Southern Hemisphere Cenozoic floras (e.g. Berry, 1913). The third contribution focused on the fossil wood from the Cretaceous, Paleocene and Eocene from Snow Hill and Seymour islands (Gothan, 1908; Figure 1.2C). Together these monographs illustrated Antarctica not as a frozen continent, but as with the deposits from the Arctic, point to the land having once been warm and vegetated.

Caption for Figure 1.2 (cont.)
floras in the northern Antarctic Peninsula. **B**, *British Antarctic Expedition (Terra Nova Expedition) 1911–1913* base with Mount Erebus in background. The polar party carried back 35 lb of rock that included specimens of *Glossopteris*. **C–G**, examples of material collected by members of the Swedish South Polar Expedition. C, thin section of fossil wood described by Gothan (1908) as *Araucarioxylon pseudoparenchymatosum*. S4061. D, leaf of *Araucaria imponens* described by Dusén (1908) from the Paleocene of Seymour Island. S132021_01. E, *Elatocladus heterophylla* S132020 from the Upper Cretaceous of Snow Hill Island. F, *Otozamites hislopii* described by Halle (1913b) from Hope Bay, S020114. G, *Pagiophyllum feistmanteli* described by Halle (1913b) from Hope Bay, S020135a_01. Scale bars C = 50 μm, D–G = 1cm. D–G courtesy of S. McLoughlin and Department of Palaeobotany, Swedish Museum of Natural History.

Whilst Nordenskjöld and the *Swedish South Polar Expedition* had been overwintering on the Antarctic Peninsula, on the other side of the continent, Captain Robert Falconer Scott was leading the *British National Antarctic Expedition 1901–1904*. The young geologist on this expedition, Hartley Ferrar, established the basic stratigraphy of the Transantarctic Mountains. He described them as a basement complex of igneous and metamorphic rocks overlain by a horizontally bedded sedimentary sequence, which was in turn intruded by dolerite sills (Ferrar, 1907). Traces of plant life were recorded (Arber, 1907) but the significance was not recognised for a further 20 years when Edwards (1928) split the samples to reveal fragments of the large, woody, seed-bearing arboreal seed plant, *Glossopteris*, confirming a Permian age for the strata. Perhaps the most well-known discoveries were made on Scott's second expedition the *Terra Nova Expedition* (1911–1913) based out of Cape Evans (Figure 1.2B). On the return journey from the South Pole the ill-fated polar party discovered beautifully preserved leaves at Mount Buckley. These were examined by Seward (1914) and again identified as *Glossopteris*. One can only speculate, had the *Glossopteris* leaves from Scott's first expedition been discovered prior to Scott's second expedition, whether the polar party would have carried the 35 lb of rock specimens from Mount Buckley. If the polar party had not carried the specimens they may have reached One Ton depot and survived, instead of succumbing to the cold some 11 miles short. The importance of the fossils from both of Scott's expeditions were realised in subsequent studies by Du Toit (1937) and the formulation of his ideas on continental drift.

With the increase in interest in the Antarctic, coupled with the increase in the number of expeditions to this region, further discoveries that included abundant fossil plant material were being made by other National Expeditions (e.g. *British Antarctic Expedition 1907–1909*; Truswell, 1991). Examples include the wood discoveries on South Georgia during Shackleton's last expedition (Gordon, 1930) and the plant fossils on Alexander Island reported by the *British Graham Land Expedition 1934–1937* (Fleming, 1938; Stephenson and Fleming, 1940). The Alexander Island material was thought to be Middle Jurassic based on similarities with the floras described from Hope Bay until recent research showed that they were from the Cretaceous (Aptian) part of the succession (Taylor, Thomson and Willey, 1979).

The next major advance in knowledge of the Antarctic fossil floras was made during the *Trans-Antarctic Expedition 1955–1958*. An extensive geological programme covering a broad geographic area made further discoveries of plant fossils of Devonian, Permian, Triassic and Jurassic. These extensive collections coupled with a review of the earlier discoveries enabled Plumstead to provide the first synthesis of Antarctic fossil floras (Plumstead, 1962) and this remained the only synthesis for nearly 30 years.

The establishment of permanent bases since the late 1940s has led to ongoing research and discovery, and thus to a dramatic increase in our knowledge of the natural history of Antarctica. The International Geophysical Year (IGY, 1957–1958) marked the beginning of this permanent presence in the Antarctic for many nations. The IGY acted as a watershed for Antarctic research as it shifted national focus from territorial claims to scientific purposes. This ultimately resulted in the formation of the Antarctic Treaty which put territorial claims into abeyance and made Antarctica a region for scientific research. International committees such as the Scientific Committee for Antarctic Research (SCAR) played, and continue to play, a role in coordinating and setting collaborative research directions. Palaeobotanical research was very much a by-product of geological reconnaissance and mapping efforts with new discoveries being passed on to international experts for research and publication. For example, British Antarctic Survey geologists made several discoveries in the 1960s and 1970s in the course of geological mapping. These included Permian floras from Dronning Maud Land (Plumstead, 1975) and the Theron Mountains (Lacey and Lucas, 1981a), a Mesozoic flora from the South Shetland Islands (Lacey and Lucas, 1981b; Orlando, 1968), and several Cenozoic floras from King George Island (e.g. Lucas and Lacey, 1981). The collaborative nature of research from the IGY continued so that it was not unusual for countries to lend material for study to scientists from other nations who had also been involved in the IGY (e.g. United Kingdom to Argentina; Orlando, 1968).

In principle, those discoveries deemed to be the most significant at the time were passed on for further research and there are many other reports of plant fragments scattered within publications focusing on the geology of specific regions (e.g. Thomson, 1975). It is only in recent decades that the British Antarctic Survey has initiated in-house palaeobotanical expertise in the field through studentships (e.g. Jefferson, 1981; Rees, 1990), and more recently through employees and joint research projects within certain British universities. These studies have resulted in a significant advancement in our understanding of the palaeobotany of Antarctica. In parallel with the development of a palaeobotanical capability was an expansion into the field of palynology, firstly through the commissioning of a pilot study (Dettmann and Thomson, 1987), followed by the appointment of staff palynologists (e.g. Duane, 1996) and joint collaborations with the British Geological Survey (e.g. Pirrie and Riding, 1988; Riding *et al.*, 1998). Other nationalities, such as Australia, have also made significant palaeobotanical and palynological collections, particularly whilst undertaking field mapping studies, which were then passed on to experts to document. These include the discoveries of Permian plant macrofossils (White, 1962, 1973) and pollen (e.g. Kemp, 1973) from the Prince Charles Mountains.

In contrast, research institutes such as the Byrd Polar Research Center (formerly Institute for Polar Studies) located within Ohio State University have had direct access to palaeobotanical expertise. James M. Schopf reported on the fossil plant findings of expeditions (Schopf, 1962, 1968) and participated in several Antarctic field seasons. Schopf also realised the significance of the discoveries made by Mercer and Gunning of permineralised peats (Schopf, 1970a) and took part in several expeditions to collect material including Triassic peats (Schopf, 1970b). Permineralised peats are now known to be from several localities in both the Permian and Triassic of the Transantarctic Mountains. These deposits have formed the focus of palaeobotanical research by researchers from the United States of America in the Transantarctic Mountains since the 1970s, firstly through the work of J. M. Schopf, and more recently through T. N. and E. L. Taylor and members of their research group. The unique permineralised peats of Permian and Triassic ages have yielded copious amounts of information both on the morphology and anatomy of the plants themselves (e.g. Schopf, 1978; Smoot and Taylor, 1986a; Smoot, Taylor and Delevoryas, 1985), as well as the biology (e.g. Smoot and Taylor, 1986b) of the plants growing in such high latitude ecosystems. Recent discoveries of peat in the Prince Charles Mountains (Neish, Drinnan and Cantrill, 1993) further extended our knowledge of the occurrence of these palaeobotanically rich deposits.

While palaeobotanical material is known from many regions of the Antarctic it will become clear that two areas, namely the Transantarctic Mountains and the Antarctic Peninsula (particularly the northern Antarctic Peninsula region) are of prime importance due to the extensive outcrop and geological history. The expansion of Antarctic Treaty nations, and the desire for a number of nationalities to have a physical presence on the continent, saw an increase in the number of bases, particularly during the 1980s, in the northern Antarctic Peninsula. Countries such as Argentina, Chile, Poland and Russia who originally had bases in this region were joined by countries such as Brazil, Uruguay, Peru, China and Korea, making the northern Antarctic Peninsula one of the most densely occupied and cosmopolitan regions of the continent. The increased presence went hand in hand with an expansion in palaeobotanical discovery and research. This phase of investigation began in the 1980s with geological mapping of the volcano-sedimentary strata of King George Island resulting in rediscovery and documentation of Late Cretaceous (e.g. Birkenmajer and Zastawniak, 1986, 1989a, 1989b; Zastawniak, 1994) and Cenozoic floras (e.g. Li, 1994; Torres, 1984; Zastawniak, 1981). Interpretation of the stratigraphy and age of these Late Cretaceous and Cenozoic sequences was not without controversy due to problems with resetting of radiometric dates (Soliani and Bonhomme, 1994). The succession of floras has now been synthesised (Dutra and Batten, 2000). Whilst numerous publications began to appear focusing on the King George Island floras other regions were also being investigated. The initial work on the Early Cretaceous flora of Byers Peninsula, Livingston Island (Fuenzalida, Araya and Hervé, 1972) and nearby Snow Island was expanded to include palynology (Askin, 1983; Duane, 1996) and macrofossils (Cantrill, 2000; Cesari *et al.*, 1998, 1999; Torres *et al.*, 1997, 1998) including wood (Torres, Valenzuela and Gonzalez, 1982). The presumed Triassic flora from Williams Point (Livingston Island) was revisited and shown to be Cretaceous (e.g. Chapman and Smellie, 1992; Rees and Smellie, 1989). Other regions, such as the flora described by Halle from Mount Flora at Hope Bay, have also been revised (Gee, 1989) based on new collections (e.g. Ociepa, 2007; Ociepa and Barbacka, 2011; Rees and Cleal, 2004). Interestingly, whilst new material and further discoveries continue to be made within existing floras, relatively few new fossiliferous regions have been found and studies have largely built upon the pioneering work of previous researchers.

Perhaps the most dramatic, and certainly the most controversial fossil discovery in recent years were those of *Nothofagus* wood (Carlquist, 1987) and leaves (Hill, Harwood and Webb, 1996; Hill and Truswell, 1993; Webb and Harwood, 1993) from glacial deposits (Sirius Formation) in the Beardmore Glacier region. These were assigned a Pliocene age based on diatoms from the base of the sequence (Webb *et al.*, 1984). The Pliocene age coupled with the interpreted growth form and implied growing conditions suggested a much warmer Antarctica, and consequently a decreased ice volume and less stable ice sheet (see Chapter 9). This created controversy and considerable debate within the Earth Science community regarding the stability of the ice sheet (Burckle and Pokras, 1991) especially in light of the current global warming. A consequence of this scientific debate was renewed interest in the onset of glaciation and the glacial history of Antarctica through the Neogene with several international research programmes arising as a result (e.g. Cape Roberts Drilling Project, ANDRILL), and including palynological investigations to determine changes in vegetation.

Antarctica was one of the last frontiers on Earth to be explored and consequently much of our knowledge of this region has accumulated over the last fifty to one hundred years. The history of palaeobotanical research in the Antarctic has seen it develop through an initial discovery stage to a culmination in knowledge of the flora through both time and space. Particular deposits have contributed enormously to our understanding of the morphology and biology of the plants that grew on this landmass. The dramatic improvement in our understanding of the palaeobotanical record of Antarctic through the 1960s to 1980s led to two important publications that synthesised knowledge to date. The first, *Antarctic Paleobiology: its Role in the Reconstruction of Gondwana* (Taylor and Taylor, 1990), is an edited volume that examined various distinct aspects of the fossil floras of Antarctica from the Devonian to Cenozoic in a Gondwana context. The second publication is by Truswell (1991) who provided a comprehensive overview of the Antarctic fossil plant record based on publications up to that time. More recently the extent of the current knowledge base has allowed the integration of changes in the plant fossil record to help interpretations of patterns of climate and environmental change. Although the nature of the region is inhospitable, and the logistics required to work in Antarctica are expensive and complex, significant discoveries will continue to add to the wealth of information already accrued regarding the geological framework of the continent and its palaeobotanical history. Given the ever increasing number of studies on Antarctic palaeobotany, the time is now right for a comprehensive overview of the palaeobotanical evidence, and the hypotheses put forward detailing the palaeoecology of the Antarctic continent through geological time.

Geological framework of Antarctica

Although oceanographic features, such as the Antarctic Circumpolar Current, are a convenient way to delimit the Antarctic region today it is a feature that only developed after Antarctica became isolated in late Oligocene times. In a geological context the continent of Antarctica is today a single entity sitting on one of the Earth's seven major global tectonic plates, but this has not always been the case. The Proterozoic core of the continent is made up of many crustal blocks that acted independently until they amalgamated into the Gondwana continent in Cambrian times. This book is largely concerned with the gondwanan and post-gondwanan history of the continent (Ordovician to recent) and consequently the Mesozoic crustal structure of Antarctica is the most relevant.

While accretionary processes fused and translated crustal fragments along the palaeo-Pacific margin of Antarctica, the break-up of Gondwana in Jurassic through to Cretaceous times saw the assembly of much of the current continental configuration. Through this

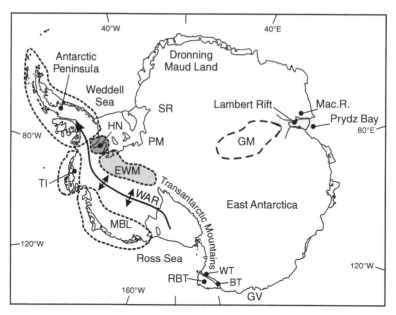

Figure 1.3 Geological framework for Antarctica showing the main crustal blocks. Antarctic Peninsula, HN = Haag Nunataks crustal block, EWM = Ellsworth-Whitmore Mountains crustal block, TI = Thurston Island crustal block, MBL = Marie Byrd Land crustal block, and East Antarctica. WAR = West Antarctic Rift System. On the East Antarctic Crustal block tectonic features mentioned include: GM = Gamburtsev Mountains; RBT = Robertson Bay Terrane; WT = Wilson Terrane. BT = Bowers Terrane. Additional geographic features mentioned in text as follows: Mac.R. = Mac. Robertson Land; GV = George V Coast; SR = Shackleton Range; PM = Pensacola Mountains.

period Antarctica can be best regarded as being composed of at least five separate crustal blocks (Storey, 1995; Figure 1.3) that by the Late Cretaceous or early Cenozoic had more or less fused into the single landmass that today we call Antarctica. This is important as each block has a separate history resulting in uncertainty in tectonic reconstructions of the region as a whole. This means that the post-Late Cretaceous fossil record can be used with confidence in determining the biotic history of Antarctica but that caution is needed when considering evidence from the small crustal blocks with unconstrained histories.

Traditionally geological syntheses separate Antarctica into East Antarctica (one crustal block) and West Antarctica (comprised of Marie Byrd Land, Ellsworth-Whitmore Mountains, Haag Nunataks, Thurston Island and the Antarctic Peninsula) (Figure 1.3). The crustal blocks that make up West Antarctica were rifted from East Antarctica at various stages and this forms a convenient geographic separation for outlining the geological framework of the Antarctic region.

East Antarctica

East Antarctica is best regarded as a Precambrian Shield composed of several Archean cratons separated by mobile belts of Proterozoic to early Paleozoic age (Boger, 2011; Tingey, 1991). The culmination of the Pan-African orogenic events in the early Cambrian saw the final assembly of Gondwana (Bodger, 2011) and a shift from convergent tectonics between the crustal blocks that compose Gondwana to convergent processes on the margin of Gondwana (Cawood *et al.*, 2009). In the Transantarctic Mountains region basement metamorphic complexes are overlain unconformably by late Proterozoic to Cambrian strata. These strata were largely deposited in marine settings on the passive margin of East Antarctica. Sedimentary settings include deep water turbidites that grade up into shallow marine sands and limestone. The Early Cambrian reorganisation of the global tectonic plates is manifested as a mountain building episode (Ross Orogen) along the present day Ross Sea margin of Antarctica. This orogenic activity affected the whole Gondwana margin extending from northern Queensland in eastern Australia, through Antarctica to South Africa and South America, and has been described as the Terra Australis Orogen (Cawood, 2005). The timing of onset varies slightly from region to region (Bodger, 2011) but with the main phase of deformation in the Early to Middle Cambrian.

The shift to convergent margin tectonics and orogenesis during the Cambrian saw the accretion of several terranes to the Antarctic margin that represent island arcs, continental fragments or back-arc basins (Federico, Capponi and Crispini, 2006; Federico *et al.*, 2009; Rocchi *et al.*, 2011). These include the Wilson, Bowers and Robertson Bay terranes in Victoria Land (Figure 1.3) that were accreted by the late Cambrian or early Ordovician (Federico, Capponi and Crispini, 2006; Federico *et al.*, 2009; Rocchi *et al.*, 2011). Intrusive granitoids (370–350 Ma) of the Robertson Bay Terrane (Stump, 1995) are interpreted as part of the widespread granitoid suite intruded into the Devonian to Carboniferous margin of Gondwana (Boger, 2011).

In East Antarctica the Cambrian Ross Orogen resulted in the creation of a widespread erosional surface, referred to as the Kukri Erosional Surface, which has been the subject of varying interpretations (see Chapter 2 for further details). One view is that the Kukri Erosional Surface has limited relief and is essentially planar (Barrett, 1991; Grindley, 1963; McElroy and Rose, 1987; Woolfe and Barrett, 1995). In contrast, other workers (Anderson, 1979; Bradshaw, 1981; Isaac *et al.*, 1996; Laird, Mansergh and Chappell, 1971; Skinner, 1965) have described considerable relief on the Kukri Erosional Surface. The exposure of plutonic rocks suggests considerable uplift and denudation (Isbell, 1999). The observations are important for understanding the palaeogeographic settings of the overlying deposits since the former implies that they have formed a sheet-like deposit over a planar surface, while the latter implies that they infill the surface topography (Isbell, 1999). Initially sedimentation infilled relief on the Kukri Erosion Surface before becoming widespread, and behind the orogenic zone a large and long-lived (Devonian to Jurassic) foreland basin system developed (Beacon Supergroup). The foreland basin extended along the Panthalassan margin from Queensland, through Antarctica, and into southern Africa and South America (Powell and Li, 1994; Vaughan and Pankhurst, 2008) and is one of the reasons for the similarities seen today in rock sequences across Australia, Antarctica, Africa and South America.

The Beacon Supergroup is divided into the Taylor Group and the overlying Victoria Group that are separated by a late Carboniferous to early Permian unconformity (Maya Erosion Surface). At least two depositional centres can be identified, one extending from the Beardmore Glacier to southern Victoria Land, and the other from the Beardmore Glacier through the Ohio Range to the Ellsworth Mountains[1] and possibly as far as the Shackleton Range (Barrett, 1991; Collinson *et al.*, 1994; Laird, 1991) (Figure 1.3). These have been referred to as the McMurdo and Ellsworth basins respectively (Barrett, 1991). Within the McMurdo Basin the Taylor Group is best known from the central Transantarctic Mountains. Rocks from the Taylor Group accumulated in fluvial and lacustrine settings. The Maya Erosion Surface merges into the Kukri surface over basement highs indicating that these highs were erosional for long periods (Ordovician to late Permian). The Maya surface has extensive terrestrial and, in places, marine glacial deposits of Upper Carboniferous to early Permian age that mark the beginning of a period of widespread sedimentation from the early Permian to Jurassic.

The sequence of rocks begins with glacial sedimentary environments (Metschel Tillite; Buckeye, Pagoda, MacKellar, Fairchild formations) that grade upwards into coal measures (Mount Glossopteris Formation, Weller Coal Measures, Buckley Formation). While the Ellsworth and McMurdo basins were separate up until the Late Permian, changes in tectonism saw the basins merge and thus a different terminology is used for the Late Permian and Mesozoic sequences (e.g. Transantarctic Basin, Victoria Land Basin; Collinson *et al.*, 1994). Whatever basin terminology is applied, the coal measures of this region were followed by red-bed sedimentation in the Early Triassic (Feather Conglomerate, Fremouw Formation, Lashly Formation). The Early Triassic strata accumulated under seasonally dry conditions and are overlain by sequences that include thin coals and dark mudstones, indicating a return to wetter conditions (Falla Formation, Upper Lashly Formation). Sedimentation is terminated in the late Early to early Middle Jurassic by widespread volcanism (Kirkpatrick Basalt) associated with the initial break-up of Gondwana. Sediments of the Mawson Formation underlie, and are associated with, these volcanics, and in places contain fossil plants (e.g. Jefferson, Siders and Haban, 1983; Yao, Taylor and Taylor, 1991) and animals (Ball *et al.*, 1979) (Chapter 5).

In contrast to the extensive Paleozoic and Mesozoic sedimentation on the Ross Sea margin, the East Antarctic margin, from Coats Land to northern Victoria Land, largely exposes Proterozoic strata. During the Paleozoic through to the Jurassic this region was in an intra-cratonic setting as a central part of Gondwana. Paleozoic strata are exposed in only a few regions. In the Lambert Glacier area a large rift sag system developed in Permian or earlier times (Lambert Graben; Harrowfield *et al.*, 2005; Veevers, 1988) and can be traced into Indian rift basins (e.g. the Krishna-Godvari Basin). This late Paleozoic rift accumulated a thick Permian to Triassic terrestrial sequence of rocks but also acted as the future site for Jurassic break-up between Antarctica and India. Further west in Dronning Maud Land, late Paleozoic strata of Early to Late Permian age comprise fluvial sandstones that formed in a similar fluvial environment. However, these sequences are more closely related to those seen in southern Africa such as the Karoo Basin (Bauer, 2009). In Dronning Maud Land break-up of Gondwana resulted in the emplacement of extensive Jurassic dolerite sills, and the eruption of basaltic lavas with the formation of thin interbeds between flows. In contrast in the Lambert Graben

---

[1] Ellsworth Mountains and the Whitmore Mountains are treated as a tectonic unit (Ellsworth-Whitmore Mountains) which is the proposed part of the Ellsworth-Whitmore Crustal Block.

region and along much of the associated margin, the break-up history is not exposed with Jurassic to Cretaceous (and Paleogene) strata largely being in offshore basins.

The onset of glaciation in the Neogene saw a shift in sedimentation patterns. Sediments continued to accumulate in offshore basins. Onshore glacial deposits drape the landscape but correlation and dating have proved difficult.

West Antarctica

West Antarctica is made up of a number of crustal blocks namely Marie Byrd Land, Thurston Island, Haag Nunataks, Ellsworth-Whitmore Mountains and the Antarctic Peninsula. A number of these are closely allied to East Antarctica: The Ellsworth-Whitmore Mountains crustal block preserves sequences of rocks that can be correlated with the foreland basin sequences of East Antarctica. This crustal fragment represents the part of East Antarctica that was detached during the break-up of Gondwana and development of the Weddell Sea and the West Antarctic Rift System. The same is also true of Marie Byrd Land which, in part, represents the margin of the Gondwana supercontinent. Pankhurst *et al.* (1998) recognised two tectonic provinces in Marie Byrd Land, namely the Ross Province, and the Amundsen Province. The Ross Province is similar to the Robertson Bay Terrane, containing Ordovician turbidites that can be interpreted as either a back-arc or a marginal marine basin, and these have been intruded by Devonian to Carboniferous granitoids supporting the placement of this region at the Gondwana margin (Pankhurst *et al.*, 1998). It appears that Ross Province was accreted to the Gondwana margin in the Ordovician or early Silurian (Boger, 2011). This province includes the Rupert Coast Volcanics that contain Devonian plant fossils. In contrast the Amundsen Province has a different Paleozoic history (Pankhurst *et al.*, 1998) and can be allied to the Median Tectonic Zone in New Zealand or the Central Domain in the Antarctic Peninsula (Vaughan and Storey, 2000). Thurston Island and Haag Nunataks also represent crustal blocks of Neoproterozoic age but are poorly constrained in terms of their palaeopositions and origins.

The Antarctic Peninsula is particularly important for examining the history of Antarctic vegetation due to the extensive Jurassic, Cretaceous and Cenozoic fossil record. This region has been traditionally interpreted as a long-lived continental island arc system (Storey and Garrett, 1985) and treated as a single geological block, but recently this region has been interpreted within a terrane accretion model (Vaughan and Storey, 2000). This has important implications for the palaeontological record because, under the traditional interpretation of the Antarctic Peninsula, the region has a single history including palaeogeographic location. However, if the terrane model is accepted then interpretation of the fossil floras depends on the age and location of the floras, and whether the terranes had been amalgamated at the time. The terrane model hinges on the interpretation of large-scale shear zones (e.g. Palmer Land Shear Zone), sutures between crustal fragments and differing geological histories on either side of the shear (Vaughan and Storey, 2000). Three domains are recognised with the eastern domain being closest to the Gondwana margin.

Much of the Paleozoic history has been recycled by later tectonic activity but the presence of early Paleozoic quartzites (Fitzgerald Beds; Laudon, 1991), Permian sandstones (Erewhon Beds) that contain *Glossopteris* in the southern part of the Antarctic Peninsula, and recycled Permian to Triassic palynomorphs in the Larsen Basin (Askin, 1990; Askin and Elliot, 1982) attest to this history. Strong support for a Gondwana margin position for this domain is the presence of Devonian to Carboniferous granites (Millar, Pankhurst and Fanning, 1999, 2002) in the Target Hill area. This suggests that the eastern domain is equivalent to the Ross Province in Marie Byrd Land or the terranes in northern Victoria Land, and thus was accreted to the Gondwana margin during the Ross Orogen.

The timing of separation of this crustal fragment from Gondwana is still unclear, but the eastern domain is separated from East Antarctica by the Weddell Sea that began forming in the Middle Jurassic. Although the precise location of the eastern domain crustal fragment is not clear, from the Middle Jurassic it had a separate history from that of East Antarctica. Whether it remains physically connected in the Jurassic is conjectural as the relationship to other crustal blocks, such as Haag Nunataks, Ellsworth-Whitmore Mountains or Thurston Island, are not well-constrained. This means that the Permian vegetation of the eastern domain was connected to the rest of Antarctica at one time but the post-Jurassic vegetational history is less certain.

The central domain of the Antarctic Peninsula contains continental crust of early Permian age (Millar, Pankhurst and Fanning, 1999, 2002) and the magmatic history is similar to that seen in the Amundsen Province of Marie Byrd Land and the Median Tectonic Zone of New Zealand. This includes Late Jurassic to middle Cretaceous arc-related intrusive bodies that imply subduction-related processes and the presence of a Jurassic to Cretaceous arc. Deformation of eastern domain rocks (Latady Group) peaked in Late Jurassic to earliest Cretaceous times (Vaughan and Storey, 2000) but may well have continued into the Cretaceous based on sequence boundaries in the Larsen Basin (Hathway, 2000). This marks collision of the eastern and central domains. From a palaeobiological perspective this means that both regions can be considered as one from the Late Jurassic onwards but the implication for floras older than Late Jurassic is that they may have had separate histories. Early and Middle Jurassic floras are only known from the eastern domain.

The western domain (Alexander Island) consists largely of accretionary complex (LeMay Group) and fore-arc basin (Fossil Bluff Group) sequences juxtaposed against the eastern domain (Vaughan and Storey, 2000). The two domains are connected through King George VI Sound, but it remains to be determined whether these two domains were separate, or whether the western domain is simply the fore-arc region of the central domain magmatic arc (Vaughan and Storey, 2000). The Alexander Island region is unusual in several respects: the accretionary prism is wide, an effect that may be due to the geometry of the margin, and subduction was oblique which means that this region acted as a tectonic trap for terranes transported along the Pacific margin. The capture of the Phoenix Plate by the Antarctic Plate resulted in successive ridge trench collisions that in the Late Cretaceous saw cessation of subduction along this part of the margin. It may well be that King George VI Sound formed as a result of this process in much the same way as Bransfield Strait to the north. Conversely,

King George VI Sound may be an extension of the West Antarctic Rift System as suggested by Eagles *et al.* (2009).

In this treatment we regard the western and central domains as part of the same magamatic arc system that docked with the eastern domain in the Late Jurassic. This means that any pre-Late Jurassic floras of the eastern domain should be considered in separate palaeogeographic contexts from those found in the western and central domains. Jurassic floras of the western and central domains should be considered as part of an arc system that sat outboard of the eastern domain, in much the same way as the Japanese arc sits outboard of the Asian landmass today. However, for the post-Late Jurassic history all floras can be considered as part of a single biogeographic unit.

Summary

- For much of its existence Antarctica has been ice-free and supporting a wealth of plant and animal life. Yet only relatively recently with the upsurge of interest in palaeobotany and palynology has it become apparent how this region, now defined by the Antarctic Circumpolar Current, played an important role in developing and maintaining biogeographic patterns across the Southern Hemisphere.

- Antarctica was the core of the Gondwana supercontinent and so maintained connections between what are now widely separated landmasses of India, South America, Africa, Australia and New Zealand. The lack of extant vegetation makes fossils from this region critical for understanding the evolution of the Southern Hemisphere biota.

- East Antarctica, a craton, and West Antarctica, comprising a number of crustal blocks, are separated by the West Antarctic Rift System, and have evolved within very different geological and climatic frameworks and provide different facets of the evolution of terrestrial plant life in the southern high latitudes. These are the themes that will be explored in detail in subsequent chapters.

References

Aagaard, B. (1930). *Fangst og Forskning i Sydishavet*. Oslo: Gyldendal Norsk Forlag, pp. 1–354.

Anderson, J. M. (1979). The geology of the Taylor Group, Beacon Supergroup, Byrd Glacier area, Antarctica. *New Zealand Antarctic Record*, **2**, 6–11.

Arber, E. A. N. (1907). Appendix to Chapter VI–Report on the plant-remains from the Beacon Sandstone. In *National Antarctic Expedition, 1901–1904 volume 1 Geology*. London: The British Museum (Natural History), pp. 1–48.

Arntz, W. E., Gutt, J. and Klages, M. (1997). Antarctic marine biodiversity: an overview. In *Antarctic Communities: Species, Structure and Survival*, eds B. Battaglia, J. Valencia and D. W. H. Walton. Cambridge: Cambridge University Press, pp. 3–14.

Askin, R. A. (1983). Tithonian (Uppermost Jurassic)–Barremian (Lower Cretaceous) spores, pollen and microplankton form the South Shetland Island, Antarctica. In *Antarctic Earth Science – Proceedings of the Fourth International Symposium on*

Antarctic Earth Sciences, Adelaide, South Australia, 16–20 August 1982, eds R. L. Oliver, P. R. James and J. B. Jago. Canberra: Australian Academy of Science/ Cambridge: Cambridge University Press, pp. 295–297.

Askin, R. A. (1990). Cryptogam spores from the Upper Campanian and Maastrichtian of Seymour Island, Antarctica. *Micropaleontology*, **36**, 141–156.

Askin, R. A. and Elliot, D. H. (1982). Geologic implications of recycled Permian and Triassic palynomorphs in Tertiary rocks of Seymour Island, Antarctic Peninsula. *Geology*, **10**, 547–551.

Ball, H. W., Borns Jr, H. W., Hall, B. A., *et al.* (1979). Biota, age, and significance of lake deposits, Carapace Nunatak, Victoria Land, Antarctica. In *Fourth International Gondwana Symposium: Papers 1*, eds B. Laskar and C. S. R. Rao. Delhi: Hindustan Publishing Corporation, pp. 166–175.

Barrett, P. J. (1991). The Devonian to Jurassic Beacon Supergroup of the Transantarctic Mountains and correlatives in other parts of Antarctica. In *The Geology of Antarctica*, *Oxford Monographs on Geology and Geophysis*, **17**, ed. R. J. Tingey. Oxford: Clarendon Press, pp. 120–152.

Bauer, W. (2009). Permian sedimentary cover, Heimefrontfjella, western Dronning Maud Land (East Antarctica). *Polar Forschung*, **79**, 39–42.

Berry, E. W. (1913). Some paleontological results of the Swedish South Polar Expedition under Nordenskiold. *Science*, **38**, 656–661.

Birkenmajer, K. and Zastawniak, E. (1986). Plant remains of the Dufayel Island Group (early Tertiary?), King George Island, South Shetlands (West Antarctic). *Acta Palaeobotanica*, **26**, 33–53.

Birkenmajer, K. and Zastawniak, E. (1989a). Late Cretaceous–early Tertiary floras of King George Island, West Antarctica: their stratigraphic distribution and palaeoclimatic significance. In *Origins and Evolution of the Antarctic Biota*, *Geological Society of London Special Publication*, **47**, ed. J. A. Crame. London: The Geological Society, pp. 227–240.

Birkenmajer, K. and Zastawniak, E. (1989b). Late Cretaceous–early Neogene vegetation history of the Antarctic Peninsula sector. Gondwana break-up and Tertiary glaciations. *Bulletin of the Polish Academy of Sciences, Earth Sciences*, **37**, 63–88.

Boger, S. D. (2011). Antarctica – before and after Gondwana. *Gondwana Research*, **19**, 335–371.

Bradshaw, M. A. (1981). Paleoenvironmental interpretations and systematics of Devonian trace fossils from the Taylor Group (lower Beacon Supergroup), Antarctica. *New Zealand Journal of Geology and Geophysics*, **24**, 615–652.

Burckle, L. H. and Pokras, E. M. (1991). Implications of a Pliocene stand of *Nothofagus* (southern beech) within 500 kilometres of the South Pole. *Antarctic Science*, **3**, 389–403.

Cantrill, D. J. (2000). A Cretaceous (Aptian) flora from President Head, Snow Island, Antarctica. *Palaeontographica B*, **253**, 153–191.

Carlquist, S. (1987). Pliocene *Nothofagus* wood from the Transantarctic Mountains. *Aliso*, **11**, 571–583.

Cawood, P. A. (2005). Terra Australis Orogen: Rodinia breakup and development of the Pacific and Iapetus margins of Gondwana during the Neoproterozoic and Paleozoic. *Earth-Science Reviews*, **69**, 249–279.

Cawood, P. A., Kröner, A., Collins, *et al.* (2009). Accretionary orogens through Earth history. In *Earth Accretionary Systems in Space and Time*, *Geological Society of London Special Publication*, **318**, eds P. A. Cawood and A. Kröner. London: The Geological Society, pp. 1–36.

Césari, S. N., Parcia, C. A., Remesal, M. B. and Salani, F. M. (1998). First evidence of Pentoxylales in Antarctica. *Cretaceous Research*, **19**, 733–743.

Césari, S. N., Parcia, C., Remesal, M. and Salani, F. (1999). Paleoflora del Cretácico Inferior de península Byers, Islas Shetland del Sur, Antártida. *Ameghiniana*, **36**, 3–22.

Chapman, J. L. and Smellie, J. L. (1992). Cretaceous fossil wood and palynomorphs from Williams Point, Livingston Island, Antarctic Peninsula. *Review of Paleobotany and Palynology*, **74**, 163–192.

Collinson, J. W., Isbell, J. L., Elliot, D. H., Miller, M. F. and Miller, J. M. G. (1994). Permian-Triassic Transantarctic Basin. In *Permian-Triassic Pangean Basins and Foldbelts along the Panthalassan Margin of Gondwanaland, Geological Society of America Memoir*, **184**, eds J. J. Veevers and C. McA. Powell. Boulder: The Geological Society of America, pp. 173–222.

Conrad, L. J. (1999). *Bibliography of Antarctic Exploration: Expedition Accounts from 1768 to 1960*. Washougal: Conrad.

Dettmann, M. E. and Thomson, M. R. A. (1987). Cretaceous palynomorphs from the James Ross Island area, Antarctica – a pilot study. *British Antarctic Survey Bulletin*, **77**, 13–59.

Duane, A. M. (1996). Palynology of the Byers Group (Late Jurassic-Early Cretaceous) of Livingston and Snow islands, Antarctic Peninsula: its biostratigraphical and palaeo-environmental significance. *Review of Paleobotany and Palynology*, **91**, 241–281.

Dusén, P. (1899). Über die Tertiäre Flora der Magellansländer. In *Wissenschaftliche Ergebnisse der Schwedischen Expedition nach den Magellansládern, 1895–1897. Volume 1(4)*, ed. O. Nordenskjöld. Stockholm: P. A. Norstedt & Söner, pp. 87–107.

Dusén, P. (1908). Über die Tertiäre Flora der Seymour-Insel. In *Wissenschaftliche Ergebnisse der Schwedischen Südpolar-Expedition 1901–1903, Volume 3(3), Geologie und Paläontologie*, ed. O. Nordenskjöld. Stockholm: P. A. Norstedt & Söner, pp. 1–27.

Du Toit, A. L. (1937). *Our Wandering Continents*. Edinburgh: Oliver and Boyd.

Dutra, T. L. and Batten, D. J. (2000). Upper Cretaceous floras of King George Island, West Antarctica, and their palaeoenvironmental and phytogeographic implications. *Cretaceous Research*, **21**, 181–209.

Eagles, G., Larter, R. D., Gohl, K. and Vaughan, A. P. M. (2009). West Antarctic Rift System in the Antarctic Peninsula. *Geophysical Research Letters*, **36**, L21305, 1–4.

Edwards, W. N. (1921). Fossil coniferous wood from Kerguelen Island. *Annals of Botany*, **35**, 609–617.

Edwards, W. N. (1928). The occurrence of *Glossopteris* in the Beacon Sandstone of Ferrar Glacier, South Victoria Land. *Geological Magazine*, **65**, 323–327.

Elliot, D. H. and Hanson, R. E. (2001). Origin of widespread, exceptionally thick basaltic phreatomagmatic tuff breccia in the Middle Jurassic Prebble and Mawson formations, Antarctica. *Journal of Volcanology and Geothermal Research*, **111**, 183–201.

Eights, J. (1833). Description of a new crustaceous animal found on the shores of the South Shetland Islands. *Transactions of the Albany Institute*, **2**, 183–201.

Engelhardt, H. (1891). Über Tertiärpflazen von Chile. *Abhandlungen der Senckenbergischen Naturforschenden Gesellschaft Band*, **16**, 629–692.

Engelhardt, H. (1895). Über neue Tertiärpflanzen Süd Amerikas. *Abhandlungen der Senckenbergischen Naturforschenden Gesellschaft Band*, **19**, 1–47.

Federico, L., Capponi, G. and Crispini, L. (2006). The Ross orogeny of the Transantarctic Mountains: a northern Victoria Land perspective. *International Journal of Earth Sciences*, **95**, 759–770.

Federico, L., Crispini, L., Capponi, G. and Bradshaw, J. D. (2009). The Cambrian Ross Orogeny in northern Victoria Land (Antarctica) and New Zealand: a synthesis. *Gondwana Research*, **15**, 188–196.

Ferrar, H. T. (1907). Report on the field-geology of the region explored during the "Discovery" Antarctic Expedition, 1901–1904. In *National Antarctic Expedition 1901–1904. Geology Volume 1 (Field Geology: Petrography)*. London: British Museum (Natural History), pp. 1–100.

Fleming, W. L. S. (1938). Geology and glaciology. In *Notes on the Scientific Work of the British Graham Land Expedition, 1934–1937, The Geographical Journal*, **91**, eds W. L. S. Fleming, A. Stephenson, B. B. Roberts and G. C. L. Bertram. London: Royal Geographical Society, pp. 508–512.

Fogg, G. E. (1992). *A History of Antarctic Science*. Cambridge: Cambridge University Press.

Fuenzalida, H., Araya, R. and Hervé F. (1972). Middle Jurassic flora from north-eastern Snow Island, South Shetland Islands. In *Antarctic Geology and Geophysics*, ed. R. J. Adie. Oslo: Universitetsforlaget, pp. 173–180.

Gee, C. T. (1989). Revision of the Late Jurassic/Early Cretaceous flora from Hope Bay Antarctica. *Palaeontographica B*, **213**, 149–214.

Gordon, W. T. (1930). A note on *Dadoxylon (Araucarioxylon)* from the Bay of Isles. In *Report on the Geological Collections made during the Voyage of the* Quest *on the Shackleton-Rowett Expedition to the South Atlantic and Weddell Sea in 1921–22*. London: British Museum of Natural History, pp. 2–24.

Gothan, W. (1908). Die fossil Hölzer von der Seymour und Snow Hill Insel. In *Wissenschaftliche Ergebnisse der Schwedischen Südpolar-Expedition 1901–1903, Volume 3(8), Geologie und Paläontologie*, ed. O. Nordenskjöld. Stockholm: P. A. Norstedt & Söner, pp. 1–33.

Grindley, G. W. (1963). The geology of the Queen Alexandra Range, Beardmore Glacier, Ross Dependency, Antarctica; with notes on the correlation of Gondwana sequences. *New Zealand Journal of Geology and Geophysics*, **6**, 307–347.

Halle, T. G. (1913a). Om de Antarktiska trakterna Juraflora. *Geologiska Föreningens Förhandlingar*, **35**, 105–106.

Halle, T. G. (1913b). The Mesozoic flora of Graham Land. In *Wissenschaftliche Ergebnisse der Schwedischen Südpolar-Expedition 1901–1903, Volume 3(4), Geologie und Paläontologie*, ed. O. Nordenskjöld. Stockholm: P.A. Norstedt & Söner, pp. 3–124.

Harrowfield, M., Holdgate, G. R., Wilson, C. J. L. and McLoughlin, S. (2005). Tectonic significance of the Lambert Graben, East Antarctica: reconstructing the Gondwanan rift. *Geology*, **33**, 197–200.

Hathway, B. (2000). Continental rift to back-arc basin: Jurassic–Cretaceous stratigraphical and structural evolution of the Larsen Basin, Antarctic Peninsula. *Journal of the Geological Society of London*, **157**, 417–432.

Heer, O. (1868). *Flora Fossilis Arctica. Die Fossil Flora der Polarländer vol. 1*. Zurich: Druck und Verlag von Friedrich Schulthess.

Heer, O. (1871). *Flora Fossilis Arctica. Die Fossil Flora der Polarländer vol. 2*. Zurich: Verlag von J. Wurster & comp.

Heer, O. (1874). Nachträge zur Miocenen flora Grönlands. *Kungliga Svenska Vetenskaps Akademiens Handlingar*, **13**, 1–29.

Heer, O. (1875). *Flora Fossilis Arctica. Die Fossil Flora der Polarländer vol. 3*. Zurich: Verlag von J. Wurster & comp.

Heer, O. (1877). *Flora Fossilis Arctica. Die Fossil Flora der Polarländer vol. 4.* Zurich: Verlag von J. Wurster & comp.

Heer, O. (1878). *Flora Fossilis Arctica. Die Fossil Flora der Polarländer vol. 5.* Zurich: Verlag von J. Wurster & comp.

Heer, O. (1880). *Flora Fossilis Arctica. Die Fossil Flora der Polarländer vol. 6.* Zurich: Verlag von J. Wurster & comp.

Heer, O. (1883). *Die Fossil Flora der Polarländer vol. 7*, Zurich: Wurster, pp. 1–275.

Hill, R. S. and Truswell, E. M. (1993). *Nothofagus* fossil in the Sirius Group Transantarctic Mountains. In *The Antarctic Paleoenvironment: a Perspective on Global Change. Part 2, Antarctic Research Series*, **60**, eds J. P. Kennett and D. Warnke. Washington: American Geophysical Union, pp. 63–73.

Hill, R. S., Harwood, D. M. and Webb, P. N. (1996). *Nothofagus beardmorensis* (Nothofagaceae), a new species based on leaves from the Pliocene Sirius Group, Transantarctic Mountains, Antarctica. *Review of Paleobotany and Palynology*, **94**, 11–24.

Hunter, M. A., Cantrill, D. J., Flowerdew, M. and Millar, I. L. (2005). Middle Jurassic age for the Botany Bay Group: implications for Weddell Sea Basin creation and Southern Hemisphere biostratigraphy. *Journal of the Geological Society of London*, **162**, 745–748.

Isaac, M. J., Chinn, T. J., Edbrokoke, S. W. and Forsythe, P. J. (1996). *Geology of the Olympus Range area, southern Victoria Land, Antarctica, scale 1:50,000, geological map 20, 1–60*. Wellington: Institute of Geological and Nuclear Sciences.

Isbell, J. L. (1999). The Kukri Erosion Surface; a reassessment of its relationship to rocks of the Beacon Supergroup in the central Transantarctic Mountains, Antarctica. *Antarctic Science*, **11**, 228–238.

Jefferson, T. H. (1981). Palaeobotanical contributions to the geology of Alexander Island, Antarctica. Ph.D. thesis, University of Cambridge.

Jefferson, T. H., Siders, M. A. and Haban, M. A. (1983). Jurassic trees engulfed by lavas of the Kirkpatrick Basalt Group, northern Victoria Land. *Antarctic Journal of the United States*, **18(5)**, 14–16.

Jones, A. E. G. (1982). *Antarctica Observed: Who Discovered the Antarctic Continent?* Whitby: Caedmon of Whitby.

Kemp, E. M. (1973). Permian flora from the Beaver Lake area, Prince Charles Mountains, Antarctica. 1. Palynological examination of samples. *Bureau of Mineral Resources, Geology and Geophysics Bulletin*, **126** (Palaeontological papers for 1969), 7–12.

Lacey, W. S. and Lucas, R. C. (1981a). A Lower Permian flora from the Theron Mountains, Coats Land. *British Antarctic Survey Bulletin*, **53**, 153–156.

Lacey, W. S. and Lucas, R. C. (1981b). The Triassic flora of Livingston Island, South Shetland Islands. *British Antarctic Survey Bulletin*, **53**, 157–173.

Laird, M. G. (1991). The Late Proterozoic–Middle Paleozoic rocks. In *The Geology of Antarctica, Oxford Monographs on Geology and Geophysics*, **17**, ed. R. J. Tingey. Oxford: Clarendon Press, pp. 74–119.

Laird, M. G., Mansergh, G. D. and Chappell, J. M. A. (1971). Geology of the central Nimrod Glacier area, Antarctica. *New Zealand Journal of Geology and Geophysics*, **14**, 427–468.

Larsen, C. A. (1894). The voyage of the "Jason" to the Antarctic region. *Geographical Journal*, **4**, 333–344.

Laudon, T. S. (1991). Petrology of sedimentary rocks from the English Coast, eastern Ellsworth Land. In *Geological Evolution of Antarctica*, eds M. R. A. Thomson, J. A. Crame and J. W. Thomson. Cambridge: Cambridge University Press, pp. 455–460.

Li, H. (1994). Early Tertiary Fossil Hill flora from Fildes Peninsula of King George Island, Antarctica. In *Stratigraphy and Palaeontology of Fildes Peninsula, King George Island, Antarctica, State Antarctic Committee Monograph*, 3, ed. Y. Shen. Beijing: China Science Press, pp. 133–171.

Linse, K., Griffiths, H. J., Barnes, D. K. A. and Clarke, A. (2006). Biodiversity and biogeography of Antarctic and sub-Antarctic mollusca. *Deep-Sea Research II*, **53**, 985–1008.

Lucas, R. C. and Lacey, W. S. (1981). A permineralized wood flora of probable early Tertiary age from King George Island, South Shetland Islands. *British Antarctic Survey Bulletin*, **53**, 147–151.

McElroy, C. T. and Rose, G. (1987). *Geology of the Beacon Heights area, southern Victoria Land, Antarctica*. New Zealand Geological Survey, Miscellaneous Series Map 15 and notes 1–47. Wellington: Government Printer.

Millar, I. L., Pankhurst, R. J. and Fanning, C. M. (1999). U-Pb zircon evidence for the nature of the Antarctic Peninsula basement. In *Abstracts Volume, 8th International Symposium on Antarctic Earth Sciences, Wellington, New Zealand, July 1999*, ed. D. N. B. Skinner. Wellington: The Royal Society of New Zealand, pp. 210.

Millar, I. L., Pankhurst, R. J. and Fanning, C. M. (2002). Basement chronology of the Antarctic Peninsula: recurrent magmatism and anatexis in the Paleozoic Gondwana margin. *Journal of the Geological Society of London*, **159**, 145–157.

Nathorst, A. G. (1884). Grönlands forntida växtvärld. *Nordisk Tidskrift*, 7, 344–363.

Nathorst, A. G. (1904). Sur le fossile des regions antarctiques. *Comptes Rendus de l'Academie des Sciences de Paris*, **138**, 1447–1450.

Nathorst, A. G. (1906). On the Upper Jurassic flora of Hope Bay, Graham Land. *Comptes Rendus 10th International Geological Congress Mexico 1906*, **10**, 1269–1270.

Neish, P. G., Drinnan, A. N. and Cantrill, D. J. (1993). Structure and ontogeny of *Vertebraria* from silicified Permian sediments in East Antarctica. *Review of Paleobotany and Palynology*, **79**, 221–244.

Nordenskjöld, O. and Andersson, J. G. (1905). *Antarctica, or two years amongst the ice of the South Pole*. London: Hurst and Blackett Ltd.

Ociepa, A. M. (2007). Jurassic liverworts from Mount Flora, Hope Bay, Antarctic Peninsula. *Polish Polar Research*, **28**, 31–36.

Ociepa, A. M. and Barbacka, M. (2011). *Spesia antarctica* gen. et sp. nov. – a new fertile fern spike from the Jurassic of Antarctica. *Polish Polar Research*, **32**, 59–66.

Orlando, H. (1968). A new Triassic flora from Livingston Island, South Shetland Islands. *British Antarctic Survey Bulletin*, **16**, 1–13.

Pankhurst, R. J., Weaver, S. D., Bradshaw, J. D., Storey, B. C. and Ireland, T. R. (1998). Geochronology and geochemistry of pre-Jurassic superterranes in Marie Byrd Land, Antarctica. *Journal of Geophysical Research*, **103**, 2529–2547.

Pirrie, D. and Riding, J. B. (1988). Sedimentology, palynology and structure of Humps Island, northern Antarctic Peninsula. *British Antarctic Survey Bulletin*, **80**, 1–19.

Plumstead, E. P. (1962). Fossil floras of Antarctica. In *Trans-Antarctic Expedition 1955–1958, Scientific Reports* 9 (Geology). London: The Trans-Antarctic Expedition Committee, pp. 1–154.

Plumstead, E. P. (1975). A new assemblage of plant fossils from Milorgfjella, Dronning Maud Land. *British Antarctic Survey Scientific Reports*, **83**, 1–30.

Powell, C. McA. and Li, Z. X. (1994). Reconstruction of the Panthalassan margin of Gondwana. In *Permian-Triassic Pangean Basins and Foldbelts along the Panthalassan Margin of Gondwanaland, Geological Society of America Memoir*,

184, eds J.J. Veevers and C.McA. Powell. Boulder: The Geological Society of America, pp. 5–9.

Rees, P.M. (1990). Palaeobotanical contributions to the Mesozoic geology of the northern Antarctic Peninsula region. Ph.D. thesis, Royal Holloway and Bedford New College, University of London.

Rees, P.M. (1993). Caytoniales in Early Jurassic floras from Antarctica. *Geobios*, **26**, 33–42.

Rees, P.M. and Cleal, C.J. (2004). Lower Jurassic floras from Hope Bay and Botany Bay, Antarctica. *Special Papers in Paleontology*, **72**, 1–96.

Rees, P.M. and Smellie, J.L. (1989). Cretaceous angiosperms from an allegedly Triassic flora at Williams Point, Livingston Island, South Shetland Island. *Antarctic Science*, **1**, 239–248.

Riding, J.B., Crame, J.A., Dettmann, M.E. and Cantrill D.J. (1998). The age of the base of the Gustav Group in the James Ross Basin, Antarctica. *Cretaceous Research*, **19**, 87–105.

Rocchi, S., Bracciali, L., di Vincenzo, G., Gemelli, M. and Ghezzo, C. (2011). Arc accretion to the early Paleozoic Antarctic margin of Gondwana in Victoria Land. *Gondwana Research*, **19**, 594–607.

Schopf, J.M. (1962). A preliminary report on plant remains and coal of the sedimentary section in the central range of the Horlick Mountains, Antarctica. *Ohio State University Institute of Polar Studies Report*, **2**, 1–61.

Schopf, J.M. (1968). Studies in Antarctic paleobotany. *Antarctic Journal of the United States*, **3(5)**, 176–177.

Schopf, J.M. (1970a). Petrified peat from a Permian coal bed in Antarctica. *Science*, **169**, 274–277.

Schopf, J.M. (1970b). Antarctic collections of plant fossils 1969–1970. *Antarctic Journal of the United States*, **5(4)**, 89.

Schopf, J.M. (1978). An unusual osmundaceous specimen from Antarctica. *Canadian Journal of Botany*, **56**, 3083–3095.

Seward, A.C. (1914). Antarctic plant fossils. In *British Antarctic ("Terra Nova") Expedition, 1910. Natural History Reports. Geology. No. 1*. London: British Museum (Natural History), 1–49.

Seward, A.C. (1919). *Fossil Plants Vol 4*. Cambridge: Cambridge University Press.

Sharman, G. and Newton, E.T. (1894). Notes on some fossils from Seymour Island, in the Antarctic regions, obtained by Dr Donald. *Transactions of the Royal Society of Edinburgh*, **37**, 707–709.

Sharman, G. and Newton, E.T. (1898). Notes on some additional fossils collected at Seymour Islands, Graham Land, by Dr Donald and Captain Larsen. *Proceedings of the Royal Society of Edinburgh*, **22**, 58–61.

Skinner, D.N.B. (1965). Petrographic criteria of the rock units between Byrd and Starshot Glaciers, South Victoria Land, Antarctica. *New Zealand Journal of Geology and Geophysics*, **8**, 292–303.

Smoot, E.L. and Taylor, T.N. (1986a). Structurally preserved fossil plants from Antarctica II. A Permian moss from the Transantarctic Mountains. *American Journal of Botany*, **73**, 1683–1691.

Smoot, E.L. and Taylor, T.N. (1986b). Evidence for simple polyembryony in Permian seeds from Antarctica. *American Journal of Botany*, **73**, 1077–1079.

Smoot, E.L., Taylor, T.N. and Delevoryas, T. (1985). Structurally preserved fossil plants from Antarctica 1. *Antarcticycas*, gen. n., a Triassic cycad stem from the Beardmore Glacier area. *American Journal of Botany*, **71**, 1410–1423.

Soliani, E. and Bonhomme, M. G. (1994). New evidence for Cenozoic resetting of K-Ar ages in volcanic rocks of the northern portion of the Admiralty Bay, King George Island, Antarctica. *Journal of South American Earth Sciences*, **7**, 85–94.

Stephenson, A. and Fleming, W. L. S. (1940). King George the Sixth Sound. *Geographical Journal*, **96**, 153–164.

Stipanicic, P. M. and Bonetti, M. I. R. (1970). Posciones estratigráficas y edades de las principales floras Jurásicas Argentinas. II. Floras Doggerianas y Málmicas. *Ameghinana*, **7**, 101–118.

Storey, B. C. (1995). The role of mantle plumes in continental breakup – case histories from Gondwanaland. *Nature*, **377**, 301–308.

Storey, B. C. and Garrett, S. W. (1985). Crustal growth of the Antarctic Peninsula by accretion, magmatism and extension. *Geological Magazine*, **122**, 5–14.

Stump, E. (1995). *The Ross Orogen of the Transantarctic Mountains*. Cambridge: Cambridge University Press.

Taylor, B. J., Thomson, M. R. A. and Willey, L. E. (1979). Geology of the Ablation Point-Keystone Cliffs area, Alexander Island. *British Antarctic Survey Scientific Report*, **82**, 1–65.

Taylor, T. N. and Taylor, E. L. (1990). *Antarctic Paleobiology: its Role in the Reconstruction of Gondwana*. New York: Springer-Verlag.

Thomson, M. R. A. (1975). New palaeontological and lithological observations on the Legoupil Formation, north-west Antarctic Peninsula. *British Antarctic Survey Bulletin*, **41**, 169–185.

Tingey, R. J. (1991). The regional geology of the Archaean and Proterozoic rocks in Antarctica. In *The Geology of Antarctica, Oxford Monographs on Geology and Geophysics*, **17**, ed. R. J. Tingey. Oxford: Clarendon Press, pp. 1–73.

Torres, T. (1984). *Nothofagoxylon antarcticus* n. sp., madera fósils del Terciario de la isla Rey Jorge, islas Shetland del Sur, Antártica. *Serie Científica Instituto Antárctico Chileno*, **31**, 39–52.

Torres, T., Valenzuela, E. A. and Gonzalez, I. M. (1982). Paleoxilología de Península Byers, Isla Livingston, Antártica. In *Actas 3rd Congreso Geológico Chileno: Concepción, Chile*. A321–A342.

Torres, T., Barale, G., Thévernard, F., Philippe, M. and Galleguillos, H. (1997). Morfología y sistemática de la flora del Cretácico Inferior de President Head, isla Snow, archipiélago de las Shetland del Sur, Antártica. *Serie Científica Instituto Antártico Chileno*, **47**, 59–86.

Torres, T., Barale, G., Galleguillos, H. and Atala, C. (1998). Hallazgo de *Neozamites* (Bennettitales) en el Cretácico Inferior de la isla Snow, Shetland del Sur, Antártica. *Serie Científica Instituto Antártico Chileno*, **48**, 9–18.

Truswell, E. M. (1991). Antarctica: a history of terrestrial vegetation. In *The Geology of Antarctica, Oxford Monographs on Geology and Geophysics*, **17**, ed. R. J. Tingey. Oxford: Clarendon Press, pp. 499–537.

Vaughan, A. P. M. and Pankhurst, R. J. (2008). Tectonic overview of the West Gondwana margin. *Gondwana Research*, **13**, 150–162.

Vaughan, A. P. M. and Storey, B. C. (2000). The eastern Palmer Land shear zone: a new terrane accretion model for the Mesozoic development of the Antarctica Peninsula. *Journal of the Geological Society of London*, **157**, 1243–1256.

Veevers, J. J. (1988). Gondwana facies started when Gondwanaland merged in Pangea. *Geology*, **16**, 732–734.

Webb, P. N. and Harwood, D. M. (1993). Pliocene fossil *Nothofagus* (Southern Beech) from Antarctica: phytogeography, dispersal strategies, and survival in high latitude glacial-deglacial environments. In *Forest Development in Cold Climates, NATO ASI Series A: Life Sciences*, **244**, eds J. Alden, J. L. Mastrantonio and S. Dum. New York: Plenum Press, pp. 135–163.

Webb, P. N., Harwood, D. M., McKelvey B. C. and Stott, L. D. (1984). Cenozoic marine sedimentation and ice-volume variation on the East Antarctic craton. *Geology*, **12**, 287–291.

White, M. E. (1962). Permian plant remains from Mount Rymill, Antarctica. Report on 1961 plant fossil collections. *Bureau of Mineral Resources, Geology and Geophysics Record*, **1962/114**, 1–15.

White, M. E. (1973). Permian flora from the Beaver Lake area, Prince Charles Mountains, Antarctica, 2: Plant fossils. *Bureau of Mineral Resources, Geology and Geophysics Bulletin*, **126** (Paleontological papers for 1969), 13–18.

von Ettingshausen, C. (1887). Beitrage zur Kenntniss der Fossilen Flora Neuseelands. *Denkschriften der Akademie der Wissenschaften Wien*, **53**, 143–194.

von Ettingshausen, C. (1888). Contributions to the Tertiary flora of Australia. *Memoirs of the Geological Survey of New South Wales, Paleontology*, **2**, 1–189.

von Ettingshausen, C. (1891). Contributions to the knowledge of the fossil flora of New Zealand. *Transactions of the New Zealand Institute*, **23**, 237–310.

Woolfe, K. J. and Barrett, P. J. (1995). Constraining the Devonian to Triassic tectonic evolution of the Ross Sea sector. *Terra Antarctica*, **2**, 7–21.

Yao, X., Taylor, T. N. and Taylor, E. L. (1991). Silicified dipterid ferns from the Jurassic of Antarctica. *Review of Paleobotany and Palynology*, **67**, 353–362.

Zastawniak, E. (1981). Tertiary leaf flora from Point Hennequin Group of King George Island (South Shetland Islands, Antarctica). Preliminary report. *Studia Geologica Polonica*, **72**, 97–108.

Zastawniak, E. (1994). Upper Cretaceous leaf flora from Blaszyk Moraine (Zamek Formation), King George Island, South Shetland Islands, West Antarctica. *Acta Paleobotanica*, **34**, 119–163.

Zinsmeister, W. J. (1988). Early geological exploration of Seymour Island. In *Geology and Paleontology of Seymour Island, Antarctic Peninsula, Geological Society of America Memoir*, **169**, eds R. M. Feldmann and M. O. Woodburne. Boulder: The Geological Society of America, pp. 1–16.

2

Early to middle Paleozoic climates and colonisation of the land

Introduction

Society is becoming increasingly concerned about global climate change and the impact that it will have on our planet. Yet climate change has always been a natural part of the Earth system dynamics and over geological timescales there is much evidence for dramatic swings from greenhouse to icehouse conditions and back again. These large-scale climatic changes have been preserved in the rock record since long-term changes in conditions, such as rainfall and temperature, all influence rock weathering, the type of sediments and patterns of sedimentation. Such changes can be illustrated by one classic example, namely the sequence of rocks in the large gondwanan foreland basin. From the Devonian to the Triassic, a time period covering some 200 Ma, the same types of rocks were being deposited over South America and Africa and across the continents of India, Australia and Antarctica. The resultant stratigraphic successions in all of these now distinct landmasses are similar and characterised by Carboniferous glacial rocks overlain by Permian coal measures, which in turn are overlain by Triassic red-beds. This example reflects global climate change on a grand scale from a glacial world in the Carboniferous through Permian temperate coal-forming climates to warm Triassic climates with seasonal rainfall.

Geographically widespread change in environmental conditions interpreted from the rock record raises several important questions: how did the vegetation respond to these changes? What influence did vegetation have on accelerating or mitigating the climate? What role did the change in climate play in plant evolution? And perhaps most fundamentally of all, how did the global environment change once plants began to colonise the land?

Plants are inextricably linked to the process of photosynthesis which uses sunlight energy to extract carbon dioxide from the atmosphere to generate useable energy (glucose) and the by-product oxygen. Since the absence of plants results in a significant reduction in the drawdown of carbon dioxide and release of oxygen into the atmosphere, proxies for carbon dioxide and oxygen in the rock record can be used to infer changes in vegetation. At no time has the influence of terrestrial plant life on the atmosphere been greater than during the earliest colonisation of land by the plants. The early to middle Paleozoic is the period of geological time associated with the rise of terrestrial vegetation from the simplest vascular land plants, which appeared in the latest Silurian to Early Devonian, to more complex floras comprising plants of greater stature. Fossilised plant remains provide direct evidence of their presence and they document changes in vegetation through geological time. But in addition plants also leave indirect evidence of their impact on the Earth system. During the Devonian as plant life became more complex, geochemical modelling has identified a rapid decrease in carbon dioxide levels coupled with increases in oxygen levels (Berner, 2004, 2009). Moreover, with increasing plant complexity through the Paleozoic not only was the atmosphere affected, but further modifications to Earth system processes impacted on the hydrosphere and lithosphere. Consequently, these modifications drove forward the evolutionary process of the plants themselves due to the link between the composition of the atmosphere, plant gas exchange mechanisms and plant evolution. This feedback between plant evolutionary processes and Earth system processes has also resulted in major changes in the carbon cycle over time with dramatic shifts in the allocation of carbon to sediments, atmosphere, hydrosphere and biosphere. In short, it is clear that globally there was (and still is) a strong link between the evolution of the Earth's climate system and plant evolution such that an understanding of the feedbacks between vegetation and climate may be informative for interpretations of the fossil record and climatic predictions for the future.

Global palaeoclimate

Earth history is characterised by periods of time when climates have been both warmer and cooler than today (e.g. Frakes, Francis and Skytus, 1992). While researchers have long recognised the variation from cool modes to warm modes, the definitions and boundaries between one mode and another differ, as do the mechanisms for transition between modes. This is perhaps not surprising as climate is complex, driven by planetary processes that interact in different ways, and is associated with a range of forcing mechanisms that can form both positive and negative feedback loops. A recent study by Kidder and Worsley (2010) proposed three climate states for the Earth – greenhouse, icehouse and hothouse – based on the dominance of differing climate processes. In practice, the boundaries between the various climate modes are to some extent gradational but with key thresholds that tip the system from one state into another. Nevertheless, the conceptual model of icehouse, greenhouse and hothouse and associated forces driving the climate from one state to another provides a good framework for discussing the Earth's climate, which impacts on subsequent Antarctic vegetation, and therefore these states are briefly outlined below.

Icehouse periods characterise one of the Earth's major climate modes (Frakes, Francis and Skytus, 1992) representing some 20 to 25% of Phanerozoic Earth time (Kidder and Worsley, 2010). In this state the Earth's climate is driven by thermohaline circulation in the oceans, with polar regions providing the driving mechanism through the creation of cold, salty, bottom water. This climate mode results in oceans that are well-oxygenated due to the ability of water to hold greater concentrations of oxygen at lower temperatures. This strongly influences the ocean hydrosphere as oxygen is a major requirement for organisms and a greater availability of oxygen, particularly on the ocean floor, results in more aerobically-based biological activity.

This in turn directly affects the global carbon cycle since the higher the oxygen level the lower the amount of carbon captured on the ocean floor.

While icehouse modes represent a significant period of Phanerozoic time, the greenhouse state represents the norm for the Earth amounting to ~70% of Phanerozoic time. In this state the Earth's climate is characterised by low pole to equator temperature gradients and consequent sluggish circulation in the oceans and atmosphere. Although the thermohaline circulation is weaker there may be alpine glaciation and some polar sea ice (Kidder and Worsley, 2010). Oceans are still considered to be well-oxygenated but this diminishes as the Earth moves towards a hothouse state.

The hothouse mode, in contrast to both the icehouse and greenhouse modes, is rare representing only ~5% of Phanerozoic time (Kidder and Worsley, 2010). Although hothouse conditions have only occurred for periods of short duration during Phanerozoic Earth history they are associated with some of the largest extinction events recorded. In a hothouse world oceanic circulation is driven by evaporative processes in the equatorial regions that create warm, dense, saline water that sinks in the mid-latitudes creating warm saline bottom waters (Kidder and Worsley, 2010; Zhang *et al.*, 2001). The lower oxygen content of warm water favours anoxia on the ocean bottom resulting in drawdown of carbon dioxide and significant carbon capture on the ocean floor.

One fundamental question that scientists are still trying to fully understand pertains to the drivers forcing Earth's climate from the normal greenhouse state to either icehouse or hothouse conditions. The converse question is also important, namely what are the processes that enable the Earth to return from either icehouse or hothouse to a greenhouse state?

Kidder and Worsley (2010) suggest that the greenhouse state can be forced into icehouse conditions by global tectonic processes – large-scale continental collisions that alter both the flux of global geochemical processes and, in particular, the carbon cycle. A review of the carbon cycle through geological time shows a strong link between global temperature and atmospheric carbon dioxide (Royer, 2006). The reasoning behind the suggestion of Kidder and Worsley (2010) is that continental collision, and consequent mountain-building events, result in the climate being forced into an icehouse state as a result of increased exposure to weathering of unaltered crustal rocks (silicates) at the Earth's surface. For every two moles of carbon dioxide needed to weather silicates one mole of carbon dioxide is back-produced and re-enters the atmosphere, whilst the second is consumed, which in turn serves to reduce pCO_2 in the atmosphere. The availability of unweathered silicates over geological timescales, combined with high moisture, favours the drawdown of carbon dioxide and global cooling. Mountains formed in arid zones will have less impact on carbon dioxide drawdown than those created in tropical areas with higher rainfall. This illustrates some of the complexity in the Earth system.

Recent estimates of global weathering and carbon dioxide drawdown suggest as much as 50% of the current drawdown occurs in active mountain belts with nearly half of that accounted for by the Himalayan mountain range (Hilley and Porder, 2008). These observations provide support for an important link between climate and mountain building through forcing of the carbon cycle (Dettman *et al.*, 2003; France-Lanord and Derry, 1997; Harris,

2006). This further reinforces the role of the uplift of the Himalayas on Cenozoic climate cooling (Raymo and Ruddiman, 1992).

Other feedbacks within the weathering process can also be influential. For example, weathering also results in more fertile soils and thus higher nutrient availability resulting in greater primary plant productivity that further draws down atmospheric carbon dioxide. One reinforcing feedback loop is that once in icehouse mode, the increased pole to equator temperature gradient promotes higher planetary wind speeds, an increase of windblown material into the oceans and stronger upwelling zones. Terrestrial windblown material acts as an iron fertilizer on the ocean surface and increases primary productivity, while stronger upwelling zones create areas of high productivity so furthering the drawdown of carbon dioxide from the atmosphere and forcing global cooling (Falkowski, 1997; Martin, 1990; Sigman and Boyle, 2000). The influence of these processes on the carbon cycle is complicated, and the actual magnitude of any effect on atmospheric carbon dioxide is still widely debated (Sigman and Boyle, 2000).

During icehouse modes the influence of ice sheets on the Earth system is profound. The freezing of sea water generates freshwater ice with salt being driven off during the freezing process. This results in increased concentration of salt, changing sea water salinity and density. The dense cold, salty water sinks to the ocean floor to create cold bottom water. The creation of cold bottom water is reinforced by ice shelves that also produce cold dense water at their base from basal melting processes. Both of these processes drive oceanic circulation today through the thermohaline system that interconnects polar water masses into a global oceanic conveyor belt (Broecker, 1987, 1991a, 1991b; Lozier, 2010; Richardson, 2008). Primarily, this system acts as a giant heat pump moving energy around the globe. The influence of the oceanic conveyor belt on continental rainfall patterns can be significant, for example, along the west coast of Africa today upwelling of cold Antarctic bottom water (Benguela Current) results in cold surface waters that influence evaporation creating dense coastal fogs. The strength of the Benguela Current, flowing along the southwest coast of Africa, and interaction with the Aghulas Current, flowing down the east coast of Africa, strongly influence rainfall patterns in southern Africa (Partridge, Scott and Schneider, 2004).

While the oceanic system controls heat fluxes around the globe it also has a major effect on the global ecosystems influencing the distribution of nutrient upwelling areas and consequently the distribution of organisms. The integral nature of ice sheets to Earth system processes means that processes forcing change in those ice sheets are important over both longer time frames and a wider geographical area than just the southern (or northern) polar regions. Orbital forcing parameters (eccentricity, obliquity, precession)[1] determine the amount of radiation the Earth receives, and where it is directed over time. These Milankovitch cycles[2] are more attenuated and influential in periods when polar ice caps

---

[1] The *eccentricity* of the Earth's orbit is its departure from a circular orbit and has a frequency of about 100,000 years, whereas the *obliquity* is the tilt of the Earth's axis away from a vertical drawn to the plane of the planet's orbit, it has a frequency of 41,000 years. The *precession* is the wobble of the Earth's axis defined as the change in the orientation of the rotational axis of the Earth.

[2] That is, the collective effects of changes in the Earth's movements, namely the eccentricity, obliquity and precession of the Earth's orbit, upon the climate.

exist, resulting in alternating glacial–interglacial cycles. Glacial and interglacial cycles strongly determine ice volumes that in turn control sea levels, and thus play an important role in the dynamics of lowland habitat at evolutionary scales.

Icehouse worlds with polar ice sheets have characterised a number of periods in Paleozoic Earth history and can be broken into two groups. The first group is marked by a rapid cooling and rapid warming and are of relatively short duration. This includes the Late Ordovician icehouse (Hambrey, 1985), which is currently suggested to have been as short as 1 Ma in duration (Brenchley *et al.*, 1994), as well as the glaciations of the Late Devonian (Fammenian; Veevers and Powell, 1987), early Carboniferous (Visean) and periods in the Jurassic and Cretaceous (Frakes and Francis, 1988) that all lasted only a few million years (Veevers and Powell, 1987). This contrasts with the long-lived glaciations of the Phanerozoic, with that of the Carboniferous and Permian estimated at 65 Ma in duration (Fielding *et al.*, 2008; Veevers and Powell, 1987; Visser 1990, 1997) as well as the recent Cenozoic glaciation that began in the late Eocene (Zachos, Stott and Lohmann, 1994; Zachos *et al.*, 2001) or early Oligocene, some 40 to 35 Ma ago and continues today.

Our understanding of the complex interactions between biosphere, atmosphere, hydro-sphere and geological processes generates new insights into the dynamics of icehouse modes where the role of orbital parameters can overprint warmer (interglacial) and cooler (glacial) periods on longer-term trends. Continued study of the dynamics within the present icehouse phase is furthering understanding of past icehouse phases.

Researchers have suggested that the onset of glaciations can be forced by the combined influence of weathering associated with mountain building (e.g. Finlay, Selby and Gröcke, 2010; Kidder and Worsley, 2010; Kump *et al.*, 1999) whilst others argue that this only applies to long-lived global glaciation events (e.g. Gibbs, Barron and Kump, 1997) and that the short-term glaciations such as those of the Late Ordovician can be driven by other processes. For the Late Ordovician, modelling experiments suggest that the glaciations were the response to short-term changes in atmospheric carbon dioxide. Ordovician atmospheric carbon dioxide levels are thought to have been up to 14 times present day levels and the climate remained ice-free (Gibbs, Barron and Kump, 1997). Yet modelling experiments suggest that a drop in carbon dioxide to 10 to 12 times present day levels resulted in the Earth cooling, and this cooling was reinforced by strong feedback from ice albedo effects resulting in glaciation even though pCO_2 was high (Cowley and Baum, 1995; Gibbs, Barron and Kump, 1997). Thus rapid changes in carbon dioxide, even at atmospheric levels 10 times those of the present day, could result in sudden onset of glaciation for a short period of time followed by rapid termination.

Gibbs, Barron and Kump (1997) argue that other short-term forcing processes can also result in glacial events. Volcanism (Buggisch *et al.*, 2010; Young *et al.*, 2009) can force large quantities of ash into the atmosphere and so lead to cooling of the Earth. Volcanism has also been used to explain a number of perturbations in the Earth's climate state through geo-logical time, and certainly large-scale volcanic events are often coincidental with extinctions and rapid climate shifts. However, the causal effect of such eruptions is yet to be fully demonstrated. This is not surprising given the complexity of response by the Earth system to even small-scale volcanic eruptions. Well-studied recent examples include the eruption of Pinatubo (in 1991) that resulted in global cooling for several years (Christiansen, 2007). Although the globe cooled overall, the effect of the Pinatubo eruption was highly regional with winter warming in some areas and summer cooling in others. Other effects included modified rainfall patterns (Spencer *et al.*, 1998; Trenberth and Dai, 2007). Yet even though by comparison with the geological past, when, for example, volcanism associated with the emplacement and eruption of the large igneous provinces were several orders of magnitude greater in both quantity and duration, the Pinatubo eruption was relatively minor and yet the Earth still took several years to return to normal (Hansen *et al.*, 1996). The effects of volcanic eruptions are complex and depend on a number of factors such as the scale of the eruption, height of the eruptive column, location of the eruption (tropical *versus* high latitude), chemistry of the volcanic ash (e.g. high sulphur content) and precursor state of the climate system. The precursor state of the climate system could be critical since an eruptive event may force the climate system through a tipping point and result in a cascade of feedbacks that result in a shift to a new climate state.

Over geological timescales polar ice cap formation has also been influenced by continental configurations and the positions of continents both latitudinally and in relation to each other. A classic example can be seen in the break-up of Gondwana where the final stages of fragmentation resulted in the isolation of Antarctica in the high southern latitudes, and the development of a circumpolar oceanic current (Chapter 9). Both the position of Antarctica relative to the other gondwanan continental fragments, coupled with its atmospheric and oceanic isolation, resulted in a reduction in heat flow from the tropics to the poles allowing Antarctica to cool and an ice cap to form[3] (Zachos *et al.*, 2001) with wide-scale knock-on effects including those affecting the terrestrial ecology. In contrast to a greenhouse mode devoid of ice caps, periods of ice cap presence probably have different evolutionary dynamics from periods without ice caps, due to sea level rise and fall leading to rapid fluctuations in lowland habitat resulting in expansion and contraction of ecological space. Moreover, fluctuations in continental rainfall resulting in geographic expansion and contraction of rainfall regimes (as illustrated by those seen today in southern Africa) rapidly influence environments, and therefore act as ecological drivers. Finally, the greater amplitude in pole to equator temperature gradients results in more biomes and thus globally greater niche partitioning.

Early Paleozoic climate

Critical to understanding early terrestrial ecosystems and environments is an accurate understanding of climate change during the Ordovician and Silurian (Munnecke *et al.*, 2010). During the earliest Paleozoic the climate was gradually warming to become strongly zonal with warm oceans (e.g. Pruss *et al.*, 2010) by the end of the Cambrian. In the

---

[3] Research also suggests that a drop in greenhouse gases past a critical threshold may have been more important in initial formation of the ice sheet on Antarctica (DeConto and Pollard, 2003; DeConto *et al.*, 2008). These ice sheets initially waxed and waned but a positive feedback loop, due to the reflection of heat back into space from the ice sheet, reinforces the cycle such that the globe cools and, ultimately, Antarctica passes into an icehouse mode. See Chapter 9 for further details.

Ordovician, conditions were generally considered to have been warm with the warmer conditions of the early Ordovician interrupted by a short-lived cool interval (Frakes, Francis and Skytus, 1992). Geochemical proxies are now confirming an overall decrease in Ordovician temperatures (Trotter *et al.*, 2008) with cooler waters perhaps being relatively hospitable and allowing a major diversification of marine life in the Ordovician (Sepkoski, 1981; Trotter *et al.*, 2008). At the Ordovician–Silurian boundary this generally warm period terminated in a short but severe icehouse that resulted in a drop in sea level and mass extinction in the oceans (Saltzman and Young, 2005; Sheehan, 2001; Trotter *et al.*, 2008).

Climate development during the Silurian, following the early Paleozoic icehouse, is still a matter for debate although periods of relative humidity alternating with relative aridity have been suggested (Jeppsson, 1990; see also Aldridge, Jeppsson and Dorning, 1993). This model has been used to explain the Silurian stable carbon and oxygen isotope record that indicates widespread organic carbon burial resulting in global positive isotopic excursions during this time (e.g. Cramer, Saltzman and Kleffner, 2006 and references therein), although not without criticism (Munnecke *et al.*, 2010 and references therein). Nevertheless sea level certainly changed significantly during this time but the interactions of sea level change, stable isotope geochemistry and extinction events are still poorly understood (Cramer and Munnecke, 2008; Munnecke *et al.*, 2010). Tectonic movements resulted in relatively warm shallow seas, which in turn helped plants in their colonisation of land around the continental margins. The presence of plants (see below) resulted in changes to the atmosphere with an increase in atmospheric oxygen levels and the beginnings of a decrease in pCO$_2$ that would continue on until the Carboniferous.

By the Devonian, the climate was exhibiting relatively warm climate conditions. Classical views suggest that the Middle Devonian was characterised by a hothouse climate (Copper, 2002; Copper and Scotese, 2003), whereas more recent data supports a relatively cooler interval in the late Early and Middle Devonian, possibly brought about by variations in pCO$_2$ levels (Joachimski *et al.*, 2009). Atmospheric CO$_2$ levels are suggested to have decreased from ~2000 ppmv in the Early Devonian to 900 ppmv in the Middle Devonian (Simon *et al.*, 2007). In the Late Devonian, palaeontological and geochemical data support a global warming as a consequence of tectonic events and enhanced volcanic CO$_2$ degassing (van Geldern *et al.*, 2006) that allowed tropical floras to spread into higher latitudes (Streel *et al.*, 2000).

Palaeogeography

The formation of by far the largest palaeocontinent up until this time (i.e. Gondwana) during the lower Paleozoic (Cambrian) (Cocks, 2001) saw the destruction of the Mawson Sea through convergent tectonic processes that resulted in collision between West Gondwana (the cratons including those of present day Africa, India and South America) and East Gondwana (the Australian and Antarctic cratons). The final amalgamation of Gondwana took place during the Cambrian such that at about 540 Ma the East Antarctic craton lay between the equator and 30° S, with Antarctica forming part of the Palaeopacific margin. Through the Cambrian, Gondwana moved south and by the Ordovician most of West Gondwana lay directly over the South Pole. In East Gondwana, Antarctica was marked by a coastal range that formed through accretion of terranes and subduction processes, while offshore continental island arcs with volcanoes formed as a result of subduction. During this time the terrestrial realm comprised mainly low-lying coastal alluvial plains. As sea level rose it encroached upon these areas forming vast shallow epicontinental seas widely distributed between continents of low relief (Munnecke *et al.*, 2010). In tropical latitudes limestones accumulated. During the Ordovician the sea level subsided before flooding the continents again in the Middle Ordovician to reach their highest point ever recorded.

In the Ordovician, Antarctica continued to lie between the equator and 30° S still forming part of the recently amalgamated Gondwana with the main landmasses of South America and Africa lying at mid-palaeolatitudes (Torsvik and Cocks, 2004). Beginning in the Silurian, Gondwana began to drift southward towards the South Pole so that by the end of the Silurian the supercontinent extended from equatorial latitudes to the South Pole (Munnecke *et al.*, 2010) whilst Antarctica lay between 30° and 60° S. In more equatorial regions other continental masses (Laurentia and Baltica and later Avalonia) drifted towards one another to form the second supercontinent, Euramerica[4].

The Devonian was a time of great tectonic activity. Near the equator, Gondwana and the relatively minor supercontinent, Euramerica, rapidly drew closer together to begin the early stages of assemblage of Pangea as the oceans of the early Paleozoic began to close (Cocks and Torsvik, 2002). High sea levels swamped much of the low-lying land to form shallow seas. Throughout the early Paleozoic the general combination of relatively warm climate and low-lying land provided the perfect conditions for plant life to flourish.

Palaeovegetation

Plant colonisation of the land

While there is an extensive fossil record that can be linked to geochemical changes in the Earth system, as the land turned green, first with the primitive land plants of the Ordovician and later with vascular plants of the Silurian, evidence for early colonisation of the land is sparse. Evidence for terrestrial colonisation has undergone a series of paradigm shifts that have pushed the terrestrial plant fossil record earlier and earlier. Prior to the development of palynological techniques the earliest known land plants were macrofossils such as *Cooksonia* of the latest Silurian age (Edwards, Feehan and Smith, 1983). Largely based on Northern Hemisphere successions, a picture emerged of increasingly more complex life forms evolving through the Devonian lending support to the emerging hypotheses of plant–climate feedback mechanisms. However, with the advent of palynology as a tool to investigate the fossil record, the origin of terrestrial vegetation was pushed back to the earliest

---

[4] Also known as Laurussia.

Silurian (Hoffmeister, 1959) – a discovery that was met by a fair degree of scepticism at the time. This was followed by further discoveries in the early Silurian (e.g. Gray and Boucot, 1971) and an even more controversial claim for an Ordovician origin of terrestrial plants (Gray, 1985, 1988). Today, however, it is accepted that the earliest evidence of colonisation of the land is Middle Ordovician (Rubenstein *et al.*, 2010; Steemans and Wellmann, 2003; Wellman, Osterloff and Mohiuddin, 2003), but controversial claims for Middle Cambrian colonisation have also been made (Strother, 2000; Taylor and Strother, 2009).

To date, the global evidence for the colonisation of land begins in the Middle Ordovician and is marked by the appearance of fossilised primitive plant spores or cryptospores (Rubenstein *et al.*, 2010; Steemans and Wellmann, 2003; Wellman, Osterloff and Mohiuddin, 2003). Cryptospores are small permanent tetrads, dyads or monads that may be naked, or enclosed within an envelope-like structure. They are generally found in non-marine strata, but when found in marine rocks their abundance significantly decreases with distance from shore. Such observations strongly indicate that cryptospores originated from a terrestrial source (Wellman, Osterloff and Mohiuddin, 2003). Support for a terrestrial origin for the spores is found in the degradation-resistant casing composed of sporopollenin. As sporopollenin is considered a synapomorphy of all land plants (Wellman, 2003a), the composition of these fossil spores suggests that they were derived from terrestrial plants. Moreover the anatomy of the spores revealed by transmission electron microscopy reveals strong similarities to extant liverworts (Wellman, Osterloff and Mohiuddin, 2003) also suggesting that they are derived from plants at this evolutionary grade.

Sporopollenin is a chemically very stable organic macromolecule that preserves well in soils and sediments. The evolution of a sporopollenin casing would have increased the effectiveness of land plants in drawing down carbon dioxide from the atmosphere by providing a mechanism for increasing the amount of carbon in the sediment. The dramatic advances in our knowledge regarding the colonisation of the land are based on the micro-fossil record, yet the macrofossil record is still remarkably poor with the earliest significant plant macrofossils not appearing until the late Silurian and earliest Devonian (Edwards, Feehan and Smith, 1983). This is probably due in part to the earliest plants lacking extensively lignified tissues and thick cuticles that would favour preservation. So our knowledge of the early phases of colonisation of the land is built on changes in the spore record. In the Late Ordovician, cryptospores become more widespread globally (e.g. Richardson and Ausich, 2007) and were joined in the latest Ordovician by new spore types, hilate and trilete spores that testify to an increasing plant diversity and adaptation to the terrestrial environment (Steemans *et al.*, 2009). By mid-Silurian times trilete spore types had become dominant, whilst cryptospore-producing plants adapted and evolved to terrestrial environments where they would persist until the latest Silurian and on into the earliest Devonian (e.g. Edwards and Richardson, 2004).

Our knowledge of these Ordovician and Silurian fossils is also limited in terms of the rock record and regions examined to date. Indeed the recent discoveries by Rubenstein *et al.* (2010) of cryptospores in the Early Ordovician of Argentina have not only changed the view that these plants had already appeared in western Gondwana some 480 million years ago, but also increased understanding regarding the timing of their appearance since the Argentinian material predates the records from the other side of the continent by some 8 to 12 Ma (Rubenstein *et al.*, 2010; Wellman, 2010). Unravelling the early history of colonisation of the land remains difficult. Lower Paleozoic rocks are often more thermally altered and so more difficult to process successfully for evidence of plant fossil remains. Furthermore, palaeobotanical research has been largely concentrated in the Northern Hemisphere, due to the geographical location of researchers, with the consequence that many Southern Hemisphere sequences have only been examined cursorily.

The geographical origin of these early terrestrial land plants is thought to be Gondwana largely due to the concentration of localities rich in cryptospores from sequences in present day Saudia Arabia (Strother, Al-Harji and Traverse, 1996), Oman (Wellman, Osterloff and Mohiuddin, 2003), Turkey (Steemans, le Hérissé and Bozdogan, 1996), Czech Republic (Vravrdová, 1990; Wellman and Gray, 2000) and Argentina (Rubenstein *et al.*, 2010) – regions located on the eastern and western margins of Gondwana (Figure 2.1). Since Antarctica during the Ordovician lay geographically between the two regions where the earliest land plants have been discovered, it raises the question as to whether the rocks in Antarctica would also yield important evidence of the early stages of land colonisation. Yet, to date, no detailed search for cryptospores from Cambrian to Ordovician rocks from

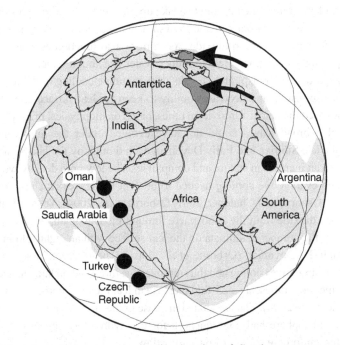

Figure 2.1 Ordovician palaeogeography showing location of Gondwana cryptospore sites in the Middle Ordovician (black dots) and regions of Antarctica with similar age sediments (dark grey area arrowed).

Antarctica has been made even though Cambrian to Ordovician sequences are widespread, and extend from Victoria Land through the Transantarctic Mountains to the Ellsworth and Pensacola Mountains and Shackleton Range.

Before a thorough exploration of any potential site in Antarctica can be undertaken some knowledge is needed regarding the depositional environments from which cryptospores of this age have already been recovered elsewhere. The cryptospore-rich middle Ordovician sequences in Oman (Wellman, Osterloff and Mohiuddin, 2003), Saudia Arabia (Strother, Al-Harji and Traverse, 1996), Turkey (Steemans, le Hérissé and Bozdogan, 1996) and Argentina (Rubenstein *et al.*, 2010) all accumulated on a passive continental margin in near-shore environments including deltas, tidal estuaries, marginal marine and sub-aerial environments. The Middle Cambrian occurrences (Strother, 2000) of putative cryptospores also occur in a similar environment. Moreover, a number of the sequences where crypto-spores have been extracted are thermally immature and, although spores are fairly resistant, the higher the degree of thermal alteration of the rock the more difficult it is to extract well-preserved palynomorphs. Therefore the ideal search profile for Antarctic occurrences would be non-marine to marginal marine settings that are little altered thermally.

The rise of vascular plants: increasing stature and status

Today vascular plants (and some bryophytes) take up carbon dioxide through stomata on leaf and stem surfaces. In a high carbon dioxide environment fewer pores are required to satisfy the photosynthetic demand for carbon dioxide. As stomata also function as sites for water loss through transpiration, fewer pores help to reduce water loss, but a deficit of stomata has an unexpected consequence. Transpiration results in evaporative cooling of the leaf surface and so fewer pores also results in a decreased ability to keep the leaf surface cool. While plants undeniably colonised the land in Ordovician times, based on biophysical principles it has been demonstrated that broad leaves could not have evolved under the high carbon dioxide conditions of the late Silurian and Early Devonian even though by the late Silurian the plant record testifies to an increase in stature and complexity. The low pore density would not have provided enough evaporative cooling needed to prevent the leaves reaching lethal temper-atures (Beerling, Osborne and Chaloner, 2001; Osborne *et al.*, 2004). Yet as plants continued to remove carbon dioxide from the atmosphere through the Devonian, this in turn drove the plants to increase the number of stomata on the leaf surface with an eight-fold increase being recorded through the Devonian (Osborne *et al.*, 2004). The increase in stomatal densities allowed greater evaporative cooling and thus removed the threshold on large leaf development such that over the same timescale a 25-fold increase in leaf area is recorded in the plant fossil record (Osborne *et al.*, 2004). The associated increase in the amount of water transpired from plants into the atmosphere had consequences for the hydrological cycle with an increase cycling of water reinforcing carbon dioxide drawdown.

As leaf size and stomatal densities increased so too did the complexity of the plants. Large leaves supporting high numbers of stomata require a transpiration stream to avoid wilting, and so plants need a well-developed root system to transport water to the leaves and increased structural tissues to support leaves of increasing size. This increase in stature is associated with the evolution of vascular tissues and the development of secondary growth that allowed water to be transported more efficiently through the plant. Vascular tissue is lignified, and the origin of the lignin macromolecule had several important implications for the interactions between plants and the Earth system. The evolution of lignin provided land plants with the necessary tool to create vascular tissue and secondary growth. Indeed the origin of secondary growth and the increase in stature of plants is seen as one of the major drivers in drawing down atmospheric carbon dioxide through the Phanerozoic (Beerling and Berner, 2005; Berner, 1998, 2003, 2006). This was achieved both through the increased biomass in the above ground parts and also the development of roots. The development of secondary growth further increased the depth of the soil profile (the rhizosphere) and in turn rock weathering reinforced carbon dioxide drawdown (Beerling and Berner, 2005). Furthermore, the evolution of resistant carbon molecules such as lignin (and cuticles) would have favoured an increased capture of carbon within sediments thus providing a mechanism for fixing atmospheric carbon and locking it up thereby slowing the return of carbon to the atmosphere.

Evidence from Antarctica

As the final amalgamation of Gondwana took place during the Cambrian, it was accompanied by a shift from passive margin to convergent margin tectonics (Ross Orogen) in that part of Antarctica covering Victoria Land through the Transantarctic Mountains to the Ellsworth and Pensacola mountains and Shackleton Range. At a continental scale the dynamic nature of the tectonics had a varying influence on sedimentation patterns from region to region. This makes it difficult to generalise about the palaeoenvironments. However, the creation of island arc systems saw Cambrian limestones replaced by shallow marine sandstones accompanied by deep marine turbidite sequences as back-arc basins deepened and large quantities of sediment were eroded from the continent in late Cambrian and Ordovician times. Ultimately the continental and island arc systems and associated marine basins were accreted to the Antarctic margin during the activity associated with the Ross Orogeny. In most regions this manifested as packets of Early to late Cambrian sediments that are often regressive in nature and are usually separated unconformably from overlying packets of sediments ranging from Ordovician up into the Devonian. These overlying sediments are generally poorly age constrained since they are largely unfossiliferous, overlain unconformably by Devonian or younger strata, and lack rocks suitable for radiometric dating. This is further compounded by the discontinuous nature of the outcrops such that the relationships between the units are often inferred rather than observed. In spite of these limitations, recent dating studies are starting to elucidate the age range of the strata (e.g. Goodge, Williams and Myrow, 2004). The sequence of rocks extending from Victoria Land to the central Transantarctic Mountains that unconformably overlies Cambrian strata is

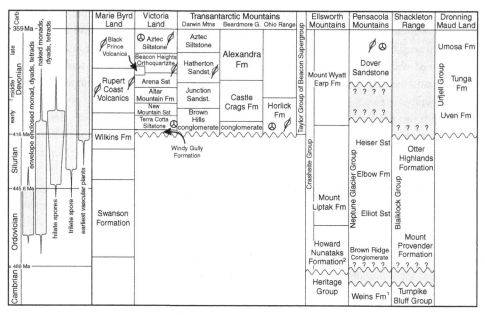

Figure 2.2 Early Paleozoic stratigraphic units in Antarctica with key events in the colonisation of the land marked by the appearance and radiation of cryptospores (left panel). Grey shaded area in right-hand columns indicates gap due to erosion or non-deposition. The presence of fossils is marked by a ⊕ for palynological assemblages or ∅ for macrofossils. Stratigraphic scheme for the central Transantarctic Mountains (Bryd Group) follows Goodge, Williams and Myrow (2004) and Myrow *et al.* (2002); Ellsworth Mountains follows Curtis and Lomas (1999). Abbreviations as follows: G. = Glacier; Sst = Sandstone. 1 = The Weins Formation is a fluvial to shallow marine unit that unconformably overlies a volcanic to volcanoclastic unit (Gambacorta Formation) the Gambacorta Formation overlies Middle Cambrian Limestones (Nelson Limestone Formation). 2 = Note the Howard Nunataks Formation is subdivided into the Linder Peak Member, Mount Twiss Member and Landmark Peak Member (Spörli, 1992).

dated as Early Ordovician and so is unlikely to contain a history of the early colonisation of the land. However, in the Ellsworth and Pensacola mountains sedimentation seems to be more continuous and the identification of potential cryptospore-rich sequences more likely.

In contrast to Victoria Land and the Transantarctic Mountains, the Ellsworth Mountains region records more or less continuous sedimentation and gentle subsidence from the Cambrian to Permian. The Crashsite Group (Howard Nunataks Formation) conformably to locally unconformably overlies the Heritage Group (Cambrian) (Curtis and Lomas, 1999) (Figure 2.2). The Howard Nunataks Formation is a complex packet of sediments that can be split into at least three members (Landmark Peak Member, Mount Twiss Member, Linder Peak Member; Spörli, 1992). There are several other unnamed wedge-shaped sandstone packages that interfinger with the Mount Twiss Member, and sedimentary features include cross bedding, mudcracks, ripple marks and thin pebble or mudflake conglomerates. These sedimentary features along with the geometry of the beds indicate accumulation in a fluvial to very shallow marine setting that was tidally influenced (Goldstrand *et al.*, 1994) with areas subject to subaerial emergence (Spörli, 1992). The Howard Nunataks Formation is overlain by the Mount Liptak Formation (Figure 2.2), a unit that also accumulated in shallow marine to fluvial settings. The clean nature of the quartzites indicates a more mature source and suggests a more stable and possibly subdued landscape. The Mount Liptak Formation is in turn overlain by the Mount Wyatt Earp Formation, deposited in a similar environment but from a more immature source (Spörli, 1992) suggesting reactivation of mountain building processes on the basin margins. The shallow marine, tidally influenced and terrestrial environments seen in the basal Crashsite Group are similar to sequences where cryptospores have been recorded elsewhere in Gondwana, and apart from the thermal alteration of the sequences would be good candidates for further investigation for evidence of land plant colonisation.

Similar sequences of rocks are recorded from the Pensacola Mountains where Middle Cambrian limestones (Nelson Limestone), late Cambrian volcanics (Gambacorta Formation) and overlying clastic sediments (Weins Formation) are unconformably overlain by the Neptune Glacier Group (Figure 2.2). The age constraints on the Neptune Glacier Group are poor: the upper surface is disconformably overlain by Dover Sandstone of Late Devonian age (Schmidt and Ford, 1969) while it rests unconformably on late Cambrian strata implying an Ordovician to Devonian age range. The Neptune Glacier Group contains a diverse range of lithologies, and the four stratigraphic units recognised are thought to represent a high-energy alluvial fan conglomerate (Brown Ridge Conglomerate) with flanking alluvial deposits (Elliot Sandstone, Elbow Formation, Heiser Sandstone) representing quieter sedimentary conditions (Laird, 1991).

Whilst Antarctic sequences have not yet revealed information about the early colonisation of land, the potential nevertheless exists in several regions where the right sedimentary environments occur. Given the discoveries of cryptospores on the western and eastern margins of Gondwana, knowledge from the Antarctic regions may be important in understanding how the plants that produced these spores spread across the continent.

The increase in size and complexity of plants in the Devonian took place as Ross orogenic activity in the Antarctic waned. By the Early Devonian an unconformity surface (Kukri Erosional Surface) had developed with considerable relief (Isbell, 1999). Topography and landscapes varied but were strongly influenced by the underlying geology as attested to by the diversity of sedimentary settings. Where Cambrian limestones were exposed, well-developed karst landscapes formed with significant relief. Anderson (1979) observed over 700 m of relief on the Shackleton limestone in the Byrd Glacier area (Figure 1.3). In other areas of high relief sediment was shed from the mountainous hinterland as evidenced by braided stream environments (Anderson, 1979). Areas of low relief were also present and basement highs often have highly weathered granite beneath the overlying sediment suggesting that they were exposed and weathered for a considerable period of time. These may have formed granite monadocks and other similar landforms. Marine sediments are also present in areas such as the Ohio Range (Bradshaw, 2002, 2010) and extending towards the Pensacola Mountains. More controversially deposits in southern Victoria Land have been interpreted as marine in parts (Bradshaw, 2010; Bradshaw, Newman and Aitchison, 2002).

The initial sites of sedimentary accumulation ultimately formed precursor sedimentary basins prior to the development of a large foreland basin system that extended from Victoria

Land and the Transantarctic Mountains to the Ellsworth and Pensacola mountains, Dronning Maud Land and into Africa along the Palaeopacific margin of Gondwana (Isbell, 1999). An analysis of the sedimentation patterns suggests the Taylor Group, the basal part of the Beacon Supergroup, accumulated in two separate precursor basins, one extending from the Beardmore Glacier into Victoria Land (McMurdo Basin; Barrett 1991), the other from the Beardmore Glacier towards the Ellsworth Mountains (Ellsworth Basin; Barrett, 1991) (Figure 1.3).

The McMurdo Basin is the more terrestrial of the two basins with sedimentation developing over the pre-existing topography in Early Devonian times. The nature of the sedimentary sequences has been the subject of considerable debate with several workers regarding the sequence as shallow marine based on the trace fossil assemblages (e.g. Bradshaw, 1981; Gevers and Twomey, 1982), whilst others regard the sequence as largely fluvial with lacustrine units (e.g. McPherson, 1978, 1979; Woolfe, 1993; Woolfe *et al.*, 1990). The classic succession is from southern Victoria Land and consists of series of sandstone and siltstone units. The lower units include trough cross-bedded sandstones that have unidirectional currents with the sheet-like nature of the sandstones suggesting deposition on an unconstrained braidplain (Woolfe *et al.*, 1990). However, elsewhere in the Transantarctic Mountains, such as in the Darwin Mountains (Brown Hills Conglomerate; Woolfe, 1993) and on the Starshot Glacier (Bradshaw, Newman and Aitchison, 2002), the basal deposits consist of conglomerates that are overlain by sandstones (e.g. Castle Crags Formation; Bradshaw, Newman and Aitchison, 2002 and Figure 2.2; Junction Sandstone; Woolfe, 1993) can, in part, be correlated with the New Mountain Sandstone and Altar Mountain Sandstone of southern Victoria Land. The outcrop of these Devonian units extends over hundreds of kilometres and relationships between the sandstone bodies at this scale are difficult to establish. Nevertheless, all the units represent deposition in fluvial environments (Bradshaw, Newman and Aitchison, 2002) by braided streams (Woolfe, 1993). These braided-stream sandstone cycles are often terminated by mudstones that show evidence of desiccation cracks and attest to standing pools of water (Woolfe, 1993). In contrast to the lower part of the Taylor Group, in Victoria Land, the upper part (Aztec Siltstone) was formed by high sinuosity meandering river systems that periodically formed lake systems (McPherson, 1978, 1979). Well-developed soils suggest a heavily vegetated landscape in distinct contrast to the lower part of the sequence, and it is likely that the shift from braided streams to meandering rivers is in some way due to increased vegetation binding the sediment and reducing sediment supply. Palaeosol horizons have been widely reported in the Aztec Siltstone (Barrett *et al.*, 1971; McKelvey, Webb and Kohn, 1977; McPherson, 1978, 1979; Retallack, 1997) with soils recorded as being >1.5 metres thick in places with extensive and wide root traces (up to several centimetres) that may contain evidence of woody tissue (McPherson, 1979). Retallack (1997) interpreted the palaeosols as being similar to modern Alfisols and Luvisols and suggested that they formed in a warm subtropical to tropical climate with annual rainfall of 779 ± 141 mm. The climate was probably seasonal as indicated by the types of soil concretions (Retallack, 1997). The amount of rainfall is also supported by animal fossils since the Aztec Siltstone is best known for its diverse fossil fish fauna (Ritchie, 1975; Young, 1988, 1991; Young, Long and Ritchie, 1992).

When this fish fauna is compared with Australian sequences a late Middle Devonian (Upper Givetian) age is suggested for the Aztec Siltstone (Young, 1993).

Sedimentology and palaeoclimatology suggest a shift in sedimentation and environments through time within the McMurdo Basin (Victoria Land). Unfortunately, evidence for changes in vegetation is sparse with relatively few macrofossil (Edwards, 1990) and palynofloral records (Playford, 1990). Within the lower part of the Devonian sequence a poorly preserved palynoflora is recorded in the Terra Cotta Siltstone (Kyle, 1977) that includes the spores *Emphanisporites* and putative *Perotriletes*, a cavate form with a circumpolar annular thickening (the application of the name *Perotriletes* to the Antarctic material may be incorrect as this genus is defined as zonate not cavate; Balme, 1995). These spore types are common elsewhere in the Early Devonian and have been recorded in association with macrofossils with *Emphanisporites* being found in the sporangia of *Horneophyton* (Gensel, 1980), and *Perotriletes* being found in rhyniophytes (Gensel 1980). This limited evidence suggests a flora dominated by primitive rhyniophytes, a situation similar to other parts of the world at this stage of Earth history. Slightly younger plant macrofossils are known from the Beacon Heights Orthoquartzite (Victoria Land) (Harrington and Speden, 1962; McKelvey *et al.*, 1972; McKelvey, Webb and Kohn, 1977; McLoughlin and Long, 1994; Plumstead, 1962) and equivalent sandstone units such as the Hatherton Sandstone (Transantarctic Mountains) (Woolfe *et al.*, 1990). These fossils are mainly lycopods. Plumstead (1962) referred to 50 mm-long axes with circular or oval spirally inserted scars as cf. *Protolepidodendron lineare* and *Haplostigma irregularis*, although Edwards (1990) questioned the assignment of this material and suggested that it would be better left as some sort of indeterminate lycopod stem. New material from both the Beacon Heights Orthoquartzite and the overlying Aztec Siltstone prompted a revision of the material (McLoughlin and Long, 1994). These authors recognised three taxa from the Beacon Heights Orthoquartzite including a new combination *Haplostigma lineare* (Figure 2.3A, D) based on both the material described by Plumstead (1962, 1964) as cf. *Protolepidodendron lineare* and new material, a second lycophyte taxon, *Malanzania*, (Figure 2.3C), and a larger form referred to *Archaeosigillaria* sp. as cf. *A. caespitosa* (equivalent to indeterminate lycopod axes of Grindley, Mildenhall and Schopf, 1980). The youngest part of the Devonian sequence (Aztec Siltstone) has both macrofossils (McLoughlin and Long, 1994) and palynomorphs (Helby and McElroy, 1969). The macroflora consists of *Praeramunculus alternatiramus* (Figure 2.3B), a taxon that was compared with psilophyte material described from Marie Byrd Land (Grindley, Mildenhall and Schopf, 1980) and *Haplostigma lineare*, a lycophyte of moderate stature with stems up to 25 mm wide and 88 mm long. The palynoflora is dominated by *Geminospora lemurata*, a taxon that has been related to the progymnosperm *Archaeopteris* (Playford, 1983). The palynoflora also contains *Apiculatisporites* sp., *Emphanisporites* sp., *Leiotriletes* spp., *Rugualtisporites* sp., *Verrucosisporites* sp. and a large spinose form *Ancyrospora* sp., all of probable Frasnian age. These spore types have been associated with rhyniophytes, zosterophylls, ferns and progymnosperms (Balme, 1995; Gensel, 1980; Traverse, 1988) pointing to considerable diversity within the vegetation some 380 million years ago.

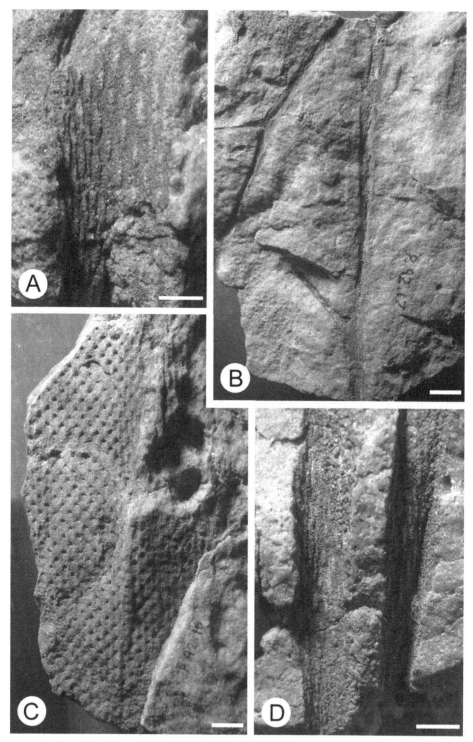

Figure 2.3 Middle Devonian fossil plants. **A**, *Haplostigma lineare* from Beacon Heights Orthoquartzite, Cook Mountains, WAM P.92.20. **B**, *Praeramunculus alternatiramus* from Mount Ritchie in the upper Aztec Siltstone, WAM P.92.17. **C**, *Malanzania* sp. from the uppermost Beacon Heights Orthoquartzite, Cook Mountains, WAM P.92.16. **D**, *Haplostigma lineare* from the uppermost Beacon Heights Orthoquartzite, Cook Mountains, WAM P.92.19. Photos courtesy of S. McLoughlin. Scale bars = 1 cm. WAM P = Western Australian Museum Palaeontological Collection number.

In the second basin, the Ellsworth Basin, the basal Horlick Formation of Early Devonian age (Boucot *et al.*, 1963; Bradshaw and McCartan, 1991; Doumani *et al.*, 1965) overlies weathered granite. In places the granite forms rounded domes that have thick weathering rinds and probably represent granite boulders on a palaeoshoreline (Bradshaw, Newman, Aitchinson, 2002). The Horlick Formation formed in nearshore to subtidal inner shelf environments (McCartan and Bradshaw, 1987) with sediments being derived from a mature landscape in Marie Byrd Land (Bradshaw, 2002, 2010). A mature landscape is further supported by sedimentary studies on terrestrial sediments that have demonstrated some of the earliest and deepest weathered soil profiles globally in this period of Earth history (Retallack, 1997). This basin has been inferred to extend to the Ellsworth and Pensacola mountains where it is equivalent in part to the Upper Crashsite Group (Ellsworth) and Heiser and Dover sandstones (Pensacola) (Figure 2.2). The Upper Crashsite Group (Mount Wyatt Earp Formation) accumulated in shallow marine settings and contains an Early Devonian fauna (Webers *et al.*, 1992). In the Pensacola Mountains the Heiser Sandstone, a shallow marine unit, is overlain unconformably by terrestrial fluvial deposits (Dover Sandstone). Elsewhere, Devonian sandstones occur in the Pauxtent Range (Pensacola) (Schmidt *et al.*, 1964). Evidence for vegetation is sparse with a few lycopod stems being recorded from the Dover Sandstone and equivalents (Edwards, 1990; Rigby and Schopf, 1969; Schmidt and Ford, 1969). These have been referred to as *Haplostigma* and *Cyclostigma*-type stems (Schopf, 1968), but Edwards (1990) pointed out differences and thought that they might be closer to *Colpodexylon cashiriense* from Venuzuela (Edwards and Benedetto, 1985). Other evidence for the vegetation comes from a single macrofloral record of smooth stemmed species (Long, 1965), later referred to as *Hostimella* (Rigby and Schopf, 1969), and from palynological records within the Horlick Formation (Kemp, 1972) of the Ohio Range. The miospore assemblage is of limited diversity that may be due in part to the thermal maturity of the rocks and the difficulty in extraction and preparation. Kemp (1972) described 16 taxa, largely trilete spores but several monolete grains. Three further spore types, which were too poorly preserved and known only from broken or single specimens, were also recorded indicating greater diversity within the assemblage.

Reconstructing Devonian vegetation based on limited palynological knowledge is diffi-cult especially as some spore taxa are known from more than one type of plant, and it is also apparent that there is considerable convergence in morphology (Gensel, 1980). In spite of the limitations in inferring the plant groups present from the palynological record, some general conclusions can be drawn with regard to the composition of the Antarctic vegeta-tion, and the types of plants that grew in the hinterland. Kemp (1972) recorded several significant spore taxa including *Reticuloidosporites antarcticus* (26%), *Stenozonotriletes* cf. *inequaemarginalis* (21%), *Dibolisporites* sp. A, (17%), *Acinosporites horlickensis* (7.5%), *Retusotriletes* cf. *R. concinnus* (7%), *Granulatisporites* (7.5%), along with several minor forms (*Dibolisporites* cf. *echinaceus*, *Emphanisporites rotatus*, *Anapiculatisporites* cf. *eifeliensis*, *Anuerospora* sp., ?*Samarisporites*). The affinity of the spores can be determined from studies of specimens with *in situ* palynomorphs (Balme, 1995; Gensel, 1980; Traverse,

1988). *Retusisporites* is known from a range of plants including the small thalloid *Protosalvinia* (Melo and Loboziak, 2003), a plant of problematic affinity. The spore types are also known from plants at the base of the Polysporangiophytes such as *Aglaophyton* (*Retusotriletes;* Wellman, 2003b), *Horneophyton* (*Emphanisporites*; Edwards and Richardson, 2000; Wellman, 2003b) and basal Tracheophyta and Eutracheophyta. Several spore types can be related to the lycopsids (e.g. *Nothia* and *Asteroxylon*, which both produce *Retusotriletes* type spores; Balme, 1995) and other unplaced taxa such as *Trichopherophyton* and *Ventarura* (Wellman, 2003b). The Cladoxylales are represented by spores such as *Dibolisporites*, a taxon found in *Calamophyton* (Bonamo and Banks, 1966; Schweitzer, 1973), and *Reticulatisporites*, which is recorded from sphenophytes (Beck and Libertin, 2010) and presumed filicalean (Andrews and Agashe, 1962) plants. Heterosporous lycopods are also represented by the spores of *Acinosporites* (Hao *et al.*, 2007; Richardson, Bonamo and McGregor, 1993) and *Reticulatisporites* (Libertin, Bek and Daskova, 2005). Overall the palynoflora suggests low vegetation of thalloid to herbaceous plants during the deposition of the Horlick Formation some 407 million years ago.

Several other regions of Antarctica have sparse records of Devonian flora. Findlay and Jordan (1984) recorded small impressions with diamond-shaped scars from the Black Prince Volcanics. Edwards (1990) discussed these stems that had been identified by Leistikow and Schweitzer (personal communication) as *Protolepidodendropsis pulchra* and concluded that, while they differed from other Devonian lycopods described from Antarctica, they could only be considered as a lycopod stem with swollen and extended leaf bases (Edwards, 1990). The Black Prince Volcanics overlie the Robertson Bay Group that is part of an exotic terrain and represents strata that formed as part of an arc system that was accreted to the Antarctic margin in Late Ordovician times.

In Marie Byrd Land two localities from the Rupert Coast Volcanics contain abundant plant fossils (Grindley and Mildenhall, 1981; Grindley, Mildenhall and Schopf, 1980). Grindley, Mildenhall and Schopf (1980) described *Drepanophycus schopfii*, a microphyllous plant lacking true leaves with cauline sporangia. Edwards (1990) raised the possibility that these plants belonged to a different genus and re-examination of the material by Xu and Berry (2008) established the presence of sagittate leaves thus allowing the material to be firmly placed within *Haskinsia colophylla*, a widespread Middle Devonian plant (Xu and Berry, 2008). *Haskinsia* has sporangia borne adaxially (Berry and Edwards, 1996) and is placed within the lycopsids. This is a small herbaceous plant and the Antarctic material consists of stems that are several centimetres long, up to 1 cm wide, and clothed in small petiolate and sagittate leaves up to 1.8 mm long. Amongst this material were several other taxa including the widespread *Haplostigma irregularis*, ?*Protolepidodendron* and some indeterminate psilophytalean axes.

While the Devonian flora of Antarctica is sparse (Table 2.1) several generalisations can, however, be made. The composition of the macroflora is similar to other areas of Gondwana and includes several widespread taxa such as *Haplostigma*. Early Devonian floras are from thalloid or herbaceous plants allied to lycopsids. This low vegetation probably lacked the ability to bind the soil strongly and so sedimentation was dynamic. An increase in

Table 2.1 *Species list for the Devonian of Antarctica.*

Taxa Lycophyta	Location/Formation	Age	References
Archaeosigillaria sp. cf. *A. caespitosa*[1]	Cook Mountains, Beacon Heights Orthoquartzite	Middle Devonian	Grindley, Mildenhall and Schopf, 1980; McLoughlin and Long, 1994
Cyclostigma sp.[2]	Dover Sandstone	Middle to Late Devonian	Schopf 1968
Haskinsia colophylla[3]	Mt Hartkopf, Marie Byrd Land, Rupert Coast Volcanics	Late Devonian	Edwards, 1990; Grindley and Mildenhall, 1981; Grindley, Mildenhall and Schopf, 1980; Xu and Berry, 2008
Haplostigma lineare[4]	Beacon Heights, Beacon Heights Orthoquartzite; southern Cook Mountains, Beacon Heights Orthoquartzite	late Middle Devonian	Edwards, 1990; McLoughlin and Long, 1994; Plumstead, 1962
Haplostigma irregularis	Beacon Heights Orthoquartzite; Mt Hartkopf, Marie Byrd Land, Rupert Coast Volcanics	Middle to Late Devonian	Grindley and Mildenhall, 1981; Grindley, Mildenhall and Schopf, 1980; Plumstead, 1962
Haplostigma sp.[2]	Dover Sandstone	Middle to Late Devonian	Schopf, 1968
Hostimella sp.	Horlick Formation, Ohio Range	Emsian, Early Devonian	Rigby and Schopf, 1969
Malanzania sp.	Cook Mountains, Beacon Heights Orthoquartzite	Middle Devonian	McLoughlin and Long, 1994

Table 2.1 (cont.)

Taxa Lycophyta	Location/Formation	Age	References
Praermunculus alternatiramus	Mount Ritchie, Aztec Siltstone	Upper Givetian, late Middle Devonian	McLoughlin and Long, 1994
Protolepidodendropsis pulchra	North Victoria Land, Black Prince Volcanics	Middle Devonian	Edwards, 1990; Findlay and Jordan, 1984
Rhynia-like plants	Aztec Siltstone	late Middle Devonian	McPherson, 1979
Indet. psilophyte axes[5]	Mt Hartkopf, Marie Byrd Land, Rupert Coast Volcanics	Late Devonian	Grindley, Mildenhall and Schopf, 1980

[1] McLoughlin and Long (1994) included the indeterminate lycopods axes from Grindley, Mildenhall and Schopf (1980) in this taxon.

[2] Note Edwards (1990) regarded these as similar to *Colpodexylon cashiriense* but given the poor preservation thought that they were best left undetermined.

[3] Originally determined as *Drepanophycus schopfii* but more recently demonstrated to belong to *Haskinsia*.

[4] Originally described by Plumstead (1962) as cf. *Protolepidodendron lineare* but regarded as *Haplostigma* by Edwards (1990) and formalised by McLoughlin and Long (1994).

[5] Regarded as indeterminate plant axes by Edwards (1990) but McLoughlin and Long (1994) regarded them as conspecific with *Praermunculus alternatiramus*.

stabilisation of the landscape and development of deep soil profiles would only come about with the development of secondary growth and the increase in stature of the plants, such as the tree-sized progymnosperms (*Archaeopteris*), that would eventually preside over an understorey of lycopods and ferns during the Upper Devonian (see Chapter 3).

It is clear that our knowledge of the Devonian floras of Antarctica is still sparse yet much potential exists. Most plant fossil collections have been made by non-palaeobotanists and discovered either in the course of field mapping exercises or while collecting other fossil groups (e.g. fish). Equally, knowledge of the palynofloras is sparse and hampered by the thermal alteration associated with younger Jurassic intrusive and volcanic rocks. Nevertheless, limited palynofloras have been recovered and more intensive sampling and different processing techniques may reveal more details of the flora and vegetation from this time period.

Summary

- The earliest phase of plant colonisation of the land that took place from the Ordovician and through the Silurian is not known from the Antarctic, predominantly due to suitable rocks not yet having been fully investigated. Although most of the potential sequences are metamorphosed by Ross orogenic tectonism, those deposits removed from the orogenic front such as in the Ellsworth and Pensacola mountains have the potential to yield important fossil material. Sedimentation in these regions is more continuous and palaeo-environments are similar to areas elsewhere in Gondwana where the earliest record of colonisation of the land has been documented.

- The Devonian record on Antarctica is equally sparse and again has not been the subject of a targeted field investigation by palaeobotanists. Much of the material collected to date has been by geological mapping parties or vertebrate palaeontologists. The limited material that has been collected suggests similarities with other gondwanan Devonian floras and seems to include a number of widespread taxa although the material is generally poorly preserved. The macrofloras are dominated by lycophytes. Sparse wood has been recorded but this may belong to younger strata since it was recovered from moraines rather than rock outcrop. If the wood is Devonian it supports the presence of arborescent progymnosperms.

- Devonian palynofloras are also present but investigations have only been preliminary in nature. These floras are rich in trilete spores that are allied to lycophytes, but the presence of *Geminospora* in abundance indicates that progymnosperms such as *Archaeopteris* were also becoming an important component of the vegetation. In the two decades since the last review of Antarctic Silurian–Devonian palaeobotany (Edwards, 1990), relatively few additional discoveries have been added (e.g. McLoughlin and Long, 1994). The Devonian would thus benefit from a systematic field study by palaeobotanists.

References

Aldridge, R. J., Jeppsson, L. and Dorning, K. J. (1993). Early Silurian oceanic episodes and events. *Journal of the Geological Society of London*, **150**, 501–513.

Anderson, J. M. (1979). The geology of the Taylor Group, Beacon Supergroup, Byrd Glacier area, Antarctica. *New Zealand Antarctic Record*, **2**, 6–11.

Andrews, H. N. and Agashe, S. N. (1962). A new sporangia from the American Carboniferous. *The Palaeobotanist*, **11**, 46–48.

Balme, B. E. (1995). Fossil *in situ* spores and pollen grains: an annotated catalogue. *Review of Palaeobotany and Palynology*, **87**, 81–323.

Barrett, P. J. (1991). The Devonian to Triassic Beacon Supergroup of the Transantarctic Mountains and correlatives in other parts of Antarctica. In *The Geology of Antarctica*, *Oxford Monographs on Geology and Geophysics*, **17**, ed. R. J. Tingey. Oxford: Clarendon Press, pp. 120–152.

Barrett, P. J., Kohn, B. P., Askin, R. A. and McPherson, J. G. (1971). Preliminary report on Beacon Supergroup studies between Hatherton and Mackay glaciers, Antarctica. *New Zealand Journal of Geology and Geophysics*, **14**, 605–614.

Beck, J. and Libertin, M. (2010). *In situ* reticulate sphenophyllalean spores, Pennsylvanian (Bolsovian) of the Czech Republic. *Review of Palaeobotany and Palynology*, **159**, 56–61.

Beerling, D. J. and Berner, R. A. (2005). Feedbacks and the coevolution of plants and atmospheric CO_2. *Proceedings of the National Academy of Sciences of the United States of America*, **104**, 1302–1305.

Beerling, D. J., Osborne, C. P. and Chaloner, W. G. (2001). Evolution of leaf-form in land plants linked to atmospheric CO_2 decline in the Late Palaeozoic era. *Nature*, **410**, 352–354.

Berner, R. A. (1998). The carbon cycle and CO_2 over Phanerozoic time: The role of land plants. *Philosophical Transactions of the Royal Society of London Series B*, **353**, 75–82.

Berner, R. A. (2003). The rise of trees and their effects on Paleozoic atmospheric CO_2 and O_2. *Comptes Rendus Geoscience*, **335**, 1173–1177.

Berner, R. A. (2004). *The Phanerozoic Carbon Cycle: CO_2 and O_2*. Oxford: Oxford University Press.

Berner, R. A. (2006). GEOCARBSULF: A combined model for Phanerozoic atmospheric O_2 and CO_2. *Geochimica et Cosmochimica Acta*, **70**, 5653–5664.

Berner, R. A. (2009). Phanerozoic atmospheric oxygen: new results using the GEOCARBSULF model. *American Journal of Science*, **309**, 603–606.

Berry, C. M. and Edwards, D. (1996). The herbaceous lycophyte *Haskinsia* Grierson and Banks from the Devonian of western Venezuela, with observations on leaf morphology and fertile specimens. *Botanical Journal of the Linnean Society*, **122**, 103–122.

Bonamo, P. M. and Banks, H. P. (1966). *Calamophyton* in the Middle Devonian of New York State. *American Journal of Botany*, **53**, 755–768.

Boucot, A. J. Caster, K. E., Ives, D. and Talent, J. A. (1963). Relationships of a new Lower Devonian terebratuloid (Brachiopoda) from Antarctica. *Bulletins of American Paleontology*, **46**, 81–151.

Bradshaw, M. A. (1981). Paleoenvironmental interpretations and systematics of Devonian trace fossils from the Taylor Group (lower Beacon Supergroup), Antarctica. *New Zealand Journal of Geology and Geophysics*, **24**, 615–652.

Bradshaw, M. A. (2002). A new ichnogenus *Catenarichnus* fron the Devonian of the Ohio Range, Antarctica. *Antarctic Science*, **14**, 422–424.

Bradshaw, M. A. (2010). Devonian trace fossil of the Horlick Formation, Ohio Range, Antarctica: systematic description and palaeoenvironmental interpretation. *Ichnos*, **17**, 58–114.

Bradshaw, M. A. and McCartan, L. (1991). Palaeoecology and systematics of Lower Devonian bivalves from the Horlick Formation, Ohio Range, Antarctica. *Alcheringa*, **15**, 1–42.

Bradshaw, M. A., Newman, J. and Aitchison, J. C. (2002). The sedimentary geology, palaeo-environments and ichnocoenoses of the Lower Devonian Horlick Formation, Ohio Range, Antarctica. *Antarctic Science*, **14**, 395–411.

Brenchley, P. J., Marshall, J. D., Carden, G. A., *et al.* (1994). Bathymetric and isotopic evidence for a short-lived Late Ordovician glaciation in a greenhouse period. *Geology*, **22**, 295–298.

Broecker, W. S. (1987). The biggest chill. *Natural History Magazine*, **96**, 74–82.

Broecker, W. S. (1991a). The great ocean conveyor. *Oceanography*, **4**, 79–89.

Broecker, W. S. (1991b). The great ocean conveyor. In *Global Warming: Physics and Facts*, *AIP Conference Proceedings*, **247**, eds B. G. Levi, D. Hafemeister and R. Scribner. pp. 129–161.

Buggisch, W., Joachimski, M. M., Lehnert, O., *et al.* (2010). Did intense volcanism trigger the first Late Ordovician icehouse? *Geology*, **38**, 327–330.

Christiansen, B. (2007). Volcanic eruptions, large-scale modes in the Northern Hemisphere, and the El Niño–Southern Oscillation. *Journal of Climate*, **21**, 910–922.

Cocks, L. R. M. (2001). Ordovician and Silurian global geography. *Journal of the Geological Society of London*, **158**, 197–210.

Cocks, L. R. M. and Torsvik, T. H. (2002). Earth geography from 500 to 400 million years ago: a faunal and palaeomagnetic review. *Journal of the Geological Society of London*, **159**, 631–644.

Copper, P. (2002). Silurian and Devonian reefs: 80 million years of greenhouse between two ice ages. In *Phanerozoic Reef Patterns*, *SEPM Special Publication*, **72**, eds W. Kiessling, E. Fluegel, and J. Golonka. Tulsa: SEPM (Society for Sedimentary Geology), pp. 181–238.

Copper, P. and Scotese, C. R. (2003). Megareefs in Middle Devonian supergreenhouse climates. In *Extreme Depositional Environments: Mega End Members in Geologic Time*, *Geological Society of America Special Publications*, **370**, eds M. A. Chan and A. W. Archer. Boulder: The Geological Society of America, pp. 209–230.

Cowley, T. J. and Baum, S. K. (1995). Reconciling Late Ordovician (440 Ma) glaciation with very high (14x) CO_2 levels. *Journal of Geophysical Research*, **100**, 1093–1101.

Cramer, B. D. and Munnecke, A. (2008). Early Silurian positive $\delta^{13}C$ excursions and their relationship to glaciations, sea-level changes and extinction events: discussion. *Geological Journal*, **43**, 517–519.

Cramer, B. D., Saltzman, M. R. and Kleffner, M. A. (2006). Spatial and temporal variability in organic carbon burial during global positive carbon isotopic excursions: new insight from high resolution $\delta^{13}C_{carb}$ stratigraphy from the type area of the Niagaran (Silurian) Provincial Series. *Stratigraphy*, **2**, 327–340.

Curtis, M. L. and Lomas S. A. (1999). Late Cambrian stratigraphy of the Heritage Range, Ellsworth Mountains: implications for basin evolution. *Antarctic Science*, **11**, 63–77.

DeConto, R. M. and Pollard, D. (2003). Rapid Cenozoic glaciation of Antarctica induced by declining atmospheric CO_2. *Nature*, **421**, 245–249.

DeConto, R. M., Pollard D., Wilson P. A., Pälike, H., Lear, C. H. and Pagani, M. (2008). Thresholds for Cenozoic bipolar glaciations. *Nature*, **455**, 625–656.

Dettman, D. L., Fang, X. M., Garzione, C. N. and Li, J. J. (2003). Uplift-driven climate change at 12 Ma: a long δ^{18}O record from the NE margin of the Tibetan Plateau. *Earth and Planetary Science Letters*, **214**, 267–277.

Doumani, G. A., Boardman, R. S., Rowell, A. J., *et al.* (1965). Lower Devonian fauna of the Horlick Formation, Ohio Range, Antarctica. In *Geology and Paleontology of the Antarctic, Antarctic Research Series*, **6**, ed. J. B. Hadley. Washington: American Geophysical Union, pp. 241–281.

Edwards D. (1990). Silurian–Devonian paleobotany: problems, progress, and potential. In *Antarctic Paleobiology: its Role in the Reconstruction of Gondwana*, eds T. N. Taylor and E. L. Taylor. New York: Springer-Verlag, pp. 89–101.

Edwards, D. and Benedetto, L. (1985). Two new species of herbaceous lycopods from the Devonian of Venezuela with comments on their taphonomy. *Palaeontology*, **28**, 599–618.

Edwards, D. and Richardson, J. B. (2000). Progress in reconstructing vegetation on the Old Red Sandstone continent: two *Emphanisporites* producers from the Lochkovian sequence of the Welsh Borderland. In *New Perspectives on the Old Red Sandstone, Geological Society of London Special Publication*, **180**, eds P. F. Friend and B. P. J. Williams. London: The Geological Society, pp. 355–370.

Edwards, D. and Richardson, J. B. (2004). Silurian and Lower Devonian plant assemblages from the Anglo-Welsh Basin: a palaeobotanical and palynological synthesis. *Geological Journal*, **39**, 375–402.

Edwards, D., Feehan, J. and Smith D. G. (1983). A late Wenlock flora from Co. Tipperary, Ireland. *Botanical Journal of the Linnean Society*, **86**, 19–36.

Falkowski, P. G. (1997). Evolution of the nitrogen cycle and its influence on biological sequestration of CO_2 in the ocean. *Nature*, **387**, 272–275.

Fielding, C. R., Frank, T. D., Birgenheier, L. P., *et al.* (2008). Stratigraphic imprint of the Late Paleozoic Ice Age in eastern Australia: a record of alternating glacial and non-glacial climate regime. *Journal of the Geological Society of London*, **165**, 129–140.

Findlay, R. H. and Jordan, H. (1984). The volcanic rocks of Mt. Black Prince and Lawrence Peaks, North Victoria Land, Antarctica. *Geologisches Jahrbuch*, **B60**, 143–151.

Finlay, A. J., Selby, D. and Gröcke, D. R. (2010). Tracking the Hirnantian glaciation using Os isotopes. *Earth and Planetary Science Letters*, **293**, 339–348.

Frakes, L. A. and Francis, J. E. (1988). A guide to Phanerozoic cold polar climates from high-latitude ice-rafting in the Cretaceous. *Nature*, **333**, 547–549.

Frakes, L. A., Francis J. E. and Skytus J. I. (1992). *Climate Modes of the Phanerozoic*. Cambridge: Cambridge University Press.

France-Lanord, C. and Derry, L. A. (1997). Organic carbon burial forcing of the carbon cycle from Himalayan erosion. *Nature*, **390**, 65–67.

Gevers, T. W. and Twomey, A. (1982). Trace fossils and their environments in Devonian (Silurian?) lower Beacon sediments, in the Asgaurd Range, Victoria Land, Antarctica. In *Antarctic Geoscience*, ed. C. Craddock. Madison: University of Wisconsin Press, pp. 639–648.

Gensel, P. G. (1980). Devonian *in situ* spores: a survey and discussion. *Review of Palaeobotany and Palynology*, **30**, 101–132.

Gibbs, M. T., Barron, E. J. and Kump, L. R. (1997). An atmospheric pCO_2 theshold for glaciation in the Late Ordovician. *Geology*, **25**, 447–450.

Goldstrand, P. M., Fitzgerald, P. G., Redfield, T. F., Stump, E. and Hobbs, C. (1994). Stratigraphic evidence for the Ross orogeny in the Ellsworth Mountains, West Antarctica: implications for the evolution of the paleo-Pacific margin of Gondwana. *Geology*, **22**, 427–430.

Goodge, J. W., Williams, I. S. and Myrow, P. (2004). Provenance of Neoproterozoic and lower Paleozoic siliciclastic rocks of the central Ross orogen, Antarctica: Detrital record of rift-, passive-, and active-margin sedimentation. *Geological Society of America Bulletin*, **116**, 1253–1279.

Gray, J. (1985). The microfossil record of early land plants: advances in understanding of early terrestrialization, 1970–1984. *Philosophical Transactions of the Royal Society of London Series B*, **309**, 167–195.

Gray, J. (1988). Land plant spores and the Ordovician-Silurian boundary. *Bulletin of the British Museum of Natural History (Geology)*, **43**, 351–358.

Gray, J. and Boucot, A. J. (1971). Early Silurian spore tetrads from New York: earliest new world evidence for vascular plants? *Science*, **173**, 918–921.

Grindley, G. W. and Mildenhall, D. C. (1981). Geological background to a Devonian plant fossil discovery, Rupert Coast, Marie Byrd Land, West Antarctica. In *Gondwana Five, Fifth International Gondwana Symposium Wellington, New Zealand*, eds A. J. Cresswell and P. Vella. Rotterdam: A. A. Balkema, pp. 23–30.

Grindley, G. W., Mildenhall, D. C. and Schopf, J. M. (1980). A mid-late Devonian flora from the Rupert Coast, Marie Byrd Land, West Antarctica. *Journal of the Royal Society of New Zealand*, **10**, 271–285.

Hambrey, M. J. (1985). The Late Ordovician–Early Silurian glacial period. *Palaeogeography, Palaeoclimatology, Palaeoecology*, **51**, 273–289.

Hansen, J., Ruedy, R., Sato, M. and Reynolds, R. (1996). Global surface air temperature in 1995: return to pre-Pinatubo level. *Geophysical Research Letters*, **23**, 1665–1668.

Hao, S. G., Xue, J. Z., Wang, Q. and Liu, Z. F. (2007). *Yuguangia ordinata* gen. et sp. nov., a new lycopsid from the Middle Devonian (late Givetian) of Yunnan, China. *International Journal of Plant Sciences*, **168**, 1161–1175.

Harrington, J. H. and Speden, I. G. (1962). Section through the Beacon Sandstone and Beacon Heights West Antarctica. *New Zealand Journal of Geology and Geophysics*, **5**, 707–717.

Harris, N. (2006). The elevation history of the Tibetan Plateau and its implications for the Asian Monsoon. *Palaeogeography, Palaeoclimatology, Palaeoecology*, **241**, 4–15.

Helby, R. A. and McElroy, C. T. (1969). Microfloras from the Devonian and Triassic of the Beacon Group, Antarctica. *New Zealand Journal of Geology and Geophysics*, **12**, 376–382.

Hilley, G. E. and Porder, S. (2008). A framework for predicting global silicate weathering and CO_2 drawdown rates over geologic time-scales. *Proceedings of the National Academy of Sciences of the United States of America*, **105**, 16855–16859.

Hoffmeister, W. S. (1959). Lower Silurian plant spores from Libya. *Micropaleontology*, **5**, 331–334.

Isbell, J. L. (1999). The Kukri Erosion Surface; a reassessment of its relationship to rocks of the Beacon Supergroup in the central Transantarctic Mountains, Antarctica. *Antarctic Science*, **11**, 228–238.

Jeppsson, L. (1990). An oceanic model for lithological and faunal changes tested on the Silurian record. *Journal of the Geological Society of London*, **147**, 663–674.

Joachimski, M. M., Breisig, S., Buggisch, W., *et al.* (2009). Devonian climate and reef evolution: insights from oxygen isotopes in apatite. *Earth and Planetary Science Letters*, **284**, 599–609.

Kemp, E. M. (1972). Lower Devonian palynomorphs from the Horlick Formation, Ohio Range, Antarctica. *Palaeontographica B*, **139**, 105–124.

Kidder, D. L. and Worsley, T. R. (2010). Phanerozoic Large Igneous Provinces (LIPs), HEATT (Haline Euxinic Acidic Thermal Transgression) episodes, and mass extinctions. *Palaeogeography, Palaeoclimatology, Palaeoecology*, **295**, 162–191.

Kump, L. R., Arthur, M. A., Patzkowsky, *et al.* (1999). A weathering hypothesis for glaciation at high atmospheric pCO_2 during the Late Ordovician. *Palaeogeography, Palaeoclimatology, Palaeoecology*, **152**, 173–187.

Kyle, R. A. (1977). Devonian paylnomorphs from the basal Beacon Supergroup of South Victoria Land, Antarctica. *New Zealand Journal of Geology and Geophysics*, **20**, 1147–1150.

Laird, M. G. (1991). The late ProterozoicMiddle Palaeozoic rocks of Antarctica. In *The Geology of Antarctica, Oxford Monographs on Geology and Geophysics*, **17**, ed. R. J. Tingey. Oxford: Clarendon Press, pp. 74–119.

Libertin, M., Bek, J. and Daskova, J. (2005). Two new species of *Kladnostrobus* nov. gen. and their spores from the Pennsylvanian of the Kladno-Rakovnik Basin (Bolsovian, Czech Republic). *Geobios*, **38**, 467–476.

Long, E. E. (1965). Stratigraphy of the Ohio Range, Antarctica. In *Geology and Paleontology of the Antarctic, Antarctic Research Series*, **6**, ed. J. B. Hadley. Washington: American Geophysical Union, pp. 71–115.

Lozier, M. S. (2010). Deconstructing the conveyor belt. *Science*, **328**, 1507–1511.

Martin, J. H. (1990). Glacial/interglacial CO_2 change: The iron hypothesis. *Paleoceanography*, **5**, 1–13.

McCartan, L. and Bradshaw, M. A. (1987). Petrology and sedimentology of the Horlick Formation (Lower Devonian), Ohio Range, Transantarctic Mountains. *United States Geological Survey Bulletin*, **1780**, 1–31.

McKelvey, B. C., Webb, P. N., Gorton, M. P. and Kohn, B. P. (1972). Stratigraphy of the Beacon Supergroup between the Olympus and Boomerang ranges, Victoria Land, Antarctica. In *Antarctic Geology and Geophysics*, ed. R. J. Adie. Oslo: Universitetesforlaget, pp. 345–352.

McKelvey, B. C., Webb, P. N. and Kohn, B. P. (1977). Stratigraphy of the Taylor and lower Victoria groups (Beacon Supergroup) between the Mackay Glacier and Boomerang Range, Antarctica. *New Zealand Journal of Geology and Geophysics*, **20**, 813–863.

McLoughlin, S. and Long, J. A. (1994). New records of Devonian plants from southern Victoria Land, Antarctica. *Geological Magazine*, **131**, 81–90.

McPherson, J. G. (1978). Stratigraphy and sedimentology of the Upper Devonian Aztec Siltstone, Victoria Land, Antarctica. *New Zealand Journal of Geology and Geophysics*, **21**, 667–683.

McPherson, J. G. (1979). Calcrete (Caliche) palaeosols in fluvial redbeds of the Aztec siltstone (Upper Devonian) southern Victoria Land, Antarctica. *Sedimentary Geology*, **22**, 267–285.

Melo, J. H. G. and Loboziak, S. (2003). Devonian–Early Carboniferous miospore biostratigraphy of the Amazon Basin, Northern Brazil. *Review of Palaeobotany and Palynology*, **124**, 131–202.

Munnecke, A., Calner, M., Harper, D. A. T. and Servais, T. (2010). Ordovician and Silurian sea-water chemistry, sea level, and climate: a synopsis. *Palaeogeography, Palaeoclimatology, Palaeoecology*, **296**, 389–413.

Myrow, P. M., Pope, M. C., Goodge, J. W., Fischer, W. and Palmer, A. R. (2002). Depositional history of pre-Devonian strata and timing of Ross orogenic tectonism in the central Transantarctic Mountains, Antarctica. *Geological Society of America Bulletin*, **114**, 1070–1088.

Osborne, C. P., Beerling, D. J., Lomax, B. H. and Chaloner, W. G. (2004). Biophysical constraints on the origin of leaves inferred from the fossil record. *Proceedings of the National Academy of Sciences of the United States of America*, **101**, 10360–10362.

Partridge, T. C., Scott, L. and Schneider, R. R. (2004). Between Agulhas and Benguela: responses of Southern African climates of the late Pleistocene to current fluxes, orbital precession and the extent of the circum-Antarctic vortex. In *Developments in Palaeoenvironmental Research, Volume 6 Past Climate Variability through Europe and Africa*, eds R. W. Batterbee, F. Gasse and C. E. Stickley. Amsterdam: Springer-Verlag, pp. 45–68.

Playford, G. (1983). The Devonian miospore genus *Geminospora* Balme 1962; a reappraisal based on topotypic *G. lemurata* (type species). *Association of Australasian Palaeontologists Memoir*, **1**, 311–325.

Playford, G. (1990). Proterozoic and Paleozoic palynology of Antarctica: A review. In *Antarctic Paleobiology: its Role in the Reconstruction of Gondwana*, eds T. N. Taylor and E. L. Taylor. New York: Springer-Verlag, pp. 51–70.

Plumstead, E. P. (1962). Fossil floras of Antarctica. *In Trans-Antarctic Expedition 1955–1958, Scientific Reports, 9 (Geology)*. London: The Trans-Antarctic Expedition Committee, pp. 1–154.

Plumstead, E. P. (1964). Palaeobotany of Antarctica. In *Antarctic Geology: proceedings of the First International Symposium on Antarctic Geology, Cape Town, 16–21 September 1963*, ed. R. J. Adie. Amsterdam: North-Holland Publishing Company, pp. 637–654.

Pruss, S. B., Finnegan, S., Fischer, W. W. and Knoll, A. H. (2010). Carbonates in skeleton-poor seas: new insights from Cambrian and Ordovician strata of Laurentia. *Palaios*, **25**, 73–84.

Raymo, M. E. and Ruddiman, W. F. (1992). Tectonic forcing of Cenozoic climate. *Nature*, **359**, 117–122.

Retallack, G. J. (1997). Early forest soils and their role in Devonian global change. *Science*, **276**, 583–585.

Richardson, J. B., Bonamo, P. M. and McGregor, D. C. (1993). The spores of *Leclerquia* and the dispersed spore morphon *Acinosporites lindlarensis* Reigel: a case of gradualistic evolution. *Bulletin of the Natural History Museum, Geology Series*, **49**, 121–155.

Richardson, J. G. and Ausich W. I. (2007). Late Ordovician–Early Silurian cryptospore occurrences on Anticosti Island (Île d'Anticosti), Quebec, Canada. *Canadian Journal of Earth Science*, **44**, 1–7.

Richardson, P. L. (2008). On the history of meridional overturning circulation schematic diagrams. *Progress in Oceanography*, **76**, 466–486.

Rigby, J. F. and Schopf, J. M. (1969). Stratigraphic implications of Antarctic paleobotanical studies. In *Gondwana Stratigraphy. 1st IUGS Gondwana Symposium*, ed. A. J. Amos. Paris: UNESCO, pp. 91–106.

Ritchie, A. (1975). *Groenlandaspis* in Antarctica, Australia and Europe. *Nature*, **254**, 569–573.

Royer, D. L. (2006). CO_2-forced climate thresholds during the Phanerozoic. *Geochimica et Cosmochimica Acta*, **70**, 5665–5675.

Rubenstein, C. V., Gerrienne, P., de la Puente, G. S., Astini, R. A. and Steemans, P. l. (2010). Early Middle Ordovician evidence for land plants in Argentina (eastern Gondwana). *New Phytologist*, **188**, 365–369.

Saltzman, M. R. and Young, S. Y. (2005). Long-lived glaciation in the Late Ordovician? Isotopic and sequence-stratigraphic evidence from western Laurentia. *Geology*, **33**, 109–112.

Schmidt, D. L. and Ford, A. B. (1969). Geology of the Pensacola and Thiel Mountains. In *Geologic Maps of Antarctica, Antarctic Folio Map Series Folio 12 Sheet 5, Plate V*, eds V. C. Bushnell and C. Craddock. New York: American Geographical Society.

Schmidt, D. L., Dover, J. H., Ford, A. B. and Brown, R. D. (1964). Geology of the Patuxent Mountains. In *Antarctic Geology: Proceedings of the First International Symposium on Antarctic Geology, Cape Town, 16–21 September 1963*, ed. R. J. Adie. Amsterdam: North-Holland Publishing Company, pp. 276–283.

Schopf, J. M. (1968). Studies in Antarctic paleobotany. *Antarctic Journal of the United States*, **3**(5), 176–177.

Schweitzer, H.-J. (1973). Die mitteldevon-Flora von Lindlar (Rheinland). 4. Filicinae *Calamophyton primaevum* Kräusel and Weyland. *Palaeontographica B*, **140**, 117–150.

Sepkoski Jr, J. J. (1981). A factor analytical description of the Phanerozoic marine fossil record. *Paleobiology*, **7**, 36–53.

Sheehan, P. M. (2001). The late Ordovician mass extinction. *Annual Review of Earth and Planetary Sciences*, **29**, 331–364.

Sigman, D. M. and Boyle, E. A. (2000). Glacial/interglacial variations in atmospheric carbon dioxide. *Nature*, **407**, 859–869.

Simon, L., Godderis, Y., Buggisch, W., Strauss, H. and Joachimski, M., (2007). Modeling the carbon and sulfur isotope composition of marine sediments: climate evolution during the Devonian. *Chemical Geology*, **146**, 19–38.

Spencer, R. W., LaFontaine, F. J., DeFelice, T. and Wentz, F. J. (1998). Tropical oceanic precipitation changes after the 1991 Pinatubo eruption. *Journal of Atmospheric Sciences*, **55**, 1707–1713.

Spörli, K. B. (1992). Stratigraphy of the Crashsite Group, Ellsworth Mountains, West Antarctica. In *Geology and Paleontology of the Ellsworth Mountains West Antarctica, Geological Society of America Memoir*, **170**, eds G. F. Webers, C. Craddock, and J. F. Splettstoesser. Boulder: The Geological Society of America, pp. 21–35.

Steemans, P. and Wellman, C. H. (2003). Miospores and the emergence of land plants. In: *The Great Ordovician Biodiversity Event*, eds B. D. Webby, M. L. Droser and I. G. Percival. New York: Columbia University Press, pp. 361–368.

Steemans, P., Le Hérissé, A. and Bozdogan, N. (1996). Ordovician and Silurian crypto-spores and miospores from Southeastern Turkey. *Review of Palaeobotany and Palynology*, **93**, 35–76.

Steemans, P., Le Hérissé, A., Melvin, J., *et al.* (2009). Origin and radiation of the earliest vascular land plants. *Science*, **324**, 353.

Streel, M., Caputo, M. V., Loboziak, S. and Melo, J. H. G. (2000). Late Frasnian–Famennian climates based on palynomorph analyses and the question of the Late Devonian glaciations. *Earth-Science Reviews*, **52**, 121–173.

Strother, P. K. (2000). Cryptospores: the origin and early evolution of the terrestrial flora. *Paleontological Society Papers*, **6**, 1–20.

Strother, P. K., Al-Harji, S. and Traverse, A. (1996). New evidence for land plants from the lower Middle Ordovician of Saudi Arabia. *Geology*, **24**, 55–58.

Taylor, W. A. and Strother, P. K. (2009). Ultrastructure, morphology, and topology of Cambrian palynomorphs from the Lone Rock Formation, Wisconsin, USA. *Review of Palaeobotany and Palynology*, **153**, 296–309.

Torsvik, T. H. and Cocks, L. R. M. (2004). Earth geography from 400 to 250 Ma: A palaeomagnetic, faunal and facies review. *Journal of the Geological Society of London*, **161**, 555–572.

Traverse, A. (1988). *Paleopalynology*. Boston: Unwin and Hyman.

Trenberth, K. E. and Dai, A. (2007). Effects of Mount Pinatubo volcanic eruption of the hydrological cycle as an analog of geoengineering. *Geophysical Research Letters*, **34**, L15702, doi: 10.1029/2007GL030524.

Trotter, J. A., Williams, I. S., Barnes, C. R., Lécuyer, C. and Nicoll, R. S. (2008). Did cooling oceans trigger Ordovician biodiversification? Evidence from conodont thermometry. *Science*, **321**, 550–554.

van Geldern, R., Joachimski, M. M., Day, J., *et al.* (2006). Carbon, oxygen and strontium isotope records of Devonian brachiopod shell calcite. *Palaeogeography, Palaeoclimatology, Palaeoecology*, **240**, 47–67.

Veevers, J. J. and Powell, M. (1987). Late Paleozoic glaciation in Gondwanaland reflected in transgressive-regressive depositional sequences in Euamerica. *Geological Society of America Bulletin*, **98**, 475–487.

Visser, J. N. J. (1990). The age of the late Palaeozoic glacigene deposits in southern Africa. *South African Journal of Geology*, **93**, 366–375.

Visser, J. N. J. (1997). Deglaciation sequences in the Permo-Carboniferous Karoo and Kalahari basins of southern Africa: a tool in the analysis of cyclic glaciomarine basin fills. *Sedimentology*, **44**, 507–521.

Vravrdová, M. (1990). Early Ordovician acritarchs from the locality Myto near Rokycany (late Arenig, Czechoslovakia). *Casopis pro Mineralogii a Geologii*, **35**, 239–250.

Webers, G. F., Glenister, B., Pojeta Jr, J. and Young, G. (1992). Devonian fossils from the Ellsworth Mountains, West Antarctica. In *Geology and Paleontology of the Ellsworth Mountains West Antarctica, Geological Society of America Memoir*, **170**, eds G. F. Webers, C. Craddock and J. F. Splettstoesser. Boulder: The Geological Society of America, pp. 269–278.

Wellman, C. H. (2003a). Dating the origin of land plants. In *Telling Evolutionary Time: Molecular Clocks and the Fossil Record*, eds P. C. J. Donoghue and M. P. Smith. London: CRC Press, pp. 119–141.

Wellman, C. H. (2003b). Palaeoecology and palaeophytogeography of the Rhynie chert plants: evidence from integrated analysis of *in situ* and dispersed spores. *Proceedings of the Royal Society of London B*, **271**, 985–992.

Wellman, C. H. (2010). The invasion of the land by plants: when and where. *New Phytologist*, **188**, 306–309.

Wellman, C. H. and Gray, J. (2000). The microfossil record of early land plants. *Philosophical Transactions of the Royal Society of London Series B*, **355**, 717–732.

Wellman, C. H., Osterloff, P. L. and Mohiuddin, U. (2003). Fragments of the earliest land plants. *Nature*, **425**, 282–285.

Woolfe, K. J. (1993). Devonian depositional environments in the Darwin Mountains: marine or non-marine. *Antarctic Science*, **5**, 211–220.

Woolfe, K. J., Long, J. A., Bradshaw, M. A., Harmsen, F. and Kirkbridge, M. (1990). Fish-bearing Aztec Siltstone (Devonian) in Cook Mountains, Antarctica. *New Zealand Journal of Geology and Geophysics*, **33**, 511–514.

Xu, H.-H. and Berry, C. A. (2008). The Middle Devonian lycopod *Haskinsia* Grierson et Banks from the Rupert Coast, Marie Byrd Land, West Antarctica. *Review of Palaeobotany and Palynology*, **150**, 1–4.

Young, G. C. (1988). Antiarchs (Placoderm fishes) from the Devonian Aztec Siltstone, southern Victoria Land, Antarctica. *Palaeontographica A*, **202**, 1–125.

Young, G. C. (1991). Fossil fishes from Antarctica. In *The Geology of Antarctica*, *Oxford Monographs on Geology and Geophysics*, **17**, ed. R. J. Tingey. Oxford: Clarendon Press, pp. 538–567.

Young, G. C. (1993). Middle Palaeozoic macrovertebrate biostratigraphy of eastern Gondwana. In *Palaeozoic Vertebrate Biostratigraphy and Biogeography*, ed. J. A. Long. London: Belhaven Press, pp. 208–251.

Young, G. C., Long, J. A. and Ritchie, A. (1992). Crossopterygian fishes from the Devonian of Antarctica: systematics, relationships and biogeographic significance. *Records of the Australian Museum Supplement*, **14**, 1–77.

Young, S. A., Saltzman, M. R., Foland, K. A., Linder, J. S. and Kump, L. R. (2009). A major drop in seawater $^{87}Sr/^{86}Sr$ during Middle Ordovician (Darriwilian): links to volcanism and climate? *Geology*, **37**, 951–954.

Zhang, R., Fellows, M. J., Grotzinger, J. P. and Marshall, J. (2001). Could the Late Permian deep ocean have been anoxic? *Paleoceanography*, **16**, 317–319.

Zachos, J., Pagani, M., Sloan, L., Thomas, E. and Billups, K. (2001). Trends, rhythms, and aberrations in global climate 65 Ma to present. *Science*, **292**, 686–693.

Zachos, J. C., Stott, L. D. and Lohmann, K. C. (1994). Evolution of early Cenozoic marine temperatures. *Paleoceanography*, **9**, 353–387.

3

Collapsing ice sheets and evolving polar forests of the middle to late Paleozoic

Introduction

Throughout the Devonian a remarkable transformation of the land was under way. The vegetation, which had comprised small, probably streamside plants only a few centimetres high in the earliest Devonian, changed dramatically. The evolution of secondary growth (wood) paved the way for an increase in stature and the origin of the tree habit (such as that exhibited by the progymnosperm, *Archaeopteris*). By the Late Devonian forests were growing across the landscape creating new niches for understorey plants, resulting in an increase in diversity within terrestrial ecosystems globally. This transformation paved the way for animal groups to follow the plants onto land and begin to colonise the new niches created by the plants. Colonisation of terrestrial environments took place during a period of warmth, albeit under a cooling trend (Nardin *et al.*, 2011) before the terrestrial biota had to face the problems associated with deteriorating climate in the late Carboniferous. This manifested in the Late Paleozoic Ice Age, a period of between 60 and 80 Ma duration that began in the Carboniferous (Visean), crossed the Carboniferous–Permian boundary and continued into the late Early Permian (Shi and Waterhouse, 2010), with some regions experiencing glaciation until the early Late Permian (Fielding, Frank and Isbell, 2008; Fielding *et al.*, 2008a, 2008b).

Global palaeogeography and palaeoclimate

The majority of the Earth's surface during the middle Paleozoic was covered by the vast Panthalassa Ocean. Sea levels were generally high and much of the land surface lay submerged beneath tropical seas that supported extensive expanses of reefs. The palaeogeography of the Devonian was dominated by three supercontinents namely Gondwana to the south, the continent of Eurasia[1] to the north, and in between the smaller supercontinent, Euramerica. Euramerica was situated in the Southern Hemisphere in close proximity to Gondwana and later amalgamated with Gondwana to form the beginnings of Pangea. During the late Paleozoic the distribution of the landmasses was highly asymmetrical relative to the palaeo-equator, with Gondwana outsizing both Euramerica and Eurasia. Subsequent northward

---

[1] The supercontinent comprising Europe and Asia.

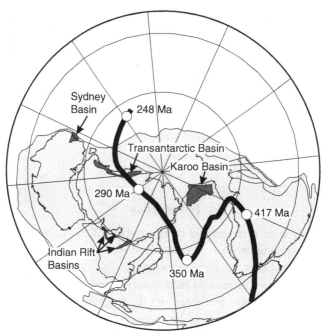

Figure 3.1 Polar projection illustrating the polar wander path (solid black line) from the Silurian to latest Permian, adapted from Scotese, Boucot and McKerrow (1999) and Torsvik and Van der Voo (2002) with key ages given. The position of the palaeopole implies movement from mid-latitudes in the early Paleozoic to high latitudes in the Permian. Sedimentary basins mentioned in Chapter 3 are marked and labelled.

movement of Pangea during the late Paleozoic progressively reduced this geographical asymmetry. Within Gondwana, northern Africa and northern South America drifted into the mid- and low-latitudes whilst other regions such as India, southern Africa, Antarctica and Australia moved into high latitudes. Large-scale continental reorganisation resulted in profound changes to the global ocean circulation patterns and in turn the climate, although how these effects manifested themselves is still under debate (Shi and Waterhouse, 2010). As Pangea continued to move northwards the South Pole migrated across the continent. For example, during the Late Devonian to early Carboniferous the palaeopole was situated over South America and Africa, but by the late Carboniferous the palaeopole was centred over Antarctica before moving to the edge of the Antarctic continent in the Permian (Scotese, Boucot and McKerrow, 1999) (Figure 3.1).

The coalescence of all major southern continents around the South Pole during the mid-Paleozoic probably created strong latitudinal contrasts in temperature as well as triggering an early and incipient episode of Gondwana glaciation (e.g. Crowell, 1999). However, other authors have favoured mechanisms such as mountain building (orogenic activity) in eastern Gondwana (Powell and Veevers, 1987; Veevers and Powell, 1987) to explain the glaciations. Chemical weathering increased and plant life on land drew down vast amounts of carbon dioxide from the atmosphere leading to a significant cooling of the Earth.

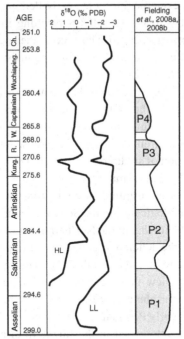

Figure 3.2 Climate curve for the Permian based on Korte *et al.* (2008) illustrating the oxygen isotope variation in high latitude (HL) *versus* low latitude (LL) brachiopod shells. Note strong congruence between HL *versus* LL indicating the global nature of the climatic variation. P1 to P4 glacial events recognised by Fielding *et al.* (2008a, 2008b).

The main period of glacial activity appears to have taken place between the late Carboniferous to the late Early Permian (Fielding, Frank and Isbell, 2008) (Figure 3.2). By the Late Devonian (Fammenian) ice sheets were centred over present day northern South America (Bolivia and Peru), which lay in the high palaeolatitudes of between 60° and 90° S (Caputo *et al.*, 2008; Melo and Loboziak, 2003; Wicander *et al.*, 2011). These ice sheets are thought to have resulted from glacial ice build-up in mountain environments and their subsequent expansion. South American Late Devonian glacial deposits are coeval with those of North Africa (Isaacson *et al.*, 2008), and more recently mid-latitude glaciations in North America have also been correlated with this interval (Brezinski, Cecil and Skema, 2010). This suggests that the Late Devonian glaciation may have been more widespread than previously thought implying greater climatic cooling, sea level change, and providing a more plausible mechanism to explain a number of widespread extinction events. These extinction events were concentrated mainly in marine (tropical) and freshwater environments, and took place during the last 20 Ma of the Devonian (e.g. Hallam and Wignall, 1997; Sepkoski, 1996; Wallister, 1996).

Climatic zonation resulting from strong latitudinal contrasts in temperature was also the result of asymmetry in Pangean continental distributions (Barron and Fawcett, 1995; Shi and Waterhouse, 2010). For example during the early Permian glacial stage the southern polar zone extended from the pole to 30–40° S resulting in narrowing temperate and tropical zones (Shi and Waterhouse, 2010). Thus during the late Paleozoic the temperature contrast between the equator and pole was large resulting in distinct tropical and temperate floras and faunas (Gibbs *et al.*, 2002; Rees *et al.*, 2002). Throughout the late Paleozoic provincialism was marked as southern and northern temperate zones became distinct. The impact of the climate gradient on the biota has had a significant impact on our ability to integrate the rock record through this time interval since few organisms occur across all climate zones and critical fossil groups, such as conodonts, do not occur in southern high latitude sites making correlation of strata from this region difficult. Furthermore, at times of marked provincialism and strong climate gradients, first and last appearances are often diachronous over long distances resulting in conflicting signals when compared with radiometric or other geochronologic ties. In the Southern Hemisphere the biostratigraphic zonation is coarse compared with that in the Northern Hemisphere. Moreover, the lack of geochronology coupled with the deficit of organisms occuring in discrete time intervals in more than one biogeographic region, make it difficult to correlate over wide areas. These issues further reinforced the view that the wide distribution of glacial rocks over the continents of the Southern Hemisphere suggest an extensive long-lived ice sheet for much of the late Carboniferous and early Permian (e.g. Veevers and Powell, 1987; Visser, 1987, 1993) – a view that became entrenched in the literature through the 1980s and 1990s (Fielding, Frank and Isbell, 2008).

In the early Carboniferous, at about 320 Ma, the continental assembly processes had finally resulted in the formation of Pangea. Increased chemical weathering, which was associated with these tectonic movements, accompanied the subsequent consolidation of Pangea. This chemical weathering has also been suggested to have played an important role in the Late Paleozoic Ice Age (e.g. Link, 2009; Tajika, 2007) since the drawdown of carbon dioxide from the atmosphere significantly cooled the Earth. Atmospheric carbon dioxide levels fell from ~5000 to 3000 ppmv in the early to mid-Paleozoic to around 400 ppmv at the Carboniferous–Permian boundary (Shi and Waterhouse, 2010), but chemical weathering may not have been the only driver.

The early Paleozoic witnessed the arrival of plants on to the land followed by their subsequent colonisation of vast terrestrial expanses albeit confined to the banks of water courses (Chapter 2). Throughout the Devonian the plants continued to evolve both in terms of diversity and number but also in stature. As plants increased in height and complexity so too did their need for mechanical support. In time plants evolved lignin and secondary growth to reinforce this increasing stature, and bark to protect these arboreal plants. Such an increase in plant size had an ultimate effect on their environment. Plants continued to use atmospheric carbon dioxide in greater and greater quantities and thereby contributed to increasing carbon dioxide drawdown in the mid- to late Paleozoic. This was coupled with an increasing amount of carbon dioxide becoming sequestered. The development of lignin and bark were new innovations. Microbes and animals were unable (as yet) to break down the new biomacromolecules residing in these structural tissues, which led to ever increasing quantities of atmospheric carbon dioxide becoming locked up in the geosphere. These undecomposed plants accumulated throughout the Devonian and Carboniferous heralding the development of great tropical coal swamps that were to typify the Carboniferous and ultimately lead to the extensive coal deposits of North America and Eurasia.

Within the Carboniferous coal measures of North America and Europe are records of alternating sea level rise and fall thought to represent glacial and interglacial events. The coal measures are distinctly cyclical in sedimentation pattern (cyclothems) with coal sequences (interglacial sequences) alternating with sandstone sequences, which are indicative of drier, cooler (glacial) periods (DiMichele *et al.*, 2010; Heckel, 1994). Those deposits that occur near sea level record changes in the base level and provide evidence for eustatic sea level change, driven by waxing and waning of the Southern Hemisphere ice sheets in response to Milankovitch forcing (Veevers and Powell, 1987). Yet, by their very nature, the Northern Hemisphere sequences record the total global signature and provide little detail on location and extent of the ice sheets. For example, the decay of an ice sheet in one area, and the growth in another, would result in a neutral signal in far flung records such as North America or Europe, whilst sea level change implied by the base level changes in the Euramerican coal deposits is estimated to have been in the order of 50 to 60 m (Gibling and Rygel, 2008; Kammer and Matchen, 2008; Stemmerik, 2008). Throughout the Carboniferous–Permian interval global sea level change varied significantly with such changes being ~10–20 m at the beginning of the Late Paleozoic Ice Age and as high as 100–120 m at the peak of the Late Paleozoic Ice Age at the Carboniferous–Permian boundary (Rygel *et al.*, 2008).

Evidence for the Late Paleozoic Ice Age is recorded from fluvial to glacial sequences often associated with uplift (Powell and Veevers, 1987). These records can be found in large sedimentary basins such as the Karoo Basin (Africa), Transantarctic Basin (Antarctica) and Sydney Basin (Australia), together with the intracontinental rift basins of India (e.g. Krishna-Godvari Basin). Indeed late Paleozoic glacial rocks are widespread across the Southern Hemisphere in the continents of Antarctica (e.g. Buckeye and Darwin tillites, Pagoda Formation; Barrett, 1991), Africa (e.g. Dwyka Group; Visser and Loock, 1987; Visser, Loock and Collinston, 1987; Visser, van Niekerk and van der Merwe, 1997), Australia (e.g. Cullens Dimictite, Perth Basin; Eyles, Mory and Backhouse, 2002; Mory, Redfern and Martin, 2008), India (Talchir Formation; Tiwari and Tripathi, 1992), South America (e.g. Itararé Group; Rocha-Campos, dos Santos and Canuto, 2008) and the Falkland Islands (Bluff Cove Formation, Fitzroy Tillite Formation; Trewin, MacDonald and Thomas, 2002). These glacial strata form part of the Gondwanide sequence of rocks that led to the view that much of the Southern Hemisphere was glaciated in the Carboniferous and Permian. Detailed investigations of the timing and extent of the Gondwana glaciation (Crowell and Frakes, 1971, 1972; Frakes and Crowell, 1969, 1970; Frakes, Matthews and Crowell, 1971) suggested a large continental ice sheet in the latest Carboniferous and Early Permian. Subsequent studies identified three phases of development: the late Carboniferous localised alpine glaciation, followed by an earliest Permian continental ice sheet that then decayed through the remainder of the Permian (Crowell, 1978; Frakes, 1979). This contrasts with the interpretation by Veevers and Powell (1987) that identified an early Carboniferous precursor followed by extensive continental-wide ice sheets in the late Carboniferous and Early Permian. The lack of late Carboniferous sediments in Gondwana was interpreted as evidence for extensive continental-scale ice sheets under which sediment could not accumulate as sediments were only deposited once the ice sheet had started to collapse (Veevers and Powell, 1987).

Recent reviews of the Late Paleozoic Ice Age across Gondwana have revealed a complex history of glaciation across southern Pangea (Fielding, Frank and Isbell, 2008; Isbell *et al.*, 2003). Since our knowledge of the timing of ice formation in different parts of Gondwana has improved, researchers are beginning to question the presence of a long-lived, large ice sheet model (e.g. Veevers and Powell, 1987) and are replacing it with a model of a more dynamic ice sheet with multiple centres (Fielding, Frank and Isbell, 2008). Therefore the emerging picture is one of regional ice centres and several discrete intervals of glaciation separated by periods of reduced ice sheets (Fielding, Frank and Isbell, 2008; Isbell *et al.*, 2003; Visser, 1989). For example, during the early Carboniferous several glacial episodes are recognised in northern South America (Caputo *et al.*, 2008), whilst at the same time sedimentary basins in eastern Australia record largely continental sedimentation in alternating marine and fluvial to lacustrine environments (Talent *et al.*, 2002). In contrast, by the late Carboniferous and Permian the sedimentary histories in several basins across New South Wales (e.g. Sydney Basin) and Queensland (e.g. Cooper and Bowen basins) have an extensive record of the late Paleozoic glaciation in a variety of settings ranging from continental through to marine environments (Eyles, Eyles and Gostin, 1997, 1998; Jones and Fielding, 2004, 2008). Compare this with coeval sequences in northern South American basins that record mid-Carboniferous deglaciation sequences, which in turn are overlain by strata that lack evidence of glaciation (Henry *et al.*, 2010). These differences can in part be attributed to the drift of the supercontinent Pangea across the high southern latitudes from the Devonian to Permian (Figure 3.1), yet the factors driving climate change in the late Carboniferous and Permian are complicated and still under discussion (see Stephenson, Angiolini and Leng, 2007; Tabor and Poulsen, 2008).

The Gondwana polar glaciation has its origins in the Late Devonian, but does not become fully established until the late early Carboniferous. The glacial period extends to at least the Middle Permian with localised ice sheets, possible montane, persisting into the Late Permian in areas such as eastern Australia. In piecing together the global record four discrete glacial intervals are recognised from the late early Carboniferous through to the late Carboniferous that reflect the growth and decay of several discrete ice sheets (Fielding *et al.*, 2008a, 2008b). In the Permian four further glacial intervals can be distinguished between the Early Permian and the Late Permian (Fielding *et al.*, 2008a, 2008b) (Figure 3.2). These Early Permian glacial intervals are the most pronounced and widespread suggesting that a large ice sheet was in place, and this is reinforced by evidence for Northern Hemisphere polar glaciation. Within each glacial interval short-term fluctuations are evident (e.g. varved sediments representing annual variation) and cyclical sedimentation patterns suggest a sequence of glacial–interglacial cycles (Milankovitch cycles) with probable longer-term fluctuations related to tectonic drivers such as organic matter burial (Birgenheier *et al.*, 2010; Cleal and Thomas, 2005).

Towards the end of the Early Permian the ice sheet collapsed dramatically, which in turn opened up new environments for plants to colonise in the high southern latitudes. The collapse of the ice sheet has strong parallels with climate scenarios suggested for current climate warming, although it is unclear what factors triggered the collapse of the Permian ice sheet. Pangea continued to occupy high southern latitudes, but some workers have suggested that the drift of Pangea northwards, resulting in the pole moving from the centre of the present day Antarctic continent towards the coast, set up oceanic circulation patterns that brought warm tropical water into polar regions. However, other scenarios have suggested a combination of sea level change (Visser, 1997) and global warming (Wopfner, 1999) driven by development of the Northern Hemisphere monsoon (Parrish, 1993). Although the exact mechanism, or series of events, that triggered the final collapse of the ice sheet is unknown, what remains clear is that, based on oxygen isotopes of marine shells, sea water became isotopically lighter indicating a release of light oxygen ($\delta^{16}O$) tied up in ice sheets, and that sea water temperatures also rose by several degrees (Korte *et al.*, 2008; Scheffler, Hoerness and Schwark, 2003; Veizer *et al.*, 1999).

Whilst the extent of the ice sheet rapidly reduced in the late Early Permian from its peak of development in the Early Permian, parts of Gondwana still experienced glacial environments. In the Late Permian the glacial intervals are less extensive with reduced severity, typically lasting 1–8 Ma, and are separated by non-glacial periods of similar length (Fielding *et al.*, 2008a, 2008b). Within eastern Australia glaciation is recorded well into the Late Permian (Fielding *et al.*, 2008a) yet coeval with the presence of glacial ice in some parts of Gondwana other areas were ice-free. For example, based on palaeolatitude alone eastern Australia lay at ~60° S and was glaciated whilst regions that were closer to the pole (e.g. Transantarctic Basin) remained ice-free. Although there is currently no explanation for this, and it may indeed seem counter intuitive, similar conditions are seen on Earth today: Greenland with its ice cap lies at the same latitude as Norway, Sweden and Finland, all regions that are unglaciated and experience milder climates as a result of the warm Gulf Stream current.

In the aftermath of the Late Paleozoic Ice Age the global climate entered a transitional phase (with episodic and sporadic glaciations mostly restricted to high latitudes) through the late Early and Middle Permian, before its transition into the greenhouse state towards the end Permian (Shi and Waterhouse, 2010).

Global palaeovegetation

Elements of the low-growing, simple and shallow-rooted plants that had comprised the global terrestrial vegetation during the early Paleozoic (Chapter 2) were now evolving to become more widespread both in structure and extent. By the Middle Devonian a number of lineages had evolved the arborescent habit and these included the lycopsids, cladoxylopsids and progymnosperms. As the plants became increasingly substantial above ground so their root systems were also becoming more substantive below ground. They were able to penetrate deep into the substrate stabilising stream banks and enhancing the formation of soil (Algeo and Scheckler, 1998; Davies and Gibling, 2010). In time this enabled the plants to become less restricted to water courses and thus able to invade drier areas away from river banks and lake sides. By the Late Devonian vegetation resembling modern forest is recorded for the first time as the primitive vascular plants, which had characterised the earlier Devonian, were beginning to disappear (e.g. Gensel and Berry, 2001). The progymnosperm *Archaeopteris* was dominant in these new ecosystems and has been recorded from virtually all known Devonian landmasses (including Euramerica, Gondwana, China and Siberia) ranging from tropical to sub-polar latitudes. Progymnosperms were characterised by secondary-thickened, gymnosperm-like axes with many exhibiting the arborescent habit (e.g. Scheckler and Galtier, 2003). *Archaeopteris* reached estimated heights of ~30 m with trunks in excess of 1 m and were formed of wood (i.e. *Callixylon* that resembles wood of modern conifers). With its lateral buds, *Archaeopteris* is thought to have been perennial with some individuals recorded to have lived for at least 40–50 years. Webbed, planar, deciduous leaves giving a frond-like appearance would have produced a relatively dense shade on the forest floor. Below ground their root system was extensive (>1m in depth) and able to accelerate paedogenesis. This resulted in the creation of new terrestrial habitats where temperature, light and humidity became moderated, for part of the year at least, and organic material was returned to the biosphere on an annual basis. These advances had significant impacts on the terrestrial environment and also freshwater and marine ecosystems (Algeo and Scheckler, 1998). By the Devonian–Carboniferous transition these first forest biomes had all but disappeared as *Archaeopteris* suddenly became extinct.

Throughout the Late Paleozoic Ice Age, from the Carboniferous and into the Early Permian, further diversification of the existing groups of vascular plants was underway (Cascales-Miñana, 2011). An increasing assemblage of seed plants, zygopterids (ancient 'ferns'), sphenopsids and arborescent lycopsids appear in the early Carboniferous fossil record. Since these plants were much larger than those of the Early Devonian, the isolated fragments and organs coupled with their increasing structure and complexity make it difficult to reconstruct them. Nevertheless, the vegetation of the early Carboniferous was characterised by the Equisetales (e.g. *Calamites*), Sphenophyllales (vine-like plants), Lycopodiales (club moss), Lepidodendrales (e.g. the arborescent club moss *Lepidodendron* and *Sigillaria*), Filicales (ferns), Medullosales and the Cordaitales (e.g. *Cordaites*), and were joined in the late Carboniferous by other groups that include the Voltziales, the Cycadophyta and Callistophytales.

The global vegetation at this time can be divided into three broad palaeogeographic provinces (Chaloner and Lacey, 1973; DiMichele, Pfefferkorn and Gastaldo, 2001). The Euramerica province was tropical in location and comprised at least two major biomes, the wetlands and the seasonally dry environments. Coal swamp floras of peat-forming mires and organic-rich swamps characterised the wetlands. Arborescent lycopods dominated the swamps and mires with lycopsid trees in ecologically distinct habitats. *Lepidodendron* and

Sigillaria were widespread, along with *Calamites*, forming a tall (up to and exceeding 20 m), yet open, vegetation with an understorey of smaller pteridosperms and *Cordaites*. These thick peats and extensive swamp forests would later become the Carboniferous coal measures that can be found across Europe and North America today. The seasonally dry environments were more advanced, relative to the wetlands, and were dominated by seed plants including conifers, peltasperms, ginkgophytes, pteridosperms, cordaites and complemented by minor amounts of ferns and sphenopsids but virtually no Lycopodiales. The second palaeogeographic province, the north temperate Angaran province, was dominated by a variety of lycopsids mostly of sub-tree stature. This was later replaced by a pteridosperm-dominated assemblage probably as a response to climate cooling and a decrease in seasonal rainfall. In turn this pteridosperm-dominated assemblage was followed by a succession of cordaitean gymnosperm vegetation types that prevailed until the later Permian. The third palaeogeographic province, the Gondwanan province was south temperate in location and of low diversity. In the late Carboniferous pteridosperms along with lycopsids, *Cordaites* and ginkgophytes prevailed, but by the Permian these floras had become more diversified and were characterised by *Glossopteris* along with conifers and ferns, as well as sphenopsids and lycopsids. Even though each province was characterised by distinctly different floras, overlaps between these floral provinces are known from several geographic areas and the provinces may have changed in composition over time (see DiMichele, Pfefferkorn and Gastaldo, 2001 and references therein for further details).

These biotic communities would have responded to the onset, demise and internal glacial–interglacial cyclicity that took place during the Late Paleozoic Ice Age, yet published information on continental to global scale studies is limited and those that are available have focused mainly on terrestrial ecosystems in palaeotropics (e.g. Cleal and Thomas, 2005; DiMichele, Pfefferkorn and Gastaldo, 2001; DiMichele *et al.*, 2009; Gastaldo, DiMichele and Pfefferkorn, 1996). The considerable research effort expended on these coal measures flora is probably due to the exceptional preservation and also, in part, to their economic importance. From these studies it is clear that even in the tropical regions the influence of glacial events on the vegetation was pronounced with glacial–interglacial cycling resulting in changes in the vegetation. During the glacials (sandstone sequences), the interglacial floras (typified by the late Carboniferous coal swamp vegetation, and dominated by aborescent lycopods along with ferns and seed plants) are thought to have retreated to refugia only to expand again during the subsequent interglacial (coal measure floras) as the glacial floras retreated to (possibly upland) refugia. The vegetation of the glacial interval was no less complex even though this vegetation is only just beginning to be understood (Falcon-Lang, 2003, 2004; Falcon-Lang and Bashforth, 2004, 2005). These glacial floras comprised ecosystems rich in xeromorphic elements (DiMichelle *et al.*, 2010; Falcon-Lang, 2003, 2004), and were probably confined to upland areas during the warmer periods (Falcon-Lang, 2003, 2004) when the warm coal swamp floras reassembled. Each cycle resulted in the disassembly of one floral type and the reassembly of another, but the composition of the reassembled vegetation differed from its previous cycle and so too did the dominant elements (DiMichelle *et al.*, 2010). So although extinction was low during the ice age, the impact on ecosystems and the success of groups was quite pronounced since certain taxa were only dominant at certain times even though they were ever-present in the vegetation.

In the mid-Permian as climates became drier the global vegetation became more regional. Arborescent lycopods were progressively replaced by more advanced seed plants, cycads and early conifers including the ancestors to many of the conifer families still growing today. The conifers characterised the northern regions of Pangea, whilst Gondwana was dominated by the broad-leafed *Glossopteris* that came to dominate the Southern Hemisphere including Antarctica in the high southern latitudes.

Geological setting and Antarctic palaeogeography

Evidence for the Antarctic Carboniferous and Permian glacial world is exposed in both the Victoria Land Basin and the large Transantarctic Mountains Basin (Collinson *et al.*, 1994), and extends into the Ellsworth and Whitmore mountains (Collinson, Vavra and Zawiskie, 1992; Matsch and Ojakangas, 1992), Pensacola Mountains (Schmidt and Ford, 1969; Storey *et al.*, 1996), Shackleton Range (Tessensohn, Kleinschmidt and Buggisch, 1999), Theron Mountains (Stephenson, 1966) and Dronning Maud Land (Figure 1.1).

In the Victoria Land Basin a basal ?late Carboniferous to Early Permian tillite (Darwin or Metschel) is overlain by fluvial coal measures (Weller Coal Measures) deposited by braided river systems (Figure 3.3). The coal measures are in turn overlain by terrestrial deposits of Triassic age (e.g. Feather Conglomerate; Barrett and Fitzgerald, 1985). In the Transantarctic Mountains ?late Carboniferous to Early Permian glacial rocks (e.g. Pagoda Tillite) are overlain by glaciomarine/glaciolacustrine sequences (e.g. MacKellar, Weaver and Discovery Ridge formations) that evidence a decreasing influence of ice and the development of fluvial systems and associated coal swamp floras (e.g. Buckley Formation, Queen Maud Formation, Mount Glossopteris Formation) (Figure 3.3). These Permian fluvial sequences pass conformably or disconformably upwards into fluvial Triassic deposits (Collinson *et al.*, 1994) although these deposits are not present in all regions. While the Victoria Land and central Transantarctic Mountains sequences show a similar pattern of development, there is a clear trend towards lacustrine or perhaps even brackish marine environments in outcrops further along the range (e.g. Ohio Range).

The sequences extending from Victoria Land through the central Transantarctic Mountains contrast with those in the Ellsworth Mountains and Pensacola Mountains where glaciomarine rocks indicate their deposition close to sea level (Figure 3.3). The ?late Carboniferous to Permian sequences from this part of Antarctica are some of the thickest, with the Whiteout Conglomerate estimated at >1000 metres thick (Matsch and Ojakangas, 1992). Here the movement of ice across the land surface is evidenced by glacial striated boulder pavements, whilst thick diamictites indicate a range of environments with deposition from glaciers and floating ice close to grounding lines, to deep water sub-ice shelf conditions (Matsch and Ojakangas, 1992). These all indicate the presence of an ice sheet in this region flowing into a marine basin (Matsch and Ojakangas, 1992). The overlying

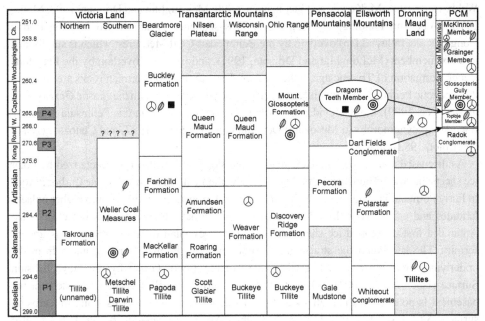

Figure 3.3 Permian stratigraphic chart for Antarctica broken up by regions and showing main lithostratigraphic units. Dark grey bars on the left represent glaciation events P1 to P4 recognised by Fielding *et al.* (2008a, 2008b). Note in Dronning Maud Land that several sections have been measured but the units have not been formally named. While outcrop have yielded Early to mid and Late Permian palynoassemblages, isolated morainal boulders contain palynoassemblages from time periods not represented in the outcrops suggest that a complete Permian sequence may be present under the ice. The presence of fossils is marked by a ⊗ for palynological assemblages, ∅ for macroflora, ◉ for wood, and ■ for petrified peat. PCM = Prince Charles Mountains, Light grey areas = no cutcrop.

Polarstar Formation consists of a deglaciation sequence beginning with deep water pro-delta mudstones that progressively shallow upwards into delta top sediments – a sequence that records infilling of the marine basin and development of terrestrial fluvial deposition (Collinson, Vavra and Zawiskie, 1992) (Figure 3.3). Whilst the sequence lacks the coal seams seen in the Transantarctic Mountains, cyclic deposition is present and abundant carbonaceous matter, including macrofossil floras, have been recovered from the Polarstar Formation (T. N. Taylor and E. L. Taylor, 1992).

Additional Permian sequences are found around the margin of present day East Antarctica with indirect evidence of the extent of these deposits being recorded in reworked material (Truswell, 1982). At this time intracontinental rift valleys had formed between Antarctica and Africa (e.g. Dronning Maud Land), and Antarctica and India (e.g. Lambert Graben, Prince Charles Mountains; Arne, 1994; Harrowfield *et al.*, 2005). The sequences in Dronning Maud Land are similar to those of the Transantarctic Mountains and Ellsworth Mountains sequences with an older basal tillite and younger fluvial strata. But the outcrop is discontinuous thus making correlation difficult. The Permian sequence exposed in the Prince Charles Mountains is slightly younger, ranging from late Early Permian through to Late Triassic (Fielding and

Webb, 1995; 1996; McKelvey and Stephenson, 1990; McLoughlin and Drinnan, 1997; Mond, 1972; Webb and Fielding, 1993a, 1993b). This sequence is subdivided into the Radok Conglomerate (alluvial fan) overlain by the Bainmedart Coal Measures, which is subdivided into six members (McLoughlin and Drinnan, 1997), and in turn is overlain by the Flagstone Bench Formation of Triassic age. Other scattered occurrences of Permian rocks are found in the Antarctic Peninsula (Erehwon Nunatak) where sandstones contain diagnostic *Glossopteris* leaves (Gee, 1989). Indirect evidence for Permian rocks in the Antarctic Peninsula is seen in reworked palynomorphs in Mesozoic (Dettmann and Thomson, 1987) and Cenozoic (Askin and Elliot, 1982) sequences.

As discussed above, the Late Devonian to mid-Permian world was characterized by polar ice sheets that waxed and waned, but peak ice development occurred in the late Carboniferous to Early Permian. During this time period Antarctica progressively drifted into higher palaeolatitudes and came under the influence of a polar climate. However, it is not clear when Antarctica first developed ice sheets due to a Late Devonian to late Carboniferous unconformity. The late Paleozoic strata (late Carboniferous to Permian) are separated from the underlying lower Paleozoic strata (Devonian) of the Taylor Group by the Maya Erosion Surface (Isbell *et al.*, 2003). Where this erosion surface overlies crystalline basement the basement is polished with striations and glacial deposits occur in places along this erosion surface. As a result not only is it unclear when Antarctic glaciation began, but it is also uncertain how much sediment might have been removed by the Carboniferous ice sheets.

This unconformity surface was inferred to have been the result of extensive Carboniferous glaciation (Barrett and Kyle, 1975; Veevers, 2000), and the lack of deposition thought to be the result of an extremely cold ice sheet that was frozen to the bedrock (Powell and Veevers, 1987). However, this interpretation has been challenged since the Devonian sediments lack the type of deformation typical of ice sheets having overridden pre-existing sediments, and thus implies a lack of both extensive and thick ice sheets in the Carboniferous (Isbell *et al.*, 2003). Palaeoflow directions of the ice, as evidenced by glacial striations and palaeocurrent indicators in the glacial sediments, suggest two centres of ice accumulation (Isbell, Gelhar and Seegers, 1997) – one in the central Transantarctic Mountains and the other in the Horlick Mountains – with flow towards the Ellsworth Mountains. Glacial rocks were deposited in the narrow basin that developed behind a coastal range and is thought to have been ice marginal (Isbell *et al.*, 2001, 2003) rather than having been deposited under an extensive ice sheet as postulated by other researchers (Scotese, Boucot and McKerrow, 1999; Veevers, 2000, 2001; Veevers and Powell, 1987).

While the extent of the pre-late Carboniferous ice sheet (Figure 3.4) has been questioned (Isbell *et al.*, 2003), the expansion of glacial conditions throughout the late Carboniferous and into the early Permian is also problematic. The main problem is determining the age range of the glacial deposits since they lack rocks suitable for geochronological dating. Palynofloras from the upper part of the Darwin Tillite and Buckeye Tillite contain palynomorphs referred to the *Parasaccites* assemblage (Kyle, 1977). These palynofloras show greatest similarity to the earliest Permian (Asselian) Stage 2 microfloras from Australia (Mory and Backhouse, 1997). However, Kyle (1977) suggested that these palynofloras

Figure 3.4 Reconstruction at 295 Ma illustrating the range of hypotheses for the extent of the Permian Ice Sheet. White area delineated by the dashed line represents the extent of the ice sheet after Scotese, Boucot and McKerrow (1999); light grey ice sheet extent follows that of Zeigler *et al.* (1997); dark grey shows the extent of land.

might also be late Carboniferous in age, a date that has been widely cited in the literature, but has not been demonstrated. Further work is needed to correlate the oldest palynofloras to accepted international stages. While early Permian spore-pollen assemblages are known from the upper parts of the tillites (Barrett and Kyle, 1975; Kyle, 1977) no age control is present at the base of the tillite units making it unclear when exactly these glacial sediments started to be deposited.

The onset of deglaciation is equally difficult to tie down due to inconsistent preservation and the lack of continuous vertical sampling to provide palynostratigraphic resolution. Perhaps the best-dated sequence occurs in Dronning Maud Land where a tillite to glacial fluvial sequence records ice retreat and ultimately grades up into non-glacial strata (Larsson, Lindström and Guy-Ohlsen, 1990). Several samples from this sequence contain palynofloras that can be dated as Asselian to Sakmarian (Larsson, Lindström, Guy-Ohlsen, 1990; Lindström, 1995a), and so providing some evidence for the timing of deglaciation. This is supported by evidence from the Transantarctic Mountains where fluvial deposits overlying the tillite (e.g. Weller Coal Measures) contain Australian Stage 3 floras (Sakmarian to Artinskian). However, no micro-floras have been obtained from the base of the fluvial unit. This suggests rapid deglaciation in the Early Permian – a process envisaged to have been fairly rapid through a series of feedbacks although it is still unclear what mechanism actually drove the deglaciation in Antarctica. In southern Africa, however, the Early Permian deglaciation appears to be the result of destabi-lisation of a grounded ice sheet (Haldorsen *et al.*, 2001). The collapse of the grounded marine ice sheet resulted in the retreat of the ice sheet by several hundreds of kilometres. Whatever the cause of the deglaciation in Antarctica, it resulted in a cascade of effects that saw the retreat of the ice sheet, the development for a short period of glacial landscapes with mountain glaciers that influenced the hydrology, apparent sea level fall as a result of isostatic uplift, and ultimately the development of high latitude fluvial systems with coal mires.

During the early to mid-Permian the Victoria Land Basin was a narrow (100 to 150 km wide) elongate basin that possibly formed as a result of glacial scouring to form a trough-like depression in the landscape (Isbell and Cúneo, 1996). Emerging from the highlands on either side of the basin were broad, braided rivers with the channel belt up to a kilometre wide. These graded downstream into more meandering systems with narrow channels (a few hundred metres wide) adjacent to floodplains. The rivers ultimately flowed into the centre of the basin which was occupied by a long, linear lake system, with the larger rivers forming distinct Gilbert-type river deltas (Isbell and Cúneo, 1996). The adjacent floodplains developed mire environments as indicated by thin to thick coal seams. The morphology of at least some of the coals indicates that they developed on frozen ground and the freezing process produced distinct mounds typical of palsa mires[2] today (Krull, 1999). To the east a second elongate basin (Transantarctic Basin) extended to the coast (Ellsworth Mountains). Deglaciation saw the retreat of the ice shelf and development of glaciomarine environments followed by lacustrine conditions. Large river systems prograded into this basin, and through the Middle and Late Permian braided river systems with extensive floodplains occurred. These were dynamic systems with channels eroding into the adjacent floodplain deposits often resulting in parts of the peat mire collapsing into the river channel and being buried (Taylor, Taylor and Collinson, 1989). In places these river and lake systems contain evidence of animal life with extensive trace fossils (Fitzgerald and Barrett, 1986; Miller and Collinson, 1994) and fossils such as conchostracans (Babcock *et al.*, 2002) and non-marine bivalves (Bradshaw, 1984). Adjacent to the rivers and lakes, floodplains developed mires with extensive vegetation.

The slightly younger (late Early Permian to Triassic) sequences of the Prince Charles Mountains record the presence of large post-glacial river systems carrying sediments from the Gamburtsev Highlands in the south. While cyclicity in the sequence suggests orbital forcing influenced sedimentation in the older parts of the sequence (Fielding and Webb, 1996), there is also evidence that tectonic influences on sedimentation, and reactivation of uplift in the adjacent highlands, may have been more important in the younger part of the sequence (McLoughlin and Drinnan, 1997). In this part of Antarctica alluvial fans developed on the uplifted range flanks while the central part of the basin was occupied by low-sinuosity braided river channels that were kilometres wide. Avulsion of the river onto the floodplain resulted in crevasse splay environments near river channels, whilst more distally in quieter settings, small lakes and ponds formed adjacent to peat mires (McLoughlin and Drinnan, 1997). Soil surfaces were also well-developed in places (Turner, 1983).

In summary, the late Carboniferous and Early Permian (Asselian) saw extensive glaciation across much of Antarctica. This may have begun as montane glaciations that formed

---

[2] Palsa mires are subarctic mire complexes with permanently frozen peat hummocks. They are marginal permafrost features occurring uniquely at high latitudes in parts of Fennoscandia, Russia, Canada and Alaska.

local ice caps that amalgamated into a large ice sheet. Rapid deglaciation in the early Permian (latest Asselian to Sakmarian) saw the retreat of the ice sheet and the initial deposition of a large sheet of till. Summer melting of local glaciers provided high discharge into rivers and resulted in braided stream environments as sediment was flushed rapidly onto coastal plains. In more distal areas where sediment supply was low, peat mires were able to form over still frozen ground (i.e. palsa mires). Narrow sedimentary basins existed and the centres of these were occupied by long linear lake systems. As the landscape matured and erosion lowered the topographic relief, river systems became more stable and meandering rivers developed along with peat swamps that accumulated thick coal seams. Fluctuations in climate influenced the sedimentation patterns and resulted in cyclic sedimentation from higher energy river systems to low-energy peat deposition that can be found in Antarctica today.

Biases associated with fossil plant material

Before outlining the changes in biodiversity in the forests of the southern high latitudes we must stipulate that, as with all fossil evidence, interpretation of fossil biodiversity comes from many sources each with their own strengths and weaknesses. As plants are disaggregated into component parts (such as leaves, trunks, spores and pollen) different types of assemblages result, and subsequent methods of preservation can further complicate the picture. In addition, taphonomic processes such as the mode of transport, environment and entombing sediments (sandstones, lacustrine) can also differentially bias the fossil record (see Gee and Gastaldo, 2005 and references cited therein). Since these concepts are relevant to the fossil assemblages throughout this book, we provide a brief overview here.

The small size and resistant composition renders spores and pollen suited to long-distance transport by wind and water. Consequently micro- or palynofloras represent a wider aspect of the vegetation probably capturing better information about the composition of the regional vegetation. However, caution is needed as some plant groups produce more pollen than others leading to biases in the abundance signal; others have different preservation potentials that can bias the biodiversity signal. The more local vegetation is represented by those organs not suited to long-distance travel, such as wood and leaves, but these organs are not immune from bias.

Wood is often permineralised or petrified preserving a three-dimensional cast of the original material, whereas leaves are more commonly found in impression or compression floras. Permineralised wood floras provide insights into the forest component of the vegetation, but at the same time miss the herbaceous or non-woody component. In addition, wood anatomy is often conservative with little variation at the generic or even family level thus tending towards taxonomic clumping and an under-representation of the floral diversity present. Permineralised material can, nevertheless, also provide greater insights into the anatomy, biology and ecology of the parent plants (e.g. Niklas, 1994). However, as permineralised material only samples a very local part of the vegetation it can be limited in the amount of biodiversity it records.

Leaves are often preserved as impression and compression floras. Impression floras in general have a lower diversity when compared with compression floras, since critical information about the cuticular features are lost thus reducing the ability to discriminate between morphotaxa.

All floras can be biased by selection processes that fail to preserve critical characters whilst preserving others. Therefore, ideally, information obtained from different types of floras exhibiting different types of preservation should be used together to constrain biodiversity and enable taxonomic segregation for that time period (e.g. different types of *Glossopteris* leaves; Gould and Delevoryas, 1977; Pigg, 1987, 1990; Pigg and Nishida, 2006; Pigg and Taylor, 1990).

In addition to the biases recognised above, there are known nomenclatural issues with Permian taxonomy that over-represent groups such as the glossopterids. For example, Lindström, McLoughlin and Drinnan (1997) demonstrated that within one microsporangia of *Arberiella*[3], there were as many as four genera and five species of previously described dispersed pollen taxa suggesting that the Permian palynological record may actually over-estimate the true diversity of this group of plants.

In spite of these issues an integration of all the information gleaned from as many palaeofloras as possible enable generalisations regarding the palaeoecology. Here (and in subsequent chapters) we combine the published evidence from impression, compression, palynoassemblages, wood and permineralised material to consider the floristic biodiversity present in these, and other, early ecosystems bearing in mind associated biases.

Biodiversity and ecology in the late Paleozoic southern high latitude boreal forests

Throughout the Permian, the regional patterns within the vegetation that could be seen globally (Figure 3.5) are characterised by relatively low diversity high latitude floras and relatively high diversity mid-latitude floras (Rees, 2002) (although this to some extent reflects our lack of detailed knowledge). The highest southern latitude macrofloras of the Gondwanan province are no exception. These are dominated mainly by *Glossopteris*, but with a significant contribution from *Paracalamites* (e.g. Rigby and Schopf, 1969). Generally, leaf material produced by the *Glossopteris* plant group overwhelms fossils derived from other groups making these other groups appear exceedingly rare. *Glossopteris* was obviously an important contributor to the Permian boreal forests, extending from within a few degrees of the pole (Taylor, Taylor and Cúneo, 1992) to the mid-latitudes (e.g. Høeg and Bose, 1960), but the composition of the forests that covered these latitudes did vary. Detailed examination of palynofloras (e.g. Lindström and McLoughlin, 2007) is beginning to provide a greater understanding of these high latitude floras. The findings of these authors illustrate a more diverse Permian vegetation, particularly in the Antarctic region, during the Late Permian. In contrast, floras from lower latitudes are richer

---

[3] A morphogenus of dispersed Permian microsporangiate reproductive organs thought to be of glossopterid affinity.

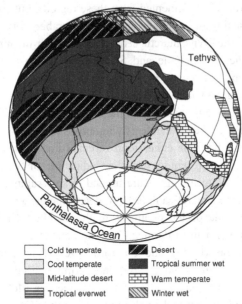

Figure 3.5 Early Permian reconstruction illustrating the inferred biomes based on analysis of leaf physiognomic features. Biome boundaries adapted from Rees *et al*. (2002).

in other groups such as ferns, ginkgos and conifers (e.g. Høeg and Bose, 1960). A global compilation of Permian floras based on morphological categories highlights these regional differences and allows the recognition of several distinct biomes or climate zones (Rees *et al*., 2002). Using this approach the Antarctic floras are considered to be equivalent to the cool to cold temperate biome.

Evidence for the biodiversity of the late Paleozoic vegetation in Antarctica comes from three main areas of outcrop. Firstly, sequences in Dronning Maud Land are the most poorly known – they are the least continuous with several of the sections being discontinuous (Guy-Ohlsen and Lindström, 1994; Hjelle and Winsnes, 1972; Juckes, 1972). These discontinuous sequences do, however, contain macrofloras (McLoughlin, Larsson and Lindström, 2005; Plumstead, 1975) and have been dated, using palynofloras, as ranging from early (Lindström, 1995a) to early Late Permian in age (Lindström, 1995b). Isolated clasts from moraines indicate the former (or sub-glacial) presence of Late Permian strata (Lindström, 1996, 2005). The discontinuous nature of these outcrops makes it difficult to build a chronology, but the short sections and the multiple palynological samples examined and counted provide important contributory evidence for the vegetation that grew in this part of Antarctica.

Further evidence regarding the palaeoecology of the late Paleozoic comes from outcrops that extend from Victoria Land through the Transantarctic Mountains and into the Ellsworth and Pensacola Mountains. These sequences span the late Carboniferous up to the Permian–Triassic boundary. Impression floras (e.g. Plumstead, 1962, 1964, 1975; T. N. Taylor and E. L. Taylor, 1992) and petrified material (e.g. Schopf, 1965, 1970, 1971; Smoot and Taylor,

1985, 1986a, 1986b) are well-documented from these regions, but considerable effort has been expended in extracting the palynoassemblages (Farabee, Taylor and Taylor, 1990, 1991; Kyle, 1977) from the Transantarctic sequences often for the main purpose of dating (e.g. Kemp, 1973; Kyle, 1977; Kyle and Schopf, 1982). This has not been entirely straightforward as several challenges have presented themselves that first needed to be overcome. Such challenges include the thermal alteration of the sequences by Jurassic Dolerite intrusions that have largely removed the organic matter. This has made it difficult to predict which samples will yield suitable palynomorphs for reliable dating. Consequently there are relatively few studies with either good vertical (time) representation, or counts of the abundance of grains that would give an indication of vegetation composition. Consequently most palynological studies are limited to spot samples that have then been linked to sequences outside Antarctica in order to determine the age.

Finally, on the other side of the continent in the Lambert Graben (Prince Charles Mountains) a late Early Permian to Permian–Triassic boundary sequence, with relatively well-preserved palynofloras (Balme and Playford, 1967; Lindström and McLoughlin, 2007), impression floras (White, 1962, 1969) and permineralised material (McLoughlin and Drinnan, 1996; Neish, Drinnan and Cantrill, 1993; Weaver, McLoughlin and Drinnan, 1997), has also been studied to elucidate clues regarding the palaeovegetation. The vegetation changes are outlined below.

?late Carboniferous to late Early Permian glacial and ice-marginal environments

The earliest record of vegetation within the Permian glacial environments of Antarctica comes from palynomorphs recovered from various sedimentary settings. These range from recent deglaciated areas of alluvial plains with braided streams, to the edges of glacial lakes within the Transantarctic Mountains in the upper parts of the Darwin Tillite, Buckeye Tillite, Pagoda Formation and the overlying Roaring Formation (an equivalent of the MacKellar Formation) (Askin, 1998; Barrett and Kyle, 1975; Kyle, 1977; Kyle and Schopf, 1982), and the Mackellar Formation (Masood *et al.*, 1994) (Figure 3.3). In addition pollen assemblages were recovered from samples from Milorgfjella in Dronning Maud Land in a sequence where a basal tillite is overlain by glacially influenced sediments (Larsson, Lindström and Guy-Ohlsen, 1990; Lindström, 1995a).

Initial palynological investigations recognised that these environments contain a differ-ent, and older, pollen assemblage referred to as the *Parasaccites* Zone (Kyle, 1977). This assemblage is likely to be Asselian in age based on its compositional similarity to Australian palynozones (e.g. Mory and Backhouse, 1997) although some samples are been considered to be slightly younger (i.e. Sakmarian; Larsson, Lindström and Guy-Ohlsen, 1990). In the Transantarctic Mountains the diversity in the *Parasaccites* assemblage is low (21 species) with a small but significant component of bisaccate (non-taeniate) grains (up to 5% frequency) and a large proportion of monosaccate (*Parasaccites*) pollen grains (33%) (Kyle, 1977). In the late Carboniferous and Early Permian deposits of the

Southern Hemisphere monosaccate grains dominate coinciding with a peak in abundance and diversity of the seed fern *Gangamopteris* and the progymnosperm *Noeggerathiopsis*. In the Late Permian monosaccate pollen were also at a peak at a time when the *Noeggerathiopsis* plant was an important component of the flora. It is not surprising therefore that it has been suggested that the monosaccate grains may be linked to the *Noeggerathiopsis* group of plants (McLoughlin and Drinnan, 1996).

The bulk of the remainder of the *Parasaccites* assemblage is made up of abundant trilete (50%) and monolete (2%) spores, and monolete pollen (*Cycadopites cymbatus*, 9%). This implies that the vegetation was largely composed of spore-bearing plants including sphenopsids (e.g. *Calamospora*), ferns and probably lycopods.

The only other significant contributor to the *Parasaccites* assemblage, in terms of later changes to the diversity and composition of the microfloras, are taeniate saccate pollen grains, but these are present in low abundances in this zone. This pollen type is indicative of the glossopterids and indicates that these seed plants formed a minor component in the vegetation. The glossopterids are first represented by *Gangamopteris*, a leaf genus that is recorded only in the earliest post-ice sheet sediments across Gondwana, dating back to the earliest Permian (Archangelsky and Cúneo, 1984; Cúneo, 1996; Pant, 1996).

A palynoflora described from the MacKellar Formation in the Transantarctic Mountains provides a unique glimpse of high palaeolatitude conditions during the transition from icehouse to greenhouse during the Early Permian (Miller and Isbell, 2010). Even though the assemblage was only determined to generic level (Masood *et al.*, 1994) an insight into the prevailing vegetation can still be gleaned. Despite the lower taxonomic resolution monosaccates were more abundant (*Cannanoropollis* 30.5 to 68%; *Plicatipollenites* 0 to 50%; *Potonieisporites* 0 to 6%; '*Parasaccites*' 3 to 7%)[4] indicating that seed plants were in abundance at this time. Whilst this formation has lower frequencies of non-taeniate bisaccates (2 to 7%), as well as trilete spores (6 to 10%) and rare *Cycadopites*, relative to other Early Permian floras, this may simply reflect a difference in sedimentary setting rather than a difference in biodiversity.

The presence of low diversity assemblages of terrestrial palynomorphs suggests that the earliest vegetation that established during the last phases of glaciation in this part of Antarctica was dominated by ferns and lycophytes with subsidiary sphenopsids and scattered glossopterid plants. The presence of rich monosaccate assemblages indicates the importance of either *Gangamopteris* or *Noeggerathiopsis*. The lack of macrofloras including wood precludes further reconstructions of this earliest post-glacial vegetation.

A similar age sequence was described from Milorgfjella in Dronning Maud Land (Larsson, Lindström and Guy-Ohlsen, 1990; Lindström, 1995a). Here a basal diamictite is overlain by mudstones and sandstones with evidence of ice rafting (Guy-Ohlsen and Lindström, 1994; Larsson, Lindström and Guy-Ohlsen, 1990). The palynoflora at Milorgfjella has low abundances of taeniate bisaccate grains (up to 3%) indicating that glossopterids also formed a minor component of the flora. In contrast, trilete spores are dominant, up 50 to 70% of the

---

[4] It should be noted that *Cannanoropollis* and *Plicatipollenites* probably equate to the *Parasaccites* of Kyle (1977).

palynoresidue, with monosaccate grains contributing between 15 and 35% (Larsson, Lindström and Guy-Ohlsen, 1990). Amongst the spores, ferns are dominant and most diverse, but lycophytes and sphenopsids also form a significant component. Strong parallels can be seen within the few samples examined from the Transantarctic Mountains all suggesting similar flora existing at this time. Interestingly, pollen of *Cycadopites cymbatus* is abundant in the Transantarctic Mountains but is rare in Dronning Maud Land, testifying to local variation within an otherwise relatively homogenous vegetation.

The oldest macrofloras of ?late Carboniferous to late Early Permian age are associated with palynofloras, and are found in both the Milorgfjella sequence (Plumstead, 1962, 1975) and slightly younger floras known from southern Victoria Land (Barrett and Kyle, 1975). Although Plumstead (1962, 1975) recorded reasonable macrofloral diversity in the Milorgfjella sequence (Table 3.1) this has been reinterpreted by McLoughlin, Larsson and Lindström (2005). This flora occurs above the tillite and a glacially influenced sequence, in unit 5, and is typical of the earliest post-glacial floras that include the presence of lycophytes, possible fern fragments, a diversity of *Gangamopteris* seed plants, rare *Glossopteris*, and possible conifer axes (cf. *Walkomiella transvaalensis*). In contrast the older floras in southern Victoria Land is dominated by *Gangamopteris* and *Noeggerathiopsis* with low numbers of *Glossopteris* (Taylor *et al.*, 1989a). In these floras the rising importance of glossopterid plants can be seen by their increasing abundance and the obvious lack of fern, and lycophyte, foliage.

The best-documented record of environmental change associated with the early Permian deglaciation is represented in the Transantarctic Mountains where fluvial deposits of the Weller Coal Measures outcrop. As with the sequences in the Dronning Maud Land the oldest part of this sequence contains evidence of ice, including small dropstones in lacustrine settings, and striated pebbles, which are suggestive of a glacially influenced palaeoclimate (Barrett, 1991; Francis *et al.*, 1993). The sequences are interpreted to have been formed as the result of small local ice caps with heavy summer melting that resulted in high seasonal runoff that formed high-energy braided-type rivers (Smith, Barrett and Woolfe, 1998). Within these settings, and probably in more stable parts of the landscape, large trees up to 20 m tall have been observed (Francis *et al.*, 1993). The trees are rooted in sediment with shallow root systems and soils containing features consistent with the presence of permafrost (Krull, 1999).

The base of the Weller Coal Measures consists of braided river systems on the basin margins with more meandering stream facies towards the basin axis (Isbell and Cúneo, 1996). Upwards the sequence retains its braided systems but grades into meandering systems with coal seams and lacustrine units in the axis of the basin (Isbell and Cúneo, 1996). This suggests that the landscape became more stable and was accompanied by the development of soil surfaces and standing vegetation for longer periods under an ameliorating climate. The coals, although thin, often have features such as dome-shaped structures that suggest formation on a frozen subsoil (permafrost) and have been interpreted as palsa mires (Krull, 1999). Associated with these environments are leaves and trunks of *Gangamopteris, Glossopteris* and *Noeggerathiopsis* (Isbell and Cúneo, 1996; Taylor *et al.*, 1989a, 1989b). Although *Noeggerathiopsis* is apparently more abundant in the lower part of the succession, *Glossopteris* and *Gangamopteris* predominate in the upper parts. Based on the sedimentology of sites within the Transantarctic

Table 3.1 *Species list for the Permian of Antarctica.*
Note: Taxonomy follows McLoughlin, Larssen and Lindström (2005) for the material
identify by Plumstead (1975)

Taxa	Location	Formation/Age	Reference
Bryophyta			
Mercia angustica	Skaar Ridge	Buckley Formation, Late Permian	Smoot and Taylor, 1985
Lycophyta			
Lycopodiopsis pedroanus	Milorgfjella, Droning Maud Land	Early Permian	Plumstead, 1975
Collinsonites schopfii	Mt Achernar, central Transantarctic Mountains	Buckley Formation, Late Permian	Schwendemann *et al.*, 2010
Sphenophyta			
Paracalamites australis	Transantarctic, Ellsworth and Theron mountains	Polarstar Formation, Early to Late Permian	Lacey and Lucas, 1981; Rigby, 1969; T. N. Taylor and E. L. Taylor, 1992
Phyllotheca australis	Prince Charles Mountains	Bainmedart Coal Measures, Late Permian	McLoughlin, Larsson and Lindström, 2005
P. cf. *P. australis*	Theron Mountains; Milorgfjella, Dronning Maud Land	Early to Middle Permian	Plumstead, 1962, 1975
Phyllotheca sp. 1	central Transantarctic Mountains, Bowden Neve	Buckley Formation, Late Permian	Rigby, 1969
Phyllotheca sp. 2	Tillite Ridge, Wisconsin Range	Upper Member, Weaver Formation	Rigby, 1969
Schizonuera gondwanensis	Ohio Range on Mt Schopf, Mercer Ridge, and Leaia Ledge	Mount Glossopteris Formation, Late Permian	Rigby, 1969
Umbellaphyllites ivini	Sentinel Range, Polarstar Peak	Polarstar Formation, Late Permian	Rigby, 1969
Raniganija bengalensis	Ellsworth Mountains	Polarstar Formation, Late Permian	Rigby, 1969
Pteridophyta			
Skaaripteris minuta	Skaar Ridge	Upper Buckley Formation, Late Permian	Galtier and Taylor, 1994

Table 3.1 (*cont.*)

Taxa	Location	Formation/Age	Reference
Spermatophyta			
Cordaitales			
Noeggerathiopsis sp.	Prince Charles Mountains	Bainmedart Coal Measures, Late Permian	McLoughlin and Drinnan, 1996
Coniferales			
cf. *Walkomiella transvaalensis*	Milorgfjella, Dronning Maud Land	Early to Middle Permian	Plumstead, 1975
Buradia heterophylla	Mount Gran, southern Victoria Land	Early to Middle Permian	Rigby and Schopf, 1969
Glossopteridales			
Glossopteris ampula	Ohio Range	Mount Glossopteris Formation, Late Permian	Cridland, 1963
G. angustifolia	Ellsworth Mountains; Ohio Range; Theron Mountains	Polarstar Formation; Mount Glossopteris Formation, Early to Late Permian	Cridland, 1963; Lacey and Lucas, 1981; T. N. Taylor and E. L. Taylor, 1992
G. browniana	Ellsworth Mountains; Ohio Range; Theron Mountains	Polarstar Formation; Mount Glossopteris Formation, Late Permian	Cridland, 1963; Lacey and Lucas, 1981; T. N. Taylor and E. L. Taylor, 1992
G. communis *G.* cf. *G. communis*	Ellsworth Mountains; Theron Mountains	Polarstar Formation, Early to Late Permian	Lacey and Lucas, 1981; T. N. Taylor and E. L. Taylor, 1992
G. cf. *conspicua*	Ellsworth Mountains; Theron Mountains	Polarstar Formation, Early to Late Permian	Lacey and Lucas, 1981; T. N. Taylor and E. L. Taylor, 1992
G. cf. *G. cordata*	Ellsworth Mountains	Polarstar Formation, Late Permian	T. N. Taylor and E. L. Taylor, 1992
G. damudica	Ellsworth Mountains; Ohio Range	Polarstar Formation; Mount Glossopteris Formation, Late Permian	Cridland, 1963; T. N. Taylor and E. L. Taylor, 1992
G. indica	Ellsworth Mountains; Ohio Range; Theron Mountains	Polarstar Formation; Mount Glossopteris Formation, Late Permian	Cridland, 1963; Lacey and Lucas, 1981; T. N. Taylor and E. L. Taylor, 1992
G. linearis	Ellsworth Mountains	Polarstar Formation, Late Permian	T. N. Taylor and E. L. Taylor, 1992

Table 3.1 (*cont.*)

Taxa	Location	Formation/Age	Reference
G. stricta	Ellsworth Mountains; Theron Mountains	Polarstar Formation, Late Permian	Lacey and Lucas, 1981, T. N. Taylor and E. L. Taylor, 1992
cf. *Gangamopteris*			Lacey and Lucas, 1981
cf. *Gangamopteris angustifolia*	Milorgfjella, Dronning Maud Land; Theron Mountains	Early to Middle Permian	Lacey and Lucas, 1981, Plumstead, 1975
Gangamopteris sp. cf. *G. angustifolia*	Milorgfjella, Dronning Maud Land	Early to Middle Permian	Plumstead, 1975
Gangamopteris sp. cf. *G. douglasii*	Milorgfjella, Dronning Maud Land	Early to Middle Permian	Plumstead, 1975
Gangamopteris sp. cf. *G. obovata*	Milorgfjella, Dronning Maud Land	Early to Middle Permian	Plumstead, 1975
Gangamopteris sp. cf. *spatulata*	Milorgfjella, Dronning Maud Land	Early to Middle Permian	Plumstead, 1975
Arberiella	Ohio Range	Mount Glossopteris Formation, Late Permian	Cridland, 1963
Arberiella inflectada	Horlick Mountains	Mount Glossopteris Formation	Ryberg, 2009
Eretmonia singulia	Horlick Mountains	Mount Glossopteris Formation	Ryberg, 2009
Scale leaf (*Ertemonia?*)	Theron Mountains	Early Permian	Lacey and Lucas, 1981
Lakkosia kerasata	Skaar Ridge, peat	Buckley Formation	Ryberg, 2010
Plumsteadia ovata	Allan Hills, Mount Feather, Victoria Land; Mt Achernar, central Transantarctic Mountains	Weller Coal Measures; upper Buckley Formation, Early to Late Permian	Cúneo *et al.,* 1993; Kyle, 1974; Ryberg, 2009; Schopf, 1976; Taylor, Taylor and Isbell, 1989
Plumsteadia sp.	Ellsworth Mountains	Polarstar Formation, Late Permian	T. N. Taylor and E. L. Taylor, 1992
Rigbya arberoides	Allan Hills	Weller Coal Measures	Schopf, 1976; Townrow, 1967
Rigbya chtenia	Horlick Mountains	Mount Glossopteris Formation	Ryberg, 2009
Samaropsis	Theron Mountains	Early Permian	Lacey and Lucas, 1981
Samaropsis longii	Ohio Range	Mount Glossopteris Formation, Late Permian	Cridland, 1963

Table 3.1 (*cont.*)

Taxa	Location	Formation/Age	Reference
Australoxylon bainii	Prince Charles Mountains	Bainmedart Coal Measures, early Late Permian	Weaver, McLoughlin and Drinnan, 1997
Australoxylon mondii	Prince Charles Mountains	Bainmedart Coal Measures, early Late Permian	Weaver, McLoughlin and Drinnan, 1997
Vertebraria indica	Prince Charles Mountains, Transantarctic mountains, Droning Maud Land	Early to Late Permian	e.g. Lacey and Lucas, 1981, McLoughlin, Larsson, Lindström, 2005

Mountains, such as the Weller Coal Measures, studies suggest that the plants were growing in periglacial environments with strong seasonal amplitude, perhaps similar to those in the high northern latitudes today. Importantly, they were probably also influenced by glacial cycles similar to those seen today.

In spite of the limited data, it appears that plants rapidly colonised the land as ice sheets retreated from Antarctica. The first phase was dominated by ferns, lycophytes, sphenophytes and *Noeggerathiopsis* with rare glossopterids (*Gangamopteris*). It is envisaged that these communities colonised an environment with continuous permafrost, which meant that trees had to be shallowly rooted (Krull, 1999). Only in the warmest microclimates were glossopterids able to survive as their aerenchymous roots were probably not suited to seasonal freezing. These open communities were replaced by ones with a more prominent tree element as indicated by the increase in wood, with *Noeggerathiopsis* and *Gangamopteris* dominating, and *Glossopteris* forming the subsidiary component. As ice retreated to the hinterland *Gangamopteris* and *Glossopteris* became increasingly important forming substantial forests at the expense of all other groups. It is likely that a mosaic of vegetation existed and that trees were concentrated in areas influenced by milder conditions. Communities with similar *Gangamopteris* – *Glossopteris* compositions in the Sydney Basin of Australia were interpreted to represent *Gangamopteris* taiga (Retallack, 1980). Given the sedimentological evidence for the climate and environments, the low diversity polar forests of Antarctica were probably similar (Cúneo *et al.*, 1993).

Post-glacial floras of the late early Permian to latest Permian

Evidence for climate amelioration from the late Early Permian through to the latest Permian can be found within the Transantarctic Mountains in the transition from glacial-influenced fluvial systems to fluvial systems lacking evidence for glacial influence. Glacio-lacustrine units overlie glacial tillites and these in turn are overlain by sandstone-rich units (e.g. Fairchild Formation). These sandstone-rich units are overlain by sandstones with coal seams (e.g. Buckley Formation). While evidence for ice is lacking, the coal measure sequences all provide evidence for strong cyclicity in the climate. Analysis of the frequency of coal–sandstone cycles in the lower part of the main coal forming sequence (Bainmedart Coal Measures) of the Prince Charles Mountains implies that the cyclicity is consistent with Milankovitch cycles (Fielding and Webb, 1996) – a factor consistent with glaciation on other parts of Gondwana (Fielding *et al.*, 2008a, 2008b).

Sequences that provide evidence of the vegetation that grew at this time include those in the Transantarctic Mountains (e.g. Buckley Formation) and disconnected outcrops in Dronning Maud Land (e.g. Larsson, Lindström and Guy-Ohlsen, 1990). In the Transantarctic Mountains there are abundant macrofossil impressions and scattered palynological samples together with unique permineralised peat floras. The macrofossil impression floras (Cridland, 1963; Edwards, 1928; Lacey and Lucas, 1981; Lambrecht, Lacey and Smith, 1972; Plumstead, 1962; Rigby, 1969; Rigby and Schopf, 1969; Schopf, 1962, 1967, 1976; Seward, 1914; T. N. Taylor and E. L. Taylor, 1992; White, 1969) are generally of low diversity. The assemblages are dominated by broad, lanceolate to spathulate leaves with mesh-like veins that have been attributed to *Glossopteris*. Macrofloras lower in the sequence have both *Gangamopteris* and *Glossopteris*, whereas those higher in the sequence are dominated by *Glossopteris*. In the coal-bearing sequences *Glossopteris* fossils typically occur at the tops of the coal seams, whilst *Vertebraria*, the segmented roots of the *Glossopteris* plant, occur at the base. To date over 15 species of *Glossopteris* leaves have been recorded from Antarctic deposits with the most common being *Glossopteris indica* and *G. browniana*. However, the circumspection species within *Glossopteris* is in urgent need of re-examination. Fertile material is also present, with scale leaves (Cridland, 1963) and seeds (Cridland, 1963; Lacey and Lucas, 1981) being widely reported. Other fructifications include *Plumsteadia* (Kyle, 1974; Taylor *et al.*, 1989a) and *Samaropsis*, which are medium sized seeds similar to those attached to the glossopterid *Lidgettonia lidgettonioides* (McLoughlin, Larsson and Lindström, 2005). The second most commonly encountered foliage impression is that of the sphenophyte *Paracalamites australis* (Rigby, 1969; Rigby and Schopf, 1969). Other sphenophytes also occur, such as *Schizoneura*, *Umbellaphyllites* and *Phyllotheca* (McLoughlin, Larsson and Lindström, 2005; Rigby, 1969; Rigby and Schopf, 1969) and *Leslotheca* (McLoughlin and Drinnan, 1996). *Collinsonites schopfii*, a slender branching axis with eligulate serrated leaves, has been described from the Late Permian of the Transantarctic Mountains (Schwendemann *et al.*, 2010). Rare impressions of other seed plants such as *Cordaites* (Isbell and Cúneo, 1996) or *Noeggerathiopsis* (Taylor, Taylor and Isbell, 1989) have also been reported.

From the impression flora it can be seen that the Permian Antarctic floras are of low diversity and comprise two main groups: the glossopterids and sphenophytes. However, the sheer dominance of glossopterid foliage tends to obscure those plants that are present in low abundance within the vegetation.

Palynofloras give additional insights into the biodiversity of these Permian ecosystems and a more complete picture of the regional vegetation. Several studies have recovered relatively well-preserved palynofloras from different stratigraphic levels within the Transantarctic Mountains (e.g. Kyle, 1976, 1977 from the Weller Coal Measures; Farabee, Taylor and Taylor, 1990, 1991 from the Buckley Formation), the Prince Charles Mountains (Balme and Playford, 1967; Dibner, 1978; Kemp, 1973; Lindström and McLoughlin, 2007 from the Amery Group), and Dronning Maud Land (Lindström, 1994, 1995a, 1995b, 2005). In contrast

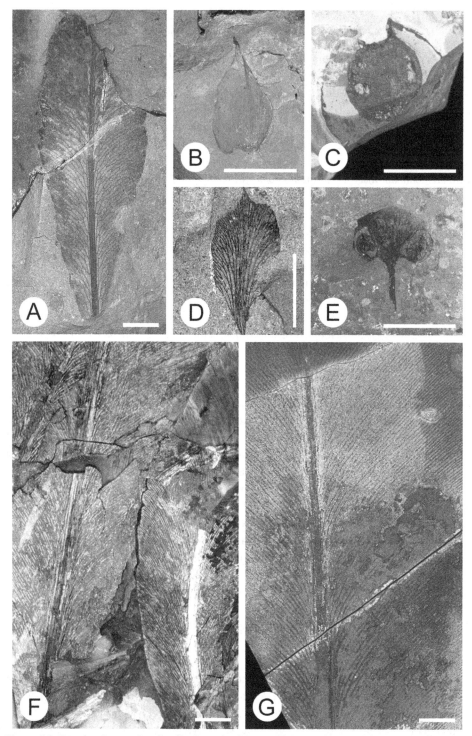

Figure 3.6 Permian impression fossils from the Mount Glossopteris Formation, Ohio Range. **A**, *Glossopteris* cf. *browniana*, AF355. **B**, bract scale, Mercer Ridge, Mt Schopf, AF778. **C**, *Samaropsis* sp. Mercer Ridge, Mt Schopf. **D**, bract scale Museum Ledge, Mt Schopf, AF337. **E**, ovulate fructification, *Rigbya arberoides*, Mercer Ridge, Mt Schopft, AF767A. **F**, *Glossopteris* cf. *indica* leaf mat, Ohio Range, AF358. **G**, large *Glossopteris* cf. *indica* with fine vein mesh, AF831. Scale bars A–G = 1 cm.

to the macrofloras, the palynofloras suggest much greater diversity but also point to the dominant role of glossopterids in the vegetation. In the older sequences (e.g. Weller Coal Measures; Kyle, 1977), bisaccate grains can form between 47 and 55% of the assemblages with half of these being taeniate and attributed to *Glossopteris*. This contrasts dramatically with the relative abundance in the post-glacial floras and points to dramatic changes in the vegetation. The increase in taeniate bisaccate grains continues up section until the Late Permian where it reaches abundances >60% (Lindström and McLoughlin, 2007), signifying the increasing importance of glossopterids in the ecosystems through the Permian.

Several deposits significant to studies on Permian ecosystems of the Antarctic have also been preserved. In the Transantarctic Mountains permineralised peat horizons occur on Skaar Ridge (Schopf, 1970; Taylor *et al.*, 1986) and Collinson Ridge (McManus *et al.*, 2002). Permineralisation of these peats occurred in at least three phases (Schopf, 1971) and served to preserve the anatomical structure of the plants in fine detail (Taylor and Taylor, 1990). It has been suggested that these permineralised peats were rafts of peat mire that were eroded into river channels. Once entombed in sand, a porous medium, they were then exposed to fluids rich in silica. An initial permineralising phase of chalcedony was followed by a second phase that infilled cell lumens and, in places, the peat was then recrystallised (Schopf, 1971). A second permineralised peat is known from the Prince Charles Mountains (see below). Both permineralised peat deposits have not only provided unique insights into the biodiversity of the Permian ecosystems, but also important information on the ecology and anatomy of the plants.

In the Transantarctic Mountains, Permian peat deposits are pervasively dominated by fragments of glossopterid plants including *Vertebraria* roots (Schopf, 1970), *Glossopteris* leaves (Pigg, 1990), wood (Taylor and Ryberg, 2007) and fertile material including ovulate structures (E. L. Taylor and T. N. Taylor, 1987; Zhao, Taylor and Taylor, 1995). Within the Skaar Ridge locality two anatomically distinct types of *Glossopteris* leaf taxa have been recognised (Pigg, 1990; Pigg and Taylor, 1985, 1987, 1990, 1993). The first leaf taxon has a well-developed fibrous bundle sheath with a two to three cell hypodermal layer, whereas the second taxon lacks the fibrous bundle sheath but has a prominent hypodermis (Pigg, 1990). Differences also extend to the vein meshes, which are fine in the former and similar to those seen in *G. angustifolia*, and coarse in the latter, comparable with *G. conspicua*. Moreover, there is also evidence of leaf attachment to stems (Pigg and Taylor, 1993) providing an important link associating leaf type with wood type. Ovulate organs are also preserved highlighting the manner in which seeds were attached to the glossopterid fruiting bodies (E. L. Taylor and T. N. Taylor, 1992). There is also a diversity of dispersed seeds (Klavins *et al.*, 2001; Smoot and Taylor, 1985, 1986b; Taylor, 1996; T. N. Taylor and E. L. Taylor, 1987) with some exhibiting exquisite detail including evidence of early development of polyembryos (Smoot and Taylor, 1986a). Although much has been learnt about the anatomy and biology of the *Glossopteris* plant through studies of these peat deposits, the emerging picture of other groups is equally important.

Smoot and Taylor (1986a) described the moss *Mercia angustica* (Bryales) from the peat. The leaves consist of several layers of cells in the central portion of the leaf forming a distinct midrib. Mosses are only rarely preserved in the fossil record and although this material lacks capsules, it provides important information on the anatomy of mosses at this period in Earth history. Ferns are also represented in the deposit by small, presumably creeping, stems of *Skaaripteris minuta* that are dichotomously branched (Galtier and Taylor, 1994). The stems are almost naked bearing only a few pinnules. Associated sporangia have been found that contain spores referred to *Horriditriletes* (Galtier and Taylor, 1994) — a spore type widely recognised from Permian deposits of Antarctica (Lindström and McLoughlin, 2007), including deposits in the Prince Charles Mountains, and elsewhere in Gondwana (Gould and Delevoryas, 1977). The anatomy of the stem and the type of sporangia suggest a relationship within the Filicales.

Whilst most studies of fossil material have concentrated on the macrofloral elements, only a few studies have focused on fungal material. In part this is due to the low preservation potential of fungi in general, but can also be attributed to the fact that fungi break down the material more usually studied. As such they are either ignored or under-collected since the host material would appear degraded and thus deemed not worthy of detailed investigation. The petrified peat material of the Transantarctic Mountains represents the forest floor, where most breakdown of plant material occurs, and so it is not surprising that the peat contains a diversity of fungi (García Massini, 2007a, 2007b; Stubblefield and Taylor, 1986; Taylor and Stubblefield, 1987). The fungi give further insights into the biodiversity and biology of these high latitude mires. Pocket-rot fungi have been observed in *Glossopteris* wood where hyphae with clamp connections were observed (Stubblefield and Taylor, 1985, 1986; Taylor and Stubblefield, 1987). In addition, García Massini (2007a) recognised an endoparasitic thallus-forming fungus, within roots, and assigned it to the extant fungal genus *Synchytrium permicus*. This material is important as it provides a link between extant chytrids – today a root-inhabiting fungus causing heterotrophy of the host plant often resulting in death (García Massini, 2007a) – and those described from the Carboniferous and Devonian (Taylor, Remy and Hass, 1992). In addition to the parasitic fungi, endomycorrhizal fungi have also been described: *Glomorphites intercalaris* chlamydospores and hyphae have been recorded in fossilised decaying root tissue. The hyphae have the distinct branching morphologies typical of endomycorrhiza, a pore-like attachment to the terminal hyphae, that points to it being a type of endomycorrhiza related to the symbiotic fungus Glomerales (García Massini, 2007b). The Transantarctic Mountains sequences have thus yielded considerable insights into the biodiversity and biology of the Permian ecosystems.

In contrast the Prince Charles Mountains provide a somewhat different picture of the Late Permian vegetation. This sequence has not been thermally altered in the same way as the Transantarctic Mountains sequence and so yields palynological assemblages more consistently. Detailed stratigraphic logging and sampling have been used to deduce the pattern of floristic change (Lindström and McLoughlin, 2007). This is augmented by macroflora records (McLoughlin and Drinnan, 1996; Ravich *et al.*, 1977; White, 1962, 1969) and a stratigraphically constrained permineralised peat.

Palynology from the Late Permian of the Prince Charles Mountains (McKinnon Member) reveals considerable diversity when compared with older sequences with at least 98 species identified so far. This may in part be due to better sampling than in either the Transantarctic Mountains or Dronning Maud Land sequences where diversity is recorded as being lower.

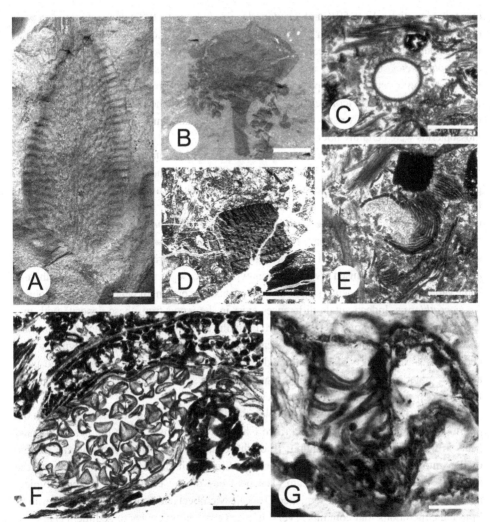

Figure 3.7 Permian plants from Antarctica. **A**, ovulate fructification *Plumsteadia ovata* PB KU Pm3–25–1. **B**, *Eretmonia singulia* from Mount Glossopteris Formation, Horlick Mountains, Late Permian, PBKU PM2046b. **C–G**, permineralised material from the Dragons Teeth Member, Prince Charles Mountains, early Late Permian. C, lycopod megaspore. D, blocky fragment of charcoal preserved in peat. E, isolated microsporangia of *Arberiella* with numerous taeniate bisaccate pollen inside. F, section through fern sporangia attached to pinnule, spore referable to dispersed spore *Horriditriletes*. G, section through *Noeggerathiopsis* stomatal furrow with numerous hairs both overarching and within furrow. Scale bars A, B = 10 mm, C, D, E = 1 mm, F = 50 μm, G = 100 μm. Photos A, B, courtesy of P. Ryberg and the Paleobotanical Collections of the University of Kansas.

Diversity in any one sample from the Prince Charles Mountains varied between 31 and 69 species, but if the stratigraphic ranges of species were incorporated this figure would rise to 81 to 85 species per stratigraphic level (Lindström and McLoughlin, 2007). In spite of the diversity present only a few taxa make up the majority of grains recovered. Two taeniate bisaccate genera

(*Protohaploxypinus*; 12 to 50% and *Striatopodocarpidites*; up to 10%) dominate together with one non-taeniate bisaccate pollen type (*Scheuringipollenites*; 10 to 15%). Fern spores occur in low abundance, with the most frequent being *Osmundacidites, Baculatisporites, Horriditriletes* and *Lophotriletes*, along with rare sphenopsid and lycopsid spores. This all points to a vegetation dominated by glossopterid plants with other rare components (*Noeggerathiopsis*) and an understorey rich in ferns relative to the lycopsids and sphenopsids. The rarity of so many taxa also suggests that diversity may have been higher than generally appreciated, but non-glossopterids were certainly not co-dominant.

The permineralised peat preserved in the Prince Charles Mountains again provides evidence for the types of plants growing at this time. While the majority of the biomass can be attributed to glossopterids, small lycophyte axes are present but remain, as yet, undescribed (McLoughlin and Drinnan, 1996). Amongst these axes are dichotomising stems together with permineralised megaspores. Other groups include seed plants such as *Noeggerathiopsis* that co-occur in specific horizons (McLoughlin and Drinnan, 1996). This plant has leaves with distinct parallel veins and sunken stomatal grooves covered with hairs (McLoughlin and Drinnan, 1996). Neish, Drinnan and Cantrill (1993) recognised several different types of glossopterid roots and, although there is some variation in developmental stages, at least two taxa are present of which one may belong to *Noeggerathiopsis*. Indeed subtle differences exist in the wood from these deposits with two species of *Australoxylon* present (Weaver, McLoughlin and Drinnan, 1997). One, *Australoxylon mondii*, has identical anatomy to that seen in *Vertebraria* roots and so is assigned to the Glossopteridales (Weaver, McLoughlin and Drinnan, 1997). Within the permineralised peats are many other organs including isolated sporangia and several different seed types reported from this deposit that still await further investigation.

Insights into the environmental and ecological dynamics can also be determined from the permineralised peat deposits in the Prince Charles Mountains. Unlike the permineralised peats in the Transantarctic Mountains (thought to have been formed from blocks of peat eroded into river channels) the peat from the Prince Charles Mountains represents the silicified top of an *in situ* coal seam. The peat lies on the boundary between a typical alternating sandstone and coal sequence (Topolje Member) and a shale unit with siderite or limonite concretions that contains several upward coarsening cycles of sediment (Dragons Teeth Member). The Dragons Teeth Member has been interpreted as a lacustrine unit with a delta building into, and filling, a lake (McLoughlin and Drinnan, 1997). The abrupt transition at the base of the Dragons Teeth Member suggests rapid formation of the lake and it is thought that injection of silica-rich waters into the base of the lake resulted in rapid precipitation of silica, and consequent permineralisation of the underlying coal. This deposit extends for about 3 km along the strike before wedging out into silicified sediments.

Not surprisingly the peat is almost entirely composed of the disaggregated parts of *Glossopteris* with only minor material from other taxa. A number of different facies can be recognised that point to the ecological processes. A typical succession consists of an undisturbed leaf mat overlain by disturbed and fragmented leaf material, which in turn is overlain by stem and root-rich material. This suggests seasonal leaf fall and subsequent break down in water of the leaves at the top of the deposit. Open water is further indicated by facies with oxidised and fluffy organic matter (freshwater palynomorphs are also present).

Still water is indicated by open fabrics and fungal material. Dynamics show the mechanical breakdown of the leaf litter into small leaf fragments followed by colonisation by roots. The root material subsequently breaks down the leaf material releasing the biomass, which is then used by the roots of the plants. In turn, after death, the roots themselves can be consumed by fungi and assimilated by other roots.

Perhaps the most interesting feature of these peats is the individual layers within the peats themselves. Charcoal is a common component of these layers and often occurs in distinct horizons. Indeed Lindström and McLoughlin (2007) record small but consistent amounts of charcoal throughout the sequence. This points to the presence of fire within the ecosystem and indicates periods of dryness. Permian coals from Australia and South Africa also contain fusain indicating that fire played an active role in this system (Glasspool, 2000, 2003a, 2003b). Indeed, in some instances layers within Permian coals from Australia and Africa are almost entirely composed of inertinite derived from burning vegetation (Glasspool, 2000, 2003a, 2003b). The permineralised peat from the Prince Charles Mountains has some horizons with concentrations of charcoal that testifies to wildfire events having occurred in this ecosystem as well. Support for dry conditions can be seen in other layers where the peat is compacted and cracked, and the cracks subsequently having been infilled by wind-blown silt. Recent geochemical analysis of the coals show an increasing ash content up section (Holdgate *et al.*, 2005), which also suggests a trend towards drier environmental conditions through the late Permian in the Prince Charles Mountains.

Other layers within the peats are rich in coprolites, indicating periods of insect activity, or contain megaspores signifying periods of colonisation by lycophytes, as well as fern spore-rich layers. Moreover leaf mats preserved within these deposits contain layers of *Arberiella*, which indicates seasonality at this time.

The Late Permian sequences in the Transantarctic Mountains (Buckley Formation) and the Prince Charles Mountains (Bainmedart Coal Measures) all have a thickening of coal horizons up sequence when compared with the lower glacial deposits. This provides evidence for climate warming and landscape stabilisation. The thickening of coal seams is also associated with a change in the style of deposition of the river systems. This indicates a more stable environment, with more even year-round runoff in the Transantarctic Mountains, suggesting that the local ice sheets had become less important in the annual hydrological regime. Compositionally the coals are largely comprised of *Glossopteris* as indicated by the soils below each coal seam being dominated by *Vertebraria* emphasising the importance of this group to the Late Permian ecosystems.

Although similar patterns can be seen in the Prince Charles Mountains and Transantarctic Mountains, the river systems of the Prince Charles Mountains differ as they are character-ised by high discharge and low sinuosity resulting in broad, braided channels that did not migrate. Adjacent to the braid channel were extensive floodplains that accumulated peat at various times. The majority of the fossil material provides evidence of floodplain (coal-forming) vegetation dominated by glossopterids (several species) with areas co-mixed with *Noeggerathiopsis*. The understorey, although diverse, had few plants that contributed significantly to the biomass — these included several types of ferns and lycopsids.

The Permian vegetation history of Antarctica can be divided into two distinct phases: an initial glacial phase supported depauperate floras, with few glossopterids, and rapid colonisation by low diversity mixed *Glossopteris* and *Gangamopteris* communities. As the climate ameliorated the mixed *Glossopteris* and *Gangamopteris* communities became firmly established giving rise to the second phase. This resulted in the first coal mires of the Permian. Although of apparently low diversity, this may be an artefact of the sheer amount of biomass produced by the glossopterid plants. *Gangamopteris* became extinct by the Late Permian leaving *Glossopteris* and *Noeggerathiopsis* as the two main overstorey elements in the vegetation. These Late Permian communities were remarkably stable and persisted until the Permian–Triassic extinction event when coal mires collapsed and were replaced by the more warm-loving plants of the Triassic.

Summary

- Glaciation in the Antarctic and the presence of ice sheets excluded vegetation from ?late Carboniferous to the earliest Permian. It is unclear if Antarctica was glaciated prior to the late Carboniferous or whether vegetation is absent as a result of the lack of appropriate aged sediments.

- The Early Permian collapse of the ice sheet created new habitats that were rapidly colonised by an early post-glacial flora. The first vegetation was dominated by ferns, lycopods and sphenophytes, with minor seed plants such as *Gangamopteris* and *Noeggerathiopsis*.

- Glaciations continued in the alpine areas through the Early Permian influencing the river dynamics with high seasonal discharge. Forest colonised the landscape even though the soil remained frozen for part of the year. Palsa mires developed on the floodplains and these were dominated by *Gangamopteris* and *Noeggerathiopsis*, but with the rare occurrence of *Glossopteris*.

- Post-glacial landscapes in the early Late Permian saw further change in the vegetation with the demise of *Gangamopteris* and the rise to dominance of *Glossopteris*. Coal swamps became widespread. The biomass was dominated by *Glossopteris* and sphenopsids such as *Paracalamites* or *Neocalamites*.

- Late Permian floras were also dominated by *Glossopteris*, but with small stands of *Noeggerathiopsis*. Pollen floras indicate an increase in diversity compared with early and mid-Permian floras. However, whilst diversity increased the species remained rare, as indicated by low pollen percentages of each taxon. The climate began to dry and fire became an important part of the landscape.

- The role of climatic change in the history of the Permian glossopterid flora cannot be underestimated. Rising out of one of the Earth's great ice ages, they firstly colonised ice-marginal environments, and held onto those ecological niches as the ice sheet retreated to ultimately occupy a broad latitudinal belt extending from within a few degrees of the South Pole to the mid-latitudes.

Case Study: The Antarctic glossopterid plants

It is apparent from the abundance of *Glossopteris* leaf fossils that this plant was dominant in Permian ecosystems of the Southern Hemisphere. The habit, biology and ecological interactions of *Glossopteris* have been built up through fortuitous discoveries of connected organs and careful analysis of the fossil deposits in which they occur (Gould and Delevoryas, 1977; Pigg and Trivett, 1994). Leaves of *Glossopteris*, and the closely related *Gangamopteris*, are diverse with more than 70 species recognised in Indian deposits alone (Chandra and Surange, 1979). The genus occurs across Gondwana (Cúneo, 1996) with conservative estimates of more than 200 described species (Taylor and Taylor, 1993). The leaves all have a similar morphology with a compound midvein and secondary veins, which arch gracefully towards the margins and anastomose to form a net-like mesh. Several species have been found attached to shoot systems (Pant and Singh, 1974; Rigby, 1967), which indicate, in at least some species, that the leaves were borne on long-shoot short-shoot systems, and leaf phyllotaxis has been determined from permineralised shoots (Pigg, 1990). Permineralised material (Gould and Delevoryas, 1977; Pigg, 1987, 1990; Pigg and McLoughlin, 1997; Pigg and Taylor, 1987, 1990) have revealed the compound nature of the midvein and the vascular system, which consist of a fibrous bundle sheath around the vascular trace. Whilst there is some variation in the density of the vein mesh and the level of development of the bundle sheath, these are all features that characterise the group.

The wood anatomy provides sound evidence to further link the leafy shoots to branch and trunk wood. Wood is of the *Araucarioxylon* type (Maheshwari, 1972), but has also been placed in other genera such as *Australoxylon* (Weaver, McLoughlin and Drinann, 1997). In the Antarctic, deposits of *Araucarioxylon* trunks up to 20 m long and 50 cm in diameter indicate that these plants formed large trees (Francis *et al.*, 1993). The long unbranched trunks support other interpretations (e.g. Gould and Delevoryas, 1977) that suggest a well-developed unbranched trunk terminating in a canopy of long shoots and short shoots. Growth rings are often wide with little cell variation across the ring, and late wood is limited and not thick-walled. These types of patterns have been attributed to a fast growing plant with cessation of wood growth at the end of the season being light-limited rather than water-limited — a scenario which is more common today for trees growing in temperate environments (Ryberg and Taylor, 2007; Taylor and Ryberg, 2007).

The large trunks in the Prince Charles Mountains have been linked anatomically to roots of *Vertebraria* (Weaver, McLoughlin and Drinann, 1997). Soil horizons beneath *Glossopteris*-dominated swamp assemblages are characterised by the distinctive segmented *Vertebraria* roots. Permineralised *Vertebraria* roots are exarch with secondary growth forming distinctive xylem arms separated by air spaces. Along the length of the root the xylem arms are connected by transverse septa of xylem (Decombeix, Taylor and Taylor, 2009; Neish, Drinnan and Cantrill, 1993) that give the root the distinct segmented appearance when compressed. This preservational feature is the result of large air spaces (aerenchyma) in the roots that allowed them to grow in water-logged environments (Gould, 1975; Schopf, 1986). Often these aerenchymatous roots go on to develop a solid xylem cylinder and this presumably reflects a shift from playing an active role in nutrient and water uptake, to a more structural supporting role.

Reproductive organs are also diverse (McLoughlin, 1990a, 1990b; Ryberg, 2009) and have been found attached to either *Glossopteris* leaves (Pant and Singh, 1974; Plumstead, 1952), or to modified leaves with the distinctive net-like venation. Most are known from impression or, more rarely, compression material and this has hampered interpretations of the morphology (Prevec, McLoughlin and Bamford, 2008). However, rare permineralised material is known from Australia (Gould and Delevoryas, 1977; Nishida *et al.*, 2007) and from several localities in Antarctica (Klavins *et al.*, 2001; McLoughlin and Drinnan, 1996; McManus *et al.*, 2002; Neish, Drinnan and Cantrill,1993; Ryberg, 2010; Zhao, Taylor and Taylor, 1995). The ovulate organs are either borne in the axil of a *Glossopteris* leaf (e.g. *Scutum*; Plumstead, 1952), or on a pedicel arising from the midrib of modified leaves (e.g. *Rigbya*), or superficially on the adaxial leaf surface of the modified reproductive leaves (e.g. *Homevalia*; Nishida *et al.*, 2007). These structures bear numerous small seeds that are shed at maturity (e.g. Nishida *et al.*, 2007) with the result that there are several isolated seed taxa described (e.g. *Plectiospermum*; E. L. Taylor and T. N. Taylor, 1987). The ovules are pollinated by distinctive bisaccate taeniate grains that have been observed in the pollination chamber; these grains produced motile sperm (Nishida, Pigg and Rigby, 2003; Nishida *et al.*, 2004). Bisaccate taeniate grains have been recovered from microsporangia (e.g. Lindström, McLoughlin and Drinnan, 1997) of the *Arberiella* type. *Arberiella* is widespread in Southern Hemisphere Permian deposits and it has been suggested that they were borne on glossopterid reproductive organs (Gould and Delevoryas, 1977).

Glossopteris formed a dominant element of the Permian high latitude vegetation of Antarctica. Dense stands of trees up to an estimated 10 m tall have been reported from Mount Achernar (Cúneo *et al.*, 1993; Taylor, Taylor, Cúneo, 1992) where up to 2134 trees per hectare grew. This contrasts with the taller trees reported by Francis *et al.* (1993) from the Allan Hills — a difference perhaps attributed to maturity of the forest with the former being a cohort of colonisers, whilst the latter formed a more mature climax community. It is believed that the trees underwent a seasonal loss of leaves as indicated by concentrations within autumn–winter varve cycles of lacustrine deposits (Retallack, 1980) and other deposits (Plumstead, 1958). Winters were harsh enough to cause the plants to become dormant (Francis *et al.*, 1994) and this is supported by the seasonal tree rings seen in trunks of the plants (Ryberg and Taylor, 2007).

As an integral element of the ecosystem, *Glossopteris* apparently not only out-competed other plants, resulting in low diversity communities, but also supported a diversity of biota. Distinctive cavities occur in the trunks of some trees and have features consistent with their formation having arisen as a result of pocket rot fungi (Stubblefield and Taylor, 1986; Taylor, 1993). However, other cavities of similar size contain frass and are consistently found extending from the late early wood to the late wood boundary (Weaver, McLoughlin and Drinnan, 1997). These are thought to be formed by invertebrates that have over-wintered beneath the bark. The lack of wound response to these cavities might evidence the fact that they occurred while the plants were dormant over winter. However, there is evidence of the ability of *Glossopteris* to respond to wounding as epicormic buds have been documented (Decombeix, Taylor and Taylor, 2010). This provides a mechanism to respond to damage such as fire or insect attack. Frass-filled cavities in a range of organs from the Transantarctic Mountains were interpreted as the result of attack by orobatid mites (Kellogg and Taylor, 2004). Indeed insect damage to *Glossopteris* leaves is widely reported (e.g. India; Srivastava and Agnihotri, 2011; Argentina; Cariglino and Gutierrez, 2011) and are also present on Antarctic material. Considerable information is now known about the morphology and biology of the glossopterids and the ecological interactions. However, important questions still remain. There are still uncertainties regarding the morphology of the reproductive structures, levels of diversity present, and how similar these plants were to other seed plants growing at the time.

References

Algeo, T. J. and Scheckler, S. E. (1998). Terrestrial–marine teleconnections in the Devonian: links between the evolution of land plants, weathering processes, and marine anoxic events. *Philosophical Transactions of the Royal Society of London Series B*, **353**, 113–130.

Archangelsky, S. and Cúneo, R. (1984). Zonacíon del Pérmico continental Argentina sobre la base de sus plantas fósiles. In *Memoria 3rd Congreso Latinoamerica de Paleontologia*. México City: Universidad Nacional Autónoma México, pp. 143–153.

Arne, D. C. (1994). Phanerozoic exhumation history of the northern Prince Charles Mountains, East Antarctica. *Antarctica Science*, **6**, 69–84.

Askin, R. A. (1998). Floral trends in the Gondwana high latitudes: palynological evidence from the Transantarctic Mountains. *Journal of African Earth Sciences*, **27**, 12–13.

Askin, R. A. and Elliot, D. H. (1982). Geologic implications of recycled Permian and Triassic palynomorphs in Tertiary rocks of Seymour Island, Antarctic Peninsula. *Geology*, **10**, 547–551.

Babcock, L. E., Isbell, J. L., Miller, M. F. and Haiostis, S. T. (2002). New late Paleozoic conchostracan (Crustacea: Brachiopoda) from the Shackleton Glacier area, Antarctica: age and paleoenvironmental implications. *Journal of Paleontology*, **76**, 70–75.

Balme, B. E. and Playford, G. (1967). Late Permian plant microfossils from the Prince Charles Mountains, Antarctica. *Revue Micropaléontologie*, **10**, 179–192.

Barrett, P. J. (1991). The Devonian to Triassic Beacon Supergroup of the Transantarctic Mountains and correlatives in other parts of Antarctica. In *The Geology of Antarctica, Oxford Monographs on Geology and Geophysics*, **17**, ed. R. J. Tingey. Oxford: Clarendon Press, pp. 120–152.

Barrett, P. J. and Fitzgerald, P. G. (1985). Deposition of the lower Feather Conglomerate, a Permian braided river deposit in southern Victoria Land, Antarctic, with notes on the regional palaeogeography. *Sedimentary Geology*, **45**, 189–208.

Barrett, P. J. and Kyle, R. A. (1975). The early Permian glacial beds of south Victoria Land and the Darwin Mountains, Antarctica. In *Gondwana Geology*, ed. K. S. W. Campbell. Canberra: Australian National University Press, pp. 333–346.

Barron, E. J. and Fawcett, P. J. (1995). The climate of Pangaea: a review of climate model simulations of the Permian. In *The Permian of Northern Pangaea, Paleogeography, Paleoclimates, Stratigraphy Vol 1*, ed. P. A. Scholle. New York: Springer, pp. 37–52.

Birgenheier, L. P., Frank, T. D., Fielding, C. R. and Rygel, M. C. (2010). Coupled carbon isotopic and sedimentological records from the Permian system of eastern Australia reveal the response of atmospheric carbon dioxide to glacial growth and decay during the late Palaeozoic ice age. *Palaeogeography, Palaeoclimatology, Palaeoecology*, **286**, 178–193.

Bradshaw, M. A. (1984) Permian nonmarine bivalves from the Ohio Range, Antarctica. *Alcheringa*, **8**, 305–309.

Brezinski, D. K., Cecil, C. B. and Skema, V. W. (2010). Late Devonian glacigenic and associated facies from the central Appalachian Basin, eastern United States. *Geological Society of America Bulletin*, **122**, 265–281.

Caputo, M. V., Melo, G., Streel, M. and Isbell, J. L. (2008). Late Devonian and Early Carboniferous glacial records of South America. In *Resolving the Late Paleozoic Ice Age in Time and Space, Geological Society of America Special Paper*, **441**, eds C. R. Fielding, T. D. Frank and J. L. Isbell. Boulder: The Geological Society of America, pp. 161–173.

Cariglino, B. and Gutierrez, P. R. (2011). Plant-insect interactions in a *Glossopteris* flora from the La Golondrina Formation (Guadalupian-Lopingian), Santa Cruz Province, Patagonia, Argentina. *Ameghiniana*, **48**, 103–112.

Cascales-Miñana, B. (2011). New insights into the reading of Paleozoic plant fossil record discontinuities. *Historical Biology*, **23**, 115–130.

Chaloner, W. G. and Lacey, W. S. (1973). The distribution of late Palaeozoic floras. *Special Papers in Palaeontology*, **12**, 271–289.

Chandra, S. and Surange, K. R. (1979). Revision of the Indian species of *Glossopteris*. *Birbal Sahni Institute of Palaeobotany Monographs*, **2**, 1–291.

Cleal, C. J. and Thomas, B. A. (2005). Palaeozoic tropical rainforests and their effect on global climates: is the past the key to the present? *Geobiology* **3**, 13–31.

Collinson, J. W., Vavra, C. L. and Zawiskie, J. M. (1992). Sedimentology of the Polarstar Formation (Permian), Ellsworth Mountains, West Antarctica. In *Geology and Paleontology of the Ellsworth Mountains West Antarctica, Geological Society of America Memoir*, **170**, eds G. F. Webers, C. Craddock and J. F. Splettstoesser. Boulder: The Geological Society of America, pp. 63–80.

Collinson, J. W., Isbell, J. L., Elliot, D. H., Miller, M. F. and Miller, J. M. G. (1994). Permian-Triassic Transantarctic Basin. In *Permian-Triassic Pangean Basins and Foldbelts along the Panthalassan Margin of Gondwanaland, Geological Society of America Memoir*, **184**, eds J. J. Veevers and C. McA. Powell. Boulder: The Geological Society of America, pp. 173–222.

Cridland, A. A. (1963). *Glossopteris* flora from the Ohio Range, Antarctica. *American Journal of Botany*, **50**, 186–195.

Crowell, J. C. (1978). Gondwana glaciation, cyclothems, continental positioning, and climate change. *American Journal of Science*, **278**, 1345–1372.

Crowell, J. C. (1999). Pre-Mesozoic ice ages: their bearing on understanding the climate system. *Geological Society of America Memoirs*, **192**, 1–106.

Crowell, J. C. and Frakes, L. A. (1971). Late Paleozoic glaciation: part IV Australia. *Geological Society of America Bulletin*, **82**, 2515–2540.

Crowell, J. C. and Frakes, L. A. (1972). Late Paleozoic glaciation. Part V: Karoo Basin, South Africa. *Geological Society of America Bulletin*, **83**, 2887–2912.

Cúneo, N. R. (1996). Permian phytogeography in Gondwana. *Palaeogeography, Palaeoclimatology, Palaeoecology*, **125**, 75–104.

Cúneo, R., Isbell, J., Taylor, E. L. and Taylor, T. N. (1993). The *Glossopteris* flora from Antarctica: taphonomy and paleocology. In *Comptes Rendu 12th International Congress of Carboniferous and Permian Stratigraphy and Geology Volume 2*, ed. S. Archangelsky. Buenos Aires, pp. 13–40.

Davies, N. S. and Gibling, M. R. (2010). Cambrian to Devonian evolution of alluvial systems: The sedimentological impact of the earliest land plants. *Earth-Science Reviews*, **98**, 171–200.

Decombeix, A.-L., Taylor, E. L. and Taylor, T. N. (2009). Secondary growth in *Vertebraria* roots from the Late Permian of Antarctica: a change in developmental timing. *International Journal of Plant Sciences*, **170**, 644–656.

Decombeix, A.-L., Taylor, E. L. and Taylor, T. N. (2010). Epicormic shoots in a Permian gymnosperm from Antarctica. *International Journal of Plant Sciences*, **171**, 772–782.

Dettmann, M. E. and Thomson, M. R. A. (1987). Cretaceous palynomorphs from the James Ross Island area, Antarctica — a pilot study. *British Antarctic Survey, Bulletin*, **77**, 13–59.

Dibner, A. F. (1978). Palynocomplexes and the age of the Amery Group deposits, East Antarctica. *Pollen et Spores*, **20**, 405–422.

DiMichele, W. A., Pfefferkorn, H. W. and Gastaldo, R. A. (2001). Response of Late Carboniferous and Early Permian plant communities to climate change. *Annual Review of Earth and Planetary Sciences*, **29**, 461–487.

DiMichele, W. A., Montañez, I. P., Poulsen, C. J. and Tabor, N. J. (2009). Climate and vegetational regime shifts in the late Paleozoic ice age earth. *Geobiology*, **7**, 200–226.

DiMichele, W. A., Cecil, C. B., Montanez, I. P. and Falcon-Lang, H. J. (2010). Cyclic changes in Pennsylvanian paleoclimate and effects on floristic dynamics in tropical Pangaea. *International Journal of Coal Geology*, **83**, 329–344.

Edwards, W. N. (1928). The occurrence of *Glossopteris* in the Beacon Sandstone of Ferrar Glacier, South Victoria Land. *Geological Magazine*, **65**, 323–327.

Eyles, C. H., Eyles, N. and Gostin, V. A. (1998). Facies and allostratigraphy of high-latitude, glacially influenced marine strata of the Early Permian southern Sydney Basin, Australia. *Sedimentology*, **45**, 121–161.

Eyles, N., Eyles, C. H. and Gostin, V. A. (1997). Iceberg rafting and scouring in the early Permian Shoalhaven Group, New South Wales: evidence of Heinrich-like events. *Palaeogeography, Palaeoclimatology, Palaeoecology*, **136**, 1–17.

Eyles, N., Mory, A. J. and Backhouse, J. (2002). Carboniferous-Permian palynostratigraphy of west Australian marine rift basins: resolving tectonic and eustatic controls during Gondwanan glaciations. *Palaeogeography, Palaeoclimatology, Palaeoecology*, **184**, 305–319.

Falcon-Lang, H. J. (2003). Response of Late Carboniferous tropical vegetation to transgressive-regressive rhythms, Joggins, Nova Scotia. *Journal of the Geological Society of London*, **160**, 643–648.

Falcon-Lang, H. J. (2004). Pennsylvanian tropical rainforests responded to glacial-interglacial rhythms. *Geology*, **32**, 689–692.

Falcon-Lang, H. J. and Bashforth, A. R. (2004). Pennsylvanian uplands were forested by giant cordaitalean trees. *Geology*, **32**, 417–420.

Falcon-Lang, H. J. and Bashforth, A. R. (2005). Morphology, anatomy, and upland ecology of large cordaitalean trees from the Middle Pennsylvanian of Newfoundland. *Review of Palaeobotany and Palynology*, **135**, 223–243.

Farabee, M. J., Taylor, E. L. and Taylor, T. N. (1990). Correlation of Permian and Traissic palynomorph assemblages from the Central Transantarctic Mountains, Antarctica. *Review of Palaeobotany and Palynology*, **65**, 257–265.

Farabee, M. J., Taylor, E. L. and Taylor, T. N. (1991). Late Permian palynomorphs from the Buckley Formation, central Transantarctic Mountains, Antarctica. *Review of Palaeobotany and Palynology*, **69**, 353–368.

Fielding, C. R. and Webb, J. A. (1995). Sedimentology of the Permian Radok Conglomerate in the Beaver Lake area of MacRobertson Land, East Antarctica. *Geological Magazine*, **132**, 51–63.

Fielding, C. R. and Webb, J. A. (1996). Facies and cyclicity of the Late Permian Bainmedart Coal Measures in the Northern Prince Charles Mountains, MacRobertson Land, Antarctica. *Sedimentology*, **43**, 295–322.

Fielding, C. R., Frank, T. D. and Isbell, J. L. (2008). The late Paleozoic ice age – a review of current understanding and synthesis of global climate patterns. In *Resolving the Late Paleozoic Ice Age in Time and Space, Geological Society of America Special Paper*, **441**, eds C. R. Fielding, T. D. Frank and J. L. Isbell. Boulder: The Geological Society of America, pp. 343–354.

Fielding, C. R., Frank, T. D., Birgenheier, L. P., *et al.* (2008a). Stratigraphic imprint of the Late Paleozoic Ice Age in eastern Australia: a record of alternating glacial and non-glacial climate regime. *Journal of the Geological Society of London*, **165**, 129–140.

Fielding, C. R., Frank, T. D., Birgenheier, L. P., *et al.* (2008b). Stratigraphic record and facies associations of the late Paleozoic ice age in eastern Australia (New South Wales and Queensland). In *Resolving the Late Paleozoic Ice Age in Time and Space, Geological Society of America Special Paper*, **441**, eds C. R. Fielding, T. D. Frank and J. L. Isbell. Boulder: The Geological Society of America, pp. 41–58.

Fitzgerald, P. G. and Barrett, P. J. (1986). *Skolithos* in Permian braided river deposit, southern Victoria Land, Antarctica. *Palaeogeography, Palaeoclimatology, Palaeoecology*, **52**, 237–247.

Frakes, L. A. (1979). *Climates through Geological Time*. Amsterdam: Elsevier.

Frakes, L. A. and Crowell, J. C. (1969). Late Paleozoic glaciation: part 1. South America. *Geological Society of America Bulletin*, **80**, 1007–1042.

Frakes, L. A. and Crowell, J. C. (1970). Late Paleozoic glaciation: part II. Africa exclusive of the Karoo Basin. *Geological Society of America Bulletin*, **81**, 2261–2286

Frakes, L. A., Matthews J. L. and Crowell, J. C. (1971). Late Paleozoic glaciation: part III. Antarctica. *Geological Society of America Bulletin*, **82**, 1581–1604.

Francis, J. E., Woolfe, K. J., Arnott, M. J. and Barrett, P. J. (1993). Permian forests of Allan Hills, Antarctica: the palaeoclimate of Gondwanan high latitudes. *Special Papers in Palaeontology*, **49**, 75–83.

Francis, J. E., Arnott, M. J., Woolfe, K. J. and Barrett, P. J. (1994). Permian climates of the southern margins of Pangea: evidence from fossil wood in Antarctica. In *Pangea: Global Environments and Resources, Canadian Society of Petroleum Geologists Memoir*, **17**, eds A. F. Embry, B. Beauchamp and D. J. Glass. Calgary: Canadian Society of Petroleum Geologists (GSPG), pp. 275–282.

Galtier, J. and Taylor, T. N. (1994). The first record of ferns from the Permian of Antarctica. *Review of Palaeobotany and Palynology*, **83**, 227–239.

García Massini, J. L. G. (2007a). A possible endoparasitic chytridomycete fungus from the Permian of Antarctica. *Paleontologia Electronica*, **10(3)**, 16A, 1–14.

García Massini, J. L. G. (2007b). A glomalean fungus from the Permian of Antarctica. *International Journal of Plant Sciences*, **168**, 673–678.

Gastaldo, R. A., DiMichele, W. A. and Pfefferkorn, H. W. (1996). Out of the icehouse into the greenhouse: a Late Paleozoic analogue for modern global vegetational change. *GSA Today*, **10**, 1–7.

Gee, C. T. (1989). Permian *Glossopteris* and *Elatocladus* from the megafossil floras from the English Coast, eastern Ellsworth Land, Antarctica. *Antarctic Science*, **1**, 35–44.

Gee, C. T. and Gastaldo, R. A. (2005). Sticks and mud, fruits and nuts, leaves and climate: plant taphonomy comes of age. *Palaios*, **20**, 415–417.

Gensel, P. G. and Berry, C. M. (2001). Early lycophyte evolution. *American Fern Journal* **91**, 74–98.

Gibbs, M. T., Rees, P. M., Kutzbach, J. E., *et al.* (2002). Simulations of Permian climate and comparisons with climate-sensitive sediments. *The Journal of Geology*, **110**, 33–55.

Gibling, M. R. and Rygel, M. C. (2008). Late Paleozoic cyclic strata of Euramerica: recognition of Gondwanan glacial signatures during periods of thermal subsidence. In *Resolving the Late Paleozoic Ice Age in Time and Space, Geological Society of America Special Paper*, **441**, eds C. R. Fielding, T. D. Frank and J. L. Isbell. Boulder: The Geological Society of America, pp. 219–234.

Glasspool, I. (2000). A major fire event recorded in the mesofossils and petrology of the Late Permian, Lower Whybrow coal seam, Sydney Basin. *Palaeogeography, Palaeoclimatology, Palaeoecology*, **164**, 357–380.

Glasspool, I. (2003a). Hypoautochtonous-allochthonous coal deposition in the Permian, South African, Witbank Basin No. 2 seam; a combined approach using sedimentology, coal petrography and palaeontology. *International Journal of Coal Geology*, **53**, 81–135.

Glasspool, I. (2003b). Palaeoecology of selected South Africa export coals from the Vryheid Formation, with emphasis on the role of heterosporous lycopods and wildfire inertinite. *Fuel*, **82**, 959–970.

Gould, R. E. (1975). A preliminary report on petrified axes of *Vertebraria* from the Permian of eastern Australia. In *Gondwana Geology: Papers Presented at the Third Gondwana Symposium, Canberra, Australia*, ed. K. S. W. Campbell. Canberra: Australian National University Press, pp. 109–115.

Gould, R. E. and Delevoryas, T. (1977). The biology of *Glossopteris*: evidence from petrified seed-bearing and pollen-bearing organs. *Alcheringa*, **1**, 387–399.

Guy-Ohlsen, D. and Lindström, S. (1994). Palaeoecology of the early Permian strata at Heimefrontfjella, Dronning Maud Land, Antarctica. *Antarctic Science*, **6**, 507–514.

Hallam, A. and Wignall, P. B. (1997). Mass extinctions and sea-level changes. *Earth-Science Reviews*, **48**, 217–250.

Haldorsen, S., Von Brunn, V., Maud, R. and Truter, E. D. (2001). A Weichselian deglaciation model applied to the Early Permian glaciation in the northeast Karoo Basins, South Africa. *Journal of Quaternary Science*, **16**, 583–593.

Harrowfield, M., Holdgate, G. R., Wilson, C. L. and McLoughlin, S. (2005). Tectonic significance of the Lambert Graben, East Antarctica: reconstructing the Gondwana rift. *Geology*, **33**, 197–200.

Heckel, P. H. (1994). Evaluation of evidence for glacio-eustatic control over marine Pennsylvanian cyclotherms in North America and consideration of possible tectonic effects. In *Tectonic and Eustatic Controls on Sedimentary Cycles, Concepts in Sedimentology and Paleontology*, **4**, eds J. M. Dennison and F. R. Ettensohn. Tulsa: SEPM (Society of Sedimentary Geology) pp. 65–87.

Henry, L. C., Isbell, J. L., Limarino, C. O., McHenry, L. J. and Fraiser, M. L. (2010). Mid-Carboniferous deglaciation of the Protoprecordillera, Argentina recorded in the Agua de Jagüel palaeovalley. *Palaeogeography, Palaeoclimatology, Palaeoecology*, **298**, 112–129.

Hjelle, A. and Winsnes, T. S. (1972). The sedimentary and volcanic sequence of Vestfjella, Dronning Maud Land. In *Antarctic Geology and Geophysics*, ed. R. J. Adie. Oslo: Universitetsforlaget, pp. 539–546.

Høeg, O. A. and Bose, M. N. (1960). The *Glossopteris* flora of the Belgian Congo: with a note on some fossil plants from the Zambesi Basin (Mozambique). *Annals Musée Royal du Congo Belge Série 8*, **32**, 1–107.

Holdgate, G. R., McLoughlin, S., Drinnan, A. N., Finkelman, R. B. and Willett, J. C. (2005). Inorganic geochemistry, petrography and palaeobotany of Permian coals in the Prince Charles Mountains, East Antarctica. *International Journal of Coal Geology*, **63**, 156–177.

Isaacson, P. E., Díaz-Martínez, E., Grader, G. W., *et al.* (2008). Late Devonian-earliest Mississippian glaciation in Gondwanaland and its biogeographic consequences. *Palaeogeography, Palaeoclimatology, Palaeoecology*, **268**, 126–142.

Isbell, J. L. and Cúneo, N. R. (1996). Depositional framework of Permian coal-bearing strata, southern Victoria Land, Antarctica. *Palaeogeography, Palaeoclimatology, Palaeoecology*, **125**, 217–238.

Isbell, J. L., Gelhar, G. A. and Seegers, G. M. (1997). Reconstruction of preglacial topography using a postglacial flooding surface: upper Paleozoic deposits, central Transantarctic Mountains, Antarctica. *Journal of Sedimentary Research*, **67**, 264–273.

Isbell, J. L., Miller, M. F., Babcock, L. E. and Hasiotis, S. T. (2001). Ice-marginal environment and ecosystem prior to initial advance of the late Palaeozoic ice sheet in the Mount Butters area of the central Transantarctic Mountains, Antarctica. *Sedimentology*, **48**, 953–970.

Isbell, J. L., Leneaker, P. A., Askin, R. A., Miller, M. F. and Babcock, L. E. (2003). Reevaluation of the timing and extent of the late Paleozoic glaciation in Gondwana: role of the Transantarctic Mountains. *Geology*, **31**, 977–980.

Jones, A. T. and Fielding, C. R. (2004). Sedimentological record of the late Paleozoic glaciation in Queensland, Australia. *Geology*, **32**, 153–156.

Jones, A. T. and Fielding, C. R. (2008). Sedimentary facies of a glacially influenced continental succession in the Pennsylvanian Jericho Formation, Galilee Basin, Australia. *Sedimentology*, **55**, 531–556.

Juckes, L. M. (1972). The geology of north-eastern Heimfrontfjella, Dronning Maud Land. *British Antarctic Scientific Reports*, **65**, 1–44.

Kammer, T. W. and Matchen, D. L. (2008). Evidence for eustasy at the Kinderhookian-Osagean (Mississippian) boundary in the United States: response to Tournaisian glaciation? In *Resolving the Late Paleozoic Ice Age in Time and Space, Geological Society of America Special Paper*, **441**, eds C. R. Fielding, T. D. Frank and J. L. Isbell. Boulder: The Geological Society of America, pp. 261–274.

Kellogg, D. W. and Taylor, E. L. (2004). Evidence of oribatid mite detritivory in Antarctica during the Late Paleozoic and Mesozoic. *Journal of Paleontology*, **78**, 1146–1153.

Kemp, E. M. (1973). Permian flora from the Beaver Lake area, Prince Charles Mountains, Antarctica. I. Palynologial examination of samples. *Bulletin of the Bureau of Mineral Resources Geology and Geophysics Australia*, **126** (Palaeontological papers for 1969), 7–12.

Klavins, S. D., Taylor, E. L., Krings, M. and Taylor, T. N. (2001). An unusual structurally preserved ovule from the Permian of Antarctica. *Review of Palaeobotany and Palynology*, **115**, 107–117.

Korte, C., Jones, P. J., Brand, U., Mertmann, D. and Veizer, J. (2008). Oxygen isotope values from high-latitudes: clues for Permian sea-surface temperature gradients and late Palaeozoic deglaciation. *Palaeogeography, Palaeoclimatology, Palaeoecology*, **269**, 1–16.

Krull, E. S. (1999). Permian palsa mires as palaeoenvironmental proxies. *Palaios*, **14**, 530–544.

Kyle, R. A. (1974). *Plumsteadia ovata* n. sp., a glossopterid fructification from south Victoria Land, Antarctica. *New Zealand Journal of Geology and Geophysics*, **17**, 719–721.

Kyle, R. A. (1976). Palaeobotanical studies of the Permian and Triassic Victoria Group (Beacon Supergroup) of South Victoria Land, Antarctica. Ph.D. thesis, Victoria University.

Kyle, R. A. (1977). Palynostratigraphy of the Victoria Group, south Victoria Land, Antarctica. *New Zealand Journal of Geology and Geophysics*, **20**, 1081–1102.

Kyle, R. A. and Schopf, J. M. (1982). Permian and Triassic palynostratigraphy of the Victoria Group, Transantarctic Mountains. In *Antarctic Geoscience*, ed. C. Craddock. Madison: University of Wisconsin Press, pp. 649–659.

Lacey, W. S. and Lucas, R. C. (1981). A Lower Permian flora from the Theron Mountains, Coats Land. *British Antarctic Survey Bulletin*, **53**, 153–156.

Lambrecht, L. L., Lacey, W. S. and Smith C. S. (1972). Observations on the Permian flora of the Law Glacier area, central Transantarctic Mountains. *Bulletin de la Sociéte Belge de Géologie, de Paléontologie et d'Hydrologie*, **81**, 161–167.

Larsson, K., Lindström, S. and Guy-Ohlsen, D. (1990). An Early Permian palynoflora from Milorgfjella, Dronning Maud Land, Antarctica. *Antarctic Science*, **2**, 331–344.

Lindström, S. (1994). Late Palaeozoic palynology of western Dronning Maud Land, Antarctica. *Lund Publications in Geology*, **121**, 1–33.

Lindström, S. (1995a). Early Permian palynostratigraphy of the northern Heimefrontfjella mountain range, Dronning Maud Land, Antarctica. *Review of Palaeobotany and Palynology*, **89**, 359–415.

Lindström, S. (1995b). Early Late Permian palynostratigraphy and palaeobiogeography of Vestfjella, Dronning Maud Land, Antarctica. *Review of Palaeobotany and Palynology*, **86**, 157–173.

Lindström, S. (1996). Late Permian palynology of Fossilryggen, Vestfjella, Dronning Maud Land, Antarctica. *Palynology*, **20**, 11–44.

Lindström, S. (2005). Palynology of Permian shale, clay and sandstone clasts from the Basen till in northern Vestfjella, Dronning Maud Land. *Antarctic Science*, **17**, 87–96.

Lindström, S. and McLoughlin, S. (2007). Synchronous palynofloristic extinction and recovery after the end-Permian event in the Prince Charles Mountains, Antarctica: implications for palynofloristic turnover across Gondwana. *Review of Palaeobotany and Palynology*, **145**, 89–122.

Lindström, S., McLoughlin, S. and Drinnan, A. N. (1997). Intraspecific variation of taeniate bisaccate pollen within Permian glossopterid sporangia from the Prince Charles Mountains, Antarctica. *International Journal of Plant Sciences*, **158**, 673–684.

Link, P. K. (2009). Icehouse (cold) climates. In *Encyclopedia of Paleoclimatology and Ancient Environments*, ed. V. Gornitz. Berlin, Heidelberg: Springer, pp. 463–471.

Maheshwari, H. K. (1972). Permian wood from Antarctica and revision of some Lower Gondwanan wood taxa. *Palaeontographica B*, **138**, 1–43.

Masood, K. R., Taylor, T. N., Horner, T. and Taylor, E. L. (1994). Palynology of the Mackellar Formation (Beacon Supergroup) of East Antarctica. *Review of Palaeobotany and Palynology*, **83**, 329–337.

Matsch, C. L. and Ojakangas, R. W. (1992). Stratigraphy and sedimentology of the Whiteout Conglomerate; an upper Paleozoic glacigenic unit, Ellsworth Mountains, West Antarctica. In *Geology and Paleontology of the Ellsworth Mountains West Antarctica, Geological Society of America Memoir*, **170**, eds G. F. Webers, C. Craddock and J. F. Splettstoesser. Boulder: The Geological Society of America, pp. 37–62.

McKelvey, B. C. and Stephenson, N. C. N. (1990). A geological reconnaissance of the Radok Lake area, Amery Oasis, Prince Charles Mountains. *Antarctic Science*, **2**, 53–66.

McLoughlin, S. (1990a). Late Permian glossopterid fructifications from the Bowen and Sydney basins, eastern Australia. *Geobios*, **23**, 283–297.

McLoughlin, S. (1990b). Some Permian glossopterid frutifications and leaves from the Bowen Basin, Queensland, Australia. *Review of Palaeobotany and Palynology*, **62**, 11–40.

McLoughlin, S. and Drinnan, A. N. (1996). Anatomically preserved *Noeggerathiopsis* leaves from East Antarctica. *Review of Palaeobotany and Palynology*, **92**, 207–227.

McLoughlin, S. and Drinnan, A. N. (1997). Revised stratigraphy of the Permian Bainmedart Coal Measures, northern Prince Charles Mountains, East Antarctica. *Geological Magazine*, **134**, 335–353.

McLoughlin, S., Larsson, K. and Lindström, S. (2005). Permian plant macrofossils from Fossilryggen, Vestfjella, Dronning Maud Land. *Antarctic Science*, **17**, 73–86.

McManus, H. A., Taylor, E. L., Taylor, T. N. and Collinson, J. W. (2002). A petrified *Glossopteris* flora from Collinson Ridge, central Transantarctic Mountains: Late Permian or Early Triassic. *Review of Palaeobotany and Palynology*, **120**, 233–246.

Melo, J. H. G. and Loboziak, S. (2003). Devonian–Early Carboniferous miospore biostratigraphy of the Amazon Basin, Northern Brazil. *Review of Palaeobotany and Palynology*, **124**, 131–202.

Miller, M. F., and Collinson, J. W. (1994). Trace fossils from Permian and Triassic sandy braided-stream deposits, central Transantarctic Mountains. *Palaios*, **9**, 605–610.

Miller, M. F. and Isbell J. L. (2010) Reconstruction of a high-latitude, postglacial lake: Mackellar Formation (Permian), Transantarctic Mountains. *Geological Society of America Special Papers*, **468**, 193–207

Mond, A. (1972). Permian sediments of the Beaver Lake area, Prince Charles Mountains. In *Antarctic Geology and Geophysics*, ed. R. J. Adie. Oslo: Universitetforlaget, pp. 585–589.

Mory, A. J. and Backhouse, J. (1997). Permian stratigraphy and palynology of the Carnarvon Basin, Western Australia. *Geological Survey of Western Australia Report*, **51**, 1–42.

Mory, A. J., Redfern, J. and Martin, J. R. (2008). A review of Permian-Carboniferous glacial deposits in Western Australia. In *Resolving the Late Paleozoic Ice Age in Time and Space, Geological Society of America Special Paper*, **441**, eds C. R. Fielding, T. D. Frank and J. L. Isbell. Boulder: The Geological Society of America, pp. 29–40.

Nardin, E., Goddéris, Y., Donnadieu, Y., *et al.* (2011). Modeling the early Paleozoic long-term climatic trend. *Geological Society of America Bulletin*, **123**, 1181–1192.

Neish, P. G., Drinnan, A. N. and Cantrill, D. J. (1993). Structure and ontogeny of *Vertebraria* from silicified Permian sediments in East Antarctica. *Review of Palaeobotany and Palynology*, **79**, 221–224.

Niklas, K. J. (1994). Predicting the height of fossil plant remains: an allometric approach to an old problem. *American Journal of Botany*, **81**, 1235–1242.

Nishida, H., Pigg, K. B. and Rigby, J. F. (2003). Swimming sperm in an extinct Gondwanan plant. *Nature*, **422**, 396–397.

Nishida, H., Pigg, K. B., Kudo, K. and Rigby, J. F. (2004). Zooidogamy in the Late Permian genus *Glossopteris*. *Journal of Plant Research*, **117**, 323–328.

Nishida, H., Pigg, K. B., Kudo, K. and Rigby, J. F. (2007). New evidence of reproductive organs of *Glossopteris* based on permineralized fossils from Queensland. I. Ovulate organ *Homevaleia* gen. nov. *Journal of Plant Research*, **120**, 539–549.

Pant, D. D. (1996). The biogeography of late Palaeozoic floras of India. *Review of Palaeobotany and Palynology*, **90**, 79–98.

Pant, D. D. and Singh, R. S. (1974). On the stem and attachment of *Glossopteris* and *Gangamopteris* leaves. Part 2 – structural features. *Palaeontographica B*, **147**, 42–73.

Parrish, J. T. (1993). Climate of the supercontinent Pangea. *The Journal of Geology*, **101**, 215–233.

Pigg, K. B. (1987). Anatomically preserved glossopterid remains from the Permian of Antarctica. *American Journal of Botany*, **74**, 687.

Pigg, K. B. (1990). Anatomically preserved *Glossopteris* foliage from the central Transantarctic Mountains. *Review of Palaeobotany and Palynology*, **66**, 105–127.

Pigg, K. B. and McLoughlin, S. (1997). Anatomically preserved *Glossopteris* leaves from the Bowen and Sydney basins, Australia. *Review of Palaeobotany and Palynology*, **97**, 339–359.

Pigg, K. B. and Nishida, H. (2006). The significance of silicified plant remains to the understanding of *Glossopteris*-bearing plants: a historical review. *Journal of the Torrey Botanical Society*, **133**, 46–61.

Pigg, K. B. and Taylor, T. N. (1985). Anatomically preserved *Glossopteris* from the Beardmore Glacier area of Antarctica. *Antarctic Journal of the United States*, **20(5)**, 8–10.

Pigg, K. B. and Taylor, T. N. (1987). Anatomically preserved *Glossopteris* from Antarctica. In *Actas VII Simposio Argentino Paleobotánica y Palinología*. Buenos Aires: Associación Palaeontólogica Argentino, pp. 177–180.

Pigg, K. B. and Taylor, T. N. (1990). Permineralized *Glossopteris* and *Dicroidium* from Antarctica. In *Antarctic Paleobiology: its Role in the Reconstruction of Gondwana*, eds T. N. Taylor and E. L. Taylor. New York: Springer-Verlag, pp. 164–172.

Pigg, K. B. and Taylor, T. N. (1993). Anatomically preserved *Glossopteris* stems with attached leaves for the central Transantarctic Mountains, Antarctica. *American Journal of Botany*, **80**, 500–516.

Pigg, K. B. and Trivett, M. L. (1994). Evolution of the glossopterid gymnosperms from Permian Gondwana. *Journal of Plant Research*, **107**, 461–477.

Plumstead, E. P. (1952). Description of two new genera and six species of fructifications borne on *Glossopteris* leaves. *Transactions of the Geological Society of South Africa*, **55**, 281–328.

Plumstead, E. P. (1958). The habit and growth of Glossopteridae. *Transactions of the Geological Society of South Africa*, **61**, 81–96.

Plumstead, E. P. (1962). Fossil floras from Antarctica. In *Trans-Antarctic Expedition 1955–1958, Scientific Reports, 9 (Geology)*. London: The Trans-Antarctic Expedition Committee, pp. 1–154.

Plumstead, E. P. (1964). Palaeobotany of Antarctica. In *Antarctic Geology: proceedings of the First International Symposium on Antarctic Geology, Cape Town, 16–21 September 1963*, ed. R. J. Adie. Amsterdam: North-Holland Publishing Company, pp. 637–654.

Plumstead, E. P. (1975). A new assemblage of fossil plants from the Milorgfjella, Dronning Maud Land. *British Antarctic Survey Scientific Reports*, **83**, 1–30.

Powell, C. McA. and Veevers, J. J. (1987). Namurian uplift in Australia and South America triggered the main Gondwanan glaciation. *Nature*, **326**, 177–179.

Prevec, R., McLoughlin, S. and Bamford, M. K. (2008). Novel double wing morphology revealed in a South African ovuliferous glossopterid fructification: *Bifariala intermittens* (Plumstead 1958) comb. nov. *Review of Palaeobotany and Palynology*, **150**, 22–36.

Ravich, G. M., Gor, Y. G., Dibner, A. F. and Dobanova, O. V. (1977). Stratigrafiya verkhne-paleozoiskikh uglenocynkh otlozheniy vostochnoy Antarktidy (rayon ozera Biver). *Antarktica*, **16**, 62–75.

Rees, P. M. (2002). Land-plant diversity and the end-Permian mass extinction. *Geology*, **30**, 827–830.

Rees, P. M., Ziegler, A. M., Gibbs, M. T., *et al.* (2002). Permian phytogeographic patterns and climate. *The Journal of Geology*, **110**, 1–31.

Retallack, G. J. (1980). Late Carboniferous to Middle Triassic megafossil floras from the *Sydney Basins*. In *A Guide to the Sydney Basin, Geological Survey of New South Wales Bulletin*, **26**, eds C. Herbert and R. Helby. Sydney: O. West, Government Printer, pp. 384–430.

Rigby, J. F. (1967). On *Gangamopteris walkomii* sp. nov. *Records of the Australian Museum*, **27**, 175–182.

Rigby, J. F. (1969). Permian sphenopsids from Antarctica. *United States Geological Survey Professional Paper*, **613F**, 1–13.

Rigby, J. F. and Schopf, J. M. (1969). Stratigraphic implications of Antarctic palaeobotanical studies. In *Gondwana Stratigraphy. 1st IUGS Gondwana Symposium*, ed. A. J. Amos. Paris: UNESCO, pp. 91–106.

Rocha-Campos, A. C., dos Santos, P. R. and Canuto, J. R. (2008). Late Paleozoic glacial deposits of Brazil: Paraná Basin. In *Resolving the Late Paleozoic Ice Age in Time and Space, Geological Society of America Special Paper*, **441**, eds C. R. Fielding, T. D. Frank and J. L. Isbell. Boulder: The Geological Society of America, pp. 97–114.

Ryberg, P. E. (2009). Reproductive diversity of Antarctic glossopterid seed-ferns. *Review of Palaeobotany and Palynology*, **158**, 167–179.

Ryberg, P. E. (2010). *Lakkosia kerasata* gen. et sp. nov., a permineralized megasporangiate glossopterid structure from the central Transantarctic Mountains, Antarctica. *International Journal of Plant Sciences*, **171**, 332–344.

Ryberg, P. E. and Taylor, E. L. (2007). Silicified wood from the Permian and Triassic of Antarctica: tree rings from polar climates. In *Online Proceedings of the 10th International Symposium on Antarctic Earth Sciences*, eds A. K. Cooper, C. R. Raymond and the 10th ISAES Editorial team. U. S. Geological Survey Open-File Report 2007–1047, Version 2. Short Research Paper 082; doi: 10.3133/of2007–1047.srp082. {Cited 2011}.

Ryberg, P. E., Hermsen, E. J., Taylor, E. L., Taylor, T. N. and Osborn, J. M. (2008). Development and ecological implications of dormant buds in the high latitude Triassic sphenophyte *Spaciinodum* (Equisetaceae). *American Journal of Botany*, **95**, 1443–1453.

Rygel, M. C., Fielding, C. R., Frank, T. D. and Birgenheier, L. (2008). The magnitude of late Paleozoic glacioeustatic fluctuations: a synthesis. *Journal of Sedimentary Research*, **78**, 500–511.

Scheckler, S. E. and Galtier, J. (2003). Tyloses and ecophysiology of the early Carboniferous progymnosperm tree *Protopitys buchiana*. *Annals of Botany*, **91**, 739–747.

Scheffler, K., Hoerness, S. and Schwark, L. (2003). Global changes during Carboniferous-Permian glaciation of Gondwana: linking polar and equatorial climate evolution by geochemical proxies. *Geology*, **31**, 605–608.

Schmidt, D. C. and Ford, A. B. (1969). Geology of the Pensacola and Thiel Mountains. In *Folio 12, sheet Geologic Maps of Antarctica, Antarctic folio map series 5, Plate v*, eds V. C. Bushnell and C. Craddock. New York: American Geographical Society.

Schopf, J. M. (1962). A preliminary report on plant remains and coal of the sedimentary section in the central range of the Horlick Mountains, Antarctica. *Ohio State University Institute in Polar Studies Report*, **2**, 1–61.

Schopf, J. M. (1965). Anatomy of the axis in *Vertebraria*. In *Geology and Paleontology of the Antarctic, Antarctic Research Series*, **6**, ed. J. B. Hadley. Washington: American Geophysical Union, pp. 217–228.

Schopf, J. M. (1967). Antarctic fossil plant collecting during the 1966–1967 season. *Antarctic Journal of the United States*, **2**(5), 114–116.

Schopf, J. M. (1970). Petrified peat from a Permian coal bed in Antarctica. *Science*, **169**, 274–277.

Schopf, J. M. (1971). Notes on plant tissue preservation and mineralization in a Permian deposit of peat from Antarctica. *American Journal of Science*, **271**, 522–543.

Schopf, J. M. (1976). Morphologic interpretation of fertile structures in glossopterid gymnosperms. *Review of Palaeobotany and Palynology*, **21**, 25–64.

Schopf, J. M. (1986). Forms and facies of *Vertebraria* in relation to Gondwana coal. In *Geology of the central Transantarctic Mountains, Antarctic Research Series*, **36**, eds M. D. Turner and J. E. Splettstoesser. Washington: American Geophysical Union, pp. 37–62.

Schwendemann, A. B., Decombeix, A.-L., Taylor, E. L. and Taylor, T. N. (2010). *Collinsonites schopfi* gen. et sp. nov., a herbaceous lycopsid from the Upper Permian of Antarctica. *Review of Palaeobotany and Palynology*, **158**, 291–297.

Scotese, C. R., Boucot, A. J. and McKerrow, W. S. (1999). Gondwanan palaeogeography and palaeoclimatology. *Journal of African Earth Sciences*, **28**, 99–114.

Sepkoski Jr, J. J. (1996). Patterns of Phanerozoic extinction: a perspective from global data bases. In *Global Events and Event Stratigraphy in the Phanerozoic*, ed. O. H. Walliser. Heidelberg, Berlin, New York: Springer-Verlag, pp. 35–52.

Seward, A. C. (1914). Antarctic fossil plants. In *British Antarctic ("Terra Nova") Expedition 1910. Natural History Reports Geology. Volume 1, No. 1.* London: British Museum (Natural History), pp. 1–49.

Shi, G. R. and Waterhouse, J. B. (2010). Late Palaeozoic global changes affecting high-latitude environments and biotas: an introduction. *Palaeogeography, Palaeoclimatology, Palaeoecology*, **298**, 1–16.

Smith, N. D., Barrett, P. J. and Woolfe, K. J. (1998). Glacier-fed(?) sandstone sheets in the Weller Coal Measures (Permian), Allan Hills, Antarctica. *Palaeogeography, Palaeoclimatology, Palaeoecology*, **141**, 35–51.

Smoot, E. L. and Taylor, T. N. (1985). Ovules from the Permian of Antarctica. *American Journal of Botany*, **72**, 900.

Smoot, E. L. and Taylor, T. N. (1986a). Structurally preserved fossil plants from Antarctica. II. A Permian moss from the Transantarctic Mountains. *American Journal of Botany*, **73**, 1683–1691.

Smoot, E. L. and Taylor, T. N. (1986b). Evidence of simply polyembryony in Permian seeds from Antarctica. *American Journal of Botany*, **73**, 1077–1079.

Srivastava, A. K. and Agnihotri, D. (2011). Insect traces on early Permian plants of India. *Paleontological Journal*, **45**, 200–206.

Stemmerik, L. (2008). Influence of the late Paleozoic Gondwana glaciations on the depositional evolution of the northern Pangean shelf, North Greenland, Svalbard, and the Barents Sea. In *Resolving the Late Paleozoic Ice Age in Time and Space, Geological Society of America Special Paper*, **441**, eds C. R. Fielding, T. D. Frank and J. L. Isbell. Boulder: The Geological Society of America, pp. 205–218.

Stephenson, M. H., Angiolini, L. and Leng, M. J. (2007). The Early Permian fossil record of Gondwana and its relationship to deglaciation: a review. In *Deep-Time Perspectives on Climate Change: marrying the Signal from Computer Models and Biological Proxies*, eds M. Williams, A. M. Haywood, F. J. Gregory and D. N. Schmidt. London: The Micropalaeontological Society Special Publications, pp. 169–189.

Stephenson, P. J. (1966). Geology 1. Theron Mountains, Shackleton Range, and Whichaway Nunataks. In *Trans-Antarctic Expedition 1955–1958. Scientific Reports, 8 (Geology)*. London: The Trans-Antarctic Expedition Committee, pp. 1–79.

Storey, B. C., Macdonald, D. I. M., Dalziel, I. W. D., Isbell, J. L. and Millar, I. L. (1996). Early Paleozoic sedimentation, magmatism, and deformation in the Pensacola Mountains, Antarctica: the significance of the Ross Orogeny. *Geological Society of America Bulletin*, **108**, 685–707.

Stubblefield, S. P. and Taylor, T. N. (1985). Fossil fungi in Antarctic wood. *Antarctic Journal of the United States*, **20**(5), 7–8.

Stubblefield, S. P. and Taylor, T. N. (1986). Wood decay in silicified gymnosperms from Antarctica. *Botanical Gazette*, **147**, 116–125.

Tabor, N. J. and Poulsen, C. J. (2008). Palaeoclimate across the Late Pennsylvanian-Early Permian tropical palaeolatitudes: a review of climate indicators, their distribution, and relation to palaeophysiographic climate factors. *Palaeogeography, Palaeoclimatology, Palaeoecology*, **268**, 293–310.

Tajika, E. (2007). Long-term stability of climate and global glaciations throughout the evolution of the earth. *Earth Planets Space*, **59**, 293–299.

Talent, J. A., Mawson, R., Simpson, A. J. and Brock, G. A. (2002). Ordovician–Carboniferous of Townsville Hinterland: Broken River and Camel Creek regions, Burdekin and Clarke River basins. *Macquarie University Centre for Ecostratigraphy and Palaeobiology Special Publication*, **1**, 1–82.

Taylor, E. L. (1996). Enigmatic gymnosperms? Structurally preserved Permian and Triassic seed ferns from Antarctica. *Review of Palaeobotany and Palynology*, **90**, 303–318.

Taylor, E. L. and Ryberg, P. E. (2007). Tree growth at polar latitudes based on fossil tree ring analysis. *Palaeogeography, Palaeoclimatology, Palaeoecology*, **255**, 246–264.

Taylor, E. L. and Taylor, T. N. (1987). An ovule bearing reproductive organ from the Permian of Antarctica. *American Journal of Botany*, **74**, 69–19

Taylor, E. L. and Taylor, T. N. (1990). Structurally preserved Permian and Triassic floras of Antarctica. In *Antarctic Paleobiology: its Role in the Reconstruction of Gondwana*, eds T. N. Taylor and E. L. Taylor. New York: Springer-Verlag, pp. 149–163.

Taylor, E. L. and Taylor, T. N. (1992). Reproductive biology of the Permian Glossopteridales and their suggested relationships to flowering plants. *Proceedings of the National Academy of Sciences of the United States of America*, **89**, 11495–11497.

Taylor, E. L., Taylor, T. N., Collinson, J. W. and Elliot, D. H. (1986). Structurally preserved Permian plants from Skaar Ridge, Beardmore Glacier region. *Antarctic Journal of the United States*, **21**(5), 27–28.

Taylor, E. L., Taylor, T. N. and Collinson, J. W. (1989). Depositional setting and paleobotany of Permian and Triassic permineralized peat from the central Transantarctic Mountains, Antarctica. *International Journal of Coal Geology*, **12**, 657–679.

Taylor, E. L., Taylor, T. N., Isbell, J. L. and Cúneo, N. R. (1989a). Fossil floras of southern Victoria Land: 2. Kennar Valley. *Antarctic Journal of the United States*, **24**(5), 26–28.

Taylor, E.L., Taylor, T.N., Isbell, J.L. and Cúneo, N.R. (1989b). Fossil floras of southern Victoria Land: 1. Aztec Mountain. *Antarctic Journal of the United States of America*, **24**(5), 24–26.

Taylor, E. L., Taylor, T. N. and Cúneo, N. R. (1992). The present is not the key to the past: a polar forest from the Permian of Antarctica. *Science*, **257**, 1675–1677.

Taylor, T. N. (1993). The role of Late Paleozoic fungi in understanding the terrestrial paleoecosystem. In *Comptes Rendu 12th International Congress of Carboniferous and Permian Stratigraphy and Geology Volume 2*, ed. S. Archangelsky. Buenos Aires, pp. 147–154.

Taylor, T. N. and Stubblefield, S. P. (1987). A fossil mycoflora. *Actas VII Simposio Argentino Paleobotánica y Palinologia*. Buenos Aries: Asociación Paleontológica Argentina, pp. 193–197.

Taylor, T. N. and Taylor, E. L. (1987). Structurally preserved fossil plants from Antarctica. III. Permian seeds. *American Journal of Botany*, **74**, 904–913.

Taylor, T. N. and Taylor, E. L. (1992). Permian plants from the Ellsworth Mountains, West Antarctica. In *Geology and Paleontology of the Ellsworth Mountains West Antarctica, Geological Society of America Memoir*, **170**, eds G. F. Webers, C. Craddock and J. F. Splettstoesser. Boulder: The Geological Society of America, pp. 285–289.

Taylor, T. N. and Taylor, E. L. (1993). *The Biology and Evolution of Fossil Plants*. New Jersey: Prentice Hall.

Taylor, T. N., Remy, W. and Hass, H. (1992). Fungi from the lower Devonian Rhynie Chert: chtridiomycetes. *American Journal of Botany*, **79**, 1233–1241.

Tessensohn, F., Kleinschmidt, G. and Buggisch, W. (1999). Permo-Carboniferous glacial beds in the Shackleton Range. *Terra Antarctica*, **6**, 317–325.

Tiwari, R. S. and Tripathi, A. (1992). Marker assemblage-zones of spore and pollen species through Gondwana Palaeozoic and Mesozoic sequences in India. *The Palaeobotanist*, **40**, 194–236.

Torsvik, T. H. and Van der Voo, R. (2002). Refining Gondwana and Pangea palaeogeography: estimates of Phanerozoic non-dipole (octupole) fields. *Geophysics Journal International*, **151**, 771–794.

Townrow, J. T. (1967). Fossil plants from Allan Hills and Carapace Nunataks, and from the Upper Mill and Shackleton glaciers, Antarctica. *New Zealand Journal of Geology and Geophysics*, **10**, 456–473.

Trewin, N. H., MacDonald, D. I. M. and Thomas, C. G. C. (2002). Stratigraphy and sedimentology of the Permian of the Falkland Islands: lithostratigraphic and palaeoenvironmental links with South Africa. *Journal of the Geological Society of London*, **159**, 5–19.

Truswell, E. M. (1982). Palynology of seafloor samples collected by the 1911–14 Australasian Antarctic Expedition: implications for the geology of coastal East Antarctica. *Journal of the Geological Society of Australia*, **29**, 343–356.

Turner, B. R. (1983). Paleosols in Permo-Triassic continental sediments from Prydz Bay, East Antarctica. *Journal of Sedimentary Petrology*, **63**, 694–706.

Veevers, J. J. (2000). Permian–Triassic Pangean basins and foldbelts along the Panthalassan margin of Gondwanaland. In *Billion-year Earth History of Australia and Neighbours in Gondwanaland*, ed. J. J. Veevers. Sydney: GEMOC Press, pp. 292–308.

Veevers, J. J. (2001). *Atlas of Billion-year Earth History of Australia and Neighbours in Gondwanaland*. Sydney: GEMOC Press.

Veevers, J. J. and Powell, M. (1987). Late Paleozoic glaciation in Gondwanaland reflected in transgressive-regressive depositional sequences in Euamerica. *Geological Society of America Bulletin*, **98**, 475–487.

Veizer, J., Ala, D., Azmy, K., *et al.* (1999). ^{87}Sr/^{86}Sr, δ^{13}C and δ^{18}O evolution of Phanerozoic seawater. *Chemical Geology*, **161**, 59–88.

Visser, J. N. J. (1987). The paleogeography of part of the southwestern Gondwana during the Permo-Carboniferous glaciation. *Palaeogeography, Palaeoclimatology, Palaeoecology*, **61**, 205–219.

Visser, J. N. J. (1989). Episodic Paleozoic glaciation in the Cape-Karoo Basin, South Africa. In *Glacier Fluctuations and Climatic Change: Proceedings of the Symposium on Glacier Fluctuations and Climatic Change, held in Amsterdam, 1–5 June 1987, Glaciology and Quaternary Geology Series*, **6**, ed. J. Overlemans. Boston, Dordrecht: Kluwer Academic Publisher, pp. 1–12.

Visser, J. N. J. (1993). A reconstruction of the late Paleozoic ice-sheet on southwestern Gondwana. In *Gondwana Eight: Assembly, Evolution and Dispersal. Proceedings of* the 8th Gondwana Symposium Hobart, eds R. H. Findlay, R. Unrug, M. R. Banks and J. J. Veevers. Netherlands: A. A. Balkema Press, pp. 449–458.

Visser, J. N. J. (1997). Deglaciation sequences in the Permo-Carboniferous Karoo and Kalahari basins of southern Africa: a tool in the analysis of cyclic glaciomarine basin fills. *Sedimentology*, **44**, 507–521.

Visser, J. N. J. and Loock, J. C. (1987). Ice margin influence on glaciomarine sedimentation in the Permo-Carboniferous Dwyka Formation and from the southwestern Karoo, South Africa. *Sedimentology*, **34**, 929–941.

Visser, J. N. J., Loock, J. C. and Collinston, W. P. (1987). Subaqueous outwash fan and esker sandstones in the Permo-Carboniferous Dwyka Formation of South Africa. *Journal of Sedimentary Petrology*, **57**, 467–478.

Visser, J. N. J., van Niekerk, B. N. and van der Merwe, S. W. (1997). Sediment transport of the late Paleozoic glacial Dwkya Group in the southwestern Karoo Basin. *South African Journal of Geology*, **100**, 223–236.

Wallister, O. H. ed. (1996). *Global Events and Event Stratigraphy*. Berlin: Springer-Verlag.

Weaver, L. S., McLoughlin, S. and Drinnan, A. N. (1997). Fossil woods from the Permian Bainmedart Coal Measures, northern Prince Charles Mountains, East Antarctica. *Australian Geological Survey Organisation Journal of Geology and Geophysics*, **16**, 655–676.

Webb, J. A. and Fielding, C. R. (1993b). Revised stratigraphical nomenclature for the Permo-Triassic Flagstone Bench Formation, northern Prince Charles Mountains, East Antarctica. *Antarctic Science*, **5**, 409–410.

Webb, J. A. and Fielding, C. R. (1993a). Permo-Triassic sedimentation within the Lambert Graben, northern Prince Charles Mountains, East Antarctica. In *Gondwana Eight: Assembly, Evolution and Dispersal. Proceedings of the 8th Gondwana Symposium Hobart*. eds R. H. Findlay, R. Unrug, M. R. Banks and J. J. Veevers. Netherlands: A. A. Balkema Press, pp. 357–369.

White, M. E. (1962). Permian plant remains from Mount Rymill, Antarctica. Report on 1961 plant fossil collections. *Bureau of Mineral Resources, Geology and Geophysics, Australia, Record*, **1962/114**, 1–14.

White, M. E. (1969). Report on plant fossil from the Beaver Lake area, Prince Charles Mountains, Antarctica. *Bureau of Mineral Resources, Geology and Geophysics, Australia, Record*, **1969/100**, 1–18.

White, M. E. (1973). Permian flora from the Beaver Lake area, Prince Charles Mountains, Antarctica. *Bureau of Mineral Resources, Geology and Geophysics Bulletin*, **126** (Palaeontological papers for 1969), 13–18.

Wicander, R., Clayton, G., Marshall, J. E. A., Troth, I. and Racey, A. (2011). Was the latest Devonian glaciation a multiple event? New palynological evidence from Bolivia. *Palaeogeography, Palaeoclimatology, Palaeoecology*, **305**, 75–83.

Wopfner, H. (1999). The early Permian deglaciation event between East Africa and north-western Australia. *Journal of African Earth Sciences*, **29**, 77–99.

Zhao, L., Taylor, T. N. and Taylor, E. L. (1995). Cupulate glossopterid seeds from the Permian Buckley Formation, central Transantarctic Mountains. *Antarctic Journal of the United States*, **30**(5), 54–55.

Ziegler, A. M., Hulver, M. L. and Rowley, D. B. (1997). Permian world topography and climate. In *Late Glacial and Post Glacial Environmental Changes: Quaternary, Carboniferous-Permian, and Proterozoic*, ed. I. P. Martini. Oxford: Oxford University Press, pp. 111–146.

4

Icehouse to hothouse: floral turnover, the Permian–Triassic crisis and Triassic vegetation

Introduction

The Permian–Triassic transition marks the turnover from ancient palaeophytic to more modern mesophytic floras and provides a striking example of the direct influence of the physical environment on plant communities (Iglesias, Artabe and Morel, 2011). The dieback of woody swamp-forest vegetation (Looy *et al.*, 1999) at the Permian–Triassic transition resulted in the disappearance of the extensive peat deposits that characterised both northern and southern cool, humid climate zones of Pangea during the Permian. Across the southern part of the Pangean supercontinent (Gondwana) the swamp forests comprised *Glossopteris*-dominated vegetation, which was deciduous and adapted to the cold winters. The disappearance of this vegetation evidences one of the most dramatic changes seen in the terrestrial fossil record of the Southern Hemisphere. The glossopterids that had dominated the southern high latitudes for nearly 50 Ma were replaced by two new groups of plants that would be as important in Southern Hemisphere Triassic ecosystems as the glossopterids had been to Permian ecosystems. The Peltaspermaceae, and a small, woody, pinnately-leaved gondwanan group, the Corystospermaceae, were two families of seed plants that rose to dominance. Ecosystem recovery through the Permian–Triassic transition, initially thought to have been unusually slow with a duration possibly at least twice that following other major extinctions (Erwin, 1998a, 1998b; Hallam, 1991), now appears to have been fairly rapid (Metcalfe and Isozaki, 2009). In the Southern Hemisphere complex coal-forming communities did not reappear until the Middle Triassic and did not become extensively developed until some 23 Ma after the end Permian crisis (Retallack, Veevers and Morante, 1996).

The changes in sedimentology and coal petrography observed across gondwanan sequences indicate that the climate became both warmer and drier in the Late Permian, reflecting the reorganisation of the Earth system through the Late Permian and into the Triassic. These environmental changes across the Permian–Triassic transition are thought to be responsible for the third, and largest, of the five big mass extinctions in the Phanerozoic (Raup and Sepkoski, 1982; Sepkoski, 1984, 1996), where it is estimated that >95% of life on Earth became extinct (Benton, 2005; Benton and Twitchett, 2003). As a consequence understanding the dramatic reorganisation of biological systems both in the oceans and on land, and also the causes of these global changes, have been the subjects of extensive research (see

Korte and Kozur, 2010; Metcalfe and Isozaki, 2009; Wignall, 2007 for recent perspectives) such that the Permian–Triassic transition is probably one of the most studied intervals in the Phanerozoic.

Global palaeogeography and palaeoclimate of the late Permian and Triassic

The convergence of almost all of the Earth's landmasses (including Gondwana) that took place during the later stages of the Paleozoic ultimately formed the single supercontinent, Pangea, in the Late Permian. Pangea straddled the globe extending from 85° N to 90° S (Ziegler, Scotese and Barrett, 1983), and the distribution of this landmass had major implications for both the climate and oceanic circulation. A vast arid region formed in the continental interior with seasonal monsoons developing near the coasts (Parrish, 1993). The Panthalassa Ocean surrounded Pangea yet very little is known about the open ocean at this time since the deep-ocean sediments laid down during the Triassic have largely disappeared through subduction of oceanic plates. In the equatorial region a broad gulf, an extension of the Palaeotethys Ocean, extended in from the east towards the centre of Pangea serving to partly separate Gondwana in the south from Laurasia[1] in the north. The southeast portion of Gondwana (including Australia, India and Antarctica) was attached to South America and Africa, and served to separate the Palaeotethys Ocean to the north and northwest from the Panthalassa Ocean to the south and southeast (Shi, Waterhouse and McLoughlin, 2010; Torsvik and Cocks, 2004).

As Pangea migrated northwards through 10° of latitude (Torsvik and Cocks, 2004) during the Permian, Antarctica moved from the polar position (above 60° S) that it had occupied at the beginning of the Permian into palaeolatitudes spanning 60° to 30° S. This northward migration continued on through the Early Triassic (Iglesias, Artabe and Morel, 2011; Li and Powell, 2001). The Panthalassan margin of Pangea that extended from northern Australia to southern Africa and South America was at this time a convergent margin with a long-lived subduction zone and associated retro-arc foreland basin (Figure 4.1). In the Late Permian the first phase of the gondwanide orogenesis occurred along this coastline due to subduction related tectonism (Veevers, Conanghan and Shaw, 1994). This phase of mountain building extended into the Triassic and was accompanied by a shift in both sedimentation and rainfall patterns (Collinson, 1990). In the Antarctic region of Pangea these shifts reflect the closer proximity of both arc environments, with their associated increase in volcanism, and mountains to the sedimentary basins.

More or less coeval with gondwanide orogenesis a sliver of land, which had begun rifting at ~300 Ma from the northern Gondwana margin (e.g. Australia) in the southern Palaeotethys, split off to form the Cimmeria microcontinent (a plate that today comprises parts of present day Anatolia, Iran, Afghanistan, Tibet, Indochina and Malaya). Between this microcontinent and the northern Gondwana margin the rift formed a new ocean, the Neotethys (Figure 4.1). The ocean gradually widened and so pushed Cimmeria northwards

---

[1] The 'new' supercontinent comprising the old continents of Laurentia and Eurasia.

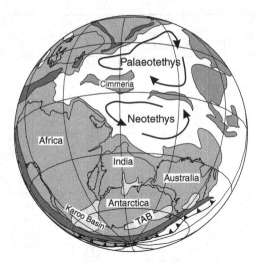

Figure 4.1 Early Triassic reconstruction at ~245 Ma. Black barbs mark subduction zone, light shading = sedimentary basins, mid grey = land, dark grey = mountain belts, TAB = Transantarctic Basin. Note the Tethys (Palaeo and Neo) is closed from the large Panthalassa Ocean (back of globe). The Tethys is subdivided into the Neotethys which is opening up behind the Cimmerian microcontinent. As the Cimmerian microcontinent is transported northwards the Palaeotethys closes with subduction and mountain building on the northern tethyean margin.

towards Euramerica, whilst at the same time causing the Palaeotethys Ocean to shrink. The Cimmerian continent ultimately collided with the northern Palaeotethys margin and was accreted to Euramerica resulting in a mountain building episode (Indosinian Orogeny) from Early Triassic times. This was followed by collisions between the South China Plate and the North China Plate in the Late Triassic. Further to the west, closure occurred along the Urals resulting in mountain building in this region.

By the Permian–Triassic boundary the (Neo)Tethys was well-developed (Stampfli and Borel, 2002) and extended from 30° N to 30° S. This sea played an important role in heat transport and oceanic circulation resulting in maximal heating in the circum-Tethys region (Kutzbach and Gallimore, 1989).

Global palaeogeographic changes through the Triassic were brought about by the onset of rifting in the Early Triassic between Euramerica and Gondwana (Golonka, 2007). This resulted in the development of uplifted highlands and rift valleys in this region of Pangea. Sedimentation in these rifts passed from red-bed sedimentation, formed under fluvial conditions, in the Early Triassic through to shallow marine carbonates and evaporates, and ultimately to open marine conditions in the Jurassic.

During the Late Permian and into the Triassic, continental movement was also causing massive volcanic activity in the north of Pangea in the Russian region of Siberia and across China. This activity lasted about 700 kyr and produced large magmatic flows, the Siberian Traps. At the same time it would have been responsible for the release of large quantities of carbon dioxide and sulphur dioxide into the atmosphere (Kiehl and Shields, 2005). This

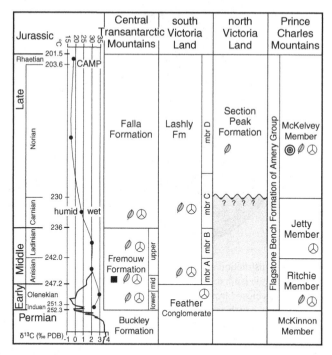

Figure 4.2 Triassic stratigraphic framework for Antarctic rock units and climate curve. Climate curve derived from Kiessling (2010) using oxygen isotopes to determine the average temperature plotted (black dots) at the mid-point for each stage. Carbon isotope curve derived from Korte and Kozur (2010). Note the sharp negative excursion in carbon isotopes with several minor perturbations on the curve just prior to the Permian–Triassic boundary. CAMP is Central Atlantic Magmatic Province, an event thought to be responsible for the Triassic–Jurassic extinction event. Symbols follow Figure 3.3.

activity was coeval with the largest mass extinction of the Phanerozoic and major pertuba-tions recorded in the global carbon cycle.

Change in the carbon cycle through the Permian to Triassic is recorded in a wide range of marine and continental deposits, such as platform carbonates, pelagic carbonates, organic-rich shales, terrestrial palaeosols and lacustrine deposits (Korte and Kozur, 2010), and is manifested in negative excursions in $\delta^{13}C$ (Figure 4.2). In the absence of biostratigraphic or radiometric data these excursions are commonly used to locate the Permian–Triassic boundary (Lindström and McLoughlin, 2007). Carbon isotope ratios were initially relatively constant in the Late Permian attesting to the stability of the global carbon cycle (Korte and Kozur, 2010). However, during the Late Permian there was a gradual negative shift in the carbon isotope record, with the $\delta^{13}C$ of organic matter recording a 10–11‰ shift and the $\delta^{13}C$ values of marine carbonates recording a shift of 8–9‰ (Erwin, 1993; Twitchett *et al.*, 2001). This reflects destability in the carbon cycle resulting from the major loss of terrestrial biomass into the ocean. This gradual negative shift then began to accelerate towards the

Permian–Triassic boundary (Kiehl and Shields, 2005; Korte and Kozur, 2010) (Figure 4.2). Superimposed upon this trend are several small negative–positive fluctuations before maximum depletion is reached at the Permian–Triassic boundary. A short recovery followed (+1 to +2‰ shift) before a second (greater) negative isotope excursion occurred in the Early Triassic (Korte and Kozur, 2010) (Figure 4.2). Following this second major excursion, values gradually become more positive until they returned once more to levels more typical of the preceding Late Permian.

A number of scenarios have been put forward to explain this isotopic excursion at the Permian–Triassic boundary, these include (i) a collapse in primary productivity resulting in the Permian–Triassic biotic crisis (Figure 4.3) (e.g. Wang, Geldsetzer and Krouse, 1994), (ii) a massive input of ^{12}C from volcanic activity in the large Siberian Traps (e.g. Reichow *et al.*, 2009; Svenson *et al.*, 2009), and (iii) marine gas hydrate release (Erwin, 1993; Krull and Retallack, 2000; Twitchett *et al.*, 2001). These scenarios are discussed in more detail below.

During the Middle to Late Permian a gradual warming trend is evident from the western and eastern parts of Gondwana. In southern and central Africa, coal deposition ended during the Middle Permian (Cairncross, 2001), and warm semi-arid conditions prevailed from about 265 Ma (Visser, 1995). In eastern Australia the reduction in ice-rafted dropstones and other cryogenic features also evidence a widespread warming climate (Draper, 1983).

Through the latest Permian, as the $\delta^{13}C$ became more negative the climate was slowly warming and the latitudinal zonation in climate zones was breaking down. This was caused in part by the amalgamation of the Earth's continents into the single large landmass (Pangea) that extended over such a great latitudinal range. Atmospheric CO_2 levels rose to between 3000 and 4000 ppmv, resulting in a warming of 4–8 °C across the Permian–Triassic boundary (Korte *et al.*, 2010; Royer, 2006; Royer *et al.*, 2004). Global temperatures now ranged from 35 °C at the equator to 15 °C at the poles (Kidder and Worsley, 2004).

In the aftermath of the end-Permian event a generally warmer and less seasonal climate prevailed in southern Gondwana (Lindström and McLoughlin, 2007). This marked the passage from the global icehouse that characterised the earlier Permian to the hothouse state of the Early Triassic, followed by warm greenhouse climates persisting throughout the rest of the Triassic (Kidder and Worsley, 2004; Preto, Kustatscher and Wignall, 2010) – a climate change scenario supported by computer generated models for this interval (Kidder and Worsley, 2004; Kiehl and Shields, 2005). While Triassic climates are thought to have been warm with ice-free poles and relatively little variation, this may not have been the case and may simply reflect a lack of investigation (Preto, Kustatscher and Wignall, 2010). Extensive carbonate deposition provides evidence for warm waters in the Neotethys, the sea that was developing in the equatorial regions, and thereby closing off connections to the high latitudes. This, together with the distribution of land, is thought to have resulted in maximal summer heating (Kutzbach and Gallimore, 1989) causing extreme continental climates with hot summers, cold winters and a strong monsoon system along the coast (Kiehl and Shields, 2005; Kutzbach and Gallimore, 1989; Parrish, 1993). This hypothesis is supported by the distribution of continental red-bed fluvial deposits across both central

Europe, which in places also includes evidence for aeolian processes (Hounslow and Ruffell, 2006), and the mid-latitude regions of the Southern Hemisphere.

Peak global red-bed development occurred in the Early Triassic and this is often interpreted as an indication of prevailing arid climates (e.g. Hounslow and Ruffell, 2006). However, palaeosol horizons and plant fossil evidence suggest that this time was a period of significant weathering and moisture availability. This implies that the formation of these red-beds, in some regions at least, was the result of a highly seasonal climate with extensive wet periods alternating with dry oxidising periods (e.g. Retallack, Veevers and Morante, 1996). In high southern latitudes red-beds are lacking from Early Triassic sedimentary sequences and only appear later. This suggests that the highly seasonal and relatively warm conditions spread poleward through the Early Triassic, with a general shift from more humid conditions in the earliest Triassic to more arid conditions in the mid-Early Triassic (Smithian/Spathian boundary) (Galfetti *et al.*, 2007). Indeed, the distribution of fossil soils suggests that the poleward shift for each climate zone may have been in the region of 20 to 25° of latitude (e.g. Retallack, 1997a, 1999). In southeastern Gondwana, Early Triassic warming was responsible for the replacement of the Permian cold temperate climate by a cool temperate climate across southeast Australia and Antarctica. This is evidenced by the change in the composition of both the vegetation and palaeosols (Chumakov and Zharkov, 2003; Retallack, 1999; Retallack and Krull, 1999).

With these shifting climatic zones, conditions became increasingly inhospitable both in the oceans and on land. In the marine realm, oceans were either anoxic to shallow depths and acidified, or euxinic (Wignall and Twitchett, 1996). As conditions deteriorated, warm water taxa moved from the low latitudes into high latitudes (Wignall, Morante and Newton, 1998) where sea surface temperatures had also warmed to reach 8 °C to great depths (as compared with 0 °C today) (Kiehl and Shields, 2005). On land the once low latitude, and presumably warm loving, plants now begin to appear in high latitude records (such as Greenland; Looy *et al.*, 2001) where high latitude palaeosols indicate warm moist conditions (Retallack, 1997a, 1999; Retallack and Alonso-Zarza, 1998). Sedimentation patterns in the terrestrial realm also changed. At the beginning of the Triassic the coal deposits that had characterised the Permian were now globally absent (Retallack, Veevers and Morante, 1996; Veevers, Conanghan and Shaw, 1994).

By the Middle Triassic the extension of the climate zones was reaching its maximum extent. The distribution of fossil floras suggest that the equatorial belt was now arid (Ziegler *et al.*, 2003) and warm temperate to temperate zones had extended into the high latitudes (Cúneo *et al.*, 2003; Ziegler *et al.*, 1993). Conditions in these high latitudes remained strongly seasonal as evidenced by the pronounced growth rings in fossil trees (Cúneo *et al.*, 2003; Taylor, 1989; Taylor and Ryberg, 2007). In the Middle Triassic coal deposition resumed, testifying to a vegetational recovery resulting from a wetter climate where precipitation exceeded evaporation (Retallack, Veevers and Morante, 1996).

Although there has been some discussion regarding the extent of the climatic zonation during the Late Triassic (e.g. Kent and Olsen, 2000; Ziegler *et al.*, 2003), owing to the distribution of the land, the early Late Triassic interval is thought to have marked the ultimate breakdown in zonal climate patterns (Sellwood and Valdes, 2006) and signifies developing monsoonal atmospheric circulation (Parrish, 1993). A region of summer wet tropical climate prevailed along the eastern side of Pangea, extending from the equator around the tethyean margin to about 30° N and 35° S. Along this eastern margin, cool temperate climate extended towards the tropics (Sellwood and Valdes, 2006), whilst on the western margin a warm temperate climate extended into the high latitudes (Cúneo *et al.*, 2003; Ziegler *et al.*, 1993).

In the early part of the Late Triassic (Carnian) a distinct wetter interval occurred that is hypothesised to have been global in extent (Colombi and Parrish, 2008; Parrish, 1993; Prochow *et al.*, 2006). This was short-lived and was thought to represent the peak of monsoon development (Parrish, 1993). It was followed by a return to relative climatic stability for the remainder of the Triassic. Although the gross pattern of climate change has been established for the Triassic there is still debate about smaller-scale fluctuations that may be in part the result of a lack of investigation (Preto, Kustatscher and Wignall, 2010). Such high frequency cycles can be found in Late Triassic deposits of North America (Newark Supergroup) (Olsen, 1986; Olsen and Kent, 1996) and similar patterns are seen in European sequences (Reinhardt and Ricken, 2000; Vollmer *et al.*, 2008). So while the Late Triassic climate has been considered uniform with no large-scale trends, a more complex picture of high-frequency fluctuations driven by orbital forcing processes is becoming evident (Haas, Gotz and Pálfy, 2010; Whiteside *et al.*, 2011a, 2011b).

In summary, the Early Triassic was a time of extreme warmth and seasonality marked by a hyperthermal event in the earliest Triassic that saw a breakdown in climate zonation, the expansion of arid areas in equatorial regions, and warming into the high latitudes. Cooling from the peak warmth in the Early Triassic, although not well-documented, records several humid episodes testifying to a wetter climate globally probably relating to fluctuations in the strength of monsoonal circulation, which reached a peak in the early part of the Late Triassic (Parrish, 1993). The Late Triassic climate became more stable but not without high-frequency fluctuations. The temporal and spatial patterns of climate change through the Triassic still form the focus of much research and would benefit from a multidisciplinary approach to develop a coherent global climate framework (Preto, Kustatscher and Wignall, 2010) needed to constrain the palaeovegetation.

Global palaeovegetation

As the Earth moved from the icehouse phase of the Permian to the greenhouse phase of the Triassic through the hothouse of the Permian–Triassic boundary so the terrestrial biomes underwent a radical reorganisation. In the Permian three floristic provinces (or realms) have been recognised (Chapter 3), but by the Late Permian these had increased to four and were associated with ten biomes: tropical ever-wet, tropical seasonal dry (summer wet), desert, subtropical seasonal dry (winter wet), warm temperate, cool temperate, mid-latitude desert, cold temperate, tundra and glacial (Rees *et al.*, 1999; Scotese, Boucot and McKerrow, 1999;

Ziegler, 1990). The floristic realms associated with these biomes were (i) the Angaran Flora in the northern middle and high palaeolatitudes dominated by Cordaitales and Equisetales, (ii) the Euramerican Flora dominated by other groups of seed plants, along with (iii) the Cathaysian Flora, characterised by the *Gigantopteris* Flora, both of which lay in low, mainly tropical palaeolatitudes, and finally (iv) the Gondwana Flora was dominated by the *Glossopteris* in the southern middle and high palaeolatitudes (Hilton and Cleal, 2007; Iglesias, Artabe and Morel, 2011; Srivastava and Agnihotri, 2010). An intermixing of these different floras resulted in the presence of so-called 'mixed' floras. One distribution pattern shows mixed floras along the marginal basins of the Tethyan region during the Middle and on into the Late Permian. Some of the gondwanan floras contain a number of extra-gondwanan elements, whereas Euramerican, Cathaysian and Angaran floras show the possible presence of glossopterid related elements (see Srivastava and Agnihotri, 2010 for details). Other floras show combined elements of high and low latitudinal vegetation (e.g. Berthelin *et al.*, 2003). The Gondwana and Angara floras, occupying the middle and high latitudes of the Southern and Northern hemispheres respectively, were generally distinct and easily recognisable (Hilton and Cleal, 2007). The recognition of the Gondwanan Realm relied on the presence of the *Glossopteris* Flora. Here *Glossopteris* represents the main constituent during and after the late Paleozoic Gondwana glaciations, and is associated with cold or cool-temperate conditions (see Chapter 3). This flora was responsible for the coal measures that typify gondwanan lithostratigraphy in the Permian.

The Gondwana supercontinent during the Permian was a highly heterogeneous landmass in terms of its vegetation cover and distribution (Cúneo, 1996). Phytogeographic differentiation through the Permian shows both latitudinal and longitudinal components, with variations from polar to mid-latitudes and from west to east (e.g. Archangelsky and Arrondo, 1975). Archangelsky and Arrondo (1975) suggest that eastern Gondwana (India, Australia and Antarctica) was vegetationally distinct from western Gondwana (Africa and South America). Western Gondwana Floras have affinities with floras to the north and share a number of taxa including sphenophylls (e.g. *Annularia*, *Sphenophyllum*) and ferns (e.g. *Asterotheca*, *Pecopteris*). In contrast these so-called 'northern' taxa are absent from eastern Gondwana. Positioned predominantly at polar to subpolar latitudes at this time, Antarctica appears to have been an area of severe physical and climatic conditions. Minimum winter temperatures would have controlled the poleward movement of vegetation and thus the flora would have been highly adapted (Cúneo *et al.*, 1993).

At the end of the Permian, the intensifying greenhouse conditions in combination with the strong light regime could no longer sustain the remaining coal forest swamps across southern Gondwana even in polar latitudes (Kidder and Worsley, 2004). The resultant cessation of coal deposition varied temporally with cessation occurring earlier in the west and east relative to the polar regions (Lindström and McLoughlin, 2007). This cessation of coal generally characterised the Permian–Triassic transition across southeastern Gondwana (Retallack, Veevers and Morante, 1996; Taylor, Taylor and Collinson, 1989).

The biotic crisis at the end of the Late Permian occurred in two phases. The initial extinction was stepwise predominantly affecting the taxa typically associated with the

Glossopteris Flora. Following the initial biotic crisis, floristic similarity did increase across Gondwana for a time (Lindström and McLoughlin, 2007).

The second, or end-Permian, extinction was not a unique event. It now appears to have been characterised by four perturbations separated by transient greenhouse warming events (Retallack *et al.*, 2005) that were particularly striking at high palaeolatitudes. Palynological assemblages[2] indicate that by the latest Permian, major floral differences were once again evident. For example, in cold-cool temperate regions (Antarctica) a stable gymnospermous glossopterid community was overwhelmingly dominant in the latest Permian, whereas following the mass extinction, a period of constant and dramatic floristic change allowed the lycopods to proliferate. Conversely, in semi-arid areas (e.g. Kenya) the lycophytes were prominent constituents of the latest Permian palynoflora but gave way to gymnosperms after the initial crisis. These differences may have been a response to once humid areas becoming drier, and at least some dry areas becoming more humid. Thus the Gondwana-wide reorganisation of the terrestrial ecosystem reflects dramatic changes taking place in the atmosphere (Kidder and Worsley, 2004; Lindström and McLoughlin, 2007). Contemporaneous stepwise introduction of new taxa and Gondwana-wide reorganisation of the terrestrial ecosystem show that the effects of the end-Permian crisis continued to affect the biota for at least another 2 Ma (Lindström and McLoughlin, 2007).

The preceding Mesozoic was characterised by particularly warm climates with less climatic differentiation (Iglesias, Artabe and Morel, 2011) and more equable, sub-humid to semi-arid climates prevailing across southern Gondwana (Lindström and McLoughlin, 2007). The number of biomes in the Triassic consequently decreased and comprised tropical seasonal dry (summer wet), desert, subtropical seasonal dry (winter wet), warm temperate, and cool temperate biomes. Notable is the absence of glacial, tundra, cold temperate, steppe (arid cold temperate) and tropical ever-wet zones from the Triassic (Scotese, Boucot and McKerrow, 1999; Sellwood and Valdes, 2006). As a result, greater floristic uniformity characterised the earliest Mesozoic as the four Paleozoic floristic realms were reduced to two, namely the Lauraic Realm dominated by conifers and the Gondwanic Realm (Artabe, Morel and Spalletti, 2003; Srivastava and Agnihotri, 2010). The Gondwanic Realm could be subdivided longitudinally into the Southwest Gondwana Province (southern South America and southern Africa) and the Southeast Gondwanan Province (Antarctica and Australasia) (Iglesias, Artabe and Morel, 2011).

During the Triassic the Gondwana floristic realm extended across three climatic belts – the subtropical seasonal dry belt, the warm temperate belt and the cool temperate climatic belt, with only those continents that made up southeast Gondwana lying within the cool temperate climate belt (Iglesias, Artabe and Morel, 2011). Following the Permian–Triassic transition the typical *Glossopteris* Flora was replaced by the *Dicroidium* (foliage attributed to the Corystospermaceae) Flora, which extended from Australasia and Antarctica through South Africa to India and South America (Iglesias, Artabe and Morel, 2011; Li, 1986; Srivastava and

---

[2] Sections across Gondwana reveal discrepancies between the palynologically inferred P–Tr boundary and the negative $\delta^{13}C$ excursion.

Agnihotri, 2010). Within the *Dicroidium* Flora the corystosperm, *Dicroidium*, is associated with a number of other plants that include other seed plants, such as the peltasperm *Lepidopteris*, ferns and tree ferns, ginkgophytes, cycadophytes, conifers and sphenopsids. Pleuromeian/isoetalean lycophytes appear to have diversified globally (Kovach and Batten, 1993; Pigg, 1992). *Dicroidium* is thought to have migrated southwards, from tropical lowland settings of northern Gondwana in the Early Triassic, along with *Lepidopteris* (Kerp *et al.*, 2006; Patterson, 1999). These taxa may well have been more adapted to better-drained, and perhaps higher elevated, areas (Lindström and McLoughlin, 2007). The increasingly warm, dry climate would have aided the radiation of these taxa with their adaptations to aridity (such as thick cuticles and a reduction in leaf surface area) into higher palaeolatitudes.

Permian–Triassic transition: the crisis, provincialism and global correlation

The cause of the end-Permian extinction event is conjectural with proposed scenarios including (i) an asteroid impact (Basu *et al.*, 2003; Becker *et al.*, 2004), (ii) flood basalt volcanism (Courtillot and Renne, 2003; Renne *et al.*, 1995), (iii) release of methane from clathrates (Ryskin, 2003), and (iv) extreme global warming during the Permian initiated by the waning of the orogenies associated with the unification of Gondwana and Laurasia and further intensified by volcanism and methane release (Kidder and Worsley, 2004; Lindström and McLoughlin, 2007). Whatever the cause(s), the Permian–Triassic boundary marks the occurrence of the largest extinction event preserved in the geological record (Figure 4.3). It lasted between 40 and

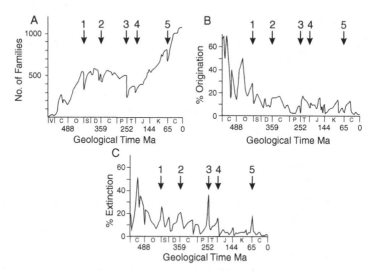

Figure 4.3 Global compilation of the fossil record to reveal diversity, origination and extinction. A, diversity curve for the Phanerozoic. B, originations. C, extinctions. Numbers arrowed 1 to 5 mark the five largest extinction events at the Ordovician/Silurian boundary (1); Devonian/Carboniferous (2); Permian/Triassic (3); Triassic/Jurassic (4); Cretaceous/Cenozoic (5). Adapted from Sepkoski (1984).

500 kyr (Bowring *et al.*, 1998; Twitchett *et al.*, 2001) and occurred some 252 Ma ago. Here 90–95% of all marine biota and 70% of terrestrial vertebrate species disappeared (Erwin, 1994). Significant impacts were also experienced by the insect population (Erwin, 1994; Labandeira and Sepkoski, 1993), and large changes occurred in land plant species abundance as well as in community structure (McElwain, 2007). This was the largest disruption in Earth's history and the mechanisms driving this crisis are still not fully understood.

Understanding of the Permian–Triassic transition has been influenced by three factors, namely (i) the nature of both the transition and resultant biological provincialism, (ii) absolute dating and definition of the Permian–Triassic boundary, and (iii) the length of the Triassic stages and substages. In Gondwana, understanding of the Permian–Triassic transition is further complicated by the sequences being uncommon and, when present, they are often palynologically barren. Gaps have occurred in the sampling, and data analysis has not always focused on taxon appearances, extinctions and changes in abundance (Lindström and McLoughlin, 2007). This has obscured the identification of any signal relating to short-term ecological changes.

The nature of the Permian–Triassic transition has rendered global correlation of this interval problematic. The transition appears to have been diachronous in nature with consequent warming and aridification appearing earlier at low latitudes than at high latitudes (e.g. McLoughlin, Lindström and Drinnan, 1997). Therefore biostratigraphic zonations, although global, may not be time equivalent. For example, vertebrates of Late Permian age in South Africa are still recorded in the Early Triassic of Antarctica, and typical Triassic plants such as *Dicroidium* are known from Late Permian strata in low latitudes (Abu Hamad *et al.*, 2008; Kerp *et al.*, 2006). Furthermore, extreme provincialism of the biota has made it difficult to correlate from one region to another as key fossil taxa are often missing. Consequently the criteria for recognising the Permian–Triassic boundary differ from one region to another[3]. This in turn has made it difficult to correlate rock sequences from region to region and thereby unravel the global chronology of events that occurred through the Permian–Triassic transition (see Lucas, 2010a, 2010b). In order to reconcile these differences, a globally agreed reference section from Zhejiang Province in China has been adopted (Global Boundary Stratotype Section and Point; GSSP; Bowring *et al.*, 1998; Walsh, Gradstein and Ogg, 2004; Yin *et al.*, 2001). With improved absolute (radiometric) dating (U-Pb) the age of the Permian–Triassic boundary has been constrained to 252.3 Ma (Mundil *et al.*, 2010). Although such constraints have undoubtedly helped in the correlation and chronology of specific changes, long distance correlation back to that reference frame still remains problematic particularly in high southern latitudes.

In order to help correlate sequences between different geographical regions, carbon isotope chemostratigraphy, using patterns of carbon isotope excursions, has been employed.

[3] For example, the Permian–Triassic boundary in Southern Hemisphere sequences has been variously taken as being either the last coal seam (see Foster, 1982; Retallack, 2002), or the last record of a certain taxa such as *Vertebraria* (Retallack *et al.* 2005) or the vertebrate *Dicynodon* (Smith and Ward, 2001), or first appearances such as the spore *Aratrisporites* (Foster, Logan and Summons, 1998), plant macrofossils *Lepidopteris* (Retallack, 2002), or animals *Lystrosuarus* (Lucas, 2010a).

The global marine negative $\delta^{13}C$ excursion at the Permian–Triassic boundary has now been found to be the first of at least four profound excursions (Payne *et al.*, 2004). All four excursions are found both on land and in the sea, and thereby confirm correlation of the Permian–Triassic boundary sequences on land with the boundary–stratotype marine sequence in China (Retallack *et al.*, 2005). Using this technique, it can be seen that the global carbon isotope perturbations of the Lopingian (252 Ma) of -4% $\delta^{13}C_{org}$ (Berner, 2002), relating to high atmospheric CO_2, were preceded by a similar global perturbation 9 Ma earlier in the Guadalupian (Retallack *et al.*, 2006). Further studies of marine extinction horizons now suggest that there were two extinction phases at the end of the Permian, one at the boundary itself (e.g. Twichett *et al.*, 2001), and one that preceded it by approximately 120 kyr (Korte and Kozur, 2010). This testifies to the occurrence of more than one geo-logically-abrupt mass extinction event making ecosystem demise less gradual than was previously envisaged (Benton, Tverdokhlebov and Sukhov, 2004; Ward *et al.*, 2005). Evidence suggests that these extinction events were separated by stable greenhouse con-ditions (Retallack *et al.*, 2006). These transient warm conditions were particularly striking at high palaeolatitudes. In southeast Australian sequences, Middle Permian periglacial palaeo-sols, marine glendonites and glacial dropstones are followed by the end-Guadalupian warm spike represented by a marine transgression. On land, kaolinitic palaeosols (Retallack, 2005a), a spike of lycopsids, fungi and algae, marked soil erosion, soil stagnation and the dominance of braided streams characterise the high latitude arid lands of South Africa and the polar wetlands of Antarctica (Retallack *et al.*, 2006).

Estimates regarding post-extinction recovery have also been confounded by inferences regarding the length of stages that make up the Early Triassic (Mundil *et al.*, 2004, 2010; Ovtcharova *et al.*, 2007). The recovery phase from the mass extinction event has been described as being prolonged (Looy *et al.*, 1999), with estimates of the duration of the crisis and environmental perturbation ranging from <60 kyr (Rampino, Prokoph and Adler, 2000) to 5–10 Ma (Wignall, 2007; Yin *et al.*, 2007). However, refinements in stage level resolution and their subsequent radiometric dating (Mundil *et al.*, 2004, 2010) suggest that the 'delayed' recovery may not have been particularly protracted considering the severity of the extinction (Metcalfe and Isozaki, 2009). Moreover, an independently dated section in East Greenland also demonstrates that terrestrial ecosystem collapse and subsequent extinc-tion of typical Late Permian gymnosperms preceded the Permian–Triassic boundary (Looy *et al.*, 2001; Twitchett *et al.*, 2001), and that floristic turnover was gradual rather than instantaneous (Lindström and McLoughlin, 2007).

This change in view of the Permian–Triassic transition from an extended period of time, to an almost instantaneous event, albeit with two extinction episodes, saw a change in the mechanisms proposed as the main cause of extinction. One hypothesis describes the crisis resulting from a meteorite impact (Becker *et al.*, 2001, 2004; Jin *et al.*, 2000). Evidence for the impact event at this time, however, postdates the onset of the extinctions (Yin *et al.*, 2007) and also appears to be several orders of magnitude smaller than the impact that resulted in the end Cretaceous extinction (Retallack *et al.*, 1998).

It is agreed that the Permian–Triassic boundary is marked by a large and rapid negative excursion of the oceanic $\delta^{13}C$ record, and that extinction occurred at the same time as large-scale volcanic eruptions (Montenegro *et al.*, 2011). Yet the causal factors are still hotly debated (see Wignall, 2007) and there is much less consensus on other aspects of the event (Montenegro *et al.*, 2011).

Suggested mechanisms have linked the crisis to the extensive volcanism in China and the formation of the Siberian Traps in Russia. Recent studies have demonstrated a strong correlation between volcanic pulses and extinction events (Kozur and Weems, 2011), yet large flood basalt provinces have occurred throughout geological time and none appear to have caused the scale of disruption to the global carbon cycle, oceanic circulation and extinction similar to that seen at the Permian–Triassic transition. This suggests that other mechanisms were also important in causing the mass ecosystem disruption in both the marine and terrestrial carbon cycle.

Other mechanisms have included ocean acidification caused by the volcanic emission of sulphur compounds and CO_2 (Heydari *et al.*, 2003; Montenegro *et al.*, 2011; Payne *et al.*, 2007), oceanic anoxia extending to shallow waters (e.g. Grice *et al.*, 2005; Wignall and Hallam, 1992; Wignall and Twitchett, 1996), acid rain and global cooling (Campbell *et al.*, 1992), as well as a cascade of environmental changes such as volcanic methane release (Heydari, Arzani and Hassanzadeh, 2008), methane clathrate dissociation on the ocean floor (Heydari and Hassanzadeh, 2003; Retallack, Smith and Ward, 2003), and H_2S outgassing from the euxinic oceans leading to direct poisoning and damage to the ozone layer (Kump, Pavlov and Arthur, 2005).

What is known for sure is that there was a change in the global carbon cycle that resulted in a negative isotope excursion at the Permian–Triassic boundary. This excursion could have been the result of either vast quantities of methane released by tectonic activity (Heydari, Arzani and Hassanzadeh, 2008), or warming of the deep ocean triggering dissociation of methane clathrates produced chemosynthetically on the ocean floor (e.g. Krull and Retallack, 2000; Krull *et al.*, 2000). Regardless of origin gas hydrate release is usually characterised by a rapid shift in isotopes as extremely light carbon is released into the terrestrial pool resulting in a dramatic negative isotope excursion (Korte and Kozur, 2010). The methane is then rapidly converted to CO_2 and absorbed back into the organic carbon pool through photosynthesis, such that the residence time of this isotopically light carbon in the atmosphere is short, and thus the isotope excursion itself is short-lived (i.e. a few tens of thousands of years). Over the Permian–Triassic transition, however, although the carbon isotope record becomes lighter it is a gradual process, taking place over several millions of years, quite unlike other methane release events. Unless, that is, the methane release was driven by more gradual processes such as a regression, whereby the confining pressure of the overlying water column was only released gradually or, alternatively, the reservoir was permafrost rather than marine clathrates, which would also result in a gradual release of methane into the atmosphere. Even though the changes in $\delta^{13}C$ have been linked with the causes of the biotic crisis (Berner, 2002), it has been suggested that these changes were associated with its consequences and are thus not considered to be the cause of the crisis at all (Korte and Kozur, 2010; Twitchett *et al.*, 2001).

Whatever the exact mechanism, the effects were catastrophic for both marine (Wagner, Kosnik and Lidgard, 2006) and terrestrial biota. As atmospheric levels of CO_2 rose to 10 times current levels, excessively high temperatures prevailed over the land (Kiehl and Shields, 2005; Ward *et al.*, 2005) and ocean waters warmed. Given the sensitivity of ocean circulation to high latitude warming, Kiehl and Shields (2005) hypothesise that some high latitude warming threshold was reached whereby connection of surface water to the deep ocean was dramatically reduced leading to a shutdown of marine biological activity. This in turn would have led to increased atmospheric CO_2 and accelerated warming (Kiehl and Shields, 2005) driving the Earth into a hothouse state.

On land, the globally diverse and highly provincial vegetation of the earlier Permian, comprising largely the *Cordaites* Flora in the Northern Hemisphere and the *Glossopteris* Flora of Gondwana (e.g. Retallack, 1995), succumbed to changing physical environments. These floras included ginkgos, sphenopsids and ferns that dominated the cold swampy bogs. Both floras were replaced by a low diversity cosmopolitan vegetation of conifers and lycopsids (Retallack, Veevers and Morante, 1996). Extensive peat swamps responsible for the coal measure deposits in Siberia and China, as well as in South Africa, Australia (Bowen and Sydney basins) and Antarctica (Transantarctic Mountains and Prince Charles Mountains), also disappeared at the Permian–Triassic boundary (Retallack, Veevers and Morante, 1996). But the peat swamp demise was not instantaneous. In India, the coal mire vegetation disappeared slightly below the Permian–Triassic transition, whereas those from South Africa (Karoo Basin) had been replaced by red-beds well before the transition (Ward, Montgomery and Smith, 2000). In South America, the uppermost Permian strata is dominated by red- or varicoloured mudstones with scarce *Glossopteris* leaves (Rohn and Rösler, 1989). With the loss of the arboreal vegetation from the coal mires (peat bogs), coal formation ceased resulting in the gap in coal deposits observed in continental boundary strata of the Early Triassic. The effects were wide-ranging and impacted on all aspects of the terrestrial ecosystem resulting in destabilisation and subsequent collapse of terrestrial ecosystems.

Although many studies have been undertaken on gondwanan Permian–Triassic palynofloras, typically only general patterns have been reported and quantitative data are limited and sometimes absent. In South Africa, Steiner *et al.* (2003) report a 100% taxonomic turnover in the Karoo Basin, with the post-transition sediments being 100% dominated by the fungal palynomorph, *Reduviasporonites chalastrus*, and woody plant remains, which is then succeeded by a palynological barren zone. In, and adjacent to, Africa the transitional sediments evidence the disappearance of ~30% of the latest Permian taxa [27% in Kenya (Hankel, 1992) and 35% in Madagascar (Wright and Askin, 1987)], and in Madagascar 42% of the taxa appear for the first time (Wright and Askin, 1987). The extinction and disappearance of taxa is also associated with dramatic shifts in abundance. For example, in Kenya 24% of the latest Permian assemblage is *R. chalastrus*, which subsequently drops to 4–10% in the earliest Triassic (Hankel, 1992). Here, as in many Permian–Triassic boundary sequences, there is a 24 m palynologically barren interval that separates the latest Permian sediments from those of the earliest Triassic (Hankel, 1992). In India, Late Permian assemblages are dominated by glossopterid pollen that decrease markedly in the earliest Triassic (Srivastava and Jha,

1990). In eastern Australia, the earliest Triassic assemblage, separated from the preceding productive sample by a ~42 m barren interval, records a loss of 25% of the taxa along with 19% of the taxa being registered for the first time (de Jersey, 1979).

Associated with this mass extinction is the unparalleled abundance of the fungal remains of *Reduviasporonites chalastrus*, which is found in all depositional environments (marine, lacustrine, fluvatile), floral provincialities and climatic zones (Visscher *et al.*, 1996). This global fungal spike is considered to reflect the excessive dieback of the prevailing arboreal vegetation (Looy *et al.*, 1999, 2001). Causes of this dieback cannot include tectonic activity or climatic explanations since deposits relating to fluctuating sea levels, and sedimentary facies and palaeosols, commonly found in coal-bearing sequences, are also present in Early Triassic rocks when coal deposits are absent (Retallack, Veevers and Morante, 1996).

Recovery from the extinction events of the latest Permian and earliest Triassic was varied. In some regions post-extinction recovery was patchy. Some high latitude regions witnessed an increase in diversity in the Early Triassic (e.g. Australia; Holmes and Ash, 1979; Antarctica; Lindström and McLoughlin, 2007), whereas in other regions the recovery was slow. For example, Looy *et al.* (1999) record the extensive dieback of forest systems in the Northern Hemisphere taking ~4 Ma to fully recover (Eshet, Rampino and Visscher, 1995; Looy *et al.*, 1999). The recovery phase saw the appearance of a new suite of plants including lycophytes, which colonised disturbed habitats, and other groups of new peat-producing plants that had to re-evolve some sort of tolerance to the moist, acidic, dysaerobic wet lands (Retallack, Veevers and Morante, 1996). It was not until the Middle Triassic that plant diversity and peat thickness had recovered to levels similar to those of the Permian, and the coal mire vegetation started forming coal deposits once more. These Middle Triassic coal deposits are characteristically thin, and of low grade (Retallack, Veevers and Morante, 1996), reflecting the vegetational change that had occurred in the flora. The new plant lineages of the Early Triassic were dissimilar to those found among their Paleozoic ancestors, but would continue growing until the end of the Late Triassic in a world exhibiting greater floristic diversity.

Permian–Triassic geology and palaeoenvironments of Antarctica

During the late Paleozoic, Antarctica was attached to Australasia, India and Africa and lay in the high southern latitudes extending from the tundra climate zone into the cold temperate zone. By the Early Triassic, Antarctica, along with the rest of southeast Gondwana, had moved north-wards into the cool temperate zone (Iglesias, Artabe and Morel, 2011). Within Antarctica, two main areas provide direct information on the Triassic history of Antarctic vegetation, namely (i) strata near Pyrdz Bay on the east side of East Antarctica in the Prince Charles Mountains (Cantrill and Drinnan, 1994; Cantrill, Drinnan and Webb, 1995; Lindström and McLoughlin, 2007; McLoughlin and Drinnan, 1997a, 1997b), and (ii) the sequences in Victoria Land and the Transantarctic Mountains (e.g. Bomfleur and Kerp, 2010; Bomfleur *et al.*, 2011a, 2011b; Boucher, Taylor and Taylor, 1993, 1995; Plumstead, 1962, 1964; Taylor, Taylor and del Fueyo, 1993; Townrow, 1967) (Figure 4.2). The Prince Charles Mountains lay at lower palaeolatitudes (between 35° and 50° S) than either the Transantarctic Mountains or Victoria

Land (~60° S) during the Triassic (Torsvik and Van der Voo, 2002). In both the Prince Charles Mountains and the Transantarctic Mountains, Permian fluvial sequences generally pass conformably or disconformably into the Triassic, accompanied by a change in style of deposition.

Within the northern Prince Charles Mountains, sediments recording the Permian–Triassic boundary can be found in the Amery Group (Figure 4.2) located within a narrow fault-bounded depression called the Lambert Graben (Lindström and McLoughlin, 2007), which was fed by a vast drainage system one fifth the area of East Antarctica (Adamson and Darragh, 1991). Within the graben alluvial fan deposits of conglomerates and coarse-grained sandstones and siltstones were derived from uplifted areas to the west and formed the Radok Conglomerate (Fielding and Webb, 1995; Ravich *et al.*, 1977). The succeeding Bainmedart Coal Measures incorporate repetitive fining-upward cycles of sandstone, siltstone and coal, which were deposited predominantly in northerly to northeasterly flowing high-energy braided rivers, alternating with low-energy forest mires and floodplains (Fielding and Webb, 1996; McLoughlin and Drinnan, 1997a; Webb and Fielding, 1993). The Permian–Triassic boundary is placed in the 24 m interval between the top of the last coal seam of the coal-bearing Bainmedart Coal Measures and the overlying coal-lacking Flagstone Bench Formation (Lindström and McLoughlin, 2007). The coarse sampling to date and the increase in sediments (sandstones) that do not yield palynofloras across this interval have precluded detailed analysis of short-term changes associated with the end-Permian extinction (Lindström and McLoughlin, 2007). It has been estimated, based on rates of sedimentation, that the sampling gap probably represents a time gap of only 96,000 years (Lindström and McLoughlin, 2007).

The uppermost unit of the Bainmedart Coal Measures is the McKinnon Member (Figure 4.2), which was deposited by low sinuosity rivers flowing in a north to northeasterly direction with floodplain sequences developing thin to thick coal seams (McLoughlin and Drinnan, 1997a). The lower and middle parts of the member are characterised by thicker and more abundant coal seams relative to the upper part where the coals are thinner and less abundant. These thick coal seams indicate extended periods of lower sedimentation rates and higher water tables (McLoughlin and Drinnan, 1997a). The thinner, less abundant coal seams in the upper part of the member suggests that the water table was lower and aridity was increasing. This interpretation is supported by the coal geochemistry which indicates greater weathering and aridity upsection (Holdgate *et al.*, 2005).

The lower part of the Flagstone Bench Formation, the Ritchie Member (Figure 4.2), comprises sandstones and siltstones deposited by predominantly northwest to north northeastly directed braided rivers under the influence of increasing aridity (McLoughlin and Drinnan, 1997b). Although lacking any evidence of coal formation, sedimentation is still cyclical in deposition, with sandstone facies being overlain by thin carbonaceous siltstones. The relative proportion of sandstone to siltstone that increases through the upperpart of the McKinnon Member, continues into the lower part of the Ritchie Member (McLoughlin and Drinnan, 1997b). The carbonaceous siltstones give way upsection to more ferrugineous siltstones, and the sequences are interpreted as accumulating under increasingly arid conditions (McLoughlin and Drinnan, 1997b; McLoughlin, Lindström and Drinnan, 1997).

The change across the Permian–Triassic boundary is interpreted as a shift from large, low-sinuosity braided and minor meandering rivers alongside poorly drained forest-mire environments to medium-sized, low-sinuosity, braided rivers with progressively more episodic discharge (McLoughlin and Drinnan, 1997a, 1997b).

The Ritchie Member is in turn overlain by the Jetty Member (Figure 4.2), a red-bed unit that records deposition in alluvial fan environments formed under episodic discharge events. The Jetty Member is ultimately replaced by the McKelvey Member of Late Triassic (Norian) age, with a return to wetter conditions. The river systems that deposited this unit were again northerly directed and of low-sinuosity (McLoughlin and Drinnan, 1997b).

The Amery Group stratigraphy hints at changes in the environment leading up to the extinction event. Sediment cyclicity, which is well-developed in the lower part of the sequence (Fielding and Webb, 1996), decreases in the upper part of the succession along with an increase in sandstone to siltstone ratios (McLoughlin and Drinnan, 1997b). This change in cyclicity indicates a shift to more episodic river flows (Lindström and McLoughlin, 2007). Ash contents in the coals also increase upsection (Holdgate *et al.*, 2005) indicating increased weathering and aridity resulting in greater terrigenous input into the coal mires. The changes in sedimentology and coal petrography all point towards a warming and drying climate in the Late Permian of Antarctica, paralleling those seen in other Gondwana sequences. These changes all heralded the reorganisation of the Earth system through the Permian–Triassic transition.

Further south and to the east, two main regions record the Permian–Triassic transition and subsequent Triassic sedimentation – one in southern Victoria Land and the other in the central Transantarctic Mountains. The alluvial sediments, deposited by large river systems within this region, suggest that during the latest Permian and Early Triassic two depositional basins existed (the central Transantarctic Basin and the Victoria Land Basin) separated by a region of highland (Ross High) (Collinson *et al.*, 1994). The Ross High isolated the Victoria Land Basin from the volcanic arc so that cratonic quartz-rich sediments dominated. However, as the Ross High was buried and the arc migrated closer to the continent, these two basins had merged into one fluvial system by the late Early Triassic when an increase in volcanogenic input is seen in the Victoria Land Basin (Collinson, 1996; Collinson *et al.*, 1994).

In the central Transantarctic Mountains, the Late Permian Buckley Formation is overlain by the Fremouw Formation, which in turn is overlain by the Falla Formation (Figure 4.2). At Coalsack Bluff the Buckley Formation terminates in a thick composite coal seam that is overlain by a distinctive claystone breccia unit (Retallack, Greaver and Jahren, 2007). The claystone breccia contains a thick pyrite layer and is interpreted as a mass wasting event that followed deforestation at the Permian–Triassic boundary (Retallack, Greaver and Jahren, 2007). The Permian–Triassic boundary is again marked by a cessation of coal at, or a few metres below, the boundary together with generally thicker packages (10 m thick) of trough cross-bedded sandstones at the base of the overlying Fremouw Formation that point to high discharge events (Retallack *et al.*, 2005). This suggests a lack of vegetation at this time. The Fremouw Formation is interpreted as being deposited by river systems that formed broad river channels with low-sinuosity, braided streams (Isbell and Macdonald, 1991). Overbank deposits are represented by siltstones and fine-grained sandstones and, in places, thin coal seams. This is followed by a return to thinner sandstone units and overbank deposits including palaeosols. However, the palaeosols are unique, containing distinctive berthierine nodules and being rich in unoxidised iron, which is thought to indicate severe hypoxia in the ground water. These soils occur over a ten metre interval before the palaeosols return to the highly oxidised types of soils more typical of the Early Triassic (Retallack, Greaver and Jahren, 2007).

The placement of the Permian–Triassic boundary in the central Transantarctic Mountains is complicated by the various criteria used for recognising the boundary in Antarctica (Collinson *et al.*, 2006). For convenience it has often been considered to be at the contact between the Buckley and Fremouw formations (Figure 4.2), but criteria for defining the boundary occur both above and below this contact. This is further complicated by the suggestion that the Buckley/Fremouw contact is diachronous across the basin (McManus *et al.*, 2002a).

Biostratigraphic criteria can also be used to help resolve the uncertainty as to the placement of the boundary. Australian Permian and Triassic strata have been the focus of intensive palynological investigations for many years. Hence, the well-established Australian palynozonation has become the *de facto* standard with which many other gondwanan assemblages are compared (Lindström and McLoughlin, 2007). Based on the Australian palynozonation, the *Protohaploxypinus microcorpus* zone (glossopterid pollen type) is regarded as latest Permian (Lindström and McLoughlin, 2007) with some workers suggesting that the Permian–Triassic boundary occurs within this zone (e.g. Morante, 1996) even though definition of this zone can be difficult. In the central Transantarctic Mountains the *P. microcorpus* zone occurs below the Buckley/Fremouw contact, but the top of the zone has not yet been recorded. On this basis the Permian–Triassic boundary is considered to lie in the uppermost coal seam of the Buckley Formation, or higher.

Using carbon isotope stratigraphy a strong negative excursion occurs below the last coal seam of the Buckley Formation and could represent the carbon isotope shift seen just prior to the boundary in global stratotype sections. However, strong negative shifts are also seen in soil horizons above the Buckley/Fremouw contact (Retallack *et al.*, 2005). Whether these actually represent the carbon isotope minima recorded in other Permian–Triassic sequences still remains to be determined especially since carbon isotope compositions can be strongly influenced by soil processes such as anaerobic breakdown resulting in lighter carbon isotope signatures.

The change in soil types from carbonaceous soils to grey-green soils has also been proposed to denote the Permian–Triassic boundary (Retallack *et al.*, 2005), as have biostratigraphical markers such as the disappearance of *Glossopteris*. Evidence of *Glossopteris* occurs just below the Buckely/Fremouw contact, but on Collinson Ridge (in the Queen Maud Mountains of the Transantarctic Mountains) *Glossopteris* is recorded about 15 m above the contact suggesting that the boundary lies somewhat higher (Collinson *et al.*, 2006). Abundant fossil wood, attributed to glossopterids, occurs above the contact and this zone overlaps with records of key Triassic vertebrates. Just above the Buckley/Fremouw contact, vertebrate burrows have been recorded (Collinson *et al.*, 2006) and several vertebrate species are known from the Fremouw Formation. The presence of the dicynodont

Lystrosaurus is taken as an indicator of the Triassic, yet in the central Transantarctic Mountains the lowest record of *Lystrosaurus* is 28 m above the top of the Buckley Formation (Collinson *et al.*, 2006). This leaves a significant gap in which to place the Permian–Triassic boundary.

Defining the Permian–Triassic boundary in Antarctica is therefore difficult, but it can be narrowed down to an interval between the last few metres of the Buckley Formation and the bottom of the Fremouw Formation (Collinson *et al.*, 2006) in the central Transantarctic Mountains, or between the last few metres of the McKinnon Member and the bottom 24 m of the Ritchie Member (Lindström and McLoughlin, 2007) in the Prince Charles Mountains. When comparisons are made between these sequences, similarities can be seen both across the Permian–Triassic transition and in the subsequent Triassic.

The precise location of the Permian–Triassic boundary in Victoria Land is also problematic. The uppermost age range of the Weller Coal Measures is constrained by palynofloras from the middle and upper parts that are correlated with Early Permian (Artinskian) assemblages of Australia (Askin, 1997; Kyle, 1977). This is supported by the presence of *Gangamopteris* (leaves of a Carboniferous–Early Permian glossopterid) at many localities. The overlying Feather Conglomerate largely lacks biostratigraphic age control although an Early Triassic palynoassemblage is known from the upper part of the unit (Kyle, 1976, 1977). Furthermore, the contact between the Weller Coal Measures and the Feather Conglomerate is disconformable to unconformable with a well-developed ferricrete that probably represents a significant time gap (Isbell and Cúneo, 1996). The nature of the contact has several important implications as it implies an unknown amount of erosional removal of the Weller Coal Measures and consequently the preserved top of the Weller Coal Measures may be different ages in different places depending on how much erosion has taken place. Moreover, the unknown time gap between erosion and the start of deposition may also not be time equivalent across the basin.

Despite these problems the Permian–Triassic boundary in Antarctica has been recognised in several places (e.g. Retallack, Greaver and Jahren, 2007; Retallack *et al.*, 1998, 2005, 2006) although not without controversy (Collinson *et al.*, 2006; Isbell and Askin, 1999). On Portal Mountain in southern Victoria Land, palaeobotanical, palaeopedological and carbon isotope studies are thought to provide evidence for the mid-Late Permian biotic crisis that took place at the end of the Guadalupian at ~260 Ma (Retallack, Greaver and Jahren, 2007). Sediments of a distinctive sandstone marker bed can be found preserved on a prominent bluff in the middle Weller Coal Measure and lie directly above a carbon isotope excursion interpreted as the end-Guadalupian event. The sandstone is remarkably similar to those of the Feather Conglomerate which lies directly above the end-Permian carbon isotope excursion found at other localities in southern Victoria Land.

Two further localities in southern Victoria Land are thought to record the Permian–Triassic transition. At Mount Crean, $\delta^{13}C$ isotope analyses have identified an excursion relating to the transition in the upper part of the Weller Coal Measures 1 m below the base of the overlying Feather Conglomerate (Krull, 1998). This is close to the last coal (2 m below the base of the Feather Conglomerate), the last *Vertebraria* root (3 m below) and the last *Glossopteris* leaf (13 m below).

By contrast, carbon isotope analyses of the Weller Coal Measures in the third locality in southern Victoria Land, the central Allan Hills, have not proved helpful in identifying the Permian–Triassic boundary (Retallack and Krull, 1999). No negative excursion was found between the last Permian coal, the last *Glossopteris* or *Vertebraria* some 15 m below the top of the Weller Coal Measures, the overlying Feather Conglomerate and the Middle Triassic lower Lashley Formation with preserved *Dicroidium* (Krull, 1998; Retallack and Alonso-Zarza, 1998).

Correlation of the Allan Hills deposits with those at Mount Crean supports a placement of the Permian–Triassic boundary near the Weller Coal Measures/Feather Conglomerate contact (Collinson *et al.*, 1994). The Weller Coal Measures are separated in places from the overlying Feather Conglomerate by either a distinct palaeosol at Mount Crean (Retallack *et al.*, 2005), or a well-developed ferricrete at Allan Hills (Isbell and Cúneo, 1996). Even though a lack of appropriate rock types and preservation of palynomorphs has precluded the recognition of the Permian–Triassic boundary in southern Victoria Land to date, Early Triassic palynoassemblages have been recovered from the upper Feather Conglomerate (Kyle, 1976, 1977). Moreover, if the well-developed boundary palaeosol at Mount Crean can be correlated with those seen in the central Transantarctic Mountains (Retallack *et al.*, 2005, 2006) (see above) then the Permian–Triassic boundary might indeed be taken as the boundary between the Weller and Feather formations.

The overlying Feather Conglomerate in southern Victoric Land (Figure 4.2) was deposited by low-sinuousity, braided rivers. River channels were wide and composite consisting of multiple stacked units indicating episodic and high flow discharge with little overbank represented by thin mud drapes in the lower part (Barrett and Fitzgerald, 1985; Collinson, Pennington and Kemp, 1983, 1986; Collinson *et al.*, 1994). The upper part (Fleming Formation of some workers) has a greater proportion of grey-green sandstones and ferruginous palaeosols (Barrett and Fitzgerald, 1985). Overlying the Feather Conglomerate is the Lashly Formation of Middle to Late Triassic age. This formation has been divided into four informal members (A–D) based on abundance of sandstone and mudstone composition. The trough cross-bedded sandstone sheets, deposited by low-sinuosity braided rivers, are overlain by grey-green mudstones with well-developed soils of the floodplain (Collinson *et al.*, 1994). The lower A and B members contain late Early to early Middle Triassic palynoassemblages, whilst members C and D contain Late Triassic *Alisporites* assemblages (Kyle, 1977; Kyle and Schopf, 1982). This dating is supported by the plant macrofossil assemblages (Retallack and Alonso-Zarza, 1998).

Vegetation change across the Permian–Triassic transition and the subsequent Triassic vegetation of Antarctica

Although the definition of the Permian–Triassic boundary in Antarctica is problematic and, given the presence of barren intervals making the determination of vegetation change through the Permian and into the Triassic is difficult, information regarding the Permian–Triassic crisis in Antarctica can only be inferred from several key floras. Palaeobotanical

Table 4.1 *Taxonomic listing of Antarctic Triassic species. SVL = southern Victoria Land, CTAMS = central Transantarctic Mountains, VAM = vescular arbuscular mycorrhiza.*

Taxa	Location	Age	Reference
Bryophyta/Algae			
Bryophyta cuticles	Timber Peak, SVL	Late Triassic	Bomfleur and Kerp, 2010
Lithothallus ganovex	Timber Peak, SVL; Beardmore Glacier area, CTAMS	Late Triassic	Bomfleur *et al.*, 2009
Lycophyta			
Isolated pleuromeian reproductive organs	Alfies Elbow	Late Triassic	Bomfleur *et al.*, 2011a
Mesenteriophylllum serratum	Alfies Elbow	Late Triassic	Bomfleur *et al.*, 2011a
Sphenophyta			
Spaciinodum collinsonii	Fremouw Peak	early Middle Triassic	Osborn and Taylor, 1989; Osborn *et al.*, 2000; Ryberg *et al.*, 2008; Schwendemann *et al.*, 2010a
Neocalamites	Gordon Valley, Horseshoe Mountain	Middle to Late Triassic	Cúneo *et al.*, 2003; Plumstead, 1962
Schizoneura	Horseshoe Mountain	Middle to Late Triassic	Plumstead, 1962
Phyllotheca brookvalensis	Shackleton Glacier area	Early Triassic	Retallack, 2005a
Pteridophyta			
Gleicheniaceae			
Antarctipteris sclericaulis	Fremouw Peak	early Middle Triassic	Millay and Taylor, 1990
Gleichenipteris antarcticus	Fremouw Peak	early Middle Triassic	Phipps *et al.*, 2000
Osmundaceae			
Ashicaulis beardmorensis	Fremouw Peak	early Middle Triassic	Schopf, 1978
Ashicaulis woolfei	Fremouw Peak	early Middle Triassic	Rothwell, Taylor and Taylor, 2002
Cladophlebis sp.	Timber Peak, SVL	Late Triassic	Bomfleur and Kerp, 2010
Osmunda claytonites	Allan Hills	Late Triassic	Phipps *et al.*, 1998

Table 4.1 (*cont.*)

Taxa	Location	Age	Reference
Marrattiaceae			
Scolecopteris antarctica	Fremouw Peak	early Middle Triassic	Delevoryas,Taylor and Taylor, 1992
Matoniaceae			
Soloropteris rupex	Fremouw Peak	early Middle Triassic	Millay and Taylor, 1990
Tomaniopteris katonii	Fremouw Peak	early Middle Triassic	Klavins, Taylor and Taylor, 2004
?Cyatheaceae/Pteridaceae			
Schopfiopteris repens	Fremouw Peak	early Middle Triassic	Millay and Taylor, 1990
Dipteridaceae			
Dictyophyllum sp.	Shackleton Glacier area	late Middle to early Late Triassic	Axsmith *et al.*, 2000
Unplaced ferns			
Fremouwa inaffecta	Fremouw Peak	early Middle Triassic	Millay and Taylor, 1990
Hapsidoxylon terpsichorum	Fremouw Peak	early Middle Triassic	McManus *et al.*, 2002b
Schleporia incarcerata	Fremouw Peak	early Middle Triassic	Millay and Taylor, 1990
Unidentified filicalean foliage	Mt Falla	Early to Middle Triassic	Bomfleur *et al.*, 2011c
Spermatophyta			
Cycadales			
Antarcticycas schopfi	Fremouw Peak	early Middle Triassic	Hermsen, Taylor and Taylor, 2009; Hermsen *et al.*, 2006, 2007; Smoot, Taylor and Delevoryas, 1985
Delemaya spinulosa	Fremouw Peak	early Middle Triassic	Klavins *et al.*, 2003; Schwendemann *et al.*, 2009a
Yelchophyllum omegapetiolaris	Fremouw Peak	early Middle Triassic	Hermsen *et al.*, 2007
Corystospermales			
Dicroidium sp.	Horseshoe Mountain, Shapeless Mountain, Mt Fleming	Triassic	Plumstead, 1962

Table 4.1 (*cont.*)

Taxa	Location	Age	Reference
Dicroidium crassinervis	Gordon Valley, Timber Peak, Prince Charles Mountains	Late Triassic	Bomfleur and Kerp, 2010; Boucher, 1995; Cantrill, Drinnan and Webb, 1995; Cúneo *et al.*, 2003
Dicroidium dubium	Prince Charles Mountains, Mt Falla, Horseshoe Mountain, Timber Peak	Late Triassic	Bomfleur and Kerp, 2010; Boucher, Taylor and Taylor, 1993; Rigby, 1985; Taylor, Boucher and Taylor, 1992
Dicroidium dutoitii	Allan Hills, Mt Bumstead, Misery Peak	Late Triassic	E. L. Taylor *et al.*, 1990; Townrow, 1967
Dicroidium elongatum	Allan Hills, Mt Bumstead, Horseshoe Mountain, Timber Peak	Late Triassic, late Middle to early Late Triassic	Bomfleur and Kerp, 2010; Gabites, 1985; Rigby, 1985, Townrow, 1967
Dicroidium feistmantelii	Allan Hills, Mt Bumstead		Rigby and Schopf, 1969; Townrow, 1967
Dicroidium fremouwensis	Fremouw Peak	early Middle Triassic	Pigg, 1990
Dicroidium lancifolium	Mt Falla, Portal Mountain	early Middle Triassic, Late Triassic	Boucher, Taylor and Taylor, 1993; Gabites, 1985; Taylor, Boucher and Taylor, 1992
Dicroidium odontopteroides	Mt Falla, Allan Hills, Horseshoe Mountain, Mt Bumstead, Portal Mountain, Timber Peak	Late Triassic	Bomfleur and Kerp, 2010; Boucher, Taylor and Taylor, 1993; Gabites, 1985; Rigby, 1985; Rigby and Schopf, 1969; Taylor, Boucher and Taylor, 1992; E. L. Taylor *et al.*, 1990; Townrow, 1967
Dicroidium spinifolium	Allan Hills, Timber Peak	Late Triassic	Bomfleur and Kerp, 2010; Boucher, Taylor and Taylor, 1995

Table 4.1 (*cont.*)

Taxa	Location	Age	Reference
Dicroidium stelznerianum	Allan Hills, Gordon Valley, Prince Charles Mountains	Late Triassic	Boucher, 1995; Cantrill, Drinnan and Webb, 1995; Cúneo *et al.*, 2003; Gabites, 1985; Webb and Fielding, 1993;
Dicroidium trilobita	Allan Hills	Late Triassic	Boucher, Taylor and Taylor, 1993; Gabites, 1985; Townrow, 1967;
Dicroidium zuberi	Horseshoe Mountain, Allan Hills, Prince Charles Mountains	Late Triassic	Cantrill, Drinnan and Webb, 1995; Gabites, 1985; Rigby, 1985
?Diplasiophyllum acutum	Allan Hills, Mt Bumstead	early Middle Triassic, Late Triassic	Townrow, 1967
Dejerseya lobata	Mt Falla	Late Triassic	Bomfleur *et al.*, 2011c
Matatiella dejerseyi	Mt Falla	Late Triassic	Bomfleur *et al.*, 2011c
Townrovia polaris	Mt Falla	Late Triassic	Bomfleur *et al.*, 2011c
Pteruchus fremouwensis	Fremouw Peak	early Middle Triassic	Yao, Taylor and Taylor, 1995
Pteruchus dubius	Prince Charles Mountains	Late Triassic	Cantrill, Drinnan and Webb, 1995
Umkomasia sp.	Mt Falla	Late Triassic	
Umkomasia resinosa	Fremouw Peak	early Middle Triassic	Klavins, Taylor and Taylor, 2002
Umkomasia uniramia	Schroeder Hill, Shackleton Glacier area	late Middle to early Late Triassic	Axsmith *et al.*, 2000
Kykloxylon fremouwensis	Fremouw Peak	early Middle Triassic	Meyer-Berthaud, Taylor and Taylor, 1992, 1993
Rhexoxylon (Antarcticoxylon) priestleyi	Priestley Glacier		
Rhexoxylon sp.	Fremouw Peak	early Middle Triassic	Taylor, 1992
Jeffersonioxylon gordonense	Gordon Valley	early to Middle Triassic	del Fueyo *et al.*, 1995
Petriellaea triangulata	Fremouw Peak		

Table 4.1 (*cont.*)

Taxa	Location	Age	Reference
		early Middle Triassic	Taylor, del Fueyo and Taylor, 1994
Peltaspermales			
Lepidopteris sp.	Prince Charles Mountains	Early Triassic	McLoughlin, Lindström and Drinnan, 1997
Lepidopteris langlohensis	Timber Peak	Late Triassic	Bomfleur and Kerp, 2010
?Ginkgoales			
Baiera sp.	Horseshoe Mountain, Shapeless Mountain		Rigby, 1985
Sphenobaiera schenkii	Horseshoe Mountain, Shapeless Mountain		Escapa *et al.*, 2011
Coniferales			
Podocarpaceae			
Notophyhtum krauselii	Fremouw Peak	early Middle Triassic	Axsmith, Taylor and Taylor, 1998; Decombeix, Taylor and Taylor, 2011; Meyer-Berthaud and Taylor, 1991; Schwendemann *et al.*, 2010b
Taxodiaceae			
Parasciadopitys aequata	Fremouw Peak	early Middle Triassic	Yao, Taylor and Taylor, 1997
Voltziaceae			
Heidiphyllum elongatum	Mt Falla, Timber Peak	Late Triassic	Bomfleur and Kerp, 2010; Bomfleur *et al.* 2011b; Yao, Taylor and Taylor, 1993
Leastrobus fallae	Mt Falla	Late Triassic	Hermsen, Taylor and Taylor, 2007
Pagiophyllum papillatus	Prince Charles Mountains	Late Triassic	Cantrill, Drinnan and Webb, 1995
Switzianthus sp.	Mt Falla	Late Triassic	Bomfleur *et al.*, 2011b
Telemachus elongatus	Mt Falla, Allan Hills	Late Triassic	Axsmith *et al.*, 1995; Escapa *et al.*, 2010;

Table 4.1 (*cont.*)

Taxa	Location	Age	Reference
			Yao, Taylor and Taylor, 1993
Conifer cone indet.	Prince Charles Mountains	Late Triassic	Cantrill, Drinnan and Webb, 1995
Bennettitales			
Nilssonia	Upper Taylor Glacier	Triassic	Plumstead, 1962
Zamites sp. A	Shapeless Mountain	early Middle Triassic	Plumstead, 1962
Zamites sp. B	Shapeless Mountain	early Middle Triassic	Plumstead, 1962
Zamites sp. C	Shapeless Mountain	early Middle Triassic	Plumstead, 1962
Unplaced			
Ignotospermum monilii	Fremouw Peak	early Middle Triassic	Perovich and Taylor, 1989
Linguifolium spp.	Victoria Land	Late Triassic	Gabites, 1985; Rigby, 1985
Probolosperma antarcticum	Fremouw Peak	early Middle Triassic	Decombeix *et al.* 2010
Taeniopteris sp.	Shackleton Glacier area	late Middle to early Late Triassic	Axsmith *et al.*, 2000
Fungi			
Combresomyces cornifer	Fremouw Peak	early Middle Triassic	Schwendemann *et al.*, 2009b
Endochaetophora antarctica	Fremouw Peak	early Middle Triassic	White and Taylor, 1988, 1989b
Gigasporites myriamyces	Fremouw Peak	early Middle Triassic	Phipps and Taylor, 1996
Glomites cycestris	Fremouw Peak	early middle Triassic	Phipps and Taylor, 1996
?*Glomus*	Fremouw Peak	early Middle Triassic	White and Taylor, 1991
Mycocerpon asterineum	Fremouw Peak	early Middle Triassic	Taylor and White, 1989
Palaeofibulus antarctica	Fremouw Peak	early Middle Triassic	Osborn, Taylor and White, 1989
?*Sclerocystis*	Fremouw Peak	early Middle Triassic	Stubblefield, Taylor and Seymour, 1987

Table 4.1 (*cont.*)

Taxa	Location	Age	Reference
Mychorriza VAM	Fremouw Peak	early Middle Triassic	Stubblefield, Taylor and Trappe, 1987a, 1987b; White and Taylor, 1991
Trichomycete	Fremouw Peak	early Middle Triassic	White and Taylor, 1989a
White Rot, Pocket Rot	Fremouw Peak	early Middle Triassic	Stubblefield and Taylor, 1986

data suggest that although the Permian–Triassic transition is associated with the demise of the *Glossopteris* boreal swamp forests, subtle changes were already underway throughout the Permian. The Prince Charles Mountains provide clues as to the changes in vegetation during the end-Permian crisis whereas younger floras, dating to the Early and Middle Triassic, are recorded from the central Transantarctic Mountains and southern Victoria Land (Table 4.1). Together these sequences provide a snapshot of vegetative turnover in Antarctica during the Late Permian and into the Triassic.

The end-Guadalupian crisis and the start of ecosystem turmoil

Evidence for the mid-Late Permian (end-Guadalupian) crisis is sparse, yet palaeobotanical data found at Portal Mountain in southern Victoria Land track the changes in the flora and are compatible with a biotic crisis and palaeoclimatic warming (Retallack *et al.*, 2006). Retallack *et al.* (2006) have undertaken a thorough investigation of this sequence and their findings are summarised here. A break in the ecosystem is marked by a slight, but not profound, change in the palaeosols and thus unlike those seen at the end Permian–Triassic boundary in other Antarctic sites. The palaeosols are silty and sandy with siderite nodules denoting an immature sandy soil with evidence of chemically reduced conditions (Sheldon, 2006; Sheldon and Retallack, 2002). The soils are accompanied by root traces and stem impressions of equisetaleans. Stagnation of these soils may have played a role in the end-Guadalupian lowland plant extinctions.

The fossil plants of the middle Weller Coal Measures (i.e. pre end-Guadalupian crisis) differ from those in the upper coal measures (post end-Guadalupian crisis), but similarities include the equisetalean *Paracalamites* and glossopterid *Vertebraria*[4]. The differences

---

[4] *Vertebraria*, the roots belonging to *Glossopteris*, are characteristically shallow-rooting and mainly horizontally-orientated in the waterlogged Permian soils, but although still present after the Guadalupian crisis they now penetrate to depths of nearly a metre in the younger, drier soils (Retallack and Krull, 1999).

pre- and post-Guadalupian crisis are greater. Seed plants pre-Guadalupian crisis are characterised by leaves of large size and lacking midribs (i.e. *Gangamopteris* and *Palaeovittaria*), as well as those exhibiting low-density venation (i.e. *Glossopteris formosa* and *G. damudica*). These are replaced by species with high density venation (i.e. *G. decipiens*) just below the level of the carbon isotope excursion. Just above the excursion isoetalean lycopsid leaves can be found. These are usually rare in Antarctica, but here are locally common. Lycopsid leaves were mainly tropical and like modern palms had a frost sensitive terminal meristem (Retallack, 1997b). The change to the warm wet conditions of the transient greenhouse phase at the end of the Guadalupian may have been responsible for these vegetation changes. Atmospheric changes served to alter the physical environment exploited by the vegetation. Stagnating lowland swampy habitats would have affected soil respiration leading to the mass deforestation and braided streams (Retallack, 2005b; Ward *et al.*, 2005), all evidenced in the rock record.

The glossopterids that appeared during the late Paleozoic glaciations and steadily diversified in the ameliorating post-glacial temperate climate during the Early and Middle Permian, proliferated markedly in the humid Late Permian climate (Lindström and McLoughlin, 2007). They were middle to high latitude deciduous trees with roots that were adapted to semi-aquatic conditions (Neish, Drinnan and Cantrill, 1993). The lack of glossopterid cuticles may reflect their thin nature and, in turn, be a function of their deciduousness and/or reflect the high humidity and temperate climate at this time (Lindström and McLoughlin, 2007). The presence of glossopterid wood with marked growth rings from the Prince Charles Mountains (see below) indicates that these plants were subject to strong seasonal variations, primarily controlled by the photoperiod (Weaver, McLoughlin and Drinnan, 1997).

The Permian–Triassic transition and the demise of the boreal forests

The Permian–Triassic biotic crisis by comparison is identified by a marked break in the ecosystem as represented by palaeosols (Retallack *et al.*, 2005) and changes in the vegetation. Although there are extensive exposures of Permian–Triassic strata in Antarctica, the definition of the Permian–Triassic boundary, coupled with only sporadic localities containing fossil assemblages through the transition zone, has made it difficult to reconstruct the chronology of events.

In the Transantarctic Mountains high resolution palynological sampling across the Permian–Triassic boundary has not yet been undertaken, but such an approach may well be hampered by the known problems of thermal destruction caused by Jurassic intrusions. Studies of macrofossils have shown sparse occurrence of glossopterid leaves above the traditional placement for the Permian–Triassic boundary at the transition between the Buckley and Fremouw formations (McManus *et al.*, 2002a). This suggests that only few elements of the *Glossopteris* Flora persisted into the very earliest Triassic, presumably in isolated humid refugia, an observation supported by the sparse occurrence of Permian pollen taxa in basal Triassic strata (Lindström and McLoughlin, 2007).

The best-documented floristic assemblage that records vegetative changes across the Permian–Triassic transition in Antarctica can be found in the Prince Charles Mountains. Here well-preserved palynofloras occur. Yet even though the record is relatively complete and there is still a significant gap (24 m) between the latest Permian and the earliest Triassic floras, a succession of changes have been described by Lindström and McLoughlin (2007) that provide the best sequence for understanding vegetation composition and dynamics in Antarctica at this time.

Lindström and McLoughlin (2007) document not only the changing composition but have also identified several distinct phases of vegetational change across the Permian–Triassic transition. Firstly, a precursor extinction event witnessed a loss of 11% of the taxa. This was followed by a second extinction event 2 m below the boundary where 19% of the taxa were lost, and a third extinction phase recorded in the last coal seam where 33% of the taxa disappeared. This transition marks the demise of the glossopterid forests and the start of their replacement by the characteristic Triassic peltasperm/corystosperm vegetation. A further extinction phase follows with both a taxonomic loss of 14% and the appearance of a range of new species. These new species are inferred as being colonisers of the new ecosystems that were being assembled. A fourth extinction phase records a further 35% loss of taxa coupled with the introduction of additional new species into the system. This was followed by an increase in the diversification of new taxa. Ironically the greatest diversity in the flora appears in sediments associated with the fourth extinction phase where many of the typical Permian taxa are recorded for the last time alongside all the new Triassic species. This suggests that the landscape was a mosaic of communities that included both the old Permian survivors and the new Triassic colonisers, coexisting side by side until the latter gained competitive advantage over the former.

Studies of palynoassemblages from the McKinnon Member, also undertaken by Lindström and McLoughlin (2007), indicate that the Late Permian vegetation was quite stable and not greatly different from that represented in the preceding members of the Bainmedart Coal Measures. The palynoflora is dominated by 98 typical Permian taxa, including gymnospermous pollen of the taeniate bisaccate glossopterid type (i.e. *Protohaploxypinus* and *Striatopodocarpidites*) and a non-taeniate bisaccate pollen type (*Scheuringispendens*). This is consistent with the leaf floras that are dominated by *Glossopteris* and *Noeggerathiopsis* (McLoughlin and Drinnan, 1997a; McLoughlin, Lindström and Drinnan, 1997), and the presence of several different wood types (Weaver, McLoughlin and Drinnan, 1997). The glossopterid and non-taeniate bisaccate pollen taxa remain in high abundance, attesting to their continued dominance in the vegetation, until just prior to the Permian–Triassic boundary (Lindström and McLoughlin, 2007).

Differences in palynological composition provide further evidence for the distribution of plants within the latest Permian ecosystem. The generally low spore/pollen ratio indicates that ferns, sphenophytes, bryophytes and herbaceous lycopsids played a subordinate role in the vegetation (Lindström and McLoughlin, 2007). Other seed plant taxa, such as those in the Peltaspermaceae and Corystospermaceae, were present but in low abundance suggesting that they played a subordinate role in the *Glossopteris* Flora, perhaps occupying drier sites.

Palynodebris is dominated by brown wood, and cuticle fragments are rare. Increased spore/pollen ratios are associated with the large coal seams where trilete fern spores (*Osmundacidites*, *Horriditriletes* and *Lophotriletes*) increase in abundance along with ?sphenophytes (*Laevigatosporites*) (Balme, 1995). Following this stable period the terrestrial ecosystem then underwent a dramatic change manifesting itself as a period of considerable turnover.

The change at the end of the Permian is marked by a sudden increase in diversity from about 30–70 taxa to 70–>90 taxa through the late Permian (Lindström and McLoughlin, 2007) with a general decline in certain seed plant pollen[5], including an especially dramatic decrease in glossopterid taeniate bisaccate pollen types. The three presumed overstorey groups at this time, represented by the pollen taxa, also decrease: *Protohaploxypinus* from ~25% to <1%, *Striatopodocarpidites* from ~6% to <1% and *Scheuringipollenites* decreasing from 12% to <1% (Lindström and McLoughlin, 2007). Glossopterids retain their <1% abundance in the Early Triassic but this virtual disappearance can be linked directly to the cessation of coal formation. Nevertheless, the minor occurrence of glossopterid pollen (<1%) preserved in these sediments testifies to the fact that glossopterids managed to survive the crisis albeit in extremely low numbers. They remained at these low levels until later in the Triassic where they finally became extinct. The uppermost coal seam of the McKinnon Member yielded the last typically Permian glossopterid-dominated palynoflora. The disappearance of the peat-forming mire that supported the glossopterids is not only conspicuous in Antarctica but also across southeast Gondwana. The cessation of coal is the most apparent lithological and sedimentological change across the Permian–Triassic transition in the Prince Charles Mountains. Concomitant with the decline in *Glossopteris* was a steady increase in pollen belonging to the new group of seed plants, the Peltaspermaceae, and the Corystospermaceae. Moreover the fern and lycopod diversity underwent a radical increase indicating that a floral turnover was underway. Between the uppermost coal seam and the next palynosample retrieved from the lower Ritchie Member of the Flagstone Bench Formation is a 24 m gap preventing further understanding of fine-scale floral dynamics at the Permian–Triassic boundary.

The Permian–Triassic extinction event lasted ~325 kyr and was stepwise with each level of extinction corresponding to the appearance of a new suite of taxa (Lindström and McLoughlin, 2007). Thirty three per cent of the typical Permian taxa disappeared followed by an initial increase in diversity of earliest Triassic taxa as the lingering Permian taxa continued to become extinct. By the earliest Triassic (Induan) the terrestrial ecosystem was already on its way to recovery with only 26% of the typical Permian taxa remaining.

There are several differences between the floras that dominated the earliest Triassic vegetation and those of the Late Permian. Seed plant pollen is dominant in the Triassic but does not reach the abundance seen in the Permian. This suggests that the communities

---

[5] Fossil spore and pollen taxa do not necessarily equate to true plant taxa (Lindström, McLoughlin and Drinnan, 1997) but are the closest indication of diversity in the parent vegetation.

were more open and diversity was greater in the understorey. The dominance of spore-producing plants in the earliest Triassic communities also suggests low-open forms of vegetation. Lycophytes underwent a rapid radiation in Antarctica at this time and became important components of the vegetation. While the taxa recorded are different from those found elsewhere in Gondwana, they do attest to similar patterns of succession. Although there is a good pollen record covering the transition from the Permian into the Triassic, there are relatively few macrofossils namely only small pinnules of the corystosperm *Lepidopteris* having been found from the Prince Charles Mountains.

Early to Middle Triassic floras and vegetation

By the Early Triassic palynoassemblages were dominated by gymnosperms but not to the same extent as in the Late Permian (e.g. the McKinnon Member). The peltaspermous and corystospermous seed plants that increased in abundance in the Early Triassic were already present in the Late Permian, although low in numbers. These were able to proliferate once the glossopterids and their ecological associates began to disppear from the landscape (Lindström and McLoughlin, 2007). Ferns were common along with a high diversity and abundance of lycophytes (McLoughlin, Lindström and Drinnan, 1997). Several fern taxa (including *Lophotriletes* and *Osmundacidites*) and seed plant taxa (e.g. *Guttulapollenites hannonicus*) appear not to have been affected by the biotic crisis (Lindström and McLoughlin, 2007). Bryophytes, which were a minor constituent in the earliest Triassic (e.g. lowermost Ritchie Member) assemblages, increase in abundance upsection. The spore/pollen ratio also increases dramatically upsection indicating that spore-producing plants played a proportionally greater role in the earliest Triassic plant communities. This is further evidenced by the decrease in woody plant debris, relative to the coal of the late Permian, where non-woody plant remains, including thick cuticles, are the most common form of palynodebris. Thick cuticles coupled with the small size and low abundance of stomata across fossilised *Lepidopteris* leaves also evidence high pCO_2 (Retallack, 2002).

Floras increase in abundance and diversity into the early Middle Triassic where a well-preserved permineralised peat is preserved in the central Transantarctic Mountains (Upper Fremouw Formation) (Schopf, 1978; Smoot, Taylor and Delevoryas, 1985; Taylor and Taylor, 1983, 1990), along with several impression floras (Lashly Formation, members A and B) (Kyle and Schopf, 1982).

In the central Transantarctic Mountains forests have been documented from the Middle Triassic (upper part of the Fremouw Formation) (Cúneo *et al.*, 2003). In Gordon Valley an *in situ* forest horizon with nearly 100 tree stumps, some >60cm in diameter and identified as *Jeffersonioxylon gordonense* (del Fueyo *et al.*, 1995), have been found exposed on a single bedding plane surface (Cúneo *et al.*, 2003). This forest has been interpreted as growing near and on a bank, or river levee, of a braided channel as evidenced by being rooted in levee bank and crevasse splay sandstones (Cúneo *et al.*, 2003). Stump densities indicate a forest

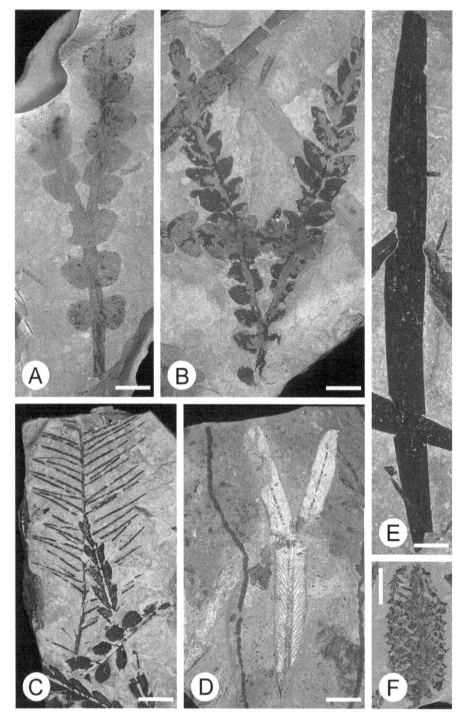

Figure 4.4 Triassic plant fossils from Antarctica. **A,** *Dicroidium odontopteroides*, AF19. **B,** *Dicroidium dubium*, AF9_1. **C,** *Rissikia media*, Mt Bumstead, B822_2. **D,** *Dicroidium dutoitii*, B1022_3, Allan Hills. **E,** *Heidiphyllum elongatum*, AF13_1. **F,** cone of *Telemachus antarcticus*, PBKU T11–411b. Scale bars A–E = 1 cm, F = 5 mm. F courtesy of R. Serbet.

density of between 300 and 2000 trees per ha and basal area of 16 to 80 m^2 per hectare. This is similar to densities observed in other polar forests (Creber and Francis, 1987; Francis, 1991; Jefferson, 1982; Pole, 1999). The trees were inferred to have been between 14 and 31 m tall and the oldest measured ring sequence gave an age of 86 years. Together with the height and age data the stumps suggest a mature forest comprising a suite of older more widely spaced trees, with younger trees set between. Around the base of the trees and across the bedding planes are fossil leaves, dominated by *Dicroidium*, that provide information on the taxonomic composition of the forest floor leaf litter. This suggests an establishment phase with some environmental disturbance (periodically inundated by flood events) that opened up the forest and enabled a second, or several, cohorts to colonise the disturbed areas. Migration of the channel left the forest buried by floodplain sediments that were in time colonised by the sphenopsid *Neocalamites* (Cúneo *et al.*, 2003).

Mire environments are also present as indicated by thin coal seams. Exceptional preservation of the mire flora occurs in rafts of permineralised peat present in river channels. It is thought that migration of the river, and subsequent erosion into the mire, saw rafts of peat eroded into the river and entombed in the base of the river channel. Within the peat considerable additional taxa have been recognised that point to a diverse early Middle Triassic vegetation. For example, Equisetales are represented by small herbaceous plants of *Spaciinodum collinsonii* (Osborn and Taylor, 1989; Ryberg *et al.*, 2008). Much of this material is preserved as young elongating buds suggesting a species with seasonal (winter) dormancy. Moreover these testify to the fact that the peat blocks were eroded into the rivers during a spring flood event (Ryberg *et al.*, 2008). Pteridophytes attained considerable diversity and included a range of small creeping rhizomes attributed to Gleicheniaceae (Millay and Taylor, 1990; Phipps *et al.*, 2000), Cyatheaeceae (Millay and Taylor, 1990) and Matoniaceae (Klavins, Taylor and Taylor, 2004; Millay and Taylor, 1990) as well as larger trunk-forming taxa belonging to the Osmundaceae (Rothwell, Taylor and Taylor, 2002; Schopf, 1978). The Marattiales are represented by pecopteroid pinnules with well-developed synangia (Delevoryas, Taylor and Taylor, 1992). There are several unplaced ferns (Millay and Taylor, 1990), including the unusual *Hapsidoxylon terpsichorum* (McManus *et al.*, 2002b), which bears some similiarities with Paleozoic cladoxylalean ferns.

Within the swamps a small slender cycad also grew (*Antarcticycas schopfi*; Smoot, Taylor and Delvoryas, 1985). Permineralised leaves (*Yelchophyllum omegapetiolaris*; Hermsen *et al.*, 2007) have been linked to pollen cones (*Delemaya spinulosa*; Klavins *et al.*, 2003) and stems (Hermsen *et al.*, 2006, 2007; Smoot, Taylor and Delevoryas, 1985) by anatomical similarities (Hermsen, Taylor and Taylor, 2009). The *Antarcticycas* plant is reconstructed as a small, possibly subterreanean, stem 3 to 5 cm in diameter with a terminal crown of cataphylls surrounding the compound leaves. Pollen cones were produced (Klavins *et al.*, 2003) and have some evidence for pollen feeding by insects that might suggest a possible insect pollination mechanism (Klavins *et al.*, 2005). Lateral bulbils of *Antarcticycas* seem to have been produced in response to damage.

Other seed plants within the same deposit include the corystosperms (*Dicroidium fremouwensis*; Pigg, 1990) and associated pollen organs (Yao, Taylor and Taylor, 1995).

Stems of the *Kykloxylon fremouwensis* (Meyer-Berthaud, Taylor and Taylor, 1992, 1993) have also been related to this group. Ovulate organs such as *Petriellaea triangulata* (Petriellales; Taylor, del Fueyo and Taylor, 1994) evidence further seed plant diversity. Modern seed plant groups are represented by coniferous plants, including forms allied to modern families such as Taxodiaceae (*Parasciadopitys aequata*; Yao, Taylor and Taylor, 1997) and Podocarpaceae (*Nothophytum krauselii*; Axsmith, Taylor and Taylor, 1998). Interestingly, associated with this peat deposit is an abundance and diversity of fungi (White and Taylor, 1989a, 1989b, 1989c, 1991) including pathogens (Stubblefield and Taylor, 1986), mycorrhizal fungi (Phipps and Taylor, 1996; Stubblefield, Taylor and Seymour, 1987; White and Taylor, 1991), including some associated with *Antarcticycas* roots, and decomposers associated with invertebrate detritivores (White and Taylor, 1989c).

The presence of these permineralised peats has given a unique insight into the plants, plant communities, diversity and ecology of the Middle Triassic mire envrionments. Forest communities developed on levee banks and were dominated by *Dicroidium* trees with a subsidiary, but as yet undocumented, understorey. Adjcent to the levee forests and moving into the swamps were sphenopsid communities (Cúneo *et al.*, 2003), while the mires contained a diversity of conifers and other seed plants. Overall while diversity was high, the diversity of *Dicroidium* appears low.

Late Triassic floras and vegetation

The warm temperate climate and abundant water supply in the polar latitudes during the Late Triassic provided favourable growing conditions resulting in high species richness in these high latitude environments. This contrasts with floras today where the species richness gradient is high in low latitudes and decreases into the high latitudes (e.g. Hillebrand, 2004). Evidence for this species-rich microthermal (polar) vegetation during the Late Triassic comes from a number of localities, including floras found in the central Transantarctic Mountains within the Falla Formation in the Queen Alexandra Range (Axsmith *et al.*, 2000; Boucher, Taylor and Taylor, 1993, 1995; Taylor and Taylor, 1988; Townrow, 1967), from southern Victoria Land in the Lashly Formation C and D and Allan Hills (Boucher, Taylor and Taylor, 1995; Townrow, 1967), from northern Victoria Land in the Section Peak Formation at Timber Peak (Bomfleur and Kerp, 2010), and from the Prince Charles Mountains in the Flagstone Bench Formation (Cantrill and Drinnan, 1994; Cantrill, Drinnan and Webb, 1995).

Lying at a palaeolatitude of 70° and 75° S (Lawver, Gahagan and Dalziel, 1998; Torsvik, Gaina and Redfield, 2008; Veevers, 2004) on the northern flank of the western ridge of Mount Falla in the central Transantarctic Mountains, the early or middle Late Triassic (Carnian or Norian) Falla Formation unconformably overlies the Middle Triassic Fremouw Formation and is in turn capped by Early Jurassic volcaniclastics. The Falla Formation is composed of sandstone and carbonaceous shale that developed in a series of fining-upward cycles with a total thickness of 282 m (Barrett, 1969). The formation represents deposits of a distal braided river system (Barrett, Elliot and Lindsay, 1986) but with a range of environments preserved. The regional vegetation as evidenced by the palynofloras is diverse with at least 44 species of dispersed spores and pollen (Farabee, Taylor and Taylor, 1989, 1990; Kyle and Fasola, 1978). Over 75% of the assemblages are made up of spore-producing plants such as sphenophytes, lycophytes and pteridophytes with the seed-producing plants including corystosperms (*Alisporites*), cycadophytes (*Cycadopites*) and probable coniferous taxa.

Several localities give unique insights into the spatial heterogeneity of the plant communities. At Mount Falla a packet of overbank mudstones and crevasse splay sandstones point to a poorly drained floodplain with occasional flood events avulsing sandstone onto the floodplain (Bomfleur *et al.*, 2011c). At one level a low diversity megafossil flora has been preserved (Bomfleur *et al.*, 2011c). The assemblage is a (par)autochthonous compression assemblage of usually complete leaves or large leaf fragments, comprising leaf litter deposited under quiescent conditions. The assemblage is thought to represent the local semi-mature, low diversity wetland community vegetation that colonised poorly drained floodplain settings (Bomfleur *et al.*, 2011c). The vegetation comprises one type of low-growing deciduous peltaspermaceous seed fern, *Dejerseya* (Bomfleur *et al.*, 2011c), a voltzialean conifer *Heidiphyllum elongatum* (Axsmith, Taylor and Taylor, 1998; Bomfleur *et al.*, 2011b; Escapa *et al.*, 2010; Taylor and Taylor, 1988; Yao, Taylor and Taylor, 1993) and a corysto-sperm represented by *Dicroidium* foliage, *Umkomasia* reproductive organs, dispersed seeds and wood (Boucher, Taylor and Taylor, 1993; Decombeix, Taylor and Taylor, 2010; Taylor and Taylor, 1988). The voltzialean conifers may have formed monodominant thickets in the poorly drained, possibly peat-forming floodplain environments along with the matatiellacean peltasperms with their spreading habit and shrubby or small tree stature (Bomfleur *et al.*, 2011c). The emergent *Dicroidium* plants, possibly reaching heights of up to 30 m and sporting a *Ginkgo*-like habit (Taylor, 1996; Taylor *et al.*, 2006), would have grown in better drained floodplain environments (Bomfleur and Kerp, 2010; Cúneo *et al.*, 2003), with filicalean ferns forming a monotonous herbaceous understorey (Bomfleur *et al.*, 2011c). The regional climate is interpreted as being mesothermal with high rainfall, no evidence of frosts or subzero temperatures, even during the prolonged period of polar twilight, and strong seasonality in photoperiod (Bomfleur and Kerp, 2010; Cúneo *et al.*, 2003; Hammer, Collinson and Ryan, 1990; Hermsen, Taylor and Taylor, 2009; Parrish, 1990; Scotese, Boucot and McKerrow, 1999; Taylor and Ryberg, 2007; Taylor, Taylor and Cúneo, 2000). Interestingly, there is little evidence for any plant–insect interactions (Bomfleur *et al.*, 2011c) but this may be explained in part by the high palaeolatitude setting. It has been suggested that insect diversity is strongly influenced by seasonality with environments that have low seasonal contrast supporting greater diversity (Archibald *et al.*, 2010).

In southern Victoria Land, Triassic plant fossils have been recorded or described from several locations including Mount Fleming and Shapeless Mountain (Gunn and Warren, 1962; Plumstead, 1962) and Allan Hills (Boucher, Taylor and Taylor, 1995; Townrow, 1967). At Allan Hills, Triassic strata of the Lashly Formation crop out. The formation was deposited by low-sinuosity, braided and meandering rivers (Ballance, 1977; Gabites, 1985). Boucher, Taylor and Taylor (1995) reported assemblages from four levels within Member C of the

Lashly Formation. Three assemblages were dominated by a diversity of *Dicroidium* with *D. dutoitii* and *D. odontopteroides* forming an association within two horizons, while the third horizon had a greater diversity. The fourth assemblage was dominated by *Cladophlebis*, now referred to *Osmunda claytonites* based on reproductive material (Phipps *et al.*, 1998; T. N. Taylor *et al.*, 1990), in association with conifers (*Heidiphyllum*) (Boucher, Taylor and Taylor, 1995). These localities point to spatial heterogeneity in the vegetation.

In northern Victoria Land the Section Peak Formation at Timber Peak, lying at a palaeo-latitude of ~70° S, preserves plant assemblages of Late Triassic age based on associated palynomorphs (Norris, 1965). The Section Peak Formation occurs between two diabase sills and is composed of a series of trough-cross-bedded quartzose sandstones with minor inter-calations of siltstones, shales and coals (Bomfleur and Kerp, 2010). Abundant silicified logs and blocks of silicified peat occur on the exposed bedding planes at the base of the profile (Gair, Norris and Ricker, 1965). Two horizons have been found with preserved plant com-pressions, the first 150 cm below, and the second 80 cm above, a coal seam. Bomfleur and Kerp (2010) undertook an investigation of these horizons and their results are presented below.

The lowermost plant horizon occurs in sediments above a well-developed soil that is overlain by a coal seam. The sedimentology suggests back swamp mire with high water tables and waterlogged soils (Bomfleur and Kerp, 2010). The flora comprises fragmentary remains of the Mesozoic fern *Cladophlebis* as well as *Heidiphyllum elongatum*. The *Cladophlebis* frag-ments are more similar to *Cladophlebis* species found from the Upper Triassic of Argentina and Australia, or *Osmunda claytoniites* from the Upper Triassic of Allan Hills (southern Victoria Land), than to the most common *Cladophlebis* species (*C. antarctica* and *C. denticulata*) found in other parts of Antarctica. This flora is therefore similar in composition to one reported from the Lashly Formation at Allan Hills (Boucher, Taylor and Taylor, 1995; Phipps *et al.*, 1998; T. N. Taylor *et al.*, 1990) and is assumed to represent a low diversity back swamp community.

The uppermost horizon is developed within a crevasse splay sandstone series indicating that the flora was derived from more elevated, less waterlogged sites on the river levee. The horizon has yielded a well-preserved cuticle-bearing plant compression assemblage domin-ated by *Dicroidium*. The dominant species is *D. elongatum* but six other species are also present. Cuticle fragments and pinnules of the peltaspermaceous *Lepidopteris* are also present (Bomfleur and Kerp, 2010). A thalloid structure described as a freshwater algae (Bomfleur *et al.*, 2009) and poorly preserved sheets of cells thought to represent thalloid bryophytes, were also recovered from this horizon (Bomfleur and Kerp, 2010). In addition, one specimen of *Heidiphyllum* was recorded. The *Dicroidium* component of the vegetation was probably tree-like, large-fronded and deciduous and represents both *Dicroidium* lineages that had evolved during the early Middle Triassic (Bomfleur and Kerp, 2010). From leaf morphology, anatomy and preservation it would appear that *Dicroidium* and *Lepidopteris* grew under different conditions, suggesting that the environment was heterogeneous. *Lepidopteris* probably formed a shrubby understorey near watersides in lowland environments in nutrient-deficient (rather than water-stressed) habitats at some distance from the *Dicroidium* Flora (Bomfleur and Kerp, 2010; Retallack and Dilcher, 1988). This flora is one of the most diverse *Dicroidium* associations known to date (Bomfleur and Kerp, 2010).

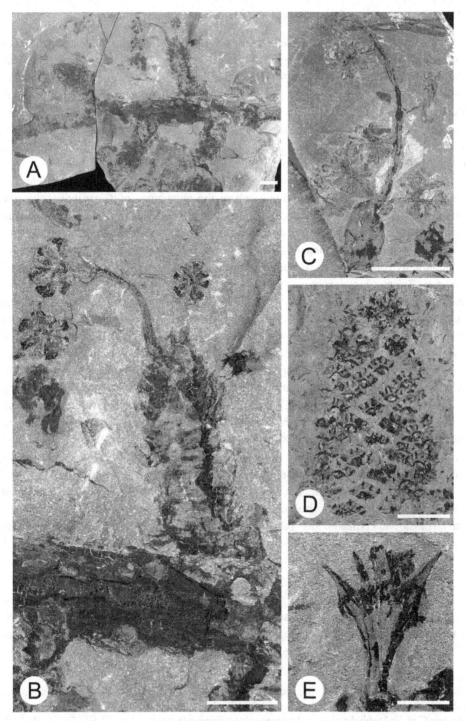

Figure 4.5 **A**, long-shoot short-shoot system bearing *Umkomasia uniramia*, PBKU T12–990, Schroeder Hill, Upper Fremouw. **B**, *Umkomasia uniramia*, PBKU T12–990 close up of short shoot with attached fructification bearing ovulate structures. **C**, *Townrovia polaris*, consisting of peltate heads bearing numerous pollen sacs with bisaccate pollen (Bomfleur *et al.*, 2011c), Mt Falla, Queen

On the east side of East Antarctica in the Prince Charles Mountains, the Flagstone Bench Formation (Amery Group) also record facets of the Late Triassic vegetation. The Jetty Member (Cantrill and Drinnan, 1994) contains megaspores referred to *Cabochonicus sinuosus* and *Minerisporites triangulatus* with affinities to the lycophytes. The overlying McKelvey Member (Cantrill, Drinnan and Webb, 1995) contains a small *Dicroidium*-dominated macroflora that includes the corystosperms *Dicroidium zuberi*, *D. crassinervis* forma *stelznerianum* and *Pteruchus dubius*, along with the conifer *Pagiophyllum papillatus* and a small charcoalified pollen cone.

Together these floras provide evidence for the composition of the *Dicroidium* Flora of Antarctica. As with the Permian floras where the abundance of *Glossopteris* material swamps the less abundant elements, the dominance of *Dicroidium* in the Late Triassic has a similar effect. While *Dicroidium* appears to be the major seed plant present in the Late Triassic floras probably forming the forest overstorey, other groups also occur. Conifers such as the shrubby voltzialean *Heidiphyllum elongatum* (Bomfleur *et al.*, 2011b), and associated seed cone *Telemachus elongatus* (Yao, Taylor and Taylor, 1993), were present along with the peltasperm, *Dejerseya lobata*, with its long-shoot short-shoot branching system and clusters of deciduous leaves on the shoots (Bomfleur *et al.*, 2011c). Fertile material includes branching fructifications referred to *Matatiella dejerseyi* (Bomfleur *et al.*, 2011c). Algae and bryophytes were present in Late Triassic floras suggesting that they were widespread (e.g. Bomfleur and Kerp, 2010; Bomfleur *et al.*, 2009). Liverworts have also been recognised as an important component of high latitude Triassic vegetation (Townrow, 1959) colonising bare ground after flood deposition (Cantrill, 1997). Ferns, such as the rhizomatous *Osmunda claytonites* that crept across the substrate producing fronds at least 21 cm long (Bomfleur and Kerp, 2010; Phipps *et al.*, 1998), and lycophytes, including small herbaceous pleruomeian lycopods, were also present although the lycopods were not as diverse as in the preceding Middle Triassic floras (Bomfleur *et al.*, 2011a).

The lack of permineralised peats in the Late Triassic floras result in a prevailing vegetation of relatively low diversity (although absolute diversity levels can vary) compared with those of the Early Triassic and Permian floras. Yet with increased collecting effort the numbers of taxa are sure to increase and then the full diversity of these Late Triassic ecosystems will be recognised.

Summary

- The Permian–Triassic transition was marked by drying and warming in the Antarctic that saw the demise of the mire-forming vegetation and ultimately the disappearance of mire forests. This was accompanied by a change in depositional style indicating more episodic rainfall.

Caption for Figure 4.5 (cont.) Alexandra Range, PBKU T5–103b. **D**, distil surface of *Telemachus antarcticus* cone illustrating the bract and lobed ovuliferous scale, PBKU T11–409b, Allan Hills, Member C of the Lashly Formation. **E**, cone scale of *Telemachus antarcticus*, PBKU T11–411b. Scale bars A–C = 1 cm, D = 5 mm, E = 2.5 mm. A–E images from Paleobotanical Collections of the University of Kansas; C, courtesy of B. Bomfleur; A, B, D , E courtesy of R. Serbet.

- The stable Permian ecosystems dominated by *Glossopteris* and *Noeggerathiopsis* dramatically collapsed and were replaced by mixed ecosystems with a mosaic of typical Triassic and typical Permian communities for a short period, resulting in an increase in diversity despite extensive extinction.

- The climate continued to warm through the Triassic resulting in the development of unique deeply weathered palaeosol horizons. Typical Triassic forms appeared and the vegetation was replaced by *Lepidopteris*-dominated assemblages, although in more humid and wetter areas reduced diversity glossopterid vegetation survived.

- *Lepidopteris* was replaced by *Dicroidium*-rich vegetation, and preserved forest floors provide information about Middle and Late Triassic ecosystems. Although *Dicroidium* was the dominant element in vegetation there was greater diversity within these forest ecosystems compared with the Permian floras.

- By the Late Triassic climate conditions were once again wetter and suitable for coal formation. *Dicroidium* further diversified and conifers such as the Voltziales became more important in the vegetation. Spatial heterogeneity of vegetation is present but still poorly understood.

Case Study: the Antarctic *Dicroidium* plant

The permineralised peat deposits from the Transantarctic Mountains (Fremouw Formation), together with localities containing impressions (Axsmith *et al.*, 2000) and compressions (Bomfleur and Kerp, 2010), have provided considerable insight into the nature of the *Dicroidium* plant. *Dicroidium* is a morphogenus for foliage leaves that typically bifurcate into two halves and range from entire to tripinnately dissected with thick leaf cuticles particularly in Late Triassic species. To date, the cuticles of only a few species have been described in any detail (Anderson and Anderson, 1983; Archangelsky, 1968; Bomfleur and Kerp, 2010; Boucher, Taylor and Taylor, 1993; Cantrill, Drinnan and Webb, 1995). Leaves are amphistomatic although stomatal density may vary between the upper and lower leaf surface (Anderson and Anderson, 1983; Bomfleur and Kerp, 2010). Stomatal complexes are longitudinally orientated, consist of polar cells that are similar to the epidermal cells and lateral subsidiary cells (2 to 5) that form butterfly-like flanges. Epidermal cells are aligned longitudinally and vary in anticlinal wall thickness and buttressing, with some species also being papillate. Pinnules have distinct palisade and spongy mesophyll, and the vascular bundles are surrounded by a bundle sheath (Pigg, 1990; Pigg and Taylor, 1987, 1990). The base of the frond also has a distinct palisade mesophyll and scattered resinous cells. The vascular tissue in the frond rachis consists of five to eight bundles in an abaxial ring and four to six in an adaxial line (Pigg and Taylor, 1990). The arrangement of the vascular bundles, the presence of resinous cells and sclerotic nests provided the anatomical link with stems preserved in the same sequence described as *Kykloxylon fremouwensis* (Meyer-Berthaud, Taylor and Taylor, 1992, 1993).

Kykloxylon fremouwensis has small stems 10–15 mm in diameter with leaf bases attached. These stems range from one to five-years-old, and within the youngest stems are abundant leaf traces with several periderm bands at the base (Meyer-Berthaud, Taylor and Taylor, 1993), suggesting that leaves were shed after only one year's growth. The secondary xylem forms a solid ring but the pitting differs from wood that has been associated with the *Dicroidium* plant (Meyer-Berthaud, Taylor and Taylor, 1993). Another wood type, *Rhexoxylon*, has been described from many sequences across Gondwana and is widely believed to be the wood produced by *Dicroidium* (Archangelsky, 1968; Archangelsky and Brett, 1961; Petriella, 1981; Retallack and Dilcher, 1988; Taylor, 1992). Although the anatomy of the small stems of *Kykloxylon fremouwensis* differs from *Rhexoxylon*, it should be noted that most *Rhexoxylon* species are based on mature wood from trunks up to a metre in diameter. Indeed the wood taxa of both *Rhexoxylon* and *Kykloxylon* share a number of similarities including sclerotic nests in the pith (Decombeix, Taylor and Taylor, 2010; Meyer-Berthaud, Taylor and Taylor, 1993) supporting a close relationship with the differences between these two genera, explained by the relative maturity of the stems.

Large tree stumps, identified as *Jeffersonioxylon gordonense*, surrounded by a forest floor litter entirely composed of *Dicroidium* (*Dicroidium odontopteroides*) foliage can be found in *in situ* forest horizons within the upper Fremouw Formation at Gordon Valley. This suggests that the stumps and foliage probably originated from the same taxon (Cúneo *et al.*, 2003). The anatomy of the trunks is poorly preserved but opposite pits in the radial walls and pitting of the cross fields suggest similarities to the Podocarpaceae (del Feuyo *et al.*, 1995). This wood is quite different in anatomy from *Rhexoxylon*, with its trunk split into distinct wedges, suggesting that there was a diversity of wood anatomy within this group. Both *Rhexoxylon* and *Jeffersonioxylon* point to the *Dicroidium* plant being a substantial tree.

Long-shoot short-shoot branching systems are also associated with *Dicroidium* leaves (Retallack and Dilcher, 1988) and several specimens illustrated in the literature support that interpretation (e.g. Anderson and Anderson, 1983). This Antarctic material is the only unequivocal evidence for the branching morphology of *Dicroidium*. Long-shoot short-shoot branching systems are often seen in deciduous plants today, and the presence of abscission scars in some *Dicroidium* species and extensive, presumably seasonal, leaf mats further supports the interpretation of a deciduous habit (Axsmith *et al.*, 2000; Retallack and Dilcher, 1988). Impression material from southeast of Schroeder Hill within the upper Fremouw Formation or lower Falla Formation illustrates the growth form of these small terminal stems further. Stems are no bigger than 2.5 cm in diameter and bear short shoots up to 3.5 cm long, with one such stem terminating in a fertile shoot (*Umkomasia uniramia*). *Dicroidium odontopteroides* foliage is attached to the long shoot, and leaf scars on the short shoot are of the same dimensions. The long shoots are over 14 cm long with no signs of tapering and are interpreted as forming large branches with short shoots, as seen in the extant *Ginkgo biloba* today. The fertile shoots are pedicellate with two lateral bracts and terminate in a whorl of cupulate organs referred to *Umkomasia uniramia*. *Umkomasia* and *Dicroidium* have long been associated elsewhere in Gondwana and have been linked through similar cuticular features (Thomas, 1933) but the Antarctic material provides the first evidence of organic connection to *Dicroidium* (Axsmith *et al.*, 2000).

In the same deposits are leaf-like microsporophylls that bear numerous large sporangia on the abaxial surface. These pollen-bearing organs, *Pteruchus*, have also been widely associated with *Dicroidium* based on similarities in cuticular morphology and consistent association at a number of localities (Anderson and Anderson, 1983; Cantrill, Drinnan and Webb, 1995; Retallack and Dilcher, 1988). Within the mircosporangia are large bisaccate pollen grains referred to the dispersed spore genus, *Alisporites*.

Interpretations of the growth form of *Dicroidium* suggest that the plant was unbranched, medium sized, with an apical crown of leaves borne on long-shoot short-shoot systems (Axsmith *et al.*, 2000). This contrasts with other interpretations that have suggested a liane-like habit (Archangelsky, 1968; Walton, 1923) or large forest trees (Retallack and Dilcher, 1988).

References

Abu Hamad, A., Kerp, H., Vörding, B. and Bandel, K. (2008). A Late Permian flora with *Dicroidium* from the Dead Sea region, Jordan. *Review of Palaeobotany and Palynology*, **149**, 83–130.

Adamson, D. and Darragh, A. (1991). Field evidence on Cainozoic history and landforms in the northern Prince Charles Mountains, East Antarctica. In *Quaternary Research in Australian Antarctica: Future Directions, The Australian Defence Force Academy, Special Publication*, **3**, eds D. Gillieson and S. Fitzsimons. Canberra: Department of Geography and Oceanography, University College, pp. 5–14.

Anderson, J. M. and Anderson, H. M. (1983). *Paleoflora of Southern Africa: Molteno Formation Vol. 1 Part 1 Introduction/Part 2* Dicroidium. Rotterdam: A. A. Balkema.

Archangelsky, S. (1968). Studies on Triassic fossil plants from Argentina. IV. The leaf genus *Dicroidium* and its possible relation to *Rhexoxylon* stems. *Palaeontology*, **11**, 500–512.

Archangelsky, S. and Arrondo, O. G. (1975). Paleogeografia y plantas fósiles en el Permico inferior Austrosudamericano. In *Actas 1st Congreso Argentino Paleontología y Bioestratigrafía*, **1**, 479–496.

Archangelsky, S. and Brett, D. W. (1961). Studies on Triassic fossil plants from Argentina. I. *Rhexoxylon* from the Ischigualasto Formation. *Philosophical Transactions of the Royal Society of London Series B*, **244**, 1–19.

Archibald, S. B., Bossert, W. H., Greenwood, D. R. and Farrell, B. D. (2010). Seasonality, the latitudinal gradient of diversity, and Eocene insects. *Paleobiology*, **36**, 374–398.

Artabe, A. E., Morel, E. M. and Spalletti, L. A. (2003). Caracterización de las provincias fitogeográficas triásicas del Gondwana extratropical. *Ameghiniana*, **40**, 387–405.

Askin, R. A. (1997). Permian palynomorphs from southern Victoria Land, Antarctica. *Antarctic Journal of the United States*, **30(5)**, 47–48.

Axsmith, B. J., Taylor, T. N., Taylor, E. L. and Boucher, L. D. (1995). Triassic conifer seed cones from the Lashly Formation, southern Victoria Land. *Antarctic Journal of the United States of America*, **30(5)**, 44–46.

Axsmith, B. J., Taylor, T. N. and Taylor, E. L. (1998). Anatomically preserved leaves of the conifer *Notophytum krauselii* (Podocarpaceae) from the Triassic of Antarctica. *American Journal of Botany*, **85**, 704–713.

Axsmith, B. J., Taylor, E. L., Taylor, T. N. and Cúneo, N. R. (2000). New perspectives on the Mesozoic seed fern order Corystospermales based on attached organs from the Triassic of Antarctica. *American Journal of Botany*, **87**, 757–768.

Ballance, P. F. (1977). The Beacon Supergroup in the Allan Hills, central Victoria Land, Antarctica. *New Zealand Journal of Geology and Geophysics*, **20**, 1003–1016.

Balme, B. E. (1995). Fossil in situ spores and pollen grains: an annotated catalogue. *Review of Palaeobotany and Palynology*, **87**, 81–323.

Barrett, P. J. (1969). Stratigraphy and petrology of the mainly fluviatile Permian and Triassic Beacon rocks, Beardmore Glacier area, Antarctica. *Ohio State University Institute of Polar Studies Report*, **34**, 1–132.

Barrett, P. J. and Fitzgerald, P. G. (1985). Deposition of the lower Feather Conglomerate, a Permian braided river deposit in southern Victoria Land, Antarctica, with notes on the regional paleogeography. *Sedimentary Geology*, **45**, 189–208.

Barrett, P. J., Elliot, D. H. and Lindsay, J. F. (1986). The Beacon Supergroup (Devonian–Triassic) and Ferrar Group (Jurassic) in the Beardmore Glacier area, Antarctica. In *Geology of the Transantarctic Mountains, Antarctic Research Series*, **36**, eds

M. D. Turner and J. E. Splettstoesser. Washington: American Geophysical Union, pp. 339–428.

Basu, A. R., Petaev, M. I., Poreda, R. J., Jacobsen, S. B. and Becker, L. (2003). Chondritic meteorite fragments associated with the Permian–Triassic boundary in Antarctica. *Science*, **302**, 1388–1392.

Becker, L., Poreda, R., Hunt, H. G., Bunch, T. E. and Rampino, M. (2001). Impact event at the Permian–Triassic boundary: evidence from extraterrestrial noble gases in fullerenes. *Science*, **291**, 1530–1533.

Becker, L., Poreda, R. J., Basu, A. R., *et al.* (2004). Bedout: A possible end-Permian impact crater offshore northwestern Australia. *Science*, **304**, 1469–1476.

Benton, M. J. (2005). *When Life Nearly Died: The Greatest Mass Extinction of All Time.* London: Thames & Hudson.

Benton, M. J. and Twitchett, R. J. (2003). How to kill (almost) all life: the end-Permian extinction event. *Trends in Ecology and Evolution*, **18**, 358–365.

Benton, M. J., Tverdokhlebov, V. P. and Sukhov, M. V. (2004). Ecosystem remodelling among vertebrates at the Permian–Triassic boundary in Russia. *Nature*, **432**, 97–100.

Berner, R. A. (2002). Examination of hypotheses for the Permo–Triassic boundary extinction by carbon cycle modelling. *Proceedings of the National Academy of Sciences of the United States of America*, **99**, 4172–4177.

Berthelin, M., Broutin, J., Kerp, H., *et al.* (2003). The Oman Gharif mixed paleoflora: a useful tool for testing Permian Pangea reconstructions. *Palaeogeography, Palaeoclimatology, Palaeoecology*, **196**, 85–98.

Bomfleur, B. and Kerp, H. (2010). *Dicroidium* diversity in the Upper Triassic of north Victorian Land, East Antarctica. *Review of Palaeobotany and Palynology*, **160**, 67–101.

Bomfleur, B., Krings, M., Kastovsky, J. and Kerp, H. (2009). An enigmatic non-marine thalloid organism from the Triassic of East Antarctica. *Review of Palaeobotany and Palynology*, **157**, 317–325.

Bomfleur, B., Krings, M., Taylor, E. L. and Taylor, T. N. (2011a). Macrofossil evidence for pleuromeialean lycophytes from the Triassic of Antarctica. *Acta Palaeontologica Polonica*, **56**, 195–203.

Bomfleur, B., Serbet, R., Taylor, E. L. and Taylor, T. N. (2011b). The possible pollen cone of the Late Triassic conifer *Heidiphyllum/Telemachus* (Voltziales) from Antarctica. *Antarctic Science*, **23**, 379–385.

Bomfleur, B., Taylor, E. L., Taylor, T. N., *et al.* (2011c). Systematics and paleocology of a new peltaspermalean seed fern from the Triassic polar vegetation of Gondwana. *International Journal of Plant Sciences*, **172**, 807–835.

Boucher, L. D., Taylor, E. L. and Taylor, T. N. (1993). *Dicroidium* from the Triassic of Antarctica. In *The Nonmarine Triassic, New Mexico Museum of Natural History and Science Bulletin*, **3**, eds S. G. Lucas and M. Morales. Albuquerque: New Mexico Museum of Natural History, pp. 39–46.

Boucher, L. D. (1995). Morphometirc and paleobiogeographic analyses of *Dicroidium* from the Triassic of Gondwana. Ph.D. thesis Ohio State University Colombus, Ohio. 1–209.

Boucher, L. D., Taylor, E. L. and Taylor, T. N. (1995). *Dicroidium* compression floras from southern Victoria Land. *Antarctic Journal of the United States of America*, **30**(5), 40–41.

Bowring, S. A., Erwin, D. H., Jin Y. G., *et al.* (1998). U/Pb Zircon geochronology and tempo of the end-Permian mass extinction. *Science*, **280**, 1039–1045.

Cairncross, B. (2001). An overview of the Permian (Karoo) coal deposits of southern Africa. *Journal of African Earth Sciences*, **33**, 529–562.

Campbell, I. H., Czamanske, G. K., Fedorenko, V. A., Hill, R. I. and Stepanov, V. (1992). Synchronism of the Siberian Traps and the Permian–Triassic boundary. *Science*, **258**, 1760–1763.

Cantrill, D. J. (1997). Hepatophytes from the Early Cretaceous of Alexander Island, Antarctica: systematic and palaeoecology. *International Journal of Plant Sciences*, **158**, 476–488.

Cantrill, D. J. and Drinnan, A. N. (1994). Late Triassic megaspores from the Amery Group, Prince Charles Mountains, East Antarctica. *Alcheringa*, **18**, 71–78.

Cantrill, D. J., Drinnan, A. N. and Webb, J. A. (1995). Late Triassic plant fossils from the Prince Charles Mountains, East Antarctica. *Antarctic Science*, **7**, 51–62.

Chumakov, N. M. and Zharkov, M. A. (2003). Climate during the Permian–Triassic biosphere reorganizations. Article 2. Climate of the Late Permian and Early Triassic: general inferences. *Stratigraphy and Geological Correlation*, **11**, 361–375.

Collinson, J. W. (1990). Depositional setting of late Carboniferous to Triassic biota in the Transantarctic Basin. In *Antarctic Paleobiology: its Role in the Reconstruction of Gondwana*, eds T. N. Taylor and E. L. Taylor. New York: Springer-Verlag, pp. 1–14.

Collinson, J. W. (1996). Depositional and tectonic setting of the Gondwana sequence in Antarctica. In *Gondwana Nine: Ninth International Gondwana Symposium vols 1 and 2*. Rotterdam: A. A. Balkema, pp. 881–894.

Collinson, J. W., Pennington, D. C. and Kemp, N. R. (1983). Sedimentary petrology of Permian–Triassic fluvial rocks in Allan Hills, central Victoria Land. *Antarctic Journal of the United States of America*, **18**(5), 20–22.

Collinson, J. W., Pennington, D. C. and Kemp, N. R. (1986). Stratigraphy and petrology of Permian–Triassic fluvial deposits in northern Victoria Land, Antarctica. In *Geological Investigations in Northern Victoria Land, Antarctic Research Series*, **46**, ed. E. Stump. Washington: American Geophysical Union, pp. 211–242.

Collinson, J. W., Isbell, J. L., Elliot, D. H., Miller, M. F. and Miller, J. M. G. (1994). Permian–Triassic Transantarctic Basin. In *Permian–Triassic Pangean Basins and Fold-Belts along the Panthalassan Margin of Gondwanaland, Geological Society of America Memoir*, **184**, eds J. J. Veevers and C. McA. Powell. Boulder: The Geological Society of America, pp.173–222.

Collinson, J. W., Hammer, W. R., Askin, R. A. and Elliot, D. H. (2006). Permian–Triassic boundary in the central Transantarctic Mountains. *Geological Society of America Bulletin*, **118**, 747–763.

Colombi, C. E. and Parrish, J. T. (2008). Late Triassic environmental evolution in southwestern Pangea: plant taphonomy of the Ischigualasto Formation. *Palaios*, **23**, 778–795.

Courtillot, V. E. and Renne, P. R. (2003). On the ages of flood basalt events. *Comptes Rendus Géoscience*, **335**, 113–140.

Creber, G. T. and Francis, J. E. (1987). Productivity in fossil forests. In *Proceedings of an International Symposium on Ecological Aspects of Tree-Ring Analysis*, ed. G. C. Jacoby. Washington DC: United States of America Department of Energy, pp. 319–326.

Cúneo, N. R. (1996). Permian phytogeography in Gondwana. *Palaeogeography, Palaeoclimatology, Palaeoecology*, **125**, 75–104.

Cúneo, N. R., Isbell, J., Taylor, E. L. and Taylor, T. N. (1993). The *Glossopteris* flora from Antarctica: taphonomy and paleocology. In *Comptes Rendu 12th International*

Congress of Carboniferous and Permian Stratigraphy and Geology volume 2, ed. S. Archangelsky. Buenos Aires, pp. 13–40.

Cúneo, N. R., Taylor, E. L., Taylor, T. N. and Krings, M. (2003). *In situ* fossil forest from the upper Fremouw Formation (Triassic) of Antarctica, paleoenvironmental setting and paleoclimate analysis. *Palaeogeography, Palaeoclimatology, Palaeoecology*, **197**, 239–261.

Decombeix, A.-L., Klavins, S. D., Taylor, E. L. and Taylor, T. N. (2010). Seed plant diversity in the Triassic of Antarctica: A new anatomically preserved ovule from the Fremouw Formation. *Review of Palaeobotany and Palynology*, **158**, 272–280.

Decombeix, A.-L., Taylor, E. L. and Taylor, T. N. (2010). Anatomy and affinities of permineralized gymnospermous trunks with preserved bark from the Middle Triassic of Antarctica. *Review of Palaeobotany and Palynology*, **163**, 26–34.

Decombeix, A.-L., Taylor, E. L. and Taylor, T. N. (2011). Root suckering in a Triassic conifer from Antarctica: paleoecological and evolutionary implications. *American Journal of Botany*, **98**, 1222–1225.

de Jersey, N. J. (1979). Palynology of the Permian–Triassic transition in the western Bowen Basin. *Geological Survey of Queensland Publication*, **374** (Palaeontological paper 46), 1–61.

Delevoryas, T., Taylor, T. N. and Taylor, E. L. (1992). A marattialean fern from the Triassic of Antarctica. *Review of Palaeobotany and Palynology*, **74**, 101–107.

del Fueyo, G. M., Taylor, E. L., Taylor, T. N. and Cúneo, N. R. (1995). Triassic wood from the Gordon Valley, central Transantarctic Mountains, Antarctica. *International Association of Wood Anatomists Journal*, **16**, 111–126.

Draper, J. J. (1983). Origin of pebbles in mudstones in the Denison Trough. In *Permian Geology of Queensland*. Brisbane: Geological Society of Australia, Queensland Division, pp. 305–316.

Erwin, D. H. (1993). *The Great Paleozoic Crisis: Life and Death in the Permian*. New York: Columbia University Press.

Erwin, D. H. (1994). The Permo-Triassic extinctions. *Nature*, **367**, 231–236.

Erwin, D. H. (1998a). After the end: recovery from extinctions. *Science*, **279**, 1324–1325.

Erwin, D. H. (1998b). The end and the beginning: recoveries from mass extinctions. *Trends in Ecology and Evolution*, **13**, 344–349.

Escapa, I. H., Decombeix, A.-L., Taylor, E. L. and Taylor, T. N. (2010). Evolution and relationships of the conifer seed cone *Telemachus*: evidence from the Triassic of Antarctica. *International Journal of Plant Sciences*, **171**, 560–573.

Escapa, I. H., Taylor, E. L., Cúneo, N. R., *et al.* (2011). Triassic floras of Antarctica: plant diversity and distribution in high paleolatitude communities. *Palaios*, **26**, 522–544.

Eshet, Y., Rampino, M. R. and Visscher, H. (1995). Fungal event and palynological record of ecological crisis and recovery across the Permian–Triassic boundary. *Geology*, **23**, 967–970.

Farabee, M. J., Taylor, T. N. and Taylor, E. L. (1989). Pollen and spore assemblages from the Falla Formation (Upper Triassic), central Transantarctic Mountains, Antarctica. *Review of Palaeobotany and Palynology*, **61**, 101–138.

Farabee, M. J., Taylor, E. L. and Taylor, T. N. (1990). Correlation of Permian and Triassic palynomorph assemblages from the central Transantarctic Mountains, Antarctica. *Review of Palaeobotany and Palynology*, **65**, 257–265.

Fielding, C. R. and Webb, J. A. (1995). Sedimentology of the Permian Radok Conglomerate in the Beaver Lake area of MacRobertson Land, East Antarctica. *Geological Magazine*, **132**, 51–63.

Fielding, C. R. and Webb, J. A. (1996). Facies and cyclicity of the Late Permian Bainmedart Coal Measures in the northern Prince Charles Mountains, MacRobertson Land, Antarctica. *Sedimentology*, **43**, 295–322.

Foster, C. B. (1982). Spore-pollen assemblages of the Bowen Basin, Queensland (Australia): their relationship to the Permian/Triassic boundary. *Review of Palaeobotany and Palynology*, **36**, 165–183.

Foster, C. B., Logan, G. A. and Summons, R. E. (1998). The Permian–Triassic boundary in Australia: where is it and how is it expressed? *Proceedings of the Royal Society of Victoria*, **110**, 247–266.

Francis, J. E. (1991). Arctic Eden. *Natural History*, **1**, 56–63.

Gabites, H. I. (1985). Triassic paleoecology of the Lashly Formation, Transantarctic Mountains, Antarctica. M.Sc. thesis, Victoria University.

Gair, H. S., Norris, G. and Ricker, J. (1965). Early Mesozoic microfloras from Antarctica. *New Zealand Journal of Geology and Geophysics*, **8**, 231–235.

Galfetti, T., Bucher, H., Brayard, A., *et al.* (2007). Late Early Triassic climate change: insights from carbonate carbon isotopes, sedimentary evolution and ammonoid paleo-biogeography. *Palaeogeography, Palaeoclimatology, Palaeoecology*, **243**, 394–411.

Golonka, J. (2007). Late Triassic and Early Jurassic palaeogeography of the world. *Palaeogeography, Palaeoclimatology, Palaeoecology*, **244**, 297–307.

Grice K., Cao, C. Q., Love, G. D., *et al.* (2005). Photic zone euxinia during the Permian–Triassic superanoxic event. *Science*, **307**, 706–709.

Gunn, B. M. and Warren, G. (1962). Geology of Victoria Land between Mawson and Mullock Glaciers, Ross Dependency, Antarctica. *New Zealand Geological Survey Bulletin*, **71**, 1–157.

Haas, J., Gotz, A. E. and Pálfy, J. (2010). Late Triassic to Early Jurassic palaeogeography and eustatic history in the NW Tethyan realm: New insights from sedimentary and organic facies of the Csővár Basin (Hungary). *Palaeogeography, Palaeoclimatology, Palaeoecology*, **291**, 456–468.

Hallam, A. (1991). Why was there a delayed radiation after the end-Palaeozoic extinctions? *Historical Biology*, **5**, 257–455.

Hammer, W. R., Collinson, J. W. and Ryan III, W. J. (1990). A new Triassic vertebrate fauna from Antarctica and its depositional setting. *Antarctic Science*, **2**, 163–167.

Hankel, O. (1992). Late Permian to Early Triassic microfloral assemblages from the Maji ya Chumvi Formation, Kenya. *Review of Palaeobotany and Palynology*, **72**, 129–147.

Hermsen, E. J., Taylor, T. N., Taylor, E. L. and Stevenson, D. W. (2006). Cataphylls of the Middle Triassic cycad *Antarcticycas schopfii* and new insights into cycad evolution. *American Journal of Botany*, **93**, 724–738.

Hermsen, E. J., Taylor, T. N. and Taylor, E. L. (2007). A voltzialean pollen cone from the Triassic of Antarctica. *Review of Palaeobotany and Palynology*, **144**, 113–122.

Hermsen, E. J., Taylor, T. N., Taylor, E. L. and Stevenson, D. W. (2007). Cycads from the Triassic of Antarctica: permineralized cycad leaves. *International Journal of Plant Sciences*, **168**, 1099–1112.

Hermsen, E. J., Taylor, E. L. and Taylor, T. N. (2009). Morphology and ecology of the *Antarcticyas* plant. *Review of Palaeobotany and Palynology*, **153**, 108–123.

Heydari, E. and Hassanzadeh, J. (2003). Deev Jahi model of the Permian–Triassic boundary mass extinction: a case for gas hydrates as the main cause of biological crisis on Earth. *Sedimentary Geology*, **163**, 147–163.

Heydari, E., Hassanzadeh, J., Wade, W. and Ghazi, A. M. (2003). Permian-Triassic boundary interval in the Abadeh section of Iran with implications for mass extinction: Part 1. Sedimentology. *Palaeogeography, Palaeoclimatology, Palaeoecology*, **193**, 405–423.

Heydari, E., Arzani, N. and Hassanzadeh, J. (2008). Mantle plume: the invisible serial killer. Application to the Permian–Triassic boundary mass extinction. *Palaeogeography, Palaeoclimatology, Palaeoecology*, **264**, 147–162.

Hillebrand, H. (2004). On the generality of the latitudinal diversity gradient. *The American Naturalist*, **163**, 192–211.

Hilton J. and Cleal, C. J. (2007). The relationship between Euramerican and Cathaysian tropical floras in the Late Palaeozoic: palaeobiogeographical and palaeogeographical implications. *Earth-Science Reviews*, **85**, 85–116.

Holdgate, G. R., McLoughlin, S., Drinnan, A. N., Finkelman, R. B. and Willett, J. C. (2005). Inorganic chemistry, petrography and palaeobotany of Permian coals in the Prince Charles Mountains, East Antarctica. *International Journal of Coal Geology*, **63**, 156–177.

Holmes, W. B. K. and Ash, S. R. (1979). An Early Triassic megafossil flora from the Lorne Basin, New South Wales. *Proceedings of the Linnean Society of New South Wales*, **103**, 47–70.

Hounslow, M. W. and Ruffell, A. (2006). Triassic – seasonal rivers, dusty deserts and salty lakes. In *The Geology of England and Wales*, eds P. J. Brenchley and P. F. Rawson. London: The Geological Society, pp. 295–324.

Iglesias, A., Artabe, A. E. and Morel, E. M. (2011). The evolution of Patagonian climate and vegetation from the Mesozoic to the present. *Biological Journal of the Linnean Society*, **103**, 409–422.

Isbell, J. L. and Askin, R. A. (1999). Search for evidence of impact at the Permian–Triassic boundary in Antarctica and Australia: comment. *Geology*, **27**, 859.

Isbell, J. L. and Cúneo, N. R. (1996). Depositional framework of Permian coal-bearing strata, southern Victoria Land, Antarctica. *Palaeogeography, Palaeoclimatology, Palaeoecology*, **125**, 217–238.

Isbell, J. L. and Macdonald, D. I. M. (1991). Lithofacies analysis of the Triassic Fremouw Formation at the Gordon Valley vertebrate site. *Antarctic Journal of the United States*, **26(5)**, 15–16.

Jefferson, T. H. (1982). Fossil forests from the Lower Cretaceous of Alexander Island, Antarctica. *Palaeontology*, **25**, 681–708.

Jin, Y. G., Wang, Y., Wang, W., *et al.* (2000). Pattern of marine mass extinction near the Permian–Triassic boundary in South China. *Science*, **289**, 432–436.

Kent, D. V. and Olsen, P. E. (2000). Magnetic polarity stratigraphy and paleolatitude of the Triassic–Jurassic Blomidon Formation in the Fundy basin (Canada): implications for early Mesozoic tropical climate gradients. *Earth and Planetary Science Letters*, **179**, 311–324.

Kerp, H., Abu Hamad, A., Vording, B. and Bandel, K. (2006). Typical Triassic Gondwana floral elements in the Upper Permian of the paleotropics. *Geology*, **34**, 265–268.

Kidder, D. L. and Worsley, T. R. (2004). Causes and consequences of extreme Permo-Triassic warming to global equable climate and relation to the Permo-Triassic extinction and recovery. *Palaeogeography, Palaeoclimatology, Palaeoecology*, **203**, 207–237.

Kiehl, J. T. and Shields, C. A. (2005). Climate simulation of the latest Permian: implications for mass extinction. *Geology*, **33**, 757–760.

Kiessling, W. (2010). Reef expansion during the Triassic: spread of photosymbiosis balancing climatic cooling. *Palaeogeography, Palaeoclimatology, Palaeoecology*, **290**, 11–19.

Klavins, S. D., Taylor, T. N. and Taylor, E. L. (2002). Anatomy of *Umkomasia* (Corystospermales) from the Triassic of Antarctica. *American Journal of Botany*, **89**, 664–676.

Klavins, S. D., Taylor, E. L., Krings, M. and Taylor, T. N. (2003). Gymnosperms from the Middle Triassic of Antarctica: the first structurally preserved cycad pollen cone. *International Journal of Plant Sciences*, **164**, 1007–1020.

Klavins, S. D., Taylor, T. N. and Taylor, E. L. (2004). Matoniaceous ferns (Gleicheniales) from the Middle Triassic of Antarctica. *Journal of Paleontology*, **78**, 211–217.

Klavins, S. D., Kellogg, D. W., Krings, M., Taylor, E. L., and Taylor, T. N. (2005). Coprolites in a Middle Triassic cycad pollen cone: evidence for insect pollination in early cycads. *Evolutionary Ecology Research*, **7**, 479–488.

Korte, C. and Kozur, H. W. (2010). Carbon isotope stratigraphy across the Permian–Triassic boundary: a review. *Journal of Asian Earth Sciences*, **39**, 215–235.

Korte, C., Pande, P., Kalia, P., *et al.* (2010). Massive volcanism at the Permian–Triassic boundary and its impact on the isotopic composition of the ocean and atmosphere. *Journal of Asian Earth Sciences*, **37**, 293–311.

Kovach, W. L. and Batten, D. J. (1993). Diversity changes in lycopsid and aquatic fern megaspores through geologic time. *Paleobiology*, **19**, 28–42.

Kozur, H. W. and Weems, R. E. (2011). Detailed correlation and age of continental late Changhsingian and earliest Triassic beds: implications for the role of the Siberian Trap in the Permian–Triassic biotic crisis. *Palaeogeography, Palaeoclimatology, Palaeoecology*, **308**, 22–40.

Krull, E. S. (1998). Paleoenvironmental and carbon isotopic studies ($\delta^{13}C_{org}$) from terrestrial and marine strata across the Permian–Triassic boundary in Antarctica and New Zealand. Ph.D. thesis, University of Oregon.

Krull, E. S. and Retallack, G. J. (2000). $\delta^{13}C$ depth profile from paleosols across the Permian–Triassic boundary: evidence for methane release. *Geological Society of America Bulletin*, **112**, 1459–1472.

Krull, E. S., Retallack, G. J., Campbell, H. J. and Lyon, G. L. (2000). $\delta^{13}C_{org}$ chemostratigraphy of the Permian–Triassic boundary in the Maitai Group, New Zealand: evidence for high-latitudinal methane release. *New Zealand Journal of Geology and Geophysics*, **43**, 21–32.

Kump, L. R., Pavlov, A. and Arthur, M. A. (2005). Massive release of hydrogen sulfide to the surface ocean and atmosphere during intervals of oceanic anoxia. *Geology*, **33**, 397–400.

Kutzbach, J. E. and Gallimore, R. G. (1989). Pangean climates: megamonsoons of the megacontinent. *Journal of Geophysical Research*, **94**, 3341–3357.

Kyle, R. A. (1976). Palaeobotanical studies of the Permian and Triassic Victoria Group (Beacon Supergroup) of South Victoria Land, Antarctica. Ph.D. thesis, Victoria University.

Kyle, R. A. (1977). Palynostratigraphy of the Victoria Group, south Victoria Land, Antarctica. *New Zealand Journal of Geology and Geophysics*, **20**, 1081–1102.

Kyle, R. A. and Fasola, A. (1978). Triassic palynology of the Beardmore Glacier area of Antarctica. *Palinologia*, **1**, 313–319.

Kyle, R. A. and Schopf, J. M. (1982). Permian and Triassic palynostratigraphy of the Victoria Group, Transantarctic Mountains. In *Antarctic Geoscience*, ed. C. Craddock. Madison: University of Wisconsin Press, pp. 649–659.

Labandeira, C. C. and Sepkoski, J. J. (1993). Insect diversity in the fossil record. *Science*, **261**, 310–315.

Lawver, L. A., Gahagan, L. M. and Dalziel, I. W. D. (1998). A tight-fit Early Mesozoic Gondwana: a plate reconstruction perspective. In *Origin and Evolution of the Continents, Memoir of the National Institute for Polar Research Special Issue*, **53**, eds Y. Motoyoshi and K. Shiraishi. Tokyo: National Institute for Polar Research, pp. 214–229.

Li, Z. X. (1986). The mixed Permian Cathaysia-Gondwana flora. *The Palaeobotanist*, **35**, 611–635.

Li, Z. X. and Powell, C. McA. (2001). An outline of the palaeogeographic evolution of the Australasian region since the beginning of the Neoproterozoic. *Earth-Science Reviews*, **53**, 237–277.

Lindström, S. and McLoughlin, S. (2007). Synchronous palynofloristic extinction and recovery after the end-Permian event in the Prince Charles Mountains, Antarctica: implications for palynoflorisitc turnover across Gondwana. *Review of Palaeobotany and Palynology*, **145**, 89–122.

Lindström, S., McLoughlin, S. and Drinnan, A. N. (1997). Intraspecific variation of taeniate bisaccate pollen within Permian glossopterid sporangia from the Prince Charles Mountains, Antarctica. *International Journal of Plant Sciences*, **158**, 673–684.

Looy, C. V., Brugman, W. A., Dilcher, D. L. and Visscher, H. (1999). The delayed resurgence of equatorial forests after the Permian–Triassic ecologic crisis. *Proceedings of the National Academy of Sciences of the United States of America*, **96**, 13857–13862.

Looy, C. V., Twitchett, R. J., Dilcher, D. L., van Konijnenburg-van Cittert, J. H. A. and Visscher, H. (2001). Life in the end-Permian dead zone. *Proceedings of the National Academy of Sciences of the United States of America*, **98**, 7879–7883.

Lucas, S. G. (2010a). The Triassic timescale based on nonmarine tetrapod biostratigraphy and biochronology. In *The Triassic Timescale, Geological Society of London Special Publication*, **334**, ed. S. G. Lucas. London: The Geological Society, pp. 447–500.

Lucas, S. G. (2010b). The Triassic chronostratgraphic scale: history and status. In *The Triassic Timescale, Geological Society of London Special Publication*, **334**, ed. S. G. Lucas. London: The Geological Society, pp.17–39.

McElwain, J. C. and Pungasena, S. W.(2007). Mass extinction events and the plant fossil record. *Trends in Ecology and Evolution*, **22**, 548–557.

McLoughlin, S. and Drinnan, A. N. (1997a). Revised stratigraphy of the Permian Bainmedart Coal Measures, northern Prince Charles Mountains, East Antarctica. *Geological Magazine*, **134**, 335–353.

McLoughlin, S. and Drinnan, A. N. (1997b). Fluvial sedimentology and revised stratigraphy of the Triassic Flagstone Bench Formation, northern Prince Charles Mountains, East Antarctica. *Geological Magazine*, **134**, 781–806.

McLoughlin, S., Lindström, S. and Drinnan, A. N. (1997). Gondwanan floristic and sedimentological trends during the Permian–Triassic transition: new evidence from the Amery Group, northern Prince Charles Mountains, East Antarctica. *Antarctic Science*, **9**, 281–298.

McManus, H. A., Boucher, L., Taylor, E. L. and Taylor, T. N. (2002b). *Hapsidoxylon terpsichorum* gen. et sp. nov., a stem with unusual anatomy from the Triassic of Antarctica. *American Journal of Botany*, **89**, 1958–1966.

McManus, H. A., Taylor, E. L., Taylor, T. N. and Collinson, J. W. (2002a). A petrified *Glossopteris* flora from Collinson Ridge, central Transantarctic Mountains: Late Permian or Early Triassic? *Review of Palaeobotany and Palynology*, **120**, 233–246.

Metcalfe, I. and Isozaki, Y. (2009). Current perspectives of the Permian–Triassic boundary and the end-Permian mass extinction: preface. *Journal of Asian Earth Sciences*, **36**, 407–412.

Meyer-Berthaud, B. and Taylor, T. N. (1991). A probable conifer with podocarpaceaen affinities from the Triassic of Antarctica. *Review of Palaeobotany and Palynology*, **67**, 179–198.

Meyer-Berthaud, B., Taylor, T. N. and Taylor, E. L. (1992). Reconstructing the Gondwana seed fern *Dicroidium*: evidence from the Triassic of Antarctica. *Geobios*, **25**, 341–344.

Meyer-Berthaud, B., Taylor, T. N. and Taylor, E. L. (1993). Petrified stems bearing *Dicroidium* leaves from the Triassic of Antarctica. *Palaeontology*, **36**, 337–356.

Millay, M. A. and Taylor, T. N. (1990). New fern stems from the Triassic of Antarctica. *Review of Palaeobotany and Palynology*, **62**, 41–64.

Montenegro, A., Spence, P., Meissner, K. J., *et al.* (2011). Climate simulations of the Permian–Triassic boundary: ocean acidification and the extinction event. *Paleoceanography*, **26**, PA3207, 1–19. DOI: 10.1029/2010PA002058

Morante, R. (1996). Permian and early Triassic isotope records of carbon and strontium in Australia and a scenario of events about the Permian–Triassic boundary. *Historical Biology*, **11**, 289–310.

Mundil, R., Ludwig, K. R., Metcalfe, I. and Renne, P. R. (2004). Age and timing of the Permian mass extinctions: U/Pb dating of closed system zircons. *Science*, **305**, 1760–1763.

Mundil, R., Pálfy, J., Renne, P. R. and Brack, P. (2010). The Triassic timescale: new constraints and a review of geochronological data. In *The Triassic Timescale, Geological Society of London Special Publication*, **334**, ed. S. G. Lucas. London: The Geological Society, pp. 41–60.

Neish, P. G., Drinnan, A. N. and Cantrill, D. J. (1993). Structure and ontogeny of *Vertebraria* from silicified Permian sediments in East Antarctica. *Review Palaeobotany Palynology*, **79**, 221–244.

Norris, G. (1965). Triassic and Jurassic miospores and acritarchs from the Beacon and Ferrar Groups, Victoria Land, Antarctica. *New Zealand Journal of Geology and Geophysics*, **8**, 236–277.

Olsen, P. E. (1986). A 40-million-year lake record of early Mesozoic orbital climatic forcing. *Science*, **234**, 842–848.

Olsen, P. E. and Kent, D. V. (1996). Milankovitch climate forcing in the tropics of Pangaea during the Late Triassic. *Palaeogeography, Palaeoclimatology, Palaeoecology*, **122**, 1–26.

Osborn, J. M. and Taylor, T. N. (1989). Structurally preserved sphenophytes from the Triassic of Antarctica: vegetative remains of *Spaciinodum* gen. nov. *American Journal of Botany*, **76**, 1594–1601.

Osborn, J. M., Taylor, T. N. and White, J. F. (1989). *Palaeofibulus* gen. nov., a clamp-bearing fungus from the Triassic of Antarctica. *Mycologia*, **81**, 622–626.

Osborn, J. M., Phipps, C. J., Taylor, T. N. and Taylor, E. L. (2000). Structurally preserved sphenophytes from the Triassic of Antarctica: reproductive remains of *Spaciinodum*. *Review of Palaeobotany and Palynology*, **111**, 225–235.

Ovtcharova, M., Bucher, H., Schaltegger, U., *et al.* (2007). New early to middle Triassic U-Pb ages from South China: calibration with ammonoid biochronozones and implications for the timing of the Triassic biotic recovery. *Earth and Planetary Science Letters*, **243**, 463–475.

Parrish, J. T. (1990). Gondwanan paleogeography and paleoclimatology. In *Antarctic Paleobiology: its Role in the Reconstruction of Gondwana*, eds T. N. Taylor and E. L. Taylor. New York: Springer-Verlag, pp. 15–26.

Parrish, J. T. (1993). Climate of the supercontinent Pangea. *The Journal of Geology*, **101**, 215–233.

Patterson, C. (1999). *Evolution* (3rd edn). London: The Natural History Museum.

Payne, J. L., Lehrmann, D. J., Wei, J. Y., *et al.* (2004). Large perturbations of the carbon cycle during recovery from the end-Permian extinction. *Science*, **305**, 506–509.

Payne, J. L., Lehrmann, D. J., Follet, D., *et al.* (2007). Erosional truncation of uppermost Permian shallow-marine carbonates and implications for Permian–Triassic boundary events. *Geological Society of America Bulletin*, **119**, 771–784.

Perovich, N. E. and Taylor, E. L. (1989). Structurally preserved fossil plants from Antarctica. IV. Triassic ovules. *American Journal of Botany*, **76**, 657–679.

Petriella, B. (1981). Sistemática y vinculaciones de las Corystospermaceae H. Thomas. *Ameghiniana*, **18**, 221–234.

Phipps, C. J. and Taylor, T. N. (1996). Mixed arbuscular mycorrhizae from the Triassic of Antarctica. *Mycologia*, **88**, 707–714.

Phipps, C. J., Taylor, T. N., Taylor, E. L., *et al.* (1998). *Osmunda* (Osmundaceae) from the Triassic of Antarctica: an example of evolutionary stasis. *American Journal of Botany*, **85**, 888–895.

Phipps, C. J., Axsmith, B. J., Taylor, T. N. and Taylor, E. L. (2000). *Gleichenipteris antarcticus* gen. et sp. nov. from the Triassic of Antarctica. *Review of Palaeobotany and Palynology*, **108**, 75–83.

Pigg, K. B. (1990). Anatomically preserved *Dicroidium* foliage from the central Transantarctic Mountains. *Review of Palaeobotany and Palynology*, **66**, 129–145.

Pigg, K. B. (1992). Evolution of isoetalean lycopsids. *Annals of the Missouri Botanical Garden*, **79**, 589–612.

Pigg, K. B. and Taylor, T. N. (1987). Anatomically preserved *Dicroidium* from the Transantarctic Mountains. *Antarctic Journal of the United States*, **22(5)**, 28–29.

Pigg, K. B. and Taylor, T. N. (1990). Permineralized *Glossopteris* and *Dicroidium* from Antarctica. In *Antarctic Paleobiology: its Role in the Reconstruction of Gondwana*, eds T. N. Taylor and E. L. Taylor. New York: Springer-Verlag, pp. 164–172.

Plumstead, E. P. (1962). Fossil floras from Antarctica. *In Trans-Antarctic Expedition 1955–1958, Scientific Reports, 9 (Geology)*. London: The Trans-Antarctic Expedition Committee, pp. 1–154.

Plumstead, E. P. (1964). Palaeobotany of Antarctica. In *Antarctic Geology: Proceedings of the First International Symposium on Antarctic Geology, Cape Town, 16–21 September 1963*, ed. R. J. Adie. Amsterdam: North-Holland Publishing Company, pp. 637–654.

Pole, M. (1999). Structure of a near-polar latitude forest from the New Zealand Jurassic. *Palaeogeography, Palaeoclimatology, Palaeoecology*, **147**, 121–139.

Preto, N., Kustatscher, E. and Wignall, P. B. (2010). Triassic climates – state of the art and perspectives. *Palaeogeography, Palaeoclimatology, Palaeoecology*, **290**, 1–10.

Prochow, S. J., Nordt, L. C., Atchley, S. C. and Hudec, M. R. (2006). Multi-proxy paleosol evidence for Middle and Late Triassic climate trends in eastern Utah. *Palaeogeography, Palaeoclimatology, Palaeoecology*, **232**, 73–72.

Rampino, M., Prokoph, A. and Adler, A. (2000). Tempo of the end-Permian event: high resolution cyclostratigraphy at the Permian–Triassic boundary. *Geology*, **28**, 643–646.

Raup, D. M. and Sepkoski Jr, J. J. (1982). Mass extinction in the marine fossil record. *Science*, **215**, 1501–1503.

Ravich, G. M., Gor, Y. G., Dibner, A. F. and Dobanova, O. V. (1977). Stratigrafiya verkhnepaleozoiskikh uglenocynkh otlozheniy vostochnoy Antarktidy (rayon ozera Biver). *Antarktica*, **16**, 62–75.

Rees, P. M., Gibbs, M. T., Ziegler, A. M., Kutzbach, J. E. and Behling, P. J. (1999). Permian climates: evaluating model predictions using global palaeobotanical data. *Geology*, **27**, 891–894.

Reichow, M. K., Pringle, M. S., Al'Mukhamedov, A. I., *et al.* (2009). The timing and extent of the eruption of the Siberian Traps large igneous province: implications for the end-Permian environmental crisis. *Earth and Planetary Science Letters*, **277**, 9–20.

Reinhardt, L. and Richan, W. (2000). The stratigraphic and geochemical record of playa cycles: monitoring a Pangean monsoon-like system (Triassic, Middle Keuper, S. Germany). *Palaeogeography, Palaeoclimatology, Palaeoecology*, **161**, 205–227.

Renne, P. R., Zhang, Z., Richards, M. A., Black, M. T. and Basu, A. R. (1995). Synchrony and causal relations between Permian–Triassic boundary crises and Siberian flood volcanism. *Science*, **269**, 1413–1416.

Retallack, G. J. (1995). Permian–Triassic life crisis on land. *Science*, **267**, 77–80.

Retallack, G. J. (1997a). Palaeosols in the upper Narrabeen Group of New South Wales as evidence of Early Triassic palaeoenvironments without exact modern analogues. *Australian Journal of Earth Sciences*, **44**, 185–201.

Retallack, G. J. (1997b). Earliest Triassic origin of *Isoetes* and quillwort evolutionary radiation. *Journal of Paleontology*, **71**, 500–521.

Retallack, G. J. (1999). Postapocalyptic greenhouse paleoclimate revealed by earliest Triassic paleosols in the Sydney Basin, Australia. *Geological Society of America Bulletin*, **111**, 52–70.

Retallack, G. J. (2002). *Lepidopteris callipteroides*, an earliest Triassic seed fern from the Sydney Basin, southeastern Australia. *Alcheringa*, **26**, 475–500.

Retallack, G. J. (2005a). Permian greenhouse crises. In *The Nonmarine Permian, New Mexico Museum of Natural History and Science Bulletin*, **3**, eds S. G. Lucas and M. Morales. Albuquerque: New Mexico Museum of Natural History, pp. 256–269.

Retallack, G. J. (2005b). Earliest Triassic claystone breccias and soil erosion crisis. *Journal of Sedimentary Research*, **75**, 679–695

Retallack, G. J. and Alonso-Zarza, A. (1998). Middle Triassic paleosols and paleoclimate in Antarctica. *Journal of Sedimentary Research*, **68**, 169–184.

Retallack, G. J. and Dilcher, D. L. (1988). Reconstructions of selected seed ferns. *Annals of the Missouri Botanical Garden*, **75**, 1010–1057.

Retallack, G. J. and Krull, E. S. (1999). Landscape ecological shift at the Permian–Triassic boundary in Antarctica. *Australian Journal of Earth Sciences*, **46**, 786–812.

Retallack, G. J., Veevers, J. J. and Morante, R. (1996). Global coal-gap between Permian–Triassic extinction and Middle Triassic recovery of peat-forming plants. *Geological Society of America Bulletin*, **108**, 195–207.

Retallack, G. J., Seyedolali, A., Krull, E. S., *et al.* (1998). Search for evidence of impact at the Permian–Triassic boundary in Antarctica and Australia. *Geology*, **26**, 979–982.

Retallack, G. J., Smith, R. M. H. and Ward, P. D. (2003). Vertebrate extinction across the Permian–Triassic boundary in Karoo Basin, South Africa. *Geological Society of America Bulletin*, **115**, 1133–1152.

Retallack, G. J., Jahren, A. H., Sheldon, N. D., *et al.* (2005). The Permian–Triassic boundary in Antarctica. *Antarctic Science*, **17**, 241–258.

Retallack, G. J., Metzger, C. A., Greaver, T., *et al.* (2006). Middle-Late Permian mass extinction on land. *Geological Society of America Bulletin*, **118**, 1398–1411.

Retallack, G. J., Greaver, T. and Jahren, A. H. (2007). Return to Coalsack Bluff and the Permian–Triassic boundary in Antarctica. *Global and Planetary Change*, **55**, 90–108.

Rigby, J. F. (1985). Some Triassic (Middle Gondwana) floras from south Victoria Land, Antarctica. In *Hornibrook Symposium Abstracts*. Christchurch: Department of Scientific and Industrial Research, pp. 78–79.

Rigby, J. F. and Schopf, J. M. (1969). Stratigraphic implications of Antarctic paleobotanical studies. In *Gondwana stratigraphy. 1st IUGS Gondwana Symposium*, ed. A. J. Amos, Paris: UNESCO, pp. 91–106.

Rohn, R. and Rösler, O. (1989). Novas ocorrencias *Glossopteris* na Formação Rio do Rasto (Bacia do Paraná, Permiano Superior). *Boletim Instituto de Geociências –Universidade de São Paulo, Publicação Especial*, **7**, 101–125.

Rothwell, G. W., Taylor, E. L. and Taylor, T. N. (2002). *Ashicaulis woolfei* n. sp.: additional evidence for the antiquity of osmundaceous ferns from the Triassic of Antarctica. *American Journal of Botany*, **89**, 352–361.

Royer, D. L. (2006). CO_2-forced climate thresholds during the Phanerozoic. *Geochimica et Cosmochimica Acta*, **70**, 5665–5675.

Royer, D. L., Berner, R. A., Montanez, I. P., Tabor, N. J. and Beerling, D. J. (2004). CO_2 as a primary driver of Phanerozoic climate. *GSA Today*, **14(3)**, 4–10.

Ryberg, P. E., Hermsen, E. J., Taylor, E. L., Taylor, T. N. and Osborn, J. M. (2008). Development and ecological implications of dormant buds in the high latitude Triassic sphenophyte *Spaciinodum* (Equisetaceae). *American Journal of Botany*, **95**, 1443–1453.

Ryskin, G. (2003). Methane-driven oceanic eruptions and mass extinctions. *Geology*, **31**, 741–744.

Schopf, J. M. (1978). An unusual osmundaceous specimen from Antarctica. *Canadian Journal of Botany*, **56**, 3083–3095.

Schwendemann, A. B., Taylor, T. N., Taylor E. L. and Krings, M. (2009a). Pollen of the Triassic cycad *Delemaya spinulosa* and implications on cycad evolution. *Review of Palaeobotany and Palynology*, **156**, 98–103.

Schwendemann, A. B., Taylor, T. N., Taylor E. L., Krings, M. and Dotzler, N. (2009b). *Combresomyces cornifer* from the Triassic of Antarctica: evolutionary stasis in the Peronosporomycetes. *Review of Palaeobotany and Palynology*, **154**, 1–5.

Schwendemann, A. B., Taylor, T. N., Taylor, E. L., Krings, M. and Osborn, J. M. (2010a). Modern traits in early Mesozoic sphenophytes: the *Equisetum*-like cones of *Spaciinodum collinsonii* with in situ spores and elaters from the Middle Triassic of Antarctica. In *Plants in Mesozoic Time: morphological Innovation, Phylogeny, Ecosystems*, ed. C. T. Gee. Bloomington: Indiana University Press, pp. 15–34.

Schwendemann, A. B., Taylor, T. N., Taylor E. L. and Krings, M. (2010b). Organization, anatomy, and fungal endophytes of a Triassic conifer embryo. *American Journal of Botany*, **97**, 1873–1883.

Scotese, C. R., Boucot, A. J. and McKerrow, W. S. (1999). Gondwanan palaeogeography and palaeoclimatology. *Journal of African Earth Sciences*, **28**, 99–114.

Sellwood, B. W. and Valdes, P. J. (2006). Mesozoic climates: General circulation models and the rock record. *Sedimentary Geology*, **190**, 269–287.

Sepkoski Jr, J. J. (1984). A kinetic model of Phanerozoic taxonomic diversity. III. Post-Paleozoic families and mass extinctions. *Paleobiology*, **10**, 246–267.

Sepkoski Jr, J. J. (1996). Patterns of Phanerozoic extinction: a perspective from global databases. In *Global Events and Event Stratigraphy in the Phanerozoic*, ed. O. H. Walliser. Heidelberg, Berlin, New York: Springer-Verlag, pp. 35–52.

Sheldon, N. D. (2006). Abrupt chemical weathering increase across the Permian–Triassic boundary. *Palaeogeography, Palaeoclimatology, Palaeoecology*, **231**, 315–321.

Sheldon, N. D. and Retallack, G. J. (2002). Low oxygen levels in earliest Triassic soils. *Geology*, **30**, 919–922.

Shi, G. R., Waterhouse, J. B. and McLoughlin, S. (2010). The Lopingian of Australasia: a reivew of biostratigraphy, correlations, palaeogeography and palaeobiogeography. *Geological Journal*, **45**, 230–263.

Smith, R. M. H. and Ward, P. D. (2001). Pattern of vertebrate extinctions across an event bed at the Permian–Triassic boundary in the Karoo Basin of South Africa. *Geology*, **29**, 1147–1150.

Smoot, E. L., Taylor, T. N. and Delevoryas, T. (1985). Structurally preserved fossil plants from Antarctica I. *Antarcticycas* gen. nov., a Triassic cycad stem from the Beardmore Glacier area. *American Journal of Botany*, **72**, 1410–1423.

Srivastava, A. K. and Agnihotri, D. (2010). Dilemma of late Palaeozoic mixed floras in Gondwana. *Palaeogeography, Palaeoclimatology, Palaeoecology*, **298**, 54–69.

Srivastava, S. C. and Jha, N. (1990). Permian–Triassic palynofloral transition in Godavari Graben, Andra Pradesh. *The Palaeobotanist*, **38**, 92–97.

Stampfli, G. M. and Borel, G. D. (2002). A plate tectonic model for the Paleozoic and Mesozoic constrained by dynamic plate boundaries and restored synthetic oceanic isochrons. *Earth and Planetary Science Letters*, **196**, 17–33.

Steiner, M. B., Eshet, Y., Rampino, M. R. and Schwindt, D. M. (2003). Fungal abundance spike and the Permian–Triassic boundary in the Karoo Supergroup (South Africa). *Palaeogeography, Palaeoclimatology, Palaeoecology*, **194**, 405–414.

Stubblefield, S. P. and Taylor, T. N. (1986). Wood decay in silicified gymnosperms from Antarctica. *Botanical Gazette*, **147**, 116–125.

Stubblefield, S. P., Taylor, T. N. and Seymour, R. L. (1987). A possible endogonaceous fungus from the Triassic of Antarctica. *Mycologia*, **79**, 905–906.

Stubblefield, S. P., Taylor, T. N. and Trappe, J. M. (1987a). Fossil mycorrhizae: a case for symbiosis. *Science*, **237**, 59–60.

Stubblefield, S. P., Taylor, T. N. and Trappe, J. M. (1987b). Vesicular-arbuscular mycorrhizae from the Triassic of Antarctica. *American Journal of Botany*, **74**, 1904–1911.

Svenson, H., Planke, S., Polozov, A. G., *et al.* (2009). Siberian gas venting and the end-Permian environmental crisis. *Earth and Planetary Sicence Letters*, **277**, 490–500.

Taylor, E. L. (1989). Tree-ring structure in woody axes from the central Transantarctic Mountains, Antarctica. In *Proceedings of the International Symposium on Antarctic Research (Hangzhou, P.R. China, May, 1989)*. Tianjin: China Ocean Press, pp. 109–113.

Taylor, E. L. (1992). The occurrence of a *Rhexoxylon*-like stem in Antarctica. *Courier Forschungsinstitutut Senckenberg*, **147**, 183–189.

Taylor, E. L. (1996). Enigmatic gymnosperms? Structurally preserved Permian and Triassic seed ferns from Antarctica. *Review of Palaeobotany and Palynology*, **90**, 303–318.

Taylor, E. L. and Ryberg, P. E. (2007). Tree growth at polar latitudes based on fossil tree ring analysis. *Palaeogeography, Palaeoclimatology, Palaeoecology*, **255**, 246–264.

Taylor, E. L. and Taylor, T. N. (1988). Late Triassic flora from Mt Falla, Queen Alexandra Range. *Antarctic Journal of the United States*, **23**(5), 2–3.

Taylor, E. L. and Taylor, T. N. (1990). Structurally preserved Permian and Triassic floras from Antarctica. In *Antarctic Paleobiology: its Role in the Reconstruction of Gondwana*, eds T. N. Talyor and E. L. Taylor. New York: Springer-Verlag, pp. 149–163.

Taylor, E. L., Taylor, T. N. and Collinson, J. W. (1989). Depositional setting and paleo-botany of Permian and Triassic permineralized peat from the central Transantarctic Mountains, Antarctica. *International Journal of Coal Geology*, **12**, 657–679.

Taylor, E. L., Taylor, T. N., Meyer-Berthaud, B., Isbell, J. L. and Cúneo, N. R. (1990). A Late Triassic flora from Allan Hills, southern Victoria Land. *Antarctic Journal of the United States of America*, **25(5)**, 20–21.

Taylor, E. L., Boucher, L. D. and Taylor, T. N. (1992). *Dicroidium* foliage from Mount Falla, central Transantarctic Mountains. *Antarctic Journal of the United States*, **27(5)**, 2–3.

Taylor, E. L., Taylor, T. N. and Cúneo, N. R. (2000). Permian and Triassic high-latitude paleoclimates: evidence from fossil biotas. In *Warm Climates in Earth History*, eds B. T. Huber, K. G. MacLeod and S. L. Wing. Cambridge: Cambridge University Press, pp. 321–350.

Taylor, E. L., Taylor, T. N., Kerp, H. and Hermsen, E. J. (2006). Mesozoic seed ferns: old paradigms, new discoveries. *Journal of the Torrey Botanical Society*, **133**, 62–82.

Taylor, T. N. and Taylor, E. L. (1983). Structurally preserved plants from the Beardmore Glacier area. *Antarctic Journal of the United States*, **18(5)**, 57–58.

Taylor, T. N. and White Jr, J. F. (1989). Fossil fungi (Endogonaceae) from the Triassic of Antarctica. *American Journal of Botany*, **76**, 389–396.

Taylor, T. N., Taylor, E. L., Meyer-Berthaud, B., Isbell, J. L. and Cúneo, N. R. (1990). Triassic osmundaceous ferns from the Allan Hills, southern Victoria Land. *Antarctic Journal of the United States of America*, **25(5)**, 18–19.

Taylor, T. N., Taylor, E. L., del Fueyo, G. (1993). Permineralized Triassic plants from Antarctica. In *The Nonmarine Triassic, New Mexico Museum of Natural History and Science Bulletin*, **3**, eds S. G. Lucas and M. Morales. Albuquerque: New Mexico Museum of Natural History, pp. 457–460.

Taylor, T. N., del Fueyo, G. M. and Taylor, E. L. (1994). Permineralized seed fern cupules from the Triassic of Antarctica: implications for cupule and carpel evolution. *American Journal of Botany*, **81**, 666–677.

Thomas, H. H. (1933). On some pteridospermous plants from the Mesozoic rocks of South Africa. *Philosophical Transactions of the Royal Society of London Series B*, **222**, 193–256.

Torsvik, T. H. and Cocks, L. R. M. (2004). Earth geography from 400 to 250 Ma: a palaeomagnetic, faunal and facies review. *Journal of the Geological Society of London*, **161**, 555–572.

Torsvik, T. H. and van der Voo, R. (2002). Refining Gondwana and Pangea palaeogeography: estimates of Phanerozoic non-dipole (octupole) fields. *Geophysical Journal International*, **151**, 771–794.

Torsvik, T. H., Gaina, C. and Redfield, T. F. (2008). Antarctica and global palaeogeography: from Rodinia, through Gondwanaland and Pangea, to the birth of the Southern Ocean and the opening of gateways. In *Antarctica: a Keystone in a Changing World. Proceedings of the 10th International Symposium on Antarctic Earth Sciences*, eds A. K. Cooper, P. J. Barrett, H. Stagg, B. C. Storey, E. Stump, W. Wise, and the 10th ISAES editorial team. Washington: National Academies Press, pp. 125–140.

Townrow, J. A. (1959). Two Triassic bryophytes from South Africa. *Journal of South African Botany*, **25**, 1–22.

Townrow, J. A. (1967). Fossil plants from Allan and Carapace Nunataks, and from the upper Mill and Shackleton glaciers, Antarctica. *New Zealand Journal of Geology and Geophysics*, **10**, 456–473.

Twitchett, R. J., Looy, C. V., Morante, R., Visscher, H. and Wignall, P. B. (2001). Rapid and synchronous collapse of marine and terrestrial ecosystems during the end-Permian biotic crisis. *Geology*, **29**, 351–354.

Veevers, J. J. (2004). Gondwanaland from 650–500 Ma assembly through 320 Ma merger in Pangea to 185–100 Ma breakup: Supercontinental tectonics via stratigraphy and radiometric dating. *Earth-Science Review*, **68**, 1–132.

Veevers, J. J., Conanghan, P. J. and Shaw, S. E. (1994). Turning point in Pangean environmental history of the Permo-Triassic (P-Tr) boundary. In *Pangea: Paleoclimate, Tectonics, and Sedimentation during Accretion, Zenith and Breakup of a Supercontinent, Geological Society of America Special Paper*, **288**, ed. G. D. Klein. Boulder: The Geological Society of America, pp. 187–196.

Visscher, H., Brinkhuis, H., Dilcher, D. L., *et al.* (1996). The terminal Paleozoic fungal event: evidence of terrestrial ecosystem destabilization and collapse. *Proceedings of the National Academy of Sciences of the United States of America*, **93**, 2155–2158.

Visser, J. N. J. (1995). Post-glacial Permian stratigraphy and geography of southern and central Africa: boundary conditions for climate modelling. *Palaeogeography, Palaeoclimatology, Palaeoecology*, **118**, 213–243.

Vollmer, T., Ricken, W., Weber, M., Tougiannidis, N., Röhling, H. -G. and Hambach, U. (2008). Orbital control on Upper Triassic playa cycles of the Steihnmergel-Kueper (Norian): a new concept for ancient playa cycles. *Palaeogeography, Palaeoclimatology, Palaeoecology*, **267**, 1–16.

Wagner, P. J., Kosnik, M. A. and Lidgard, S. (2006). Abundance distributions imply elevated complexity of post-Paleozoic marine ecosystems. *Science*, **314**, 1289–1292.

Walsh, S. L., Gradstein, F. M. and Ogg, J. G. (2004). History, philosophy, and application of the Global Stratotype Section and Point (GSSP). *Lethaia*, **37**, 201–218.

Walton, J. (1923). On *Rhexoxylon* Bancroft – a Triassic genus of plants exhibiting a liane-type of vascular organisation. *Philosophical Transactions of the Royal Society of London B*, **212**, 79–109.

Wang, K., Geldsetzer, H. H. J. and Krouse, H. R. (1994). Permian–Triassic extinction: Organic δ^{13}C evidence from British Columbia, Canada. *Geology*, **22**, 580–584.

Ward, P. D., Montgomery, D. R. and Smith, R. (2000). Altered river morphology in South Africa related to the Permian-Triassic extinction. *Science*, **289**, 1740–1743.

Ward, P. D., Botha, J., Buick, R., *et al.* (2005). Abrupt and gradual extinction among Late Permian vertebrates in the Karoo Basin, South Africa. *Science*, **307**, 709–714.

Webb, J. A. and Fielding, C. R. (1993). Permo-Triassic sedimentation within the Lambert Graben, northern Prince Charles Mountains, East Antarctica. In *Gondwana 8 – Assembly, Evolution and Dispersal, Proceedings of the 8th Gondwana Symposium, Hobart*, eds R. H. Findlay, R. Unrug, H. R. Banks and J. J. Veevers. Rotterdam: A. A. Balkema, pp. 357–369.

Weaver, L., McLoughlin, S., Drinnan, A. N. (1997). Fossil woods from the Upper Permian Bainmedart Coal Measures, northern Prince Charles Mountains, East Antarctica. *Australian Geological Survey Organisation Journal of Geology and Geophysics*, **16**, 655–676.

White Jr, J. F. and Taylor, T. N. (1988). Triassic fungus from Antarctica with possible ascomycetous affinities. *American Journal of Botany*, **75**, 1495–1500.

White Jr, J. F. and Taylor T. N. (1989a). A trichomycete-like fossil from the Triassic of Antarctica. *Mycologia*, **81**, 643–646.

White Jr, J. F. and Taylor T. N. (1989b). An evaluation of sporocarp structure in the Triassic fungus *Endochaetophora*. *Review of Palaeobotany and Palynology*, **61**, 341–345.

White Jr, J. F. and Taylor T. N. (1989c). Triassic fungi with suggested affinities to the Endogonales (Zygomycotina). *Review of Palaeobotany and Palynology*, **61**, 53–61.

White Jr, J. F. and Taylor T. N. (1991). Fungal sporocarps from Triassic peat deposits in Antarctica. *Review of Palaeobotany and Palynology*, **67**, 229–236.

Whiteside, J. H., Olsen, P. E., Eglinton, T. I., *et al.* (2011a). Pangean great lake paleoecology on the cusp of the end-Triassic extinction. *Palaeogeography, Palaeoclimatology, Palaeoecology*, **301**, 1–17.

Whiteside, J. H., Grogan, D. S., Olsen, P. E. and Kent, D. V. (2011b). Climatically driven biogeographic provinces of Late Triassic tropical Pangea. *Proceedings of the National Academy of Sciences of the United States of America*, **108**, 8972–8977.

Wignall, P. B. (2007). The end-Permian mass extinction—how bad did it get? *Geobiology*, **5**, 303–309. doi:10.1111/j.1472–4669.2007.00130.x.

Wignall, P. B. and Hallam, A. (1992). Anoxia as a cause of the Permian/Triassic extinction: facies evidence from northern Italy and the western United States. *Palaeogeography, Palaeoclimatology Palaeoecology*, **93**, 21–46.

Wignall, P. B. and Twitchett, R. J. (1996). Oceanic anoxia and the end Permian mass extinction. *Science*, **272**, 1155–1158.

Wignall, P. B., Morante, R. and Newton, R. (1998). The Permo-Triassic transition in Spitsbergen: $\delta^{13}C_{org}$ chemostratigraphy, Fe and S geochemistry, facies, fauna and trace fossils. *Geological Magazine*, **135**, 47–62.

Wright, R. P. and Askin, R. A. (1987). The Permian–Triassic boundary in the Southern Morondava Basin of Madagascar as defined by plant microfossils. In *Gondwana Six: Stratigraphy, Sedimentology, and Palaeontology*, ed. G. D. McKenzie. *Geophysical Monographs*, **41**, 157–166.

Yao, X., Taylor, T. N. and Taylor, E. L. (1993). The Triassic seed cone *Telemachus* from Antarctica. *Review of Palaeobotany and Palynology*, **78**, 269–276.

Yao, X., Taylor, T. N. and Taylor, E. L. (1995). The corystosperm pollen organ *Pteruchus* from the Triassic of Antarctica. *American Journal of Botany*, **82**, 535–546.

Yao, X., Taylor, T. N. and Taylor, E. L. (1997). A taxodiaceous seed cone from the Triassic of Antarctica. *American Journal of Botany*, **84**, 343–354.

Yin, H., Zhang, K. X., Tong, J. N., Yang, Z. Y. and Wu, S. B. (2001). The global stratotype section and point (GSSP) of the Permian–Triassic boundary. *Episodes*, **24**, 102–114.

Yin, H., Feng, Q., Lai, X., Baud, A. and Tong, J. (2007). The protracted Permo-Triassic crisis and multi-episode extinction around the Permian–Triassic boundary. *Global and Planetary Change*, **55**, 1–20.

Ziegler, A. M. (1990). Phytogeographic patterns and continental configurations during the Permian period. In *Palaeozoic Palaeogeography and Biogeography, Geological Society of London Memoir*, **12**, eds W. S. McKerrow and C. R. Scotese. London: The Geological Society, pp. 363–379.

Ziegler, A. M., Scotese, C. R. and Barrett, S. F. (1983). Mesozoic and Cenozoic paleogeographic maps. In *Tidal Friction and the Earth's Rotation II. Proceedings of a workshop held at the Centre for Interdisciplinary Research (ZiF) of the University of Bielefled, September 28–October 3, 1981*, eds P. Broche and J. Sundermann. Berlin: Springer-Verlag, pp. 240–252.

Ziegler, A. M., Parrish, J. M., Yao, J., *et al.* (1993). Early Mesozoic phytogeography and climate. *Philosophical Transactions of the Royal Scoiety of London Series B*, **341**, 297–305.

Ziegler, A. M., Eshel, G., Rees, P. M., *et al.* (2003). Tracing the tropics across land and sea: Permian to present. *Lethaia*, **36**, 227–254.

5

Gondwana break-up and landscape change across the Triassic–Jurassic transition and beyond

Introduction

Events associated with the transition from the Triassic to Jurassic ~200 Ma ago have been regarded either as an artifact or as marking the fourth largest extinction in the Earth History (Tanner, Lucas and Chapman, 2004) (Figure 4.3). This contradiction highlights the importance for investigations in both the ocean and on land to determine the sequence of events that took place over a period of time spanning <10 kyr, immediately prior to the start of the break-up of Pangea. This biotic crisis is evidenced both on land and in the sea where at least half, and probably up to 80%, of the species now known to have been living on Earth became extinct (Raup and Sepkoski, 1982; Sepkoski, 1996). On land, the vegetation underwent a sudden and severe crisis freeing up numerous ecological niches, and so allowing the dinosaurs that had first appeared in the fossil record some 30 Ma earlier to assume dominant roles through the Jurassic. In the terrestrial realm the typical *Dicroidium* Flora of the Triassic were replaced by Jurassic floras rich in bennettites, conifers and ferns. In spite of the severity of this mass extinction, associated environmental changes have been poorly documented relative to other extinction events (Wignall, 2001), and as such the causal mechanisms driving the end-Triassic extinction continue to remain enigmatic (Hallam and Wignall, 1997). Changes in global climate brought about by major perturbations in the biogeochemical cycle are often cited as possible drivers, but the details still need to be fully elucidated (cf. Beerling and Berner, 2002).

Globally the Jurassic is considered to be an exotic world – a period of warm, wet greenhouse conditions (e.g. Huber, MacLeod and Wing, 2000) with high levels of atmospheric CO_2 (Sellwood and Valdes, 2008). Compared with today, Jurassic Earth has been modelled to have been warmer by about 5–10 °C causing higher atmospheric humidity and greatly enhancing the hydrological cycle (e.g. Sellwood and Valdes, 2008). Superimposed on this period of relative equable warmth were several extremely warm periods or hyperthermal events (Kidder and Worsley, 2010; Suan *et al.*, 2010), or multiscale perturbations with dynamic polar ice (Dera *et al.*, 2011). Latitudinal climate variations were strong and this is reflected in the distribution of the biota (Rees, Zeigler and Valdes, 2000). Fossil plant data can be used to infer vegetation patterns, but determining the climatic parameters under which the vegetation grew is more problematic. It is still unclear how cold the polar regions actually were, and whether regions of more tundra-like vegetation existed as suggested by climate models (Sellwood and Valdes, 2008). Across the Southern Hemisphere the lack of data is one of the main problems associated with these general inferred vegetation belts (Rees, Ziegler and Valdes, 2000) particularly in the high latitude regions of Antarctica where cold climates are thought to have been in existence.

Global palaeogeography

For much of the Jurassic, the supercontinent Pangea (Laurasia plus Gondwana), surrounded by the Panthalassa Ocean, continued to straddle the globe extending from the high northern latitudes (80° N) across the equator, and south into the high southern latitudes (80° S). The western margin extended from the present day eastern Australia seaboard to Antarctica and up the west coast of South and North America. This region was dominated by an active margin on the western side that bordered the Panthalassa Ocean and was characterised by subduction processes forming a coastal range of mountains (Ziegler, Scotese and Barrett, 1983). In contrast, the eastern side of Pangaea had a large embayment that formed the Tethys Ocean, a region bounded on the north by Laurasia and to the south by Gondwana (Golonka, 2007). Continental blocks that had rifted from the Gondwana margin during the Permian then drifted across the Tethys during the Triassic and Jurassic to become accreted onto the Laurasian landmass (Golonka, 2007). These continental blocks have complicated reconstructions of the palaeogeography of the Tethys region during the Jurassic since the timing of accretion and drifting of these blocks is in general poorly constrained. Nevertheless it is likely that there were island continents in the Tethys Sea during the Jurassic, which in turn would have influenced oceanic circulation patterns.

Using conceptual models, based on an understanding of primary oceanic circulation principles (Arias, 2008), and Global Circulation Model simulations (e.g. Kutzbach, Guetter and Washington, 1990; Smith, Dubois and Marotzke, 2004), it has been possible to recreate palaeoceanographic circulation patterns for both the Panthalassa and Tethys oceans. In the Early Jurassic there was no connection between the western Tethys Ocean, on the east side on Pangea, and the Panthalassa Ocean on the west so that the only connection was through the eastern Tethys. In eastern Pangea, continental monsoonal conditions drove surface wind direction that influenced the development of circulation within the Tethys. With time, as Pangaea began to break up, narrow and shallow marine connections developed (namely the Hispanic Corridor) to link the Tethys Sea in the east to Panthalassa in the west and thus separating Laurasia in the north from Gondwana to the south in the Early Jurassic (Arias, 2008). Evidence derived from the mixing of boreal and tethyean faunas suggests that the development of connections between these two oceans occurred in the Pliensbachian (196 Ma) (Arias, 2006) and so marks the beginning of the separation of Laurasia and Gondwana. The Laurasian landmass went on to develop shallow warm epicontinental seas in Europe and a series of island archipelagos (Dercourt *et al.*, 2000).

As the Jurassic advanced, the seaway continued to widen so increasing the separation of Laurasia from Gondwana until they finally separated at 150 Ma. The separation of these two great landmasses and the progressive creation of a seaway between them profoundly altered oceanic circulation. Currents that were previously deflected north or south of the equator, to flow into mid and high latitudes when the continents were connected, were replaced by equatorial through-flow between the now separating landmasses (Scotese, 1991). This effective isolation of Laurasia from Gondwana limited the ability for floral and faunal exchange between the northern and southern landmasses.

While the palaeogeographic configuration in the Northern Hemisphere was changing, dramatic changes were also taking place in Gondwana. The long-lived active margin along the western coastline shifted from convergent to extensional tectonics, and associated with this was widespread volcanism. The change in tectonic regime has been associated with mantle plume activity, and these mantle plumes have been suggested to have been a major mechanism in the break-up of Gondwana (Dalziel, Lawver and Murphy, 2000; Storey, 1995). However, the break-up mechansim has remained controversial. Although it has been demonstrated that mantle plumes with voluminous magmas were in existence at this time, some researchers maintain that they are a consequence of break-up rather than the cause (e.g. Collins, 2003; Eagles and Vaughan, 2009). Whatever the mechanism that led to the break-up of Gondwana, tholeitic magmatic rocks are widespread from southern Africa (Karoo Dolerite, volcanic Drakensberg Group; see Riley *et al.*, 2006 and references therein) into Dronning Maud Land (Riley *et al.*, 2005) and along the length of the Transantarctic Mountains (Ferrar Dolerite, Kirkpatrick Basalt; Elliot and Fleming, 2008), through southern and northern Victoria Land (Ferrar Dolerite, Kirkpatrick Basalt; Elliot and Fleming, 2008), and finally extending into southern Australia (Bromfield *et al.*, 2007; Hergt and Brauns, 2001).

The emplacement and eruption of these rocks marked a change in Antarctic landscapes along the Palaeopacific gondwanan margin. This transformation from a mature coastal plain to dynamic landscape occurred over a relatively short period of time, beginning at around 183 Ma, and continuing for some 8 million years such that by ~175 Ma, widespread changes to terrestrial environments had taken place. These changes were associated not only increased volcanism but also the initiation of inversion and uplift of the long-lived foreland basin on this margin of Gondwana. The previous stable land surface in this area became covered in small explosive volcanic events, and in places, the development of flood basalt lava fields. Today these are well-exposed in both northern and southern Victoria Land (Ross, White and McClintock, 2008), but exposures in the Queen Alexandra Range and Otway Massif (Hanson and Elliot, 1996) suggest that this change in landscape was widespread throughout the Transantarctic Mountains and possibly extended into Dronning Maud Land. Analogies can be seen in other parts of the world that suggest the landscape would have been flat with small scattered volcanic cones with lava flows infilling palaeovalleys to create a new surface with new drainage patterns. In places, small lakes developed on the basalt flow tops (Stigall *et al.*, 2008) and these were periodically buried by further volcanism with lava flowing into the lakes (Ross, White and McClintock, 2008).

Initially, Gondwana fragmented into an eastern fragment (Australia, Antarctica, India, New Zealand) and a western fragment (Africa, South America), with connections maintained between the two via the Antarctic Peninsula. As the rift system between Africa and

Antarctica, India and Antarctica, and Australia and Antarctica developed, previously relatively subdued topography was first thermally uplifted and subsequently dissected by rift valleys with associated volcanism. Whilst evidence of these processes is not well-preserved in onshore Antarctic rock exposures, it can be inferred from sedimentary and volcanic strata preserved on the conjugate rift margin (e.g. Bunbury Basalt for Australia/India/Antarctic rift, and Rajmahal Traps for India/Antarctic rift) (Sushchevskaya *et al.*, 2009).

Although the timing and pattern of break-up of Gondwana is relatively well-understood for the large continental fragments, it is less so for the smaller microplates (Storey *et al.*, 1988). During the Jurassic all the major fragments of the supercontinent (namely South America, Africa, India, Australia, New Zealand) were attached at some time to Antarctica. Therefore, in order to understand biogeographic connections across the Southern Hemisphere through time, an insight into the development and palaeogeographical history of Antarctica is essential. The present day Antarctic region can be divided into two areas that are now separated by the West Antarctic Rift System (initiated in the late Early Cretaceous; see Chapter 6). East Antarctica is a craton whilst West Antarctica is composed of a number of microcontinental fragments (namely the Antarctic Peninsula, Thurston Island-Eights Land, Marie Byrd Land, Ellsworth-Whitmore Mountains, Haag Nunataks), which have had a long and complicated history (Storey *et al.*, 1988), and represent one of the major uncertainties in Gondwana reconstructions. Such changes in palaeogeography had implications not only for the Jurassic palaeoclimate, but also for the biota in the surrounding sea and across the land.

Global palaeoclimates

The warm, wet greenhouse conditions that prevailed during the Jurassic (Frakes, Francis and Syktus, 1992; Huber, MacLeod and Wing, 2000) were preceded by the Triassic–Jurassic transition, which marks one of the five largest extinctions in the Earth history (Raup and Sepkoski, 1982; Sepkoski, 1996) (Figure 4.3). This biotic crisis is observed both in the sea and on land. In the oceans profound changes occurred that included the demise of coral reefs (e.g. Kiessling *et al.*, 2007, 2009) and extinction of pelagic groups (Thorne, Ruta and Benton, 2011). This extinction event paved the way for radiations within other groups (e.g. sharks and rays; Kriwet, Kiessling and Klug, 2009). In the terrestrial realm, the characteristic Triassic floras (see Chapter 4) were replaced by Jurassic floras rich in bennettites, conifers and ferns. These biotic changes at the Triassic–Jurassic boundary are thought to be the result of marked climate shifts resulting from geologically rapid global warming events.

Geologically rapid climate changes require some unusual trigger resulting in a major input of greenhouse gases into the atmosphere (Twitchett, 2006). Indeed the climatic changes seen at the Triassic–Jurassic transition are marked by evident shifts in the carbon isotope record and atmospheric CO_2 is estimated to have risen from 600 to 2100–2400 ppmv across the Triassic–Jurassic boundary (McElwain, Beerling and Woodward, 1999). This increase in CO_2 is inferred to have led to a 3–4 °C warming effect relative to the latest Triassic, and was accompanied by an increase in lightning activity resulting in prolific wildfires (Belcher *et al.*, 2010). The source of the CO_2 is thought to have been volcanic outpouring of the Central Atlantic Magmatic Province (CAMP), which pumped vast amounts of CO_2 into the oceans and atmosphere. Even though the amount of CO_2 generated by the CAMP was not enough to produce the inferred warming of 3–4 °C (Beerling and Berner, 2002), it would have acted as a driver for raising the oceanic bottom water temperatures. This would have led to the dissociation of gas hydrates in the shallow shelf seas and/or in high latitude permafrost, both resulting in methane release, a gas several times more potent than CO_2 in terms of greenhouse effect (e.g. Pálfy *et al.*, 2001; Shindell *et al.*, 2009). This in turn exacerbated the subsequent global warming leading to further temperature rise, further methane release and a runaway greenhouse effect (e.g. Benton and Twitchett, 2003). With this runaway greenhouse state, the Earth entered a hothouse (hyperthermal) state during the early Middle Jurassic (Toarcian) (Kidder and Worsley, 2010). Following this hothouse condition was a period of global cooling as the Earth returned to its normal greenhouse state. The Early Jurassic climate is thus regarded as a period of warmth, with average global temperatures significantly warmer than today, and much wetter (e.g. Huber, MacLeod and Wing, 2000).

The greenhouse world of the Early Jurassic is evidenced by geological proxies that testify to relative warmth extending further polewards when compared with current day distributions. For example, Jurassic coral reefs extended by at least 10° of the latitude (Leinfelder *et al.*, 2002) of their present day distribution, which equates to approximately 30° either side of the equator. On land indications for a warmer world come from widespread evidence for bauxites, resulting from extreme leaching of soils in a warm, wet climate (Price, Valdes and Sellwood, 1997), as well as extensive aeolian desert deposits such as sediments of the Clarens Formation in Southern Africa (Bordy and Catuneanu, 2002; Holzförster, 2007) and the Navajo Formation in North America (e.g. Kocurek, 2003; Porter, 1987). This is in addition to interpretations based on isotopes from palaeosols (Retallack, 2009). Moreover other proxies, based on palaeotemperature estimates of oxygen isotopes calculated from marine biogenic carbonates (e.g. Gómez *et al.*, 2009; Price and Page, 2008), can also provide evidence for Jurassic warmth but interpretations of isotope data can be confounded by other environmental conditions (see Bice, Huber and Norris, 2003; Price and Sellwood, 1997), poor preservation, diagenetic alteration and other factors leading to possible overestimates of warmth from using this method (McArthur *et al.*, 2007; Pearson *et al.*, 2001; Williams *et al.*, 2005). Nevertheless, Dera *et al.* (2011) have recently demonstrated two significant negative $\delta^{18}O$ excursions (of –2‰), one during the Early Jurassic (Toarcian) and a second in the Late Jurassic (Oxfordian–early Tithonian).

These two $\delta^{18}O$ excursions, separated by steady high values of 0‰ during the Middle Jurassic, coincide with intensive volcanism in large igneous provinces (Jourdan *et al.*, 2007; Wang *et al.*, 2006). This supports a strong influence of repeated volcanic pulses on pCO_2, temperature and polar ice cap volumes over projected periods (Dera *et al.*, 2011). On shorter time scales, thirteen relatively rapid (0.5–1 Ma) and significant warming and cooling events have now been identified between the Early Jurassic (Pliensbachian) and the Late Jurassic (Tithonian). These are associated with fluctuations in greenhouse gas concentrations [e.g. destabilisation of methane gas hydrates (Jenkyns, 2003) or volcanic activity (Dera *et al.*,

2011), with all causing warming] but the actual mechanisms are still under debate (Dera *et al.*, 2011). A review of the evidence for polar ice in the Mesozoic suggests that glacial conditions existed in the Northern Hemisphere (Price, 1999), but it is unclear if this was the case in the south. Indeed the case for polar glaciation has been contested by some researchers, and some of the sedimentological sequences that had been interpreted as glacially derived have been shown to be formed by other processes (e.g. Mawson Formation; Elliot, 2000; Elliot and Fleming, 2008). According to some models high summer temperatures at the poles would have prevented the formation of long-term, large-scale ice sheets, a conclusion supported by the lack of direct geological evidence including more negative oxygen isotopic ratios of marine shells relative to later times (Sellwood and Valdes, 2008), even though dynamic polar ice caps may well have been present (Dera *et al.*, 2011).

Throughout the Early Jurassic, although the Earth was in a greenhouse mode the climate still underwent a number of warming and cooling phases that may have resulted in shifting climatic belts, coupled with several episodes of extreme warming that resulted in hothouse conditions (*sensu* Kidder and Worsley, 2010). More recent synthetic studies have enhanced this view suggesting that the Jurassic climate may have been more complex with multiscale perturbations that included both short-term and longer-term changes (Dera *et al.*, 2011). Yet as a broad generalisation, the climate of the Early Jurassic was strongly monsoonal across eastern Pangea as a result of the continental effects of the extensive low latitude landmass (e.g. Kutzbach and Gallimore, 1989). The equatorial region was characterised by summer wet conditions and to both the north and south of this zone were deserts with extensive sand seas (Rees, Ziegler and Valdes, 2000). Further out, a narrow belt of winter wet climate separated the low latitudes from the warm temperate mid and high latitudes, and in turn these warm temperate belts were separate from the cool temperate polar latitudes (Rees, Ziegler and Valdes, 2000). This pattern of climatic belts persisted throughout the Jurassic.

This greenhouse state lasted until the Pliensbachian–Toarcian (~183 Ma) when several warm events culminated in the Toarcian Ocean Anoxic Event (T-OAE; Suan *et al.*, 2010) – a warm excursion that lasted 900 kyr (Suan *et al.*, 2008). This period of Earth history was characterised by a number of events, namely the first negative $\delta^{18}O$ excursion (Dera *et al.*, 2011), a negative carbon isotope excursion with an associated ocean anoxic event (OAE, Suan *et al.*, 2010), mass extinction and global warming estimated to be in the order of 10 °C (Suan *et al.*, 2010). The causal mechanism for the T-OAE is still under debate (see Beerling and Brentall, 2007), but suggestions include the release of methane clathrates (Hesselbo *et al.*, 2000; Kemp *et al.*, 2005) and release of carbon dioxide associated with the extensive Karoo-Ferrar Large Igneous Province that intruded into the extensive Permian coal sequences in Gondwana (Kidder and Worsley, 2010; McElwain, Wade-Murphy and Hesselbo, 2005). However, Dera *et al.* (2011) suggest that the destabilisation of clathrates may represent a consequence of previous warming, rather than its driving force.

Following the early Toarcian warm excursion the Earth once again returned to greenhouse conditions. From the late Toarcian through the early Middle Jurassic (Aalenian and into the Bajocian), the climate continued to fluctuate and although these fluctuations saw several warmer periods (Gómez *et al.*, 2009), none triggered global ocean anoxia. From the

Bajocian into the Bathonian (~172–168 Ma) there was a distinct warming trend followed by a subsequent cooling through the later Bathonian and into the Callovian (Malchus and Steuber, 2002) (~165–161 Ma). During this episode of volcanic quiescence, high latitude glacial deposits such as tillites, dropstones and glendonites have been documented (Price, 1999) coinciding with the less negative $\delta^{18}O$ values all indicating possible waxing of polar ice caps over long periods (Dera *et al.*, 2011).

Another warming phase followed in the Late Jurassic (Oxfordian) and reached a peak in middle to late Oxfordian where the second negative $\delta^{18}O$ excursion was identified (Dera *et al.*, 2011). A period of cooling then followed through the Kimmeridgian and Tithonian (155–145 Ma) to the end of the Jurassic (see Weissert and Ebra, 2004).

Superimposed on these general warming–cooling trends are short-term fluctuations. Indeed, detailed studies of carbonate platforms in both the Early (Bonis, Ruhl and Kürschner, 2010; Ruhl *et al.*, 2010) and Late (Strasser, 2007; Védrine and Strasser, 2009; Weedon, Coe and Gallois, 2004) Jurassic have revealed Milankovitch scale cyclicity, suggesting orbital forcing of climate and sedimentation. Today, orbital forcing triggers glacial–interglacial cycles that drive sea level change and influence sedimentation patterns. However, the scale of sea level change in the Jurassic is small and has led to the hypothesis that the Jurassic was relatively ice-free (Husinec and Read, 2007). Any orbitally forced sea level change that did take place can be attributed to thermal expansion of the ocean and influences on precipitation that locked water up in rift basin lakes, aquifers and mountain glaciers (Jacobs and Sahagian, 1993). Nevertheless, the idea of a relatively ice-free world (e.g. Husinec and Read, 2007) implies some high latitude glaciation. Indeed some researchers have interpreted high latitude Jurassic sequences as ice-derived (Price, 1999), and this has been supported by climate model studies that suggest extreme cold in the polar regions (Sellwood, Valdes and Price, 2000). However, some of the high latitude sequences initially thought to have been glacial have since been demonstrated to have had quite a different origin. For example, the Jurassic Mawson Formation in the Transantarctic Mountains was originally thought to be a glacial diamictite (Gunn and Warren, 1962), but is in fact a breccia formed by phreatomagmatic eruptions (Elliot and Fleming, 2008).

In spite of variations in the climate, the Jurassic was characterised by very strong latitudinal zonation in both temperature and precipitation. Rainfall was convective in character and focused largely under the Intertropical Convergence Zone over the oceans, resulting in major desert expanses on the continents (e.g. Sellwood and Valdes, 2006, 2008). This strong climatic zonation is reflected in the distribution of the biota, including both animals (e.g. dinosaurs; Rees *et al.*, 2004) and vegetation, including specific flora types (Rees, Ziegler and Valdes, 2000; Vakhrameev, 1991) and supported by models of Jurassic biomes (e.g. Sellwood and Valdes, 2008).

Global palaeovegetation

During the Jurassic several distinct global floral associations have been recognised (Vakhrameev, 1991) based on associations of key taxa such as ferns (*Dictyophyllum*),

Bennettitales (*Ptilophyllum*) and Ginkgoales. The close relationship between the plant distributions and interpreted climate can be also be related to leaf morphology (Rees, Ziegler and Valdes, 2000), with microphyllous taxa often with thick cuticles, occurring in drier and more arid regions, whilst large-leafed taxa occur in higher latitudes that are inferred to have been wetter. Using this variation in leaf physigonomic features, Rees, Ziegler and Valdes (2000) suggested the presence of several different biomes. This work is supported by sedimentological evidence and the distribution of climate sensitive sedimentary strata (Chandler, Rind and Ruedy, 1992).

Jurassic vegetation biomes (Figure 5.1) are generally thought to have consisted of an equatorial tropical biome but there are few fossil floras from this region to support this interpretation (Rees, Ziegler and Valdes, 2000). The low equatorial latitudes were dominated by desert systems (Rees, Ziegler and Valdes, 2000) with large sand-seas (Bordy and Catuneanu, 2002; Eriksson, 1986; Holzförster, 2007; Kocurek, 2003; Porter, 1987) and thus presumed limited vegetation. Desert oases, created by upwelling ground water that formed carbonate spring deposits, supported coniferous forests within these arid equatorial environments (Parrish and Falcon-Lang, 2007). A narrow belt of winter wet climate either side of the equatorial deserts served to separate the low latitude ecosystems from the warm temperate mid and high latitudes that were dominated by large-leafed vegetation (Rees, Ziegler and Valdes, 2000).

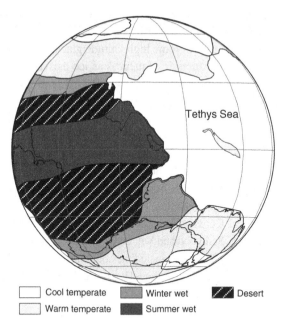

Figure 5.1 Simplified diagram of inferred Jurassic biomes. Biome boundaries adapted from Rees, Ziegler and Valdes (2000) and based on morphological analysis of leaf characters. Note that recent coupled global general circulation and vegetation models have provided a finer level of detail and more inferred biomes (e.g. Sellwood and Valdes, 2008) but testing these biomes against palaeontological data requires a greater number of fossil localities and detail of the floras than currently available.

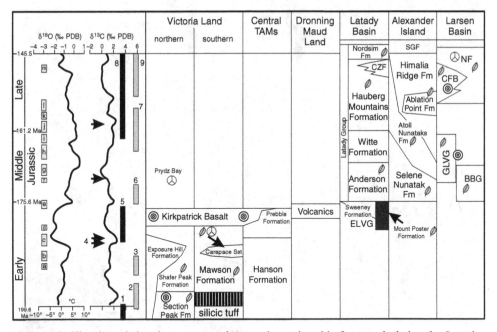

Figure 5.2 Climatic variation, key events and Antarctic stratigraphic framework during the Jurassic. Abbreviations as follows: ELVG = Ellsworth Land Volcanic Group; CZF = Cape Zumberge Formation; BBG = Botany Bay Group; GLVG = Graham Land Volcanic Group; CFB = Cape Frammes Beds; NF = Nordenskjöld Formation; SGF = Spartan Glacier Formation. Note that there is an age discrepancy between the palynological age (late Sinemurian to earliest Pliensbachian; ~191 Ma) of the Carapace Sandstone (Sst) Formation and Kirkpatrick Basalts (Pliensbachian, ~189 Ma; Ribecai, 2007) and the radiometric ages of 177 Ma (Fleming *et al.*, 1997; Heimann *et al.*, 1994). a–m, times of rapid climate variation events identified as important in the climate history of the Jurassic. Arrows indicate periods of rapid negative carbon isotope excursions not represented in the smoothed curve. 1 = eruption of Central Atlantic Magmatic Province with injection of CO_2 into the atmosphere. 2 = opening of oceanic gateway Viking Corridor. 3 = opening of the oceanic gateway between Gondwana and Laurasia. 4 = early Toarcian Oceanic Anoxic Event. 5 = Eruption of Ferrar/Karoo magmas associated with the early phase of Gondwana break-up and thought to have injected large quantities of CO_2 into the atmosphere through both the magma and metamorphism of coal-rich sequences. 6 = opening of central Atlantic Ocean. 7 = closure of the Mongol-Okhotsk Ocean. 8 = northeast Asian orogeny and magmatism. 9 = opening of the Indian Ocean. Black bars represent eruptive periods, grey bars represent tectonic events. Climatic curves and events adapted from Dera *et al.* (2011) using belemnite $\delta^{18}O$ and $\delta^{13}C$ data. Note Botany Bay Group also represents a number of isolated outcrops with sequences referred to the Tower Peak Formation, Camp Hill Formation, Mount Flora Formation that are in turn overlain by volcanics of the Graham Land Volcanic Group (e.g. Kenny Glacier Formation, Mapple Formation; see Riley *et al.*, 2010). Stratigraphic summaries derived from Bomfleur, Pott and Kerp (2011), Elliot and Fleming (2008), Hunter and Cantrill (2006), Luttinen, Leat and Furnes (2010), Miller and Macdonald (2004), Riley *et al.* (2010). ∅ = leaf macroflora, ◎ = wood flora, Ⓐ = palynoflora.

These regions had a distinct biota that supported numerous dinosaurs but were largely confined to the mid-latitude regions (Rees *et al.*, 2004). The warm temperate mid-latitude vegetation belts were in turn separated from the cool temperate polar latitudes with their own distinct flora (Rees, Ziegler and Valdes, 2000).

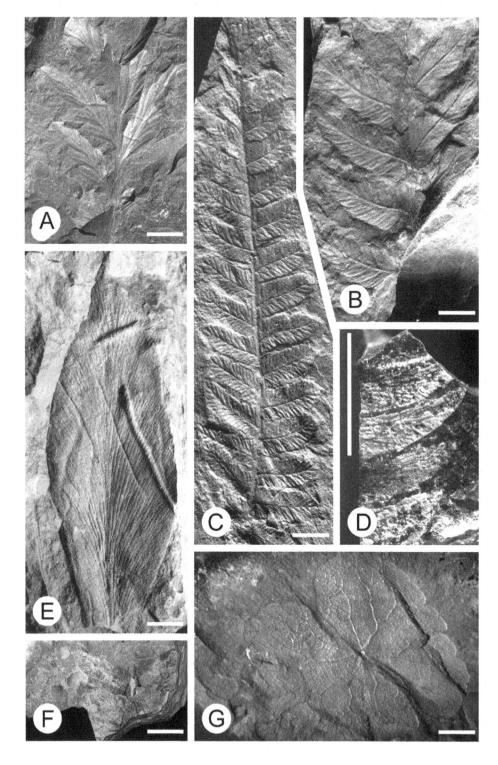

Early to Middle Jurassic floras and environments of Antarctica

The Jurassic plant fossil record of Gondwana is generally meagre and has thus rendered phytogeographic interpretations difficult (Bomfleur, Pott and Kerp, 2011). Fossil material preserved on Antarctica, however, is helping us to understand more fully the homogeneity of the vegetation that grew across southern Gondwana at this time of environmental instability (Bomfleur, Pott and Kerp, 2011). The record of Early to Middle Jurassic vegetation of Antarctica is limited to macrofossils from three sites. Two come from the Transantarctic Mountains, the first in southern Victoria Land (Garland *et al.*, 2007; Plumstead, 1962; Townrow, 1967a, 1967b) and more recently a second from northern Victoria Land (Bomfleur and Kerp, 2010; Bomfleur, Pott and Kerp, 2011; Bomfleur *et al.*, 2007; Yao, Taylor and Taylor, 1991). A third flora has been documented from the Antarctic Peninsula region on West Antarctica (e.g. Halle, 1913; Rees, 1990, 1993a, 1993b; Rees and Cleal, 1993, 2004).

East Antarctica

In the Transantarctic Mountains, the Early Jurassic marked the end of the Beacon Supergroup sedimentation that took place in the long-lived (Devonian to Jurassic) foreland basin (Collinson *et al.*, 1994). Silicic tuffaceous rocks and interbedded sandstones of the Hanson Formation (Elliot, 1996, 2000) contain an Early to middle Early Jurassic vertebrate fauna (Hammer and Hickerson, 1994; Hammer and Smith, 2008; Smith and Pol, 2007; Smith *et al.*, 2007) and rafts of the Hanson Formation, reworked into the Middle Jurassic Mawson Formation, have yielded a limited macroflora. This flora testifies to the presence of a vegetation comprising ferns (*Cladophlebis oblonga*), conifers (*Pagiophyllum, Nothodacrium*) and marchantialean liverworts (*Marchantites mawsonii*) dominating the region during the middle Early Jurassic.

The widespread magmatism that marked the beginning of Gondwana break-up and dispersal resulted in both the emplacement of large dolerite sills (Ferrar Group) in the Transantarctic Mountains at 183 Ma (Fleming *et al.*, 1997) and widespread volcanism at the surface (Kirkpatrick Basalt, 183 Ma; Heimann *et al.*, 1994). This event was synchronous with the emplacement of the Karoo Dolerites in Southern Africa (Duncan *et al.*, 1997;

Caption for Figure 5.3 Antarctic Jurassic plant fossils. **A**, *Komlopteris indica*, S 021010_04, Middle Jurassic, Mount Flora Formation, Botany Bay Group, Hope Bay, northern Antarctic Peninsula. **B**, *Cladophlebis antarctica*, BAS KG 4703.2, Middle Jurassic, Selene Nunatak Formation, Zebra Ridge, Alexander Island. **C**, *Cladophlebis oblonga* BAS KG 4703.35, Middle Jurassic, Selene Nunatak Formation, Zebra Ridge, Alexander Island. **D**, *Dictyozamites* cf. *minisculus*, BAS R. 7801.1, Nordsim Formation, Latady Group, Cantrill Nunataks, Ellsworth Land, southern Antarctic Peninsula. **E**, *Sagenopteris nilssoniana*, BAS KG. 4703.3. Middle Jurassic, Selene Nunatak Formation, Zebra Ridge, Alexander Island. **F**, *Otozamites antarcticus* with numerous conchostrachans, B697_3, Carapace Sandstone, Allan Hills, southern Victoria Land. **G**, *Hausmannia papilio*, V63620, Middle Jurassic, Mount Flora Formation, Botany Bay Group, Hope Bay, northern Antarctic Peninsula. Scale bars = 1 cm.

Encarnación *et al.*, 1996) and tholeitic volcanism in Dronning Maud Land. Indeed, basaltic rocks of early Middle Jurassic age (183 Ma) extend from Dronning Maud Land (e.g. Faure, Bowman and Elliot, 1979) to the Transantarctic Mountains (e.g. Barrett, Elliot and Lindsay, 1986), into northern Victoria Land (Elliot, Haban and Siders, 1986), and even into Australia (Bromfield *et al.*, 2007; Hergt and Brauns, 2001; Tingey, 1991). Interbedded with these volcanic units are strata that record deposition in lacustrine, fluvial and terrestrial environments (Ball *et al.*, 1979; Tasch, 1987).

In northern Victoria Land, lava flows in the Mesa Range entombed coniferous forests with individual trees becoming buried and silicified (Jefferson, Siders and Haban, 1983). These trees range from 0.5 to 1 m in diameter with preserved lengths of over four metres that attest to substantial forests growing across this area. Diameters of one metre would imply that trees grew up to 64.5 m in height (cf. allometric approach of Niklas, 1994) and these have been identified as *Protocupressinoxylon* (Jefferson, Siders and Haban, 1983) – a wood type characteristic of cheirolepidiacean conifers (Francis, 1983). More recently permineralised dipterid fern (*Clathropteris*) frond material has also been described from this region (Bomfleur and Kerp, 2010) together with a small macroflora including *Isoetites, Equisetites, Matonidium, Coniopteris murrayana, C. hymenophylloides, Spiropteris*, Bennettitales (*Otozamites linearis, O. sanctae-crucis, Zamites* spp., *Cycadolepis*) as well as other conifers (*Schizolepis*, cf. *Allocladus*, cf. *Elatocladus*, cf. *Pagiophyllum*) (Bomfleur and Kerp, 2010; Bomfleur, Pott and Kerp, 2011; Bomfleur *et al.*, 2007).

Unlike most floras from the Antarctic, the deposits within the Shafer Peak Formation in northern Victoria Land are well-preserved and have yielded cuticles. The cuticles are papillate and stomatal distributions are often hypostomatic — possible adaptations to the warm climate. The dominant plants occurring in the deposits of the Shafer Peak Formation that can be correlated with parts of the Hanson Formation in the central Transantarctic Mountains (Schöner *et al.*, 2007), have been identified as belonging to sphenopsids (*Equisetites*), ferns (*Coniopteris*) and bennettites (*Otozamites*), and in this respect the flora was similar to Jurassic floras growing elsewhere in Antarctica. In fact, small-leafed bennettites were generally found in most floras from this part of the world during the Jurassic, although in West Antarctica at this time some large-leafed (>50 cm) bennettites (*Zamites*) have been found.

Palynofloras from Section Peak in northern Victoria Land are dominated by *Classopollis* (53–64%) with lesser amounts of *Exesipollenites* (10–16%) and bisaccate coniferous pollen comprising up to 9% of the flora (Norris, 1965). The rest of the assemblage is composed of bryophytes (up to 2.6%), ferns (<5%) and other seed plants. Interestingly, relatively high levels of monosulcate type pollen are recorded (~5% abundance) that probably reflect a higher contribution of bennettitaleans present in the vegetation of this area. Given the considerable variation in the abundance of different taxa between samples from this locality, the local vegetation of this area was probably fairly heterogeneous. Nevertheless generalisations can be made regarding the prevailing vegetation: the microfloras all suggest the vegetation was dominated by Cheirolepidiaceae and Araucariaceae conifers with an understorey of bryophytes, ferns and other seed plants. The only ferns identified to date are dipterids, a group that appears to have been common and relatively diverse in Early and

Middle Jurassic assemblages of Antarctica (Rees, 1993a). The palynoflora at Section Peak can be compared with palynofloras from a similar stratigraphic position at Carapace Nunatak, located in southern Victoria Land, which also reveal an assemblage dominated by *Classopollis* (22–61%) and *Exesipollenites* (1.5–51%), but also dominant is *Araucariacites* (1.8–18%) (Ribecai, 2007; Shang, 1997; Tasch, 1977; Tasch and Lammons, 1978). Bisaccate grains account for ~7% of the abundance with the rest of the flora being composed of a sparse cryptogam assemblage.

In southern Victoria Land, leaf floras from the Carapace Sandstone, a unit that underlies basaltic flows (Kirkpatrick Basalt equivalents), contain benettitaleans (*Otozamites*) and conifers (*Brachyphyllum, Pagiophyllum, Nothodacrium, Elatocladus*) (Plumstead, 1962, 1964; Townrow, 1967a, 1967b). In this area, as in northern Victoria Land, volcanic eruptions entomb standing trees to preserve a history of the forest vegetation. Individual trees have been recorded from other localities such as the Coomb Hills (Garland *et al.*, 2007) and silicified dipterid ferns (*Polyphacelus stormensis*) have been recovered from Storm Peak (Yao, Taylor and Taylor, 1991). These ferns occur in interflow sediments and represent wet environments that were colonised between volcanic eruptions. Today dipterid ferns are known colonisers of open disturbed sites, and may provide an insight into the role played by the dipterid ferns during this time.

Other records of early to middle Jurassic strata within East Antarctica come from the Mac. Robertson Shelf. Seismic interpretation suggests initial (early–middle Jurassic) accumulation in an intracratonic sag, with rifting between India and Antarctica developing later in Callovian to Oxfordian times (~165–160 Ma) — rifting that would continue on until the middle of the Lower Cretaceous (Truswell, Dettmann and O'Brien, 1999). Palynomorphs recovered from the oldest strata indicate a late early to early middle Jurassic age. The microflora is wholly terrestrial and is made up of conifers, including *Araucariacites* (42–46%), *Classopollis* (17–19%) and *Callialasporites turbatus* (12–19%), with a minor component of bisaccate pollen belonging to *Alisporites*. A diversity of pteridophyte spores occurring only in low frequencies (*Cyathidites, Baculatisporites, Gleicheniidites*) has also been recorded along with lycopohytes and bryophytes. Slightly younger microfloras of late Middle to early Late Jurassic (Callovian–Oxfordian) also contain *Araucariacites* (65%) and *Classopollis* (8%), but with a diverse cryptogam flora of bryophytes (four species), lycopods (two species) and ferns (12 species). Compositionally these floras differ from those in the Transantarctic Mountains by the higher araucarian and lower *Classopollis* (cheirolepidiacean) conifer percentages that are thought to reflect regional differences in the vegetation across Antarctica (Truswell, Dettmann and O'Brien, 1999).

Overall in East Antarctica, along the present day Transantarctic Mountains, the subdued topography was uplifted and disturbed by extensive phreatomagmatic eruptions that in turn periodically disturbed the vegetation. Few localities are known but they record sedimentation between eruptions in a range of environments including lakes, high energy fluvial systems with air fall deposits. These conditions would have led to a patchwork of successional types. The coniferous trees belong to a range of families including the Araucariaceae, Podocarpaceae and Cheirolepidiaceae, although the latter family was the most common.

Cheirolepidiaceae (as indicated by *Classopollis*) is regarded as an indicator of warm conditions (Alvin, 1982; Francis, 1983; Watson, 1988) as it is often found in association with saline environments and evaporites (e.g. Purbeck; Francis, 1983). Only a limited diversity of understorey plants has been documented, with those present (e.g. Bomfleur, Pott and Kerp, 2011) indicating an understorey of dipterid and matoniaceous ferns, sphenophytes and Bennettitales, including large *Zamites*, as evidenced by their leaves. Together the sparse Early to Middle Jurassic floras of East Antarctica point to a humid and warm temperate vegetation, a conclusion supported by the cuticular evidence (Bomfleur, Pott and Kerp, 2011). This is in direct contrast to the climates predicted by GCM simulations that retrodict frigid high latitude regions (e.g. Sellwood and Valdes, 2006, 2008). The discrepancy between GCM models and data for greenhouse worlds has already been highlighted (e.g. Rees, Ziegler and Valdes, 2000; Sellwood and Valdes, 2008). It is not unusual for many GCM simulations, run in times when the world is in a greenhouse or hothouse state, to yield anomalously cold polar regions, suggesting that the models are not yet accurately simulating climate (Sellwood, Valdes and Price, 2000; Sellwood and Valdes, 2006, 2008).

West Antarctica

In West Antarctica, Early and Middle Jurassic sequences in the Antarctic Peninsula area contain non-marine conglomeratic sequences that accumulated in rifted half grabens developed on Permo-Triassic basement. These deposits mark the first stages of Gondwana breakup in this region (Hathway, 2000). Two sedimentary systems are identified: the Larsen Basin to the north, and the Latady Basin to the south, with synrift sedimentation and volcanism being diachronous from south to north.

The earliest recorded sedimentation in the Latady Basin is late Early Jurassic volcanic rocks of the Ellsworth Land Volcanic Group (Figure 5.2). They are estimated to be at least 2 km thick (Rowley, Schmidt and Williams, 1982), but individual outcrops rarely expose more than 600 m of silicic volcanic rocks. These have been referred to the Mount Poster Formation and have been dated at between 167 and 188 Ma from various exposures (Fanning and Laudon, 1999; Riley and Leat, 1999). At Potter Peak a 300 m-thick sequence of laminated mudstones and sandstones is interbedded with basalts and silicic pyroclastics (Hunter *et al.*, 2006), and are regarded as lateral equivalents of the thick silicic volcanics. These are referred to the Sweeney Formation (Figure 5.2) and represent extra-caldera environments including both lacustrine and terrestrial deposits. Soils occur within the Sweeney Formation together with silicified wood and leaf material, including sphenopsids (*Equisetum laterale*) and conifers such as *Brachyphyllum* and *Elatocladus confertus* (Cantrill and Hunter, 2005; Hunter *et al.*, 2006).

The Ellsworth Land Volcanic Group is succeeded by shallow marine and coastal deposits of the Anderson Formation (Aalenian–Bajocian; 175–168 Ma) (Figure 5.2) that also contain plant fragments including horsetails (*Equisetum laterale*), ferns (*Cladophlebis antarctica*, *C. galliantinensis*), bennettites (*Zamites antarcticus*), cycads (*Taeniopteris* sp.), other seed plants (*Archangelskya furcata*) and conifers (*Brachyphyllum, Pagiophyllum feistmanteli, Elatocladus confertus*). Although fragmentary, they point to considerable diversity with many of the same taxa also recovered from Hope Bay in the Larsen Basin (see below). The coastal to shallow marine sediments pass upwards into the more fully marine rocks of the Witte Formation (Callovian) (Figure 5.2), that accumulated in a marine basin with restricted circulation and anoxic conditions, and finally into shelf (Hauberg Mountains Formation, Oxfordian to Tithonian) and outer-shelf (Cape Zumberge Formation, Tithonian) environments (Hunter and Cantrill, 2006).

To the north, the Larsen Basin sequence begins with conglomeratic to fluvial sequences that include the classic Hope Bay Flora originally described by Halle (1913), along with associated macrofloras of the Botany Bay Group (Rees, 1990; Rees and Cleal, 2004) (Figure 5.2). The age of the Hope Bay flora has been variously dated as Early Jurassic (Orlando, 1971), Middle Jurassic (Halle, 1913; Rao, 1953), and even latest Jurassic to Early Cretaceous (Gee, 1989a; Stipanicic and Bonetti, 1970). However, this flora is known to underlie silicic volcanics of the Antarctic Peninsula Volcanic Group that have yielded high precision dates of 172–162 Ma (Pankhurst *et al.*, 2000) (Middle Jurassic). Recent U-Pb studies of detrital zircons have dated this synrift sedimentation at 167 Ma (Hunter *et al.*, 2005) that firmly places the Hope Bay Flora and the Botany Bay Group as pre-early Middle Jurassic in age. Hope Bay equivalent floras are widespread both to the north in the South Orkney Islands (Cantrill, 2000), and south in the Larsen and Latady basins (e.g. Aitkenhead, 1975; del Valle and Núñez, 1988). On the western (fore-arc) side of the Antarctic Peninsula, on Alexander Island, plants that are presumed to be coeval with the Hope Bay Flora occur at Zebra Ridge (D. J. Cantrill, personal observation) and in the southwestern Douglas range (B. C. Storey, personal communication), but plant material from these sites have not been studied to date.

The Early to early Middle Jurassic flora of the Botany Bay Group is dominated by *Archangelskya furcata* (54%), *Cladophlebis antarctica* (52%), *Pagiophyllum* sp. (44%), *Todites williamsonii* (36%), *Elatocladus confertus* (34%), *Pterophyllum ensiformis* (30%) and *Equisetum laterale* (28%) (frequency occurrence data from Rees, 1990). However, a number of other important taxa are present: dipterid ferns are widespread and include *Goepertella, Dictyophyllum* and *Hausmannia* (Rees, 1993a), and seed plants such as *Sagenopteris* (Rees, 1993b) and *Pachypteris* (Rees and Cleal, 2004) characterise this assemblage. Other taxa do occur (e.g. *Cladophlebis*), but relatively infrequently, and they often form the most conspicuous element of the flora in that particular bedding horizon. This suggests considerable spatial heterogeneity in the vegetation. For example, Bennettitales are represented by two genera and six species (Gee, 1989a) but are not particularly abundant; remains of Coniferales, the assumed overstorey component, can be assigned to the Araucariaceae. However, it is unclear which other groups may be present. Unfortunately there are no complementary palynofloras from this sequence to augment the macrofloral information.

The widespread silicic volcanism (Mapple Formation and correlatives; Riley and Leat, 1999) seen in the Antarctic Peninsula forms part of a large igneous province that extends from southern South America through to the Antarctic Peninsula (Pankhurst *et al.*, 2000; Riley and Leat, 1999). In Graham Land the silicic volcanic rocks overlie Botany Bay Group strata at a number of localities and represent a transition from synrift to rift sequences (Hathway, 2000). The volcanism has been dated by high precision techniques as having occurred between 172–162 Ma (Middle Jurassic) and forms part of a later phase of volcanism, the V2 event (Pankhurst *et al.*, 2000). Minor epiclastic rocks are interbedded with the volcanics (Riley and Leat, 1999), and locally these contain abundant, largely broken plant debris including bennettitalean fragments (Riley and Leat, 1999) and carbonised trees (del Valle *et al.*, 1997).

In general, pteridophytes and seed plants (*Archangelskya*, *Pterophyllum*) dominate the Early and Middle Jurassic floras of the Antarctic Peninsula with minor conifers (*Pagiophyllum*, *Araucarites*). Ferns are both abundant, occurring in 78% of the localities examined by Rees (1990), and diverse comprising between 37.5 and 41.4% of the species identified in two separate studies (Gee, 1989a; Rees, 1990). Dipterid ferns are widespread and diverse (three genera, four species) and other seed plants are well-represented (e.g. *Archangelskya*, *Pachypteris*, *Sagenopteris*, *Thinnfeldia*, *Taeniopteris*) (Rees, 1990, 1993a, 1993b; Rees and Cleal, 1993, 2004). Although the macrofloral and microfloral records are sparse, all fossil remains indicate forest vegetation including substantial trees growing in volcanically disturbed environments.

In contrast to the vegetation of East Antarctica, the differences exhibited by the vegetation growing on the Antarctic Peninsula may reflect local habitat dynamics or the environment of preservation. Here mudstone, sandstone and conglomeratic sequences record deposition in fluvial environments (e.g. Botany Bay Group) that, unlike those of East Antarctica, are not dominated by volcanic processes. Volcanic events were, however, still an important factor affecting the palaeoenvironment with large silicic cones producing dense ash clouds resulting in terrestrial (and marine) sequences thicker than those seen in East Antarctica. Soil horizons have also been identified that suggest longer periods of colonisation between events that would have resulted in blanketing the landscape in ash, or the rapid deposition of sediment. Extensive and often large wood fragments in channel deposits and rare standing trees testify to the presence of forests having once grown here (del Valle *et al.*, 1997). Again the forests comprised a range of conifers as evidenced by preserved cone scales of the Araucariaceae and foliage that is likely to be related to Podocarpaceae. The presence of other groups such as the Cheirolepidiaceae cannot be confirmed since subsequent metamorphism has rendered palynofloras absent from the Early and Middle Jurassic deposits. The understorey comprised diverse colonisers of both fresh ground and wet sites, and probably included the sphenophytes. Ferns were dominated by Osmundaceae (*Cladophlebis*, *Todites*) with other fern taxa including probable tree ferns of the Cyatheaceae/Dicksoniaceae, rare Matoniaceae and ferns of uncertain affinities. Ferns such as the Dipteridaceae are present and locally abundant with genera, such as *Dipteris*, today being a primary coloniser of disturbed sites (Cantrill, 1995). Given the sedimentary setting and frequent disturbance, and the abundance of dipterids growing together in particular localities, it suggests a similar situation-role for the dipterids at this time. Understorey seed plants are dominated by *Archangelskya furcata* and cycads, such as *Pterophyllum*, although the Bennettitales are also present. As with nearly all Southern Hemisphere floras described to date, the Bennettitales are generally small-leafed, unlike their counterparts in the Northern Hemisphere.

Late Jurassic–earliest Cretaceous (Valanginian) floras
and environments of Antarctica

From the Jurassic to earliest Cretaceous much of the East Antarctica coast between the eastern edge of George V Coast and the western end of Dronning Maud Land (Figure 1.1) was undergoing rifting, as first India and then Australia rifted from the Antarctic continent. Even though no definite Late Jurassic to earliest Cretaceous macrofloras are known from East Antarctica, recycled palynomorphs in Quaternary glacial deposits indicate strata of this age does occur along the East Antarctic margin (Truswell, 1982, 1987). Truswell, Dettmann and O'Brien (1999) reported *in situ* strata of early Middle Jurassic to Valanginian (Lower Cretaceous) age from offshore Mac. Robertson Land. Two microfloral samples of late Tithonian to early Valanginian and late Tithonian to possibly early Aptian were also identified from off shore Mac. Robertson Land (Truswell, Dettmann and O'Brien, 1999). These assemblages are dominated by *Araucariacites* (40–44%) with subsidiary cryptogam spores (25–34%). The cryptogams mainly comprise pteridophytes (21 species) but include a significant lycophyte- (five species) and bryophyte/hepatophyte component (four species). In one sample the cryptogam spores, although common (34%), are of restricted diversity and accompanied by significant numbers of algal spores. This points to heterogeneity in the vegetation and this sample probably reflects vegetation growing in wetter conditions. The conifers and other gymnosperms make up a further 12 species in the overall assemblage.

In West Antarctica the only recorded macrofloras of late Jurassic to Valanginian age occur in the Antarctic Peninsula. To the west of the Antarctic Peninsula, in the fore-arc region, the Fossil Bluff Group contains marine strata of Late Jurassic to Early Cretaceous age (Ablation Point, Himalia Ridge, and Spartan Glacier formations of the Fossil Bluff Group) (Figure 5.2). Rare plant remains of Kimmeridgian to ?Barremian are known from this succession and are dominated by bennettite fronds (*Ptilophyllum*, *Otozamites*, *Dictyozamites*, *Zamites*) and other gymnosperms (e.g. *Pachypteris*, *Pagiophyllum*, *Brachypyllum*) (D. J. Cantrill, personal observations). However, the marine setting favours the preservation of robust plants and so the absence of delicate material such as fern fronds is not so surprising. Strata are also known from the intra-arc region (e.g. Anchorage Formation, Byers Group), with plant remains not yet being described, but include *Pachypteris*, the conifer *Elatocladus*, and ferns such as *Cladophlebis* (D. J. Cantrill, personal observations).

On the east side of the Antarctic Peninsula, in the back-arc region, Late Jurassic non-marine to shallow marine strata occur from the Orville Coast in the south to the Nordenskjöld Coast in the north. The most southerly strata are referred to the Latady Group (Hunter *et al.*, 2006) and range from Bajocian to Tithonian in age (Crame, 1983; Quilty, 1977, 1983; Thomson, 1983). Within the Hauberg Mountains Formation of the Latady Group (Figure 5.2), plant fossils are present and represented by *Elatocladus planus* and *Cladophlebis antarctica* (Gee, 1989b). The uppermost unit in the Latady Group is a terrestrial sequence (Nordsim Formation) that has been dated as latest Jurassic or earliest Cretaceous (Hunter, Cantrill and Flowerdew, 2006). The juxtaposition of this unit with the deep marine Cape Zumberge

Formation suggests emergence of the Antarctic Peninsula Volcanic Arc close to the basin margin in the south, with rapid erosion and progadation of deltaic sediments over the shallow marine Latady Basin. This may, in part, be due to the failure of the Weddell Sea to undergo sea floor spreading combined with continued uplift of the arc, and it may explain the lack of younger sediments when compared with the more northerly Larsen Basin – a basin that contains a Cretaceous to Eocene sequence.

The flora preserved in the Nordsim Formation is relatively diverse with sixteen taxa (Cantrill and Hunter, 2005) including sphenopyhtes (*Equisetum laterale*), pteridophytes (*Cladophlebis, Microphyllopteris, Coniopteris*), small-leafed Bennettitales (*Otozamites boolensis, O. rowleyi, O. sanctae-crucis*) and larger forms (*Ptilophyllum, Zamites, Dictyozamites*). Conifers (*Brachyphyllum, Pagiophyllum, Elatocladus*) and several other seed plants (*Taeniopteris, Archangelskya, Pachypteris*) are also present.

Further north, upper Latady Group equivalents are exposed at Mount Hill (Meneilly *et al.*, 1987), Jason Peninsula (Riley *et al.*, 1997) and also possibly on the Sobral Peninsula, with deeper water equivalents including the Nordenskjöld Formation further north. At Jason Peninsula the Bennettitales *Otozamites* and *Williamsonia* have been described (Riley *et al.*, 1997). While further north, *Otozamites saxatilis, Pachypteris* and *Ptilophyllum* are present in the Nordenskjöld Formation and equivalents (del Valle and Núñez, 1988; Whitham and Doyle, 1989). The Late Jurassic to Valanginian record can be supplemented by information from palynofloras. No Late Jurassic terrestrial microfloras have been described but a well-documented Berriasian to mid-Valanginian palynoflora occurs in the lower part of the Byers Group (Duane, 1996). This flora is numerically dominated by pteridophytes mainly *Cyathidites, Biretrisporites* and *Ruffordiaspora*. Bryophytic grains are also common making up to 7% of the terrestrial component. Conifers are largely composed of bisaccate grains (11–35%) with minor *Classopollis* (0–5%).

In general, plant material from the Late Jurassic to earliest Cretaceous interval is poorly known for the Antarctic Peninsula region. However, considerable potential exists for new discoveries of important floras, particularly in the Latady Formation. Since most floras from this time interval tend to occur in marine strata, there is probably a strong preservational bias towards the more robust elements. In spite of this some generalisations can be made: the vegetation is largely composed of bennettites, conifers and ferns but does include other seed plants (such as *Pachypteris*). Groups that were important in Early and Middle Jurassic floras, such as the dipteridaceous ferns, are rare (only *Hausmannia* is present) and *Sagenopteris* is apparently absent. *Classopollis* declines substantially in the palynofloras and is only common in marine strata (frequency <5%; Duane, 1996).

The Late Jurassic to earliest Cretaceous vegetation of the Antarctic is similar to the early and Middle Jurassic floras, with certain groups such as *Sagenopteris*, the dipterid ferns *Goepertella* and *Dictyophyllum* all disappearing in this interval. Similar successions of floras are seen in Australia with Early Jurassic floras containing these groups (e.g. Jansson *et al.*, 2008) followed by their disappearance in either the Middle or Late Jurassic. *Cladophlebis* (Osmundaceae), however, remains important in Late Jurassic Antarctic floras. The major change in the vegetation at this time is marked by taxonomic turnover in Bennettitales with the appearance of a number of new species. These are all characterised by small leaves with small leaflets, a feature common to most southern high latitude floras including those of southern Australia (Douglas, 1969; McLoughlin and Pott, 2009).

Summary

- Information regarding the Jurassic vegetation of Antarctica is limited due to changing sedimentation patterns associated with the altered tectonic regime and the break-up of Gondwana. From the limited evidence available, the vegetation of the Jurassic was similar to that observed elsewhere in Gondwana.

- Although some climate models suggest frigid conditions prevailed over the Antarctic continent there is no evidence of this from the Early and Middle Jurassic floras, which consisted of substantial forests with a diversity of taxa. The greatest influence on the vegetation was environmental dynamics: a shift from passive margin tectonics to those dominated by rifting, uplift and the development of high energy fluvial systems in the Antarctic Peninsula region and the development of explosive, but small scale, volcanism and associated flood basalts in the Transantarctic Mountains. The burial of the pre-existing topography by basalt flows entombed, and silicified, trees and created shallow lake environments that were later colonised. Cheirolepidiaceae was a dominant element in the overstorey, whilst ferns and bennettites formed the understorey.

- Late Jurassic floras are similar to those of the Early and Middle Jurassic but with subtle differences. Forests were still present as indicated by rare wood but the composition differed. Cheirolepidiaceae became less important while the Araucariaceae and Podocarpaceae increased. Small-leafed bennettites became more important and replaced their larger-leaved relatives of the Early and Middle Jurassic. Ferns belonging to the Dipteridaceae also decreased with the disappearance of *Goepertella* and *Dictyophyllum*.

References

Aitkenhead, N. (1975). The geology of the Duse Bay – Larsen Inlet area, north-east Graham Land. *British Antarctic Survey Scientific Reports*, **51**, 1–62.
Alvin, K. L. (1982). Cheirolepidiaceae: biology, structure and paleoecology. *Review of Palaeobotany and Palynology*, **37**, 71–98.
Arias, C. (2006). Northern and Southern Hemispheres ostracod palaeobiogeography during the Early Jurassic: possible migration routes. *Palaeogeography, Palaeoclimatology, Palaeoecology*, **233**, 63–95.
Arias, C. (2008). Palaeoceanography and biogeography in the Early Jurassic Panthalassa and Tethys Oceans. *Gondwana Research*, **14**, 306–315.
Ball, H. W., Borns Jr, H. W., Hall, B. A., *et al.* (1979). Biota, age, and significance of lake deposits, Carapace Nunatak, Victoria Land, Antarctica. In *Fourth International Gondwana Symposium: papers 1*, eds B. Laskar and C. S. R. Rao. Delhi: Hindustan Publishing Corporation, pp. 166–175.
Barrett, P. J., Elliot, D. H. and Lindsay, J. F. (1986). The Beacon Supergroup (Devonian to Triassic) and the Ferrar Group (Jurassic) in the Beardmore Glacier area, Antarctica. In

Geology of the Central Transantarctic Mountains, Antarctic Research Series, **36**, eds M. D. Turner and J. F. Splettstoesser. Washington: American Geophysical Union, pp. 339–428.

Beerling, D. J. and Berner, R. A. (2002). Biogeochemical constraints on the Triassic-Jurassic boundary carbon cycle event. *Global Biogeochemical Cycles*, **16**, 1036, doi:10.1029/2001GB001637.

Beerling, D. J. and Brentall, S. J. (2007). Numerical evaluation of mechanisms driving Early Jurassic changes in global carbon cycling. *Geology*, **35**, 247–250.

Belcher, C. M., Mander, L., Rein, G., *et al.* (2010). Increased fire activity at the Triassic/Jurassic boundary in Greenland due to climate-driven floral change. *Nature Geoscience*, **3**, 426–429.

Benton, M. J. and Twitchett, R. J. (2003). How to kill (almost) all life: the end-Permian extinction event. *Trends in Ecology and Evolution*, **18**, 358–365.

Bice, K. L., Huber, B. T. and Norris, R. D. (2003). Extreme polar warmth during the Cretaceous greenhouse? Paradox of the late Turonian delta $\delta^{18}O$ record at Deep Sea Drilling Project Site 511. *Paleoceanography*, **18**, 1031, 1–11.

Bomfleur, B. and Kerp, H. (2010). The first record of the dipterid fern leaf *Clathropteris* Brongniart from Antarctica and its relations to *Polyphacelus stormensis* Yao, Taylor and Taylor. *Review of Palaeobotany and Palynology*, **160**, 143–153.

Bomfleur, B., Schneider, J., Schöner, R., Viereck-Götte, L. and Kerp, H. (2007). Exceptionally well-preserved Triassic and early Jurassic floras from North Victoria Land, Antarctica. In *Online Proceedings of the 10th International Symposium on Antarctic Earth Sciences*, eds A. K. Cooper and C. R. Raymond and the 10th ISAES Editorial team. U. S. Geological Survey Open-File Report 2007–1047, Version 2. Extended Abstract 034, available at http://pubs.usgs.gov/of/2007/1047/ea/of2007–1047ea034.pdf. {Cited 2011}.

Bomfleur, B., Pott, C. and Kerp, H. (2011). Plant assemblages from the Shafer Peak Formation (Lower Jurassic), north Victoria Land, Transantarctic Mountains. *Antarctic Science*, **23**, 188–208.

Bonis, N. R., Ruhl, M. and Kürschner, W. M. (2010). Milankovitch-scale palynological turnover across the Triassic–Jurassic transition at St Audrie's Bay, SW UK. *Journal of the Geological Society of London*, **167**, 877–888.

Bordy, E. M. and Catuneanu, O. (2002). Sedimentology and palaeontology of the upper Karoo aeolian strata (Early Jurassic) in the Tuli Basin, South Africa. *Journal of African Earth Sciences*, **35**, 301–314.

Bromfield, K., Burrett, C. F., Leslie, R. A. and Meffre, S. (2007). Jurassic volcaniclastic–basaltic andesite–dolerite sequence in Tasmania – new age constraints for fossil plants from Lune River. *Australian Journal of Earth Sciences*, **54**, 965–974.

Cantrill, D. J. (1995). The occurrence of the fern *Hausmannia* Dunker (Dipteridaceae) in the Cretaceous of Alexander Island, Antarctica. *Alcheringa*, **19**, 243–254.

Cantrill, D. J. (2000). A new macroflora from the South Orkney Islands, Antarctica: evidence of an Early to Middle Jurassic age for the Powell Island Conglomerate. *Antarctic Science*, **12**, 185–195.

Cantrill, D. J. and Hunter, M. A. (2005). Macrofossil floras of the Latady Basin, Antarctic Peninsula. *New Zealand Journal of Geology and Geophysics*, **48**, 537–553.

Chandler, M. A., Rind, D. and Ruedy, R. (1992). Pangaean climate during the Early Jurassic: GCM simulations and the sedimentary record of paleoclimate. *Geological Society of America Bulletin*, **194**, 543–549.

Collins, W. J. (2003). Slab pull, mantle convection, and Pangaean assembly and dispersal. *Earth and Planetary Science Letters*, **205**, 225–237.

Collinson, J. W., Isbell, J. L., Elliot, D. H., Miller, M. F. and Miller, J. M. G. (1994). Permian–Triassic Transantarctic Basin. In *Permian–Triassic Pangaean Basins and Foldbelts along the Panthalassan Margin of Gondwanaland, Geological Survey of America Memoir*, **184**, eds J. J. Veevers and C. McA. Powell. Boulder: The Geological Society of America, pp. 173–222.

Crame, J. A. (1983). The occurrence of the Upper Jurassic bivalve *Malayomaorica malayomaorica* (Krumbeck) on the Orville Coast, Antarctica. *Journal of Molluscan Studies*, **49**, 61–76.

Dalziel, I. W. D., Lawver, L. A. and Murphy, J. B. (2000). Plumes, orogenesis, and supercontinental fragmentation. *Earth and Planetary Science Letters*, **178**, 1–11.

del Valle, R. A. and Núñez, J. H. (1988). Estratigafía de la región comprendida entre el Cerro Torre y el cabo Longing, en el Borde ne de la Peninsula Antártida. *Instituto Antártico Contribución*, **332**, 1–83.

del Valle, R. A., Lirio, J. M., Lusky, J. C., Morelli, J. R. and Núñez, J. H. (1997). Jurassic trees at Jason Peninsula, Antactica. *Antarctic Science*, **9**, 443–444.

Dera, G., Brigaud, B., Monna, F., *et al.* (2011). Climatic ups and downs in a disturbed Jurassic world. *Geology*, **39**, 215–218.

Dercourt, J., Gaetani, M., Vrielynck, B., *et al.* eds (2000). *Peri-Tethys Atlas; Palaeogeographical Maps and Explanatory Notes*. Paris: Commission de la Carte Géologique du Monde.

Douglas, J. G. (1969). The Mesozoic flora of Victoria. Parts 1 and 2. *Geological Survey of Victoria Memoir*, **28**, 1–310.

Duane, A. M. (1996). Palynology of the Byers Group (Late Jurassic–Early Cretaceous) of Livingston and Snow islands, Antarctic Peninsula: its biostratigraphical and palaeoenvironmental significance. *Review of Palaeobotany and Palynology*, **91**, 241–281.

Duncan, R. A., Hooper, P. R., Rehacek, J., Marsh, J. S. and Duncan, A. R. (1997). The timing and duration of the Karoo igneous event, southern Gondwana. *Journal of Geophysical Research*, **102**, 18127–18138.

Eagles, G. and Vaughan, A. P. M. (2009). Gondwana breakup and plate kinematics: business as usual. *Geophysical Research Letters*, **36**, L10302, doi:10.1029/2009GL037552.

Elliot, D. H. (1996). The Hanson Formation: a new stratigraphical unit in the Transantarctic Mountains, Antarctica. *Antarctic Science*, **8**, 389–394.

Elliot, D. H. (2000). Stratigraphy of Jurassic pyroclastic rocks in the Transantarctic Mountains. *Journal of African Earth Sciences*, **31**, 77–89.

Elliot, D. H. and Fleming, T. H. (2008). Physical volcanology and geological relationships of the Jurassic Ferrar Large Igneous Province, Antarctica. *Journal of Volcanology and Geothermal Research*, **172**, 20–37.

Elliot, D. H., Haban, M. A. and Siders, M. A. (1986). Jurassic tholeiites in the region of the upper Rennick Glacier, north Victoria Land. In *Geological Investigations in Northern Victoria Land, Antarctic Research Series*, **46**, ed. E. Stump. Washington: American Geophysical Union, pp. 249–266.

Encarnación, J., Fleming, T. H., Elliot, D. H. and Eales, H. V. (1996). Synchronous emplacement of Ferrar and Karoo dolerites and the early breakup of Gondwana. *Geology*, **24**, 535–538.

Eriksson, P. G. (1986). Aeolian dune and alluvial fan deposits in the Clarens Formation of the Natal Drakensberg. *Transactions of the Geological Society of South Africa*, **89**, 389–393.

Fanning, C. M. and Laudon, T. S. (1999). Mesozoic volcanism, plutonism and sedimentation in eastern Ellsworth Land, West Antarctica. In *8th International Symposium on Antarctic Earth Sciences, Programme and Abstracts*, ed. D. N. B. Skinner. Wellington: Victoria University, p. 102.

Faure, G., Bowman, J. R. and Elliot, D. H. (1979). The initial $^{87}Sr/^{86}Sr$ ratios of the Kirwan Volcanics of Dronning Maud Land: comparison with the Kirkpatrick Basalt, Transantarctic Mountains. *Chemical Geology*, **26**, 77–90.

Fleming, T. H., Heimann, A., Foland, K. A. and Elliot, D. H. (1997). $^{40}Ar/^{39}Ar$ geochronology of Ferrar Dolerite sills from the Transantarctic Mountains, Antarctica: implications for the age and origin of the Ferrar magmatic Province. *Geological Society of America Bulletin*, **109**, 533–546.

Frakes, L. A., Francis J. E. and Syktus J. I. (1992). *Climate Modes of the Phanerozoic*. Cambridge: Cambridge University Press.

Francis, J. E. (1983). The dominant conifer of the Jurassic Purbeck Formation, England. *Palaeontology*, **26**, 277–294.

Garland, M. J., Bannister, J. M., Lee, D. E. and White, J. D. L. (2007). A coniferous tree stump of late Early Jurassic age from the Ferrar Basalt, Coombs Hills, southern Victoria Land, Antarctica. *New Zealand Journal of Geology and Geophysics*, **50**, 263–269.

Gee, C. T. (1989a). Revision of the Late Jurassic/Early Cretaceous flora from Hope Bay, Antarctica. *Palaeontographica B*, **213**, 149–214.

Gee, C. T. (1989b). Permian *Glossopteris* and *Elatocladus* megafossil floras from the English Coast, Ellsworth Land, Antarctica. *Antarctic Science*, **1**, 35–44.

Golonka, J. (2007). Late Triassic and Early Jurassic paleogeography of the world. *Palaeogeography, Palaeoclimatology, Palaeoecology*, **244**, 297–307.

Gómez, J. J., Canales, M. L., Ureta, S. and Goy, A. (2009). Palaeoclimatic and biotic changes during the Aalenian (Middle Jurassic) at the southern Laurasian Seaway (Basque-Cantabrian Basin, northern Spain). *Palaeogeography, Palaeoclimatology, Palaeoecology*, **275**, 14–27.

Gunn, B. M. and Warren, G. (1962). Geology of Victoria Land between the Mawson and Mulock glaciers, Antarctica. *Bulletin of the New Zealand Geological Survey*, **71**, 1–157.

Hallam, A. and Wignall, P. B. (1997). *Mass Extinctions and their Aftermath*. New York: Oxford University Press.

Halle, T. G. (1913). The Mesozoic flora of Graham Land. In *Wissenschaftliche Ergebnisse der Schwedischen Südpolar-Expedition 1901–1903, Volume 3(4), Geologie und Paläontologie*, ed. O. Nordenskjöld. Stockholm: P. A. Norstedt & Söner, pp. 3–124.

Hammer, W. R. and Hickerson, W. J. (1994). A new crested theropod dinosaur from Antarctica. *Science*, **264**, 828–830.

Hammer, W. R. and Smith, N. D. (2008). A tritylodont postcanine from the Hanson Formation of Antarctica. *Journal of Vertebrate Paleontology*, **28**, 269–273.

Hanson, R. E. and Elliot, D. H. (1996). Rift-related Jurassic phreatomagmatic volcanism in the central Transantarctic Mountains: precursory stage to flood-basalt effusion. *Bulletin of Volcanology*, **58**, 327–347.

Hathway, B. (2000). Continental rift to back-arc basin: Jurassic – Cretaceous stratigraphical and structural evolution of the Larsen Basin, Antarctic Peninsula. *Journal of the Geological Society of London*, **157**, 417–432.

Heimann, A., Fleming, T. H., Elliot, D. H. and Foland, K. A. (1994). A short interval of Jurassic continental flood basalt volcanism in Antarctica as demonstrated by $^{40}Ar/^{39}Ar$ geochronology. *Earth and Planetary Science Letters*, **121**, 19–41.

Hergt, J. M. and Brauns, C. M. (2001). On the origin of the Tasmanian dolerite. *Australian Journal of Earth Sciences*, **48**, 543–549.

Hesselbo, S. P., Grocke, D. R., Jenkyns, H. C., *et al.* (2000). Massive dissociation of gas hydrate during a Jurassic oceanic anoxic event. *Nature*, **406**, 392–395.

Holzförster, F. (2007). Lithology and depostional environments of the Lower Jurassic Clarens Formation in the eastern Cape, South Africa. *South African Journal of Geology*, **110**, 543–560.

Huber, B. T., MacLeod, K. G. and Wing S. L. eds (2000). *Warm Climates in Earth History.* Cambridge: Cambridge University Press.

Hunter, M. A. and Cantrill, D. J. (2006). A new stratigraphy for the Latady Basin, Antarctica. Part 2. Latady Group and basin evolution. *Geological Magazine*, **143**, 797–819.

Hunter, M. A., Cantrill, D. J., Flowerdew, M. and Millar, I. L. (2005). Middle Jurassic age for the Botany Bay Group: implications for Weddell Sea Basin creation and southern hemisphere biostratigraphy. *Journal of the Geological Society of London*, **162**, 745–748.

Hunter, M. A., Cantrill, D. J. and Flowerdew, M. (2006). Latest Jurassic–earliest Cretaceous age for a fossil flora from the Latady Basin, Antarctic Peninsula. *Antarctic Science*, **18**, 261–264.

Hunter, M. A., Riley, T. R., Cantrill, D. J., Flowerdew, M. J. and Millar, I. L. (2006). A new stratigraphy for the Latady Basin, Antarctica Part 1. Ellsworth Land Volcanic Group. *Geological Magazine*, **143**, 777–796.

Husinec, A. and Read, J. F. (2007). The Late Jurassic Tithonian, a greenhouse phase in the Middle Jurassic–Early Cretaceous 'cool' mode: evidence from the cyclic Adriatic Platform, Croatia. *Sedimentology*, **54**, 317–337.

Jacobs, D. K. and Sahagian, D. L. (1993). Climate-induced fluctuations in sea-level during non-glacial times. *Nature*, **361**, 710–712.

Jansson, I.-M., McLoughlin, S., Vadja, V. and Pole, M. (2008). An Early Jurassic flora from the Clarence-Moreton Basin, Australia. *Review of Palaeobotany and Palynology*, **150**, 5–21.

Jefferson, T. H., Siders, M. A. and Haban, M. A. (1983). Jurassic trees engulfed by lavas of the Kirkpatrick Basalt Group, northern Victoria Land. *Antarctic Journal of the United States*, **185**, 14–16.

Jenkyns, H. C. (2003). Evidence for rapid climate change in the Mesozoic-Paleogene greenhouse world. *Philosophical Transactions of the Royal Society of London, Series A*, **361**, 1885–1916.

Jourdan, F., Féraud, G., Bertrand, H., Watkeys, M. K. and Renne, P. R. (2007). Distinct brief major events in the Karoo large igneous province clarified by new ^{40}Ar/^{39}Ar ages on the Lesotho basalts. *Lithos*, **98**, 195–209.

Kemp, D. B., Coe, A. L., Cohen, A. S. and Schwark, L. (2005). Astronomically paced methane release and mass extinction in the Early Jurassic. *Nature*, **437**, 396–399.

Kidder, D. L. and Worsley, T. R. (2010). Phanerozoic Large Igneous Provinces (LIPs), HEATT (Haline Euxinic Acidic Thermal Transgression) episodes, and mass extinctions. *Palaeogeography, Palaeoclimatology, Palaeoecology*, **295**, 162–191.

Kiessling, W., Aberhan, M., Brenneis, B. and Wagner, P. J. (2007). Extinction trajectories of benthic organisms across the Triassic–Jurassic boundary. *Palaeogeography, Palaeoclimatology, Palaeoecology*, **244**, 201–222.

Kiessling, W., Roniewicz, E., Villier, L., Léonide, P. and Struck, U. (2009). An early Hettangian coral reef in southern France: implications for the end-Triassic reef crisis. *Palaios*, **24**, 657–671.

Kocurek, G. (2003). Limits on extreme aeolian systems: Sahara of Mauritania and Jurassic Navajo Sandstone examples. In *Extreme Depositional Environments: Mega End Members in Geologic Time, Geological Society of America Special Paper*, **370**, eds M. Chan and A. Archer. Boulder: The Geological Society of America, pp. 143–156.

Kriwet, J., Kiessling, W. and Klug, S. (2009). Diversification trajectories and evolutionary life-history traits in early sharks and batoids. *Proceedings of the Royal Society of London Series B*, **276**, 945–951.

Kutzbach, J. E. and Gallimore, R. G. (1989). Pangaean climates: megamonsoons of the megacontinent. *Journal of Geophysical Research*, **94**, 3341–3357.

Kutzbach, J. E., Guetter, P. J. and Washington W. M. (1990). Simulated circulation of an idealized ocean for Pangaean time. *Paleoceanography*, **5**, 299–317.

Leinfelder, R. R., Schmid, D. U., Nose, M. and Werner, W. (2002). Jurassic reef patterns – the expression of a changing globe. In *Phanerozoic Reef Patterns*, *SEPM Special Publication*, **72**, eds W. Kiessling, E. Flügel and J. Golonka. Tulsa: SEPM (Society for Sedimentary Geology), pp. 465–520.

Luttinen, A. V., Leat, P. T. and Furnes, H. (2010). Björnnutane and Sembberget basalt lavas and the geochemical provinciality of Karoo magmatism in western Dronning Maud Land. *Journal of Volcanology and Geothermal Research*, **198**, 1–18.

Malchus, N. and Steuber, T. (2002). Stable isotope records (O, C) of Jurassic aragonitic shells from England and NW Poland: palaeoecologic and environmental implications. *Geobios*, **35**, 29–39.

McArthur, J. M., Doyle, P., Leng, M. J., *et al.* (2007). Testing palaeo-environmental proxies in Jurassic belemnites: Mg/Ca, Sr/Ca, Na/Ca, δ^{18}O and δ^{13}C. *Palaeogeography, Palaeoclimatology, Palaeoecology*, **252**, 464–480.

McElwain, J. C., Beerling, D. J. and Woodward F. I. (1999). Fossil plants and global warming of the Triassic-Jurassic boundary. *Science*, **285**, 1386–1390.

McElwain, J. C., Wade-Murphy, J. and Hesselbo, S. P. (2005). Changes in carbon dioxide during an oceanic anoxic event linked to the intrusion into Gondwana coals. *Nature*, **435**, 479–482.

McLoughlin, S. and Pott, C. (2009). The Jurassic flora of Western Australia. *GFF*, **131**, 113–136.

Meneilly, A. W., Harrison, S. M., Piercy, B. A. and Storey, B. C. (1987). Structural evolution of the magmatic arc in northern Palmer Land, Antarctic Peninsula. In *Gondwana Six: Structure, Tectonics and Geophysics*, *American Geophysical Union Geophysical Monograph*, **40**, ed. G. D. McKenzie. Washington: American Geophysical Union, pp. 209–219.

Miller, S. and Macdonald, D. I. M. (2004). Metamorphic and thermal history of a fore-arc basin: the Fossil Bluff Group, Alexander Island, Antarctica. *Journal of Petrology*, **45**, 1453–1465.

Niklas, K. J. (1994). Predicting the height of fossil plant remains: an allometric approach to an old problem. *American Journal of Botany*, **81**, 1235–1242.

Norris, G. (1965). Triassic and Jurassic miospores and acritarchs from the Beacon and Ferrar Groups Victoria Land, Antarctica. *New Zealand Journal of Geology and Geophysics*, **8**, 236–277.

Orlando, H. A. (1971). Las floras fósiles de Antártida occidental y sus relaciones estratigráficas. *Instituto Antártico Argentino Contribución*, **140**, 1–12.

Pálfy, J., Demeny, A., Haas, J., *et al.* (2001). Carbon isotope anomaly and other geochemical changes at the Triassic-Jurassic boundary. *Geology*, **29**, 1047–1050.

Pankhurst, R. J., Riley, T. R., Fanning, C. M. and Kelley, S. P. (2000). Episodic silicic volcanism in Patagonia and the Antarctic Peninsula: chronology of magmatism associated with break-up of Gondwana. *Journal of Petrology*, **41**, 605–625.

Parrish, J. T. and Falcon-Lang, H. J. (2007). Coniferous trees associated with interdune deposits in the Jurassic Navajo Sandstone Formation, Utah, USA. *Palaeontology*, **50**, 829–843.

Pearson, P. N., Ditchfield, P. W., Singano, J., *et al.* (2001). Warm tropical sea surface temperatures in the Late Cretaceous and Eocene epochs. *Nature*, **413**, 481–487.

Plumstead, E. P. (1962). Fossil floras from Antarctica. In *Trans-Antarctic Expedition 1955–1958, Scientific Reports 9 (Geology)*. London: The Trans-Antarctic Expedition Committee, pp. 1–154.

Plumstead, E. P. (1964). Palaeobotany of Antarctica. In *Antarctic Geology: proceedings of the First International Symposium on Antarctic Geology, Cape Town, 16–21 September 1963*, ed. R. J. Adie. Amsterdam: North-Holland Publishing Company, pp. 637–654.

Porter, M. L. (1987). Sedimentology of an ancient erg margin: the Lower Jurassic Aztec Sandstone, southern Nevada and California. *Sedimentology*, **34**, 661–680.

Price, G. D. (1999). The evidence and implications of polar ice during the Mesozoic. *Earth-Science Reviews*, **48**, 183–210.

Price, G. D. and Page, K. N. (2008). A carbon and oxygen isotopic analysis of molluscan faunas from the Callovian-Oxfordian boundary at Redcliff Point, Weymouth, Dorset: Implications for belemnite behavior. *Proceedings of the Geologists Association*, **119**, 153–160.

Price, G. D. and Sellwood, B. W. (1997). "Warm" palaeotemperatures from high Late Jurassic palaeolatitudes (Falkland Plateau): ecological, environmental or diagenetic controls? *Palaeogeography, Palaeoclimatology, Palaeoecology*, **129**, 315–327.

Price, G. D., Valdes, P. J. and Sellwood, B. W. (1997). Prediction of modern bauxite occurrence: implications for climate reconstruction. *Palaeogeography, Palaeoclimatology, Palaeoecology*, **131**, 1–13.

Quilty, P. G. (1977). Late Jurassic bivalves from Ellsworth Land, Antarctica: their systematics and palaeobiogeographic implications. *New Zealand Journal of Geology and Geophysics*, **20**, 1033–1080.

Quilty, P. G. (1983). Bajocian bivalves from Ellsworth Land, Antarctica. *New Zealand Journal of Geology and Geophysics*, **26**, 395–418.

Rao, A. R. (1953). Some observations on the Rajmahal flora. *The Palaeobotanist*, **2**, 25–28.

Raup, D. M. and Sepkoski, J. J. (1982). Mass extinctions in the marine fossil record. *Science*, **215**, 1501–1503.

Rees, P. M. (1990). Palaeobotanical contributions to the Mesozoic geology of the northern Antarctic Peninsula region. Ph.D. thesis, Royal Holloway and Bedford New College, University of London.

Rees, P. M. (1993a). Dipterid ferns from the Mesozoic of Antarctic and New Zealand and their stratigraphical significance. *Palaeontology*, **36**, 637–656.

Rees, P. M. (1993b). Caytoniales in Early Jurassic floras from Antarctica. *Geobios*, **26**, 33–42.

Rees, P. M. and Cleal C. J. (1993). Marked polymorphism in *Archangelskya furcata*, a pteridospermous frond from the Jurassic of Antarctica. *Special Papers in Palaeontology*, **49**, 85–100.

Rees, P. M. and Cleal, C. J. (2004). Lower Jurassic floras from Hope Bay and Botany Bay, Antarctica. *Special Papers in Palaeontology*, **72**, 1–96.

Rees, P. M., Ziegler, A. M. and Valdes, P. J. (2000). Jurassic phytogeography and climates: new data and model comparisons. In *Warm Climates in Earth History*, eds B. T. Huber, K. G. Macleod and S. L. Wing. Cambridge: Cambridge University Press, pp. 297–318.

Rees, P. M., Noto, C. R., Parrish, J. M. and Parrish, J. T. (2004). Late Jurassic climates, vegetation and dinosaur distributions. *The Journal of Geology*, **112**, 643–653.

Retallack, G. J. (2009). Greenhouse crises of the past 300 million years. *Geological Society of America Bulletin*, **121**, 1441–1455.

Ribecai, C. (2007). Early Jurassic miospores from Ferrar Group of Carapace Nunatak, South Victoria Land, Antarctica. *Review of Palaeobotany and Palynology*, **144**, 3–12.

Riley, T., Crame, J. A., Thomson, M. R. A. and Cantrill, D. J. (1997). Late Jurassic (Kimmeridgian-Tithonian) macrofossil assemblage from Jason Peninsula, Graham

Land: Evidence for a significant northward extension of the Latady Formation. *Antarctic Science*, **9**, 432–440.

Riley, T. R. and Leat, P. T. (1999). Large volume silicic volcanism along the proto-Pacific margin of Gondwana: lithological and stratigraphical investigations from the Antarctic Peninsula. *Geological Magazine*, **136**, 1–16.

Riley, T. R., Leat, P. T., Curtis, M. L., *et al.* (2005). Early-Middle Jurassic dolerite dykes from western Dronning Maud Land (Antarctica): Identifying mantle sources in the Karoo Large Igneous Province. *Journal of Petrology*, **46**, 1489–1524.

Riley, T. R., Curtis, M. L., Leat, P. T., *et al.* (2006). Overlap of Karoo and Ferrar magma types in KwaZulu-Natal, South Africa. *Journal of Petrology*, **47**, 541–566.

Riley, T. R., Flowerdew, M. J., Hunter, M. A. and Whitehouse, M. J. (2010). Middle Jurassic rhyolite volcanism of eastern Graham Land, Antarctic Peninsula: age correlations and stratigraphic relationships. *Geological Magazine*, **147**, 581–595.

Ross, P.-S., White, J. D. L. and McClintock, M. (2008). Geological evolution of the Coombs–Allan Hills area, Ferrar large igneous province, Antarctica: debris avalanches, mafic pyroclastic density currents, phreatocauldrons. *Journal of Volcanology and Geothermal Research*, **172**, 38–60.

Rowley, P. D., Schmidt, D. L. and Williams P. L. (1982). Mount Poster Formation, southern Antarctic Peninsula and eastern Ellsworth Land. *Antarctic Journal of the United States*, **17** (5), 38–39.

Ruhl, M., Deenen, M. H. L., Abels, H. A., *et al.* (2010). Astronomical constraints on the duration of the early Jurassic Hettangian stage and recovery rates following the end-Triassic mass extinction (St Audrie's Bay/East Quantox Head, UK). *Earth and Planetary Science Letters*, **295**, 262–276.

Schöner, R., Viereck-Goette, L., Schneider, J. and Bomfleur, B. (2007). Triassic-Jurassic sediments and multiple volcanic events in North Victoria Land, Antarctica: A revised stratigraphic model. In *Online Proceedings of the 10th International Symposium on Antarctic Earth Sciences*, eds A. K. Cooper, C. R. Raymond and the 10th ISAES Editorial team. U. S. Geological Survey Open-File Report 2007–1047, Version 2. Short Research Paper 102, doi:10.3133/of2007–1047.srp102.

Scotese, C. R. (1991). Jurassic and Cretaceous plate tectonic reconstructions. *Palaeogeography, Palaeoclimatology, Palaeoecology*, **87**, 493–501.

Sellwood B. W. and Valdes, P. J. (2006). Mesozoic climates. General circulation models and the rock record. *Sedimentary Geology*, **190**, 269–287.

Sellwood, B. W. and Valdes, P. J. (2008). Jurassic climates. *Proceedings of the Geologists Association*, **119**, 5–17.

Sellwood, B. W., Valdes, P. J. and Price, G. D. (2000). Geological evaluation of multiple circulation model simulations of Late Jurassic palaeoclimate. *Palaeogeography, Palaeoclimatology, Palaeoecology*, **156**, 147–160.

Sepkoski Jr, J. J. (1996). Patterns of Phanerozoic extinction: a perspective from global databases. In *Global Events and Event Stratigraphy in the Phanerozoic*, ed. O. H. Walliser. Heidelberg, Berlin, New York: Springer-Verlag, pp. 35–52.

Shang, Y. (1997). Middle Jurassic palynology of Carapace Nunatak, Victoria Land, Antarctica. *Acta Palaeontologica Sinica*, **36**, 170–186.

Shindell, D. T., Faluvegi, G., Koch, D. M., *et al.* (2009). Improved attribution of climate forcing emissions. *Science*, **326**, 716–718.

Smith, N. D. and Pol, D. (2007). Anatomy of a basal sauropodomorph dinosaur from the Early Jurassic Hanson Formation of Antarctica. *Acta Palaeontologica Polonica*, **52**, 657–674.

Smith, N. D., Makovicky, P. J., Hammer, W. R. and Currie, P. J. (2007). Osteology of *Crylophosaurus ellioti* (Dinosauria: Theropoda) from the Early Jurassic of Antarctica and implications for early theropod evolution. *Zoological Journal of the Linnean Society*, **151**, 377–421.

Smith, R. S., Dubois, C. and Marotzke, J. (2004). Ocean circulation and climate in an idealized Pangaean OAGCM. *Geophysical Research Letters*, **31**, 1–18.

Stigall, A. L., Babcock, L. E., Briggs, D. E. G. and Leslie, S. A. (2008). Taphonomy of lacustrine interbeds in the Kirkpatrick Basalt (Jurassic), Antarctica. *Palaios*, **23**, 344–355.

Stipanicic, P. M. and Bonetti, M. I. R. (1970). Posciones estratigráficas y edades de las principales flores Jurásicas Argentinas. II. Floras Doggerianas y Málmicas. *Ameghinana*, **7**, 101–118.

Storey, B. C. (1995). The role of mantle plumes in continental breakup: cases histories from Gondwanaland. *Nature*, **377**, 310–308.

Storey, B. C., Dalziel, I. W. D., Garrett, S. W., *et al.* (1988). West Antarctic in Gondwanaland: crustal blocks, reconstruction and break-up processes. *Tectonophysics*, **155**, 381–390.

Strasser, A. (2007). Astronomical time scale for the Middle Oxfordian to Late Kimmeridgian in the Swiss and French Jura Mountains. *Swiss Journal of Geosciences*, **100**, 407–429.

Suan, G., Pittet, B., Bour, I., *et al.* (2008). Duration of the Early Toarcian carbon isotope excursion deduced from spectral analysis: consequence for its possible causes. *Earth and Planetary Science Letters*, **267**, 666–679.

Suan, G., Mattioli, E., Pittet, B., *et al.* (2010). Secular environmental precursors to Early Toarcian (Jurassic) extreme climate changes. *Earth and Planetary Science Letters*, **290**, 448–458.

Sushchevskaya, N. M., Belyetsky, B. V., Leichenkov, G. L. and Laiba, A. A. (2009). Evolution of the Karoo–Maud mantle plume in Antarctica and its influence on the magmatism of the early stages of Indian Ocean opening. *Geochemistry International*, **47**, 1–17.

Tanner, L. H., Lucas, S. G. and Chapman, M. G. (2004). Assessing the record and causes of Late Triassic extinctions. *Earth-Science Reviews*, **65**, 103–139.

Tasch, P. (1977). International correlation by conchostrachans and palynomorphs from Antarctica, western Australia, India and Africa. *Antarctic Journal of the United States*, **12(5)**, 121.

Tasch, P. (1987). Fossil Conchonstracha of the southern hemisphere and continental drift palaeontology, biostratigraphy and dispersal. *Geological Society of America Memoir*, **165**, 1–290.

Tasch, P. and Lammons, J. M. (1978). Palynology of some lacustrine beds of the Antarctic Jurassic. *Palinología numéro extraordinaire 1, Proceedings of the Coloquio International de Palinología, Léon*, pp. 455–460.

Thomson, M. R. A. (1983). Late Jurassic ammonites from the Orville Coast, Antarctica. In *Antarctic Earth Science – Proceedings of the Fourth International Symposium on Antarctic Earth Sciences, Adelaide, South Australia, 16–20 August 1982*, eds R. L. Oliver, P. R. James and J. B. Jago. Canberra: Australian Academy of Science, pp. 315–319.

Thorne, P. M., Ruta, M. and Benton, M. J. (2011). Resetting the evolution of marine reptiles at the Triassic-Jurassic boundary. *Proceedings of the National Academy of Sciences of the United States of America*, **108**, 8339–8344.

Tingey R. J. (1991). Mesozoic tholeiitic rocks in Antarctica: the Ferrar (Super) Group. In *The Geology of Antarctica, Oxford Monographs on Geology and Geophysics*, **17**, ed. R. J. Tingey. Oxford: Clarendon Press, pp. 153–174.

Townrow, J. A. (1967a). On a conifer from the Jurassic of East Antarctica. *The Papers and Proceedings of the Royal Society of Tasmania*, **101**, 137–147.

Townrow, J. A. (1967b). Fossil plants from Allan and Carapace Nunataks, and from the upper Mill and Shackleton glaciers, Antarctica. *New Zealand Journal of Geology and Geophysics*, **10**, 456–473.

Truswell, E. M. (1982). Palynology of sea floor samples collected by the 1911–1914 Australasian Antarctic expedition: implications for the geology of coastal East Antarctica. *Journal of the Geological Society of Australia*, **29**, 343–356.

Truswell, E. M. (1987). The palynology of core samples from the *S.P. Lee* Wilkes Land cruise. *Circum-Pacific Council for Energy and Mineral Resources Earth Science Series*, **5A**, 215–218.

Truswell, E. M., Dettmann, M. E. and O'Brien, P. E. (1999). Mesozoic palynofloras from Mac. Robertson Shelf, East Antarctica: geological and phytogeographical implications. *Antarctic Science*, **11**, 239–255.

Twitchett, R. (2006). The palaeoclimatology, palaeoecology and palaeoenvironmental analysis of mass extinction events. *Palaeogeography, Palaeoclimatology, Palaeoecology*, **232**, 190–213.

Vakhrameev, V. A. (1991). *Jurassic and Cretaceous Floras and Climates of the Earth*. Cambridge: Cambridge University Press.

Védrine, S. and Strasser A. (2009). High-frequency palaeoenvironmental changes on a shallow carbonate platform during a marine transgression (Late Oxfordian, Swiss Jura Mountains). *Swiss Journal of Geosciences*, **102**, 247–270.

Wang, F., Zhou, X.-H., Zhang, L.-C., *et al.* (2006). Late Mesozoic volcanism in the Great Xing'an Range (NE China): Timing and implications for the dynamic setting of NE Asia. *Earth and Planetary Science Letters*, **251**, 179–198.

Watson, J. (1988). The Cheirolepidiaceae. In *Origin and Evolution of Gymnosperms*, ed. C. B. Beck. New York: Columbia University Press, pp. 382–447.

Weedon, G. P., Coe, A. L. and Gallois, R. W. (2004). Cyclostratigraphy, orbital tuning and inferred productivity for the type Kimmeridge Clay (Late Jurassic), Southern England. *Journal of the Geological Society of London*, **161**, 655–666.

Weissert, H. and Ebra, E. (2004). Volcanism, CO_2 and palaeoclimate: a Late Jurassic–Early Cretaceous carbon and oxygen isotope record. *Journal of the Geological Society of London*, **161**, 695–702.

Whitham, A. G. and Doyle, P. (1989). Stratigraphy of the Upper Jurassic–Lower Cretaceous Nordenskjöld Formation of eastern Graham Land, Antarctica. *Journal of South American Earth Sciences*, **2**, 371–384.

Wignall, P. B. (2001). Large igneous provinces and mass extinctions, *Earth-Science Reviews*, **53**, 1–33.

Williams, M., Haywood, A. M., Taylor, S. P., *et al.* (2005). Evaluating the efficacy of planktonic foraminifer calcite $\delta^{18}O$ data for sea surface temperature reconstruction for the Late Miocene. *Geobios*, **38**, 843–863.

Yao, X., Taylor, T. N. and Taylor, E. L. (1991). Silicified dipterid ferns from the Jurassic of Antarctica. *Review of Palaeobotany and Palynology*, **67**, 353–362.

Ziegler, A. M., Scotese, C. R. and Barrett, S. F. (1983). Mesozoic and Cenozoic paleogeographic maps. In *Tidal Friction and the Earth's Rotation II. Proceedings of a workshop held at the Centre for Interdisciplinary Research (ZiF) of the University of Bielefeld, September 28–October 3, 1981*, eds P. Brosche and J. Sündermann. Berlin: Springer-Verlag, pp. 240–252.

6

Fern-conifer dominated Early Cretaceous (Aptian–Albian) ecosystems and the angiosperm invasion

Introduction

The palaeofloras of the Early Cretaceous gradually became more provincial relative to the earlier Jurassic to earliest Cretaceous floras. This was as a result of the development of new zonal circulation patterns in both hemispheres, more humid conditions, increasing palaeo-precipitation in the tropical regions, and expanding cold temperate climates at high latitudes (Iglesias, Artabe and Morel, 2011). Early studies (e.g. Meyen, 1987; Vakhrameev, 1991) have suggested that a relatively uniform flora existed across (southern) Gondwana during the Early Cretaceous. Indeed, a degree of uniformity was certainly apparent as some taxa were widely distributed across the supercontinent (e.g. McLoughlin, 2001). The vegetation was rich in ferns and, at high latitudes, exceptionally rich in liverworts with greater biomass quite probably allocated to non-vascular and lower vascular plant groups. However, with continued studies of the fossil assemblages coupled with taxonomic identification it has become clear that, although there was a degree of uniformity with some taxa widely distributed across the supercontinent (McLoughlin, 2001; Vakhrameev, 1991), the flora also showed signs of provincialism with eastern Gondwana Floras broadly distinguishable from western Gondwana Floras. At this time East Antarctica held a central position within Gondwana and this is critical for understanding the floristic relationships across the supercontinent. However, understanding of the vegetation in this area has been restricted to evidence from reworked palynomorphs preserved around the periphery of the continent as exposed Early Cretaceous sediments are lacking. Nevertheless floristic (sub)provinces, including regions of endemism, are still recognisable within southern Gondwana (Herngreen et al., 1996; McLoughlin, 2001). Allopatric (geographic) speciation was promoted by continued fragmen-tation of terrestrial environments through gradational environmental features, including latitudinal or climatic differences, tectonic separation of the landmass, and separation by marine incursions. Floras were further influenced by local habitat differences (e.g. topography, moisture levels, competition, fire: Askin, 1989; Dettmann, 1986a, 1986b; Douglas, 1994). Together these factors resulted in broad phytogeographical patterns evidenced by the Early Cretaceous conifers and other seed plants that dominated floras across Gondwana, which became increasingly divergent as Gondwana became increasingly segregated (McLoughlin, 2001), and was influenced both by the arrival of the angiosperms and the ameliorating climate.

189

Global palaeoclimate

The greenhouse world that characterised the southern high latitudes during the Jurassic continued into the Cretaceous. The warming trend that went on to peak in the mid- to Late Cretaceous (Albian–Turonian; 112–88 Ma) (e.g. Clarke and Jenkyns, 1999; Price and Nunn, 2010 and references therein; Steuber *et al.*, 2005) was not consistent but punctuated by cooler oscillations. Although the timing and significance of these oscillations are not well-constrained, it is possible that sub-freezing polar conditions occurred during the Early and mid-Cretaceous (e.g. Ditchfield, 1997; Koch and Brenner, 2009; Price and Mutterlose, 2004). The Cretaceous was thus a time of general global warmth with unusually warm poles and an almost total lack of polar ice caps. High latitudes exhibited cool and warm temperate climates with warmer-than-modern mean annual and winter minimum temperatures (Bice and Norris, 2002; Bornemann *et al.*, 2008; Miller *et al.*, 2005; Wilson and Norris, 2001). The cause of globally warm temperatures remains elusive but is widely attributed to high levels of atmospheric greenhouse gases including CO_2 and methane (Bice *et al.*, 2006). This is thought to have resulted from increased volcanic outgassing associated with major tectonic activity that was taking place at this time. Although estimates vary widely, concentrations of CO_2 may have reached ~2000 ppm in the Early Cretaceous, rising to >4000 ppm in the mid-Cretaceous and then dropping steadily to 800 ppm by the end of the Cretaceous (Berner and Kothavala, 2001; Bice and Norris, 2002; Bice *et al.*, 2006; Fletcher *et al.*, 2005). However, these maximum levels have been suggested to be overestimates[1] and may have only reached levels of around 1000 ppm during the mid-Cretaceous (Breecker, Sharp and McFadden, 2010).

Globally, in the marine realm this was a time of episodic, but critical, changes in sea water composition associated with worldwide anoxia resulting from enhanced burial of organic-rich sediments. Causal mechanisms for this enhanced burial are still debated, but it resulted in the locking up of vast quantities of ^{12}C leading to negative carbon isotope excursions in organic matter and positive shifts in sea water $\delta^{13}C$ values. The first of these so-called Oceanic Anoxic Events occurred during the Aptian and early Albian, with the two further events in the Cretaceous (Sellwood and Valdes, 2006). Major tectonic activity and sea floor spreading caused palaeogeographic changes that altered the global oceanic circulation. Large ocean gyres distributed heat between the equatorial tropics and the polar regions of the proto-Pacific Ocean (Panthalassa), and so established low equatorial to pole thermal gradients. These gradients helped maintain a greenhouse Earth with ocean waters at 60° S suggested to have reached ~30 °C during its peak (Bice, Huber and Norris, 2003; Huber, Hodell and Hamilton, 1995), although this again may be an overestimation (Sellwood and Valdez, 2006). Species responded by migrating into the high latitudes such that by its acme, the mid-Cretaceous is widely considered to be the archetypal ice-free greenhouse interval sporting poles with luxuriant forest ecosystems teeming with life (e.g. Askin and Spicer, 1995; Herman and Spicer, 1996). Although the time of inception of polar ice over the last 100 Ma and whether widespread glaciations were

---

[1] Quantification of atmospheric CO_2 concentrations is based primarily on the carbon isotope composition of calcium carbonate in fossil soils. Breecker, Sharp and McFadden (2010) report that previous use of too high levels of assumed soil CO_2 concentrations during carbonate formation has resulted in a significant overestimation of greenhouse atmospheric CO_2 levels.

present during the Cretaceous is a long running debate (e.g. Moriya *et al.*, 2007 and references therein), the mid-Cretaceous was undeniably a time of warmth that existed for millions of years. This warm period was followed by gradual cooling through the Late Cretaceous and into the early Cenozoic (Ditchfield, Marshall and Pirrie, 1994; Zachos *et al.*, 2001).

The Cretaceous greenhouse climate, complete with polar ecosystems, was therefore markedly different from the world today yet its importance cannot be underestimated. Understanding the composition and dynamics of these polar environments is the key to understanding the consequences of possible future global warming. Any change in the global mean surface temperature is expressed most markedly at the poles, as the poles are the ultimate heat sink for the planet, and conditions in these regions determine, to a large extent, the rate and process of global climate change (Spicer, 2003). Since there are no modern analogues for such polar ecosystems, the high latitude Cretaceous fossil record becomes an important source of information for the future Earth.

Palaeogeography of southern Gondwana

Pivotal to migrations of plants and animals across Gondwana and into the high latitudes were the terrestrial connections between what are now the widely separated landmasses of Australia, Africa, India, New Zealand, South America and Antarctica. These biotic gateways were integral to the development of biogeographic and diversity patterns seen across the Southern Hemisphere today. Therefore an understanding of the relative positions of these fragments through geological time is critical to understanding the evolution of floristic realms.

The history of Gondwana break-up is long and complicated and major uncertainties still exist with regard to timing. The long-lived Pangean supercontinent that began fragmenting in the Middle Jurassic had all but shifted in landmass balance from the Southern Hemisphere to the Northern Hemisphere by the Late Cretaceous to early Cenozoic as new oceans developed. These events were contemporaneous with some of the greatest climatic, topographic, volcanic and orogenic, oceanic and floral–faunal changes in Earth's history – events that continue to permeate and control our planet today (Blakey, 2008).

Gondwana break-up began at 180 Ma with the onset of rifting and widespread Karoo and Ferrar magmatic events. However, sea floor spreading did not begin until about 20 Ma later, at ~160 Ma, when East Gondwana (Antarctica, India, Australia, New Zealand and Madagascar) began rotating southward away from West Gondwana (Africa, Arabia and South America). By the dawn of the Cretaceous (145 Ma) the five main Gondwana continents were becoming discernible as the Central Atlantic became a mature ocean, the Somoma Ocean opened between East and West Gondwana and continental rifting was initiated between South America and South Africa (Blakey, 2008; Scotese, 1998; Stampfli and Borel, 2002; Veevers, 2004). At about 130 Ma, the former supercontinent started breaking up into four continental blocks: South America, Africa–Arabia, India–Madagascar, and Australia–Antarctica (Blakey, 2008; Li and Powell, 2001) (Figure 6.1A). By 120 Ma the Australia–Antarctica block was separated from India, by the developing Indian Ocean, and earliest rifting between Antarctica

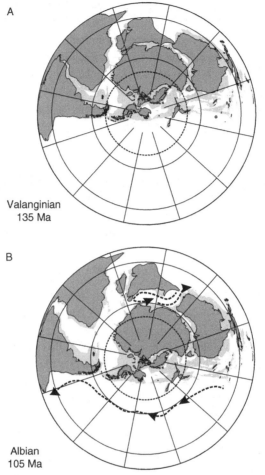

Figure 6.1 View of Gondwana break-up during the Early Cretaceous. Latitude intervals are 30°, thin dashed black line represents the palaeopolar circle. **A**, Valanginian (135 Ma) illustrating the close proximity of Africa and India to the rest of Gondwana. **B**, Albian (105 Ma), note that India and South Africa have separated from the rest of Gondwana. Arrowed, dotted lines indicate inferred direction of the palaeocurrents in existence at the time; light grey = continental shelves, dark grey = land. Reconstructions provided by R.A. Livermore, British Antarctic Survey. Reproduced from Cantrill and Poole (2002) with permission from The Geological Society of London.

and Australia had begun (Li and Powell, 2001; Scotese, 1998). The South Atlantic Ocean was also forming, separating South America from Africa (Lawver, Gahagan and Coffin, 1992). India rotated northwards at 105 Ma and started moving away from the middle southern latitudes. At the same time some of the fragments that make up New Zealand were beginning to separate from Antarctica (Figure 6.1B). The break-up between Antarctica and Australia began in the earliest Cretaceous with a narrow gulf forming between western Australia and Antarctica. This gulf propagated eastwards throughout the Cretaceous and separation only accelerated in the Cenozoic (or latest Cretaceous) – a split that took until the latest Eocene to complete (Blakey, 2008) (Figure 6.1B). By now Africa, South America, Australia, Antarctica and India–Madagascar were independent continental blocks, leaving the only connection between West and East Gondwana via the Antarctic Peninsula and Tasmania.

Uncertainties with regard to exact timings become more acute when trying to understand the method of fragmentation of the individual continents due to the lack of sea floor data and age constraints on magnetic anomalies along parts of the Antarctic margin. For example, the development of marine conditions in present day eastern Indian sedimentary basins suggests that the separation of India from Antarctica was diachronous from west to east (Truswell, Dettmann and O'Brien, 1999), although reconstructions based on sea floor magnetic anomalies indicate opening from east to west (Royer and Coffin, 1992; Veevers, Powell and Roots, 1991). Another consideration is the role played by microcontinental fragments that may have maintained some degree of terrestrial connection to Gondwana for a considerable period of time. One example is the Kerguelen Plateau between India and Antarctica (Figure 6.2). Sedimentary strata that cap the plateau are non-marine indicating a long period of emergence up until the Turonian/Coniacian (Shipboard Scientific Party, 2000). In the Coniacian these strata become overlain by marginal marine deposits that suggest nearby emergent land. One could argue that parts of the plateau may have been above sea level since the Early Cretaceous, for example the Kerguelen Islands today are an emergent land area with interesting Cenozoic fossils. Although small portions of the present day Kerguelen Plateau were sub-aerial during the Late Cretaceous, sea floor spreading continued to isolate this landmass from both Antarctica and India and large oceanic gaps exisited (Ali and Aitchison, 2009). The timing of Antarctic isolation from India, and Kerguelen Plateau isolation from both India and Antarctica, is further complicated by the oceanic hot spot that formed the Ninetyeast Ridges[2] (Carpenter, Truswell and Harris, 2010). This hot spot trail resulted in a chain of islands that progressively formed and sank as the hot spot moved past them – a situation similar to the Hawaiian Islands today. Pollen, spores, leaf material and carbonised wood fragments provide evidence of the Kerguelen Plateau vegetation in the early Albian (~110 Ma) (Francis and Coffin, 1992; Mohr and Gee, 1992a, 1992b; Mohr, Wähnert and Lazarus, 2002). The volcanic surface was probably initially colonised by ferns followed by the subsequent development of a more complex conifer- and fern-dominated forest with some angiosperms (Mohr and Gee, 1992b). By the early Late Cretaceous (~85 Ma) angiosperms had probably arrived (Carpenter, Truswell and Harris, 2010) but then the Kerguelen Plateau subsided below sea level sometime after ~90 Ma. Given its location on the Kerguelen Plateau (Figure 6.2), the Ninetyeast Ridge hot spot chain may have shared some sort of extended history with East Gondwana (Carpenter, Truswell and Harris, 2010; Shipboard Scientific Party, 2000), although the temporal as well as geographical extent is not well-constrained. Ali and Aitchison (2009) conclude that there is no physical data to support the hypothesis that the plateau ever formed part of an unbroken land route (>2000 km long) connecting South America through Antarctica to India and Madagascar, at least by 80 Ma. This has important implications for the southern

---

[2] A 5000 km linear seamount chain that today lies in the Indian Ocean and runs paralled to the 90th meridian.

Figure 6.2 Albian reconstruction of Gondwana. Polar view illustrating the microcontinents that made up West Antarctica and their importance in maintaining terrestrial connections between East and West Gondwana. (1) Antarctic Peninsula; (2) Thurston Island-Eights Land; (3) Marie Byrd Land; (4) Ellsworth-Whitmore Mountains; (5) Haag Nunataks. Note the potential importance of the Kerguelen Plateau (KP) (with which the Ninetyeast Ridge was more or less contiguous) prior to rifting and the Mozambique Rise (MR) in the early stages of rifting. Base map courtesy of R.A. Livermore, British Antarctic Survey. Reproduced from Cantrill and Poole (2002) with permission from The Geological Society of London.

continent connection hypothesis and associated dispersal routes (Ali and Aitchison, 2009). However, this does not preclude rafting of biota on successive islands in the hot spot chain.

Another example of problems arising due to paucity of magnetic anomaly data is that concerning Africa–Antarctica and the subsequent early stages of break-up of these continents (Marks and Tikku, 2001) including Madagascar. Reconstructions such as those of Lawver, Gahagan and Dalziel (1998), Muller, Mihut and Baldwin (1998) and Roeser, Fritsch and Hinz (1996) differ fundamentally from those of Livermore and Hunter (1996), Marks and Tikku (2001) and Segoufin and Patriat (1990), and consequently have markedly different implications for the timing of separation. For example, by 155 Ma sea floor spreading had started in the Mozambique Basin (Segoufin and Patriat, 1990) and off the Dronning Maud Land Coast (Roeser, Fritsch and Hinz, 1996). However, much of the initial African–Antarctic motion had a large transform component that probably allowed landmasses to be juxtaposed (e.g. through the Mozambique Ridge), with terrestrial connections present until at least 120 Ma (Reeves and de Wit, 2000). Recent discoveries of Late Cretaceous dinosaurs in Madagascar (Sampson *et al.*, 1998), South America and India, also suggest that these landmasses were still connected via Antarctica in the Late Cretaceous

(Hay *et al.*, 1999; Krause *et al.*, 1999; Sampson *et al.*, 1998). The uncertainties in the timing of terrestrial separation between the component continents of Gondwana still remain a major constraint to explaining present day disjunct biotic distribution patterns.

Antarctica occupied a unique position within Gondwana in that it was attached at some time to all of the other major components of the supercontinent (Figure 6.1 and 6.2). Thus an understanding of the geological evolution of Antarctica is essential for evaluating terrestrial connections across Gondwana through time. The timing and pattern of break-up is relatively well-understood for the large continental fragments, but for smaller microplates this is less well-constrained (Storey *et al.*, 1988). Two areas make up the present day Antarctic region: (i) East Antarctica, a craton, and (ii) West Antarctica, comprising a number of microcontinental fragments (e.g. Antarctic Peninsula, Thurston Island–Eights Land, Marie Byrd Land, Haag Nunataks, Ellsworth-Whitmore Mountains) (Figure 6.2). West Antarctica still represents one of the major uncertainties in Gondwana reconstructions, mainly as a result of its long and complicated history (Storey *et al.*, 1988). Today West Antarctica is separated from East Antarctica by the West Antarctic Rift System, a region now covered by the West Antarctic Ice Sheet.

Based on rift related sediments from the New Zealand margin, the study of plutonic rocks in Marie Byrd Land (e.g. Siddoway, 2008; Siddoway *et al.*, 2004; Storey *et al.*, 1999) coupled with geophysical studies (Luyendyk, Wilson and Siddoway, 2003), the development of the West Antarctic Rift System is thought to have taken place in the late Early Cretaceous (Lawver and Gahagan, 1994). In contrast, a Late Cretaceous age based on fission track thermochronology[3] has also been suggested (Fitzgerald *et al.*, 1986). The onset of rifting is not well-constrained, nor is the length of time during which the rift was active. Yet it is likely that there were several phases of development since there is also evidence for Cenozoic extension (Cande *et al.*, 2000; Hamilton *et al.*, 2001). It is clear that developing a framework that reconstructs the position of these microcontinental fragments, along with role of the West Antarctic Rift system through time, is critical for understanding biogeographic connections between the various Gondwana landmasses.

Cretaceous polar vegetation: function and physiology

Throughout most of the Cretaceous, Southern Gondwana was positioned mainly across the middle to high southern latitudes. The vegetation that established across these latitudes as far south as 85° S (Spicer and Parrish, 1986) was rich in bennettites and Cycadales, taxa more characteristic of the earlier Mesozoic. In time they were gradually replaced by conifers and, later still, underwent a further modernisation with the introduction of the angiosperms. This has enabled the use of analogues to help with experimental and modelling studies to further our understanding of the physiology and distribution of these relatively modern fossil polar biomes.

---

[3] Fission track dating is a type of radiometric dating based on the analyses of fission tracks formed in certain Uranium-bearing minerals (e.g. apatite) as Uranium decays. The fission track age is a function of the number of tracks that have formed and the amount of Uranium remaining as measured by induced fission events. This method enables a reconstruction of the time–temperature histories of rock samples from shallow parts of the Earth's crust.

An extraordinary feature of these polar forests was their ability to tolerate the extremely seasonal light regime. This included protracted permanent darkness between two periods of extremely low irradiances during the winter months, followed by summers with continuous low-angle and low to moderate irradiance. Assuming that the obliquity of the Earth's rotational axis has remained constant over geological time (Barron, 1984), forests growing at 70° S would have experienced up to 70 days of unbroken darkness each year, whereas those at 85° S would have been deprived of light for nearly 160 days per year (Read and Francis, 1992). Different biomes at different palaeolatitudes may have evolved different strategies to cope with this extreme seasonality. Deciduousness is one strategy that modern plants exploit to cope with seasonal conditions. Conversely, an evergreen habit is more commonly associated with relatively aseasonal conditions. Indeed the question relating to the evergreen *versus* deciduous nature of polar forests has remained elusive until fairly recently. This is not simply a piece of additional information that completes the picture of polar ecology. On the contrary, the question of habit has important implications for past (and future) climate. At high latitudes the effects of a vegetated, as opposed to a barren, land surface are most pronounced when the cover is a forest. In such cases the effect exerted on regional climate is on a scale similar to either that of doubling atmospheric pCO_2 or to those changes associated with orbital forcing (Bonan, Pollard and Thompson, 1992; Foley *et al.*, 1994). A vegetated land surface, especially with forests or woodland biomes, significantly decreases the albedo of the land surface by increasing the amount of solar radiation absorbed. As more warmth is absorbed, the efficiency with which heat and water vapour are transferred from the land back to the atmosphere and into adjacent oceans also increases and so initiates a positive feedback that further increases and maintains high latitude temperatures (Dorman and Sellers, 1989; Upchurch, Otto-Bliesner and Scotese, 1999). Therefore, vegetation with an evergreen habit would greatly enhance this positive feedback relative to one with a deciduous habit, especially with respect to winter temperatures (Upchurch, Otto-Bliesner and Scotese, 1999). The presence or absence of a winter canopy also has important implications for the terrestrial faunal ecology, such as therapod, sauropod and ornithopod dinosaurs that inhabited these Cretaceous polar latitudes (Dettmann *et al.*, 1992). The result is that leaf longevity and the seasonal timing of leaf abscission in a community is closely related to climate and provides important data needed to refine numerical models of (past) global climates and the interplay between vegetation and environments (Upchurch, Otto-Bliesner and Scotese, 1999; Wolfe and Upchurch, 1987).

The forests that covered the polar region during the late Early Cretaceous comprised long-lived trees with growth ring sequences reaching 180 years (Chapman, 1994). Their trunks reached diameters of 1 m and heights of up to 40 m, and productivity is estimated to have been similar to those of temperate deciduous forests today (Chapman, 1994; Equiza, Day and Jagels, 2006; Falcon-Lang and Cantrill, 2000). Low temperatures would not have restricted plant growth, although the light regime would differ qualitatively from that to which contemporary temperate tree species are adapted (Equiza, Day and Jagels, 2006). For decades it was simply assumed that plants growing in these polar latitudes would have coped with the long, dark and probably cold winters by being deciduous. Yet only recently have experiments been designed to test this hypothesis. Results clearly show the cost of annually shedding leaves for deciduous trees greatly exceeds the cost of respiration for an evergreen canopy. When this is scaled up to the whole forest, a deciduous habit would have been approximately twice as expensive in terms of winter carbon loss when compared with its evergreen counterpart (Brentnall *et al.*, 2005; Osborne, Royer and Beerling, 2004; Royer, Osborne and Beerling, 2003). Therefore this suggests that the polar canopy might indeed have been evergreen. However, these authors also observe distinct 'pulses' of carbon gain occurring during spring and autumn in deciduous trees that were missing from evergreen trees. This allowed deciduous trees to 'catch up' in terms of their annual carbon capture and put them on an equal footing with trees with an evergreen habit. Interestingly neither deciduous nor evergreen trees were able to capitalise fully on the continuous light of the polar summer. Continuous light for six weeks would have saturated the photosynthetic capacity resulting in suppression of the photosystem and was probably limited by root nitrogen uptake (Osborne and Beerling, 2003). Therefore functioning of ancient polar forests might not have been as dependent upon carbon balance as previously thought, and other aspects of plant physiology, such as leaf retention time, may shed further light on the functioning of these unique ecosystems (Osborne, Royer and Beerling, 2004).

The fossil record coupled with model simulations indicate a predominance of evergreen trees with leaf retention spans of up to four years in Early Cretaceous Antarctic forests (Brentnall *et al.*, 2005; Cantrill and Falcon-Lang, 2001; Falcon-Lang, 2000a, 2000b). Within the Antarctic environment (e.g. Alexander Island) volcanic disturbance, including fire, was a distinct, albeit relatively infrequent, possibility. Slow colonising trees with their long leaf retention times (~4 years) would have had enough time to reach a competitive advantage for light, in terms of height, relative to their deciduous counterpart. This tips the canopy balance in favour of evergreen trees such that with time only a small proportion (~10%) of the land surface would have supported deciduous trees (Brentall *et al.*, 2005). With an increase in disturbance frequency the balance may have tipped back in favour of deciduous trees, thus serving to decrease the albedo of the land surface with potential climatic implications.

These insights coupled with computer climate–vegetation models have begun to provide answers to other aspects of the physiological ecology of the polar biome (e.g. Beerling and Osborne, 2002; Osborne, Royer and Beerling, 2004). With increased understanding of polar plant physiology and productivity, more realistic continental-scale palaeoclimate–vegetation generalisations can be made (e.g. Brentnall *et al.*, 2005). Simulated polar biogeography, derived from data and interpretations obtained directly from fossil material, suggest that in both the Northern and Southern Hemisphere high latitudes there were alternating belts of predominantly evergreen and deciduous forest biomes (e.g. Brentall *et al.*, 2005).

In the Southern Hemisphere the models simulate predominantly evergreen forests (with relatively long leaf retention times) ranging from 40° S and giving way to a belt of deciduous forest at ~60° S. Then above 80° S this deciduous biome grades into a central core of evergreen forest, but this time with short leaf retention times (12–18 months) (Brentnall *et al.*, 2005).

Observed (but as yet sparse) fossil data from the Southern Hemisphere show initial support for these simulated patterns except for the extreme southern high latitude core, where there is no fossil evidence to test the validity of the model.

Fossil data from Antarctica is currently sparse and restricted to areas along the edge of the present day ice sheet, and the pool of deciduous taxa available to form deciduous forests is severely restricted (i.e. to *Ginkgo* and some Cupressaceae conifer types). Yet evidence from the Antarctic Peninsula clearly shows that evergreen forests occupied this region (~62° S) and extended around the eastern margins (e.g. Dettmann, 1989; Falcon-Lang and Cantrill, 2001a, 2001b; Upchurch and Askin, 1989). Moreover, a number of Early to mid-Cretaceous sites yielding both micro and macrofossils (wood and foliage), located between 40 and 60° S palaeolatitude along the Palaeopacific coastline in what are now New Zealand, India, southern South America, and the more continental setting of southeastern Australia, provide evidence for coniferous taxa such as Araucariaceae, Podocarpaceae and Taxodiaceae with long leaf retention spans (i.e. 4–8 years) (Archangelsky, 1963a; Cantrill, 1991, 1992; Cantrill and Douglas, 1988; Cantrill and Falcon-Lang, 2001; Cantrill and Webb, 1987; Dettmann, 1989; Dettmann *et al.*, 1992; Douglas and Williams, 1982; Falcon-Lang and Cantrill, 2001a, 2001b; Parrish *et al.*, 1998; Upchurch and Askin, 1989). During the Lower Cretaceous in East Gondwana (i.e. New Zealand), at a palaeolatitude of 68° S and adjacent to the East Antarctic landmass, fossils provide evidence for a mixed evergreen and deciduous forest that became increasingly evergreen throughout the Late Cretaceous, contradicting the simulated occurrence of a general deciduous biome at this latitude. Abundant araucarian fossils testify to the dominance of evergreen conifers, but both deciduous and evergreen species formed the canopy, as evidenced by the presence of deciduous taxa such as *Ginkgo* and evergreen podocarps (Parrish *et al.*, 1998; Stopes, 1914, 1916). Moreover, abundant mid-Cretaceous fossil floras lying at a palaeolatitude of 70–85° S have been described from the Otway Basin in southeastern Australia. Here the vegetation was dominated by a mixed evergreen–deciduous or predominantly evergreen biome with conifers, such as araucarians and podocarps, alongside a deciduous, or semi-deciduous, component such as *Ginkgoites* and *Elatocladus* (Cantrill, 1991, 1992; Cantrill and Douglas, 1988; Cantrill and Webb, 1987; Dettmann *et al.*, 1992; Douglas and Williams, 1982). The fossil record provides some limited support for the biome model, but with further investigations of fossil floras any biome zonation of deciduous and/or evergreen forests will become clearer and biogeographic models can then be refined.

Floristic provincialism

Continental-scale generalisations of palaeoclimate and vegetation interactions have been the focus of much research concerned with understanding Early Cretaceous polar vegetation. This has been undertaken at the expense of community-scale studies despite the fact that there is abundant fossil material available that can provide the necessary data to test large-scale generalisations. When intracontinental evidence is amassed from community studies already undertaken, the vegetation suggests that the Southern Hemisphere was marked by strong floral provincialism delimited by ecological and geographical differences (Herngreen *et al.*, 1996).

Within the pollen flora one such distinct floral province was dominated by bisaccate and trisaccate pollen grains largely allied to the Podocarpaceae and important subsidiary fern spores. This Trisaccate Province characterised microfloras of Australia, India, New Zealand, Antarctica, southern Africa and southern South America (Herngreen *et al.*, 1996 and references therein) (Figure 6.3), and was separated by a transitional zone from the more northerly African–South American Elaterates Province (characterised by certain well-known angiosperm pollens such as *Afropollis*, conifer pollen such as *Classopollis*, *Araucariacites*, fern spores such as *Gleichiniidites*, and dinoflagellates). Within this large Trisaccate Province two subprovinces can be recognised: The *Cyclusphaera–Classopollis* subprovince (Africa plus southern South America), and the *Murospora* subprovince (Australia and India) (Herngreen *et al.*, 1996). Macrofloral evidence for these large floristic provinces is hampered by the lack of well-documented floras from many regions, in particular Antarctica (although investigations of both new, and previously described, floras are significantly expanding the macrofloral and microfloral record from this region). Within these two provinces the microflora exhibits important regional variations. For example, *Microcachryidites antarcticus*, one of the elements that define the *Murospora* subprovince, typically represents up to 25% by abundance of palynoassemblages in Australian localities. In contrast, *M. antarcticus* is never more than 5–6% abundance in the Antarctic Peninsula microfloras (Cantrill and Poole, 2002). Moreover,

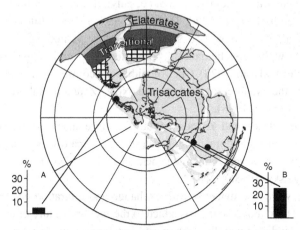

Figure 6.3 Floristic provincialism across Gondwana. Note regional differences in abundance of key trisaccate species such as *Microcachryidites antarcticus*, one of the elements that made up the *Murospora* subprovince in the Trisaccate Province. Histograms show the percentage abundance of *Microcachryidites antarcticus* in Lower Cretaceous strata. (A) Antarctic Peninsula from Dettmann and Thomson (1987); (B) southeast Australia from Dettmann (1986b) and central Australia from Burger (1980). Floristic boundaries derived from Herngreen *et al.* (1996). Reproduced from Cantrill and Poole (2002) with permission from The Geological Society of London.

along this transect from South America to Australia regionalism was also present at a taxonomic level in the subcanopy and understorey associations, and is believed to have been influenced by climatic as well as edaphic factors (Dettmann and Thomson, 1987). For example, the ferns *Cyatheacidites annulatus* (Lophosoriaceae) and *Polypodiaceoisporites elegans* (?Polypodiaceae) are present in the South American strata but absent from Australian deposits of the same age (Dettmann, 1986a).

Based on both macro- and microfossil data, McLoughlin (2001) outlined four floristic subprovinces, or regions of endemism, that can be identified within southern Gondwana. These included southeastern Australia–New Zealand, northern Australia–India, southern South America–Antarctic Peninsula, and South Africa, which by this time had separated from the supercontinent. The Early Cretaceous vegetation of Australia was characterised by ferns and lycopods. The ferns were more diverse and abundant in coastal settings whereas away from marine influence the lycopods were more common, along with sphagnalean mosses and liver-worts in low-relief, epicontinental basins (Dettmann, 1994). *Phyllopteroides* ferns (which peaked in abundance and diversity in this subprovince) characterised southeastern Australia along with bryophytes, small-leafed bennettitaleans (with <5 mm^2 pinnules), abundant pentoxylaleans and seed plants (e.g. *Komlopteris*, *Pachypteris* and '*Rienitsia*') (Douglas, 1969; Drinnan and Chambers, 1986). By the Aptian, ginkgophytes had become common yet cheirolepidiacean conifers remained in low abundance (Dettmann, 1994). New Zealand was also rich in *Phyllopteroides* ferns, small-leafed pentoxylaleans and bennettites (Arber, 1917; Edwards, 1934; Parris, Drinnan and Cantrill, 1995) although they appear to persist slightly later in the stratigraphic record here than in Australia (McLoughlin, 2001).

In the northern Australia–India subprovince, palynofloras were more diverse relative to those further south (Burger, 1990; Dettmann, 1973). Bennettites were large-leafed (commonly >100 mm^2 pinnules and some reticulately veined) and characterised the landscape along with gleicheniacean, dipteridacean and *Phyllopteroides* ferns (McLoughlin, 1996; Walkom, 1928). Cheirolepidiacean conifers were also more abundant here relative to the southeastern Australia–New Zealand subprovince, as were Araucariaceae (Dettmann, 1994).

Large-leafed bennettites and cycads of large stature also characterised the South America–Antarctic Peninsula floras, with greatest Gondwana diversity appearing in South America, which included several endemic genera (e.g. *Ticoa*, *Ruflorinia*, *Mesosingeria* and *Ktalenia*; Archangelsky, 1963a).

Floral elements were also dynamic with exchange and migration occurring within and between regions. These movements also added to the regional nature of the vegetation and occurred stepwise and latitudinally (east–west) and longitudinally (north–south). This is evidenced from both the macro- and microfloras. East to west migration is exemplified in the macroflora by *Phyllopteroides* being reported in Australia from the Aptian (Cantrill and Webb, 1987), but not occurring in Antarctica until the late Albian. This pattern extends to other taxa such as *Aculea bifida* (Cantrill, 1996; Cantrill and Poole, 2002). West to east exchange is highlighted by the taxa such as *Dicotyozamites* and *Pachypteris* that are present

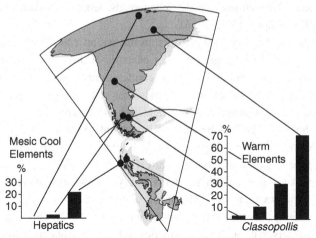

Figure 6.4 Palaeolatitudinal gradient through South America and the Antarctic Peninsula highlighting the increasing diversity of mesic elements (hepatics) and the decrease in abundance of thermophilic elements (e.g. *Classopollis*) from high to low latitudes. Diversity data derived from Pons (1988), Falcon-Lang, Cantrill and Nichols (2001) and Riccardi (1988). Abundance data for *Classopollis* derived from Dettman and Thomson (1987), Herngreen (1975), Herngreen *et al.* (1996) and Prámparo and Volkheimer (1999). Reproduced from Cantrill and Poole (2002) with permission from The Geological Society of London.

in the latest Jurassic to Early Cretaceous sediments of Antarctica and the Baqueró Formation in South America, but absent from Australia until the Aptian.

North–south floristic gradients are also present in both the micro- and macrofossil record. Many palynomorph taxa that occur in both the southeastern Australia–New Zealand sub-province and the northern Australia–India subprovince appeared slightly earlier in the stratigraphic record of the former region (Helby, Morgan and Partridge, 1987), suggesting poleward advance with climatic amelioration during the mid-Cretaceous (McLoughlin, 2001). In the northern part of Gondwana (India, northern Australia and Argentina) the conifer that produced the pollen *Hoegisporis* was restricted to the northern (and western) areas of Australia, along with cheirolepidiacean conifers that were more abundant in the north when compared with the south (Dettmann, 1994). Moreover araucarians were generally a more important contributor to the forest canopy in the north relative to the more humid south where podocarps dominated. In Australia, further to the south, the vegetation was more of a mixed evergreen–deciduous or predominantly evergreen type, with conifers such as araucarians and podocarps alongside a deciduous or semi-deciduous component such as *Ginkgoites* and *Elatocladus* (Cantrill, 1991, 1992; Cantrill and Douglas, 1988; Cantrill and Webb, 1987; Dettmann *et al.*, 1992; Douglas and Williams, 1982). Compare this with New Zealand where the forests were evergreen and dominated by araucarian conifers with fewer podocarps and *Ginkgo* contributed to the upper canopy level. The understoreys in the north and south were, however, more similar with taeniopterids, and other pentoxylaleans, bennettites, cycads, sphenopsids, ferns and liverworts all coexisting (Parrish *et al.*, 1998; Stopes, 1914, 1916).

The southern South America–Antarctic Peninsula region can also be used to illustrate the north–south palaeolatitudinal gradient. For example in the Aptian, *Classopollis* pollen (produced by dry-loving cheirolepidiacean conifers growing in a warm climate on well-drained soils of upland slopes and lowlands near coastal areas) accounts for 30–60% of certain palynofloras by abundance. However, at high latitudes this abundance drops sharply (Dettmann and Thomson, 1987) with *Classopollis* accounting for only 0–5% of the residue in the Antarctic Peninsula region (Figure 6.4). This pattern is also seen in the Elaterate (Gnetales), and later angiosperm, pollen record (Cantrill and Poole, 2002). The reverse is also true – some taxa (e.g. hepatophytes, ferns and lycopsids) were more abundant at high latitudes (e.g. lycophytes represent 18% of the within-floral diversity) than low latitudes (lycophytes are absent) (Cantrill, 1997; Cantrill and Poole, 2002). One explanation for the floristic gradients observed through South America and into the Antarctic Peninsula is a strong climatic gradient (Dettmann, 1986a). Thermal and precipitation gradients through this latitudinal transect are thought to have been extreme (Valdes, Sellwood and Price, 1996) and this is supported by plant fossil evidence. Yet with enlargement of the South Atlantic Ocean this abrupt vegetational zonation across southern South America and the Falkland Plateau and into the Antarctic Peninsula appears to have decreased (Dettmann, 1986b, 1989; Dettmann *et al.*, 1992). Interestingly, the vegetation fringing this ocean on the Falkland Plateau and in South Africa retained its cheirolepidiacean character for much of the Early Cretaceous (Dettmann, 1986b).

Further examples of this palaeolatitudinal floral gradient can be found in Aptian floras on the Antarctic Peninsula, and in southern South America (e.g. Baqueró Formation). Antarctic floras with their abundant cool- and moisture-loving plants contrast strongly with South American floras that are dominated by other desiccation-tolerant, warm-loving plants such as bennettites and other seed plants. Even when the same taxa (e.g. bennettites) are represented in both locations, their morphological and anatomical adaptations (e.g. strongly sunken stoma, thick cuticles) also testify to the presence of an extreme thermal and moisture gradient acting as an effective barrier against palaeolatitudinal plant migration at this time (Archangelsky, 1963a, 1963b; Archangelsky and Taylor, 1986; Archangelsky, Taylor and Kurman, 1986; Archangelsky and Villar de Seoane, 2004; Archangelsky *et al.*, 1995; Cantrill and Poole, 2002; Villar de Seoane, 2005).

Early angiosperms and subsequent vegetational change in the southern high latitudes

Although general regionalism was evident along both longitudinal and latitudinal transects, reflecting both climatic and geographic gradients, the taxonomic composition of the vegetation that extended across southern Gondwana during the late Early Cretaceous was essentially characterised by extensive conifer forests with important and diverse ferns, liverworts and mosses (e.g. Cantrill, 1997; Cantrill and Poole, 2002; Douglas, 1973; Palma-Heldt *et al.*, 2004, 2007). Along forest fringes *Ginkgo* grew alongside cycads, taeniopterids, bennettitaleans, and other seed plants, tree ferns and ferns. Ferns also grew across more open land forming heaths (Cantrill, 1996). In regions comprising areas of damp habitat, semi-aquatic to fully aquatic associations of mosses, liverworts, Isoetales and Equisetales colonised. Yet the appearance of a new (monoaperturate) pollen type in the late Barremian – early Aptian of the southern high latitudes heralded the arrival of a new competitor in the flora – the angiosperms. Yet in order to determine the timing of their first appearance, the recognition of angiosperms is paramount and so to this end a well-constrained stratigraphic framework and presence of unique defining characters, by which angiosperm organs can be identified, is required (see Friis, Pedersen and Crane, 2006).

The earliest unequivocal angiosperm fossils worldwide have been reported from Hauterivian[4] (~135 Ma) and younger strata where diverse and distinctive monoaperturate pollen occurs (see Hughes, 1994). Grains from Israel ascribed to angiosperms, which are of possible late Valanginian–early Hauterivian age (Brenner, 1996), are more difficult to evaluate because the available illustrations reveal few structural details (Friis, Pedersen and Crane, 2010). Some of the earliest angiosperm megafossils are found later in the lower Aptian to lower Cenomanian sequences from the United States and Portugal (Hochuli, Heimhofer and Weissert, 2006; Royer *et al.*, 2010 and references therein). To account for the substantial gap between the first microfossils and the first macrofossils appearing in the rock record, various hypotheses have been put forward. Evidence is mounting for the early angiosperms to have adopted a fast-growing, weedy, shrub or herb habit, colonising either disturbed streamside and/or dry habitats with associated low preservation potentials. These fast-growing plants with their short leaf retention time and high photosynthestic rates probably coexisted with ferns that had a similar life strategy, but the rapid flower-to-seed timespan gave the angiosperms competitive advantage in physically disturbed, nutrient-rich environments at a time of major climatic destabilisation (Heimhofer *et al.*, 2005; Royer *et al.*, 2010; Wing and Boucher, 1998). Other evidence points either to an aquatic, or marsh origin (Friis, Pedersen and Crane, 2001; Sun, Dilcher and Zheng, 2008), or to slow-growing, understorey evergreen shrubs and small trees in dimly lit, evermoist, physically-disturbed montane forests (Feild and Arens, 2005; Feild, Chatelet and Brodribb, 2009; Feild *et al.*, 2004). In the Northern Hemisphere it is well-established that angiosperms rose in the low palaeolatitudes of what is now tropical Africa and South America and, as a result of several major pulses of radiation that increased diversity exponentially, spread polewards. By the late Barremian to early Aptian (~125 Ma) the micro- and macrofossil records document a global array of different angiosperm fossil types (flowers, leaves, pollen; e.g. Friis, Pedersen and Crane, 2006; Hickey and Doyle, 1977; Li, 2003) indicating that eudicot angiosperms[5] were geographically widespread with their establishment

---

[4] Records of potential pre-Cretaceous angiosperm fossils (such as those of Sun *et al.*, 1998; Wang and Wang, 2010) are considered to be either questionable and/or, in several cases, have been demonstrated to have been misplaced chronostratigraphically or misinterpreted systematically (Friis, Pedersen and Crane, 2006; Friis *et al.*, 2003).

[5] eudicotyledonous angiosperms (or tricolpates) form a monophyletic group of the flowering plants constituting >70% of all angiosperms.

coinciding with the progressive long-term warming trend (Crane, Friis and Pedersen, 1995; Crane and Lidgard, 1989; Heimhofer *et al.*, 2005). The radiation and rise to ecological dominance thus caused major, irreversible changes in the global vegetation (Friis, Pedersen and Crane, 2006; Lidgard and Crane, 1990). However, the paucity of knowledge about Southern Hemisphere floras, and in particular those of Antarctica, means that the pattern of angiosperm radiation and diversification in the Southern Hemisphere is much less well-constrained (Drinnan and Crane, 1990). Evidence for early angiosperms in the gymnosperm–fern dominated floras can be found in a number of continents that made up southern Gondwana, and together provide an idea of timings of origins and radiation patterns across this region through the Lower Cretaceous.

In Australia, the earliest evidence of angiosperms comes from the basal part of the Aptian in the *Cyclosporites hughesii* spore-pollen Zone of the Eromanga Basin (Burger, 1988, 1993; Dettmann, 1986b; Dettmann, Clifford and Peters, 2009) (Figure 6.5) where both *Clavatipollenites* and *Asteropollis* occur. In slightly younger late Aptian strata, *Clavatipollenites* is joined by leaves and tricolpate pollen (*Tricolpites variabilis*, *Rousea* spp.), and other species indicative of eudicots appear in the early middle Albian (Burger, 1993). Across the Australian continent the mid-Cretaceous (post-late Albian) was largely a period of non-deposition and deep weathering (BMR Palaeogeographic Group, 1990) so little evidence exists as to the type of vegetation that grew at this time. Only in northeastern Australia is the vegetation from the latest Albian–Cenomanian known in any detail. Here a large but shallow inland waterway, the Eromanga Sea, covered much of central west

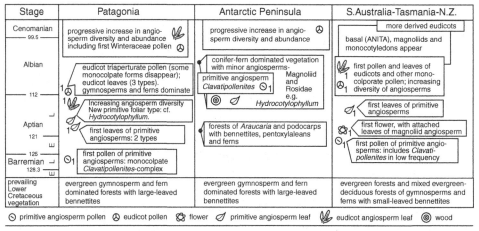

Figure 6.5 Summary of the timing of events relating to angiosperm introduction to, and radiation across southern Gondwana (N.Z. = New Zealand) along a latitudinal transect. Where the age of the first occurence (1) is known, the symbol is outside the box, when the age range of the strata in which the fossil(s) was found covers a larger time span this is indicated by the symbols placed within the box marking that range. For further details see text and references cited therein.

Queensland. Over time as the continent uplifted and the sea gradually regressed, a basin was left which subsequently became infilled with sediment. This sediment was brought into the basin by great meandering rivers, similar in size to the Amazon or Mississippi today. River plains, swamps, lakes, lagoons and coastal estuaries characterised this region and the remains of this ecosystem can be found in the Winton Formation. By virtue of its age and the prevailing environmental conditions under which the rocks were deposited, this formation has preserved one of the richest ecosystems in Australia. It supported sporadic terrestrial and aquatic vertebrate fauna such as sauropod dinosaurs (Hocknull *et al.*, 2009), crocodiles and turtles and an array of invertebrates as well as a diverse flora (Dettmann and Clifford, 2000; Dettmann, Clifford and Peters, 2009; McLouglin, Drinnan and Rozefelds, 1995; McLoughlin, Pott and Elliott, 2010; Peters and Christophel, 1978; Pole, 1999).

Lying north of the Antarctic Circle, the climate prevailing over this basin was humid and warm enabling diverse plant communities to thrive on the open floodplains. Across the uplands bordering this lower land, subalpine vegetation dominated. Scattered conifer trees dominated lowland forests comprising other seed plants. In the understorey there is the first evidence of significant, yet minor, flowering plants (Dettmann, Clifford and Peters, 2009; McLoughlin, Drinnan and Rozefelds, 1995; McLoughlin, Pott and Elliott, 2010). This disputes the suggestion that angiosperms arrived in Australia via valleys associated with the broadening of the rift in the Gippsland–Otway region aided by both the opening of the South Atlantic and Indian oceans and the associated decrease in sea level (Dettmann and Thomson, 1987). Evidence for the earliest angiosperms comes from pollen grains of *Clavatipollenites*, the earliest known angiosperm, recorded from (Barremian)– Aptian sediments around the margin of the shallow epicontinental sea in the Eromanga Basin (Burger, 1993; Dettmann, Clifford and Peters, 2009). Monocolpate forms such as *Asteropollis* appear in the early Albian and *Liliacidites* in the middle Albian (Burger, 1993) (Figure 6.5). Cuticular fragments (including dicot types believed to be of chlorantha-ceous or illiciaceous affinity and monocotyledonous, possibly Areciflorae, affinity; Pole, 1999, 2000a) and flowers (with possibly affinity to *Gomortega*, Laurales) provide incon-trovertible evidence of magnoliid clade angiosperms having been established in Australia by the late Albian times (Dettmann, Clifford and Peters, 2009). Further angiospermous macro-fossils that appear in sediments of late Aptian and Albian age in Victoria (Otway Basin) and Queensland include chloranthaceous or saururaceous flowering shoots, and leaves (of uncertain familial affinity)[6] (Douglas, 1965, 1969, 1973, 1994; McLoughlin, Drinnan and Rozefelds, 1995; Medwell, 1954; Taylor and Hickey, 1990; Walkom, 1919).

The vegetation consequently underwent a change as the angiosperms, with their herb and arboreal habit, began to encroach into new habitats. The pattern of clade representation in the fossil assemblages at this time suggests that there was indeed a causal link between angiosperm diversification and the decline of key understorey and mid-storey plants, particularly equisetaleans, Bennettitales, ginkgophytes, other non-angiospermous seed plants,

---

[6] Also described from Aptian and Albian sediments of Victoria are impressions of awned nut-like structures (of Douglas, 1963, 1969) that have been compared with *Hemitrapa*, an aquatic early Cenozoic angiosperm. However, these were more recently assessed to be of uncertain angiospermous origin (Drinnan and Chambers, 1986; Friis, Crane and Pedersen, 2011).

and some fern families (McLoughlin, Drinnan and Rozefelds, 1995; McLoughlin, Pott and Elliott, 2010; Pole, 1999, 2000a, 2000b; Pole and Douglas, 1999). These angiosperms probably formed an herbaceous or shrubby groundcover in lakeside and riverine habitats and across the broad coastal plains of southeastern Australia (Dettmann, 1994). The dominant podocarps and *Araucaria* conifers started to decrease. Equisetales, Pentoxylales, *Ginkgo* and bennettites, such as *Ptilophyllum*, became increasingly scattered across the landscape along with rare cycads (McLoughlin, Pott and Elliott, 2010 and references therein). Ferns, such as *Gleichenia*, gradually became more dominant and abundant in the understorey niches and spread out to form open fern prairies of *Alamatus*, *Phyllopteroides*, *Sphenopteris*, *Coniopteris* and *Microphyllopteris* (Cantrill, 1996). After an initial proliferation, the bryophytes also declined both in relative diversity and abundance along with the lycopods, even though *Sphagnum* remained locally abundant. Certain ferns (e.g. Schizaeaceae) then also succumbed although the diversity of other ferns, such as Osmundaceae, Matoniaceae/Dipteridaceae and Cyatheaceae/Dicksoniaceae, remained unchanged.

The angiosperms appear to have been subsidiary elements in the flora for several million years through the Aptian until the early middle Albian when the angiosperm pollen provides evidence for an explosive diversification that is matched, to a degree, by the macrofossil record (McLoughlin, Pott and Elliott, 2010). Whether the change in climate, from the seasonally frigid conditions of the earlier Cretaceous to the warmer conditions characteristic of the mid-Cretaceous climate, was linked to the decline of the gymnosperms and expansion of the angiosperms, or whether this was coincidental remains unclear (McLoughlin, Pott and Elliott, 2010).

To the west, in southern South America, under a seasonal temperate climate, similar non-angiosperm seed plant–fern vegetation dominated during the Early Cretaceous. Local environmental differences resulted in edaphic changes and microclimatic effects, which in turn gave rise to variations in vegetation assemblages (Archangelsky and Taylor, 1986; Archangelsky *et al.*, 1995). In valleys, araucarians, cycads, *Taeniopteris* and *Gleichenites* were abundant and formed the main ground cover. Around lakes and ponds, conifers such as *Brachyphyllum*, *Ginkgo*, bennettites, ferns (e.g. Osmundales), waterferns, lycopods (Selaginellales and Isoetales) dominated. Xeromorphic characters of the leaves of a number of seed plants (Archangelsky *et al.*, 1995; Villar de Seoane, 2005) suggest that the plants were living under, and thus adapting to, conditions of stress arising from volcanic disturbance and/or the relatively warm, dry climate. During the late Barremian–mid-Aptian in Santa Cruz Province (Argentina) gymnosperms, pteridophytes (with some fern families at their maximum diversification) and bryophytes dominated the vegetation (Archangelsky, 2001, 2003; Quattrocchio *et al.*, 2006). In lenses within the Springhill Formation (Austral Basin; ~126 Ma) the first angiosperm remains, i.e. *Clavatipollenites* pollen, have been found (Figure 6.5). The earliest occurrence of leaves in South America exhibiting a morphology typical of angiosperms from early Late Cretaceous floras globally[7], have been found in the late Aptian (~119 Ma) sediments of the Anfiteatro de Ticó Formation (Deseado Masif Basin) to the north (Archangelsky *et al.*, 2009 and

---

[7] Generally entire and unlobed micro- or notophyllous with a pinnate-brochidodromous or eucamptodromous venation pattern with relatively irregular intercostal areas and disorganised higher venation categories – characters that set them apart from modern angiosperm families (Cúneo and Gandolfo, 2005).

referencess therein; Romero and Archangelsky, 1986). These angiosperms are marginally represented and constitute a subordinate component of the floras (Archangelsky *et al.*, 2009). Within 1 Ma there had been an incipient increase in angiopserm diversity (Llorens, 2003). By the latest Aptian–early Albian, although ferns were still abundant and diverse, and taxodioid and cheirolepidiacean conifers thrived (Llorens and del Fueyo, 2003), the vegetation was beginning to change as the still subordinate angiosperm component became more abundant and diverse (Cúneo and Gandolfo, 2005). Floras of the earliest Albian (~110 Ma) are characterised by the presence of the first definitive fossils with affinities to eudicots. These early eudicots are represented by both pollen (i.e. *Tricolpites* and *Retitricolpites*; Medina, *et al.* 2008; Volkheimer and Salas, 1975) and leaves. As several of the early angiosperm pollen types became extinct, eudicots progressively increased in abundance and diversity through the conifer–non angiosperm seed plant–fern dominated landscape of the Albian (e.g. Archangelsky *et al.*, 2009; Prámparo, 1990, 1994; Puebla, 2004). During the mid-Cretaceous these landscapes were becoming increasingly influenced by volcanic activity (Archangelsky *et al.*, 1995) that repeatedly devastated soils and vegetation. Although the early angiosperms may have taken advantage of these continually disturbed environments, as ash fall increased, so soil development and the formation of plant-bearing horizons became increasingly rare and eventually disappeared altogether (Archangelsky *et al.*, 1995).

To the south across the northern Peninsula region of Antarctica, typical Early to mid-Cretaceous highly productive forests were thriving (Dettmann, 1989; Palma-Heldt *et al.*, 2004) under an equable, wet and temperate climate. Growth rings from wood further south in Alexander Island suggest that the climate was markedly seasonal (Jefferson, 1982a). The vegetation of the Antarctic Peninsula region reflected the prevailing climate with profuse abundant cool and moisture-loving taxa such as liverworts and ferns. Yet the earliest evidence of angiosperms in Antarctica comes from rare pollen grains found in principally podocarp-dominated floras of the early Albian, some 15 Ma after their first recorded appearance worldwide, thus postdating their arrival in Australia and southern South America. These *Clavatipollenites* pollen grains are found in the Kotick Point Formation (early to mid-Albian) assemblage on James Ross Island and are recognised as being chloranthaceous in origin, indicative of a shrubby habit exhibited by these earliest angiosperms (Dettmann and Thomson, 1987; Riding and Crame, 2002). The earliest angiosperm macrofossil (leaf) evidence comes from sediments of late Albian age on Alexander Island (see below) and it was some 10 Ma later, during the Turonian, before evidence reveals an increase in angisoperm abundance coeval with the Campanian–Turonian radiation of angiosperm taxa in northern latitudes (Askin, 1983, 1992; Cantrill and Nichols, 1996; Dettmann and Thomson, 1987; Lupia, Lidgard and Crane, 1999; see Chapter 7).

Even though our understanding regarding the timing of angiosperm introductions into the southern high latitudes is gradually becoming resolved, the question still remains as to the route(s) by which these angiosperms moved out of the low latitudes southwards towards the pole. Regionalism may have been facilitated through the Antarctic Peninsula region, which offered a major connection for floral exchange between East and West Gondwana (Hill and Scriven, 1995). However, other alternatives have been offered, such as along the widening rift valleys between the separating landmasses of Africa, India, Australia and Antarctica (Dettmann and Jarzen, 1990) with the climatic gradient acting as an important filter and a barrier to exchange (Cantrill and Poole, 2002). Establishing the timing and development of the systematic composition of these early angiosperm floras is important for understanding both the radiation and diversification of this group across Antarctica and the rest of Southern Gondwana, as well as implications in relation to the modernisation of the vegetation structure and ecology in the southern high latitudes.

Angiosperm migration patterns across the southern high latitudes

The pathway of angiosperm migration into, and their subsequent radiation across, Gondwana is underpinned by the timing of gondwanan break-up and the existence of biotic gateways, and facilitated by the ameliorating climate during the mid-Cretaceous. Terrestrial connections between the landmasses provided routes for plant migration from warmer, more tropical regions southwards towards the high southern latitudes. If Africa–South America had separated from Gondwana by at least 155 Ma this was substantially before angiosperm radiation and diversification and thus migration could not have taken place into Antarctica via Africa. Since the Antarctic Peninsula offered the only connection between West and East Gondwana at this time, this gateway has been postulated as providing the main connecting link for radiation into the high latitudes. However, alternative scenarios have also been postulated: palaeogeographic reconstructions of southeast Asia, Australia and East Antarctica during the Cretaceous indicate land connections between these areas (Metcalfe, 1990), and point to a possible invasion from Asia to the north (Burger, 1981; Hill and

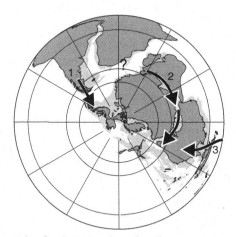

Figure 6.6 Alternative scenarios for the expansion of angiosperms across Gondwana plotted on an Albian reconstruction. (1) Invasion route via South America and Antarctic Peninsula; (2) invasion via India; (3) invasion from Asia to the north. The fossil record suggests that initial invasion occurred either through Africa? or India. Base map provided by R. A. Livermore, British Antarctic Survey. Reproduced from Cantrill and Poole (2002) with permission from The Geological Society of London.

Scriven, 1995; Truswell, Kershaw and Sluiter, 1987). Other workers have postulated that angiosperms arrived in Gondwana via India (Burger, 1990) (Figure 6.6). Although the exact route, or routes, still remains a matter for conjecture from a geological perspective, evidence from the palaeobotanical realm can help to shed further light on this matter.

Cretaceous angiosperm floras from Antarctica are critical for testing the validity of hypothesised radiation routes (Truswell, 1991). If the Antarctic Peninsula acted as the main gateway, then Cretaceous floras from this region should provide earlier records of angiosperms than elsewhere in eastern Gondwana (i.e. Australia, New Zealand, India). Alternatively, if angiosperms radiated into Gondwana via another route, then they should appear later in Antarctic floras than elsewhere in Gondwana (Cantrill and Poole, 2002). Therefore, in order to evaluate whether the Antarctic Peninsula did indeed act as a biotic gateway during the mid-Cretaceous radiation of angiosperms, the floral succession needs to be densely sampled and well age-constrained.

In spite of intensive sampling of over 250 pre-Albian sites (e.g. Duane, 1996; Hathway *et al.*, 1999; Keating, Spencer-Jones and Newham, 1992; Riding *et al.*, 1998), and numerous Albian and younger samples (e.g. Barreda, Palamarczuk and Medina, 1999) across the Antarctic Peninsula including the ?Barremian to Aptian strata in the South Shetlands, the earliest evidence of angiosperms is still *Clavatipollenites* pollen from early Albian strata on James Ross Island (i.e. West Gondwana). Further support for postulated routes and the timing of angiosperm arrival comes from the time lag between extinctions within certain taxonomic groups across southern Gondwana. Although surviving further north (Pole and Douglas, 1999) bennettites disappeared from southeastern Australia macrofloras in the late Aptian, whereas in the Antarctic Peninsula they persist until the Coniacian. This evidence also suggests that angiosperms arrived later in the Antarctic Peninsula region relative to elsewhere in Gondwana.

The early Albian arrival of angiosperms in the Antarctic Peninsula clearly postdates their occurrence elsewhere in eastern Gondwana (Burger, 1990, 1993). As discussed above, in Australia angiosperm pollen (*Clavatipollenites*) first appeared in latest Barremian–Aptian strata, and macrofossils with attached flowers were found in Aptian (~125 Ma) floras of southeastern Australia (Dettmann, 1994; Taylor and Hickey, 1990). In South America the first angiosperms were evidenced by pollen appearing in the conifer–non angiosperm seed plant–fern dominated floras also from the Barremian–mid-Aptian with the first definitive occurrence of eudicots in the latest Aptian–early Albian. In Patagonia, angiosperm macro-fossils first appear in Albian sequences (Archangelsky *et al.*, 2009; Cantrill and Nichols, 1996; Cúneo and Gondalfo, 2005; Halle, 1913; Passalía *et al.*, 2003; Rebassa, 1982). Therefore, it seems more likely that the gateway for the initial angiosperm invasion of Gondwana occurred through Africa or India. The macro- and microfloral records are not well-known for southern Africa and poorly constrained for India thus making it difficult to discriminate between these hypotheses (Cantrill and Poole, 2002). Further work in this area is needed to refine both the plate tectonic reconstructions and floristic history of this area and thereby shed further light on probable migration pathways (Cantrill and Poole, 2002). Even though land connections existed from South America into West Antarctica during the

Cretaceous, it is unlikely that the Antarctic Peninsula acted as a biotic gateway for dispersal across Gondwana, at least in the initial stages of radiation (Cantrill and Poole, 2002). The steep climatic gradient must have acted as a barrier to any potential (pre-)Aptian migration of angiosperms through South America and into Antarctica via this route. Radiation was probably blocked and further radiations prevented until later in the Cretaceous when warmer more homogeneous climatic conditions developed and the latitudinal range of the climatic belts expanded polewards (Cantrill and Poole, 2002).

These ideas are broadly supported by present day distributions. Evidence for *Nothofagus*, for example, is lacking in Africa and India but present in South America, Australia and Antarctica. *Nothofagus* is a prolific pollen producer and its pollen so distinctive that the lack of a record is probably proof of absence. At around 90 Ma this group appears in the fossil record implying that connections between Africa/India and the rest of Gondwana had been severed by this time. In contrast, other groups with older fossil records such as the Proteaceae, with pollen grains dating back to 100 Ma, are present on all Gondwana landmasses implying existing connections between these landmasses at this time.

Regardless of their route of passage into Antarctica, having arrived the angiosperms gradually began to transform the early Albian vegetation. The angiosperms initially formed part of the pioneer and/or understorey vegetation along with diverse rainforest ferns such as Osmundaceae, Hymenophyllaceae, Dicksoniaceae and *Lophosoria* (Dettmann and Thomson, 1987). The lack of angiosperm wood and the dominance of herbaceous leaves in the Albian floras of the Antarctic Peninsula (Cantrill and Nichols, 1996; Dettmann and Thomson, 1987) also suggest that these early angiosperms were shrubs rather than trees, and thus occupied the understorey. Within-flora angiosperm diversity increased initially in the colonisation niches, such as bare soil, and understorey thickets at the expense of the hepatophytes, bryophytes and ferns, and lycophytes respectively (Cantrill, 1996, 1997; Nagalingum *et al.*, 2002).

By the late Albian and early Cenomanian, angiosperms had increased in diversity and abundance as evidenced by the pollen record. Variation indicates the presence of monocots, magnoliids and eudicots (e.g. Askin, 1992). However, only a few can be compared with extant taxa and, apart from pollen of *Liliacidites* (with possible affinity to Liliaceae), it is still only the Chloranthaceae that can be recognised with confidence based on *Clavatipollenites* pollen. On the western side of the Antarctic Peninsula, on Alexander Island, younger Albian strata of the Fossil Bluff Group contain beds with the oldest known angiosperm leaves in Antarctica. These leaves comprise seven different types which were compared with Chloranthaceae, 'magnoliids' (such as Piperales, Laurales, Magnoliales) and rosids, and assigned to herbaceous (e.g. *Hydrocotylophyllum*) and shrubby (e.g. *Dicotylophyllum*) taxa (Cantrill and Nichols, 1996). Only a few leaf types represent more substantial plants, perhaps understorey trees (e.g. *Araliaephyllum*, *Ficophyllum*) (Cantrill and Nichols, 1996). The leaves were characterised by actinodromous, acrodromous and craspedodromous venation – patterns not represented in contemporaneous floras of South America (Cantrill and Nichols, 1996). The morphological differences between the leaves of southern South America and those on the Antarctic Peninsula again suggest that angiosperm radiation was blocked, through the Antarctic Peninsula at least, in the initial stages of migration. If not, then the morphotypes present in Antarctica should have been present in Patagonia prior to the early Aptian. Essentially the mid-Albian–Cenomanian vegetation of the Antarctic Peninsula was similar to that of the early Albian, distinguished only by the increasingly abundant and diverse angiospermous pollen. Podocarps, ferns and bryophytes dominated the vegetation, with other conifers bearing *Brachyphyllum* foliage, Cheirolepidiaceae, Araucariaceae, Ephedrales, cycadophytes, ginkgophytes and other seed plants were all less frequent.

Using an increased dataset derived from a decade of intensive studies focusing on macrofloras from Aptian to Maastrichtian strata, general patterns of floristic evolution driven by the angiosperms within the Antarctic vegetation have been documented (Cantrill and Poole, 2002 and references therein). These authors used within-floral diversity plots to capture patterns of change in absolute diversity. In spite of taxonomic biases in both the macrofloral and microfloral records that make it difficult to discriminate certain groups (e.g. bennettites in the microflora, lycopods in the macroflora), the two datasets show similar patterns of floristic change, indicating that the observed patterns are a real reflection of changes in vegetation composition through the Cretaceous. The general pattern of change within the Antarctic ecosystem resulting from the introduction of the angiosperm is as follows.

In the earliest Cretaceous, bryophytes and hepatophytes form an important and characteristic component of the floristic diversity of Antarctica (Figure 6.7). Not only were they at their most diverse, accounting for up to 20% of the species, but they were also ecologically abundant ranging from colonisers of fresh sediment to components of established fern thickets and forests (Cantrill, 1997). Indeed many localities are characterised by assemblages comprising monotypic stands of hepatophytes where they cover individual bedding surfaces for hundreds of metres. Other taxonomic groups forming important contributions to the Antarctic ecology at this time with their high abundances, and often high diversities, include the ferns (e.g. Dipteridaceae) and lycophytes, bennettites and other non-coniferous seed plants (e.g. *Pachypteris*, Ginkgoales).

By the mid-Cretaceous hepatophytes, ferns and lycopods underwent a steep decline, probably as a result of competition from the angiosperms, even though some fern families (e.g. Gleicheniaceae) seem to have remained relatively stable throughout. This suggests that as angiosperms increased in abundance the pteridophytes were able to colonise new niches created by the arboreal angiosperms. The bennettites remained reasonably diverse (representing 14% of the floristic diversity) and abundant (especially in fossil leaf floras) during the Aptian, but were soon to undergo a rapid decline through the Albian (3.5% of the floristic diversity) to become very rare and thus of negligible importance ecologically. A similar story is seen in southeastern Australia (Douglas, 1969) even though here bennettite history is clouded by the fact that their monosulcate pollen lacks distinguishing features making species discrimination difficult (see Dettmann, 1986b). This emphasises the importance of combining both macro- and microfossil assemblages when making such interpretations. Other non-coniferous seed plants also show a similar decline from the Aptian to Albian with a slight recovery during the Cenomanian.

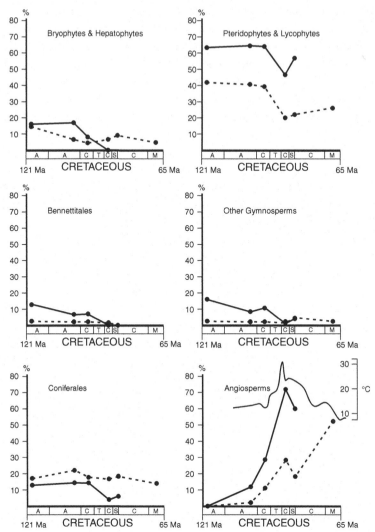

Figure 6.7 Diversity trends through the Cretaceous for macrofloral (solid line) and microfloral (dashed line) data from Antarctica with the introduction and rise of angiosperms. Note the close correspondence between angiosperm diversity and palaeotemperature (°C) estimates for the Southern Ocean (bottom right). Temperature curve modified and smoothed from Huber (1998). Macrofloral taxon diversity compiled from Banerji and Lemoigne (1987), Barale *et al.* (1995), Cantrill (2000), Césari *et al.* (1999), Falcon-Lang, Cantrill and Nichols (2001), Hayes (1999), Hernández and Azcárate (1971), Lacey and Lucas (1981), Orlando (1968), Rees and Smellie (1989). Microfloral taxon diversity compiled from Askin (1990), Barreda, Palamarczuk and Medina (1999), Cantrill (2000), Dettmann and Thomson (1987), Dolding (1992), Hathway *et al.* (1999), Keating (1992), Keating, Spencer-Jones and Newham (1992), Riding *et al.* (1998). Reproduced from Cantrill and Poole (2002) with permission from The Geological Society of London.

Throughout the Cretaceous, the conifer microfloral record indicates a stable within-flora diversity of about 10–20%, whilst the macrofloral record undergoes a distinct reduction in within-floral diversity in the mid-Cretaceous. These differences may be due to wind dispersed conifer pollen representing a wider range of plant communities from a greater geographical area relative to those represented by the more local macrofloras. This suggests that the angiosperms perhaps initially filled sun gaps in the canopy left by the conifers. The angiosperms radically changed the appearance of the vegetation across Antarctica, but the consequences of this new angiosperm dominated vegetation will be discussed in Chapter 7.

Palaeovegetation of Antarctica

Across Antarctica as a whole, Early Cretaceous macrofossil floras are relatively rare. In East Antarctica there are no macrofossils of this age, and pollen occurs only in offshore sediments where luxuriant vegetation is recorded (Mohr, 1990). ODP sites 692 and 693 have revealed the presence of an *in situ* ?Valangian to Albian palynoflora comprising bryophytes, lycophytes, several fern families (especially Gleicheniaceae, Osmundaceae, ?Lygodiaceae, and Schizaeaceae-Parkeriaceae-Lygodiaceae) along with other seed plants inculding *Cycadopites* (cycadophyte-ginkgophyte-Pentoxylales). Conifers are represented by Cheirolepidiaceae, Podocarpaceae (*Michrocachrys*, *Podocarpus*) and Araucariaceae. This composition shows strongest affinities with other floras from South Australia and the Antarctic Peninsula and suggests prevailing cool temperate rainforests, composed mainly of podocarps and araucarian conifers, with an understorey of ferns that were adapted to strong seasonality, high humidity and winter temperatures dropping to below freezing (Mohr, 1990). In contrast, the Antarctic Peninsula has extensive Lower Cretaceous outcrops complete with fossil material (Truswell, 1991). Therefore discussions on the conifer-dominated vegetation of the pre-Aptian to Albian will centre on the ecosystem preserved in the northern Antarctic Peninsula region.

Mid-Cretaceous pre-angiosperm conifer-dominated vegetation of the northern Antarctic Peninsula region

Late Early Cretaceous pre-angiosperm, conifer-fern flora has been found on the east side of the Antarctic Peninsula in the Larsen Basin. Here an extensive fossiliferous Cretaceous sedimentary sequence crops out totalling more than 5 km in thickness and has an ever-increasing significance for regional stratigraphic correlations in the Southern Hemisphere (e.g. Crame *et al.*, 2004; Crame, Pirrie and Riding, 2006). This basin, lying at palaeolatitude ~62° S, was one of a series of extensive back-arc basins that formed in the Patagonia–Antarctic Peninsula region during the mid-Mesozoic to early Cenozoic as the Palaeopacific plate subducted under the western margin of the Antarctic Peninsula (Hathway, 2000). This back-arc basin is filled with a regressive mega-sequence of arc-derived clastic and volcaniclastic marine rocks that has been subdivided into the older Gustav Group (Barremian to Coniacian, and the younger Marambio Group (Coniacian to Paleocene) (Figure 6.8).

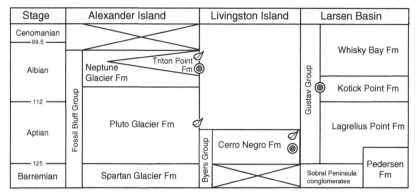

Stage	Alexander Island	Livingston Island	Larsen Basin
Cenomanian 99.5			Whisky Bay Fm
Albian	Neptune Glacier Fm — Triton Point Fm ◎		Kotick Point Fm ◎
112			
Aptian	Pluto Glacier Fm ◁	Cerro Negro Fm ◁◎	Lagrelius Point Fm
125			Pedersen Fm
Barremian	Spartan Glacier Fm		Sobral Peninsula conglomerates

Figure 6.8 Stratigraphic summary of key formations and groups for the Cretaceous of the Antarctic Peninsula. Units with fossil leaf and/or wood material are marked (after Hathway and Lomas, 1998; Poole and Cantrill, 2006). Note on Livingston Island the Byers Group Consists of the basal Late Jurassic Anchorage Formation that is overlain successively by the Devils Point Formation, President Beaches Formation and Chester Cone Formation. These Early Cretaceous units are separated from the Cerro Negro Formation by a mid-Valanginian to Barremian unconformity.

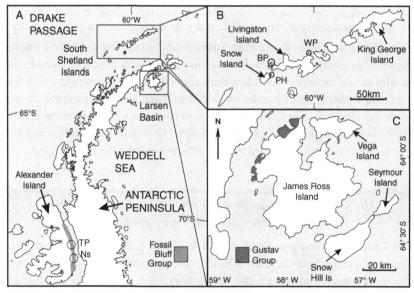

Figure 6.9 Locality map for key locations mentioned in text. **A**, Antarctic Peninsula and Alexander Island with the extent of the Fossil Bluff Group outcrop and the location of Triton Point (TP), and Citadel Bastion, Coal Nunatak and Titan Nunatak (Ns). **B**, South Shetland Islands with Byers Peninsula (BP), Presidents Head (PH) and Williams Point (WP). **C**, James Ross Island and the extent of the Gustav Group outcrop. See text for further details.

The Gustav Group is confined to the northwest coast of James Ross Island (Figure 6.9C) and to certain isolated outcrops on the adjacent margins of the Antarctic Peninsula where it reaches a maximum thickness of 2.6 km (Riding and Crame, 2002). In recent years it has become apparent that within the Gustav Group a particularly thick (~1750 m) and continuous Aptian–Albian sequence is present, but has so far only yielded sparse microfossils and macrofossils comprising *Agathoxylon* (Araucariaceae) wood, from the early Albian Kotick Point Formation (Figure 6.8; Crame, Pirrie and Riding, 2006; Feldman and Woodburne, 1988; Francis, 1986; Ottone and Medina, 1998; Pirrie *et al.*, 1997; Riding and Crame, 2002; Rinaldi, 1982).

A better record of Early Cretaceous floras occurs in the intra-arc and fore-arc basin sequences on the west side of the Antarctic Peninsula. Here the upper part of the Byers Group crops out in the South Shetland Islands (intra-arc) and in the Fossil Bluff Group on Alexander Island (fore-arc) (Figure 6.8) to the southwest (Falcon-Lang and Cantrill, 2001a). The floras along this north–south transect are very similar in composition and thus enable a reconstruction of the pre-angiospermous vegetation that grew across the Antarctic Peninsula at this time.

The Upper Jurassic–Lower Cretaceous Byers Group crops out on both Byers Peninsula on Livingston Island (62° 38' S, 61° 05' W) and nearby President Head on Snow Island (63° 45' S, 61° 14' W) within the southeast portion of the South Shetland Islands (Falcon-Lang and Cantrill, 2001b; Hathway, 1997) (Figure 6.9A, B). The Byers Group is divided into four units: the Anchorage, President Beaches, Chester Cone and Cerro Negro formations. The Upper Jurassic Anchorage Formation is overlain by the Lower Cretaceous President Beaches and Chester Cone Formations. These three units were folded gently and uplifted prior to the deposition of the Cerro Negro Formation. Radiometric ^{40}Ar/^{39}Ar dates from the base and top of the Cerro Negro Formation indicate that it was rapidly deposited in the early Aptian, between 120–119 Ma, following a depositional hiatus of ~17 Ma. This hiatus corresponds to a phase of increasing arc activity that is reflected in both the James Ross Island Basin to the west and the Alexander Island Fore Arc Basin to the south (Hathway, 1997; Hathway and Lomas, 1998; Hathway *et al.*, 1999). The Cerro Negro Formation represents a generally terrestrial sequence that developed in regions dominated by active andesitic stratovolcanoes. Palynomorphs from this formation have been used to correlate sequences from Byers Peninsula and President Head (Askin, 1983; Duane, 1996).

The volcanic-sedimentary deposits of the Cerro Negro Formation (Figure 6.8) on Byers Peninsula and President Head are complex and laterally variable (Pirrie and Crame, 1995). Seven facies can be recognised within the Cerro Negro Formation each with its own distinct megafloral assemblage (Hathway, 1997), and thus provides an insight into the temporal community scale dynamics of a high southern latitude conifer-dominated ecosystem during the Early Cretaceous. Falcon-Lang and Cantrill (2002) identified two forested communities (one lowland and one upland), and one lowland shrub community type growing in this region. The structure and composition of these communities underwent changes in response to volcanic environmental disturbance. Although there is no geological data to ascertain the exact ecological relationship between the lowland community comprising a forest and shrub vegetation, it is reasonable to speculate that volcanic disturbances may have been of high enough frequency to enable the establishment of the shrub community following an eruption event, but prevented the vegetation from developing to ecological climax. If periods of quiescence were long enough, vast tracts of forest representing the climax community would have become established. Here we are able to identify snapshots from two community associations that grew at this time.

The community evolution begins in Facies 1 (Figure 6.10), which is interpreted as the deposits of sediment laden streams that drained valleys adjacent to explosive volcanoes. An arboreal climax forest community of substantial araucarian and podocarp conifers grew alongside a relatively minor component of taxodioid conifers and other seed plants (e.g. the putative bennettite, *Sahnioxylon*). This forest vegetation covered the slopes of the active silicic volcanoes, along the valleys and over levees formed by the rivers (Cantrill, 2000; Falcon-Lang and Cantrill, 2001b, 2002; Hathway, 1997; Torres *et al.*, 1997a). These levees were undercut by the relatively high energy streams, which then transported plant material including wood, foliage, plentiful pollen, and trunks of up to 1.5 m in diameter and 6 m long downstream. Along with this debris were charred tree trunks (>1 m in diameter) preserved in pyroclastic flows (Falcon-Lang and Cantrill, 2002; Scheihing and Pfefferkorn, 1984). These logs represent the remains of conifers that had once been growing on exposed sites proximal to the volcanoes with no protection from the volcanic activity.

One upland forest community was dominated by tall (17–45 m) podocarps and arborescent bennettites. This would have represented an ecosystem growing in a period of favourable conditions, uninterrupted by volcanic disturbance, where undisturbed stands had completed the vegetational cycle from coloniser to climax vegetation. Pentoxylaleans, an enigmatic seed plant, and tree-sized ferns shaded an understorey comprised of herbaceous ferns and sphenopsids. Some seed plants (e.g. *Sahnioxylon*) were probably fast growing, producing growth rings up to 8 mm wide per year, whereas the podocarps, such as *Podocarpoxylon*, were slower with growth rates of ~3 mm per year. Together they suggest favourable climates existed across the Antarctic Peninsula region at this time. The coexistence of fast and slow growing trees is not surprising because the slow growing podocarps probably had a greater life span and, once mature, dominated the canopy for long periods relative to the quick growing, possibly short-lived plants (Falcon-Lang and Cantrill, 2002).

Asymmetry of fossilised trunk specimens indicates that the conifer trees were probably occupying the steep slopes of the volcanic cones and so were particularly vulnerable to volcanic activity. Pyroclastic flows, which probably occurred on a centennial to millennial timescale, were the main ecological disturbance in this area, immediately charring the vegetation or burying it (Cas and Wright, 1987; Falcon-Lang and Cantrill, 2002). The hot pyroclastic flows would have killed the trees and engulfed them for several weeks at a time leaving them charred to the core. These remains, along with other trees, would have been enveloped in ash as periodic air fall deposits covered vast swathes of the landscape. Vegetation further from the cone would have been engulfed by relatively cooler pyroclastic flows, which also left them charred but friable. At some time later, remains of this material would have been transported tens of kilometres from these sites and deposited in fluvial sediments within the palaeovalleys. Plants growing in the protection offered by the valley sides remained relatively unaffected by the volcanic activity, but would have succumbed to periodic flooding, which was probably

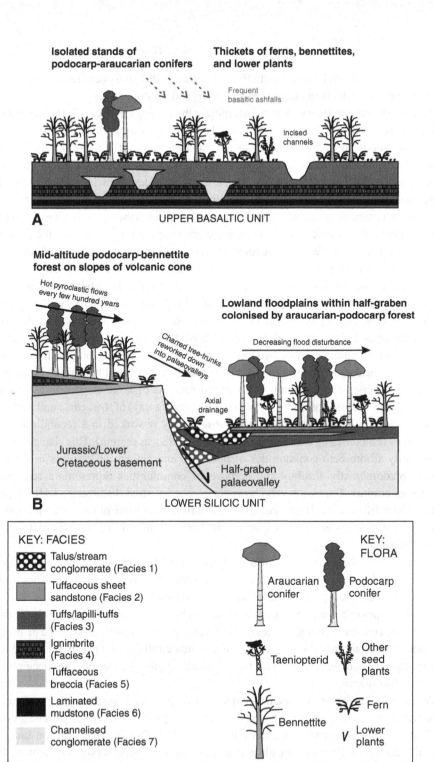

Figure 6.10 Palaeoecological reconstruction for the upper and lower units of the Cerro Negro Formation integrating data from biodiversity, facies assemblages and growth rings in wood. Reproduced from Falcon-Lang and Cantrill (2002) with permission from SEPM Society for Sedimentary Geology.

the most frequently encountered disturbance process in this ecosystem reoccurring with an intra-annual to decadal frequency (cf. Falcon-Lang and Cantrill, 2002).

This ecosystem can be widened to incorporate the findings in Facies 2, the deposits of a palaeovalley floodplain. These sandy floodplain soils also supported continuous lowland evergreen woodland or forests of *Araucaria* and podocarps (Falcon-Lang and Cantrill, 2002). Stumps of large araucarian trees preserved in growth position testify to the closely spaced nature of these forests. Two species of the fastest growing taxon, *Araucaria*, were spatially separated along the floodplain, with one species occupying the relatively unfavourable riparian niche whereas the second species inhabited a more distal, and thus drier and more favourable, setting. During the growing season the trees on the floodplain came under attack from fungi: white rot was widespread and attacked the wood resulting in irregular annual growth patterns that ultimately impacted on tree growth itself.

Facies 3 records the vegetational response to a volcanically disturbed environment with tuffs representing the product of cool pyroclastic flows and accretionary lapilli suggesting air fall deposits. These putative air fall deposits soon became colonised as evidenced by rooted horizons. Foliage remains preserved above this level suggest shrubby vegetation composed of a combination of bennettites and pentoxylaleans, along with herbaceous ferns, sphenopsids and thalloid liverworts, before being covered by further ash deposits following a subsequent eruption. The charred remains of large conifer trunks indicate relatively local stands (possibly only tens of kilometres away) of *Araucaria* and podocarps buried within hot pyroclastic flows and subsequently reworked in a second cold pyro- clastic flow (Falcon-Lang and Cantrill, 2002 and references therein). This change from the dominantly arborescent communities of podocarp and araucarian conifers in the lower unit to predominantly shrubby and herbaceous communities represents a response to increasing eruption frequency, which progressively favoured disturbance-adopted plants with shorter life cycles. Both ecological scenarios have been observed in present day araucarian-podocarp-fern forests growing in similar volcanic arc settings in Chile and New Zealand.

Volcanic disturbance also features in Facies 4 with evidence of charred trunks of podocarps and *Sahnioxylon* originating from a seed plant–podocarp forest. These trunks would have been transported tens of kilometres downstream from the forests that grew across the slopes of the active volcano after having been destroyed by hot pyroclastic flows. Disturbance continues through Facies 5 with further air fall deposits on the slope of an active volcanic cone. Plant material found here are rare, unidentifiable and poorly preserved wood fragments consistent with woody vegetation colonising this harsh terrain and succumbing to volcanic disturbance.

At times when basin subsidence outpaced volcanic sedimentation, large permanent freshwater lakes formed (Facies 6) fed by streams that ran through deeply incised valleys draining the volcanic cones that periodically flooded (Duane, 1996; Falcon-Lang and Cantrill, 2002). On Byers Peninsula a lacustrine sequence is preserved with a more distal (terrestrial) part of the same system cropping out at President Head. The plant material preserved in the mudstones records both regional and local components of the vegetation

(Cantrill, 2000). The low depositional energy environment facilitated the preservation of small delicate plants and insects alongside the more robust cycads, ferns, conifers and rare bennettitalean and pentoxylalean remains (Cantrill, 2000). The fossil flora records lake margins supporting shrubby bennettite-herbaceous fern dominated communities with a minor component of lycophytes and bryophytes (Cantrill, 2000). The absence of woody trunks in this flora provides evidence for the herbaceous and shrubby nature of the vegetation along the valleys and around the lake margins.

The streams that flowed into the lakes brought with them evidence for compositionally similar but temporally more persistent communities from further up the valleys. Relative to the leaf floras, the wood is more widespread and locally abundant, and together with abundant pollen and spores, provides evidence for substantial conifer trees and pteridophytes being present on the flanks of the volcanic edifice. These allochthonous assemblages originate from a large stratigraphic interval in the upper unit (Facies 6 and 7) and are always bennettite-dominated with rare conifer, pentoxylalean and fern remains. Here too volcanic activity was periodic providing newly exposed surfaces, which in time became colonised.

Away from the margin, the understorey vegetation around the lakes was characteristic-ally rich in pteridophytes, often dominated by a single taxon such as the lophosoriaceous tree ferns (*Cyatheacidites*) and representatives of Dicksoniaceae and Cyatheaceae (e.g. *Cyathidites australis* and *C. minor*) (cf. Césari, 2006; Palma-Heldt et al., 2007). Another important post-eruption community around the lakeside margins were the fern-dominated thickets of *Aculea* and *Lophosoria cupulatus*, characterised by its large trunk and fronds of at least 1m in length, along with *Phyllopteroides* and *Sphenopteris* (Falcon-Lang and Cantrill, 2002). This might be a reflection of local abundance and/or disturbance since regular ash falls would have ensured that these communities were short-lived. Today the Lophosoriaceae are particularly renowned for their ability to colonise open and disturbed sites as monospecific thickets (Cantrill, 1996, 1998) therefore suggesting that *Lophosoria cupulatus* probably behaved in much the same way during the Early Cretaceous. Spores of the palynoflora indicate that the fern diversity was great, even though this diversity is not reflected to the same extent in the macrofossil record (Cantrill, 2000).

In periods of volcanic quiescence diversity increased. Ferns were represented by Cyatheaceae, Lophosoriaceae, Osmundaceae, Dennstaedtiaceae, Gleicheniaceae, as well as a new genus of core tree ferns, *Alienopteris* (Cantrill, 2000; Césari, 2006; Césari, Remesal and Parica, 2001; Césari et al., 1999; Vera, 2007, 2009). Bennettites included *Ptilophyllum*, *Otozamites* and *Dictyozamites*, and pentoxylaleans (e.g. *Taeniopteris*) and/or Ginkgoales may have become present as evidenced by *Cycadopites*-type pollen grains (Araya and Hervé, 1966; Cantrill, 2000; Torres et al., 1997a, 1997b). Other members of the community included shrubby pentoxylaleans and herbaceous liverworts, sphenopsids, ferns, along with relatively abundant gymnospermous plants such as *Pachypteris* and isolated stands of conifers. The gymnosperms were more infrequent in the proximal lake areas and thus indicate that the arboreal component of the vegetation differed according to the prevailing (micro)habitats (Araya and Hervé, 1966; Cantrill, 2000; Palma-Heldt *et al.*, 2007; Torres *et al.*, 1997a, 1997b).

Many Cretaceous high latitude floras are characterised by diverse and abundant bryophytes and hepatophytes, which were adapted to the warm climate and strongly seasonal light regime (e.g. Cantrill; 1997; Douglas, 1973; Drinnan and Chambers, 1986), and the floras of Byers Peninsula and President Head are no exception. Between forested stands, the open ground and disturbed sites were colonised by pioneer species of liverworts (e.g. Sphaerocarpales), hornworts (Anthocerotales) and bryophytes. Lycophytes, represented by both lycopods and members of the Selaginellales, were present probably as both herbaceous scrambling plants and epiphytes (Cantrill, 2000). Colonies of liverworts and other bryophytes covered the forest floors forming clumps, which trapped other organs such as fern pinnules, seeds, bracts and sporangia that inadvertently became entangled. Epiphytic communities of bryophytes were also present occupying niches higher on tree trunks and canopy forming branches. The forest floors were probably very wet with boggy areas colonised by mosses (*Stereisporites*, a form allied to the extant *Sphagnum*), and true mires characterised by bryophyte species. Humidity levels were high, thus enabling the small forest-dwelling filmy fern *Hymenophyllum* to thrive.

Reconstructions of the Cerro Negro Formation plant communities indicate that their stature and composition changed in response to volcanic style (Falcon-Lang and Cantrill, 2002). The exact relationship between the shrubby bennettite-fern community and araucarian-podocarp community in the lower unit (Figure 6.10A) is not known, but it is reasonable to speculate that they may represent end members of one community association. The shrubby bennettite-fern community may represent a post-volcanic disturbance community, whilst the araucarian-podocarp community represents the climax community. Similarly, the shrubby bennettite-dominated community of the upper unit (Figure 6.10B) may be representative of the vegetation type characteristic of areas influenced by high frequency volcanic disturbances that prevented vegetation from developing to ecological climax (Falcon-Lang and Cantrill, 2002).

Analogues for the Antarctic Cretaceous araucarian-podocarp communities can be found in the compositionally similar evergreen conifer forests of New Zealand today. Here the araucarians and podocarps attain similar heights of 20–30 m with a few emergent araucarias reaching >40 m (Enright and Hill, 1995; Wardle, 1991). Trees can grow between 2 and 5 mm per year in the extant forests, (Ahmed and Ogden, 1987; Dunwiddie, 1979; Norton, Herbert and Beveridge, 1988), whereas in the Cretaceous forest growth rate is estimated to have been 2–6 mm per year (Falcon-Lang and Cantrill, 2002). Such growth comparisons suggest that a warm temperate vegetation existed in the Antarctic Peninsula at this time, an interpretation supported by coupled climate-biome models which have also predicted a warm temperate vegetation across the Antarctic Peninsula during the Cretaceous (Beerling, 2000).

Although these Aptian floras are interesting in that they represent the last fern-conifer-other seed plant–dominated vegetation in Antarctica immediately prior to the angiosperm invasion, probably the most well-known and best-documented flora of the Lower Cretaceous is the late Albian Alexander Island Flora (Cantrill, 1995, 1996, 1997; Jefferson, 1982a, 1982b). This flora, found on the western side of the Antarctic Peninsula, is considered to be of great importance as it represents one of the earliest Antarctic fossil floras that the angiosperms had begun to invade. This flora testifies to the massive and irreversible transformation in terms of floral diversity that took place during the Early Cretaceous, a change that was to affect the subsequent evolution of the Southern Hemisphere vegetation as a whole.

Alexander Island: earliest evidence of angiosperms in a fern-conifer-dominated ecosystem

The second example of a late Early Cretaceous fern-gymnosperm dominated vegetation comes from Alexander Island, lying on the western side of the Antarctic Peninsula at a palaeolatitude of ~75° S (Smith, Smith and Funnel, 1994). The flora found on Alexander Island represents not only one of very few southern high latitude floras but also one of the most complete and most southerly early angiosperm forests known to date. The vegetation type occurring here is probably representative of the flora that grew across the whole of the Antarctic Peninsula at the time, having replaced the vegetation exemplified by the Cerro Negro Formation Flora in the northern Antarctic Peninsula region, and extended from Alexander Island in the south across the James Ross Island Basin and South Shetland Islands to Patagonia in the north, and across East Antarctica towards New Zealand in the east. The Alexander Island flora postdates the Cerro Negro Formation Flora by about 10 million years, but within this window the earliest evidence of angiosperms can be seen to have reached Antarctica.

Alexander Island contains a suite of rocks deposited in a fore-arc setting to the west of the magmatic arc formed as the Pacific Plate subducted beneath Antarctica (Barker, Dalziel and Storey, 1991). The sequence of Mesozoic rocks includes the Upper Jurassic–Lower Cretaceous Fossil Bluff Group, a series of marine, fluvatile and deltaic sediments that formed as the infill of the subsiding fore-arc basin (Butterworth *et al.*, 1988). They are dominated by volcaniclastic components derived from the eroding volcanic arc. At Triton Point the Triton Point Formation (formerly Triton Point Member) of the Fossil Bluff Group (Figure 6.8) has been dated as late Albian (~105 Ma) on the basis of a bracketing molluscan flora (Moncrief and Kelly, 1993), and contains a wealth of Lower Cretaceous plant-bearing rocks complete with *in situ* fossil forests (Jefferson, 1982a,1982b, 1987), leaf floras (Cantrill, 1995, 1996, 1997, 2001; Cantrill and Falcon-Lang, 2001; Cantrill and Nagalingum, 2005; Nichols and Cantrill, 2002) and exceptionally well-preserved palaeosols (Howe, 2003; Howe and Francis, 2005).

The fossiliferous Triton Point Formation represents a prograded wedge of non-marine fluvatile sediments that thickens southward from 200 m at Triton Point through Pagoda and Phobos ridges to ~950 m at the southern nunataks, namely Citadel Bastion, Coal Nunatak, and Titan Nunatak (Moncrieff and Kelly, 1993). Two distinct facies associations can be recognised within the Triton Point Formation (Cantrill and Nichols, 1996). The lower facies association is interpreted as the product of a braided alluvial plain environment, cropping out at all localities and is called the Citadel Bastion Member. The overlying facies association has been interpreted as a coastal meandering river system with broad river belts and extensive floodplains but geographically restricted to the upper part of Coal Nunatak, and is called the Coal Nunatak Member (Cantrill and Nichols, 1996; Nichols and Cantrill, unpublished data). This river system deposition marks one of the last stages of infilling in the Fossil Bluff fore-arc basin (Moncrieff and Kelly, 1993). Evidence suggests that the climate was temperate with a high mean annual rainfall (cf. Falcon-Lang, Cantrill and Nichols, 2001).

The vegetation preserved within the Triton Point Formation represents the diverse forest communities that grew across floodplain areas. The fluvial environment matured from a braided river system with frequent floods and unstable channel banks, to a more meandering river system with fewer floods and more stable floodplains (Howe, 2003). According to Howe (2003) across the floodplain the flora was divided into well-structured communities that occupied different niches within the river floodplain. Open woodland forests comprising long-lived pioneer taxa grew on the stable areas of the floodplain and were dominated by araucarian conifers, ferns and small shrubs. Low energy floods covered the smaller plants in fine sands and silts that then enabled young saplings to take root. The river banks supported a disturbance vegetation comprising *Taeniopteris*, liverworts, ferns and the newly evolved angiosperms that were frequently covered in sediment when the river broke its banks (Figure 6.11B–E). Away from the river channels high density climax forests of podocarps, ginkgos, cycadophytes and ferns grew on the more stable areas, affected only by catastrophic flood events that covered the forest floors in coarse sand and entombing the trees (Figure 6.11A). The dynamics of this vegetation is outlined in more detail below.

The diversity of the Alexander Island flora comprises 42 genera containing approximately 70 species (Falcon-Lang, Cantrill and Nichols, 2001). This is considered to be extremely diverse when compared with contemporaneous mid-latitude floras (Cantrill, 1997), but is a feature shared with other Southern Hemisphere high latitude Albian floras such as those in southeastern Australia (~65° S) (Douglas, 1973). In general, conifers are the most abundant plant fossil represented by wood and foliage remains belonging to large trees of Araucariaceae, Podocarpaceae and Taxodiaceae (Cantrill, 1997; Cantrill and Falcon-Lang, 2001; Falcon-Lang and Cantrill, 2000). However, although they are abundant, the conifers only make up a small percentage of the total within floral species diversity (~25%). The unique nature of the Alexander Island flora includes an unusually high liverwort and fern diversity (Figure 6.12A–D). Although abundant in other Early Cretaceous floras, the ferns are the most diverse component of this flora, contributing possibly in excess of 40% of the species diversity, which is much higher than in the slightly older Aptian floras of James Ross Island and in South America (e.g. the Baquéro Formation with ~20% fern within floral diversity), but similar to those found in southeastern Australia (also 40%). Taxa preserved have been allied to extant families including the Dipteridaceae, Gleicheniaceae, ?Lophosoriaceae, Matoniaceae and Osmundaceae (Cantrill, 1995; Jefferson, 1981, 1982a), although the majority, such as *Microphyllopteris*, *Coniopteris*, *Sphenopteris* (Cantrill, 1996; Cantrill and Nagalingum, 2005), cannot be attributed with confidence to family level. Other Mesozoic taxa include members of the Bennettitales, Cycadales, Ginkgoales and Pentoxylales, collectively making up only ~10% of the species diversity. Of the lower

Figure 6.11 Fossil plants from the late Albian, Triton Point Formation, Alexander Island. **A**, entombed standing tree with pseudomonopodial habit (note hammer for scale). **B–C**, angiospermous leaves. B, *Hydrocotylophyllum alexandri* BAS KG 2815.171a; C, *Araliaephyllum quiquelobatus* BAS KG. 4693.16. **D**, *Microphyllopteris unisorus* a gleicheniaceous fern, BAS KG. 2815.206. **E**, *Zamites* sp. BAS KG. 4728.18. Scale bars B–D = 1 cm, E = 2 cm.

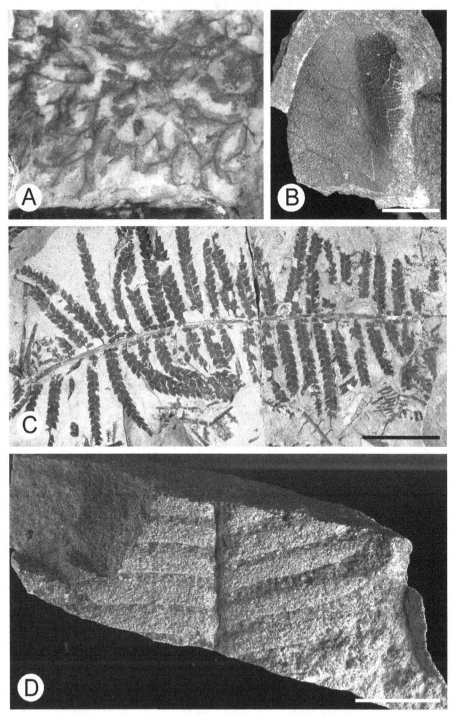

Figure 6.12 Albian plants from Alexander Island. **A**, thalloid liverwort *Hepaticites minutus*, BAS KG. 1704.6. **B–C**, pteridophytes. B, dipteridaceous fern *Hausmannia papilio*, BAS KG. 4737.80. C, gleicheniaceous fern *Gleicheniaceaephyllum acutum*, BAS KG 2817.75a/76a. **D**, bennettite *Ptilophyllum* sp., BAS KG. 4653.3. Scale bars A = 5 mm, B, C = 1 cm, D = 2 cm.

plants, lycopods and horsetails are rare when compared with other Cretaceous floras from lower palaeolatitudes (Cantrill, 1997). Liverworts contribute ~15% species diversity and include forms that can be allied to the Marchantiales with minor representatives of the Metzgeriales. The interesting component of the vegetation is the presence of angiosperms that, although only moderate in diversity (12% of species diversity), consist mainly of large arborescent plants and a few small herbaceous forms (Cantrill and Nichols, 1996). All these plant remains occur as assemblages in sedimentary facies that indicate the existence of spatially complex and heterogeneous vegetation communities.

Abiotic and biotic factors can be seen to have affected the plants within the Triton Point Formation. Charcoal is present in association with tuff layers suggesting that volcanism may have played a similar role to that in New Zealand (on Mount Taupo) today (Wilmshurst and McGlone, 1996). However, the rarity of charcoal indicates that fires were indeed rare events in this mid-Cretaceous polar biome (Falcon-Lang, Cantrill and Nichols, 2001). Within the charcoalified wood, rings of traumatic tissue may represent frost rings caused by the unusually low temperature drop following a volcanic eruption (LaMarche and Hirschboek, 1984). Attack from biotic causes was also prevalent, as evidenced by fungal bodies and arthropod frass within the latewood tracheids of the Alexander Island woods, indicating that attack occurred during the dormant phase of the dusky late autumn or dark winter when the trees' defence mechanism would have been particularly vulnerable (Falcon-Lang, Cantrill and Nichols, 2001).

The lower facies association of the Citadel Bastion Member comprises sediments deposited on an alluvial plain where braided rivers were able to cut deep (~15 m) channelised incisions hundreds of metres wide (Falcon-Lang, Cantrill and Nichols, 2001). Sediments of the lower channel-fill units within the river deposits provide evidence for mid-channel bars (Nichols and Cantrill, 2002). Large-scale flood events, represented by thick sheet sandstone units containing coarse-grained sediment locally buried the prevailing vegetation (A. C. M. Moncrieff, unpub. British Antarctic Survey Field Report AD6/2R/1988/G4 1989, in Falcon-Lang, Cantrill and Nichols, 2001) and swept around standing trees with currents scouring the substrate (Underwood and Lambert, 1974). Localities at distance from the braided river system were periodically covered in fine-grained sands and silts deposited from suspension onto the floodplain, but did not escape regular flooding as evidenced by thin sandstone beds (Jefferson, 1981). Through time, these braided river systems gradually decreased in gradient and channel size such that the initial broad channel belts formed by the migration of braided channels over the alluvial plain became a braidplain environment that graded into a lower-energy, possibly meandering, river system that continued to flood periodically. The extent of the outcrop of this member has enabled a detailed reconstruction of the ecology of this region during the late Albian.

Numerous palaeosol horizons can be found associated with this member, providing information on the development of the ecosystem and the associated dynamics of the fluvial environment. The fossil plants testify to an increasing floral complexity, grading from the frequently disturbed sandy riparian sites to more distal floodplain localities (Figure 6.13A). The palaeosols represent soil formation that took place over hundreds to thousands of years.

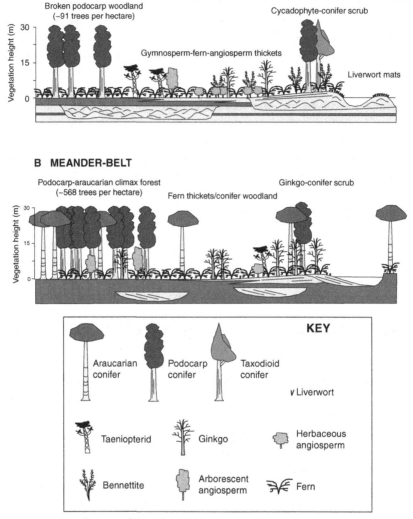

Figure 6.13 Summary diagram of the Triton Point Formation ecosystems based on facies analysis of plant assemblages. Reproduced from Falcon-Lang, Cantrill and Nichols (2001) with permission from The Geological Society of London.

They provide little evidence of regular flooding or waterlogging for long periods, but interbedded fluvial sandstones and mudstones indicate that flood events did occur, but only periodically, with sediments being deposited over relatively short periods of time i.e. days to months (Howe and Francis, 2005). Throughout this facies, overbank deposits with immature fossil soils and rootlet horizons, abundant plant debris and *in situ* plants adjacent to channel bodies provide further evidence for regular, but not frequent or long-lasting, flood events. These events may have been associated with weak riverbanks, leading to local and short-lived flooding (Cantrill and Nichols, 1996).

Mature palaeosol profiles have also been preserved. These consist of an upper O horizon (leaf litter layer), a middle, medium-brown, carbonaceous mudstone layer (~8 cm thick) corresponding to the A horizon, and a lower bleached sandstone and siltstone layer (<60 cm thick) comprising the E\C horizon. Throughout these facies major flood deposits have preserved *in situ* trunks in growth position that are up to seven metres high and rooted in the carbonaceous palaeosols (A. C. M. Moncrieff, unpub. British Antarctic Survey Field Report AD6/2R/1988/G4 1989, in Falcon-Lang, Cantrill and Nichols, 2001). The stumps have basal diameters of 8–50 cm with roots departing downwards at angles of 35–90° into the palaeosol. The roots are most abundant in the A horizon where they form a dense mat, but some have penetrated through to the weathered E/C horizon (Cantrill and Falcon-Lang, 2001). Where they penetrate sandy soils, distinctive mycorrhizal nodules characteristic of podocarp conifers today can be found associated with the roots, suggesting that these sandy soils were well-drained and nutrient-poor (cf. Cantrill and Douglas, 1988). The size of these preserved conifer stumps indicates that they were large trees with probable heights of up to about 30 m (Falcon-Lang and Cantrill, 2000) and many have lived for more than 100–200 years (Chapman, 1994). The palaeosols also suggest formation under a seasonal climate, characterised by a well-defined dry season (Howe and Francis, 2005).

Following these relatively short periods of inundation, newly emergent sandsheets near the water bodies became colonised by diverse and extensive mats of liverworts (e.g. *Thallites*; Cantrill, 1997). These communities helped to stabilise the sediment and allow immature soils to develop (Falcon-Lang, Cantrill and Nichols, 2001). Once stabilised, these sandy riparian soils could become further colonised, but this time by ferns (Figure 6.13A). Abundant herbaceous Dipteridaceae (*Hausmannia*) have been discovered in growth position with stipes attaching the plant to the substrate, forming patchy, monotypic communities (Cantrill, 1995). Further from the river channels the relatively immature palaeosols became dominated by shrubby angiosperms such as *Gnafelea* (with fagalean leaves and a possible ancestor to *Nothofagus*), together with other ferns (*Cladophlebis*, *Aculea*), *Ginkgo*, bennettites (*Ptilophyllum*) and rare conifers including *Brachyphyllum* and shrubby podocarps. In areas where flooding was less destructive and the soils were able to reach greater maturity, sporadic liverworts such as *Marchantites* grew over the surface with ferns (*Cladophlebis* and *Hausmannia*), angiosperms (*Araliaephyllum*, *Timothyia* and *Dicotylophyllum*) and the gymnospermous Pentoxylales (*Taeniopteris*) (Cantrill, 1995, 1997; Cantrill and Nichols, 1996).

Further still from channel influences, sandy soils were able to support a variety of two-tier plant communities. At one site, thickets of shrubby angiosperms and ferns occurred between widely spaced *Ginkgo* and bennettites. On a second site, the vegetation was supported by an overstorey of Pentoxylales (*Taeniopteris*) interspersed with occasional angiosperms and conifers, and a diverse understorey of ferns (*Hausmannia* and *Cladophlebis*), rare angiosperm shrubs (*Timothyia* and *Dicotylophyllum*) and liverworts (*Marchantites*).

In the interchannel areas on the distal parts of the braided alluvial plain, podzolic soils were frequently flooded and thus prevented the development of the forest climax vegetation. These disturbed interchannel sites supported low-density clumps of podocarp conifers (Figure 6.13A). Elsewhere in localities where the vegetation was able to mature, the riparian flora was mixed, and composed of large podocarps and taxodioid conifers interspersed with bennettitalean trees. Abundant *Podocarpoxylon* and *Taxodioxylon* conifer logs, up to 3 m in length, had been transported downstream by the river along with foliage of *Ptilophyllum* and *Brachyphyllum* and dumped in both the braided channel deposits and the sandstone beds between the soil horizons (Falcon-Lang, Cantrill and Nichols, 2001).

At Coal Nunatak, the lowermost part of the section contains the Citadel Bastion Member, which represents a similar braided river and floodplain environment to that described above for Citadel Bastion and Titan Nunatak. The overlying facies association, the Coal Nunatak Member (only exposed on Coal Nunatak itself), has a thicker succession of fine-grained units than those in the Citadel Bastion Member and is characterised by gravelly to coarse-grained, trough cross-bedded sandstone bodies with distinct channels (Cantrill and Nichols, 1996; Falcon-Lang, Cantrill and Nichols, 2001). The facies association represents a fluvial environment comprising a series of shallow, slow flowing, meandering river channels (Cantrill and Nichols, 1996). Mud drapes in one channel body suggest that the meander-belt developed in a coastal setting and thus was under a tidal influence (Cantrill and Nichols, 1996). Channel levees and flood deposits are evidenced by sheet sandstone units suggesting that periodic flooding also took place over this coastal meander-belt in nearby interchannel areas. Sheet flooding also occurred and these deposits preserved palaeosols complete with *in situ* plant material (Falcon-Lang, Cantrill and Nichols, 2001). With time the meandering fluvial environment developed briefly into a shallow marine setting before returning to a fluvial condition once again.

In this coastal plain meander belt environment the river channels were more confined and contain highly macerated plant material and a few drifted logs. The extensive immature sandy substrates covered by dense mats of liverworts, which characterised the braided alluvial plain further inland, were more sporadic. Further away from the channel influence, sandy substrates were colonised by continuous thickets of ferns (*Aculea* and *Alamatus*) interspersed with shrubby podocarps and scattered stands of *Ginkgo* and *Araucaria* (Figure 6.13B). Marchantioid liverworts formed a ground layer here but were comparatively rare relative to the immature sandy substrates closer to the river (Cantrill, 1996). Trees colonised the channel levees forming a canopy of podocarp and araucarian conifers with a subcanopy of *Ginkgo*, cycadophytes and an understorey of the fern *Cladophlebis* (Figure 6.13B). At greater distance from the river channels, where flood disturbance was more infrequent, palaeosols (see below) were able to mature and support a medium density (~600 trees per hectare) podocarp-araucarian climax forest. The understorey, dominated by ferns (*Cladophlebis* and *Sphenopteris*) with intermittent *Ginkgo*, angiosperms (*Ficophyllum*) and liverworts (*Hepaticites*) formed a ground cover on uncolonised sites (Falcon-Lang, Cantrill and Nichols, 2001).

As with the braided alluvial plain of the Citadel Bastion Member, at sites away from the channels, the floodplains of the coastal meander belt contain mature podzolic soils. In the most well-developed palaeosol, the upper A-horizon (topsoil) forms a dark brown, organic-rich (but not peaty) layer >25 cm in thickness. A gradation from this well-developed palaeosol to relative less well-developed palaeosols, characterised by an upper layer of high detrital (volcanic) mineral components, can also be seen. The preserved E/B (eluviated/subsoil) horizon of the mature palaeosol appears more weathered and probably encompasses a weakly developed mineral (cambic) horizon with relic bedding and features of the underlying bedrock (Howe and Francis, 2005). These fossil soils are similar to modern soils with a dark humus-rich upper horizon formed by the underground decomposition of organic residues of roots and plant material, and a lower horizon usually of fine-grained sand that has undergone alteration and weathering (Howe and Francis, 2005). Today such soils develop in the sub-humid to sub-arid regions of, typically, the mid-latitudes often under grassland but also under well-drained forest vegetation, such as those found beneath the mixed *Araucaria*, *Podocarpus* and *Agathis* forests of the Southern Hemisphere (Howe and Francis, 2005; Pritchett and Fisher, 1987). These soils most closely resemble the leached podzolic soils of New Zealand (Falcon-Lang, Cantrill and Nichols, 2001) that form on acidic volcanic terraines and support a podocarp-araucarian conifer forest vegetation growing in a humid, warm temperate climate (Wardle, 1991).

The vegetation structure growing on the alluvial floodplain (Citadel Bastion Member) can be compared with that growing nearer the coast (Triton Point Member). On the braided alluvial plain, where catastrophic flooding was a regular occurrence, over a dozen mature palaeosol horizons can be identified each bearing low density stands of podocarp (*Podocarpoxylon*) and taxodioid conifer stumps and trunks (A. C. M. Moncrieff, unpub. British Antarctic Survey Field Report AD6/2R/1988/G4 1989, in Falcon-Lang, Cantrill and Nichols, 2001). At one site on Offset Ridge, thirteen stumps occur over 1540 m^2 of palaeosol exposure where they are grouped into clumps of four or five individuals. The mean density of this forest would have been low at ~90 stumps per hectare due to either periodic catastrophic flooding (Falcon-Lang, Cantrill and Nichols, 2001) or species competition (Howe, 2003). Unfortunately, nothing is known of the leaf litter composition of these mature palaeosols. By contrast, on the more stable conditions offered by the coastal meander plain, less frequent stands of medium density *Araucaria* and podocarps conifer stumps can be found in growth position on mature palaeosols of the coastal plain meander belt association of the Coal Nunatak Member. On one site at Coal Nunatak, the tree density of the alluvial braidplain is six times greater, with 54 stumps occurring on an exposed area of palaeosol measuring 950 m^2 (equating to a high mean density of 568 stumps per hectare; Jefferson, 1981). Moreover, on the alluvial braidplain only podocarps were found; in the coastal plain both araucarian (*Araucariopitys*; ~30%) and podocarp (*Podocarpoxylon*; ~70%) conifers are found. The leaf litter layer of the palaeosol is also preserved here and dominated by conifers (*Podozamites* 46%, *Elatocladus* 41%, with rare *Pagiophyllum* and *Brachyphyllum*) together with minor components of ferns (*Cladophlebis* and *Sphenopteris*), angiosperms (*Ficophyllum*) and liverworts (*Hepaticites*).

Using the plant assemblages, more detailed ecological interpretations can be made. The widely spaced conifers of the braided river association exhibit narrower growth rings when compared with those of the coastal meander belt implying that growing conditions in the braidplain were more stressful even though the interception of sunlight and availability of soil nutrients would have been greater relative to the more densely vegetated coastal meander belt (Falcon-Lang, Cantrill and Nichols, 2001). Braided river systems are much more prone to bank bursting and overbank flooding as a result of the relative amount of energy they contain when compared with meandering systems (Reading, 1996). The narrow, false ring sequences exhibited by the braidplain conifers tell us that inundation was a frequent and regular scenario (up to as many as 14 times within a single growing season) of this locality leading to long-term suppression of growth (Falcon-Lang, Cantrill and Nichols, 2001). Further evidence for flooding in this environment can be gained from the presence of thick bark, adventitious roots and the shallow, horizontally-orientated nature of the podocarp roots (Falcon-Lang, Cantrill and Nichols, 2001).

In summary therefore, fossil plants in sedimentary facies coupled with *in situ* leaf litter layers on palaeosols, indicate that the late Albian vegetation of the Alexander Island floodplains consisted of a spatially complex flora. This flora was composed of intergrading conifer and fern dominated plant communities with widely spaced individuals, or stands of araucarian conifers being the most widespread group in relatively well-drained habitats. In the more waterlogged, swampy settings, dense stands of podocarps dominated. This vegetation grew under a humid seasonal climate characterised by cool, dark winters and warm, light summers and high year-round rainfall (Spicer and Chapman, 1990). Alluvial floodplains were dominated by podocarp and taxodioid, low density forest stands, whereas coastal belts were dominated by less frequent stands of medium density araucarian-podocarp rainforests. This more distal climax vegetation consisted of a four-tier structure comprising a ~30 m high conifer canopy, a subcanopy of tree ferns and ginkgos, bennettites, Pentoxylales and angiosperms, a herbaceous layer dominated by ferns and minor angiosperms, and a liverwort ground layer. Almost all of the arborescent vegetation on both types of floodplain possessed a broad-leafed evergreen canopy, although some elements of the subcanopy and herbaceous layer were probably deciduous (Falcon-Lang and Cantrill, 2001b; Howe, 2003). Vegetation disturbances due to unusually hard frosts, wildfires and biotic attack from wood-boring arthropods and fungi periodically occurred but with low frequency (Falcon-Lang, Cantrill and Nichols, 2001). Such open-canopied, evergreen araucarian-podocarp conifer forests were characteristic of the mid-Cretaceous Pacific margins of the Gondwana continent (Falcon-Lang and Cantrill, 2000).

Warm temperate broad-leafed araucarian-podocarp rainforests, similar to those of Alexander Island, appear to have dominated much of Gondwana above a palaeolatitude of 60° S (Spicer and Chapman, 1990). Plant growth was subject to conditions of relatively high humidity and extreme light seasonality, characterised by up to 70 days of unbroken darkness each winter (Parrish, Ziegler and Scotese, 1982; Read and Francis, 1992). Although a classical conical shape has been inferred for the trees occupying such high latitudes, in order to cope with the low-angle light and intercept as much radiation as possible (e.g. Chaloner and Creber, 1989), more recent research suggests that this was not the case (see Brodribb and Hill, 2004) since no evidence of whorled branch insertion has been found at this time (Cantrill, 2001). Where branching has been observed it appears to be towards a pseudo-monopodial habit, more consistent with the habit of extant podocarps. Stand density ranges from a median of 500–600 stems per hectare (perhaps representing colonisation stands: Falcon-Lang, Cantrill and Nichols, 2001; see below) to dispersed clumps of individuals similar to those in open woodland today (Cantrill, 2001), inferring that this may have been the maximum density possible to ensure the minimisation of mutual shading due to the low-angle radiation.

Modern analogues

The composition and structure of the Alexander Island forest associations bears some resemblance to the extant rainforests of New Zealand (Falcon-Lang, Cantrill and Nichols, 2001). These forests grow today under a humid (1500 mm per year) and warm temperate climate with mean maximum summer temperature between 16 and 22 °C, and a mean minimum winter temperature of 3–8 °C (Wardle, 1991). They are composed of a podocarp-araucarian canopy 20–30 m in height, and several understorey layers dominated by angiosperms, tree ferns, palms, ferns, mosses and liverworts (Wardle, 1991). The greatest difference between the Alexander Island flora and the New Zealand temperate rainforests is the greater abundance of angiosperms (locally up to 52% of the woody vegetation; Duncan, 1993) in the latter. This is not surprising considering the angiosperms had only recently migrated into polar regions prior to the deposition of the Triton Point Formation (Hill and Scriven, 1995). Therefore, even though further angiosperm radiation has modified the ecology of the temperate rainforests in the Southern Hemisphere, these relatively modern forests can provide interesting and important clues to the palaeoecology of Alexander Island (Falcon-Lang, Cantrill and Nichols, 2001). We have seen that the araucarian conifers of Alexander Island were restricted to the stable meander belts whereas the podocarp conifers dominated the disturbed floodplains. Within the putative closest modern analogue, the araucarian and podocarp conifers are also ecologically separated. The araucarians prefer well-drained, ultra-infertile podzolic soils (Ecroyd, 1982), whereas the podocarps prefer the flood-prone, semi-infertile alluvial substrates where periodic flooding can play an important role in forest regeneration (Duncan, 1993; Ogden and Stewart, 1995). Temperatures may have occasionally fallen to below freezing as evidenced by the rare occurrence of frost rings in the Alexander Island araucarian woods, assuming similar thermal tolerances of modern and fossil conifers. However, from comparisons with the New Zealand rainforests, modern araucarians are frost hardy to −11 °C, which therefore implies that temperatures could have fallen this low during the late Albian. Moreover, the absence of frost rings in the fossil podocarp wood indicates that, even though temperatures may have fallen to below −11 °C they did not fall to below −23 °C, the temperature at which modern podocarps produce frost rings (Falcon-Lang, Cantrill and Nichols, 2001; Sakai and Larcher, 1987). Present day podocarp-araucarian rainforests of New Zealand grow in medium density stands of 450–1400 trees per hectare similar to the density (i.e. ~600 per hectare) of the *in situ* stumps found in growth position on the alluvial braidplains (Falcon-Lang, Cantrill and Nichols, 2001). Together with mean ring increment and stump density, the stand density of the fossil wood annual productivity of the Alexander Island forests was determined to have been 5.6–7.3 m^3 per hectare per year (Falcon-Lang, Cantrill and Nichols, 2001). These values are similar to the 4.5–7.5 m^3 per hectare per year productivities of the araucarian-podocarp stands in warm temperate North Island, New Zealand (Wardle, 1991). Interestingly, these findings have been verified by numerical models (Beerling, 2000; Valdes, Sellwood and Price, 1996) where Alexander Island and much of coastal Antarctica were estimated to have been influenced by a mild climate with winter temperatures of 0–4 °C and summer temperatures of 20–24 °C enabling a productivity similar to that of the warm temperate forests of New Zealand today.

Summary

- The vegetation of southern Gondwana during the mid-Cretaceous, a period of warmth with rising global temperatures, was initially dominated by ferns along with evergreen and deciduous gymnosperms. Polar forests were highly productive and comprised long-lived evergreen conifer trees attaining heights exceeding 40 m. Within this relatively homogeneous vegetation, floristic subprovinces have been identified along both latitudinal (east–west) and longitudinal (north–south) gradients. Migration occurred between and within them facilitated by climate change and tectonic processes.

- Within southern Gondwana, Antarctica was generally characterised by cool temperate rainforests comprising abundant bryophytes, lycophytes, ferns along with seed plants and conifers. The activity of the volcanic arc that made up the Antarctic Peninsula in West Antarctica had a major influence on the vegetation with evidence of altitudinal statification. Upland communities were dominated by arborescent podocarps and bennettites, with pentoxylaleans, tree-sized ferns and other seed plants forming a canopy above an understorey of herbaceous ferns and sphenopsids. Lowlands supported continuous woodlands or forests of *Araucaria* and more podocarps. Sites near to volcanic activity were occupied by bennettite-podocarp forests with shrubby bennettite-herbaceous fern communities dominating lake margins. Disturbed ground enabled colonisation by pioneer species of liverworts, hornworts and bryophytes that also grew in more stable environments along with *Sphagnum*-like plants in boggy areas. Humidity levels were high, enabling abundant epiphytes to occupy higher niches and filmy ferns to thrive.

- Angiosperms probably initially invaded Gondwana through an African or Indian gateway and radiated out across the continent reaching South America and Australia by the Barremian–Aptian and then finally into West Antarctica. The earliest evidence of angiosperms in the fossil record of Antarctica date from the early Albian (~112 Ma), some 15 Ma after their first recorded appearance worldwide and postdating their arrival in Australia and southern South America. Therefore, the Antarctic Peninsula did not act as a biological gateway for initial dispersal between east and west Gondwana, but rather as a barrier as a result of the steep climate gradient that was in existence at this time.

- Initially the angiosperms were probably shrubby contributing in a minor way to the understorey vegetation. By the late Albian they had become more substantial and were exhibiting a more arboreal habit. Vegetation change, as a result of the diversification of the angiosperms, was initially at the expense of the ferns, hepatophytes and lycopods and only later affecting the bennettites and gymnosperms.

References

Ahmed, M. and Ogden, J. (1987). Population dynamics of the emergent conifer *Agathis australis* (D. Don) Lindl. (Kauri) in New Zealand. 1. Population structures and tree growth rates in mature stands. *New Zealand Journal of Botany*, **25**, 217–229.

Ali, J. R. and Aitchison, J. C. (2009). Kerguelen Plateau and the Late Cretaceous southern-continent bioconnection hypothesis: tales from a topographical ocean. *Journal of Biogeography*, **36**, 1778–1784.

Araya, R. and Hervé, F. (1966). Estudio Geomorphológico y geológico en las Islas Shetland del Sur, Antártico. *Publicación del Instituto Antártico Chileno*, **8**, 1–76.

Arber, E. A. N. (1917). The earlier Mesozoic floras of New Zealand. *New Zealand Geological Survey Palaeontological Bulletin*, **6**, 1–80.

Archangelsky, A., Andreis, R. R., Archangelsky, S. and Artabe, A. (1995). Cuticular characters adapted to volcanic stress in a new Cretaceous cycad leaf from Patagonia, Argentina. Considerations on the stratigraphy and depositional history of the Baqueró Formation. *Review of Palaeobotany and Palynology*, **89**, 213–233.

Archangelsky, S. (1963a). A new Mesozoic flora from Ticó Santa Cruz Province, Argentina. *Bulletin of the British Museum (Natural History) Geology*, **8**, 47–92.

Archangelsky, S. (1963b). Notas sobre la flora fósil de la zona de Ticó, Provincia de Santa Cruz 2. Tres nuevas especies de *Mesoingeria. Ameghiniana*, **3**, 113–122.

Archangelsky, S. (2001). Evidences of an early Cretaceous floristic change in Patagonia, Argentina. In *7th International Symposium on Mesozoic Terrestrial Ecosystems, Asociación Paleontológica Argentina, Publicación Especial*, **7**. Buenos Aires: APA, pp. 15–19.

Archangelsky, S. (2003). La flora Cretácica del Grupo Baqueró, Santa Cruz, Argentina. *Monografías del Museo Argentino de Ciencias Naturales*, **4**, 1–14.

Archangelsky, S. and Taylor, T. N. (1986). Ultrastructural studies of fossil plant cuticles. II. *Tarphyderma* gen. n., a Cretaceous conifer from Argentina. *American Journal of Botany*, **73**, 1577–1587.

Archangelsky, S. and Villar de Seoane, L. (2004). Cycadean diversity in the Cretaceous of Patagonia, Argentina. Three new *Androstrobus* species from the Baqueró Group. *Review of Palaeobotany and Palynology*, **131**, 1–28.

Archangelsky, S., Taylor, T. N. and Kurman, M. H. (1986). Ultrastructural studies of fossil plant cuticles: *Ticoa harrisii* from the Early Cretaceous of Argentina. *Botanical Journal of the Linnean Society*, **92**, 101–116.

Archangelsky, S., Barreda, V., Passalia, M. G., *et al.* (2009). Early angiosperm diversification: evidence from southern South America. *Cretaceous Research*, **30**, 1073–1082.

Askin, R. A. (1983). Tithonian (uppermost Jurassic) – Barremian (Lower Cretaceous) spore, pollen and microplankton from the South Shetland Islands, Antarctica. In *Antarctic Earth Science – Proceedings of the Fourth International Symposium on Antarctic Earth Sciences, Adelaide, South Australia, 16–20 August 1982*, eds R. L. Oliver, P. R. James and J. B. Jago. Canberra: Australian Academy of Sciences, pp. 295–297.

Askin, R. A. (1989). Endemism and heterochroneity in the Late Cretaceous (Campanian) to Paleocene palynofloras of Seymour Island, Antarctica: implications for origins, dispersal and palaeoclimates of southern floras. In *Origins and Evolution of the Antarctic Biota, Geological Society of London Special Publication*, **47**, ed. J. A. Crame. London: The Geological Society, pp. 107–119.

Askin, R. A., (1990). Cryptogam spores from the upper Campanian and Maastrichtian of Seymour Island, Antarctica. *Micropaleontology*, **36**, 141–156.

Askin, R. A. (1992). Late Cretaceous Early Tertiary Antarctic outcrop evidence for past vegetation and climates. In *The Antarctic Paleoenvironment: A Perspective on Global Change. Part 1. Antarctic Research Series*, **56**, eds J. P. Kennett and D. A. Warnke. Washington: American Geophysical Union, pp. 61–73.

Askin, R. A. and Spicer, R. A. (1995). The Late Cretaceous and Cenozoic history of vegetation and climate at northern and southern high latitudes: a comparison. In *Studies in Geophysics. Effects of Past Global Change on Life*, eds Panel on Effects of Past Global Change on Life, National Research Council. Washington: National Academy Press, pp. 156–173.

Banerji, J. and Lemoigne, Y. (1987). Significant additions to the Upper Triassic flora of Williams Point, Livingston Island, South Shetlands (Antarctica). *Geobios*, **20**, 469–487.

Barale, G., Philippe, M., Torres, T. and Thévenard, F. (1995). Reappraisal of the Triassic flora from Williams Point, Livingston Island (South Shetland Islands, Antarctica): systematical, biostratigraphical and paleogeographical implications. *Serie Científica Instituto Antártico Chileno*, **45**, 9–38.

Barker, P. E, Dalziel I. W. D. and Storey, B. C. (1991). Tectonic development of the Scotia arc region. In *The Geology of Antarctica, Oxford Monographs on Geology and Geophysics*, **17**, ed. R. J. Tingey. Oxford: Clarendon Press, pp. 215–248.

Barreda, V. D., Palamarczuk, S. and Medina, F. (1999). Palinología de la Formación Hidden Lake (Coniaciano e Santoniano), isla James Ross, Antártida. *Revista Española de Micropaleontología*, **31**, 53–72.

Barron, E. J. (1984). Climatic implications of the variable obliquity explanation of Cretaceous–Paleogene high-latitude floras. *Geology*, **12**, 595–598.

Beerling, D. J. (2000). The influence of vegetation cover on soil organic matter preservation in Antarctica during the Mesozoic. *Geophysical Research Letters*, **27**, 253–256.

Beerling, D. J. and Osborne, C. P. (2002). Physiological ecology of Mesozoic polar forests in a high CO_2 environment. *Annals of Botany*, **89**, 329–339.

Berner, R. A. and Kothavala, Z. (2001). GEOCARB III: a revised model of atmospheric CO_2 over Phanerozoic time. *American Journal of Science*, **301**, 182–204.

Bice, K. L. and Norris, R. D. (2002). Possible atmospheric CO_2 extremes of the Middle Cretaceous (late Albian–Turonian). *Paleoceanography*, **17**, 1070, doi:10.1029/2002PA000778.

Bice, K. L., Huber, B. T. and Norris, R. D. (2003). Extreme polar warmth during the Cretaceous greenhouse? Paradox of the late Turonian $\delta^{18}O$ record at Deep Sea Drilling Project Site 511. *Paleoceanography*, **18**, 1031, doi:10.1029/2002PA000848.

Bice, K. L., Birgel, D., Meyers, P. A, *et al.* (2006). A multiple proxy and model study of Cretaceous upper ocean temperatures and atmospheric CO_2 concentrations. *Paleoceanography*, **21**, PA2002, doi:10.1029/2005PA001203.

Blakey, R. C. (2008). Gondwana paleogeography from assembly to breakup— a 500 m.y. odyssey. In *Resolving the Late Paleozoic Ice Age in Time and Space, Geological Society of America Special Papers*, **441**, eds C. R. Fielding, T. D. Frank and J. L. Isbell. Boulder: The Geological Society of America, pp. 1–28.

BMR Palaeogeographic Group. (1990). *Australia: Evolution of a Continent*. Canberra: Australian Government Publishing Service, Bureau of Mineral Resources.

Bonan, G. B., Pollard, D. and Thompson, S. L. (1992). Effects of boreal forest vegetation on global climate. *Nature*, **359**, 716–718.

Bornemann, A., Norris, R. D., Friedrich, O., *et al.* (2008). Isotopic evidence for glaciation during the Cretaceous supergreenhouse. *Science*, **319**, 189–192.

Breecker, D. O., Sharp, Z. D. and McFadden, L. D. (2010). Atmospheric CO_2 concentrations during ancient greenhouse climates were similar to those predicted for AD 2100, *Proceedings of the National Academy of Sciences of the United States of America*, **107**, 576–580.

Brenner, G. J. (1996). Evidence for the earliest stage of angiosperm pollen evolution: A paleoequatorial section from Israel. In *Flowering Plant Origin, Evolution, and Phylogeny*, eds D. W. Taylor and L. J. Hickey. New York: Chapman and Hall, pp. 91–115.

Brentnall, S. J., Beerling, D. J., Osborne, C. P., *et al.* (2005). Climatic and ecological determinants of leaf lifespan in polar forests of the high CO_2 Cretaceous 'greenhouse' world. *Global Change Biology*, **11**, 2177–2195.

Brodribb, T. and Hill, R. S. (2004). The rise and fall of the Podocarpaceae in Australia – a physiological explanation. In *The Evolution of Plant Physiology*, eds A. R. Hemsley and I. Poole. Amsterdam: Elsevier Academic Press, pp. 381–399.

Burger, D. (1980). Palynology of the Lower Cretaceous in the Surat Basin. *Bureau of Mineral Resources, Geology and Geophysics Bulletin*, **189**, 1–106.

Burger, D. (1981). Observations on the earliest angiosperms development with special reference to Australia. In *Proceedings of the 4th International Palynological Congress Lucknow, volume 3*, Lucknow: Birbal Sahni Institute of Palaeobotany, pp. 418–428.

Burger, D. (1988). Early Cretaceous environments in the Eromanga Basin: palynological evidence from GSQ Wyandra 1 corehole. *Association of Australasian Palaeontologists Memoir*, **5**, 173–186.

Burger, D. (1990). Early Cretaceous angiosperms from Queensland, Australia. *Review of Palaeobotany and Palynology*, **65**, 153–163.

Burger, D. (1993). Early and middle Cretaceous angiosperm pollen grains from Australia. *Review of Palaeobotany and Palynology*, **78**, 183–234.

Butterworth, P. J., Crame, J. A., Howlett, P. J. and MacDonald, D. I. M. (1988). Lithostratigraphy of Upper Jurassic–Lower Cretaceous strata of eastern Alexander Island, Antarctica. *Cretaceous Research*, **9**, 249–264.

Cande, S. C., Stock, J., Müller, R. D. and Ishihara, T. (2000). Cenozoic motion between East and West Antarctica. *Nature*, **404**, 145–150.

Cantrill, D. J. (1991). Broad leafed coniferous foliage from the Lower Cretaceous Otway Group, southeastern Australia. *Alcheringa*, **15**, 177–190.

Cantrill, D. J. (1992). Araucarian foliage from the Lower Cretaceous of southern Victoria, Australia. *International Journal of Plant Sciences*, **153**, 622–645.

Cantrill, D. J. (1995). The occurrence of the fern *Hausmannia* Dunker (Dipteridaceae) in the Cretaceous of Alexander Island, Antarctica. *Alcheringa*, **19**, 243–254.

Cantrill, D. J. (1996). Fern thickets from the Cretaceous of Alexander Island, Antarctica containing *Alamatus bifarius* Douglas and *Aculea acicularis* sp. nov. *Cretaceous Research*, **17**, 169–182.

Cantrill, D. J. (1997). Hepatophytes from the Early Cretaceous of Alexander Island, Antarctica: systematic and palaeoecology. *International Journal of Plant Sciences*, **158**, 476–488.

Cantrill, D. J. (1998). Early Cretaceous fern foliage from President Head, Snow Island, Antarctica. *Alcheringa*, **22**, 241–258.

Cantrill, D. J. (2000). A Cretaceous (Aptian) flora from President Head, Snow Island, Antarctica. *Palaeontographica B*, **253**, 153–191.

Cantrill, D. J. (2001). Cretaceous high-latitude terrestrial ecosystems: an example from Alexander Island, Antarctica. In *7th International Symposium on Mesozoic Terrestrial Ecosystems, Asociación Paleontológica Argentina Publicación Especial*, **7**. Buenos Aires: APA, pp. 39–44.

Cantrill, D. J. and Douglas, J. G. (1988). Mycorrhizal conifer roots from the Lower Cretaceous of the Otway Basin, Victoria. *Australian Journal of Botany*, **36**, 257–272.

Cantrill, D. J. and Falcon-Lang, H. J. (2001). Cretaceous (Late Albian) Coniferales of Alexander Island, Antarctica. Part 2: Foliage, reproductive structures and roots. *Review of Palaeobotany and Palynology*, **115**, 119–145.

Cantrill, D. J. and Nagalingum, N. S. (2005). Ferns from the Cretaceous of Alexander Island, Antarctica: implications for Cretaceous phytogeography of the Southern Hemisphere. *Review of Palaeobotany and Palynology*, **137**, 83–103.

Cantrill, D. J. and Nichols, G. J. (1996). Taxonomy and palaeoecology of Early Cretaceous (late Albian) angiosperm leaves from Alexander Island, Antarctica. *Review of Palaeobotany and Palynology*, **92**, 1–28.

Cantrill, D. J. and Poole, I. (2002). Cretaceous patterns of floristic change in the Antarctic Peninsula. In *Palaeobiogeography and Biodiversity Change: the Ordovician and Mesozoic-Cenozoic Radiations, Geological Society of London Special Publication*, **194**, eds J. A. Crame and A. W. Owen. London: The Geological Society, pp. 141–152.

Cantrill, D. J. and Webb, J. A. (1987). A reappraisal of *Phyllopteroides* Medwell (Osmundaceae) and its stratigraphic significance in the Early Cretaceous of eastern Australia. *Alcheringa*, **11**, 59–85.

Carpenter, R. J., Truswell, E. M. and Harris, W. K. (2010). Lauraceae fossils from a volcanic Palaeocene oceanic island, Ninetyeast Ridge, Indian Ocean: ancient long-distance dispersal? *Journal of Biogeography*, **37**, 1202–1213.

Cas, R. A. F. and Wright, J. V. (1987). *Volcanic Successions: Modern and Ancient*. London: Unwin-Hyam.

Césari, S. N. (2006). Aptian ferns with in situ spores from the South Shetland Islands, Antarctica. *Review of Palaeobotany and Palynology*, **138**, 227–238.

Césari, S. N., Parica, C., Remesal, M. and Salani, F. (1999). Paleoflora del Cretácico Inferior de península Byers, Isla Shetland del Sur, Antártida. *Ameghiniana*, **36**, 3–22.

Césari, S. N., Remesal, M. and Parica, C. (2001). Ferns: a palaeoclimatic significant component of the Cretaceous flora from Livingston Island, Antarctica. In *7th International Symposium on Mesozoic Terrestrial Ecosystems, Asociación Paleontológica Argentina Publicación Especial*, **7**. Buenos Aires: APA, pp. 45–50.

Chaloner, W. G. and Creber, G. T. (1989). The phenomenon of forest growth in Antarctica: a review. In *Origins and Evolution of the Antarctic Biota, Geological Society of London Special Publication*, **47**, ed. J. A. Crame. London: The Geological Society, pp. 85–88.

Chapman, J. L. (1994). Distinguishing internal developmental characteristics from external palaeoenvironmental effects in fossil wood. *Review of Palaeobotany and Palynology*, **81**, 19–32.

Clarke, L. J. and Jenkyns, H. C. (1999). New oxygen isotope evidence for long-term Cretaceous climates in the Southern Hemisphere. *Geology*, **27**, 699–702.

Crame, J. A., Francis, J. E., Cantrill, D. J. and Pirrie, D. (2004). Maastrichtian stratigraphy of Antarctica. *Cretaceous Research*, **25**, 411–423.

Crame, J. A., Pirrie, D. and Riding, J. B. (2006). Mid-Cretaceous stratigraphy of the James Ross Basin, Antarctica. In *Cretaceous–Tertiary High-Latitude Palaeoenvironments,*

James Ross Basin, Antarctica, Geological Society of London Special Publication, **258**, eds J. E. Francis, D. Pirrie, and J. A. Crame. London: The Geological Society, pp. 7–19.

Crane, P. R. and Lidgard, S. (1989). Angiosperm diversification and paleolatitudinal gradients in Cretaceous floristic diversity. *Science*, **246**, 675–678.

Crane, P. R., Friis, E. M. and Pedersen, K. R. (1995). The origin and early diversification of angiosperms. *Nature*, **374**, 27–33.

Cúneo, R. and Gandolfo, M. A. (2005). Angiosperm leaves from the Kachaike Formation, Lower Cretaceous of Patagonia, Argentina. *Review of Palaeobotany and Palynology*, **136**, 29–47.

Dettmann, M. E. (1973). Angiospermous pollen from Albian to Turonian sediments of eastern Australia. *Special Publication of the Geological Society of Australia*, **4**, 3–34.

Dettmann, M. E. (1986a). Significance of the Cretaceous Tertiary spore genus *Cyatheacidites* in tracing the origin and migration of *Lophosoria* (Filicopsida). *Special Papers in Palaeontology*, **35**, 63–94.

Dettmann, M. E. (1986b). Early Cretaceous palynoflora of subsurface strata correlative with the Koonwarra Fossil Bed, Victoria. *Association of Australasian Palaeontologists Memoir*, **3**, 79–110.

Dettmann, M. E. (1989). Antarctica: Cretaceous cradle of austral temperate rainforests? In *Origins and Evolution of the Antarctic Biota, Geologial Society of London Special Publication*, **47**, ed. J. A. Crame. London: The Geological Society, pp. 89–105.

Dettmann, M. E. (1994). Cretaceous vegetation: the microfloral record. In *History of the Australian Vegetation: Cretaceous to Recent*. ed. R. S. Hill. Cambridge: Cambridge University Press, pp. 143–170.

Dettmann, M. E. and Clifford, H. T. (2000). Gemmae of the Marchantiales from the Winton Formation (mid-Cretaceous), Eromanga Basin, Queensland. *Memoirs of the Queensland Museum*, **45**, 285–292.

Dettmann, M. E. and Jarzen, D. M. (1990). The Antarctic/Australian rift valley: Late Cretaceous cradle of northeastern Australasian relicts? *Review of Palaeobotany and Palynology*, **65**, 131–144.

Dettmann, M. E. and Thomson, M. R. A. (1987). Cretaceous palynomorphs from the James Ross Island area, Antarctica — a pilot study. *British Antarctic Survey Bulletin*, **77**, 13–59.

Dettmann, M. E., Molnar, R. E., Douglas, J. G., *et al.* (1992). Australian Cretaceous terrestrial faunas and floras: biostratigraphic and biogeographic implications. *Cretaceous Research*, **13**, 207–262.

Dettmann, M. E., Clifford, H. T. and Peters, M. (2009). *Lovellea wintonensis* gen. et sp. nov.– Early Cretaceous (late Albian), anatomically preserved, angiospermous flowers and fruits from the Winton Formation, western Queensland, Australia. *Cretaceous Research*, **30**, 339–355.

Ditchfield, P. W. (1997). High northern palaeolatitude Jurassic–Cretaceous palaeotemperature variation: new data from Kong Karls Land, Svalbard. *Palaeogeography, Palaeoclimatology, Palaeoecology*, **130**, 163–175.

Ditchfield, P. W., Marshall, J. D. and Pirrie, D. (1994). High latitude temperature variation: new data from the Tithonian to Eocene of James Ross Island Antarctica. *Palaeogeography, Palaeoclimatology, Palaeoecology*, **107**, 79–101.

Dolding, P. J. D. (1992). Palynology of the Marambio Group (Upper Cretaceous) of northern Humps Island. *Antarctic Science*, **4**, 311–326.

Dorman, J. L. and Sellers, P. J. (1989). A global climatology of albedo, roughness length and stomatal resistance for atmospheric general-circulation models as represented by the simple biosphere model (SIB). *Journal of Applied Meteorology*, **28**, 833–855.

Douglas, J. G. (1963). Nut-like impressions attributed to aquatic dicotyledons from Victorian Mesozoic sediments. *Proceedings of the Royal Society of Victoria*, **76**, 23–28.

Douglas, J. G. (1965). A Mesozoic dicotyledonous leaf from the Yangery No.1 bore Koroit, Victoria. *Mining and Geological Journal*, **6**, 64–67.

Douglas, J. G. (1969). The Mesozoic flora of Victoria. Parts 1 and 2. *Geological Survey of Victoria Memoir*, **28**, 1–310.

Douglas, J. G. (1973). The Mesozoic flora of Victoria. Part 3. *Geological Survey of Victoria Memoir*, **29**, 1–185.

Douglas, J. G. (1994). Cretaceous vegetation: the macrofossil record. In *History of the Australian Vegetation: Cretaceous to Recent*, ed. R. S. Hill. Cambridge: Cambridge University Press, pp. 171–188.

Douglas, J. G. and Williams, G. E. (1982). Southern polar forests: the Early Cretaceous floras of Victoria and their palaeoclimatic significance. *Palaeogeography, Palaeoclimatology, Palaeoecology*, **39**, 171–185.

Drinnan, A. N. and Chambers, T. C. (1986). Flora of the Lower Cretaceous Koonwarra Fossil Bed (Korumburra Group), South Gippsland, Victoria. *Association of Australasian Palaeontologists Memoir*, **3**, 1–77.

Drinnan, A. N. and Crane, P. R. (1990). Cretaceous palaeobotany and its bearing on the biogeography of austral angiosperms. In *Antarctic Paleobiology: its Role in the Reconstruction of Gondwana*, eds T. N. Taylor and E. L. Taylor. New York: Springer-Verlag, pp. 192–219.

Duane, A. M. (1996). Palynology of the Byers Group (Late Jurassic–Early Cretaceous) of Livingston and Snow islands, Antarctic Peninsula: its biostratigraphic and palaeoenvironmental significance. *Review of Palaeobotany and Palynology*, **91**, 241–281.

Duncan, R. P. (1993). Flood disturbance and the coexistence of species in a lowland podocarp forest, south Westland, New Zealand. *Journal of Ecology*, **81**, 403–416.

Dunwiddie, P. W. (1979). Dendrochronological studies of indigenous New Zealand trees. *New Zealand Journal of Botany*, **17**, 251–266.

Ecroyd, C. E. (1982). Biological flora of New Zealand 8. *Agathis australis* (D.Don) Lindl. (Araucariaceae) Kauri. *New Zealand Journal of Botany*, **20**, 17–36.

Edwards, W. N. (1934). Jurassic plants from New Zealand. *Annals and Magazine of Natural History, Series 10*, **13**, 81–109.

Enright, N. J. and Hill, R. S. (1995). *Ecology of the Southern Conifers*. Australia: Melbourne University Press.

Equiza, M. A., Day, M. E. and Jagels, R. (2006). Physiological responses of three deciduous conifers (*Metasequoia glyptostroboides*, *Taxodium distichum* and *Larix laricina*) to continuous light: adaptive implications for the early Tertiary polar summer. *Tree Physiology*, **26**, 353–364.

Falcon-Lang, H. J. (2000a). A method to distinguish between woods produced by evergreen and deciduous coniferopsids on the basis of growth ring anatomy: a new palaeoecological tool. *Palaeontology*, **43**, 785–793.

Falcon-Lang, H. J. (2000b). The relationship between leaf longevity and growth ring markedness in modern conifer woods and its implications for palaeoclimatic studies. *Palaeogeography, Palaeoclimatology, Palaeoecology*, **160**, 317–328.

Falcon-Lang, H. J. and Cantrill, D. J. (2000). Cretaceous (Late Albian) Coniferales of Alexander Island, Antarctica. 1: Wood taxonomy: a quantitative approach. *Review of Palaeobotany and Palynology*, **111**, 1–17.

Falcon-Lang, H. J. and Cantrill, D. J. (2001a). Leaf phenology of some mid-Cretaceous polar forests, Alexander Island, Antarctica. *Geological Magazine*, **138**, 39–52.

Falcon-Lang, H. J. and Cantrill, D. J. (2001b). Gymnosperm woods from the Cretaceous (mid-Aptian) Cerro Negro Formation, Byers Peninsula, Livingston Island, Antarctica: the arborescent vegetation of a volcanic arc. *Cretaceous Research*, **22**, 277–293.

Falcon-Lang, H. J. and Cantrill, D. J. (2002). Terrestrial paleoecology of the Cretaceous (early Aptian) Cerro Negro Formation, South Shetlands Islands, Antarctica: a record of polar vegetation in a volcanic arc environment. *Palaios*, **17**, 491–506.

Falcon-Lang, H. J., Cantrill, D. J. and Nichols, G. J. (2001). Biodiversity and terrestrial ecology of a mid-Cretaceous, high-latitude floodplain, Alexander Island, Antarctica. *Journal of the Geological Society of London*, **158**, 709–725.

Feild, T. S. and Arens N. C. (2005). Form, function and environments of the early angiosperms: Merging extant phylogeny and ecophysiology with fossils. *New Phytologist*, **166**, 383–408.

Feild, T. S., Arens, N. C., Doyle, J. A., Dawson, T. E. and Donoghue, M. J. (2004). Dark and disturbed: a new image of early angiosperm ecology. *Paleobiology*, **30**, 82–107.

Feild, T. S., Chatelet, D. S. and Brodribb, T. J. (2009). Ancestral xerophobia: a hypothesis on the whole plant ecophysiology of early angiosperms. *Geobiology*, **7**, 237–264.

Feldmann, R. M. and Woodburne, M. O. (eds) (1988). Geology and paleontology of Seymour Island, Antarctic Peninsula. *Geological Society of America Memoir*, **169**, 1–555.

Fitzgerald, P. G., Sandiford, M., Barrett, P. J. and Gleadow, A. J. W. (1986). Asymmetric extension associated with uplift and subsidence in the Transantarctic Mountains and Ross Embayment. *Earth and Planetary Science Letters*, **81**, 67–78.

Fletcher, B. J., Beerling, D. J., Brentnall, S. J. and Royer, D. L. (2005). Fossil bryophytes as recorders of ancient CO_2 levels: experimental evidence and a Cretaceous case study. *Global Biogeochemical Cycles*, **19**, GB3012, doi:10.1029/2005GB002495.

Foley, J. A., Kutzback, J. E., Coe, M. T. and Levis, S. (1994). Feedbacks between climate and boreal forests during the Holocene epoch. *Nature*, **371**, 52–54.

Francis, J. E. (1986). Growth rings in Cretaceous and Tertiary wood from Antarctica and their palaeoclimatic implications. *Palaeontology*, **25**, 665–684.

Francis, J. E. and Coffin, M. F. (1992) Cretaceous fossil wood from the Raggatt Basin, Southern Kerguelen Plateau. In *Proceedings of the Ocean Drilling Program, Volume 120, Scientific Results, Part 1 Central Kerguelen Plateau, covering Leg 120 of the cruises of the Drilling Vessel JOIDES Resolution, Fremantle, Australia, to Fremantle, Australia, Sites 747–751, 20 February to 30 April 1988*, eds S. W. Wise Jr, A. A. Palmer Julson, R. Schlich and E. Thomas. doi:10.2973/odp.proc.sr.120.195.1992 available at http://www-odp.tamu.edu/publications/120_SR/VOLUME/CHAPTERS/sr120_18.pdf. {Cited 2011}.

Friis, E. M., Pedersen, K. R. and Crane, P. R. (2001). Fossil evidence of water lilies (Nymphaeales) in the Early Cretaceous. *Nature*, **410**, 357–360.

Friis, E. M., Doyle, J. A., Endress, P. K. and Leng, Q. (2003). *Archaefructus* – angiosperm precursor or specialized early angiosperm? *TRENDS in Plant Science*, **8**, 369–373.

Friis, E. M., Pedersen, K. R. and Crane, P. R. (2006). Cretaceous angiosperm flowers: innovation and evolution in plant reproduction. *Palaeogeography, Palaeoclimatology, Palaeoecology*, **232**, 251–293.

Friis, E. M., Pedersen, K. R. and Crane, P. R. (2010). Diversity in obscurity: fossil flowers and the early history of angiosperms. *Philosophical Transactions of the Royal Society Series B*, **365**, 369–382.

Friis, E. M., Crane, P. R. and Pedersen, K. R. (2011). *Early Flowers and Angiosperm Evolution*. Cambridge: Cambridge University Press.

Halle, T. G. (1913). Some Mesozoic plant-bearing deposits in Patagonia and Tierra del Fuego and their floras. *Kongliga Svenska Vetenskaps Akademiens Handlingar*, **51**, 1–58.

Hamilton, R. J., Luyendyk, B. P., Sorlien, C. C. and Bartek, L. R. (2001). Cenozoic tectonics of the Cape Roberts Rift Basin and Transantarctic Mountains Front, southwestern Ross Sea, Antarctica. *Tectonics*, **20**, 325–342.

Hathway, B. (1997). Nonmarine sedimentation in an Early Cretaceous extensional continental-margin arc, Byers Peninsula, Livingston Island, South Shetland Islands. *Journal of Sedimentary Research*, **67**, 686–697.

Hathway, B. (2000). Continental rift to back-arc basin: Jurassic–Cretaceous stratigraphical and structural evolution of the Larsen Basin, Antarctic Peninsula. *Journal of the Geological Society* of London, **157**, 417–432.

Hathway, B. and Lomas, S. A. (1998). The Upper Jurassic–Lower Cretaceous Byers Group, South Shetland Islands, Antarctica: Revised stratigraphy and regional correlations. *Cretaceous Research*, **19**, 43–67.

Hathway, B., Duane, A. M., Cantrill, D. J. and Kelley, S. P. (1999). ^{40}Ar/^{39}Ar geochronology and palynology of the Cerro Negro Formation, South Shetland Islands, Antarctica: A new radiometric tie for Cretaceous terrestrial biostratigraphy in the Southern Hemisphere: *Australian Journal of Earth Sciences*, **46**, 593–606.

Hay, W. W., de Conto, R. M., Wold, C. N., *et al.* (1999). An alternative global Cretaceous palaeogeography. In *Evolution of the Cretaceous Ocean Climate System, Geological Society of America Special Papers*, **332**, eds E. Barrera and C. Johnson. Boulder: The Geological Society of America, pp. 1–47.

Hayes, P. A. (1999). Cretaceous angiosperm floras of Antarctica. Ph.D. thesis, University of Leeds.

Heimhofer, U., Hochuli, P. A., Burla, S., Dinis, J. M. L. and Weissert, H. (2005). Timing of Early Cretaceous angiosperm diversification and possible links to major paleoenvironmental change. *Geology*, **33**, 141–144.

Helby, R., Morgan, R. and Partridge, A. D. (1987). A palynological zonation of the Australian Mesozoic. *Association of Australasian Palaeontologists Memoir*, **4**, 1–94.

Herman, A. B. and Spicer, R. A. (1996). Palaeobotanical evidence for a warm Cretaceous Arctic Ocean. *Nature*, **380**, 330–333.

Hernández, P. J. and Azcárate, V. (1971). Estudio paleobotánico preliminar sobre restos de una tafoflora de la Peninsula Byers (Cerro Negro), Isla Livingston, Islas Shetland del Sur, Antartica. *Serie Científica Instituto Antárctico Chileno*, **2**, 15–50.

Herngreen, G. F. W. (1975). Palynology of Middle and Upper Cretaceous strata in Brazil. *Mededelingen Rijks Geologische Dienst Nieuwe Serie*, **26**, 39–90.

Herngreen, G. F. W., Kedves, M., Rovina, L. V. and Smirnova, S. B. (1996). Cretaceous palynofloral provinces: a review. In *Palynology: Principles and Applications, Volume 3*, eds J. Jansonius and D. C. McGregor. College Station: Amercian Association of Stratigraphic Palynologists Foundation, pp. 1157–1588.

Hickey, L. J. and Doyle, J. A. (1977). Early Cretaceous evidence for angiosperm evolution. *Botanical Review*, **43**, 3–104.

Hill, R. S. and Scriven, L. J. (1995). The angiosperm-dominated woody vegetation of Antarctica: a review. *Review of Palaeobotany and Palynology*, **86**, 175–198.

Hochuli, P. A., Heimhofer, U. and Weissert, H. (2006). Timing of early angiosperm radiation: Recalibrating the classical succession. *Journal of the Geological Society*, **163**, 587–594.

Hocknull, S. A., White, M. A., Tischler, T. R., *et al.* (2009). New mid-Cretaceous (latest Albian) dinosaurs from Winton, Queensland, Australia. *PlosOne*, **4**, e6190, doi:10.1371/journal.pone.0006190.

Howe, J. (2003). Mid-Cretaceous fossil forests of Alexander Island, Antarctica. Ph.D. thesis, University of Leeds.

Howe, J. and Francis, J. E. (2005). Metamorphosed palaeosoils associated with Cretaceous fossil forests, Alexander Island, Antarctica. *Journal of the Geological Society of London*, **162**, 951–957.

Huber, B. T. (1988). Tropical paradise at the Cretaceous poles? *Science*, **282**, 2199–2200.

Huber, B. T., Hodell, D. A. and Hamilton, C. P. (1995). Middle-Late Cretaceous climate of the southern high latitudes: stable isotopic evidence for minimal equator-to-pole thermal gradients. *Geological Society of America Bulletin*, **107**, 1164–1191.

Hughes, N. F. (1994). *The Enigma of Angiosperm Origins*. Cambridge: Cambridge University Press.

Iglesias, A., Artabe, A. E. and Morel, E. M. (2011). The evolution of Patagonian climate and vegetation from the Mesozoic to the present. *Biological Journal of the Linnean Society*, **103**, 409–422.

Jefferson, T. H. (1981). Palaeobotanical contributions to the geology of Alexander Island, Antarctica. Ph.D. thesis, University of Cambridge.

Jefferson, T. H. (1982a). Fossil forests from the Lower Cretaceous of Alexander Island, Antarctica. *Palaeontology*, **25**, 681–708.

Jefferson, T. H. (1982b). The preservation of fossil leaves in Cretaceous volcaniclastic rocks from Alexander Island, Antarctica. *Geological Magazine*, **119**, 291–300.

Jefferson, T. H. (1987). The preservation of conifer wood: examples from the Lower Cretaceous of Antarctica. *Palaeontology*, **30**, 233–249.

Keating, J. M. (1992). Palynology of the Lachmans Crags Member, Santa Marta Formation (Upper Cretaceous) of north-west James Ross Island. *Antarctic Science*, **4**, 293–304.

Keating, J. M., Spencer-Jones, M. and Newham, S. (1992). The stratigraphical palynology of the Kotick Point and Whisky Bay formations, Gustav Group (Cretaceous), James Ross Island. *Antarctic Science*, **4**, 279–292.

Koch, J. T. and Brenner, R. L. (2009). Evidence for glacioeustatic control of large, rapid sea-level fluctuations during the Albian-Cenomanian: Dakota Formation, eastern margin of western interior seaway, USA. *Cretaceous Research*, **30**, 411–423.

Krause, D. W., Roger, R. R., Forster, C. A., Hartman, J. H. and Buckley, G. A. (1999). The Late Cretaceous vertebrate fauna of Madagascar: implications for Gondwanan paleo-biogeography. *GSA Today*, **98**, 1–7.

Lacey, W. S. and Lucas, R. C. (1981). The Triassic flora of Livingston Island, South Shetland Islands. *British Antarctic Survey Bulletin*, **53**, 157–173.

LaMarche, V. C. and Hirschboeck, K. K. (1984). Frost rings in trees as records of major volcanic eruptions. *Nature*, **307**, 121–126.

Lawver, L. A. and Gahagan, L. M. (1994). Constraints on the timing of extension in the Ross Sea region. *Terra Antartica*, **1**, 545–552.

Lawver, L. A., Gahagan, L. M. and Coffin, M. F. (1992). The development of paleoseaways around Antarctica. In *The Antarctic Paleoenvironment: A Perspective on Global Change. Part 1, Antarctic Research Series*, **56**, eds J. P. Kennett and D. A. Warnke. Washington: American Geophysical Union, pp. 7–30.

Lawver, L. A., Gahagan, L. M. and Dalziel, I. W. D. (1998). A tight fit early Mesozoic Gondwana, a plate reconstruction perspective. In *Origin and Evolution of the Continents, Memoirs of the National Institute of Polar Research*, **53**, Tokyo: National Institute for Polar Research, pp. 214–229.

Li, H. (2003). Lower Cretaceous angiosperm leaf from Wuhe in Anhui, China. *Chinese Science Bulletin*, **48**, 611–614.

Li, Z. X. and Powell, C. McA. (2001). An outline of the palaeogeographic evolution of the Australasian region since the beginning of the Neoproterozoic. *Earth-Science Reviews*, **53**, 237–277.

Lidgard, S. and Crane, P. R. (1990). Angiosperm diversification and Cretaceous floristic trends: a comparison of palynofloras and leaf macrofloras. *Paleobiology*, **16**, 77–93.

Livermore, R. A. and Hunter, R. J. (1996). Mesozoic seafloor spreading in the southern Weddell Sea. In *Weddell Sea Tectonics and Gondwana Break-up, Geological Society of London Special Publications*, **108**, eds B. C. Storey, E. C. King and R. A. Livermore, London: The Geological Society, pp. 227–241.

Llorens, M. (2003). Granos de polen de angiospermas de la Formación Punta del Barco (Aptiano), provincia de Santa Cruz, Argentina. *Revista del Museo Argentino de Ciencias Naturales, Nueva Serie*, **5**, 235–240.

Llorens, M. and del Fueyo, G. M. (2003). Coníferas fértiles de la Formación Kachaike, Cretácico medio de la provincia de Santa Cruz, Argentina. *Revista del Museo Argentino de Ciencias Naturales, Nueva Serie*, **5**, 241–244.

Lupia, R., Lidgard, S. and Crane, P. R. (1999). Comparing palynological abundance and diversity: implications for biotic replacement during the Cretaceous angiosperm radiation. *Paleobiology*, **25**, 305–340.

Luyendyk, B. P., Wilson, D. and Siddoway, C. S. (2003). The eastern margin of the Ross Sea rift in western Marie Byrd Land: crustal structure and tectonic development. *Geochemistry, Geophysics, Geosystems*, **4**, 1090, doi:10.1029/2002GC000462.

Marks, K. M. and Tikku, A. A. (2001). Cretaceous reconstructions of East Antarctica, Africa and Madagascar. *Earth and Planetary Science Letters*, **186**, 479–495.

McLoughlin, S. (1996). Early Cretaceous macrofloras of Western Australia. *Records of the Western Australian Museum*, **18**, 19–65.

McLoughlin, S. (2001). The breakup history of Gondwana and its impact on pre-Cenozoic floristic provincialism. *Australian Journal of Botany*, **49**, 271–300.

McLoughlin, S., Drinnan, A. N. and Rozefelds, A. C. (1995). The Cenomanian flora of the Winton Formation, Eromanga Basin, Queensland. *Memoirs of the Queensland Museum*, **38**, 273–313.

McLoughlin, S., Pott, C. and Elliott, D. (2010). The Winton Formation flora (Albian–Cenomanian, Eromanga Basin): implications for vascular plant diversification and decline in the Australian Cretaceous. *Alcheringa*, **34**, 303–323.

Medina, F., Archangelsky, S., Guler, V., Archangelsky, A. and Cárdenas, O. (2008). Estudio bioestratigráfico integrado del perfil La Horqueta (límite Aptiano–Albiano), Lago Cardiel, Patagonia, Argentina. *Revista del Museo Argentino de Ciencias Naturales, Nueva Serie*, **10**, 141–155.

Medwell, L. M. (1954). Fossil plants from Killara, near Casterton, Victoria. *Proceedings of the Royal Society of Victoria*, **66**, 17–23.

Metcalfe, I. (1990). Allochthonous terrane processes in Southeast Asia. *Philosophical Transactions of the Royal Society London series A*, **331**, 625–640.

Meyen, S. V. (1987). *Fundamentals of Palaeobotany*. London: Chapman and Hall.

Miller, K. G., Kominz, M. A., Browning, J. V., *et al.* (2005). The Phanerozoic record of global sea-level change. *Science*, **310**, 1293–1298.

Mohr, B. A .R. (1990). Early Cretaceous palynomorphs from ODP sites 692 and 693, the Weddell Sea, Antarctica. In *Proceedings of the Ocean Drilling Program, Volume 113. Scientific Results, Weddell Sea, Antarctica covering Leg 113 of the cruises of the Drilling Vessel JOIDES Resolution, Valparaiso, Chile, to East Cove, Falkland Islands, Sites 689–697, 25 December 1986–11 March 1987*, eds P. F. Barker, J. P. Kennett, S. O'Connell and N. G. Pisias. Online doi:10.2973/odp.proc. sr.113.207.1990, available at http://www-odp.tamu.edu/publications/113_SR/ VOLUME/CHAPTERS/sr113_29.pdf. {Cited 2011}.

Mohr, B. A. R. and Gee, C. T. (1992a). Late Cretaceous palynofloras (sporomorphs and dinocysts) from the Kerguelen Plateau, Southern Indian Ocean (sites 748 and 750). In *Proceedings of the Ocean Drilling Program, Volume 120, Scientific Results, Part 1 Central Kerguelen Plateau, covering Leg 120 of the cruises of the Drilling Vessel* JOIDES Resolution, *Fremantle, Australia, to Fremantle, Australia, Sites 747–751, 20 February to 30 April 1988*, eds S. W. Wise Jr, A. A. Palmer Julson, R. Schlich and E. Thomas. doi:10.2973/odp.proc.sr.120.196.1992, available at http://www-odp.tamu.edu/publications/120_SR/VOLUME/CHAPTERS/sr120_17.pdf. {Cited 2011}.

Mohr, B. A. R. and Gee, C. T. (1992b). An early Albian palynoflora from the Kerguelen Plateau, southern Indian Ocean (Leg 120). In *Proceedings of the Ocean Drilling Program, Volume 120, Scientific Results, Part 1 Central Kerguelen Plateau, covering Leg 120 of the cruises of the Drilling Vessel* JOIDES Resolution, *Fremantle, Australia, to Fremantle, Australia, Sites 747–751, 20 February to 30 April 1988*, eds S. W. Wise Jr, A. A. Palmer Julson, R. Schlich and E. Thomas doi:10.2973/odp.proc.sr.120.161.1992, available at http://www-odp.tamu.edu/publications/120_SR/VOLUME/CHAPTERS/sr120_17.pdf. {Cited 2011}.

Mohr, B. A. R., Wähnert, V. and Lazarus, D. (2002). Mid-Cretaceous palaeobotany and palynology of the central Kerguelen Plateau, southern Indian Ocean (ODP Leg 183, Site 1138). In *Proceedings of the Ocean Drilling Program, Volume 183 Scientific Results Kerguelen Plateau–Broken Ridge: A large igneous province, Covering Leg 183 of the cruises of the Drilling Vessel JOIDES* Resolution, *Fremantle, Australia, to Fremantle, Australia Sites 1135–1142 7 December 1998–11 February 1999*, eds F. A. Frey, M. F. Coffin, P. J. Wallace and P. G. Quilty doi:10.2973/odp.proc.sr.183.008.2002, available at http://www-odp.tamu.edu/publications/183_SR/008/008.htm. {Cited 2011}.

Moncrieff, A. C. M. and Kelly, S. R. A. (1993). Lithostratigraphy of the uppermost Fossil Bluff Group (Early Cretaceous) of Alexander Island: History of an Albian regression. *Cretaceous Research*, **14**, 1–15.

Moriya, K., Wilson, P. A., Friedrich, O., Erbacher, J. and Kawahata, H. (2007). Testing for ice sheets during the mid-Cretaceous greenhouse using glassy foraminiferal calcite from the mid-Cenomanian tropics on Demerara Rise. *Geology*, **35**, 615–618.

Muller, R. D., Mihut, D. and Baldwin, S. (1998). A new kinematic model for the formation and evolution of the west and northwest Australian margin. In *The Petroleum Basins of Western Australia*, eds P. G. Purcell and P. R. Purcell. Perth: Petroleum Exploration Society of Australia, pp. 55–71.

Nagalingum, N. S, Drinnan, A. N., Lupia, R. and McLoughlin, S. (2002). Fern spore diversity and abundance in Australia during the Cretaceous. *Review of Palaeobotany and Palynology*, **119**, 69–92.

Nichols, G. J. and Cantrill, D. J. (2002). Tectonic and climatic controls on a Mesozoic forearc basin succession, Alexander Island, Antarctica. *Geological Magazine*, **139**, 313–330.

Norton, D. A., Herbert, J. W. and Beveridge, A. E. (1988). The ecology of *Dacrydium cupressinum*: a review. *New Zealand Journal of Botany*, **26**, 37–62.

Ogden, J. and Stewart, G. H. (1995). Community dynamics of the New Zealand conifers. In *Ecology of the Southern Conifers*, eds N. J. Enright and R. S. Hill. Melbourne: Melbourne University Press, pp. 81–119.

Orlando, H. A. (1968). A new Triassic flora from Livingston Island, South Shetland Islands. *British Antarctic Survey Bulletin*, **16**, 1–13.

Osborne, C. P. and Beerling, D. J. (2003). The penalty of a long hot summer. Photosynthetic acclimation to high CO_2 and continuous light in "living fossil" conifers. *Plant Physiology*, **133**, 803–812.

Osborne, C. P., Royer, D. L. and Beerling, D. J. (2004). Adaptive role of leaf habit in extinct polar forests. *International Forestry Review*, **6**, 181–186.

Ottone, E. G. and Medina, F. A. (1998). A wood from the Early Cretaceous of James Ross Island, Antarctica. *Ameghiniana*, **35**, 291–298.

Palma-Heldt, S., Fernandoy, F., Quezada, I. and Leppe, M. (2004). Registro Palinológico de Cabo Shirreff, Isla Livingston, nueva localidad para el Mesozoico de las Shetland del Sur, In *V° Simposio Argentino y 1° Latinoamericano sobre Investigaciones Antárticas Ciudad Autónoma de Buenos Aires, Argentina 30 de agosto al 3 de septiembre de 2004 Resúmen Expandido N° 104GP*, 4pp. Buenos Aires, Argentina. Available at http://www.dna.gov.ar/CIENCIA/SANTAR04/CD/PDF/104GP.PDF. {Cited 2011}.

Palma-Heldt, S., Fernandoy, F., Henríquez, G. and Leppe, M. (2007). Palynoflora of Livingston Island, South Shetland Islands: Contribution to the understanding of the evolution of the southern Pacific Gondwana margin. In *Online Proceedings of the 10th International Symposium on Antarctic Earth Sciences*, eds A. K. Cooper, C. R. Raymond and the 10th ISAES Editorial team. U. S. Geological Survey Open-File Report 2007–1047, Version 2. Extended Abstract 100, available at http://pubs.usgs.gov/of/2007/1047/ea/of2007–1047ea100.pdf. {Cited 2011}.

Parris, K. M., Drinnan, A. N. and Cantrill D. J. (1995). *Palissya* cones from the Mesozoic of Australia and New Zealand. *Alcheringa*, **19**, 87–111.

Parrish, J. T., Ziegler, A. M. and Scotese, C. R. (1982). Rainfall patterns and the distribution of coals and evaporates in the Mesozoic and Cenozoic. *Palaeogeography, Palaeoclimatology, Palaeoecology*, **40**, 67–101.

Parrish, J. T., Daniel, I. L., Kennedy, E. M. and Spicer, R. A. (1998). Palaeoclimatic significance of mid-Cretaceous floras from the Middle Clarence Valley, New Zealand. *Palaios*, **13**, 149–159.

Passalía, M. G., Archangelsky, S., Romero, E. J. and Cladera, G. (2003). A new early angiosperm leaf from the Anfiteatro de Ticó Formation (Aptian), Santa Cruz Province, Argentina. *Revista del Museo Argentino de Ciencias Naturales Nueva Serie*, **5**, 245–252.

Peters, M. D. and Christophel, D. C. (1978). *Austrosequoia wintonensis*, a new taxodiaceous cone from Queensland, Australia. *Canadian Journal of Botany*, **56**, 3119–3128.

Pirrie, D. and Crame, J. A. (1995). Late Jurassic paleogeography and anaerobic-dysaerobic sedimentation in the northern Antarctic Peninsula region. *Journal of the Geological Society of London*, **152**, 469–480.

Pirrie, D., Crame, J. A., Lomas, S. A. and Riding, J. B. (1997). Late Cretaceous stratigraphy of the Admiralty Sound region, James Ross Basin, Antarctica. *Cretaceous Research*, **18**, 109–137.

Pole, M. (1999). Latest Albian–early Cenomanian monocotyledonous leaves from Australia. *Botanical Journal of the Linnean Society*, **129**, 177–186.

Pole, M. (2000a). Dicotyledonous leaf macrofossils from the latest Albian–earliest Cenomanian of the Eromanga Basin, Queensland, Australia. *Palaeontological Research*, **4**, 39–52.

Pole, M. (2000b). Mid Cretaceous conifers from the Eromanga Basin, Australia. *Australian Systematic Botany*, **13**, 153–197.

Pole, M. S. and Douglas, J. G. (1999). Bennettitales, Cycadales and Ginkgoales from the mid-Cretaceous of the Eromanga Basin, Queensland, Australia. *Cretaceous Research*, **20**, 523–538.

Pons, D. (1988). *Le Mésozoïque de Colombie macroflores et microflores*. Paris: CNRS.

Poole, I. and Cantrill, D. J. (2006). Cretaceous and Cenozoic vegetation of Antarctica integrating the fossil wood record. In *Cretaceous-Tertiary High-Latitude Palaeoenvironments, James Ross Basin, Antarctica, Geological Society of London Special Publication*, **258**, eds J. E. Francis, D. Pirrie and J. A. Crame. London: The Geological Society, pp. 63–81.

Prámparo, M. B. (1990). Palynostratigraphy of the Lower Cretaceous of the San Luis Basin, Argentina. Its place in the Lower Cretaceous flora provinces pattern. *Neues Jahrbuch für Geologie und Paläontologie*, **181**, 255–266.

Prámparo, M. B. (1994). Lower Cretaceous palynoflora of the La Cantera Formation, San Luis Basin: Correlation with other Cretaceous palynofloras of Argentina. *Cretaceous Research*, **15**, 193–203.

Prámparo, M. B. and Volkheimer W. (1999). Palinología del Miembro Avilé (Formación Agrio, Cretácico Inferior) en la cerro de la Parva, Neuquén. *Ameghiniana*, **36**, 217–227.

Price, G. D. and Mutterlose, J. (2004). Isotopic signals from Late Jurassic–Early Cretaceous (Volgian–Valanginian) sub-Arctic belemnites, Yatria River, western Siberia. *Journal of the Geological Society of London*, **161**, 959–968.

Price, G. D., and Nunn, E. V. (2010). Valanginian isotope variation in glendonites and belemnites from Arctic Svalbard: Transient glacial temperatures during the Cretaceous greenhouse. *Geology*, **38**, 251–254.

Pritchett, W. L. and Fisher, R. F. (1987). *Properties and Management of Forest Soils*. New York: Wiley.

Puebla, G. G. (2004). La megaflora de la Formación La Cantera (Cretácico Temprano) Sierra del Gigante, San Luis, Argentina. M.Sc. thesis, Universidad Nacional de San Luis.

Quattrocchio, M. E., Martinez, M. A., Carpinelli Pavisich, A. and Volkheimer, W. (2006). Early Cretaceous palynostratigraphy, palynofacies and palaeoenvironments of well sections in northeastern Tierra del Fuego, Argentina. *Cretaceous Research*, **27**, 584–602.

Read, J. and Francis, J. (1992). Responses of some Southern Hemisphere tree species to a prolonged dark period and their implications for high-latitude Cretaceous and Tertiary floras. *Palaeogeography, Palaeoclimatology, Palaeoecology*, **99**, 271–290.

Reading, H. G. (1996). *Sedimentary Environments: Processes, facies and stratigraphy*. 3rd edn. Oxford: Blackwell Science.

Rebassa, M. (1982). Análisis estratigráfico y paleoambiental de la Formación Kachaike, aflorante en la barranca epónima, prov. de Santa Cruz. M.Sc. thesis, University Buenos Aires.

Rees, P. M. and Smellie, J. L. (1989). Cretaceous angiosperms from an allegedly Triassic flora at Williams Point, Livingston Island, South Shetland Islands. *Antarctic Science*, **1**, 239–248.

Reeves, C. V. and de Wit, M. J. (2000). Making ends meet in Gondwana: retracing the transforms of the Indian Ocean and reconnecting continental shear zones. *Terra Nova*, **12**, 272–280.

Riccardi, A. C. (1988). The Cretaceous system of southern South America. *Geological Society of America Memoir*, **168**, 1–161.

Riding, J. B., and Crame, J. A. (2002). Aptian to Coniacian (Early–Late Cretaceous) palynostratigraphy of the Gustav Group, James Ross Basin, Antarctica. *Cretaceous Research*, **23**, 739–760.

Riding, J. B., Crame, J. A., Dettmann, M. E. and Cantrill, D. J. (1998). The age of the base of the Gustav Group in the James Ross Basin, Antarctica. *Cretaceous Research*, **19**, 87–105.

Rinaldi, C. A. (1982). Upper Cretaceous in the James Ross Island group. In *Antarctic Geoscience*, ed. C. Craddock. Madison: University of Wisconsin Press, pp. 281–286.

Roeser, H. A., Fritsch, J. and Hinz, K. (1996). The development of the crust of Donning Maud Land, East Antarctica. In *Weddell Sea Tectonics and Gondwana Break-up, Geological Society of London Special Publications*, **108**, eds B. C. Storey, E. C. King and R. A. Livermore. London: The Geological Society, pp. 243–264.

Romero, E. J. and Archangelsky, S. (1986). Early Cretaceous angiosperm leaves from southern South America. *Science*, **234**, 1580–1582.

Royer, J. V. and Coffin, M. F. (1992). Jurassic to Eocene plate tectonic reconstructions in the Kerguelen Plateau region. In *Proceedings of the Ocean Drilling Program, Volume 120, Scientific Results, Part 1 Central Kerguelen Plateau, covering Leg 120 of the cruises of the Drilling Vessel* JOIDES Resolution, *Fremantle, Australia, to Fremantle, Australia, Sites 747–751, 20 February to 30 April 1988*, eds S. W. Wise Jr, A. A. Palmer Julson, R. Schlich and E. Thomas. Online doi:10.2973/odp.proc.sr.120.200.1992, available at http://www-odp.tamu.edu/publications/120_SR/VOLUME/CHAPTERS/sr120_50. pdf. {Cited 2011}.

Royer, D. L., Osborne, C. P. and Beerling, D. J. (2003). Carbon loss by deciduous trees in a CO_2 rich ancient polar environment. *Nature*, **424**, 60–62.

Royer, D. L., Miller, I. M., Peppe, D. J. and Hickey, L. J. (2010). Leaf economic traits from fossils support a weedy habit for early angiosperms. *American Journal of Botany*, **97**, 438–445.

Sakai, A. and Larcher W. (1987). *Frost survival in plants. Ecostudies 62*. Berlin: Springer-Verlag.

Sampson, S. D., Witmer, L. M., Forster, C. A., *et al.* (1998). Predatory dinosaur remains from Madagascar: implications for the Cretaceous biogeography of Gondwana. *Science*, **280**, 1048–1051.

Scheihing, M. H. and Pfefferkorn, H. W. (1984). The taphonomy of land plants in the Orinoco Delta: A model for the incorporation of plants in clastic sediments of Late Carboniferous age of Euramerica. *Review of Palaeobotany and Palynology*, **41**, 205–240.

Scotese, C. R. (1998). Quicktime Computer Animations: PALEOMAP Project Department of Geology: Arlington, Texas, University of Texas at Arlington, CD-ROM. Available at http://www.scotese.com.

Segoufin, J. and Patriat, P. (1990). Existence d'anomalies mésozoïques dans le bassin de Somalie. Implications pour les relations Afrique–Antarctique–Madagascar. *Compte Rendu Académie Sciences Paris*, **291**, 85–88.

Sellwood, B. W. and Valdes, P. J. (2006). Mesozoic climates: General circulation models and the rock record. *Sedimentary Geology*, **190**, 269–287.

Shipboard Scientific Party (2000). Leg 183 summary: Kerguelen Plateau – Broken Ridge – A large igneous province. In *Proceedings of the Ocean Drilling Program, Volume 183, Initial Reports Kerguelen Plateau-Broken Ridge: A Large Igneous Province covering Leg 183 of the cruises of the Drilling Vessel* JOIDES Resolution, *Fremantle, Australia, to Fremantle, Australia Sites 1135–114 7 December 1988 – 11 February 1999*, eds M. F. Coffin, F. A. Frey, P. J. Wallace, *et al.* Online doi:10.2973/odp.proc. ir.183.101.2000, available from http://www-odp.tamu.edu/publications/183_IR/ 183ir.html. {cited 2011}.

Siddoway, C. S. (2008). Tectonics of the West Antarctic Rift System: new light on the history and dynamics of distributed intracontinental extension. In *Antarctica: A Keystone in a Changing World, Proceedings of the 10th International Symposium on Antarctic Earth Sciences*. eds A. K. Cooper, P. J. Barrett, H. Stagg, *et al.* and the 10th ISAES editorial team. Washington: U.S. Geological Survey and The National Academies Press, pp. 91–114.

Siddoway, C. S., Baldwin, S. L., Fitzgerald, P. G., Fanning, C. M. and Luyendyk, B. P. (2004). Ross Sea mylonites and the timing of intercontinental extension within the West Antarctic rift system. *Geology*, **32**, 57–60.

Smith, A. G., Smith, D. G. and Funnel, B. M. (1994). *Atlas of Mesozoic and Cenozoic Coastlines*. Cambridge: Cambridge University Press.

Spicer, R. S. (2003). Changing Climate and biota. In *The Cretaceous World*, ed. P. Skelton. Cambridge: Cambridge University Press, pp. 85–162.

Spicer, R. A. and Chapman, J. L. (1990). Climate change and the evolution of high-latitude terrestrial vegetation and floras. *Trends in Ecology and Evolution*, **5**, 279–284.

Spicer, R. A. and Parrish, J. T. (1986). Paleobotanical evidence for cool north polar climates in middle Cretaceous (Albian–Cenomanian) time. *Geology*, **14**, 703–706.

Stampfli, G. M. and Borel, G. D. (2002). A plate tectonic model for the Paleozoic and Mesozoic constrained by dynamic plate boundaries and restored synthetic ocean isochrones. *Earth and Planetary Science Letters*, **196**, 17–33.

Steuber, T., Rauch, M., Masse, J. P., Graaf, J. and Malkoc, M. (2005). Low-latitude seasonality of Cretaceous temperatures in warm and cold episodes. *Nature*, **437**, 1341–1344.

Storey, B. C., Dalziel, I. W. D., Garrett, S. W., *et al.* (1988). West Antarctica in Gondwanaland: crustal blocks, reconstruction and break-up processes. *Tectonophysics*, **155**, 381–390.

Storey, B. C., Leat, P. T., Weaver, S. D., Pankhurst, R. J. and Kelley, S. (1999). Mantle plumes and Antarctica-New Zealand rifting: evidence from mid-Cretaceous mafic dykes. *Journal of the Geological Society of London*, **156**, 659–671.

Stopes, M. C. (1914). A new *Araucarioxylon* from New Zealand. *Annals of Botany*, **28**, 341–350.

Stopes, M. C. (1916). The Lower Greensand flora. *Nature*, **97**, 261.

Sun, G., Dilcher, D. L., Zheng, S.-L. and Zhou, Z. K. (1998). In search of the first flower: a Jurassic angiosperm, *Archaefructus*, from northeast China. *Science*, **282**, 1692–1695.

Sun, G., Dilcher, D. L. and Zheng, S.-L. (2008). A review of recent advances in the study of early angiosperms from northeastern China. *Paleoworld*, **17**, 166–171.

Taylor, D. W. and Hickey, L. J. (1990). An Aptian plant with attached leaves flowers: implication for angiosperm origin. *Science*, **247**, 702–704.

Torres, T., Barale, G., Thévenard, F., Philippe, M. and Galleguillos, H. (1997a). Morfología y sistemática de la flora del Cretácico Inferior de President Head, isla Snow, archipiélago de las Shetland del Sur, Antártica. *Serie Científica Instituto Antárctico Chileno*, **47**, 59–86.

Torres, T., Barale, G., Méon, H., Philippe, M. and Thévnard, F. (1997b). Cretaceous floras from Snow Island (South Shetland Islands, Antarctica) and their biostratigraphic significance. In *The Antarctic Region: Geological Evolution and Processes*, ed. C. A. Ricci. Siena: Terra Antarctica Publication, pp. 1023–1028.

Truswell, E. M. (1991). Antarctica: a history of terrestrial vegetation. In *The Geology of Antarctica, Oxford Monographs on Geology and Geophysics*, **17**, ed. R. J. Tingey. Oxford: Clarendon Press, 499–537.

Truswell, E. M., Kershaw, A. P. and Sluiter, I. R. (1987). The Australian–south-east Asian connection: evidence from the palaeobotanical record. In *Biogeographical Evolution of the Malay Archipelago*, ed. T. C. Whitemore. Oxford: Oxford University Press, pp. 32–49.

Truswell, E. M., Dettmann, M. E. and O'Brien, P. E. (1999). Mesozoic palynofloras from the Mac. Robertson Shelf, East Antarctica: geological and phytogeographic implications. *Antarctic Science*, **11**, 239–255.

Underwood, J. R. and Lambert, W. (1974). Centroclinial cross strata, a distinctive sedimentary structure. *Journal of Sedimentary Petrology*, **44**, 1111–1114.

Upchurch, G. R. and Askin, R. A. (1989). Latest Cretaceous and earliest Tertiary dispersed cuticles from Seymour Island, Antarctica. *Antarctic Journal of the United States*, **24** (**5**), 7–10.

Upchurch, G. R., Otto-Bliesner, B. L. and Scotese, C. R. (1999). Terrestrial vegetation and its effects on climate during the latest Cretaceous. In *Evolution of the Cretaceous Ocean-Climate System, Geological Society of America Special Paper*, 332, eds E. Barrera and C. C. Johnson. Boulder: The Geological Society of America, pp. 407–426.

Vakhrameev, V. A. (1991) *Jurassic and Cretaceous Floras and Climates of Earth.* Cambridge: Cambridge University Press.

Valdes, P. J., Sellwood, B. W. and Price, G. D. (1996). Evaluating concepts of Cretaceous equability. *Palaeoclimates*, **2**, 139–158.

Veevers, J. J. (2004). Gondwanaland from 650–500 Ma assembly through 320 Ma merger in Pangaea to 185–100 Ma breakup: Supercontinental tectonics via stratigraphy and radiometric dating. *Earth-Science Reviews*, **68**, 1–132.

Veevers, J. J., Powell, C. McA. and Roots, S. R. (1991). Review of seafloor spreading around Australia. I. Synthesis of patterns of spreading. *Australian Journal of Earth Sciences*, **38**, 373–389.

Vera, E. I. (2007). A new species of *Ashicaulis* Tidwell (Osmundaceae) from Aptian strata of Livingston Island, Antarctica. *Cretaceous Research*, **28**, 500–508.

Vera, E. I. (2009). *Alienopteris livingstonensis* gen. et sp. nov., enigmatic petrified tree fern stem (Cyatheales) from the Aptian Cerro Negro Formation, Antarctica. *Cretaceous Research*, **30**, 401–410.

Villar de Seoane, L. (2005). New cycadalean leaves from the Anfiteatro de Ticó Formation, early Aptian, Patagonia, Argentina. *Cretaceous Research*, **26**, 540–550.

Volkheimer, W. and Salas, A. (1975). Die älteste Angiospermen-Palynoflora Argentiniens von der Typuslokalität der unterkretazischen Huitrín-Folge des Neuquén-Beckens. Mikrofloristische Assoziation und biostratigraphische Bedeutung. *Neues Jahrbuch Geologie Paläontologie, Monatshefte*, 7, 424–436.

Walkom, A. B. (1919). Mesozoic floras of Queensland. Parts 3 and 4. The floras of the Burrum and Styx River Series. *Geological Survey of Queensland Publication*, **263**, 1–77.

Walkom, A. B. (1928). Fossil plants from Plutoville, Cape York Peninsula. *Proceedings of the Linnean Society of New South Wales*, **53**, 145–150.

Wang, X. and Wang, S. J. (2010). *Xingxueanthus*: an enigmatic Jurassic seed plant and its implications for the origin of angiospermy. *Acta Geologica Sinica-English Edition*, **84**, 47–55.

Wardle, P. (1991). *Vegetation of New Zealand*. Cambridge: Cambridge University Press.

Wilmshurst, J. M. and McGlone, M. S. (1996). Forest disturbance in the central North Island, New Zealand, following the 1850 BP Taupo eruption. *The Holocene*, **6**, 399–411.

Wilson, P. A. and Norris, R. D. (2001). Warm tropical ocean surface and global anoxia during the mid-Cretaceous period, *Nature*, **412**, 425–429.

Wing, S. L. and Boucher, L. D. (1998). Ecological aspects of the Cretaceous flowering plant radiation. *Annual Review of Earth and Planetary Science*, **26**, 379–421.

Wolfe, J. A. and Upchurch Jr, G. R. (1987). North American nonmarine climates and vegetation during the Late Cretaceous. *Palaeogeography, Palaeoclimatology, Palaeoecology*, **61**, 33–77.

Zachos, J., Pagani, M., Sloan, L., Thomas, E. and Billups, K. (2001). Trends, rhythms, and aberrations in global climate 65 Ma to Present. *Science*, **292**, 686–693.

7

The origin of southern temperate ecosystems

Introduction

The explosive radiation of the angiosperms has puzzled scientists since Darwin famously referred to this as an 'abominable mystery' whilst contemplating the topic with his friend J. D. Hooker in 1879. This seemingly sudden appearance of so many angiosperms in the Late Cretaceous conflicted strongly with Darwin's gradualist perspective on evolutionary change (Friedman, 2009). Over the last 150 years, however, many new fossils have been found suggesting that this diversification may not have been quite so 'rapid' (Friis, Pedersen and Crane, 2010) as Darwin had presumed. Moreover, the use of molecular data has enabled scientists to determine this radiation more fully by furthering our understanding of evolutionary patterns and routes of dispersal taken by the early angiosperms (e.g. Doyle, 2001).

By the Early Cretaceous (~125 Ma) major plant groups including the Chloranthaceae, the magnoliids, eudicots and monocots had originated (Figure 7.1). After the divergence of the major angiosperm lines in the late Barremian there was a steady but impressive increase in diversity (e.g. Friis, Pedersen and Crane, 2006 and references therein). Thus for the first few million years angiosperms remained relatively rare in the global landscape (i.e. 5–20% in the Albian ~105 Ma), until their impressive rise to ecological dominance by the end of the Cretaceous (80–100%) some 40 Ma later (Crane and Lidgard, 1990). Following the events at the Cretaceous–Cenozoic boundary (65 Ma) many angiosperm lines, comprising a large part of the extant eudicot groups, showed a rapid diversification at the genus and even family level replacing some of the old lineages and filling empty niches (Magallón and Castillo, 2009; Wolfe, 1997). This rapid diversification and rise to ecological dominance went hand in hand with important key innovations such as pollination and seed dispersal by animals, their fast growth rates, and growth forms (Berendse and Sheffer, 2009 and references therein). In the high latitudes the angiosperms may have had an increased advantage. Their relatively large leaf surface area would have enhanced shading of competitors in an environment characterised by low-angle light (e.g. suppressing regeneration of relatively slow growing conifers after fire). Moreover, leaf size may have been a more effective sink for coping with inhibition from excess solar energy absorption and photosynthesis end-product accumulation, as well as an increased photosynthetic surface area with associated multiplicative effects on productivity (cf. Equiza, Day and Jagels, 2006).

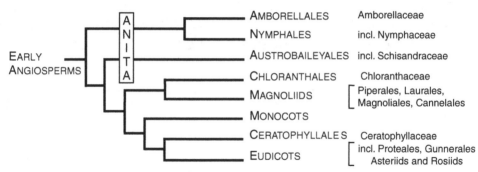

Figure 7.1 Simplified tree based on molecular phylogenetic data of the Angiosperm Phylogeny Group (2009) showing the relationship between eudicots, basal angiosperms and the ANITA grade with some of the families mentioned in the text (note Illiciaceae is now included in the Schisandraceae).

Higher growth rates, coupled with consequent increases in environmental nutrient status, are suggested to have resulted in a runaway process such that once angiosperms had reached a certain threshold of abundance (Berendse and Scheffer, 2009) they became front runners in the competition for ecological dominance with their Mesozoic counter-parts – the conifers, other seed plants and ferns. This ongoing process radically trans-formed the gymnosperm-fern dominated floras of the Mesozoic into those with which we are more familiar today.

Direct evidence for the radiation and diversification of angiosperms that resulted in the modernisation of Late Cretaceous ecosystems has advanced significantly during the past two to three decades. Findings of Cretaceous angiosperm fruits and flowers have supplemented information previously obtained from pollen, wood and leaves. Many of these flowers and fruits, although small, are three-dimensional charcoalified or lignitic compressions retaining fine anatomical detail. The overwhelming majority of Cretaceous angiosperm fossils with preserved anatomy are from the Northern Hemisphere, which contrasts strongly with the meagre record from the Southern Hemisphere (Friis, Pedersen and Crane, 2006, 2010). The same distribution applies to preserved flowers that have been found in mesofloras of mid-early (Barremian–Aptian) to Late Cretaceous (late Santonian–early Campanian) sediments (e.g. Chambers, Poinar and Buckley, 2010; Friis, 1985; Friis, Pedersen and Crane, 1999, 2006, 2010; Friis and Skarby, 1981; Friis *et al.*, 2009; Wang and Zheng, 2009). These fossil flowers typically measure a few millimetres or less in length, and are complete with sepals, petals, stamens, carpels and anthers and exquisite detail allowing detailed comparisons with extant taxa (Friis, Pedersen and Crane, 2006). Since a variety of angiosperm lineages exhibited considerable diversity in terms of organisation and structure on this scale, this confirms that primitive angiosperms did not have large flowers with numerous floral parts as once thought. Moreover, the newly recognised basal groups have simple flowers, which are reduced inflorescences rather than being primitively simple (Endress and Doyle, 2009).

In the Southern Hemisphere, contemporary mesofossils are less well-preserved, with detail often insufficient to enable systematic identification (Eklund, Cantrill and Francis, 2004). This results in a strongly biased view regarding floristic composition. Nevertheless, a few southern records of Cretaceous fruits and flowers with preserved anatomy aligned with the basal angiosperms have been found in the Southern Hemisphere, namely, southern Africa (M. Bamford, personal communication in Eklund, 2003), Australia (Dettmann, Clifford and Peters, 2009; Taylor and Hickey, 1990), New Zealand (Daniel, 1989; Daniel, Lovis and Reay, 1988; Kennedy, Lovis and Daniel, 2003; Cantrill, Wanntorp and Drinnan, 2011) and South America (Archangelsky and Taylor, 1993; Mohr and Bernardes-de-Oliveira, 2004; Mohr and Eklund, 2003; Mohr and Friis, 2000). As yet the only mesoflora described so far from Antarctica comes from an isolated outcrop of Upper Cretaceous sediments at Table Nunatak on the Antarctic Peninsula (Eklund, 2003; Eklund, Cantrill and Francis, 2004; see Case Study p. 286). These mesofossil records have provided important insights into early angiosperm history, diversification and radiation routes across southern Gondwana (Dettmann, Clifford and Peters, 2009; Eklund, Cantrill and Francis, 2004). This augments the information derived from the macroflora and microflora (Herendeen *et al.*, 1999). Data from all three records provide an understanding of the origins and subsequent evolution of the southern temperate ecosystems under the ameliorating climates of the Upper Cretaceous, before they succumbed to the biotic crisis at the Cretaceous–Tertiary transition.

Global palaeoclimate

The climate of the mid-Cretaceous and on into the Late Cretaceous was one of the most equable in Earth's history. It was marked by a reduction in seasonal contrast at mid-latitudes and a weaker (flatter) than present pole to equator latitudinal thermal gradient (Crowley and Zachos, 2000; Fluteau *et al.*, 2007). It is considered to be largely ice-free even though fluctuations in ice cap volume have been suggested to explain rapid sea level changes evidenced from the rock record (e.g. Miller, 2009). This suggests that the equable Cretaceous climate was determined by a combination of different factors, namely higher atmospheric CO_2 concentrations, which could have been responsible for a general warming, especially at high latitudes (Barron, 1983), and increased ocean heat transport, which would have resulted in flattening the latitudinal gradient. Plate movements associated with Gondwana break-up would have also influenced global oceanic circulation and caused changes in sea level including a major marine transgression of 150–200 m (Haq, Hardenbol and Vail, 1987; Sellwood and Valdes, 2006). Global fluctuations of sea level characterised the early Albian to early Coniacian period with highstands being reached during the Cenomanian–Turonian (95–80 Ma) and followed by a subsequent regression in the mid-late Turonian (Haq, Hardenbol and Vail, 1987; Hart, Joshi and Watkinson, 2001). Such major tectonic activity and sea floor spreading was also associated with increased volcanic outgassing, resulting in enhanced CO_2 levels in the atmosphere (~1000 ppm; Royer, 2006).

In the marine realm the episodic, yet critical, change in sea water composition that had led to worldwide anoxia (Oceanic Anoxic Event; OAE) in the Aptian–Albian, happened twice more in the Upper Cretaceous. The most extensive event took place at the Cenomanian–Turonian boundary (OAE-2 at 93.5 Ma) followed by another at the Coniacian–Santonian boundary (~86 Ma). Over a period of a half to one million years, the rate and magnitude of organic carbon production and preservation greatly increased above background Cretaceous levels resulting in worldwide oceanic anoxia (Sageman, Meyers and Arthur, 2006). These phases of enhanced rates of organic burial locked up ^{12}C and resulted in positive carbon isotope excursions ($\delta^{13}C$) detectable in both the global marine carbonates and organic carbon (e.g. Arthur, Dean and Pratt, 1988; Tsikos *et al.*, 2004). These events were accompanied by substantial and abrupt decreases in atmospheric CO_2 concentration (of up to 26% in OAE-2) although the quantification of this drawdown remains controversial (Arthur, Dean and Pratt, 1988; Barclay, McElwain and Sageman, 2010; Freeman and Hayes, 1992; Kuypers, Pancost and Sinninghe-Damsté, 1999). Yet sea surface temperatures across the OAE-2 interval were extremely high and punctuated by large-scale cooling of up to 11 °C (Sinninghe Damsté *et al.*, 2010).

Throughout the Late Cretaceous pCO_2 dropped steadily, as did global temperatures. When CO_2 levels drop below a certain threshold (560–1120 ppm) the formation of permanent ice sheets becomes a distinct possibility such that periods of low pCO_2 correlate with long-lived extensive continental glaciations (DeConto and Pollard, 2003; Pollard and DeConto, 2005; Royer, 2006). Therefore, superimposed on this downward trend were brief periods of globally cool climates associated with the OAEs (Haq, Hardenbol and Vail, 1987; Miller *et al.*, 2003, 2004; Royer, 2006; Steuber *et al.*, 2005; Wilson *et al.*, 2002). During the Campanian, however, CO_2 levels rose to ~3000 ppm (Royer, 2006) only to be followed by three (brief) cooling events. These events occurred near the Campanian–Maastrichtian boundary (71.6–69.6 Ma), in the mid-Maastrichtian (67.5–66.5 Ma) and at the end of the Maastrichtian (65.6–65.5 Ma), and coincided with drops in pCO_2 to <1000 ppm, before returning to values >1000 ppm after these events (Royer, 2006). The first event is supported by rapid shifts in both sea level and $\delta^{18}O$ of marine carbonates as well as temperature reconstructions. These rapid shifts also support the mid-Maastrichtian event, whilst geographically widespread marine and terrestrial temperature reconstructions support the very brief (100 kyr) end-Maastrichtian event (Barrera and Savin, 1999; Frakes, 1999; Haq, Hardenbol and Vail, 1987; Huber, Norris and MacLeod, 2002; MacLeod and Huber, 1996; Miller *et al.*, 1999, 2003, 2004; Nordt, Atchley and Dworkin, 2003; Parrish and Spicer, 1988; Pirrie and Marshall, 1990; Royer 2006; Wilf, Johnson and Huber, 2003).

Palaeogeography of southern Gondwana

Tectonic activity associated with the break-up of Gondwana and consequent readjustments of oceanic circulation contributed greatly to climatic instability

(Fluteau *et al.*, 2007). By the beginning of the Late Cretaceous, there had been considerable relative motion between Antarctica and the other continents that formed Gondwana (Dingle and Lavelle, 2000). The land connection between New Zealand–Lord Howe Rise complex through Marie Byrd Land to Gondwana that had existed through much of the Early Cretaceous (at least up until ~100 Ma; Weaver *et al.*, 1994), as evidenced from the presence of dinosaur remains on the islands (Molnar and Wiffen, 1994), had all but disappeared by ~80–70 Ma (Lawver, Gahagan and Coffin, 1992). This separation is the only major palaeogeographical change to take place in Gondwana during the latest Cretaceous, yet it was enough to generate a new, more polar, sea route along the southern edge of the Pacific Ocean, which may have brought colder surface currents to the western side of the Antarctic Peninsula (Dingle and Lavelle, 2000).

Tectonic rifting heralding the separation of Australia from East Antarctica was, however, increasing and was accompanied by volcanism, uplift and erosion. By ~80 Ma, the rifting that had begun in the west during the Jurassic now extended from the Indian Ocean eastwards towards the Pacific Ocean forming a deep sea trough between East Antarctica and Australia (Zinsmeister, 1987) (Figure 7.2A). A tongue of this proto-Southern Ocean extended eastwards through the Otway Basin towards Tasmania, which still separated the proto-Southern Ocean in the west from the advancing Tasman Sea to the east (Dettmann *et al.*, 1992). By the Turonian–Santonian, the sea had retreated from central Australia and inland lakes and swamps developed across the Eromanga Basin (Frakes *et al.*, 1987). A newly established river system drained into the embryonic Southern Ocean rendering most of the continent above sea level.

On the west side of Gondwana, East Antarctica was linked to South America via West Antarctica (made up of the Antarctic Peninsula, Marie Byrd Land, Ellsworth Land and the Ellsworth-Whitmore Mountains; Chapter 6, Figure 6.2). Throughout the Cretaceous the Antarctic Peninsula lay adjacent to, if not connected to, the southern tip of South America and thus the southern Andes (Hay *et al.*, 1999) (Figure 7.2). The Antarctic Peninsula was undergoing rapid uplift and erosion as a result of eastwards subduction of the Palaeopacific plate beneath the western side of the Antarctic Peninsula. Active volcanoes erupted vast quantities of volcanoclastic debris into a relatively shallow anoxic sea (Birkenmajer, 1981). Land comprising the Antarctic Peninsula was composed principally of a series of volcanic islands that consisted of near-continuous highlands extending its length. Small semi-isolated deep ocean basins in the South Atlantic, Weddell and southern Indian oceans had formed with shallow water connections and a seaway linking the southern Weddell region to the Tasman and east Australian seaboards (Lawver, Gahagan and Coffin, 1992). By the end of the Turonian the shallow transpolar seaway had closed (Lawver, Gahagan and Coffin, 1992) and direct surface and mid-water access for South Atlantic equatorial waters into the Weddell Basin had been initiated (Dingle, 1999). By the end of the Cretaceous, Antarctica was isolated at the bottom of the world except for the final link between the Antarctic Peninsula and South America (Figure 7.2B).

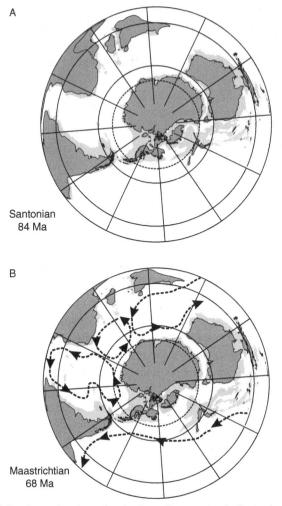

Figure 7.2 View of Gondwana break-up for the Late Cretaceous. **A**, Santonian showing extensive separation between India, Africa and the rest of Gondwana. **B**, Maastrichtian showing New Zealand and Lord Howe Rise rifting from the eastern margin of Gondwana. Note continuous connections still in existence between South America and Antarctica. Latitude intervals are 30°, narrow dashed black line = palaeoPolar Circle, dashed arrowed lines = suggested oceanic circulation, light grey = continental shelf. Reconstructions provided by R. A. Livermore, British Antarctic Survey. Reproduced from Cantrill and Poole (2002) with permission from The Geological Society of London.

Palaeovegetation of southern Gondwana

Across southern Gondwana the vegetation was in general floristically homogenous during the early mid-Cretaceous (Dettmann, 1994) with forests of araucarian and podocarp conifers, prairies of ferns, glades of tree ferns, and mosses and liverworts covering fallen logs and shaded outcrops. Over time these similarities became less distinct as angiosperms encroached further into the conifer-fern realm, such that only remnants of this Gondwana Flora survive today in parts of Australia and New Zealand (Dettmann, 1994; Dettmann and Thomson, 1987; Webb, Tracey and Jessup, 1986). Regionalism (as outlined in Chapter 6) was still evident during the Late Cretaceous, across the eastern part of southern Gondwana, both within the same latitudinal belt and across latitudes reflecting ecophysiological adaptations to variations in climatic and edaphic conditions (Dettmann, 1989, 1994; Specht, Dettmann and Jarzen, 1992). Although angiosperms had reached Australia and New Zealand by the Aptian and Albian respectively (Chapter 6), open podocarp and araucarian conifer forests continued to dominate the peneplain of New Zealand and swathes of Australia (Dettmann, Clifford and Peters, 2009; Raine, 1984; Raine, Speden and Strong, 1981).

In East Gondwana the angiosperms across Australia diversified and became more widely distributed resulting in more regionally defined vegetation (Dettmann, 1994). During the early Late Cretaceous as widening of the southern rift valley floodplain continued and the central and eastern Australian sea regressed (Dettmann, 1994), a second wave of angiosperm (this time eudicot) migration occurred from northern Gondwana. This route was via the southern South America–Antarctic corridor and was probably helped by the high level of disturbance as well as combinations of a large repetoire of traits, possibly including an early successional nature, and new innovations sported by many of these angiosperms (Feild and Arens, 2005; Feild *et al.*, 2004 and references therein; Hill and Brodribb, 2003; Wing and Tiffney, 1987). This led to an increasingly diversified angiosperm component within the understoreys of the forests (Hill and Brodribb, 2003). More subtle, short-term changes in vegetation (e.g. podocarp distribution) also took place at this time, but have been related to fluctuations in the temperate climate (wetter to drier). The causal factor put forward to account for these climatic changes is fluctuations in eccentricity, which in turn may have led to ephemeral ice sheets developing in Antarctica (Gallagher *et al.*, 2008; Miller, Wright and Browning, 2005; Miller *et al.*, 2004).

Around the Cenomanian–Turonian boundary (~94 Ma) the podocarp-araucarian forests were still abundant in Australia, but important elements of the modern austral vegetation had begun to appear including additional podocarps (such as *Lagarostrobus*, *Dacrydium* and the newly emerging *Dacrycarpus*) and Proteaceae. Angiosperms, which had occupied understorey niches, were increasing in diversity and stature such that by the Santonian some angiosperms had evolved to become arboreal components of the canopy. *Nothofagus* appeared after 84 Ma and Proteaceae further diversified to include canopy and understorey taxa, as well as scleromorphic communities on forest fringes and/or on nutrient deficient soils (Hill, 2004; Hill *et al.*, 1999). Along the coastal shorelines of Australia, cheirolepidiacean conifers still grew alongside *Brachyphyllum*, whereas the Proteaceae was becoming increasingly dominant in the central Australian vegetation (Dettmann *et al.*, 1992). In western and northern parts of Australia, angiosperms, ferns and hepatics were becoming more abundant and diverse when compared with the flora in the southeast.

Angiosperm migration in East Gondwana continued during the Turonian–Maastrichtian such that by the latest Cretaceous floral communities in southeastern Southern Gondwana were diverse and angiosperm dominated, and thus were distinct from those of the earlier Turonian–Santonian. Moreover, during the Turonian–Maastrichtian, modernisation encompassed '*in situ* evolution' of some groups, differentiation of other groups considered to be 'southern' in the modern flora, along with further introductions that pose an evolutionary problem with no satisfactory resolution (Hill, 2004). The centre for the origin and diversification of many organisms (plants and animals) during the past 100 Ma has been suggested to be a region covering southern South America, western Antarctica, southeastern Australia and New Zealand, when all these land masses were connected. Little is known about events in this region during the Late Cretaceous, but the environment would have been totally foreign to us today (Hill, 2004). This region, known as the Weddellian Biogeographic Province, seems to have been the site of many major evolutionary innovations during the Cretaceous and spawned some remarkable plant and animal groups that help characterise the living Southern Hemisphere biota (Hill, 2004). From here organisms were able to spread northwards to mid and low latitudes where their descendents can still be found today. For example, the probable cradle for *Ilex* has now been identified as the area surrounding the proto-Southern Ocean in eastern Gondwana (Dettmann, 1989; Dettmann and Jarzen 1990, 1991). Moreover, three angiosperms characteristic of modern southern ecosystems, Casuarinaceae, Nothofagaceae and Proteaceae, all first appeared in the Southern Hemisphere in the pollen record of the Weddellian Biogeographic Province albeit at different times. Yet nothing is known about the actual plants bearing the pollen since they are absent from the macrofossil record. Casuarinaceae and Proteaceae are thought to have radiated to very high latitudes relatively quickly during the Late Cretaceous, and have changed little since the late Paleocene when they first appear in the macrofossil record (Hill, 2004). Nothofagaceae, however, with its single extant genus, *Nothofagus*, is less clear. Superficially, *Nothofagus* resembles extant broad-leafed families of the Northern Hemisphere (e.g. Fagaceae, Betulaceae) and thus may not have altered its morphology, relative to its sister families, to the degree shown by Casuarinaceae and Proteaceae and their respective sister families, which are now centred in the Northern Hemisphere (Hill, 2001, 2004).

The cool temperate conditions of the latest Cretaceous supported tall (up to 30 m height), open rainforests with canopies comprising diverse conifers (*Lagarostrobus*, *Dacrydium*, *Dacrycarpus*, *Podocarpus*, *Microcachrys*, Araucariaceae) and angiosperms such as Proteaceae (e.g. *Knightia* and *Macadamia*) and the newly introduced, but still rare, *Nothofagus*. The shrubby proteaceous understorey, with elements that still occur in present day austral rainforests, had now diversified to include other angiosperm taxa such as the Winteraceae, Trimeniaceae, probable Epacridaceae and *Ilex*, *Gunnera*, *Ascarina* and a ground stratum of diverse ferns and bryophytes (Dettmann, 1994; Dettmann and Jarzen, 1991; Specht, Dettmann and Jarzen, 1992). Some Proteaceae had adapted to survive in drier regions along the forest fringes and/or on nutrient-poor sites along with Epacridaceae (now the Styphelioideae in the Ericaceae) and ferns adapted to nutrient-poor and waterlogged soils (Specht, Dettmann and Jarzen, 1992). The vegetation assemblage that fringed the estuary separating East Antarctica from southern Australia suggests a mosaic of rainforest and sclerophyll communities. Although many of these latest Cretaceous East Gondwana elements are still typical of the modern Australasian flora today, these community associations have no modern analogue (Dettmann and Jarzen 1991).

In West Gondwana, the first angiosperms had arrived in southern South America during the early Aptian where they enjoyed a humid and (warm) temperate climate (Archangelsky and Villar de Seoane, 1998; Archangelsky *et al.*, 2009; Cúneo and Gondolfo, 2005; Passalía *et al.*, 2003; Romero and Archangelsky, 1986) (see Chapter 6). But regionalism both north–south and east–west rendered the increasingly angiospermous vegetations distinct. During the mid-Cretaceous, angiosperms expanded and diversified to occupy the herbaceous, shrub and small tree strata of the South American vegetation that had been principally dominated by ferns, conifers and other seed plants (Iglesias *et al.*, 2007). This reflected their ability to adapt to different environments, and signified the probable beginnings of angiosperm domination. This diversification is reflected in both the pollen and leaf floras (Archangelsky *et al.*, 2009 and references therein). By the middle Late Cretaceous angiosperm diversity in West Gondwana also peaked coinciding with a peak in global warmth.

During the Late Cretaceous, as the climates steadily warmed, the extreme thermal and moisture gradients (Valdes, Sellwood and Price, 1996) that had been acting as an effective barrier to polar migration were slowly breaking down. Latitudinal belts of vegetation had expanded, pushing floristic boundaries southwards, dampening the steep floristic gradient that had existed during the Early Cretaceous and allowing angiosperms to migrate into West Antarctica via the Peninsula. This coincided with the progressive establishment of south Indian seaways connecting into the Weddell Basin (Dingle and Lavelle, 2000). The earliest angiosperms in Antarctica[1] were joined by this second wave of angiosperms that began modernising the vegetation in West Gondwana. The open-canopied, podocarp-rich forests of the Cenomanian that had extended across South America, into the Falkland Plateau and across Antarctica and southern Australia (Dettmann and Thomson, 1987) saw the introduction of *Lagarostrobus*, Proteaceae (Baldoni and Medina, 1989; Barreda, Palamarczuk and Medina, 1999; Dettmann and Thomson, 1987) and *Ilex* by the Turonian, and then diverse eudicots during the Coniacian (e.g. Dutra and Batten, 2000; Hayes, 1999; Poole, Gottwald and Francis, 2000; Poole, Richter and Francis, 2000; Poole *et al.*, 2000) along with the first record of Myrtaceae (Baldoni and Medina, 1989; Dettmann and Thomson, 1987). Although floras of this age are scarce from the Antarctic Peninsula (Askin, 1992; Dettmann, 1989), the increased abundance of wood fragments suggests that the angiosperms were no longer herbaceous, shrubby understorey elements, but had now become a more important component of the canopy (Poole and Cantrill, 2006). Angiosperm diversity is assumed to have peaked during the Turonian–Coniacian (although further work on the floras needs to be undertaken to establish exactly when this peak occurred since evidence from the microfossils suggests that it occurred in the Turonian, whereas the macroflora suggests the Coniacian). This change in vegetation was accompanied by an increase in polar warmth

---

[1] These records are of fagalean lineage in the South Shetland Islands (see Chapter 6) with their two main foliar morphologies (palmate and pinnate craspedodromous) and mesophyllic to microphyllic leaf sizes, acute apices and entire margins (Alves, Guerra-Sommer and Dutra, 2005; Cantrill and Nichols, 1996; Dingle and Lavelle, 2000; Philippe *et al.*, 1993).

such that during the Coniacian to early Campanian, widespread warm temperate to sub-tropical floras still extended across the Peninsula region.

With the second wave of invasion across southern Gondwana came ancestral *Nothofagus* (represented by the *Nothofagidites* pollen type) (Case, 1988; Dutra, 1997, 2000, 2001; Dutra and Batten, 2000; McLoughlin and Hill, 1996; Menéndez and Caccavari de Filici, 1975; Romero, 1978; Zastawniak, 1990). The ancestral area for the genus is thought to be South America, and together with the Antarctic Peninsula, became the most likely centre of origin, in the initial stages of differentiation and diversification of modern *Nothofagus* (Swenson, Hill and McLoughlin, 2000). Periodic environmental disturbance probably allowed the opportunistic *Nothofagus* to colonise new volcanic soils and niches. By exploiting unfavourable environments, this genus was able to characterise the austral basin assemblage from South America across to Australasia (Alves, Guerra-Sommer and Dutra, 2005). Today, it still represents the most common element, along with southern conifers, of the Southern Hemisphere temperate rainforests where it sustains its preference for disturbed environments and thin volcanic soils, particularly in South America (Hill, 1990, 1992, 1994; Veblen, Hill and Read, 1996; Veblen Schlegel and Oltremari, 1983). Throughout this modernisation the fossil record of the Antarctic Peninsula has recorded stages of the changing vegetation throughout the Late Cretaceous – these vegetational snapshots, based on relatively well-dated, well-preserved floras are detailed below. Other Cretaceous floras of uncertain date, comprising limited material and/or poor preservation are not discussed here but detailed summaries are provided in the Appendix.

Warm temperate–subtopical floras of early Late Cretaceous Antarctica (Coniacian–early Campanian)

During the Cenomanian, vegetation remained similar to that of the early Albian, distinguished only by the increasingly diverse angiosperm component in an otherwise conifer-fern-rich flora (Dettmann, 1989). Podocarps, ferns and bryophytes dominated the vegetation; less frequent were conifers, such as Araucariaceae, Cheirolepidiaceae, and other seed plants such as *Ginkgo*, Ephedrales, and cycads (Figure 7.3). The angiosperms were increasing in both abundance and diversity as a second wave of introductions reached the Antarctic Peninsula and included eight pollen taxa of monocotyledonous and dicotyledonous angiosperms, some representing eudicots. At the end of the Cenomanian, angiosperms had represented a minimum of 16% of the palynoflora but rapidly increased in importance to peak at >72% in the Coniacian–Turonian (Cantrill and Poole, 2002) (Figure 7.3). In these Coniacian sediments the abundant angiosperm wood signifies their more arboreal habit and ability to join the middle and upperstorey canopy niches, hitherto dominated by bennettites and other gymnosperms.

The bryophytes and hepatophytes which were (and still are) important colonisers of fresh sediments, and components of established fern thickets and forests, had undergone a steep decline along with both the ferns (although some, e.g. Gleicheniaceae, remained fairly

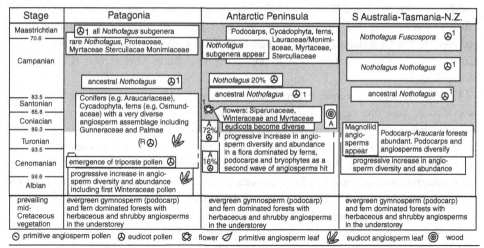

Figure 7.3 Summary of the timing of events relating to angiosperm introduction to and radiation across Southern Gondwana (N.Z. New Zealand) along a latitudinal transect updated and revised from Dutra and Batten (2000). Where the age of the first occurrence (1) is known the symbol is outside the box, when the age range of the strata in which the fossil(s) was found covers a larger time span this is indicated by the symbols placed within a box marking that range. (R) reworked; (A) angiosperm. Note that all four *Nothofagus* subgenera are not present in East Gondwana (i.e. South Australia) until some 12–15 Ma after they are recorded in South America (Swenson, Hill and McLoughlin, 2000). Note angiosperm abundance increases to a peak of 72% in the Turonian along with a marked rise in wood in the Coniacian–Santonian testifying to an increase in stature and expansion into the forest canopy. Background shading denotes when floras became dominated by angiosperms. For further details (including references) see Figure 6.7 and text.

stable) and lycopods in the earlier mid-Cretaceous as a result of increasing competition from the angiosperms (Chapter 6, Figure 6.7). This now minor component of the vegetation recovered during the latter part of the Cretaceous suggesting that the increase in arboreal angiosperms, post-Coniacian, opened up new niches that were colonised by the pterido-phytes. Although ferns were probably diverse in the latter part of the Cretaceous, ecologically they remained less important than they had been during the Aptian and Albian. This pattern of fern demise and recovery is recorded globally, which probably represents an ecological opportunistic response to the diversification of the angiosperms (Cantrill and Poole, 2002; McLoughlin *et al.*, 2008; Schneider *et al.*, 2004). Similarly, the hepatophytes and non-coniferous seed plants had also become a minor constituent of the vegetation. The bennettites, that had contributed to the understorey and had remained reasonably abundant and diverse (representing 14% of the floristic diversity) during the Aptian, rapidly declined through the Albian (to 3.5%) to become very rare and of negligible importance ecologically in the Upper Cretaceous. Finally, they had become all but extinct by the end of the Coniacian where only a few specimens have been recorded (Cantrill and Poole, 2002; Stockey and Rothwell, 2003). A similar pattern is seen elsewhere globally, although the timing of their ultimate demise spatially does vary (Alvin *et al.*, 1967; Cantrill and Poole, 2002; Pole and Douglas, 1999). The decline of other non-coniferous seed plants (such as *Pachypteris* and *Ginkgo*) in the mid-Cretaceous shows a slight, but short-lived recovery in the Cenomanian. Even though they do not become extinct at the end of the Cretaceous their contribution to the prevailing vegetation was minor (Cantrill and Poole, 2002).

The distinct drop in diversity experienced by the conifers during the mid-Late Cretaceous is evidenced from the macrofossil record, yet the microfossils indicate that within-floral diversity was maintained. These differing signals may be due to wind-dispersed conifer pollen representing a wider range of plant communities relative to those evidenced from the macrofossil record (Cantrill and Poole, 2002). Alternatively, the distribution of conifers may have become more restricted to form pockets of local abundance. This way the copious pollen produced would still have entered the fossil record whereas other organ parts may have been subject to greater taphonomic bias and thus not so greatly represented.

The recovery noted for the other plant groups such as the ferns, mosses and liverworts, and non-coniferous seed plants may be explained by the short-lived decline in angiosperms during the Santonian (Cantrill and Poole, 2002). Whilst some (e.g. the ferns) were able to adapt, the remainder, including the dominant podocarp-araucarian forest associations, succumbed as the angiosperms subsequently recovered in the latest Cretaceous. The causal factor(s) for this short-lived reversal in diversity remains unknown. The Turonian–Coniacian was an important transitional time when the established podocarp conifer forest vegetation was diversifying and new angiosperm families that henceforth typified southern vegetation were on the increase both in diversity and abundance (Askin, 1989; Cantrill and Poole, 2002, 2005; Leppe *et al.*, 2007). Palynomorph assemblages indicate the predominance of temperate podocarp-araucarian-fern forest vegetation with lycopod and fern moorland vegetation (Cantrill and Poole, 2005; Dettmann, 1986; Dettmann and Thomson, 1987; Dettmann *et al.*, 1992; Douglas and Williams, 1982; Truswell, 1990, 1991). This Antarctic region then remained a locus of evolutionary innovation from the Turonian until the end of the Cretaceous (Askin and Spicer, 1995).

The plant fossil record from the Cenomanian to Turonian of Antarctica is poorly known, but on East Antactica microfloras from Pyrdz Bay (Figure 7.4A) provide insights into the vegetation during the Turonian–Santonian. This record is an important one given that sediments of Turonian–Santonian age have not previously been reported from Antarctica (Macphail and Truswell, 2004). These authors describe a coastal plain flora with strong affinities to the vegetation of southeastern Australia during the Late Cretaceous. The vegetation was dominated by Araucariaceae and Podocarpaceae conifers, with ferns and fern allies forming diverse understoreys and perhaps heathlands in open areas. The physiological constraints imposed by low light levels suggest that the trees and taller shrubs were widely spaced to form woodland rather than forests. Two rare woody podocarp species, *Lagarostrobus* and *Microcachrys*, indicate the presence of freshwater swamps. The presence of gymnosperms points to moderate year-round humidity and relatively mild temperatures. The low relative abundance of hydrophytes indicates conditions not being sufficiently cold or uniformly wet to support extensive freshwater swamps or raised peat

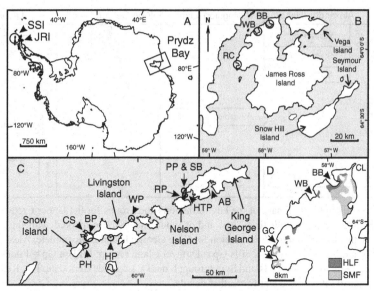

Figure 7.4 Map indicating the places mentioned in the text. **A**, Antarctica showing position of Prydz Bay, South Shetland Islands (SSI) and James Ross Island (JRI). **B**, James Ross Island with Brandy Bay (BB), Rum Cove (RC), Whisky Bay (WB). **C**, South Shetland Islands with (PH) President Head on Snow Island, Livingston Island with Cape Shirreff (CS), Byers Peninsula (BP), Hannah Point (HP) and Williams Point (WP); Nelson Island with Rip Point (RP); King George Island with Admiralty Bay (AB), Half Three Point (HTP), Price Point and Skua Bay (PP & SB). **D**, northwest James Ross Island showing the extent of the Hidden Lake Formation (HLF) and the Santa Marta Formation (SMF) and Brandy Bay (BB), Cape Lachman (CL), Gin Cove (GC), Rum Cove (RC) and WB (Whisky Bay).

bogs. This flora represents the Austral Conifer Woodland or Heath vegetation type and is so termed because of its broad resemblance to modern boreal communities that can be found close to the limits of tree growth in the Northern Hemisphere (Macphail and Truswell, 2004). The understorey would have witnessed the encroachment of angiosperms especially from Proteaceae and *Nothofagus* (Truswell and Macphail, 2009). The microfloras from Pyrdz Bay supports Askin's (1992) suggestion that this angiosperm takeover peaked in East Antarctica slightly earlier than the Campanian–Maastrichtian takeover in West Antarctica, as evidenced by relative abundance data (Cantrill and Poole, 2002; Macphail and Truswell, 2004; Truswell and Macphail, 2009).

In West Antarctica, the Whisky Bay Formation (Figure 7.5) on James Ross Island (late Albian to latest Turonian; Crame, Pirrie and Riding, 2006; Riding and Crame, 2002) is the only unit where strata of this age occur on the Peninsula. The formation outcrops both in the Brandy Bay–Whisky Bay area and the Gin Cove–Rum Cove area (Crame, Pirrie and Riding, 2006) (Figures 7.4B, D and 7.5). The complex lateral variation has made it difficult to correlate between outcrops and between the subdivisions in different parts of the basin (Crame, Pirrie and Riding, 2006). No detailed collecting for plant macrofossils has yet taken place, and broad conclusions regarding the palaeoecology are drawn from studies based on a

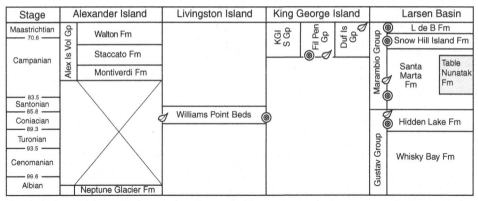

Figure 7.5 Stratigraphic summary of the key formations and groups for the Upper Cretaceous of the Antarctic Peninsula. Units with fossil leaf and/or wood material are marked with the appropriate symbol. Note that the stratigraphic succession for King George Island is incomplete. Alex Is Vol Gp = Alexander Island Volcanic Group; Duf Is Gp = Dufayel Island Group; Fil Pen Gp = Fildes Peninsula Group; KGI S Gp = King George Island Supergroup; L de B = López de Bertodano Fm. Redrawn from Poole and Cantrill (2006). See text for further details.

few terrestrial microfloras (e.g. Crame, Pirrie and Riding, 2006; Dettmann and Thomson, 1987). The microflora indicates a dominance of ferns (*Cyathidites*) and bryophytes but low angiosperm diversity (Askin, 1992; Crame, Pirrie and Riding, 2006; Dettmann, 1989).

From the Cenomanian to early Campanian, conifer forests still dominated the vegetation of West Antarctica with an understorey of bennettites, and abundant ferns and bryophytes. Evidence for this floral association comes from the fossil floras preserved at Williams Point on Livingston Island (Figures 7.4C, 7.6 and 7.7) in the South Shetland Islands, lying at a palaeolatitude of ~62° S (Chapman and Smellie, 1992; Grunow, Kent and Dalziel, 1991; Palma-Heldt *et al.*, 2007; Poole and Cantrill, 2001; Rees and Smellie, 1989; Torres and Lemoigne, 1989). Amongst Antarctic Peninsula Cretaceous floras, those found at Williams Point have attracted considerable interest because, unlike many sites around the Antarctic Peninsula, the floras are often well-preserved. The Williams Point Beds are a sequence of conglomerates, sandstones and mudstones with interbedded tuffaceous horizons that accumulated in a dominantly fluvial environment (Chapman and Smellie, 1992). This sequence is intruded by several hydroclastic vents and a thick (~50 m) dolerite sill (Rees and Smellie, 1989). The sedimentary strata are preserved either as isolated outcrops on top of the sill in squeeze-up structures between pods of intrusive sill material, or as flat-lying strata beneath the sill. The lithologies and lithofacies associations within the Williams Point Beds indicate a terrestrial environment with deposition in a fluvial, possibly braided river, system that may have been part of a large alluvial fan (Lacey and Lucas, 1981; Rees and Smellie, 1989; Smellie *et al.*, 1984) that was periodically influenced by volcanic activity. This scenario evidenced from the rock record is also reflected in the plant record.

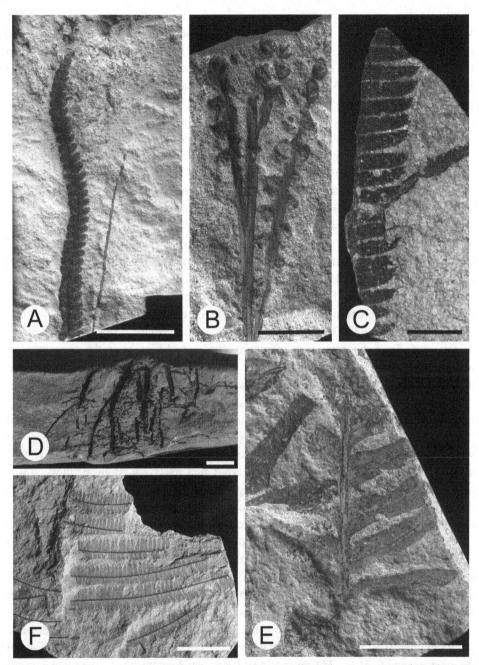

Figure 7.6 Coniacian–Santonian Williams Point Flora. **A**, *Asterotheca crassa* BAS P. 101.8. **B**, *Coniopteris distans* BAS P. 101. 24. **C**, *Zamites* sp. BAS P. 3051.A.44. **D**, conifer roots with mychorrhizal nodules in a soil horizon, BAS P. 3051.B.66. **E**, *Cladophlebis* sp. 1. BAS P. 224.38b. **F**, *Asterotheca crassa* BAS P. 3052A.79a. Scale bars A–E = 1 cm, F = 3 mm.

Initially, the fragmentary and poorly preserved megafossil plant material from the Williams Point Beds were all identified as Triassic along with the better preserved floral remains, which were described as representing Triassic groups of non-woody plants, even though these fossils were equally similar to known Jurassic and Cretaceous Gondwana genera (Orlando, 1967, 1968). This Triassic age assignment was subsequently supported by later workers (Banerji and Lemoigne, 1987; Banerji, Lemoigne and Torres, 1987; Barale *et al.*, 1995; Lacey and Lucas, 1981; Lemoigne, 1987) and thus placed the Williams Point Beds firmly within the pre-fragmentaion history of Gondwana. With their volcanic provenance and evidence for active basic-intermediate arc magnetism, they seemingly occupied a unique position in the stratigraphy of the Antarctic Peninsula (cf. Askin and Elliot, 1982; Smellie, 1981). However, the Triassic age posed serious problems in interpreting the geological evolution of an essentially Jurassic–early Cenozoic volcanic arc, and had subsequent repercussions for Gondwana reconstructions (Rees and Smellie, 1989). In 1987 the Williams Point Beds were revisited and a large palaeobotanical collection was made from both existing and new localities. These collections comprised much fragmentary material alongside a well-preserved and diverse fossil flora from a number of localities at Williams Point. Then in all but one of the alleged Triassic sites, the presence of angiosperms emerged. These leaves, wood, and pollen (Chapman and Smellie, 1992; Poole and Cantrill, 2001; Rees and Smellie, 1989; Torres and Lemoigne, 1989) thus cast doubt on the supposed Triassic age. Since the case for a Triassic age had rested heavily on the correctness of identification, these findings suggested that either the deposits were allochthonous, the angiosperms present in the beds are also Late Triassic, or the beds are, more plausibly, simply not Triassic. Evidence for a revised age came from various sources: field relations, lithological and sedimentological similarity and all indicate a Cretaceous age (Chapman and Smellie, 1992; Rees and Smellie, 1989). The angiosperm leaves have a morphology and venation patterns fully consistent with those of angiosperms growing at this time. This suggests that the beds can be no older than early Late Cretaceous. Angiosperm wood was also found in the Williams Point Beds (Chapman and Smellie, 1992; Poole and Cantrill, 2001), but angiosperm wood only occurs in floras from sequences dating from the Coniacian and younger. Preliminary palynological investigations have revealed that the distinctive pollen type of the prolific pollen-producer and characteristic austral taxon, *Nothofagus*, was still absent relative to its widespread record from Antarctic Campanian (and younger) sequences throughout the Peninsula region (e.g. Askin 1988a; Chapman and Smellie, 1992; Dettmann and Thomson, 1987; Dutra and Batten, 2000). This supports an upper age limit of Campanian. The sill that intrudes the succession has also been radiometrically dated as 81 Ma (Campanian) and supports this upper age limit to the beds (Rees and Smellie, 1989). Further Ar-Ar dating by Hunt (unpublished data) of the tuff bed[2], and fossil material returned an age of ~85 ± 1 Ma, suggesting a Coniacian–Santonian age. Therefore, considering the above, the Williams Point Beds are best considered to be Cenomanian to early

---

[2] British Anatarctic Survey sample number P.3057.1, which returned a plateau age of 84.09 ± 0.97 Ma that correlates with the K-Ar stratigraphy erected for the area (R. Hunt personal communication, 1999 and 2010).

Campanian and probably latest Coniacian, thus representing an early angiosperm-influenced vegetation. The whole assemblage, particularly the leaves, is however in need of taxonomic revision in light of this new age.

Given the age and diversity of the wood, palynomorph and leaf flora, this assemblage is important palaeobotanically and crucial to our understanding of the diversity and ecology of early angiosperm palaeocommunities (Banerji and Lemoigne, 1987; Banerji, Lemoigne and Torres, 1987; Chapman and Smellie, 1992; Lacey and Lucas, 1981; Orlando, 1967, 1968; Palma-Heldt et al., 2007; Poole and Cantrill 2001; Rees and Smellie, 1989). Some leaf specimens are very well-preserved, with epidermal cell details on the cuticles making these some of the best preserved leaves on Antarctica from either the late Mesozoic or Cenozoic age (Rees and Smellie, 1989).

Both the lithological and plant record from different localities within the Williams Point Beds suggest that the environment was overshadowed by volcanic activity. This volcanic activity coupled with ash-fall deposition would have destroyed climax plant communities and preserved them poorly in tuffaceous siltstones and sandstones. In time, the climax vegetation would have been replaced by pioneer vegetation, possibly comprising more xeric forms composed predominantly of non-angiospermous plants in a less stable environment. Subsequently, remains of these pioneer communities would also have become deposited at distance from these disturbed sites, and other communities would have been able to flourish under relatively stable conditions. Natural leaf fall near depositional environments, such as stream and lakes, would have facilitated the preservation of leaf floras. It is highly probable that during this time in Antarctica angiosperms only occurred as sporadic constituents in stream-margin facies sediments and only locally dominated the vegetation – a scenario similar to the Lower Cretaceous floras of the middle palaeolatitudes in North America (Hickey and Doyle, 1977; Rees and Smellie, 1989).

The leaf, wood and microfossils are probably all derived from assemblages that were growing in a similar riparian locality and provide a clearer picture of the ecology in this part of Antarctica (Chapman and Smellie, 1992; Poole and Cantrill, 2001). Against the backdrop of periodic volcanic activity occurring less than 20 km away (Smellie, Roberts and Hirons, 1996), the landscape would have been dominated, at least in stature, by conifers. Not surprisingly they consisted of araucarians and podocarps (Cantrill and Poole, 2002; Palma-Heldt et al., 2007), typical of Cretaceous vegetation, but alongside were other conifers, for example those bearing the *Pagiophyllum* and *Elatocladus*-type foliage (Banerji and Lemoigne, 1987). In amongst them, probably occupying both the understorey and ground cover but also reaching into the canopy, was a diverse angiosperm component with elements that had by now attained much greater height and status (as evidenced by the presence of wood). These included taxa from the Monimiaceae (cf. *Hedycarya* from Southeast Asia) and Cunoniaceae (cf. *Ackama* from New Zealand) along with other angiosperms which lie within the 'Magnoliidae–Hamamelidae–Rosidae' (Poole and Cantrill, 2001), and ?lauraceous forms with the *Cinnamomoides*-type leaf (Rees and Smellie, 1989). These angiosperms probably had small, simple leaves with entire margins and pinnate primary venation (similar to the primitive forms of Early or mid-Cretaceous age in the Northern Hemisphere; Hickey and Wolfe, 1975), some were palmate but few would have had larger (notophyllous), serrate leaves (Rees and Smellie, 1989). Moreover, they were probably insect-pollinated (Chapman and Smellie, 1992). The wood of these early woody/arboreal angioserms shows no evidence of distinct growth rings suggesting that these plants may indeed have been evergreen (Cantrill and Poole, 2005). In amongst the tree flora were bennettites and probably large tree-ferns (Palma-Heldt *et al.*, 2007; Poole and Cantrill, 2001). The understorey was made up of ferns (such as Gleicheniaceae and Osmundaceae), represented by leaves and petrified rachides that may have dominated the vegetation in terms of abundance and diversity. Other, possibly bushy, conifers and shrubby angiosperms, such as the chloranthaceous plant that produced the *Clavatipollenites* pollen, were also present (Chapman and Smellie, 1992). Ferns, along with sphenopsids, such as *Equisetites* (Cantrill, 1997; Lacey and Lucas, 1981; Palma-Heldt *et al.*, 2007), inhabited the moist forest floor and banks of streams. Thalloid liverworts probably allied to the Marchantiales (e.g. *Thallites*; Lacey and Lucas, 1981) and epiphytic ferns would have grown on moist, shady banks or clung tenaciously to damp rocks, trunks and branches that lay on the forest floor (Chapman and Smellie, 1992; Palma-Heldt *et al.*, 2007; Poole and Cantrill, 2001). All plants would have slowly succumbed to decomposition and fungal attack, either initially whilst the plant was still living (e.g. epiphyllous fungi), or whilst lying on the forest floor. This adds to the support that great changes, albeit with some temporal differentiation, were underway across southern Gondwana.

The plant fossils from the Williams Point Beds can also provide indication of the climatic conditions within the environment. Both the leaves and wood have been used to this end. Relatively little is known with regard to the relationship between these relatively few angiosperm leaves and either their original growth position (i.e. riparian, climax, etc.) or habit (shrub, tree etc.), and details regarding their depositional environment(s), precise age and rapidly evolving states. Therefore leaf-size indices employed to determine palaeotemperature (e.g. Wolfe and Upchurch, 1987) cannot be applied to these floras with any great certainty. Nevertheless, the dominance of entire margined microphyllous leaves over serrate forms (3:1 respectively) suggests that a relatively mild climate prevailed at this time (cf. Royer *et al.*, 2005; Spicer and Parrish, 1986 and references therein), possibly with a mean annual temperature range of 13–20 °C (Rees and Smellie, 1989) – a range matched by wood physiognomy and the application of Coexistence Analysis (cf. Mosbrugger and Utescher, 1997) – a warm month mean of ~25 °C and high (~3000 mm) levels of annual precipitation (Poole, Cantrill and Utescher, 2005). Evidence from wood anatomy also indicates relatively favourable growing conditions: the wide tracheid zones in the presumed bennettite wood, *Sahnioxylon*, coupled with the relative abundance of early wood tissue in the conifer woods indicate high biomass productivity. Pronounced, but narrow, growth rings in the fossil conifer wood suggest sufficient seasonality (probably annual given the palaeolatitude of 59° S; Grunow, Kent and Dalziel, 1991) to induce tree dormancy, but these growth interruptions are generally absent in the angiosperm wood, which at most exhibit indistinct rings (Cantrill and Poole, 2002; Chapman and Smellie, 1992; Torres and Lemoigne, 1989). This conifer–angiosperm growth ring phenomenon has been observed in other floras (e.g. Poole, 2000) and might reflect the degree of evergreen *versus* deciduousness (Falcon-Lang, 2000a, 2000b) and/or taxon specific responses to rapid changes in photoperiod at the end of the growing season (cf. Chapman and Smellie, 1992). Extrapolating from modern tree responses, the distinct growth rings in conifers could be taken as evidence for a deciduous habit, whereas the early arboreal angiosperms with their indistinct/absent growth rings might have had a longer leaf longevity and thus could be considered to be evergreen in habit. Therefore, perhaps the mixed conifer–angiosperm floras of the mid-Cretaceous also exhibited an interesting mix in evergreen and deciduous habits. Needless to say, however, these relatively primitive angiosperms were still evolving rapidly at this time so that the use of growth characters, such as ring markedness, for climate determination may be confounded by incomplete evolution of adaptational stategies to such extreme conditions (Poole, Cantrill and Utescher, 2005; Wheeler and Baas, 1991).

On the other side of Livingston Island at Hannah Point (Figure 7.4C), a Coniacian–Santonian–Campanian flora has been found between two igneous strata and comprises leaf imprints and some fossil trunks (Leppe *et al.*, 2007). The macro and microfloras record evidence of a vegetation dominated by ferns, mainly ground ferns (Osmundaceae and Gleicheniaceae) and tree ferns (Dicksoniaceae and Cyatheaceae), and conifers (podocarps). Interestingly, angiosperms appear to have been rare (Leppe *et al.*, 2007; Palma-Heldt *et al.*, 2007). At Cape Shirreff, a further Upper Cretaceous palynoflora has been described and also indicates a vegetation dominated by ferns as well as podocarps, but this time *Nothofagus* and proteaceous angiosperms were also represented as were fungal spores (Palma-Heldt *et al.*, 2004). These floras provide further evidence for the abundance and importance of ferns at this time.

Therefore, within the 10 or so million years between the earliest evidence of angiosperms in early Albian sediments of Antarctica and the Upper Cretaceous, floral modernisation was well underway with angiosperms having radiated and diversified both in terms of taxa and habit from the small herbaceous colonisers of newly exposed land, to obvious trees forming a dominant component of the canopy. The vegetation continued to thrive through the peak warmth of the Turonian and a snapshot of the now subtropical-warm temperate vegetation has been found preserved on the eastern side of the Peninsula in the James Ross Basin.

Subtropical floras following peak warmth: Coniacian to early Campanian floras of the James Ross Basin

Evidence for diverse subtropical floras across the Peninsula region at this time comes from the proximal marine sediments on James Ross Basin located within part of the larger Larsen Basin (Figure 7.4B). These sediments provide an unparalleled record of Cretaceous marine and terrestrial life in the southern high latitudes, as well as important information on biostratigraphy, palaeoenvironments, palaeoclimates and palaeobiology (e.g. Crame 1994; Crame *et al.*, 1991, 1996, 1999). The 6 km thick Aptian–Oligocene shallow marine sedimentary succession exposed in the James Ross Island area forms the thickest onshore sequence of upper Cretaceous–Cenozoic sediments in Antarctica, and is one of the thickest and most complete Upper Cretaceous–lower Cenozoic sedimentary successions exposed in the Southern Hemisphere (Hathway *et al.*, 1998; Pirrie, Marshall and Crame, 1998; Pirrie *et al.*, 1997; Zinsmeister, 1982). This Late Cretaceous to early Cenozoic section basin infill consists of siltstones, sandstones and conglomerates that were deposited in progressively shallowing conditions in proximal submarine-fan and slope-apron settings, shelf settings and deltaic environments (Elliot, 1988; Ineson, 1989). The succession is divided into three main units: Gustav Group (early Aptian to Coniacian), Marambio Group (late Coniacian to Paleocene) and the Seymour Island Group (Paleocene–?earliest Oligocene; see Chapter 8) (McArthur, Crame and Thirlwall, 2000; Riding *et al.*, 1998) (Figure 7.5). Within these sediments are the remains of abundant plants that were washed into the back-arc basin on the eastern side of the emergent volcanic arc. The most continuous fossil record comes from palynomorphs that, along with ammonite and bivalve faunas and isotopic studies, provide reasonably accurate dating (Askin, 1997; Barreda, Palamarczuk and Medina, 1999; Crame *et al.*, 1991; Dettmann and Thomson, 1987; Riding *et al.*, 1998; Truswell, 1990). Information pertaining to the composition of the contemporaneous vegetation flourishing on the Antarctic continent above 60° S during the Cretaceous greenhouse was scarce until a detailed study of the leaf floras in the James Ross Basin, namely those from the Gustav and Marambio groups, was undertaken by Hayes (1999). Angiosperms formed an important part of the Antarctic vegetation from the mid-Cretaceous onwards therefore it is not surprising that the majority of the leaf impressions were found to be angiospermous in origin and thus represent the oldest known angiosperm remains from the James Ross Basin, along with ferns, bennettites and conifers (Hayes, 1999).

Since these floras date from a period of peak warming in high latitudes during one of the warmest periods in Earth history, the architectural features, such as morphology, margins and venation patterns (cuticles are lacking) of the angiosperm material have provided important new data (Hayes, 1999) on possibly the most diverse high latitude flora ever to have existed. Moreover the floras provide a unique insight into the general palaeoecology of the Antarctic Peninsula at this time.

The oldest of the fossil floras is the Coniacian Hidden Lake Formation Flora (Figures 7.8 and 7.9A) within the uppermost part of the Gustav Group (Crame, Pirrie and Riding, 2006; Ineson, Crame and Thomson, 1986; McArthur, Crame and Thirlwall, 2000; Whitham, Ineson and Pirrie, 2006). This formation is exposed in the northwest of James Ross Island (Figure 7.4D) and comprises more than 400 m of coarse-grained volcaniclastic sandstones and channelled conglomerates grading into burrowed sandy siltstones. The sand-dominated sediments have been interpreted as being deposited in a shallow marine deltaic environment within fan delta shelf and slope settings (Elliot, 1988; Hathway *et al.*, 1998; Pirrie, 1991). This deposition closely relates to a major pulse of proximal volcanic activity and probable arc uplift (Dingle and Lavelle, 1998; Hathway *et al.*, 1998; Pirrie, 1991), with the distance between sediment source and site of deposition being small and sedimentation rates rapid (Pirrie, 1991). Ammonite and bivalve faunas suggest a probable age range of Coniacian–Santonian (Ineson *et al.*, 1986), but dinoflagellates from the beds containing the leaf flora and strontium

Figure 7.7 Coniacian–Santonian Williams Point Flora. **A**, *Asterotheca crassa* BAS P. 3052A. **B**, angiosperm type 2 BAS P. 1801B. **C**, cf. *Cinnamomoides* BAS P. 1801.D. **D**, angiosperm type 1 BAS P. 1801.H. **E**, *Pterophyllum dentatum* BAS P. 426.5a. Scale bars = 1 cm.

isotope analyses confirm a Coniacian (88.7–86.4 Ma) age (McArthur, Crame and Thirwall, 2000; Riding and Crame, 2002). The plant fossils preserved within this formation occur in one horizon within deltaic sediments, suggesting that they were victims of a catastrophic event, probably relating to flooding and storms with strong winds that hit the volcanic arc itself, and were then washed a short distance into the marine basin where they were deposited rapidly in a low-energy current regime. The assemblage comprises a relatively large, well-preserved mix of juvenile and mature foliage that may represent the leaf fall from one season and may explain the dominance of angiosperms, the high proportion of ferns and low amounts of conifers preserved. Bias, however, may also have introduced an overrepresentation of deciduous species and more leaves having originated from woody dicots relative to evergreen foliage of herbaceous plants (Burnham, 1989; Ferguson, Hofmann and Denk, 1999).

Overlying the Hidden Lake Formation is the younger Santa Marta Formation, also exposed on the northwest of James Ross Island (Figures 7.4D and 7.5). The Santa Marta Formation is the basal of four formations recognised within the Marambio Group (Figure 7.5). The Marambio Group, composed of finer-grained sandstones and mudstones,

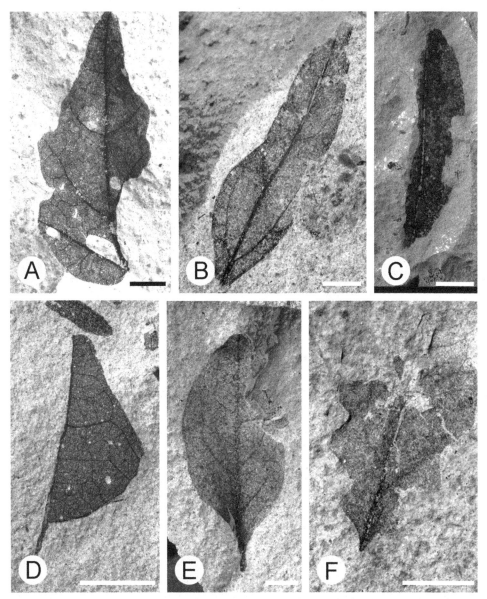

Figure 7.8 Angiosperm leaves from the Coniacian Hidden Lake Formation Flora. **A**, morphotype 2, Sterculiaceae BAS D. 8754.8.1a. **B**, morphotype 11, Lauraceae BAS D. 8754.8.57a. **C**, morphotype 28, BAS D. 8754.8.48a. **D**, morphotype 2, BAS D. 8754.8.27a. **E**, morphotype 3, BAS specimen D. 8754.8.45a. **F**, morphotype 15, BAS D 8754.8.28a. Morphotypes follows the concept of Hayes (1999) and Hayes *et al*. (2006). Scale bars = 1 cm.

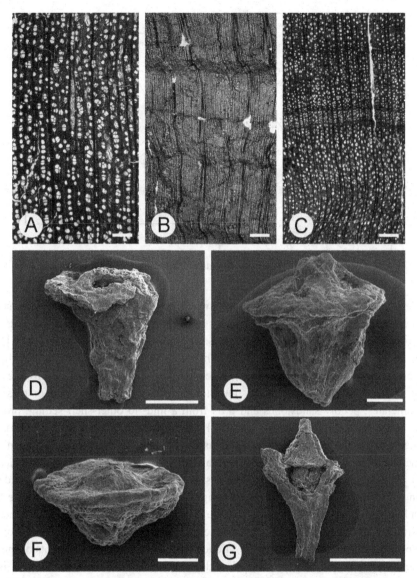

Figure 7.9 Santonian to Campanian plant fossils from the Antarctic Peninsula. **A–C**, Cretaceous woods (transverse sections) from the Antarctic Peninsula. A, wood of an unknown angiosperm from the Coniacian–Santonian Williams Point Flora with primitive wood anatomy, BAS P1806.17. B, winteraceous wood from Santonian–Campanian Santa Marta Formation, James Ross Island, BAS DJ 1051.05. C, lauraceous wood from the Maastrichtian López de Bertodano Formation, Seymour Island, BAS DJ 1056.49. **D–G**, scanning electron micrographs of late Santonian charcoalified flowers from Table Nunatak (photos courtesy of H. Eklund). D, Siparunaceae flower with cup-shaped hypanthium, BAS R 2959.8.23. E and F, myrtaceous flowers, BAS DJ 755.130 and R.2959.27.53 respectively. G, unknown eudicot flower BAS, R. 2959.27.430. Scales bars A–D, G = 1 mm, E, F = 300 μm.

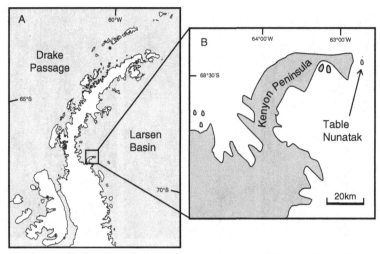

Figure 7.10 Map showing location of Table Nunatak to the east of Kenyon Peninsula.

was deposited in a shallow marine environment. The basal Santa Marta Formation comprises eroded volcanic and plutonic elements from the Antarctic Peninsula and is dated as late Coniacian–latest Campanian, based on Sr/Sr dating (Browne and Pirrie, 1995; Dingle and Lavelle, 1998; McArthur, Crame and Thirwall, 2000). The flora is found within the Lachman Crags Member (Figure 7.9B), a sequence consisting of mainly turbidite sandstones and intensely bioturbated silty mudstones and sandstones with minor mass-flow conglomerates, pebbly sandstones and tuffs from a mid- to outer-shelf setting, probably below storm wave-base (Crame *et al.*, 1991; Hathway *et al.*, 1998; Pirrie, 1989). This member has been dated using molluscan floras and dinoflagellates as early Santonian–early Campanian in age (Crame *et al.*, 1991; Dettman and Thomson, 1987; Doyle, 1990; Keating, 1992). However, more recent studies suggest that the Coniacian–Santonian boundary is within the lower 150 m of the Lachman Crags member (the lowermost part of the Santa Marta Formation), and thus extending the age to probable late Coniacian–early Campanian (McArthur, Crame and Thirwall, 2000). The Santa Marta Formation represents quieter, more offshore conditions relative to the Hidden Lake Formation, and elements of the flora are more widely distributed spatially. Slow deposition rates indicate that this flora is less affected by seasonal patterns and thus seasonally deciduous leaves are less dominant and conifer remains are more abundant (Hayes, 1999). The flora comprises predominantly angiosperm leaf impressions alongside those of ferns, bennettites and conifers, but also seeds, concretions containing large wood fragments and some leaves, and calcified logs up to several metres in length (Pirrie, 1987).

Even though these two floras were probably temporally separated by a period of several million years, a comparison of the two floras indicate that angiosperms growing in Antarctica during the Turonian–Coniacian (Dettmann, 1989) had increased rapidly such that by the mid-Coniacian they had established themselves as a diverse and dominant component of the regional Antarctic flora. Conifers (predominantly podocarps but also araucarians), ferns, bryophytes along with rarer bennettites, cycadophytes, ginkgophytes and other seed plants, were also present in the floras (Baldoni and Medina, 1989). Between the Santonian and Campanian there is an unmistakable turnover in pollen taxa, which is also reflected in the wood and other macrofloras (Askin, 1992; Dettmann, 1989; Poole and Cantrill, 2006). Within the angiosperm wood flora the more 'primitive' wood types with affinities to the 'Magnoliidae-Hamamelidae-Rosidae' (Poole and Cantrill, 2001) do not continue through to the Campanian. The only anatomically primitive types of wood occurring in the Coniacian flora that also appear in the Campanian are the cunoniaceous *Weinmannioxylon* and the monimiaceous *Hedycaryoxylon*, both of which disappear during the Campanian. *Hedycaryoxylon* is replaced by other anatomically different, monimiaceous taxa, whereas *Weinmannioxylon* xylotype continues through until the Eocene as a distinct second xylotype (Cantrill and Poole, 2005). The western and eastern sides of the Antarctic Peninsula record the changing vegetation during the Campanian and into the Maastrichtian.

Comparisons can be made between the floras preserved in the Hidden Lake Formation and the younger Santa Marta Formation. The Hidden Lake Formation macroflora is dominated by leaves of both Magnoliales and of the laurelean type, along with Annonaceae and Dilleniid types. Modern eudicot forms, for example those with affinities to Sterculiaceae, Elaeocarpaceae are common as are magnoliid forms, such as Lauraceae and Winteraceae (cf. *Bubbia*), and atherospermataceous plants (Hayes, 1999; Poole and Francis, 1999, 2000). Eudicots (e.g. Cunoniaceae, Nothofagaceae, Fagales, Urticales, Hamamelidae and Rosales) become more abundant in the younger Santa Marta Formation, although the Magnoliales still dominate alongside a strong component of laurelean types. Modern Sterculiaceae, Lauraceae, Elaeocarpaceae and Winteraceae are also common along with the first evidence of Myrtaceae, Ebenaceae, Ericaceae and Illiciaceae (Hayes, 1999; Poole *et al.*, 2000).

The Hidden Lake leaf flora contains ~70% angiosperms compared with 28% ferns and <3% bennettites and conifers. The Santa Marta leaf flora contains 71% angiosperms, 15% ferns (predominantly gleicheniaceous ferns), relatively more (8%) conifers than in the Hidden Lake Flora and only 1% bennettites (Hayes, 1999). This dominance of angiosperms in the James Ross Basin continues to increase throughout the late Santonian and into the Maastrichtian (Askin and Spicer, 1995), with increasing abundance of wood remains testifying to the increasing arboreal habit. The exact timing of the radiation into the canopy is problematic because of the paucity of data from the Cenomanian and Turonian. This vertical invasion would have drastically changed the prevailing landscape, yet in time it enabled ferns to colonise the new niches created by the angiosperms, thus heralding fern rediversification (Cantrill and Poole, 2002).

Wound reaction is also present in leaf material from both the Hidden Lake and Santa Marta floras as well as plant–animal interactions signified by the presence of galls associated with some of the leaves. Pre-abscission leaf damage (i.e. wound tissue which only forms on living plants; Scott and Titchner, 1999) surrounding holes on some of the Hidden Lake Formation leaves and traces may represent non-marginal feeding by arthropods, although such damage could have been formed from fungal attack or wind damage causing abrasions when leaves rub together (MacKerron, 1976; Wilson, 1980, 1984). Together these provide further details of the dynamic environment in which these plants grew (Hayes, 1999).

Living organisms often have restricted tolerances to amounts of temperature and rainfall, but plants are the most sensitive indicators of terrestrial climate conditions since, being immobile, they have to be able to adapt to local conditions to survive. There is little evidence for cool and cold climates from the fossil plants or their putative modern equivalents (Hayes *et al.*, 2006). However, since individual plants and the communities they form are both morphologically and physiologically adapted to their environment, plant fossils have been able to provide more precise palaeoclimate data for the terrestrial environment through statistical analyses based on the relationship between climate and their physiognomic characters. The only other direct measure of terrestrial climate is from geochemical evidence of terrestrial sediments, but even these require knowledge from palaeofloras before they can be utilised (Dingle and Lavelle, 1998). The use of plant fossils as proxies for palaeoclimatic parameters operates under certain caveats (see summary by Poole and van Bergen, 2006), but can provide meaningful data when no direct evidence for palaeoclimate is forthcoming. Leaf morphology, wood anatomy and community analyses have been undertaken by a number of workers to try to determine the climate that prevailed over the Antarctic Peninsula region during the Late Cretaceous, and these results can be compared with outputs from computer generated climate models (General Circulation Models, GCM).

Climatic indications from nearest living relative analyses and plant physiognomy indicate that during the Coniacian–Campanian period the growing conditions were not interrupted by cold spells after the growing season had been initiated by increasing light intensity. The proposed increased levels in pCO_2 would have enhanced plant sensitivity to unseasonal fluctuations in temperature (Royer, Osborne and Beerling, 2002) and thus any cold snaps would more likely be reflected in (conifer) wood physiognomy, albeit randomly, if such cold snaps had occurred (Poole, Cantrill and Utescher, 2005). Mean annual temperature and mean annual range in temperature, derived from plant physiognomy and community analyses, are suggested to have been 15 °C and 3.5 °C respectively. A maximum estimated summer temperature of ~25 °C, with winter temperatures remaining above freezing (~7 °C) is also recorded (Poole, Cantrill and Utescher, 2005). Mean annual precipitation was high (>1000 mm) suggesting that the maritime climate was warm, permanently wet with high humidity (Hayes, 1999; Hayes *et al.*, 2006; Poole, Cantrill and Utescher, 2005). These findings compare well with those derived from independent sources: coal deposits in these high latitudes also indicate a relatively high rainfall (Francis and Frakes, 1993), and climate simulations (e.g. Valdes, Sellwood and Price, 1996) predict summer surface air temperatures of >20 °C, and winter temperatures above freezing (4–8 °C) even though further south on the Antarctic continent freezing conditions probably did occur.

As discussed at the beginning of this chapter, a forested polar region would have impacted on climate. In an assessment to find the conditions required to sustain polar forests, DeConto *et al.* (1999) found that forests at the poles required pCO_2 levels of 1500 ppm, coupled with greater poleward oceanic heat transport. The warming effect would have been in the region of about 2 °C and, coupled with its low albedo, the polar vegetation would have added more water vapour to the atmosphere through evapotranspiration thereby inducing a greater warming. This would have had a knock-on effect by warming land surfaces that consequently warmed adjacent oceans, preventing sea ice formation and causing higher winter temperatures (Otto-Bliesner and Upchurch, 1997). This is reflected in marine records, which also indicate peak warmth during the Coniacian and Santonian, with isotopic analysis of shells from the Hidden Lake Formation giving a palaeotemperature estimate of between 13 °C and ~19 °C (Ditchfield, Marshall and Pirrie, 1994; Marshall, Ditchfield and Pirrie, 1993; Pirrie and Marshall, 1990). The presence of forest vegetation on the Antarctic Peninsula would have provided a positive feedback favouring the growth of more forest vegetation in southern high latitudes, and thus may have acted as a trigger for expansion of the angiosperm floras in Antarctica (Hayes *et al.*, 2006).

The biodiversity and prevailing warm climatic regime of the Turonian–Coniacian polar latitudes declined during the Santonian before recovering during the Campanian and Maastrichtian. This taxonomic turnover has been noted in both the pollen and macro (wood) floras, where 'primitive' forms tended to disappear prior to the Campanian (Askin, 1992; Dettmann, 1989; Poole and Cantrill, 2006), heralding the extensive vegetational changes that were underway.

Record of cool temperate vegetation in the Campanian–Maastrichtian from the Antarctic Peninsula

During, or just prior to, the mid-Campanian under a relatively warm, wet climate where mean annual temperatures ranged from 10 to 16 °C and mean precipitation levels were in the region of 3000 mm (Dutra and Battern, 2000; Poole, Cantrill and Utescher, 2005), profound vegetational changes were occurring that included the introduction and proliferation of several key extant angiosperm taxa such as the Nothofagaceae (Askin, 1992; Dettmann and Thomson, 1987; Dutra and Batten, 2000; Poole and Francis, 1999, 2000; Poole and Gottwald, 2001; Zastawniak, 1994). From the mid-Campanian through to the Campanian–Maastrichtian boundary the palynofloras diversified further to include new elements such as Gunneraceae, Aquifoliaceae, Loranthaceae and possibly Dilleniaceae, Palmae and Pedaliaceae (Askin, 1992). The southern podocarp conifers were also diversifying and Antarctica became a locus of evolutionary novelty for many plant groups (Askin, 1992; Dettmann, 1989). These modernisations are supported by an increase in abundance and/or diversity of other taxa including *Dacrydium* (which originated in the Coniacian), and Proteaceae, Loranthaceae, Gunneraceae, *Dacrycarpus* and Myrtaceae in the Campanian (Askin, 1989; Dettmann, 1989; Dettmann and Jarzen, 1990, 1991; Dettmann and Thomson, 1987; Dettmann *et al.*, 1990). Diversification continued throughout the Maastrichtian with many endemic angiosperm pollen taxa of unknown botanical affinity appearing in the fossil record. During the short, warm interval of the late Maastrichtian, members of the

Bombacaceae, Olacaceae and Sapindaceae (Cupanieae) are recorded for the first time from Seymour Island (Askin, 1989).

Perhaps the most dramatic change accompanied the increase in importance of the Nothofagaceae. The Nothofagaceae is a typical, and one of the most important, elements of the relictual Southern Hemisphere temperate floras today, having radiated out from its putative centre of origin within southern South America (Manos, 1997; Swenson, Hill and McLoughlin, 2000). Its pollen is extremely common such that absence can be regarded as evidence for the lack of the genus in areas where it has not been found (Swenson and Hill, 2001). *Nothofagus* first appeared, represented by the ancestral *Nothofagidites senectus* pollen type (notably distinct from extant *Nothofagus* pollen), in early Campanian (88–80 Ma) sediments over wide areas of southern Gondwana, including the Antarctic Peninsula and South Australia (Dettmann *et al.*, 1990; Swenson, Hill and McLoughlin, 2000). The Antarctic Peninsula played an important role in the differentiation and diversification of the genus such that by the late Campanian (75–71 Ma) the *Nothofagus* lineage had diversified into all four modern subgenera (Figure 7.3). The distinct subgenera first appeared in West Antarctica, shortly thereafter in South America and finally in Australia some 12–15 Ma later (Dettmann *et al.*, 1990).

On the eastern side of the Antarctic Peninsula, on Vega and Seymour Islands (Figure 7.4), Maastrichtian strata reveal the oldest known occurrences of *Nothofagus* subgenera *Fuscospora*, *Lophozonia* and *Brassospora* in the pollen record (Dettmann *et al.*, 1990) alongside wood of the *Fuscospora* type (Poole, 2002). Manos (1997) and Swenson, Hill and McLoughlin (2000) recognise the subgenus *Lophozonia* as the most basal clade within Nothofagaceae with the pollen record extending back to the late Campanian in Antarctica (Dettmann *et al.*, 1990), and appearing only slightly later in the wood flora (Poole, 2002). By the Maastrichtian all but the *Brassospora* subgenus are represented in the wood flora, whereas all four subgenera are represented in the pollen flora. The lack of wood allied to *Brassospora* is not surprising as this group, although a copious pollen producer today, is often confined to montane areas. We can confidently assume that by the mid-Campanian the Nothofagaceae had now become an important and diverse component of the Antarctic ecosystem having substantially changed the face of the flora relative to the ancestral Coniacian–Santonian vegetation.

During the Maastrichtian, the wood floras continue to increase in abundance and diversity with other angiospermous taxa appearing and Nothofagaceae diversifying still further. *Nothofagus* became increasingly important both within the flora as well as in stature, and was on route to its maximum diversity and spatial spread in the Oligocene (Hill, 2001; Poole, 2002; Poole, Mennega and Cantrill, 2003). Wood anatomical characters, coupled with good preservation of these now prolific angiosperms, provide us with important indications pertaining to the local environment as well as the vegetational composition and dynamics. The Antarctic material comprises predominantly aerial organs (branches or trunks – only one specimen could be confidently identified as having a root origin), possibly dominated by sap wood remains [cf. Wheeler and Manchester (2002) and their inferences drawn from the

Eocene Clarno Formation woods], testifying to the presence of substantial trees. The dominance of sap wood is evidenced from preserved fungal hyphae associated with the ray parenchyma and vessel lumens or tracheids, suggesting that these fungi are sap-staining fungi restricted to sap wood. These woods detached from the parent plant probably during the Antarctic polar winter, or at least during the dormant phase of the year, as many of the angiosperm woods have vessels filled with well-developed tyloses with no evidence of bud-like outgrowths of parenchyma into the vessel lumen marking the beginning of tylose formation (cf. Wheeler and Manchester, 2002; Chattaway, 1949). The woods probably resided, for a time at least, in a damp, but not waterlogged, aerobic environment. The prevailing moist conditions across the volcanic arc adjacent to the James Ross Island Basin probably encouraged fungal growth within the woody debris covering the forest floor. Then before decomposition was complete, this woody debris (e.g. Figure 7.9C) was transported, by streams or rivers during times of storms or flash floods, to the coast where the material floated in the shallow sea before becoming waterlogged and sinking into the sediments that ultimately became the López de Bertodano Formation (Poole and Cantrill, 2006).

Late Cretaceous vegetation of the South Shetland Islands, western Antarctic Peninsula region

On King George Island, mega- and microfloras are known from late Campanian to early Maastrichtian successions. The plant remains are found in thin intercalations of reworked volcanic sediments beneath andesitic flows, which vary in radiometric age from Paleocene (~58 Ma) to latest Maastrichtian (~66.7 Ma) (Fensterseifer *et al.*, 1988; Soliani and Bonhomme, 1994), and above basaltic rocks dated as 77 Ma (Birkenmajer *et al.*, 1983; Shen, 1994). The depositional characters of the tuffaceous conglomeratic sandstones, silt-stones and mudstones, together with some coals in the lower part, suggest deposition took place in an alluvial fan (mainly preserving pieces of wood), lakes and floodplains (with leaves and pollen) within a major braided river system that developed during periods of volcanic quiescence. The substantial and often extensive mud and debris flow deposits indicate the relatively close proximity of highlands (Dutra and Batten, 2000).

Plant debris accumulated in the fluvial settings not normally confined by channels and were prone to sheet flooding following heavy rain. For short periods floodplains containing bodies of fresh water developed in areas between the volcanic mountains and the sea. The plant horizons represent the vegetation that had established after periods of disturbance, and vary in composition reflecting lateral facies changes as well as taphonomic processes and climatic differences (Dutra and Batten, 2000). The fossil assemblages have been recovered from sections in the Half Three Point Formation at Half Three Point, Skua Bay and Price Point, all on Fildes Peninsula, and from the Zamek Formation[3], in the Admiralty Bay area.

---

[3] The age of this formation is under dispute and may in fact be early Cenozoic (T. Dutra personal communication, 2010).

At Half Three Point (Figure 7.4) the palynoflora is preserved in a lacustrine setting between two magmatic flows and records an assemblage representing the vegetation that dominated the low hills and lake shores near the coast during the warm humid (subtropical) conditions of the Upper Cretaceous (Cao, 1990, 1992, 1994; Dutra and Batten, 2000; Shen, 1994; Song and Cao, 1994; Zhou and Li, 1994). The plant-bearing beds comprise a short section (~4 m) of finely laminated tuffs and tuffaceous sandstones, siltstones and mudstones. An Rb-Sr age of ~71 Ma has been obtained from analysis of whole rock samples in the tuffaceous interval (Shen, 1994; Wang and Shen, 1994). Fungal, bryophyte and pterido-phyte remains suggest that the plant debris preserved in the lake is autochthonous. However, those in the braided river deposits will have been transported some distance from the source flora (Dutra and Batten, 2000; Shen, 1994). The palynological and leaf assemblages testify to a fern dominated (80% abundance) flora with representatives from the tree fern Cyatheaceae, and Gleicheniaceae being especially numerous (Dutra and Batten, 2000). Members of these families along with Adiantaceae probably dominated the hill tops with other tree ferns (Dicksoniaceae) and ferns (e.g. Polypodiaceae, Lygodiaceae Osmundaceae) whilst Selaginallaceae, mosses including *Sphagnum* and hornworts occupied the lake shores. Araucariaceae, Podocarpaceae and primitive deciduous Nothofagaceae[4], with leaves similar to those of the *N. obliqua*, *N. alessandri* and *N. glauca*, probably formed sparse mixed broad-leafed angiosperm to coniferous woodlands giving way to denser rainforest communities on more distant mountainous regions, complete with climbing lianas (Loranthaceae) and Proteaceae. Abundant fungi were present decomposing the remains of the dead and dying vegetation and testifying to the presence of humid and at least cool–warm temperate conditions (Cao, 1992, 1994; Dutra *et al.*, 1996; Zhou and Li, 1994). Seed plants also include pollen attributed to Gnetaceae–Ephedraceae. Both Gnetaceae and Ephedraceae are uncommon in high latitudes today, with Ephedraceae being xerophytic shrubs and *Gnetum* growing in tropical rainforests. If the grains have an ephedracean origin, the proximity of the sea or seasonal aridity might explain their presence and suggest a warm climate (Dutra and Batten, 2000).

Tree ferns and ferns (Dicksoniaceae) also dominate the assemblage preserved in a braided river facies at Rip Point on the northeast of neighbouring Nelson Island. This flora is also dated as Campanian (77–71 Ma; Shen, 1994) and is characterised by few conifers and angiosperms, such as primitive deciduous *Nothofagus* and others with similarities to extant Magnoliidae (with lauralean leaf architecture), Dillenidae and Rosidae (Dutra, Hansen and Fleck, 1998; Shen, 1990). Both the Half Three Point Flora and the Rip Point Flora suggest that the environment experienced a wet meso-microthermic or warm temperate climate with a relatively dry season similar to the conditions described for the Cretaceous Santa Marta Formation on James Ross Island, or the prevailing Mediterranean climate of Chile today (Baldoni, 1987; Dutra, Hansen and Fleck, 1998).

---

[4] The presence of *Nothofagidites* supports a late Campanian or Campanian–Maastrichtian age for this flora (Dutra and Batten, 2000).

The second assemblage originates from Price Point (Figure 7.4). The section comprising tuffaceous sandstones, siltstones and mudstones extends up to 10 m laterally and is approximately 6 m thick, 4 m of which is fossiliferous (Dutra and Batten, 2000). The fossil plant assemblage comprises a few poorly-preserved leaves and abundant spores and pollen grains. Angiosperms dominated the vegetation (up to 50% of the pollen assemblage), with *Nothofagus* becoming the most dominant (up to 20%; Dutra *et al.*, 1996; Dutra and Batten, 2000). Conifers such as Podocarpaceae and *Bellarinea* would also have contributed to the canopy, but in much smaller numbers (10%). Tree ferns (Cyatheaceae) and abundant (40%) ferns (e.g. Osmundaceae) made up the understorey. One noticeable difference between Price Point flora and the Half Three Point assemblage is the presence of cycads, such as *Podozamites*, which are absent from the Half Three Point assemblage, yet probably contributed to the understorey on King George Island (Dutra and Batten, 2000).

A mixed broad-leafed-conifer woodland with abundant ferns is also suggested for the third assemblage, namely that from Skua Bay (Figure 7.4). Ferns again dominate the vegetation with one type very similar to *Thyrsopteris elegans* growing in the Juan Fernández Archipelago today (700 km off the Chilean coast), others include taxa such as Cyatheaceae, Dicksoniaceae and Gleicheniaceae that are also found in the other assemblages of similar age from King George Island. Angiosperms are represented by wood, pollen and leaves. The leaves are micro- to notophyllous, simple, broad and coriaceous with entire margins derived from primitive types of Magnoliidae and palmate Dilleniidae. Rare pinnate Dilleniidae and Rosidae (with cunonioid teeth) represent less than 10% of the angiosperm remains. Nearly all of the rest (93%) can be included in the cinamomumophyll morphological group of Crabtree (1987). The leaves also exhibit characteristics of the Lauraceae type II group (see Macphail *et al.*, 1993). A representative of *Ficophyllum*, *F. skuaensis*, is very similar to another species, *F. palustris*, described from upper Albian deposits on Alexander Island (Cantrill and Nichols, 1996). Other elements common in the Upper Cretaceous deposits of King George Island show morphologies similar to those of *Magnoliidaephyllum* from the Zamek Formation (Zastawniak, 1994) and *Rhoophyllum nordenskjoeldii* (Dusén, 1899), a leaf with affinities to *Monimiophyllum antarcticum* found in the Skua Bay section (Birkenmajer and Zastawniak, 1989a). Other types resemble species of *Sterculiaephyllum* and the 'laurophyllous impressions' of Birkenmajer and Zastawniak (1986, 1989b). These are also found in younger Paleocene deposits on the island, as important components of assemblages in the late Paleocene of South America, as well as in Eocene sediments of southeast Australia and New Zealand. This suggests that the Skua Bay Flora might be slightly younger than the other Late Cretaceous plant assemblages on King George Island. Alternatively, it might (also) confirm the diachronism in first appearances of taxa in the Antarctic palaeoflora in relation to the other austral continents (Askin, 1989; Dutra, 1997). *Nothofagus* is represented by 10–20% of the palynoflora of the other two sections on Fildes Peninsula as well as by some megafossils. *Nothofagus* does not form a significant component of the megaflora of Skua Bay, which suggests that it was not part of the local vegetation. This is in marked contrast with the Admiralty Bay sites where the genus dominates both the leaf and wood flora (Dutra and Batten, 2000).

In Admiralty Bay (Figure 7.4) a number of Upper Cretaceous fossils have been recovered (e.g. *Nothofagus* wood at Paradise Cove, and in *ex situ* leaves in moraine material from Block Point), but dating remains uncertain. The material described by Zastwaniak (1990, 1994) from blocks in the Blaszyk Moraine of Admiralty Bay is supported by the discovery of identical fragments in an outcrop in the lower part of the Zamek Formation at the base of the Zamek Hill succession (Birkenmajer *et al.*, 1983; Dutra and Batten, 2000). The lithologies in the Zamek Formation are similar to those in the Half Three Point Formation on Fildes Peninsula. The sediments accumulated in quiet, shallow-water lakes that disappeared during periods of drought leaving floodplains exposed until the next rains. The large number of leaves and deficit of pollen and spores (probably destroyed by *in situ* oxidation) indicate an environment significantly different from those inferred from the section on Fildes Peninsula (Dutra and Batten, 2000). The Zamek assemblage indicates the remains of a broad-leafed forest community that included both evergreen types (with thick coriaceous leaves) and deciduous *Nothofagus* with abundant ferns. Angiosperms were dominant, whilst conifers and ferns were rare. The plant material was probably derived from the mountainous hinterland and brought downstream by dynamic rivers. This assemblage shares characteristics with each of the floras on Fildes Peninsula. Some of the *Nothofagus* leaves preserved show most similarity to the northern hemispheric Betulaceae, whilst others are more similar to *N. alessandri*, *N. glauca*, *N. obliqua* and *N. alpina*, exhibiting a plicate vernation and deciduous habit, preserved in the Half Three Point Flora. These leaves are associated with other angiosperms referable to the Magnoliidae, Asteridae, palmate Dilleniidae, and Rosidae. Other angiosperms had affinities to the magnoliids (lauraceous-magnoliaceous forms) and eudicots (those with proteaceous and myrtaceous affinities), although eudicots representing Araliaceae, Malvaceae and Sterculiaceae were the most significant elements (Dutra, 1997; Dutra and Batten, 2000; Zastawniak, 1994). The source forest was probably adapted to a micro-mesothermal climate with an estimated mean annual temperature of 10–16 °C that agrees well with the 11 °C estimate derived from wood physiognomy and 16 °C from Coexistence Analyses (Dutra and Batten, 2000; Poole, Cantrill and Utescher, 2005; Zastwniak, 1994). Similar plant associations are known today where the climate is oceanic and involves a cold and/or dry season such as between the latitudes of 40–50° S on the Chilean coast, in New Zealand, or at altitudes of 900–1300 m in the coastal mountains of eastern Brazil from 20 to 33° S where the Mata Atlantica subtropical rain forest grows (Dutra and Batten, 2000).

The presence of small basins/depositional areas at a short distance from the sea is also suggested by the plant taxa recorded in the Skua Bay and Zamek Formation assemblages, which indicate a close affinity to taxa living today on islands and continents with oceanic climates (e.g. Tasmania, eastern Australia, the west coast of Chile and New Zealand). Such taxa include representative of *Thyrsopteris*, *Dacrydium*, *Podocapus*, *Nothofagus*, Myrtaceae and Sterculiaceae. The presence of lianes and ferns in these Late Cretaceous floras suggest warmth and humidity. Leaf characters in the floras may suggest a climate cooling to one similar to the high humid altitudes of Brazil. Here, a microphyllous flora dominated by plants bearing leaves with entire margins and leathery textures are very similar to the physiognomic character of leaves in the Late Cretaceous floras of King George Island (Accordi, Dutra and Müller, 1996). This does not, however, eliminate the possibility of seasonally dry conditions, at least around Admiralty Bay (Zamek section), which was located on the eastern side of the highlands within the fore-arc mass (Dutra and Batten, 2000).

The absence of evidence of vertebrate life in the sedimentary succession, apart from the footprints of birds, led Dutra and Batten (2000) to suggest that the distinctive insular character of the vegetation may have been attained early in the evolution of the fore-arc basins on the west side of the Peninsula. This would be in marked contrast with the back-arc basins on the east side, where the remains of dinosaurs, turtles and other reptiles have been found. The source of new plant and animal taxa to both the fore-arc and back-arc setting may have been from nearby continental areas.

Similarities in vegetational composition in the South Shetland Islands are shared with those on the east side of the Antarctic Peninsula where podocarps and angiosperms formed the main canopy elements of the perhumid, tall, open forests (Askin, 1988a; Specht, Dettmann and Jarzen, 1992). Palynofloras and macrofloras suggest a diverse angiosperm component, with Myrtaceae along with diverse species of *Nothofagus*, *Gunnera*, Proteaceae, Aquifoliaceae, Olacaceae, Eucryphiaceae, Atherospermataceae, Myrtaceae, Loranthaceae, Sapindaceae and Illiciaceae, accompanied by a rich fern component including the Osmundaceae and Gleicheniaceae (e.g. Askin, 1983, 1989, 1990a, 1990b; Cantrill and Poole, 2005 and the references therein; Dettmann and Thomson, 1987). At the end of the Maastrichtian, the warmer temperate elements, such as the lauraceous *Sassafras* (*Sassafrasoxylon*) and Illiciaceae had disappeared, whilst the abundance of fern taxa also decreased (Askin, 1988a, 1988b, 1990a, 1990b; Cantrill and Poole, 2005; Poole *et al.*, 2000), possibly in response to a cooling in the climate. This changing climate is evidenced by the change in wood characters (e.g. increase in angiosperm woods exhibiting ring porosity and other character combinations; Poole, Cantrill and Utescher, 2005). Moreover, the angiosperms were probably also deciduous (Poole, Cantrill and Utescher, 2005).

The overall vegetational composition of the Antarctic Peninsula during the Late Cretaceous (and on into the early Cenozoic) has been likened to the extant low- to mid-altitudinal Valdivian rainforests (see Chapter 8 for more details; Poole, Hunt and Cantrill, 2001; Poole, Mennega and Cantrill, 2003). Indeed the similarities in terms of environmental dynamics in addition to vegetational composition are remarkable. Along the Andean margin of South America, stratovolcanic activity is a major source of disturbance along with associated events such as landslides, earthquakes and the flooding of lake systems. Glacial erosion and deposition contribute to the general disturbance, especially at altitude. Thus the vegetated active volcanic arc ecosystem of the Antarctic Peninsula region during the Cretaceous and Cenozoic would have been subject to similar ecological disturbances. There is no doubt that climate change influenced vegetational composition, although changing palaeoenvironments in the Antarctic Peninsula region probably contributed to an equal, if not greater, degree especially over shorter time scales.

Across the Cretaceous–Cenozoic boundary these rainforest communities thrived and remained dominated by abundant podocarps and diverse angiosperms with a minor fern component. Angiosperm diversity had almost doubled through the latest Cretaceous, from the latest Campanian (~33 taxa) into the latest Maastrichtian (~60), with almost half the angiosperm species endemic to the Antarctic region by the end of the Cretaceous. However, towards the end of the Cretaceous a turnover in angiosperm taxa is evident with the disappearance of certain taxa whilst new forms appeared (e.g. Palmae and a distinctive new species of *Liliacidites*). This long-term floral turnover has been attributed to cooling climate as mean annual temperatures along the Antarctic coast dropped to ~6 °C, yet probably remained frost-free at least in the coastal lowlands of the northern Antarctic Peninsula (Askin, 1988a, 1988b, 1989, 1994; Askin and Jacobson, 1996; Crame, 1994; Dutton *et al.*, 2007; Thorn, Riding and Francis, 2009; Upchurch and Askin, 1990).

The Cretaceous–Cenozoic boundary and evidence for a catastrophic event in the Southern Hemisphere

The end of the Mesozoic era is marked by a global event that led to a modernisation of terrestrial ecosystems when those plants and animals characteristic of the Cretaceous were replaced by taxa more typical of the Cenozoic. Evidence for this global event lies in a thin layer (a few cm thick) of clay covering much of the globe and sandwiched between Cretaceous and Cenozoic rocks. This clay layer was found to be rich in iridium, a rare element found in crustal rocks deep inside the Earth and primitive stony meteorites from outer space. The amount of iridium in this layer is many tens of times greater than the background on Earth. The most plausible explanation to account for such quantities of iridium to be concentrated in a thin layer across the Earth is that an asteroid, 10 km in diameter, impacted on Earth some 65 Ma ago. This resulted in the extinction of approximately 75% of the marine and terrestrial species, the most well-known being the demise of the dinosaurs. In the oceans >90% of the plankton alone were extinguished, which led to the collapse of the oceanic food chain.

The site of impact has been argued to be the ~180 km-wide Chicxulub crater in the northern Yucatan Peninsula (Mexico), although some authors believe that the Chicxulub impact occurred 300 kyr before the K-T boundary (Bralower *et al.*, 2010; Schulte *et al.*, 2010 and references therein). Perhaps a later impact, or multiple impacts, were responsible for the K-T boundary event (e.g. Keller *et al.*, 2007) – a controversy not helped by the fact that at most outcrops the K-T sediments, and thus much of the evidence, have been eroded (Stüben *et al.*, 2005). This impact event was, however, catastrophic causing an abrupt climate change, mass ecological turnover, shutdown of terrestrial and marine primary production (e.g. Maruoka, Koeberl and Bohor, 2007) and throwing the planet into a scenario similar to that of a nuclear winter.

The impact of such a meteorite would initially have triggered seismic activity and tsunamis with far-reaching consequences (e.g. Preisinger *et al.*, 2002; Scasso *et al.*, 2005).

Shortly afterwards, dust containing the iridium, toxic gases, vaporised evaporates and fine ejecta would have all contributed to the vapour-rich impact plume that was ejected into the atmosphere. The bolide's passage through the atmosphere, radiation from the impact plume and returning ejecta, and lightning strikes would have contributed to spontaneous ignition of vegetation across much of the Earth's surface. Relatively quickly, a blanketing cloud of dust and fine ejecta would have formed around the planet preventing sunlight from reaching the Earth's surface. This would have resulted in dramatic cooling of land surface temperatures and possibly triggering a short period of glaciation (cf. Twitchett, 2006). The dust cloud would have lasted for at least a few months, possibly years, slowly building up the millimetre-thick Cretaceous–Paleogene boundary clay layer with anomalously high iridium levels worldwide (Smit, 1990). Acid rain (both sulphuric and nitric), formed by the vaporised evaporites reacting with water in the atmosphere, would have washed back down to Earth causing mass defoliation of continental vegetation and aquatic plants in shallow lakes and seas. Animals would have been asphyxiated by nitrous oxides and toxic poisoning by metals leaching from the ground. Wildfires would have raged across the globe, initiated by the temperature spike igniting the vegetation. The immediate after effects of the impact, such as the temperature spike, high winds, storms and tsunamis would have lasted for a few hours; the fires, darkness and subsequent cold, for months, possibly more than a year; whereas the effects of the acid rain and toxic substances would have lasted several years; the greenhouse warming persisted for hundreds of years. Within a time span of perhaps just 10 years after the impact, freshwater environments (in contrast to marine environments) had recovered[5] with highly productive microbiota taking advantage of increased nitrogen fertilisation and/or eutrophication, induced by enhanced sulphide formation (Maruoka, Koeberl and Bohor, 2007).

However, what is also of interest is that a period of long-term biotic stress observed during the late Maastrichtian had begun prior to any impact, regardless of where the meteorite(s) hit the Earth. One phenomenon famously associated with a K-T event is the demise of the dinosaurs. At the poles, a variety of dinosaur taxa had been surviving in conditions of relative cold since the Jurassic, possibly migrating or overwintering the long periods of winter darkness. This way they were able to survive conditions too harsh for ectothermic reptiles, which to date have not been found in the high latitudes. Although the climate gradually cooled during the Maastrichtian, the dinosaurs, as opposed to other ectothermic reptiles, remained largely unaffected (Buffetaut, 2004). Yet at the boundary itself, the dinosaurs became globally extinct, whereas the ectotherms were able to survive. Therefore perhaps the extinction of the dinosaurs was related to the breakdown of the food chain, rather than directly to the impact-induced period of cold and dark which has frequently been put forward as one of the main causes of dinosaur extinction (Buffetaut, 2004).

---

[5] This is evidenced by a coeval $\delta^{13}C$ shift by 2‰ to less negative values and a negative ^{15}N excursion in the boundary clay (Maruoka, Koeberl and Bohor, 2007).

In the Southern Hemisphere, the evidence for devastation on a scale similar to that observed in the Northern Hemisphere is limited (Pole and Vajda, 2009; Vajda, Raine and Hollis, 2001). Although the K-T boundary is marked by a pronounced iridium anomaly in both marine and terrestrial sections (Brooks *et al.*, 1986; Vajda and McLoughlin, 2004; Vajda, Raine and Hollis, 2001) the evidence suggests the effects of the impact(s) were less severe.

In the east (i.e. New Zealand), the boundary records evidence of a mass-kill of the diverse flora of the latest Cretaceous, where gymnosperms (especially Araucariaceae and podocarps) dominated (57%), along with abundant ferns (25%) and increasingly common and diverse angiosperms (14%). This vegetation was abruptly replaced by one dominated by a few species of ferns (Vajda, Raine and Hollis, 2001). The thin (4 mm) layer demarking the K-T boundary yields fungal spores and hyphae, woody debris and sparse cuticle fragments, and a fern spike signalling widespread deforestation brought about by an impact winter or massive wildfires (Pole and Vajda, 2009; Vajda and McLoughlin, 2004; Vajda, Raine and Hollis, 2001). The acme of fungal spores is envisaged to represent short-term proliferation of saprotrophs on abundant organic substrates in the wake of forest destruction, before regeneration of the vegetation from rhizomes, lignotubers, spores and seed banks. The herbaceous, pioneer vegetation that rapidly established after the K-T event was floristially poor in terms of diversity and consisted predominantly of ground ferns (especially Blechnaceae and Gleicheniaceae), followed by a gradual transition to tree-fern dominance with small quantities of podocarp and araucarian conifers (Vajda and McLoughlin, 2007). Angiosperms appear to have been most severely affected by the K-T event with evidence for their presence initially sparse, but within a few thousand years, following an initial flourish of opportunistic herbaceous species, shrubs and trees – including those of proteaceous and nothofagaceous affinity – had (re)established themselves to become dominant alongside the podocarp conifers. This formed part of the complex, stratified closed-forest communities that was part of the major transition from deciduous to evergreen vegetation that took place under the moist, cool temperate climate of the earliest Cenozoic (Pole, 1992, 1993; Pole and Vajda, 2009; Vajda and McLoughlin, 2007; Vajda and Raine, 2003).

Further west, in Australia, the Late Cretaceous vegetation comprised cool-temperate, mixed podocarp–Proteaceae–*Nothofagus*-dominated open forests, the canopies of which attained heights of up to 30 m above a thriving substratum of other angiosperms, including shrubby Proteaceae, Winteraceae, Trimeniaceae and *Ilex*, and a ground storey of diverse ferns (Dettmann, 1994). Woodlands were also common in the Late Cretaceous, as were heathlands and aquatic communities. But in Australia, there are no well-exposed Cretaceous–Cenozoic boundary sections (Douglas, 1994). Because of this, Australia can contribute little to the debate on how global the effects of a catastrophic 'mass extinction' event were, and the impact on the evolution of the modern Southern Hemisphere flora, other than to say that many of the angiosperm families that came to achieve prominence in the Cenozoic were present in both the rainforest and sclerophyllous vegetation of the latest Cretaceous (Douglas, 1994; Dettmann, 1994). Palynological evidence suggests that in southeast Australia, the resultant Cenozoic vegetation once again became dominated by *Araucaria*, podocarp conifers and pteridophytes (Macphail *et al.*, 1994). Angiosperm representation in these assemblages is variable, but Proteaceae, Casuarinaceae and Callitrichaceae were all locally abundant.

In the high southern latitudes, it has been (controversially) suggested that climates cooled considerably during the Maastrichtian, at times so severely that high latitude regions suffered short-term glaciations that affected sea levels worldwide (Miller *et al.*, 2003; Miller, Wright and Browning, 2005; Thorn, Riding and Francis, 2009). This is thought to have punctuated the supposedly warm, stable climates putting terrestrial (and marine) biota under considerable stress, making them particularly susceptible to early extinction relating to the end-Cretaceous catastrophe (Thorn *et al.*, 2007). In West Gondwana, the only near-continuous outcropping record of the fossil succession during the last stages of Gondwana break-up (Askin, 1989) and across the Cretaceous–Cenozoic boundary occurs on Seymour Island on the east side of the Antarctic Peninsula. These uppermost Cretaceous and early Paleogene sediments make up the López de Bertodano Formation, a marine sequence that outcrops along the southwestern and central parts of Seymour Island, reaching 1190 m in thickness. The formation is composed of muddy sandy siltstones, muddy sandstones and occasional glauconite sandstones, deposited in nearshore to offshore environments (Crame *et al.*, 2004; Macellari, 1988) and records the final turbulent events of the Mesozoic.

The expanded section records volcanic activity continuing unabated across the northern Peninsula region throughout the late Maastrichtian and into the earliest Paleocene (Danian), coincident with a general regressive trend in sea level (Elliot and Trautman, 1982; Macellari, 1986; Pirrie, Crame and Riding, 1991). During the late Maastrichtian, a lowstand in sea level was accompanied by probable changes in sea water chemistry due to the increased volcanism (Askin, 1988a, 1988b). A subsequent short-term rise in sea level then culminated in a high stand 1 m below a significant glauconite horizon (Brizuela *et al.*, 2007). This high stand spans the Maastrichtian–Danian boundary and, at its peak, it is starved of terrigenous material that allowed the glauconite to form at the water–sediment interface (Van Wagoner *et al.*, 1990). The increase in sea level is reflected both on land and in the sea with an increase in the number and diversity of terrestrial palynomorphs and the size and abundance of phytoclasts, alongside a decrease in marine palynomorphs (Brizuela *et al.*, 2007; Thorn, Riding and Francis, 2009).

Precise placement of the K-T boundary on Seymour Island varies according to the particular fossil group examined, but has usually been placed in the interval approximating to the boundary between units 9 and 10 of the López de Bertodano Formation. Here, relative abundances of typically Maastrichtian dinoflagellate species (*Manumiella* spp.) change to a typical Cenozoic species (such as *Senegalinium obscurum*) (Upchurch and Askin, 1990). Recent palynofacies analyses (Brizuela *et al.*, 2007; Thorn *et al.*, 2007) of these units agree with earlier studies (e.g. Askin, 1988a, 1988b; Elliot *et al.*, 1994; Harwood, 1988; Huber, 1988; Zinsmeister *et al.*, 1989) documenting changes in diatoms, silicoflagellates, palynology and foraminifera such that the K-T boundary coincides with the glauconite horizon in the uppermost López de Bertodano unit of Macellari (1988) (Crame *et al.*, 2004). Nevertheless, it is interesting to note that there are only small levels of iridium enrichment associated with the K-T boundary on Seymour Island, relative to the high concentrations found further east. Moreover, analyses of the palynofacial data across this Maastrichtian–Paleogene section on Seymour Island by Brizuela *et al.* (2007) and Thorn *et al.* (2007) both support the earlier conclusions that, in spite of the disappearance of a few diagnostic pollen types at or near the boundary here and throughout the Weddellian Biogeographic Province, there was no obvious decline in floral and/or planktonic productivity (Askin, 1988a, 1988b, 1990a, 1992; Askin and Jacobson, 1989) – a phenomenon that is not unique to Antarctica (cf. Cripps *et al.*, 2005). The absence of diagnostic pollen types and gradual increase in angiosperm pollen across the boundary (Thorn *et al.*, 2007) may have been in response to increased volcanism causing relatively localised palaeoecological changes (Askin, 1992). The lack of a sudden devastating cold temperature event at the K-T boundary in the high latitudes is also evidenced by the consistent occurrence of the frost sensitive epiphyllous fungi, *Trichopeltinites*, from the Maastrichtian and into the Danian (Askin, 1992). This concurs with other southern mid to high latitude sections that record some change during the K-T transition, but no major ecological trauma on land. However, fossil wood collected from across the K-T section on Seymour Island record a 4-6 ‰ shift in both the standardised $\delta^{13}C$ signature of the bulk wood organic matter and the $\delta^{13}C$ values for specific wood compounds (see Figure 3 in Poole and van Bergen, 2006). This provides evidence for a dramatic alteration in the $\delta^{13}C$ of the atmosphere that might be linked to increased volcanism taking place at this time.

Therefore, whatever the cause of the massive extinction event in the mid and low latitudes, it exerted no instantaneous extinction effect across the southern high latitudes. Changes appear to have been relatively gradual in Antarctica when compared with the sudden changes that typify K-T sections to the north and east (i.e. New Zealand).

Case Study: Late Santonian Table Nunatak Flora – evidence in charcoal

Table Nunatak, located in the Larsen Basin (Figure 7.10), is 1 km long and 400 m wide with a total thickness of 62 m and is situated just east of Kenyon Peninsula, which lay at a palaeolatitude of about 65° S during the late Santonian. The Table Nunatak Formation includes three sedimentary facies: Facies A consisting of fine- to very fine-grained, and thick- to very thick-bedded sandstone; Facies B consisting of a rich plant debris, fine- to very fine-grained sandstone and siltstone, and Facies C a massive bioturbated mudstone (Hathway *et al.*, 1998). The charcoal layers can be found in the 0.1–0.5 cm-thick Facies B and contain plant debris consisting of wood fragments up to 10 cm long, small twigs, leaves, fruits, seeds and rare flowers (Figure 7.9D–G).

During the late Santonian, the Larsen Basin was still tectonically active with the Antarctic Peninsula volcanically active (Hathway, 2000). The highly unstable conditions of the region would have included periodic wildfires, which were probably responsible for the charring of plant parts. This charcoalification has rendered the plant tissues three-dimensional, rigid and resistant, so preventing collapse of the delicate tissues after burial. Even so, charcoalified material can be fragile and thus finding intact specimens suggests that they have probably not undergone any significant transport prior to deposition. The flora represents an interesting snapshot in time whereby a vegetation growing in a warm temperate seasonal environment was subjected to a period of wildfire (Hathway *et al.*, 1998).

Charcoalified plant remains were then washed into the marine environment and accumulated in shallow marine sands (Hathway *et al.*, 1998). Subsequent (minimal) abrading and sorting has probably resulted in the loss of some size classes, but has also rendered it difficult to determine with certainty the systematic identity of the remaining fossils (Eklund, Cantrill and Francis, 2004). The assemblage, albeit incomplete, comprises structures from lycopods (megaspores), ferns (megaspores, circinate rachides), conifers (wood, leaves, leafy shoots, pollen cones, microsporophylls, ovulate cone scales, seeds) and angiosperms (leaves, flowers, fruits, seeds).

Lycophytes are relatively rare in the Table Nunatak assemblage and are represented by four types of isolated megaspore, possibly originating from isoetalean or sellaginellalean affinities. Megaspores and fragments of circinate fronds signify the presence of ferns of possible marsileacean affinity, which are thought to have grown in waterlogged or aquatic conditions. Conifers dominate the assemblage with pollen cone axes and microsporophylls with two pollen sacs suggesting the presence of possible Podocarpaceae and/or Taxodiaceae. An ovulate bract scale complex also suggests the presence of Araucariaceae; leaves suggest the presence of Podocarpaceae and Taxodiaceae/Cupressaceae/Cheirolepidiaceae, whereas leafy shoot remains suggest affinities with Podocarpaceae/Taxodiaceae/Araucariaceae. Wood fragments also indicate the importance of Podocarpaceae in this assemblage, but, surprisingly, no araucarian wood has been identified to date. Interestingly, there is an abundance of the *Phyllocladoxylon* wood type, which is relatively rare in the Lower Cretaceous strata but becomes more important upsequence, evidencing the increasing importance of podocarps (Cantrill and Poole, 2005). Angiosperm remains are also abundant in the Table Nunatak assemblage, and it is clear that they formed an important component of the vegetation. Both reproductive structures and seven different leaf morphotypes are represented. The leaf material is fragmentary and has been broken into size classes similar to that of the conifers. This indicates strong taphonomic bias at work, which results in a bias against angiosperms both in terms of diversity and abundance. The fruit and seed flora, however, suggest angiosperms were represented in greater diversity than is suggested by the leaf flora alone. The few flowers described indicate the presence of Magnoliids and eudicots within the flora – an interpretation supported by leaf floras from the Larsen Basin to the north where Coniacian and Santonian sediments, which were quickly deposited and less strongly transported, yield a high angiosperm diversity (60–70%) and abundance (Hayes, 1999).

The flowers in the Table Nunatak assemblage include a high percentage of flowers with cup-shaped hypanthium and an epigynous to perigynous organisation similar to those found in contemporary sites in the Northern Hemisphere (Friis, 1984; Herendeen *et al.*, 1999; Takahashi, Crane and Ando, 1999a, 1999b) suggesting that similar evolutionary pressures were operating in both hemispheres during the Cretaceous (Eklund, 2003). The systematic affinites of the eleven different flower types uncovered are difficult to determine with certainty but lie within the magnoliids and eudicots. Three flowers may be related to extant Siparunaceae, Winteraceae and Myrtaceae. These taxa are clearly different from those in the Northern Hemisphere which supports previous indications that a distinctive austral flora had been established in the Southern Hemisphere by the Late Cretaceous (Eklund, 2003; Cantrill and Poole, 2002; Drinnan and Crane, 1990). The deep floral cups seen in most of the flowers uncovered to date may have evolved as a means of protecting the reproductive organs against catastrophic events such as wildfires (Eklund, 2003) – a strategy known from extant plants such as the Myrtaceae (e.g. Judd, 1993). Unfortunately, no pollen grains have been found adhering to the flowers indicating that either the pollen had been removed by abrasion and/or the plants were largely insect-pollinated and were thus low pollen producers. Several flowers show additional evidence for insect pollination: highly specialised flowers, the occurrence of nectaries and possible oil cells. Even though no insects have been found the abundance of coprolites in the assemblage provides indirect evidence for the presence of insects in and among the plants.

With their flowers and deciduous habit the angiosperms had transformed Antarctica during the latest Cretaceous into a world of riches both in terms of species and in colour, but this was relatively short-lived since the closure of the Mesozoic was a time of global devastation associated with one of the greatest mass extinctions to have taken place on Earth (Figure 4.3).

Summary

- As Gondwana was breaking up a second wave of angiosperm migration took place. They radiated out from northern Gondwana during the Cenomanian–Turonian through the South American–Antarctic Peninsula corridor. The reduced seasonal contrast coupled with the weaker equator-to-pole thermal gradient, resulting in the breakdown of latitudinal climatic barriers, allowed latitudinal belts of vegetation to expand poleward during one of the most equable periods in Earth's history. The migration was probably aided by the high levels of ecological disturbance and fortuitous combinations of new traits exhibited by these angiosperms. Angiosperm diversity is thought to have peaked in the Turonian–Coniacian as the vegetation was transformed from gymnosperm-fern dominated to angiosperm dominated by the end of the Cretaceous.
- From the Turonian to the Coniacian, the extensive podocarp-rich vegetation underwent a rapid modernisation. These forests diversified with the appearance of new taxa belonging to both conifers (e.g. the podocarps *Lagarostrobus*, *Dacrydium* and *Dacrycarpus*) and angiosperms (e.g. Casuarinaceae, Proteaceae, *Nothofagus*) – elements characteristic of modern southern ecosystems today. The Antarctic region remained a locus for evolutionary innovation until the end of the Cretaceous.
- In East Gondwana (East Antarctica and southeastern Australia) during the Turonian–Santonian the vegetation comprised Araucariaceae and Podocarpaceae conifer forests with fern and fern-allies forming open understoreys and heathland. Typical Southern Hemisphere angiosperm taxa such as Proteaceae and *Nothofagus* gradually encroached into this habitat. The peak in angiosperm abundance predated a similar peak recorded in West Antarctica floras during the Campanian–Maastrichtian.
- The record of vegetation change is most detailed from West Gondwana (i.e. West Antarctica). During the Santonian, angiosperm numbers declined briefly, allowing an initial recovery of fern numbers, even though the once dominant podocarp-*Araucaria* forest associations continued to decline. As the angiosperms recovered during the Coniacian to Maastrichtian, new niches were opened up allowing a recovery in fern abundance and diversity, even though they never attained their Early Cretaceous levels. Non-coniferous seed plants (e.g. *Ginkgo*, bennettites) and hepatophytes became minor constituents of the flora and increasingly rare. The last record of bennettites dates back to the Coniacian.

- Following the period of peak warmth in the Turonian, in West Gondwana the vegetation underwent a rapid modernisation. Primitive angiosperm morphotypes were replaced by modern types belonging to, for example, the Magnoliales, Cunoniaceae, Lauraceae, Sterculiaceae, Elaeocarpaceae and key taxa Nothofagaceae, Atherospermataceae, Monimiaceae, Proteaceae, Myrtaceae and Winteraceae. Angiosperms dominated the floras (~70%) with ferns, conifers and other gymnosperms making up the remaining 30%. Abundant wood testifies to the increasing arboreal status, which in turn led to opening of new niches that were exploited by ferns that consequently underwent a rediversification.

- The Williams Point Flora from Livingston Island provides an insight into one of the earliest angiosperm-influenced vegetations growing in Antarctica. Originally described as Triassic, radiometric dating and characteristics of the flora suggest an early Campanian age. Volcanic activity led to the preservation of various stages of vegetation succession: araucarian and podocarps dominated the forests, growing under a mild, wet climate, but angiosperms also contributed to the canopy as many had by now attained arboreal stature. Taxa represented include a mix of modern taxa including monimiaceous- and cunoniaceous types, as well as more primitive forms. Ferns still formed an important element of the understorey along with sphenopsids, bushy conifers and shrubby angiosperms (e.g. *Clavatipollenites*).

- In the latest Cretaceous (Campanian–Maastrichtian) the climate had cooled relative to the Turonian and became more oceanic in nature. Diversification continued as new elements appeared, such as the angiospermous Gunneraceae, Aquifoliaceae, Loranthaceae, and coniferous podocarp elements such as *Dacrydium*. *Nothofagus* continued to increase in importance as the Antarctica Peninsula played an integral role in its radiation. By the late Campanian all four modern *Nothofagus* subgenera had evolved and migrated into South America and across to Australia. In the South Shetland Islands differences in environmental conditions may account for differences seen in the composition of the vegetation that are either spatially or temporally separated. Areas dominated by ferns with sparse mixed conifer-*Nothofagus* woodlands contrast with the angiosperm (especially *Nothofagus*)-fern dominated vegetation, the mixed broad-leafed-evergreen (often conifer) woodlands with abundant ferns, or the angiosperm-rich rainforests occupying higher altitudes. These vegetation associations have been likened to the Valdivian Rainforest communities of southern South America, which would go on to characterise the early Cenozoic.

- Global catastrophe is recorded in a thin clay layer worldwide and marks the Cretaceous–Cenozoic boundary. In Antarctica, the boundary sequence is only evident on Seymour Island and marks both the final stage of Gondwana break-up and the end-Cretaceous mass extinction event. Although the boundary is marked by an iridium anomaly, the transitional nature of the fossil faunas and floras coupled with a relatively gradual change in the megafauna, complicate the precise placement of the boundary and the extent of the biotic turnover at the end of the Cretaceous. The seeming lack of evidence for a sudden devastating extinction makes the placement of the K-T boundary uncertain, yet this apparent lack of evidence for extinction concurs with other southern mid- to high latitude sections that record some ecological change during the K-T transition but no major extinction.

References

Accordi, I. A., Dutra, T. L. and Müller, J. (1996). Fisionomía foliar e parâmetros ambientais: um exemplo a partir das florestas úmidas do sul do Brasil. *Resumo do 47th Congresso Nacional de Botânica, 1996, Nova Friburgo*. Rio de Janeiro: Sociedade Botânica do Brasil, pp. 496.

Alves, L. S. D. R., Guerra-Sommer, M. and Dutra, T. (2005). Paleobotany and Paleoclimate. Part 1: Growth rings in fossil wood and paleoclimate. Part 2: Leaf assemblages (taphonomy, paleoclimatology and paleogeography. In *Applied Stratigraphy*, ed. E. A. M. Koutsoukos. Netherlands: Springer, pp. 179–202.

Alvin, K. L., Barnard, P. D. W., Harris, T. M., *et al.* (1967). Gymnospermopsida. In *The Fossil Record, Geological Society of London Special Publication*, **2**, eds W. B. Harland, C. H. Holland, M. R. House, N. F. Hughes, A. B. Reynolds, M. J. S. Rudwick, G. E. Satterthwaite, L. B. H. Tarlo and E. C. Willey. London: The Geological Society, pp. 247–267.

Angiosperm Phylogeny Group. (2009). An update of the Angiosperm Phylogeny Group classification for the orders and families of flowering plants: APG III. *Botanical Journal of the Linnean Society*, **161**, 105–121.

Archangelsky, S. and Taylor, T. N. (1993). The ultrastructure of *in situ Clavatipollenites* pollen from the Early Cretaceous of Patagonia. *American Journal of Botany*, **80**, 879–885.

Archangelsky, S. and Villar de Seoane, L. (1998). Estudios palinológicos de la formación Baqueró (Cretácico), Provincia de Santa Cruz, Argentina VIII. *Ameghiniana*, **35**, 7–19.

Archangelsky, S., Barreda, V., Passalia, M. G., *et al.* (2009). Early angiosperm diversification: evidence from southern South America. *Cretaceous Research*, **30**, 1073–1082.

Arthur, M. A., Dean, W. E. and Pratt, L. M. (1988). Geochemical and climatic effects of increased marine organic carbon burial at the Cenomanian–Turonian boundary. *Nature*, **335**, 714–717.

Askin, R. A. (1983). Campanian palynomorphs from James Ross and Vega islands, Antarctic Peninsula. *Antarctic Journal of the United States*, **18(5)**, 63–64.

Askin, R. A. (1988a). Campanian to Paleocene palynological successions of Seymour and adjacent islands, northwestern Antarctic Peninsula. In *Geology and Paleontology of Seymour Island, Antarctic Peninsula, Geological Society of America Memoir*, **169**, eds R. M. Feldmann and M. O. Woodburne. Boulder: The Geological Society of America, pp. 131–153.

Askin, R. A. (1988b). The palynological record across the Cretaceous-Tertiary transition on Seymour Island, Antarctica. In *Geology and Paleontology of Seymour Island, Antarctic Peninsula, Geological Society of America Memoir*, **169**, eds R. M. Feldmann and M. O. Woodburne. Boulder: The Geological Society of America, pp. 155–162.

Askin, R. A. (1989). Endemism and heterochroneity in the Late Cretaceous (Campanian) to Paleocene palynofloras of Seymour Island, Antarctica: implications for origins, dispersal and palaeoclimates of southern floras. In *Origins and Evolution of the Antarctic Biota, Geological Society of London Special Publication*, **47**, ed. J. A. Crame. London: The Geological Society, pp. 107–119.

Askin, R. A. (1990a). Campanian to Paleocene spore pollen assemblages of Seymour Island, Antarctica. *Review of Palaeobotany and Palynology*, **65**, 105–113.

Askin, R. A. (1990b). Cryptogam spores from the upper Campanian and Maastrichtian of Seymour Island, Antarctica. *Micropaleontology*, **36**, 141–156.

Askin, R. A. (1992). Late Cretaceous–early Tertiary Antarctic outcrop evidence for past vegetation and climates. In *The Antarctic Paleoenvironment: A Perspective on Global*

Change. Part 1, Antarctic Research Series, **56**, eds J. P. Kennett and D. A. Warnke. Washington: American Geophysical Union, pp. 61–73.

Askin, R. A. (1994). Monosulcate angiosperm pollen from the López de Bertodano Formation (upper Campanian–Maastrichtian–Danian) of Seymour Island, Antarctica. *Review of Palaeobotany and Palynology*, **81**, 151–164.

Askin, R. A. (1997). Eocene–?Earliest Oligocene terrestrial palynology of Seymour Island, Antarctica. In *The Antarctic Region: Geological Evolution and Processes*. ed. C. A. Ricci. Siena: Terra Antartica, pp. 993–996.

Askin, R. A. and Elliot, D. H. (1982). Geologic implications of recycled Permian and Triassic palynomorphs in Tertiary rocks of Seymour Island, Antarctic Peninsula. *Geology*, **10**, 547–551.

Askin, R. A. and Jacobson, S. R. (1989). Total Organic Carbon content and Rock Eval pyrolysis on outcrop samples across the Cretaceous/Tertiary boundary, Seymour Island, Antarctica. *Antarctic Journal of the United States*, **23(5)**, 37–39.

Askin, R. A. and Jacobson, S. R. (1996). Palynological change across the Cretaceous–Tertiary boundary on Seymour Island, Antarctica: environmental and depositional factors. In *Cretaceous–Tertiary Mass Extinctions: Biotic and Environmental Changes*, eds N. MacLeod and G. Keller. New York: Norton, pp. 7–25.

Askin, R. A. and Spicer, R. A. (1995). The Late Cretaceous and Cenozoic history of vegetation and climate at northern and southern high latitudes: a comparison. In *Studies in Geophysics. Effects of Past Global Change on Life*, eds Panel on Effects of Past Global Change on Life, National Research Council. Washington: National Academy Press, pp. 156–173.

Baldoni, A. M. (1987). Estudios palinológicos de los niveles basales de la Formación Santa Marta. Cretácico superior de Bahía Brandy (Isla James Ross, Antártida). *Paleobotánica Latinoamericana*, **8**, 5.

Baldoni, A. M. and Medina, F. (1989). Fauna y microflora del Cretácico, en Bahía Brandy, Isla James Ross, Antártida. *Serie Científica Instituto Antártico Chileno*, **39**, 43–58.

Banerji, J. and Lemoigne, Y. (1987). Significant additions to the Upper Triassic flora of Williams Point, Livingston Island, South Shetlands (Antarctica). *Geobios*, **20**, 469–487.

Banerji, J., Lemoigne, Y. and Torres, T. (1987). Nuevos registros de la paleoflora del Triásico Superior de Punta Williams, Isla Livingston, Shetland del Sur (Antártica). *Serie Científica Instituto Antártico Chileno*, **36**, 35–58.

Barale, G., Philippe, M., Torres, T. and Thévenard, F. (1995). Reappraisal of the Triassic flora from Williams Point, Livingston Island (South Shetland Islands, Antarctica): systematical, lithostratigraphical and palaeogeographical implications. *Serie Científica Instituto Antártico Chileno*, **45**, 9–38.

Barclay, R. S., McElwain, J. C. and Sageman, B. B. (2010). Carbon sequestration activated by a volcanic CO_2 pulse during Ocean Anoxic Event 2. *Nature Geoscience*, **3**, 205–208.

Barreda, V. D., Palamarczuk, S. and Medina, F. (1999). Palinología de la formación Hidden Lake (Coniaciano–Santoniano), Isla James Ross, Antártida. *Revista Española de Micropaleontología*, **31**, 53–72.

Barrera, E. and Savin, S. M. (1999). Evolution of late Campanian–Maastrichtian marine climates and oceans. In *Evolution of the Cretaceous Ocean-Climate System, Geological Society of America Special Paper*, **332**, eds E. Barrera and C. C. Johnson. Boulder: The Geological Society of America, pp. 245–282.

Barron, E. J. (1983). A warm equable Cretaceous: the nature of the problem. *Earth-Science Reviews*, **19**, 305–338.

Berendse, F. and Scheffer, M. (2009). The angiosperm radiation revisited, an ecological explanation for Darwin's 'abominable mystery'. *Ecological Letters*, **12**, 865–872.

Birkenmajer, K. (1981). Raised marine features and glacial history in the vicinity of H. Arctowski Station, King George Island (South Shetland Islands, West Antarctica). *Bulletin de l'Academie Polonaise des Sciences, Series des Sciences de la Terre*, **29**, 109–117.

Birkenmajer, K. and Zastawniak, E. (1986). Plant remains of the Dufayel Island Group (early Tertiary?), King George Island, South Shetland Islands (West Antarctic). *Acta Palaeobotanica*, **26**, 33–53.

Birkenmajer, K. and Zastawniak, E. (1989a). Late Cretaceous–early Neogene vegetation history of the Antarctic Peninsula sector. Gondwana break-up and Tertiary glaciations. *Bulletin of the Polish Academy of Sciences, Earth Sciences*, **37**, 63–88.

Birkenmajer, K. and Zastawniak, E. (1989b). Late Cretaceous–early Tertiary floras of King George Island, West Antarctica: their stratigraphic distribution and palaeoclimatic significance. In *Origins and Evolution of the Antarctic Biota, Geological Society of London Special Publication*, **47**, ed. J. A. Crame. London: The Geological Society, pp. 227–240.

Birkenmajer, K., Narębski, W., Nicoletti, M. and Petrucciani, C. (1983). Late Cretaceous through late Oligocene K-Ar ages of the King George Island Supergroup volcanics, South Shetland Islands (West Antarctica). *Bulletin de l'Académie Polonaise des Sciences, Série des Sciences de la Terre*, **30**, 133–143.

Bralower, T., Eccles, L., Kutz, J., *et al.* (2010). Grain size of Cretaceous–Paleogene boundary sediments from Chicxulub to the open ocean: Implications for interpretation of the mass extinction event. *Geology*, **38**, 199–202.

Brizuela, R. R., Marenssi, S., Barreda, V. and Santillana, S. (2007). Palynofacies approach across the Cretaceous–Paleogene boundary in Marambio (Seymour) Island, Antarctic Peninsula. *Revista de la Asociación Geológica Argentina*, **62**, 236–241.

Brooks, R. R., Hoek, P. L., Reeves, R. D. and Strong, C. P. (1986). Geochemical delineation of the Cretaceous/Tertiary boundary in some New Zealand rock sequences. *New Zealand Journal of Geology and Geophysics*, **29**, 1–8.

Browne, J. R. and Pirrie, D. (1995). Sediment dispersal patterns in a deep marine back-arc basin, James Ross Island, Antarctica; evidence from petrographic studies. In *Characterisation of Deep Marine Clastic Systems, Geological Society of London Special Publication*, **94**, eds A. J. Hartley and D. J. Prosser. London: The Geological Society, pp. 137–154.

Buffetaut, E. (2004). Polar dinosaurs and the question of dinosaur extinction: a brief review. *Palaeogeography, Palaeoecology, Palaeoclimatology*, **214**, 225–231.

Burnham, R. J. (1989). Relationships between standing vegetation and leaf litter in a paratropical forest: implications for paleobotany. *Review of Palaeobotany and Palynology*, **58**, 5–32.

Cantrill, D. J. (1997). The pteridophyte *Ashicaulis livingstonensis* (Osmundaceae) from the Upper Cretaceous of Williams Point, Livingston Island, Antarctica. *New Zealand Journal of Geology and Geophysics*, **40**, 315–323.

Cantrill, D. J. and Nichols, G. J. (1996). Taxonomy and palaeoecology of Early Cretaceous (late Albian) angiosperm leaves from Alexander Island, Antarctica. *Review of Palaeobotany and Palynology*, **92**, 1–28.

Cantrill, D. J. and Poole, I. (2002). Cretaceous patterns of floristic change in the Antarctic Peninsula. In *Palaeobiogeography and Biodiversity Change: the Ordivician and Mesozoic–Cenozoic Radiations, Geological Society of London Special Publication*, **194**, eds J. A. Crame and A. W. Owen. London: The Geological Society, pp. 141–152.

Cantrill, D. J. and Poole, I. (2005). Taxonomic turnover and abundance in Cretaceous to Tertiary wood floras of Antarctica: implications for changes in forest ecology. *Palaeogeography, Palaeoecology, Palaeoclimatology*, **215**, 205–219.

Cantrill, D. J., Wanntorp, L. and Drinnan, A. N. (2011). Mesofossil flora from the Late Cretaceous of New Zealand. *Cretaceous Research*, **32**, 164–173.

Cao, L. (1990). Discovery of a Late Cretaceous palynoflora from Fildes Peninsula, King George Island, Antarctica and its significance. *Acta Palaeontologica Sinica*, **29**, 140–146 (In Chinese, with English summary).

Cao, L. (1992). Late Cretaceous and Eocene palynofloras from Fildes Peninsula, King George Island (South Shetland Islands), Antarctica. In *Recent Progress in Antarctic Earth science*, eds Y. Yoshida, K. Kaminuma and K. Shiraishi. Tokyo: Terra Scientific Publishing Company, pp. 363–369.

Cao, L. (1994). Late Cretaceous palynoflora in King George Island of Antarctica with reference to its palaeoclimatic significance. In *Stratigraphy and Palaeontology of Fildes Peninsula, King George Island, Antarctica, State Antarctic Committee Monograph*, **3**, ed. Y. Shen. Beijing: Science Press, pp. 51–83 (In Chinese, English abstract and description of new species).

Case, J. A. (1988). Paleogene floras of Seymour Island, Antarctic Peninsula. In *Geology and Paleontology of Seymour Island, Antarctic Peninsula. Memoir of the Geological Society of America*, **169**, eds R. M. Feldmann and M. O. Woodburne. Boulder: The Geological Society of America, pp. 523–530.

Chambers, K. L., Poinar, G. and Buckley, R. (2010). *Tropidogyne*, a new genus of Early Cretaceous eudicots (Angiospermae) from Burmese amber. *Novon: A Journal for Botanical Nomenclature*, **20**, 23–29.

Chapman, J. L. and Smellie, J. L. (1992). Cretaceous fossil wood and palynomorphs from Williams Point, Livingston Island, Antarctic Peninsula. *Review of Palaeobotany and Palynology*, **74**, 163–192.

Chattaway, M. M. (1949). The development of tyloses and secretion of gum in heartwood formation. *Australian Journal of Scientific Research*, **B2**, 227–240.

Crabtree, D. R. (1987). Angiosperms of the northern Rocky Mountains: Albian to Campanian (Cretaceous) megafossil floras. *Annals of the Missouri Botanical Garden*, **74**, 707–747.

Crame, J. A. (1994). Evolutionary history of Antarctica. In *Antarctic Science*, ed. G. Hempel. Berlin, Heidelberg: Springer–Verlag, pp. 188–214.

Crame, J. A., Pirrie, D., Riding, J. B. and Thomson, M. R. A. (1991). Campanian–Maastrichtian (Cretaceous) stratigraphy of James Ross Island area, Antarctica. *Journal of the Geological Society of London*, **148**, 1125–1140.

Crame, J. A., Lomas, S. A., Pirrie, D. and Luther, A. (1996). Late Cretaceous extinction patterns in Antarctica. *Journal of the Geological Society of London*, **153**, 503–506.

Crame, J. A., McArthur, J. M., Pirrie, D. and Riding, J. B. (1999). Strontium isotope correlation of the basal Maastrichtian Stage in Antarctica to the European and US biostratigraphic schemes. *Journal of the Geological Society of London*, **156**, 957–964.

Crame, J. A., Francis, J. E., Cantrill, D. J. and Pirrie, D. (2004). Maastrichtian stratigraphy of Antarctica. *Cretaceous Research*, **25**, 411–423.

Crame, J. A., Pirrie, D. and Riding, J. B. (2006). Mid-Cretaceous stratigraphy of the James Ross Basin, Antarctica. In *Cretaceous–Tertiary High-Latitude Palaeoenvironments, James Ross Basin, Antarctica, Geological Society of London Special Publication*, **258**, eds J. E. Francis, D. Pirrie and J. A. Crame. London: The Geological Society, pp. 7–19.

Crane, P. R. and Lidgard, S. (1990). Angiosperm radiation and patterns of Cretaceous palynological diversity. In *Major Evolutionary Radiations*, eds P. D. Taylor and G. P. Larwood. Oxford: Oxford University Press, pp. 377–407.

Cripps, J. A., Widdowson, M., Spicer, R. A. and Jolley, D. W. (2005). Coastal ecosystem responses to late stage Deccan Trap volcanism: the post K–T (Danian) palynofacies of Mumbai (Bombay), west India. *Palaeogeography, Palaeoclimatology, Palaeoecology*, **216**, 303–332.

Crowley, T. J. and Zachos, J. C. (2000). Comparison of zonal temperature profiles for past warm time periods. In *Warm Climates in Earth History*, eds B. T. Huber, K. G. Macleod and S. L. Wing. Cambridge: Cambridge University Press, pp. 50–76.

Cúneo, R. and Gandolfo, M. A. (2005). Angiosperm leaves from the Kachaike Formation, Lower Cretaceous of Patagonia, Argentina. *Review of Palaeobotany and Palynology*, **136**, 29–47.

Daniel, I. L. (1989). Taxonomic investigation of elements from Early Cretaceous megaflora from the middle Clarence Valley, New Zealand. Ph.D. thesis, University of Canterbury.

Daniel, I. L., Lovis, J. D. and Reay, M. B. (1988). A brief introductory report on the mid-Cretaceous megaflora of the Clarence Valley, New Zealand. In *Proceedings of the 3rd International Organisation of Palaeobotany Conference*, eds J. G. Douglas and D. C. Christophel. Melbourne: A-Z Printers, pp. 27–28.

DeConto, R. M. and Pollard, D. (2003). Rapid Cenozoic glaciation of Antarctica induced by declining atmospheric CO_2. *Nature*, **421**, 245–249.

DeConto, R. M., Hay, W. W., Thompson, S. L. and Bergengren, J. (1999). Late Cretaceous climate and vegetation interactions: cold continental interior paradox. In *Evolution of the Cretaceous-Ocean Climate System, Geological Society of America Special Paper*, **332**, eds E. Barrera and C. C. Johnson. Boulder: The Geological Society of America, pp. 391–406.

Dettmann, M. E. (1986). Significance of the Cretaceous–Tertiary spore genus *Cyatheacidites* in tracing the origin and migration of *Lophosoria* (Filicopsida). *Special Papers in Paleontology*, **35**, 63–94.

Dettmann, M. E. (1989). Antarctica: Cretaceous cradle of austral temperate rainforests? In *Origins and Evolution of the Antarctic Biota, Geologial Society of London Special Publication*, **47**, ed. J. A. Crame. London: The Geological Society, pp. 89–105.

Dettmann, M. E. (1994). Cretaceous vegetation: the microfossil record. In *History of the Australian Vegetation: Cretaceous to Recent*, ed. R. S. Hill. Cambridge: Cambridge University Press, pp. 143–170.

Dettmann, M. E. and Jarzen D. M. (1990). The Antarctic/Australian riftvalley: Late Cretaceous cradle of northeastern Australasian relicts? *Review of Palaeobotany and Palynology*, **65**, 131–144.

Dettmann, M. E. and Jarzen D. M, (1991). Pollen evidence for Late Cretaceous differentiation of Proteaceae in southern polar forests. *Canadian Journal of Botany*, **69**, 901–906.

Dettmann, M. E. and Thomson, M. R. A. (1987). Cretaceous palynomorphs from the James Ross Island area, Antarctica—a pilot study. *British Antarctic Survey Bulletin*, **77**, 13–59.

Dettmann, M. E., Pocknall, D. T., Romero, E. J. and Zamaloa, M. C. (1990). *Nothofagidites* Erdtman ex Potonié, 1960; a catalogue of species with notes on the paleogeographic distribution of *Nothofagus* Bl. (Southern Beech). *New Zealand Geological Survey Paleontological Bulletin*, **60**, 1–79.

Dettmann, M. E., Molnar, R. E., Douglas, J. G., *et al.* (1992). Australian Cretaceous terrestrial faunas and floras: biostratigraphic and biogeographic implications. *Cretaceous Research*, **13**, 207–262.

Dettmann, M. E., Clifford, H. T. and Peters, M. (2009). *Lovellea wintonensis* gen. et sp nov.– Early Cretaceous (late Albian), anatomically preserved, angiospermous flowers and fruits from the Winton Formation, western Queensland, Australia. *Cretaceous Research*, **30**, 339–355.

Dingle, R. V. (1999). Walvis Ridge barrier: its influence on palaeoenvironments and source rock generation deduced from ostracod distributions in the early South Atlantic Ocean. In *Hydrocarbon Habitats of the South Antlantic, Geological Society of London Special Publication*, **153**, eds N. Cameron, R. H. Bate and R. Clure. London: The Geological Society, pp. 293–302.

Dingle, R. V. and Lavelle, M. (1998). Late Cretaceous–Cenozoic climatic variations of the northern Antarctic Peninsula: new geochemical evidence and review. *Palaeogeography, Palaeoclimatology, Palaeoecology*, **141**, 215–232.

Dingle, R. V. and Lavelle, M. (2000). Antarctic Peninsula Late Cretaceous–Early Cenozoic palaeoenvironments and Gondwana palaeogeographies. *Journal of African Earth Sciences*, **31**, 91–105.

Ditchfield, P. W., Marshall, J. D. and Pirrie, D. (1994). High latitude temperature variation: new data from the Tithonian to Eocene of James Ross Island, Antarctica. *Palaeogeography, Palaeoclimatology, Palaeoecology*, **107**, 79–101.

Douglas, J. G. (1994). Cretaceous vegetation: the macrofossil record. In *History of the Australian Vegetation: Cretaceous to Recent*, ed. R. S. Hill. Cambridge: Cambridge University Press, pp. 171–188.

Douglas, J. G. and Williams, G. E. (1982). Southern polar forests: the Early Cretaceous floras of Victoria and their palaeoclimatic significance. *Palaeogeography, Palaeoclimatology, Palaeoecology*, **39**, 171–185.

Doyle, J. A. (2001). Significance of molecular phylogenetic analyses for paleobotanical investigations on the origin of angiosperms. *The Palaeobotanist*, **50**, 167–188.

Doyle, P. (1990). New records of dimitobelid belemnites from the Cretaceous of James Ross Island, Antarctica. *Alcheringa*, **14**, 159–175.

Drinnan, A. N. and Crane, P. R. (1990). Cretaceous paleobotany and its bearing on the biogeography of Austral angiosperms. In *Antarctic Paleobiology: its Role in the Reconstruction of Gondwana*, eds T. N. Taylor and E. L. Taylor. New York: Springer-Verlag, pp. 192–219.

Dusén, P. C. H. (1899). Über die tertiäre Flora der Magellans länder. In *Wissenschaftliche Ergebnisse der Schwedischen Expeditionen nach den Magellansländern, 1895–1897, 1(4)*, ed. O. Nordenskjöld. Stockholm: P. A. Norstedt & Söner, pp. 87–107.

Dutra, T. L. (1997). Composicão e história da vegetacão do Cretáceo e Terciário da Ilha Rei George, Península Antártica. Ph.D. thesis, Universidade Federal do Rio Grande do Sul.

Dutra, T. L. (2000). *Nothofagus* do noroeste da Península Antártica. I. Cretáceo Superior. *Revista Universidade Guarulhos Geociências*, **5**, 102–106.

Dutra, T. L. (2001). Paleoflora da ilha 25 de Mayo, Península Antártica: contribuição à paleogeografia, paleoclima e para a evolução de *Nothofagus*. In *Proceedings XI Simposio Argentino de Paleobotánica y Palinología, Asociación Paleontológica Argentina Publicación Especial*, **8**. Buenos Aires: APA, pp, 29–37.

Dutra, T. L. and Batten, D. (2000). Upper Cretaceous floras of King George Island, West Antarctica, and their palaeoenvironmental and phytogeographic implications. *Cretaceous Research*, **21**, 181–209.

Dutra, T. L., Leipnitz, B., Faccini, U. F. and Lindenmayer, Z. (1996). A non-marine Upper Cretaceous interval in West Antarctica (King George Island, northern Antarctic Peninsula). *SAMC (South Atlantic Mesozoic Correlations, IGCP Project No. 381) News*, **5**, 21–22.

Dutra, T., Hansen, M. A. F. and Fleck, A. (1998). New evidence of wet and mild climate in northern Antarctic Peninsula at the end of the Cretaceous. In *Proceedings of the 3rd Annual Conference of IGCP Project 381, SAMC III Chubut, Argentina*, pp. 10–14.

Dutton, A., Huber, B. T., Lohmann, K. C. and Zinsmeister, W. J. (2007). High-resolution stable isotope profiles of a dimitobelid belemnite: Implications for paleodepth habitat and late Maastrichtian climate seasonality. *Palaios*, **22**, 642–650.

Eklund, H. (2003). First Cretaceous flowers from Antarctica. *Review of Palaeobotany and Palynology*, **127**, 187–217.

Eklund, H., Cantrill, D. J. and Francis, J. E. (2004). Late Cretaceous plant mesofossils from Table Nunatak, Antarctica. *Cretaceous Research*, **25**, 211–228.

Elliot, D. H. (1988). Tectonic setting and evolution of the James Ross Basin, northern Antarctic Peninsula. In *Geology and Paleontology of Seymour Island, Antarctic Peninsula. Geological Society of America Memoir*, **169**, eds R. M. Feldmann and M. O. Woodburne. Boulder: The Geological Society of America, pp. 541–555.

Elliot, D. H. and Trautman, T. A. (1982). Lower Tertiary strata on Seymour Island, Antarctic Peninsula. In *Antarctic Geoscience*, ed. C. Craddock. Madison: University of Wisconsin Press, pp. 287–297.

Elliot, D. H., Askin, R. A., Kyte, F. T. and Zinsmeister, W. J. (1994). Iridium and dinocysts at the Cretaceous–Tertiary boundary on Seymour Island, Antarctica: implications for the K-T event. *Geology*, **22**, 675–678.

Endress, P. K. and Doyle, J. A. (2009). Reconstructing the ancestral angiosperm flower and its initial specializations. *American Journal of Botany*, **96**, 22–66.

Equiza, M. A., Day, M. E. and Jagels, R. (2006). Physiological responses of three deciduous conifers (*Metasequoia glyptostroboides*, *Taxodium distichum* and *Larix laricina*) to continuous light: adaptive implications for the early Tertiary polar summer. *Tree Physiology*, **26**, 353–364.

Falcon-Lang, H. J. (2000a). A method to distinguish between woods produced by evergreen and deciduous coniferopsids on the basis of growth ring anatomy: a new palaeo-ecological tool. *Palaeontology*, **43**, 785–793.

Falcon-Lang, H. J. (2000b). The relationship between leaf longevity and growth ring markedness in modern conifer woods and its implications for palaeoclimatic studies. *Palaeogeography, Palaeoecology, Palaeoclimatology*, **160**, 317–328.

Feild, T. S. and Arens, N. C. (2005). Form, function and environments of the early angiosperms: merging extant phylogeny and ecophysiology with fossils. *New Phytologist*, **166**, 383–408.

Feild, T. S., Arens, N. C., Doyle, J. A., Dawson, T. E. and Donoghue, M. J. (2004). Dark and disturbed: a new image of early angiosperm ecology. *Paleobiology*, **30**, 82–107.

Fensterseifer, H. C., Soliani, E., Hansen, M. A. F. and Troian, F. L. (1988). Geología e estratigrafía da associação de rochas do setor centro-norte da península Fildes, Ilha Rei George, Shetland do Sul, Antártica. *Serie Científica Instituto Antártico Chileno*, **38**, 29–43.

Ferguson, D. K., Hofmann, C.-C. and Denk, T. (1999). Taphonomy: field techniques in modern environments. In *Fossil Plants and Spores: Modern Techniques*, eds T. P. Jones and N. P. Rowe. London: The Geological Society, pp. 210–213.

Fluteau, F., Ramstein, G., Besse, J., Guiraud, R. and Masse, J. P. (2007). Impacts of palaeogeography and sea level changes on mid-Cretaceous climate. *Palaeogeography, Palaeoclimatology, Palaeoecology*, **247**, 357–381.

Frakes, L. A. (1999). Estimating the global thermal state from Cretaceous sea surface and continental temperature data. In *Evolution of the Cretaceous Ocean-Climate System*,

Geological Society of America Special Paper, **332**, eds E. Barrera and C. C. Johnson. Boulder: The Geological Society of America, pp. 49–57.

Frakes, L. A., Burger, D., Apthorpe M., *et al.* (1987). Australian Cretaceous shorelines, stage by stage. *Palaeogeography, Palaeoclimatology, Palaeoecology*, **59**, 31–48.

Francis, J. E. and Frakes, L. A. (1993). Cretaceous climates. In *Postgraduate Research Institute of Sedimentology, Sedimentology Review*, ed. V. P. Wright. Oxford: Blackwell Scientific Publications, pp.17–30.

Freeman, K. H. and Hayes, J. M. (1992). Fractionation of carbon isotopes by phytoplankton and estimates of ancient CO_2 levels. *Global Biogeochemical Cycles*, **6**, 185–198.

Friedman, W. E. (2009). The meaning of Darwin's "abominable mystery". *American Journal of Botany*, **96**, 5–21.

Friis, E. M. (1984). Preliminary report on Upper Cretaceous angiosperm reproductive organs from Sweden and their level of organization. *Annals of the Missouri Botanical Garden*, **71**, 403–418.

Friis, E. M. (1985). *Actinocalyx* gen. nov. sympetalous angiosperm flowers from the Upper Cretaceous of Southern Sweden. *Review of Palaeobotany and Palynology*, **45**, 171–483.

Friis, E. M. and Skarby, A. (1981). Structurally preserved angiosperm flowers from the Upper Cretaceous of southern Sweden. *Nature*, **291**, 485–486.

Friis, E. M., Pedersen, K. R. and Crane, P. R. (1999). Early angiosperm diversification: the diversity of pollen associated with angiosperm reproductive structures in Early Cretaceous floras from Portugal. *Annals of the Missouri Botanical Garden*, **86**, 259–296.

Friis, E. M., Pedersen, K. R. and Crane, P. R. (2006). Cretaceous angiosperm flowers: innovation and evolution in plant reproduction. *Palaeogeography, Palaeoclimatology, Palaeoecology*, **232**, 251–293.

Friis, E. M., Pedersen, K. R., von Balthazar, M., Grimm, G. W. and Crane, P. R. (2009). *Monetianthus mirus* gen. et sp nov., a nymphaealean flower from the Early Cretaceous of Portugal. *International Journal of Plant Science*, **170**, 1086–1101.

Friis, E. M., Pedersen, K. R. and Crane, P. R. (2010). Diversity in obscurity: Fossil flowers and the early history of angiosperms. *Philosophical Transactions of the Royal Society Series B*, **365**, 369–382.

Gallagher, S. J., Wagstaff, B. E., Baird, J. G., Wallace, M. V. and Li, C. L. (2008). Southern high latitude climate variability in the Late Cretaceous greenhouse world. *Global and Planetary Change*, **60**, 351–364.

Grunow, A. M., Kent, D. V. and Dalziel, I. W. D. (1991). New paleomagnetic data from Thurston Island: Implications for the tectonics of West Antarctica and Weddell Sea opening. *Journal of Geophysical Research*, **96**, 17935–17954.

Haq, B. U., Hardenbol, J. and Vail, P. R. (1987). Chronology of fluctuating sea levels since the Triassic. *Science*, **235**, 1156–1167.

Hart, M. B., Joshi, A. and Watkinson, M. P. (2001). Mid-late Cretaceous stratigraphy of the Cauvery Basin and the development of the Eastern Indian Ocean. *Journal of the Geological Society of India*, **58**, 217–229.

Harwood, D. H. (1988). Upper Cretaceous and Lower Paleocene diatom and dinoflagellate biostratigraphy of Seymour Island, eastern Antarctic Peninsula. In *Geology and Paleontology of Seymour Island, Antarctic Peninsula, Memoir of the Geological Society of America*, **169**, eds R. M. Feldmann and M. O. Woodburne. Boulder: The Geological Society of America, pp. 55–129.

Hathway, B. (2000). Continental rift to back-arc basin: Jurassic–Cretaceous stratigraphical and structural evolution of the Larsen Basin, Antarctic Peninsula. *Journal of the Geological Society of London*, **157**, 417–432.

Hathway, B., Macdonald, D. I. M., Riding, J. B. and Cantrill, D. J. (1998). Table Nunatak: a key outcrop of Upper Cretaceous shallow-marine strata in the southern Larsen Basin, Antarctic Peninsula. *Geological Magazine*, **135**, 519–535.

Hay, W. W., de Conto, R. M., Wold, C. N., *et al.* (1999). An alternative global Cretaceous palaeogeography. In *Evolution of the Cretaceous Ocean Climate System, Geological Society of America Special Papers*, **332**, eds E. Barrera and C. Johnson. Boulder: The Geological Society of America, pp. 1–47.

Hayes, P. A. (1999). Cretaceous angiosperm floras of Antarctica. Ph.D. thesis, University of Leeds.

Hayes, P. A., Francis, J. E., Cantrill, D. J. and Crame, J. A. (2006). Palaeoclimate analysis of the Late Cretaceous angiosperm leaf floras, James Ross Island, Antarctica. In *Cretaceous–Tertiary High-Latitude Palaeoenvironments, James Ross Island, Antarctic, Geological Society of London Special Publications*, **258**, eds J. E. Francis, D. Pirrie and J. A. Crame. London: The Geological Society, pp. 49–62.

Herendeen, P. S., Magallon-Puebla, S., Lupia, R., Crane, P. R. and Kobylinska, J. (1999). A preliminary conspectus of the Allon flora from the Late Cretaceous (Late Santonian) of central Georgia, USA. *Annals of the Missouri Botanical Garden*, **86**, 407–471.

Hickey, L. J. and Doyle, J. A. (1977). Early Cretaceous evidence for angiosperm evolution. *Botanical Review*, **43**, 3–104.

Hickey, L. J. and Wolfe, J. A. (1975). The bases of angiosperm phylogeny: vegetative morphology. *Annals of the Missouri Botanical Garden*, **62**, 538–589.

Hill, R. S. (1990). *Araucaria* (Araucariaceae) species from Australian Tertiary sediments – a micromorphological study. *Australian Systematic Botany*, **3**, 203–220.

Hill, R. S. (1992). Evolution from a southern perspective. *Trends in Ecology and Evolution*, **7**, 190–194.

Hill, R. S. (1994). The history of selected Australian taxa. In *History of the Australian Vegetation: Cretaceous to Recent*, ed. R. S. Hill. Cambridge: Cambridge University Press, pp. 390–419.

Hill, R. S. (2001). Biogeography, evolution and palaeoecology of *Nothofagus* (Nothofagaceae): the contribution of the fossil record. *Australian Journal of Botany*, **49**, 321–332.

Hill, R. S. (2004). Origins of the southeastern Australian vegetation. *Philosophical Transactions of the Royal Society of London Series B*, **359**, 1537–1549.

Hill, R. S. and Brodribb, T. J. (2003). The evolution of Australia's living biota. In *Ecology: an Australian Perspective*, eds P. Attiwill and B. Wilson. South Melbourne: Oxford University Press, pp. 13–33.

Hill, R. S., Truswell, E. M., McLoughlin, S. and Dettmann, M. E. (1999). Evolution of the Australian flora: fossil evidence. In *Flora of Australia Volume 1, Introduction*, 2nd edition. Melbourne: CSIRO, pp. 251–320.

Huber, B. T. (1988). Upper Campanian–Paleocene foraminifera from the James Ross Island region (Antarctic Peninsula). In *Geology and Paleontology of Seymour Island, Antarctic Peninsula, Memoir of the Geological Society of America*, **169**, eds R. M. Feldmann and M. O. Woodburne. Boulder: The Geological Society of America, pp. 163–251.

Huber, B. T., Norris, R. D. and MacLeod, K. G. (2002). Deep-sea paleotemperature record of extreme warmth during the Cretaceous. *Geology*, **30**, 123–126.

Iglesias, A., Zamuner A. B., Poire, D. G. and Larriestre F. (2007). Diversity, taphonomy and palaeoecology of an angiosperm flora from the Cretaceous (Cenomanian–Coniacian) in southern Patagonia, Argentina. *Palaeontology*, **50**, 445–466.

Ineson, J. R. (1989). Coarse-grained submarine fan and slope apron deposits in a Cretaceous back-arc basin, Antarctica. *Sedimentology*, **36**, 793–819.

Ineson, J. R., Crame, J. A. and Thomson, M. R. A. (1986). Lithostratigraphy of the Cretaceous strata of west James Ross Island, Antarctica. *Cretaceous Research*, **7**, 141–159.

Judd, T. S. (1993). Seed survival in Myrtaceae capsules subjected to experimental heating. *Oecologia*, **93**, 576–581.

Keating, J. M. (1992). Palynology of the Lachman Crags Member, Santa Marta Formation (Upper Cretaceous) of north-west James Ross Island. *Antarctic Science*, **4**, 293–304.

Keller, G., Adatte, T., Berner, Z., *et al.* (2007). Chicxulub impact predates K–T boundary: New evidence from Brazos, Texas. *Earth and Planetary Science Letters*, **255**, 339–356.

Kennedy, E. M., Lovis, J. D. and Daniel, I. L. (2003). Discovery of a Cretaceous angiosperm reproductive structure from New Zealand. *New Zealand Journal of Geology and Geophysics*, **46**, 519–522

Kuypers, M. M. M., Pancost, R. D. and Sinninghe Damsté, J. S. (1999). A large and abrupt fall in atmospheric CO_2 concentration during Cretaceous times. *Nature*, **399**, 342–345.

Lacey, W. S. and Lucas, R. C. (1981). The Triassic flora of Livingston Island, South Shetland Islands. *British Antarctic Survey Bulletin*, **53**, 157–173.

Lawver, L. A., Gahagan L. M. and Coffin, M. F. (1992). The development of paleoseaways around Antarctica. In *The Antarctic Paleoenvironment: A Perspective on Global Change. Part 1*, Antarctic Research Series, **56**, eds J. P. Kennett and D. A. Warnke. Washington: American Geophysical Union, pp. 7–30.

Lemoigne, Y. (1987). Confirmation de l'existence d'une flore Triasique dans l'île Livingston des Shetland du Sud (Ouest Antarctique). *Comptes Rendu de l'Académie des Sciences*, **304**, 543–546.

Leppe, M., Michea, W., Muñoz, C., Palma-Heldt, S. and Fernandoy, F. (2007). Paleobotany of Livingston Island: The first report of a Cretaceous fossil flora from Hannah Point. In *Online Proceedings of the 10th International Symposium on Antarctic Earth Sciences*, eds A. K. Cooper, C. R. Raymond and the 10th ISAES Editorial team. U. S. Geological Survey Open-File Report 2007–1047, Version 2. Short Research Paper 081, available at http://pubs.usgs.gov/of/2007/1047/srp/srp081/of2007–1047srp081.pdf. {cited 2011}.

Macellari, C. E. (1986). Late Campanian–Maastrichtian ammonite fauna from Seymour Island (Antarctic Peninsula). *Memoir of the Palaeontolgoical Society*, **18**, 1–55.

Macellari, C. E. (1988). Stratigraphy, sedimentology and paleoecology of Upper Cretaceous/Paleocene shelf-deltaic sediments of Seymour Island. In *Geology and Paleontology of Seymour Island, Antarctic Peninsula, Memoir of the Geological Society of America*, **169**, eds R. M. Feldmann and M. O. Woodburne. Boulder: The Geological Society of America, pp. 25–53.

MacKerron, D. K. L. (1976). Wind damage to the surface of strawberry leaves. *Annals of Botany*, **40**, 351–354.

MacLeod, K. G. and Huber, B. T. (1996). Reorganization of deep ocean circulation accompanying a Late Cretaceous extinction event. *Nature*, **380**, 422–425.

Macphail, M. K. and Truswell, E. M. (2004). Palynology of Neogene slope and rise deposits from ODP Sites 1165 and 1167, East Antarctica. In *Proceedings of the Ocean Drilling Program, Volume 188 Scientific Results Pyrdz Bay-Cooperation Sea, Antarctica: Glacial History and Paleoceanography covering Leg 188 of the cruises of the Drilling Vessel* JOIDES Resolution, *Fremantle, Australia, to Hobart, Tasmania Sites 1165–1167 10 January–11 March 2000*, eds A. K. Cooper, P. E. O'Brien and C. Richter. Online doi:10.2973/odp.proc.sr.188.012.2004, available at http://www-odp.tamu.edu/publications/188_SR/012/012.htm. {Cited 2011}.

Macphail, M. K., Colhoun, E. A., Kiernan, K. and Hannan, D. (1993). Glacial climates in the Antarctic region during the late Paleogene: evidence from northwest Tasmania, Australia. *Geology*, **21**, 145–148.

Macphail, M. K., Alley, N., Truswell, E. M. and Sluiter, I. R. K. (1994). Early Tertiary vegetation: evidence from spores and pollen. In *History of the Australian Vegetation: Cretaceous to Recent*, ed. R. S. Hill. Cambridge: Cambridge University Press, pp. 189–262.

Magallón, S. and Castillo, A. (2009). Angiosperm diversification through time. *American Journal of Botany*, **96**, 349–365.

Manos, P. S. (1997). Systematics of *Nothofagus* (Nothofagaceae) based on rDNA spacer sequences (ITS): taxonomic congruence with morphology and plastid sequences. *American Journal of Botany*, **84**, 1137–1155.

Marshall, J. D., Ditchfield, P. W. and Pirrie, D. (1993). Stable isotope palaeotemperatures and the evolution of high palaeolatitude climate in the Cretaceous. In *University Research in Antarctica 1989–1992. Proceedings of the British Antarctic Survey Antarctic Special Topic Award Scheme Round 2 Symposium 30th September–1st October 1992*, ed. R. B. Heywood. Cambridge: British Antarctic Survey, pp. 71–79.

Maruoka, T., Koeberl, C. and Bohor, B. F. (2007). Carbon isotopic compositions of organic matter across continental Cretaceous–Tertiary (K–T) boundary sections: Implications for paleoevironment after the K–T impact. *Earth and Planetary Science Letters*, **253**, 226–238.

McArthur, J. M., Crame, J. A. and Thirlwall, M. F. (2000). Definition of Late Cretaceous stage boundaries in Antarctica using strontium isotope stratigraphy. *Journal of Geology*, **108**, 623–640.

McLoughlin, S. and Hill, R. S. (1996). The succession of Western Australian Phanerozoic terrestrial flora. In *Gondwanan Heritage: Past, Present and Future of the Western Australian Biota*, eds S. D. Hopper, J. A. Chappill, M. S. Harvey and A. S. George. Chipping Norton: Beatty and Sons, pp. 61–80.

McLoughlin, S., Carpenter, R. J., Jordan, G. J. and Hill, R. S. (2008). Seed ferns survived the end-Cretaceous mass extinction in Tasmania. *American Journal of Botany*, **95**, 465–471.

Menéndez, C. A. and Caccavari de Filice, M. A. (1975). Las especies de *Nothofagidites* (pólen fósil de *Nothofagus*) de sedimentos Terciarios y Cretácicos de Estancia La Sara, Norte de Tierra del Fuego, Argentina. *Ameghiniana*, **12**, 165–183.

Miller, K. G. (2009). Broken greenhouse windows. *Nature Geoscience*, **2**, 465–466.

Miller, K. G., Barrera, E., Olsson, R. K., Sugarman, P. J. and Savin, S. M. (1999). Does ice drive early Maastrichtian eustasy? *Geology*, **27**, 783–786.

Miller, K. G., Sugarman, P. J., Browning, J. V., *et al.* (2003). Late Cretaceous chronology of large, rapid sea level changes: glacioeustasy during the greenhouse world. *Geology*, **31**, 585–588.

Miller, K. G., Sugarman, P. J., Browning, J. V., *et al.* (2004). Upper Cretaceous sequences and sea level history, New Jersey Coastal Plain. *Geological Society of America Bulletin*, **116**, 368–393.

Miller, K. G., Wright, J. D. and Browning, J. V. (2005). Visions of ice sheets in a greenhouse world. *Marine Geology*, **217**, 215–231.

Mohr, B. A. R. and Bernardes-de-Oliveira, M. E. C. (2004). *Endressina brasiliana*, a magnolialean angiosperm from the Lower Cretaceous Crato Formation (Brazil). *International Journal of Plant Sciences*, **165**, 1121–1133.

Mohr, B. A. R. and Eklund, H. (2003). *Araripia florifera*, a magnoliid angiosperm from the Lower Cretaceous Crato Formation (Brazil). *Review of Palaeobotany and Palynology*, **126**, 279–292.

Mohr, B. A. R. and Friis, E. M. (2000). Early angiosperms from the Lower Cretaceous Crato Formation (Brazil), a preliminary report. *International Journal of Plant Sciences*, **161** (supplement), 155–167.

Molnar, R. E. and Wiffen, J. (1994). A Late Cretaceous polar dinosaur fauna from New Zealand. *Cretaceous Research*, **15**, 689–706.

Mosbrugger, V. and Utescher, T. (1997). The coexistence approach–a method for quantitative reconstructions of Tertiary terrestrial palaeoclimate data using plant fossils. *Palaeogeography, Palaeoclimatology, Palaeoecology*, **134**, 61–86.

Nordt, L., Atchley, S. and Dworkin, S. (2003). Terrestrial evidence for two greenhouse events in the latest Cretaceous. *GSA Today*, **13(12)**, 4–9.

Orlando, H. A. (1967). Primera florula Triásica de la Antártida Occidental. *Instituto Antártico Argentino Contribución*, **118**, 1–16.

Orlando, H. A. (1968). A new Triassic flora from Livingston Island, South Shetland Islands. *British Antarctic Survey Bulletin*, **16**, 1–13.

Otto-Bliesner, B. L. and Upchurch Jr, G. R. (1997). Vegetation-induced warming of high-latitude regions during the Late Cretaceous period. *Nature*, **385**, 804–807.

Palma-Heldt, S., Fernandoy, F., Quezada, I. and Leppe, M. (2004). Registro Palinológico de Cabo Shirreff, Isla Livingston, nueva localidad para el Mesozoico de las Shetland del Sur, In *Vº Simposio Argentino y 1º Latinoamericano sobre Investigaciones Antárticas Ciudad Autónoma de Buenos Aires, Argentina 30 de agosto al 3 de septiembre de 2004 Resúmen Expandido Nº 104GP*, 4pp. Buenos Aires, Argentina. Available at http://www.dna.gov.ar/CIENCIA/SANTAR04/CD/PDF/104GP.PDF. {Cited 2011}.

Palma-Heldt, S., Fernandoy, F., Henríquez, G. and Leppe, M. (2007). Palynoflora of Livingston Island, South Shetland Islands: Contribution to the understanding of the evolution of the southern Pacific Gondwana margin. In *Online Proceedings of the 10th International Symposium on Antarctic Earth Sciences*, eds A. K. Cooper and C. R. Raymond and the 10th ISAES Editorial team. U. S. Geological Survey Open-File Report 2007–1047, Version 2. Extended Abstract 100, available at http://pubs.usgs.gov/of/2007/1047/ea/of2007–1047ea100.pdf. {Cited 2011}.

Parrish, J. T. and Spicer, R. A. (1988). Late Cretaceous terrestrial vegetation: a near-polar temperature curve. *Geology*, **16**, 22–25.

Passalía, M. G., Archangelsky, S., Romero, E. J. and Cladera, G. (2003). A new early angiosperm leaf from the Anfiteatro de Ticó Formation (Aptian), Santa Cruz Province, Argentina. *Revista de Museo Argentino de Ciencias Naturales Nueva Serie*, **5**, 245–252.

Philippe, M., Barale, G., Torres, T. and Covacevich, V. (1993). First study of *in situ* woods from the Upper Cretaceous of Livingston Island, South Shetland Islands, Antarctica: palaeoecological investigations. *Comptes Rendus de l Académie des Sciences Series 2 Mécanique, Physique, Chimie, Sciences de la Terre et de l'Universe*, **317**, 103–108.

Pirrie, D. (1987). Orientated calcareous concretions from James Ross Island, Antarctica. *British Antarctic Survey Bulletin*, **75**, 41–50.

Pirrie, D. (1989). Shallow marine sedimentation within an active margin basin, James Ross Island, Antarctica. *Sedimentary Geology*, **63**, 69–82.

Pirrie, D. (1991). Controls on the petrographic evolution of an active margin sedimentary sequence: the Larsen Basin, Antarctica. In *Developments in Sedimentary Provenance Studies, Geological Society of London Special Publication*, **57**, eds A. C. Morton, S. P. Todd and P. D. W. Haughton. London: The Geological Society, pp. 231–249.

Pirrie, D. and Marshall, J. D. (1990). High-paleolatitude Late Cretaceous paleotemperatures: New data from James Ross Island, Antarctica. *Geology*, **18**, 31–34.

Pirrie, D., Crame, J. A. and Riding, J. B. (1991). Late Cretaceous stratigraphy and sediment-ology of Cape Lamb, Vega Island, Antarctica. *Cretaceous Research*, **12**, 227–258.

Pirrie, D., Crame, J. A., Lomas, S. A. and Riding J. B. (1997). Late Cretaceous stratigraphy of the Admiralty Sound region, James Ross Basin, Antarctica. *Cretaceous Research*, **18**, 109–137.

Pirrie, D., Marshall, J. D. and Crame, J. A. (1998). Marine high Mg calcite cements in *Teredolites*-bored fossil wood; evidence for cool paleoclimates in the Eocene La Mesata Formation, Seymour Island, Antarctica. *Palaios*, **13**, 276–286.

Pole, M. S. (1992). Cretaceous macrofloras of eastern Otago, New Zealand: angiosperms. *Australian Journal of Botany*, **40**, 169–206.

Pole, M. S. (1993). Keeping in touch: vegetation prehistory on both sides of the Tasman. *Australian Systematic Botany*, **6**, 387–397.

Pole, M. S. and Douglas, J. G. (1999). Bennettitales, Cycadales and Ginkgoales from the mid-Cretaceous of the Eromanga Basin, Queensland, Australia. *Cretaceous Research*, **20**, 523–538.

Pole, M. and Vadja, V. (2009). A new terrestrial Cretaceous–Paleogene site in New Zealand turnover in macroflora confirmed by palynology. *Cretaceous Research*, **30**, 917–938.

Pollard, D. and DeConto, R. M. (2005). Hysteresis in Cenozoic Antarctic ice sheet varia-tions. *Global and Planetary Change*, **45**, 9–21.

Poole, I. (2000). Fossil angiosperm wood: its role in the reconstruction of biodiversity and palaeoenvironment. *Botanical Journal of the Linnean Society*, **134**, 361–381.

Poole, I. (2002). Systematics of Cretaceous and Tertiary *Nothofagoxylon*: Implications for Southern Hemisphere biogeography and evolution of the Nothofagaceae. *Australian Systematic Botany* **15**, 247–276.

Poole, I. and van Bergen, P. F. (2006). Physiognomic and chemical characters in wood as palaeoclimate proxies. *Plant Ecology*, **182**, 175–195.

Poole, I. and Cantrill, D. J. (2001). Fossil woods from Williams Point Beds, Livingston Island, Antarctica: A Late Cretaceous southern high latitude flora. *Palaeontology*, **44**, 1081–1112.

Poole, I. and Cantrill, D. J. (2006). Cretaceous and Tertiary vegetation of Antarctica implications from the fossil wood record. In *Cretaceous–Tertiary High-Latitude Palaeoenvironments, James Ross Basin, Antarctica*, eds J. E. Francis, D. Pirrie and J. A. Crame. *Geological Society of London Special Publication*, **258**, 63–81.

Poole, I. and Francis, J. E. (1999). Reconstruction of Antarctic palaeoclimates using angio-sperm wood anatomy. *Acta Palaeobotanica Supplement*, **2**, 173–179.

Poole, I. and Francis, J. E. (2000). The first record of fossil wood of Winteraceae from the Upper Cretaceous of Antarctica. *Annals of Botany*, **85**, 307–315.

Poole, I. and Gottwald, H. (2001). Monimiaceae *sensu lato*, an element of Gondwanan polar forests: Evidence from the Late Cretaceous–early Tertiary wood flora of Antarctica. *Australian Systematic Botany*, **14**, 207–230.

Poole, I., Cantrill, D. J., Hayes, P. and Francis, J. E. (2000). The fossil record of Cunoniaceae: new evidence from Late Cretaceous wood of Antarctica? *Review of Palaeobotany and Palynology*, **111**, 127–144.

Poole, I., Gottwald, H. and Francis, J. E. (2000). *Illicioxylon*, an element of Gondwanan polar forests? Late Cretaceous and Early Tertiary woods of Antarctica. *Annals of Botany*, **86**, 421–432.

Poole, I., Richter, H. and Francis, J. E. (2000). Evidence for Gondwanan origins for *Sassafras* (Lauraceae)? Late Cretaceous fossil wood of Antarctica. *International Association of Wood Anatomists Journal*, **21**, 463–475.

Poole, I., Hunt, R. J. and Cantrill, D. J. (2001). A fossil wood flora from King George Island: ecological implications for an Antarctic Eocene vegetation. *Annals of Botany*, **88**, 33–54.

Poole, I., Mennega, A. M. W. and Cantrill, D. J. (2003). Valdivian ecosystems in the Late Cretaceous and early Tertiary of Antarctica further evidence from myrtaceous and eucryphiaceous fossil wood. *Review of Palaeobotany and Palynology*, **124**, 9–27.

Poole, I., Cantrill, D. J. and Utescher, T. (2005). Reconstructing Antarctic palaeoclimate from wood floras – a comparison using multivariate anatomical analysis and the Coexistence Approach. *Palaeogeography, Palaeoclimatology, Palaeoecology*, **222**, 95–121.

Preisinger, A., Aslanian, S., Brandstätter, F., *et al.* (2002). Cretaceous–Tertiary profile, rhythmic deposition and geomagnetic polarity reversals of marine sediments near Bjala, Bulgaria. In *Catastrophic Events and Mass Extinctions: Impacts and Beyond. Geological Society of America Special Paper*, **356**, eds C. Koeberl and K. G. MacLeod. Boulder: The Geological Society of America, pp. 213–229.

Raine, J. I. (1984). Outline of a palynological zonation of Cretaceous to Palaeogene terrestrial sediments in West Coast Region, South Island, New Zealand. *New Zealand Geological Survey Report*, **109**, 1–82.

Raine, J. I., Speden, L. G., Strong, C. P. (1981). New Zealand. In *Aspects of Mid-Cretaceous Regional Geology*, eds R. A. Reyment and P. Bengtson. London: Academic Press, pp. 221–267.

Rees, P. M. and Smellie, J. L. (1989). Cretaceous angiosperms from an allegedly Triassic flora at Williams Point, Livingston Island, South Shetland Islands. *Antarctic Science*, **1**, 239–248.

Riding, J. B. and Crame, J. A. (2002). Aptian to Coniacian (Early-Late Cretaceous) palyno-stratigraphy of the Gustav Group, James Ross Basin, Antarctica. *Cretaceous Research*, **23**, 739–760.

Riding, J. B., Crame, J. A., Dettmann, M. E. and Cantrill, D. J. (1998). The age of the base of the Gustav Group in the James Ross Basin, Antarctica. *Cretaceous Research*, **19**, 87–105.

Romero, E. J. (1978). Paleoecología y paleofitogeografía de las tafofloras del Cenofítico de Argentina y áreas vecinas. *Ameghiniana*, **15**, 209–227.

Romero, E. J. and Archangelsky, S. (1986). Early Cretaceous angiosperm leaves from southern South America. *Science*, **234**, 1580–1582.

Royer, D. L. (2006). CO_2-forced climate thresholds during the Phanerozoic. *Geochimica et Cosmochimica Acta*, **70**, 5665–5675.

Royer, D. L., Osborne, C. P. and Beerling, D. J. (2002). High CO_2 increases the freezing sensitivity of plants: Implications for paleoclimatic reconstructions from fossil floras. *Geology*, **30**, 963–966.

Royer, D. L., Wilf, P., Janesko, D. A., Kowalski, E. A. and Dilcher, D. L. (2005). Correlations of climate and plant ecology to leaf size and shape: potential proxies for the fossil record. *American Journal of Botany*, **92**, 1141–1151.

Sageman, B. B., Meyers, S. R. and Arthur, M. A. (2006). Orbital timescale and new C-isotope record for Cenomanian–Turonian boundary stratotype. *Geology*, **34**, 125–128.

Scasso, R. A., Concheyro, A., Kiessling, W., *et al.* (2005). A tsunami deposit at the Cretaceous/Paleogene boundary in the Nequén Basin of Argentina. *Cretaceous Research*, **26**, 283–297.

Schneider, H., Schuettpelz, E., Pryer, K. M., *et al.* (2004). Ferns diversified in the shadow of angiosperms. *Nature*, **428**, 553–557.

Schulte, P., Alegret, L., Arenillas, I., *et al.* (2010). The Chicxulub asteroid impact and mass extinction at the Cretaceous–Paleogene boundary. *Science*, **327**, 1214–1218.

Scott, A. C. and Titchner, F. R. (1999). Techniques in the study of plant-animal interactions. In *Fossil Plants and Spores: Modern Techniques*, eds T. P Jones and N. P. Rowe. London: The Geological Society, pp. 310–315.

Sellwood, B. W. and Valdes, P. J. (2006). Mesozoic climates: General circulation models and the rock record. *Sedimentary Geology*, **190**, 269–287.

Shen, Y. (1990). Progress in stratigraphy and palaeontology of Fildes Peninsula, King George Island, Antarctica. *Acta Palaeontologica Sinica*, **29**, 136–139.

Shen, Y. (1994). Subdivision and correlation of Cretaceous to Paleogene volcano-sedimentary sequence from Fildes Peninsula, King George Island, Antarctica. In *Stratigraphy and Palaeontology of Fildes Peninsula, King George Island, Antarctica, State Antarctic Committee, Monograph* **3**, ed. Y. Shen Beijing: China Science Press, pp. 1–36 (In Chinese, English abstract).

Sinninghe Damsté, J. S., van Bentum, E. C., Reichart, G. J., Pross, J. and Schouten, S. (2010). A CO_2 decrease-driven cooling and increased latitudinal temperature gradient during the mid-Cretaceous Oceanic Anoxic Event 2. *Earth and Planetary Science Letters*, **293**, 97–103.

Smellie, J. L. (1981). A complete arc-trench system recognized in Gondwana sequences of the Antarctic Peninsula region. *Geological Magazine*, **118**, 139–159.

Smellie, J. L., Pankhurst, R. J., Thomson, M. R. A. and Davies, R. E. S. (1984). The geology of the South Shetland Islands: VI. stratigraphy, geochemistry and evolution. *British Antarctic Survey Scientific Reports*, **87**, 1–85.

Smellie, J. L., Roberts, B. and Hirons, S. R. (1996). Very low- and low-grade metamorphism in the Trinity Peninsula Group (Permo-Triassic) of northern Graham Land, Antarctic Peninsula. *Geological Magazine*, **133**, 583–594.

Smit, J. (1990). Meteorite impact, extinctions and the Cretaceous–Tertiary boundary. *Geologie en Mijnbouw*, **69**, 187–204.

Soliani, E. and Bonhomme, M. G. (1994). New evidence for Cenozoic resetting of K-Ar ages in volcanic rocks of the Northern Portion of the Armiralty Bay, King George Island, Antarctica. *Journal of South American Earth Sciences*, **7**, 85–94.

Song, Z. and Cao, L. (1994). Late Cretaceous fungal spores from King George Island, Antarctica. In *Stratigraphy and Palaeontology of Fildes Peninsula, King George Island, Antarctica, State Antarctic Committee, Monograph* **3**, ed. S. Yanbin. Beijing: China Science Press. pp. 37–49 (In Chinese, English abstract).

Specht, R. L., Dettmann, M. E. and Jarzen, D. M. (1992). Community associations and structure in the Late Cretaceous vegetation of southeast Australasia and Antarctica. *Palaeogeography, Palaeoclimatology, Palaeoecology*, **94**, 283–309.

Spicer, R. A. and Parrish, J. T. (1986). Paleobotanical evidence for cool north polar climates in middle Cretaceous (Albian–Cenomanian) time. *Geology*, **14**, 703–706.

Steuber, T., Rauch, M., Masse, J. P., Graaf, J. and Malkoc, M. (2005). Low-altitude seasonality of Cretaceous temperatures in warm and cold episodes. *Nature*, **437**, 1341–1344.

Stockey, R. A. and Rothwell, G. W. (2003). Anatomically preserved *Williamsonia* (Williamsoniaceae): Evidence for bennettitalean reproduction in the Late Cretaceous of Western North America. *International Journal of Plant Sciences*, **164**, 251–262.

Stüben, D., Kramar, U., Harting, M., Stinnesbeck, W. and Keller, G. (2005). High-resolution geochemical record of Cretaceous–Tertiary boundary sections in Mexico: New constraints on the K/T and Chicxulub events. *Geochimica et Cosmochimica Acta*, **69**, 2259–2579.

Swenson, U. and Hill, R. S. (2001). Most parsimonious areagrams versus fossils: the case of *Nothofagus* (Nothofagaceae). *Australian Journal of Botany*, **49**, 367–376.

Swenson, U., Hill, R. S. and McLoughlin, S. (2000). Ancestral area analysis of *Nothofagus* (Nothofagaceae) and its congruence with the fossil record. *Australian Systematic Botany*, **13**, 469–478.

Takahashi, M., Crane, P. R. and Ando, H. (1999a). Fossil flowers and associated plant fossils from the Kamikitaba locality (Ashizawa Formation, Futaba Group, Lower Coniacian, Upper Cretaceous) of Northeast Japan. *Journal of Plant Research*, **112**, 187–206.

Takahashi, M., Crane, P. R. and Ando, H. (1999b). *Esgueiria futabensis* sp. nov., a new angiosperm flower from the Upper Cretaceous (lower Coniacian) of northeastern Honshu, Japan. *Paleontological Research*, **3**, 81–87.

Taylor, D. W. and Hickey, L. J. (1990). An Aptian plant with attached leaves and flowers: implication for angiosperm origin. *Science*, **247**, 702–704.

Thorn, V. C., Francis, J. E., Riding, J. B., *et al.* (2007). Terminal Cretaceous climate change and biotic response in Antarctica, In *Online Proceedings of the 10th International Symposium on Antarctic Earth Sciences*, eds A. K. Cooper, C. R. Raymond and the 10th ISAES Editorial team. U. S. Geological Survey Open-File Report 2007–1047, Version 2. Extended Abstract 096, available at http://pubs.usgs.gov/of/2007/1047/ea/of2007–1047ea096.pdf. {Cited 2011}.

Thorn, V. C., Riding, J. B. and Francis, J. E. (2009). The Late Cretaceous dinoflagellate cyst *Manumiella*–biostratigraphy, systematics, and palaeoecological signals in Antarctica. *Review of Palaeobotany and Palynology*, **156**, 436–448.

Torres, T. and Lemoigne, Y. (1989). Hallazgos de maderas fósiles de angiospermas y gimnospermas del Cretácico Superior en Punta Williams, Isla Livingston, Islas Shetland del Sur, Antártica. *Serie Científica Instituto Antártico Chileno*, **39**, 9–29.

Truswell, E. M. (1990). Cretaceous and Tertiary vegetation of Antarctica: a palynological perspective. In *Antarctic Paleobiology: its Role in the Reconstruction of Gondwana*, eds T. N. Taylor and E. L. Taylor. New York: Springer-Verlag, pp. 71–88.

Truswell, E. M. (1991). Antarctica: A history of terrestrial vegetation. In *The Geology of Antarctica, Oxford Monographs on Geology and Geophysics*, **17**, ed. R. J. Tingey. Oxford: Clarendon Press, pp. 499–537.

Truswell, E. M. and Macphail, M. K. (2009). Polar forests on the edge of extinction: what does the fossil spore and pollen evidence from East Antarctica say? *Australian Systematic Botany*, **22**, 57–106.

Tsikos, H., Jenkyns, H. C., Walsworth-Bell, B., *et al.* (2004). Carbon-isotope stratigraphy recorded by the Cenomanian–Turonian Oceanic Anoxic Event: correlation and implications based on three key-locations. *Journal of the Geological Society of London*, **161**, 711–719.

Twitchett, R. J. (2006). The palaeoclimatology, palaeoecology and palaeoenvironmental analysis of mass extinction events. *Palaeogeography, Palaeoclimatology, Palaeoecology*, **232**, 190–213.

Upchurch, G. R. and Askin, R. A. (1990). Latest Cretaceous and earliest Tertiary dispersed cuticles from Seymour Island, Antarctica. *Antarctic Journal of the United States*, **24** (**5**), 7–10.

Vajda, V. and McLoughlin, S. (2004). Fungal proliferation at the Cretaceous–Tertiary boundary. *Science*, **303**, 1489.

Vajda, V. and McLoughlin, S. (2007). Extinction and recovery patterns of the vegetation across the Cretaceous–Palaeogene boundary – a tool for unravelling the causes of the end-Permian mass-extinction. *Review of Palaeobotany and Palynology*, **144**, 99–112.

Vajda, V. and Raine, J. I. (2003). Pollen and spores in marine Cretaceous/Tertiary boundary sediments at mid-Waipara River, North Canterbury, New Zealand. *New Zealand Journal of Geology and Geophysics*, **46**, 255–273.

Vajda, V., Raine, J. I. and Hollis, C. J. (2001). Indication of global deforestation at the Cretaceous–Tertiary boundary by New Zealand fern spike. *Science*, **294**, 1700–1702.

Valdes, P. J., Sellwood, B. W. and Price, G. D. (1996). Evaluating concepts of Cretaceous equability. *Palaeoclimates*, **2**, 139–158.

Van Wagoner, J. C., Mitchum, R. M., Campion, K. M. and Rahamnian, V. D. (1990). *Siliciclastic Sequence Stratigraphy in Well Logs, Cores and Outcrops: concepts for High-Resolution Correlation of Time and Facies, American Association of Petroleum Geologists, Methods in Exploration Series*, **7**. Tulsa: AAPG Foundation, pp.1–55.

Veblen, T. T., Schlegel, F. M. and Oltremari, J. V. (1983). Temperate broad-leaved evergreen forests of South America. In *Temperate Broad-Leaved Evergreen Forests*, ed. J. D. Ovington. Amsterdam: Elsevier, pp. 5–31.

Veblen, T. T., Hill, R. S. and Read, J. (1996). *The Ecology and Biogeography of Nothofagus Forests*. London: Yale University Press.

Wang Y. and Shen, Y. (1994). Rb-Sr isotopic dating and trace element, REE geochemistry of Late Cretaceous volcanic rocks from King George Island, Antarctica. In *Stratigraphy and Palaeontology of Fildes Peninsula, King George Island, Antarctica, State Antarctic Committee, Monograph* **3**, ed. Y. Shen. Beijing: China Science Press, pp. 109–131 (In Chinese, English abstract).

Wang, X. and Zheng, S. (2009). The earliest normal flower from Liaoning Province, China. *Journal of Integrative Plant Biology*, **51**, 800–811.

Weaver, S. D., Storey, B. C., Pankhurst, R. J., *et al.* (1994). Antarctica-New Zealand rifting and Marie Byrd Land lithospheric magmatism linked to ridge subduction and mantle plume activity. *Geology*, **22**, 811–814.

Webb, L. J., Tracey, J. G. and Jessup, L. W. (1986). Recent evidence for autochthony of Australian tropical and subtropical rainforest floristic elements. *Telopea*, **2**, 575–589.

Wheeler, E. A. and Bass, P. (1991). A survey of the fossil record for dicotyledonous wood and its significance for evolutionary and ecological wood anatomy. *International Association of Wood Anatomists Bulletin*, **12**, 275–332.

Wheeler, E. A. and Manchester, S. (2002). Woods of the middle Eocene Nut Beds Flora, Clarno Formation, Oregon, USA. *International Association of Wood Anatomists Journal, Supplement*, **3**, 1–188.

Whitham, A. G., Ineson, J. R. and Pirrie, D. (2006). Marine volcaniclastics of the Hidden Lake Formation (Coniacian) of James Ross Island, Antarctica: an enigmatic element in the history of a back arc basin. In *Cretaceous–Tertiary High-Latitude Palaeoenvironments, James Ross Basin, Antarctica. Geological Society of London Special Publications*, **258**, eds J. E. Francis, D. Pirrie and J. A. Crame. London: The Geological Society, pp. 21–47.

Wilf, P., Johnson, K. R. and Huber, B. T. (2003). Correlated terrestrial and marine evidence for global climate changes before mass extinction at the Cretaceous–Paleogene boundary. *Proceedings of the National Academy of Sciences of the United States of America*, **100**, 599–604.

Wilson, J. (1980). Macroscopic features of wind damage to leaves of *Acer pseudoplatanus* L., and its relationship with season, leaf age, and wind speed. *Annals of Botany*, **46**, 303–311.

Wilson, J. (1984). Microscopic features of wind damage to leaves of *Acer pseudoplatanus* L. *Annals of Botany*, **53**, 73–82.

Wilson, G. S., Barron, J. A., Ashworth, A. C., *et al.* (2002). The Mount Feather Diamicton of the Sirius Group: an accumulation of indicators of Neogene Antarctic glacial and climatic history. *Palaeogeography, Palaeoclimatology, Palaeoecology*, **28**, 1–15.

Wing, S. L. and Tiffney, B. H. (1987). Interactions of angiosperms and herbivorous tetrapods through time. In *The Origins of Angiosperms and their Biological Consequences*, eds, E. M. Friis, W. G. Chaloner and P. R. Crane. Cambridge: Cambridge University Press, pp. 203–224.

Wolfe, J. A. (1997). Relations of environmental change to angiosperm evolution during the Late Cretaceous and Tertiary. In *Evolution and Diversification of Land Plants*, eds K. Iwatsuki, and P. R. Raven. Tokyo: Springer, pp. 269–290.

Wolfe, J. A. and Upchurch Jr, G. R. (1987). North American nonmarine climates and vegetation during the Late Cretaceous. *Palaeogeography, Palaeoclimatology, Palaeoecology*, **61**, 33–77.

Zastawniak, E. (1990). Late Cretaceous leaf flora of King George Island, West Antarctica. *Proceedings of the Symposium Paleofloristic and Paleoclimatic Changes in the Cretaceous and Tertiary*. Prague: Geological Survey Publisher, pp. 81–86.

Zastawniak, E. (1994). Upper Cretaceous leaf flora from the Blaszyk Moraine (Zamek Formation), King George Island, South Shetland Islands, West Antarctica. *Acta Palaeobotanica*, **34**, 119–163.

Zhou, Z. and Li, H. (1994). Some Late Cretaceous plants from King George Island, Antarctica. In *Stratigraphy and Palaeontology of Fildes Peninsula, King George Island, Antarctica, State Antarctic Committee Monograph*, **3**, ed. Shen Yanbin. Beijing: Science Press. pp. 85–96 (In Chinese, English abstract, including systematic descriptions).

Zinsmeister, W. J. (1982). Review of the Upper Cretaceous–Lower Tertiary sequence on Seymour Island, Antarctica. *Journal of the Geological Society*, **139**, 779–785.

Zinsmeister, W. J. (1987). Cretaceous paleogeography of Antarctica. *Palaeogeography, Palaeoclimatology, Palaeoecology*, **59**, 197–206.

Zinsmeister, W. J., Feldmann, R. M., Woodburne, M. O. and Elliot, D. H. (1989). Latest Cretaceous/earliest Tertiary transition on Seymour Island, Antarctica. *Journal of Paleontology*, **63**, 731–738.

8

The heat is on: Paleogene floras and the Paleocene–Eocene warm period

Introduction

Over the last 65 million years Earth's climate has undergone a significant and complex evolution with shifts from extremes of extensive warmth with ice-free poles, to periods of intense cold with massive continental ice sheets and polar ice caps (Zachos *et al.*, 2001). The long-term climate drivers, controlled largely by plate tectonics, occurred on timescales covering millions of years and were as a result, gradual and unidirectional. However, superimposed on these gradual changes were unusually rapid or extreme changes in climate (Crowley and North, 1988; Diester-Haass *et al.*, 2009; Zachos *et al.*, 1993). The evolution of climate during the Cenozoic has been studied at very high resolution because the record of deep sea oxygen and carbon isotopes has been sampled in great detail (Zachos *et al.*, 2001). These isotope records contain long-term warming and cooling trends with geologically abrupt shifts highlighting brief, yet extreme, excursions. It is becoming increasingly apparent that the Cenozoic climate includes gradual trends in warming and cooling driven by plate tectonic processes on timescales of 10^5 to 10^7 years, rhythmic or periodic cycles driven by orbital processes with a 10^4 to 10^6 year cyclicity and rare rapid aberrant shifts, and extreme climatic transients with durations of 10^3 to 10^5 years (e.g. Agnini *et al.*, 2009; Cramer *et al.*, 2003; Hays, Imbrie and Shackleton, 1976; Lourens *et al.*, 2005; Stap *et al.*, 2009; Westerhold *et al.*, 2007; Zachos *et al.*, 2001, 2010).

During this phase of Earth history the vegetation also evolved to become increasingly modern. Molecular phylogenetic studies and the use of molecular clock approaches have raised questions about when, where and how plant lineages diverged, questions that can be tested with the fossil record. Therefore reliably identified fossils in accurately dated sediments play a crucial role in providing minimum age constraints on the divergence of lineages and evidence for their geographic distribution. Although fossils cannot provide direct evidence of migration/dispersal pathways they are of special interest when such distributions are remote from extant distributions, and when they provoke new questions about possible dispersal routes in the past (Carpenter, Truswell and Harris, 2010). Only fossils can provide material evidence that certain taxa did indeed occur in a particular palaeogeographic region at a particular time. This is especially important when the past land is now submerged, but may have acted as important stepping stones for plant migration in the past (Carpenter, Truswell and Harris, 2010). Over the past couple of decades, major advances in our knowledge of the palaeobotany of Australasia, South America and Antarctica have been made, and these discoveries can help further our understanding of the role of long-distance dispersal and vicariance as means of speciation and ultimately the modernisation of the austral vegetation (cf. Carpenter, Truswell and Harris, 2010). Thus the importance of Antarctic palaeobotany remains central to our understanding of austral biogeographic patterns seen today.

Global palaeoclimate

Early in the Cenozoic, during the mid-Paleocene, the sedimentological, micropalaeonto-logical and organic geochemical records provide evidence for a short-lived cold spell (Haq, Hardenbol and Vail, 1987; Leckie *et al.*, 1995) (Figure 8.1). The climate cooled, perhaps only for a few thousand years, but this was long enough to allow the formation of sea ice around Antarctica and for sea levels to drop by ~50–125 m. This in turn had subsequent, and marked, repercussions for the oceanic biota.

Following this brief cooling event, a long-term warming trend was initiated at about 59 Ma, close to the Selandian–Thanetian boundary, which continued for several million years and finally peaked between 52 and 50 Ma at the Early Eocene Climatic Optimum (EECO) (Figure 8.1). Superimposed on the 6 Ma period of gradual global warming during the late Paleocene and early Eocene were geologically brief, but very dramatic, increases in temperature known as hyperthermal events. One such event manifested at ~55 Ma, and spanned the Paleocene–Eocene transition (Figure 8.1). This has been termed the Paleocene–Eocene Thermal Maximum[1] (PETM), and lasted for ~170 kyr (Röhl *et al.*, 2007).

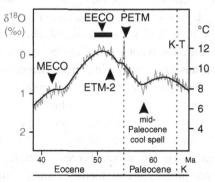

Figure 8.1 Climate change during the last 65 million years as expressed by the oxygen isotope curve based on benthic $\delta^{18}O$ and showing the computed ocean equivalent sea surface temperature (°C) for high latitudes based on an ice-free ocean. K-T = Cretaceous–Cenozoic (Paleogene) boundary event; PETM = Paleocene–Eocene Thermal Maximum; EECO = Early Eocene Climatic Optimum; MECO = Middle Eocene Climatic Optimum; ETM-2 = Eocene Thermal Maximum-2. Adapted from Zachos *et al.* (2001) (reproduced with permission from AAAS).

---

[1] Otherwise known as the Eocene Thermal Maximum 1 (ETM-1), and formerly referred to as the Late Paleocene Thermal Maximum (LPTM) or Initial Eocene Thermal Maximum (IETM).

The PETM is the most rapid and extreme global warming event recorded to date in the history of the Earth. Within less than 20 kyr, global temperatures rose by 5–8 °C and atmospheric humidity increased to 20–25% higher than present (Bowen *et al.*, 2004). At the poles, with the positive feedback effect of melting ice reducing albedo, annual average temperatures rose to 10–20 °C (Shellito, Sloan and Huber, 2003; Weijers *et al.*, 2007). Sea levels also rose and sea temperatures increased both at depth and at the surface (by 5–6 °C and ~4–8 °C respectively) (Kennett and Stott, 1991; Weijers *et al.*, 2007; Zachos *et al.*, 2001). The events of the PETM also led to rapid (<5000 years) changes in atmospheric and marine circulation patterns. Global-scale oceanic currents were reversed with the effects of changing circulation patterns, changes once again being most deeply felt at the high latitudes (Nunes and Norris, 2006). This event was marked by a perturbation in the hydrological cycle and major biotic responses that included radiations, extinctions and migrations, both on land and in the sea (Bowen *et al.*, 2006; Sluijs *et al.*, 2007a). For example, numerous deep sea benthic foraminifera became extinct as a result of changes in temperature of the bottom water and degree of bioturbation (Thomas and Shackleton, 1996). Subtropical and tropical marine and terrestrial biota were able to migrate poleward such that during the height of the PETM very high latitudes were able to support tropical life forms (e.g. Hickey *et al.*, 1983; McKenna, 1980; Shellito, Sloan and Huber, 2003; Wolfe, 1980). On land, the PETM event was characterised by shifting geographic distributions predominantly through migration, rather than extinction, of terrestrial organisms (such as plants, mammals and turtles) result- ing in new and unique taxonomic assemblages and ecosystems (e.g. Bowen *et al.*, 2002; Clyde and Gingerich, 1998; Iglesias *et al.*, 2007; Wing *et al.*, 2005).

The PETM was not one unique event, but represents the extreme example in a series of similar hyperthermal events that took place during the late Paleocene and early Eocene, superimposed on the long-term late Paleocene and early Eocene warming (Lourens *et al.*, 2005; Thomas and Zachos, 2000; Zachos, Dickens and Zeebe, 2008). Although the climatic aberration of the PETM is the most extreme ever recorded, its cause(s) is still unclear. To try and determine the mechanisms for such a rapid climate change, scientists have focused on the carbon isotope record. During the late Paleocene and early Eocene the carbon cycle was in stasis, but marine calcium carbonates and carbon locked up within the terrestrial realm record both numerous and short-lived depletions, or transient negative excursions, of between 2.5 and 6‰. This suggests that the global carbon cycle had undergone a massive perturbation with a fast injection of ^{12}C-enriched carbon in the form of CO_2 or methane into the global exogenic carbon pool[2]. Recent models estimate that this equates to up to ~3000 Pg[3] carbon released over a period of 5000 years at the start of the PETM[3], followed by a continuous release of ~1480 Pg carbon over the next 70 kyr (Zeebe, Zachos and Dickens, 2009). This would have resulted in pCO$_2$ rising from 1000 ppmv to 1700 ppmv

---

[2] i.e. the carbon contained within the atmosphere and ocean.

[3] This is equivalent to an annual increase in carbon to the atmosphere of 0.6 Pg yr^{-1} at the PETM followed by a reduction in carbon input to 0.021 Pg yr^{-1} (i.e. ~30 times less) over the subsequent 70 kyr following the PETM. Compare this with current annual increases of 3–3.5 Pg yr^{-1} (i.e. five times as much as at the PETM) when the pCO$_2$ of the atmosphere is assumed to increase by 1.5 ppm each year (cf. Sarmiento and Gruber, 2002).

over the main phase of the PETM. Although the source of this carbon is still under discussion (Panchuk, Ridgewell and Kump, 2008; Sluijs *et al.*, 2007a; Stap *et al.*, 2009), the severity in terms of magnitude and rate of carbon isotope excursion that occurred at the PETM is consistent with a sudden dissociation of clathrates, which trap methane hydrates formed in cold water at great pressure, deep below the sea floor near continental margins (Dickens *et al.*, 1995; Katz *et al.*, 1999; Sluijs *et al.*, 2007b; Storey, Duncan and Swisher, 2007; Thomas *et al.*, 2002). Even though definitive evidence for similar clathrate destabilisation is currently lacking, this scenario is probably (one of) the most likely causal factors for other recorded hyperthermal events (such as the Eocene Thermal Maximum 2 at ~53.7 Ma) (Quillévéré *et al.*, 2008; Sluijs and Brinkhuis, 2008; Sluijs *et al.*, 2009; Stap *et al.*, 2009). The trigger for methane release at the PETM could have resulted from one, or a combination of, factors including changing oceanic circulation, comet impact, seafloor erosion and/or volcanic activity (e.g. Cramer and Kent, 2005; Dickens, 2000; Lourens *et al.*, 2005; Tripati and Elderfield, 2005). Arguably, it is also possible that the long-term deep sea warming trend, which began at the very end of the Cretaceous, crossed a thermal threshold (Thomas and Shackleton, 1996) that resulted in the melting of these hydrates causing massive amounts of methane to be released, probably through multiple injections, into the atmosphere (Bains, Corfield and Norris, 1999; Dickens, 2000; Dickens *et al.*, 1995; Pagani *et al.*, 2006). The methane, a potent greenhouse gas, would have subsequently reacted with oxygen to produce huge amounts of carbon dioxide that trapped the Sun's warmth. This coupled with the consequent unbalancing of the global carbon cycle, resulted in a runaway global warming event characterised by soaring global temperatures (Zeebe, Zachos and Dickens, 2009). The increase in greenhouse gases would, in all likelihood, have been coupled with other as yet unidentified processes and/or feedbacks, which contributed to a substantial portion of the warming during the PETM (Zeebe, Zachos and Dickens, 2009). Moreover, other research has shown that some warming events and biotic changes did slightly predate the carbon isotope excursion and were therefore not caused by it (Bowen *et al.*, 2001; Thomas *et al.*, 2002; Tripati and Elderfield, 2005). For example, sea surface temperatures in both hemispheres appear to have undergone an extreme warming significantly prior to the injection of ^{12}C-enriched carbon. This implies that some initial warming, possibly by a non-carbon greenhouse gas, triggered the injection of ^{12}C-enriched carbon into the atmosphere, which resulted in global warming, rather than being the result of the carbon excursion itself (e.g. Dickens *et al.*, 1995; Thomas *et al.*, 2002, Tripati and Elderfield, 2005; Sluijs *et al.*, 2007a). Although it has been suggested that the carbon dioxide concentration in the atmosphere fluctuated from <300 to >2000 ppm during this period, evidence seems to suggest that excursions to high CO_2 did not necessarily correspond with warm pulses (e.g. Veizer, Godderis and François, 2000). This is contrary to records from the earlier Mesozoic, and then again in the late Eocene (when CO_2 re-established itself as a climate driver), where a consistent pattern between relatively low CO_2 and relatively low temperature is evidenced (Royer, 2006). Therefore, in spite of the plentiful CO_2 and temperature data available, the driver(s) for the Paleocene and early Eocene warm climate still remains somewhat enigmatic.

Regardless of cause, during the PETM water throughout the oceanic water column warmed causing a massive change in the ocean chemistry and associated oxygen depletion (Thomas, Zachos and Bralower, 2000). Sea surface temperatures rose by up to 8 °C resulting in subtropical (~23 °C) temperatures in the Arctic (Sluijs *et al.*, 2006; Zachos *et al.*, 2001). Vertical, as well as latitudinal, thermal gradients were greatly reduced leading to both the largest extinction event in the deep sea in the last 90 million years (Thomas, Zachos and Bralower, 2000), and a coeval burst of evolution in the surface oceans (Bralower, 2002; Kelly *et al.*, 1996). This resulted in a rise in new morphotypes (excursion taxa) coincident with the carbon isotope excursion, which has helped to stratigraphically correlate this event worldwide (Bralower, Kelly and Leckie, 2002).

On land, notable climatic changes occurred including an increase in global temperatures, by an average of ~5 °C, in conjunction with enhanced humidity and precipitation (Zachos, Dickens and Zeebe, 2008). Major turnovers in the terrestrial floral and faunal realms are recorded in the fossil record with widespread dispersal and radiation of new lineages (e.g. Wing *et al.*, 2005). Neo-mammalian lines appeared at this time, proliferated and migrated rapidly at the expense of the primitive mammals that had evolved during the Late Cretaceous (Gingerich, 2006; Gingerich, Rose and Krause, 1980). Increased greenhouse gases provides one explanation for Eocene warmth (Pearson and Palmer, 2000) but the concentrations necessary to sufficiently warm high latitudes and continental interiors are so great that climate models have had difficulty matching them to tropical sea surface temperature reconstructions — yet other works (e.g. Head *et al.*, 2009) also suggest very high tropical temperatures based on terrestrial fossil data. Other possible drivers include warm waters from the tropics feeding the deep oceans and high latitudes maintaining polar warmth during this time and sea ice feedback mechanisms (cf. Zachos *et al.*, 1993, 1996). However, recent studies show that neither atmosphere nor oceans transported substantially more heat in the Eocene relative to today (Huber and Nof, 2006).

At the end of the EECO, during the middle and late Eocene, global temperatures cooled once again over the ensuing 17 Ma. The driver for this downturn is thought to be falling atmospheric CO_2 (Eagles, Livermore and Morris, 2006; Pagani *et al.*, 2005). Sea temperature dropped from ~12 °C to ~4.5 °C (Zachos *et al.*, 2001) and Antarctica witnessed the inception and expansion of ice sheets (Zachos and Kump, 2005; see Chapter 9). The cause for this decrease in pCO_2 is still not well-understood, but circulation model experiments suggest that the opening of even a shallow (<1000 m) Pacific–Antarctic gateway (Drake Passage) at mid-southerly latitudes would be enough to upset global circulation and increase long-term carbon sequestration in the oceans[4] resulting in falling atmospheric CO_2 concentration (DeConto and Pollard, 2003; Eagles, Livermore and Morris, 2006; Tripati *et al.*, 2005). Moreover, the increase in biological activity recorded in the oceans at this time, stimulated by the initiation/increase of the Antarctic Circumpolar Current (ACC), might

---

[4] Prior to ~52 Ma, organic carbon was stored in coals, coal swamps and peatlands and could easily be remobilised (through wildfire for example) and thereby sustaining high atmospheric CO_2 concentrations. Later carbon burial switched to the oceans, at first in the euxinic seas for a short period and then in the deep ocean sediments where it could be locked away from the atmosphere for longer timescales (Eagles, Livermore and Morris, 2006; Kurtz *et al.*, 2003).

have indirectly contributed to lowering the atmospheric pCO_2 (Miller *et al.*, 2009; Scher and Martin, 2006).

In the southern high latitudes the early Cenozoic climate witnessed the continued decline in temperature that had started following the mid-Campanian optimum (Chapter 7) and, after undergoing minor perturbations in the late Maastrichtian, culminated in relatively cool–cold conditions that marked the short mid-Paleocene cool episode (Barrera *et al.*, 1987; Dingle and Lavelle, 1998a; Ditchfield, Marshall and Pirrie, 1994). These changing conditions were reflected in the temperature of marine waters that fluctuated between warm (~18 °C) and cool (~7 °C) conditions from the Late Cretaceous through to the early Paleogene. The mid-Paleocene cool episode was marked by the influx of cooler planktonic foraminiferal faunas into the Weddell/proto-Southern Ocean area (Barrera *et al.*, 1987; Dingle and Lavelle, 2000; Ditchfield, Marshall and Pirrie, 1994; Huber, Hodell and Hamilton, 1995; Pirrie and Marshall, 1990; Pirrie, Marshall and Crame, 1998). Then at ~59 Ma, during the late middle Paleocene, the climate began warming – an event that was to last for at least six million years.

This gradual amelioration was interrupted by the sudden PETM warming event at 55 Ma. High humidity, warmth and little or no polar ice characterised the southern high latitudes (Askin, 1992; Bijl *et al.*, 2009; Hollis *et al.*, 2009; Leckie *et al.*, 1995; Zachos *et al.*, 1993). Other research, however, has found evidence for significant sea level variations at this time (Pekar *et al.*, 2005) and continental ice may have indeed existed on Antarctica, despite the globally warm temperatures (Sluijs, 2006), even though the first ephemeral ice sheet on Antarctica would not appear for another 10 Ma (Birkenmajer *et al.*, 2005; Ivany *et al.*, 2008; Westerhold and Röhl, 2009). During this brief (270 ka) period of warmth, East Antarctica experienced an increase in the continent-to-ocean temperature gradient as Southern Ocean sea surface temperatures reached acmes of 18–22 °C. As the transport of atmospheric heat towards the pole increased, conditions similar to modern subtropical–tropical areas manifested across Antarctica, with soil temperatures remaining above a minimum of 15 °C for much of the year (Robert and Kennett, 1992, 1994). The increase in temperature and perennial rainfall (which was to become more seasonal with alternating wet and dry seasons in the ~120 ka following the peak) resulted in a major increase in chemical weathering as evidenced by stable isotopes and clay minerals (Dingle and Lavelle, 1998a, 2000; Robert and Kennett, 1992, 1994).

Recovery from the PETM was more gradual, taking between ~200 to 30 kyr from the onset of the event to return to more 'normal' conditions globally (Farley and Eltgroth, 2003; Röhl *et al.*, 2000). This recovery time was relatively rapid (given that the residence time of carbon in the atmosphere today is 100–200 kyr). Therefore, in order to explain this rapid recovery various feedback mechanisms have been invoked, which include an intensified flux of organic carbon from the ocean surface to the deep ocean, and its subsequent burial through biogeochemical feedback mechanisms, changes in boundary conditions, increased residence time of atmospheric methane and/or changes in palaeogeography (Bains *et al.*, 2000; Royer, 2006).

Warming continued until ~50 Ma followed by the onset of long-term cooling that would culminate in widespread Antarctic glaciations at the Eocene–Oligocene boundary. This cooling was most likely to have been driven by atmospheric pCO_2, as values fell from >2000 ppmv in the Paleocene and early Eocene to near modern values (<500 ppmv) in the Neogene (Pagani, Arthur and Freeman, 1999; Pearson and Palmer, 2000). Amid this cooling through the middle Eocene, at sites across the Southern Ocean, another important climatic reversal occurred. This event has been designated the Middle Eocene Climatic Optimum (MECO) and is accompanied by several other short-lived warming and cooling phases (Bohaty and Zachos, 2003). Between ~42 and 41 Ma large negative $\delta^{18}O$ excursions found as far apart as Italy (e.g. Jovane *et al.*, 2007) and Antarctica (Ivany *et al.*, 2008) indicate that this was indeed a global warming event. In the Southern Ocean, warming began at ~42 Ma and peaked 500 kyr later at 41.5 Ma, before cooling to pre-event temperatures within ~200 kyr (Bohaty and Zachos, 2003). Climate perturbations were ubiquitous high southern latitude phenomena, experienced on shelf as well as in the open oceans (Ivany *et al.*, 2008). The MECO is therefore both distinct and one of the most rapid hyperthermal events to have taken place during the Cenozoic. Although as yet not fully resolved, it is thought that changes in pCO_2 through increased volcanism or metamorphic activity associated with plate reorganisations once again played an important role in the short-term climate variability during the late middle Eocene (Boharty and Zachos, 2003). As sea temperatures of the Southern Ocean warmed by 4–5 °C, precipitation and consequent runoff increased. This climatic event dramatically affected the biological communities and coastal environments across the southern high latitudes around Antarctica and connecting landmasses (Boharty and Zachos, 2003; Boharty *et al.*, 2009; Ivany *et al.*, 2008).

Southern high latitude landmass and vegetation

By the Paleocene, the once great landmass of Gondwana that had straddled the Southern Hemisphere was now reduced to East and West Antarctica, attached to Australia and South America respectively (Figure 8.2). New Zealand had by now separated from the rest of Gondwana through the opening of the Tasman Sea and Southern Ocean, which separated the New Zealand–Lord Howe Rise complex from Australia–Antarctica. New Zealand remained within the cool to warm temperate region of the mid-high latitudes where tropical and polar air and water masses met, and sea surface temperatures were as high as ~20 °C (Huber *et al.*, 2004; Killops *et al.*, 2000; Mildenhall, 1980; Shackleton and Kennett, 1975). On land the podocarp dominated coastal plain swamps that had characterised the Late Cretaceous were being replaced by an increasing abundance of angiosperms. Subsequently during the Paleocene, the New Zealand vegetation became characterised by mesothermal conifer-dominated multistratal forest, lowland rainforests, with abundant angiosperms such as Epacridaceae and Proteaceae (Killops *et al.*, 1995; Pole, 1998a, 1998b), fern-dominated

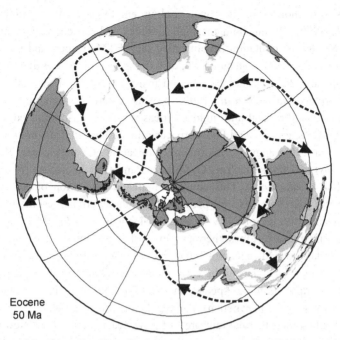

Figure 8.2 View of Gondwana break-up for the Eocene warm period. Latitude intervals are 30°. Note continuous connections still in existence between South America and Antarctica through the Peninsula region of West Antarctica. Arrowed, dotted lines indicate direction of the palaeocurrents in existence at the time.

open heathland, and *Nypa* mangroves along the coast. Interestingly, the aberrant shifts in climate associated with the PETM were not reflected in changes in the terrestrial plant communities of New Zealand (Crouch and Brinkhuis, 2005).

During the Paleocene, Australia was still joined to East Antarctica via a land bridge extending along the South Tasman Rise connecting Tasmania, at ~70° S, to northern Victoria Land. As spreading continued during the early Cenozoic, this land bridge began to disintegrate such that by the early Eocene, the gateway was open to marine waters (Stickley *et al.*, 2004). In the late Eocene, the final detachment of the western South Tasman Rise from Antarctica enabled changes in oceanic circulation patterns (Brown, Gaina and Müller, 2006) and deep marine conditions were in existence by about 35.5 Ma (1.8 Ma before the Eocene–Oligocene boundary) (Stickley *et al.*, 2004). The last remaining contact between Antarctica and Australia was thus severed by the end of the Eocene.

Whilst still in contact with Antarctica, Australia spanned a huge latitudinal extent (Veevers, Powell and Roots, 1991), which was reflected in both the climate and vegetation. The northern margin extended to ~40° S, allowing northeast Australia to enjoy sea surface temperatures characteristic of the subtropics today (Feary *et al.*, 1991). The southern margin lay at a polar position of ~70° S where the climate in the southeastern highlands of New

South Wales was moist with seasonal cooling emphasised by the high local relief (Taylor *et al.*, 1990). Within both the high latitudes and the continental hinterlands an increase in seasonal precipitation accompanied the PETM (Robert and Chamley, 1991). Even though the Paleocene vegetation of Australia is sparsely represented, evidence of a complex temperate rainforest has been found that yields support for both mesomorphic and sceleromorphic taxa (Hill *et al.*, 1999). In southeastern Australia, plant communities that occupied lowland/coastal and highland areas were conifer-dominated wet forests of predominantly Araucariaceae and Podocarpaceae (including *Lagarostrobus* swamp forest) along with broad-leafed angiosperms (mainly Proteaceae, Casuarinaceae, *Eucryphia*, but also *Nothofagus*) and ferns (such as Cyatheaceae and Gleicheniaceae) (Gonzalez *et al.*, 2007; Macphail *et al.*, 1994). This conifer-dominated vegetation of Australia has few possible modern parallels, except perhaps the microthermal/mesothermal podocarp-dominated swamp forests of New Zealand (Macphail *et al.*, 1994). Across inland Australia a shifting mosaic of mesothermal rainforest vegetation existed, similar in composition to the coastal/lowland vegetation but with Cunoniaceae and/or Proteaceae dominating and only rare *Nothofagus*. During the early Paleocene (Danian) the angiosperm element, particularly ancestral *Nothofagus* species, increased in both diversity and abundance such that by the late Paleocene (Thanetian), the vegetation was more akin to the *Nothofagus*-conifer-broad-leafed mesothermal rainforest of New Zealand and Papua New Guinea/New Caledonia. Minor differences are evident in the vegetation across the lowlands of southeastern Australia at this time, which suggests little altitudinal zonation (Macphail *et al.*, 1994).

In the late Paleocene of southeast Australia, plants with scleromorphic characters, such as thick leaves with high vein density and well-defined areoles, belonging to the Proteaceae (e.g. *Banksieaephyllum*) and Myrtaceae began to evolve (Hill and Merrifield, 1993). Considering that seasonal precipitation increased at this time and that these scleromorphic plants occurred alongside vegetation representing closed forests growing under high rainfall conditions, it is supposed that scleromorphy, especially as preserved in the Proteaceae, evolved as a response to low soil nutrients (in particular low phosphorous and nitrogen), prior to enabling adaptation to low water availability (Hill, 1998).

By the early Eocene (~55–49 Ma) as a result of increasing global warmth and high year-round humidity, coupled with repeated marine inundation of the low-lying coastal plains, the rainforest vegetation of the coastal plains and low-lying land had become distinctly megathermal in character (Macphail *et al.*, 1994). The angiosperms were highly diverse with increasingly abundant *Nothofagus*, and mangroves that are today restricted to tropical regions (Macphail *et al.*, 1994; Pole and Macphail, 1996; Truswell, 1993). Similar megathermal rainforest vegetation, which are also *Nothofagus*-poor, are today confined to the lowland tropics where mean annual temperatures are in excess of 24 °C. Away from these coastal/lowland areas, rainforest communities were more mesothermal in character although many species were shared with the megathermal forests. Strong ecological differentiation was prevalent with (local) *Nothofagus*-dominated communities largely confined to Tasmania and the southeastern Highlands, with a mosaic of rainforest communities characterising inland central Australia with dominance varying between gymnosperms (including Cupressaceae and/or Taxodiaceae), Casuarinaceae, Cunoniaceae and Myrtaceae, and an evolving proteaceous community (Macphail *et al.*, 1994).

The high moisture levels of the early Eocene probably moderated high latitude seasonal effects allowing these forests to expand further across Antarctica. This may account for fossils with similarity to Australasian taxa being found preserved in the Antarctic Peninsula. In Tasmania, the forests were conifer dominated with abundant podocarps and few *Libocedrus* and *Araucaria*. The diversity of angiosperms, however, was also relatively low and included *Nothofagus* along with elements such as *Eucryphia*, Lauraceae, Proteaceae and Casuarinaceae. The climate conditions were quite different from those of any modern environment and this was reflected in the structural make up of the forests: some elements lacked close affinities to modern plants, whilst those that do have living relatives are related to taxa growing today in microthermal environments or mesothermal–megathermal environments. The presence of taxa from these three climatic regimes, coexisting, might reflect possible altitudinal zonation as well as unfamiliar taxonomic associations. These forests have no modern structural or floristic analogue (Carpenter, Hill and Jordan, 1994).

On the other side of the landmass, West Antarctica was still connected to South America via Tierra del Fuego and the Antarctic Peninsula, thus enabling biotic interaction to take place. The climate of southern South America was predominantly warm temperate and humid with highly seasonal precipitation during the late Paleocene. Across the Paleocene–Eocene boundary, climatic changes affected the mid-latitudes as climate became more subtropical/tropical with an increase in year-round rainfall (Raigemborn *et al.*, 2009). The vegetation was markedly diverse and adapted to the wet paratropical climatic conditions of the Paleocene. Early Paleocene floras in northern and central Patagonia were still gymnosperm dominated but megathermal angiosperms were present and increasing in diversity (Iglesias *et al.*, 2007; Palazzesi and Barreda, 2007). Within the gymnosperms, the Cheirolepidiaceae was an important coniferous element, along with Araucariaceae and Podocarpaceae (*Dacrydium*, *Podocarpus*, *Phyllocladus*, *Microcachrys*), comprising the forest canopy (Barreda and Palazzesi, 2007). The understorey was dominated by ferns belonging to taxa such as *Lygodium*, Gleicheniaceae, Matoniaceae, Cyatheaceae and Dicksoniaceae (Iglesias *et al.*, 2007; Palazzesi and Barreda, 2007). Neotropical megathermal angiosperm taxa were abundant and included Arecaceae (*Nypa* and palms), Olacaceae (*Anacolosa*), Pandanaceae (*Pandanus*), Symplocaceae (*Symplocos*), alongside ?Mimosoideae, Chrysobalanaceae, Vitaceae, Menispermaceae, Lauracae, Urticaceae, Fabaceae, Malvaceae ('*Sterculia*'), Rosaceae the southern gondwanan lineage *Akania* (Akaniaceae), Proteaceae, Cunoniaceae, Nothofagaceae, Gunneraceae and Myrtaceae (Barreda and Palazzesi, 2007; Berry, 1937; Iglesias *et al.*, 2007), and even the more northern hemispheric element, Ulmaceae (Archangelsky and Zamaloa, 1986; Barreda and Palazzesi, 2007; Romero, 1968).

With increasingly tropical conditions during the late Paleocene–early Eocene, *Anacolosa*, Myrtaceae and Proteaceae were able to reach the southernmost tip of Patagonia where a high diversity of megathermal Lauraceae dominated the vegetation (Freile, 1972; Suarez, de la

Cruz and Troncoso, 2000). During the late Paleocene–Eocene, richly diverse, mixed rain-forest communities similar to those growing today in Australia, southeastern Asia and southeastern Brazil, covered central Patagonia. Austral taxa such as Nothofagaceae, Cunoniaceae and Podocarpaceae, dominated alongside more temperate and tropical elements such as Styracaceae and Araliaceae (Iglesias *et al.*, 2007; Raigemborn *et al.*, 2009). The coastal lowlands were covered by mangrove (*Nypa*) communities (Petriella and Archangelsky, 1975) and swampy areas were colonised by Restionaceae and *Sphagnum*. The presence of many taxa such as *Anacolosa*, *Gunnera* (Gunneraceae), and the diversity of ferns all suggest that rainfall was high. Interestingly *Nothofagus*, one of the most important elements in the relictual Southern Hemisphere flora today, is relatively uncommon in both the palynological and macrofossil record of southern South America during the Paleocene (Archangelsky and Romero, 1974; Iglesias *et al.*, 2007; Palazzesi and Barreda, 2007; Romero, 1973). Although this might simply represent local rather than regional abundance, being a copious pollen producer its poor fossil record might indeed suggest that it was only a small component of the overall vegetation. This in turn may shed light on possible migration pathways at this time.

By the early Eocene, the vegetation in central Patagonia had become highly diverse with abundant megathermal elements supported by warm temperatures and abundant rainfall, and winter mean temperatures staying above 10 °C (Gandolfo *et al.*, 2004; Melendi, Scafati and Volkheimer, 2003; Wilf *et al.*, 2003, 2005; Zamaloa and Andreis, 1995). In Patagonia, one flora growing during the EECO now preserved at Laguna del Hunco (47° S) testifies to a rich subtropical vegetation in existence at this time, which was undergoing *in situ* speciation and diversification in response to the warm temperatures and increased rainfall (Wilf *et al.*, 2003), with strong taxonomic links to Australasia. Fossil evidence confirms that meso-megathermal rainforests continued to be gymnosperm dominated with a diverse angiosperm component. The conifer element comprised podocarps (*Dacrydium*, *Lagarostrobus*, *Dacrycarpus*, *Microcachrys*) and Araucariaceae (Melendi, Scafati and Volkheimer, 2003) and minor Taxaceae/Cephalotaxaceae (Brea, Bellosi and Krause, 2009). Angiosperms were represented by abundant Proteaceae (*Beauprea*), Juglandaceae, Myrtaceae (*Myrcia*), Sapindaceae (*Schmidelia*, *Cupania*), Lauraceae, Rubiaceae (*Coprosma*), Casuarinaceae (*Gymnostoma*), Trimeniaceae, palms and bamboo (*Chusquea*) (Barreda and Palazzesi, 2007; Frenguelli and Parodi, 1941). The forest understorey comprised ferns and club-mosses belonging to the Cyatheaceae, Schizaeaceae, Dicksoniaceae, Polypodiaceae, Lycopodiaceae and bryophytes. Other taxa present (with broader climatic requirements) include cycads and ginkgos, further Proteaceae (*Lomatia*) and certain malvalean elements (Barreda and Palazzesi, 2007; Wilf *et al.*, 2005). The vegetation did not comprise all wet forest: locally dry habitats were exploited by more xerophytic elements such as Amaranthaceae, along with Casuarinaceae and Proteaceae (*Beauprea* sp., *Symphyonema* sp.) and Myrtaceae (*Eucalyptus*) whose relatives grow in Australia today (Palazzesi and Barreda, 2007; Gandolfo *et al.*, 2011). Low statue, more open vegetation is supported by the presence of shrub and low trees such as Anacardiaceae, Celtidaceae (*Celtis*) and Fabaceae (*Cassia*), possibly also adapted to more arid conditions in a generally wet continent (Barreda and Palazzesi, 2007; Hünicken, 1955; Wilf *et al.*, 2003, 2005). Further south, the presence of taxa related to Juglandaceae, Tiliaceae-Bombacaceae and Chloranthaceae and Arecaceae indicate that these megathermal forests were able to extend deeper into the higher latitudes (Zamaloa and Andreis, 1995). Interestingly though, throughout the early and middle Eocene, still no evidence has been found for *Nothofagus* forests in the Patagonian vegetation, even though such forests were widespread across the Antarctic Peninsula at this time (Askin, 1990). This may suggest that these forests were restricted to the areas of highland to the west that were undergoing tectonic uplift to form the Andes.

Throughout the Eocene, volcanic disturbance began to play a major role as a regional forcing factor in both the composition and structure of the vegetation. Ash soils provided drier niches that were subsequently colonised by xerophytic taxa (Barreda and Palazzesi, 2007). The climate became drier, as evidenced by evaporites and calcareous palaeosols, so that the warm, wet climate of the preceding Paleocene–Eocene was gradually replaced by a warm, semi-arid climate with mean annual temperatures above 20 °C (Melchor, Genise and Miquel, 2002). This heralded a change in ecology. By the middle Eocene, the prevailing mixture of Neotropical and Antarctic elements in the South American flora ('Palaeoflora Mixta' of Romero, 1986) and expanding grasslands (Spalletti and Mazzoni, 1978) highlight the difference between the Antarctic floras to the south (see below) and the more northern, southern South American floras.

Palaeogeography of Antarctica

The continued fragmentation of Gondwana during the Cenozoic resulted in the progressive isolation of Antarctica at the bottom of the world. This isolation brought with it the gradual establishment of a more polar oceanic circulation route along the southern edge of the Pacific Ocean, which in turn contributed to a number of factors that would influence the change from Paleocene greenhouse to late Eocene icehouse conditions in the southern high latitudes (Lawver and Gahagan, 2003; Zachos *et al.*, 2001).

Although the extent of West Antarctica's connection to East Antarctica is poorly understood, the same is true of the land bridge that connected the Antarctic Peninsula to South America. The Antarctic Peninsula was an emergent volcanic arc, similar to the Andes mountain range of South America today. The volcanic arc is thought to have consisted of several landmasses (Lawver, Gahagan and Coffin, 1992), possibly forming a series of islands separated by shallow marine basins (e.g. Huber and Webb, 1986). The connection between the land mass of the Antarctic Peninsula and that of South America has been suggested as being either a narrow (100–200 km) strip of continuous land (Shen, 1998), a series of land bridges, or indeed a chain of islands. Regardless of its extent and form, there existed a continuous geologic provenance between these two continents from the Late Cretaceous into the Paleogene (e.g. Case, 1988; Grunow, Kent and Dalziel, 1987; Woodburne and Zinsmeister, 1984). Yet in spite of this connection, evidence supports the existence of an isolating barrier – climatic, geographic or topographic – between South

America and the Antarctic Peninsula (Cantrill and Poole, 2005; Poole and Cantrill, 2006; Reguero, Marenssi and Santillana, 2002), resulting in distinct eco-latitudinal differences. Nevertheless, biotic interchange probably persisted well into the Cenozoic, although it is uncertain how far this interchange would then have continued into West Antarctica, and ultimately to East Antarctica, since *in situ* plant assemblages of known Cenozoic age from East Antarctica are, to date, still rare until the subsequent period of Eocene cooling (Chapter 9). Ocean Drilling Project (ODP) sites from around East Antarctica have at best yielded pollen and occasional plant macrofossils (Hill, 1989; Truswell, 1990)[5], whereas other such projects, for example the Cape Roberts Project, failed to locate any early Paleogene strata.

The only extensive evidence for Cenozoic Antarctic vegetation is restricted to the Antarctic Peninsula region although clasts containing rare vestiges of Cenozoic flora are also known from East Antarctica. In the Antarctic Peninsula region, the Cenozoic sequences contain well-preserved plants. Rare angiosperm remains have been found in the Elgar Uplands of Alexander Island and Adelaide Island (Thomson and Burn, 1977), but it is the 5–6 km thick Late Cretaceous to Cenozoic marine strata in the James Ross Basin and Late Cretaceous to Cenozoic terrestrial, volcaniclastic sequences from the South Shetland Islands that provide abundant palaeofloras. These floras allow us to glimpse the extensive vegetation that covered the Antarctic Peninsula during the early Cenozoic warming event that culminated in the Middle Eocene Climatic Optimum prior to the global cooling during the late Eocene.

Ecology of the Antarctic Peninsula during the Paleocene to Eocene warm period

Subarctic environments today are generally associated with coniferous vegetation, often exhibiting a stunted and deformed (*krummholz*) appearance, yet during the Paleocene to Eocene warm period, dense forests of angiosperm trees growing to full stature (at least 14 m tall; Cantrill and Poole, 2005) were able to grow in polar regions. Evidence for the extent of this arboreal angiosperm migration can be found located towards the southern part of the Antarctic Peninsula on Adelaide Island (67° S) and in the Elgar Uplands of Alexander Island (71° S) where ~53 Ma old volcanic strata has preserved facets of the vegetation (McCarron and Millar, 1997). Although these high latitude remains tend to be fragmentary, they provide enough detail for us to conclude that floras with a rich diversity of large-leafed (>10 cm) dicotyledonous angiosperms, including *Nothofagus* (Jefferson, 1980; Thomson and Burn, 1977), were able to reach far south during the late Paleocene–early Eocene (Askin, 1992). Unfortunately, since definitive evidence for the Paleocene–Eocene event is lacking and a

---

[5] Whilst this book was in press evidence from an ocean sediment core recovered off the Wilkes Land coast of East Antarctica (palaeolatitude ~70° S) was published providing the first well-dated record of EECO climate and vegetation for East Antarctica. Palynological and organic geochemical climate proxies for 53.9 to 51.9 Ma indicate coastal environments characterised by a warm climate (MAT 11–21 °C, MST 16–26 °C, MWT 6–16 °C and essentially frost-free, MAP >1000 mm yr^{-1}) supported highly diverse, paratropical, lowland rainforests with mesothermal to megathermal floral elements including ferns, tree ferns, palms (Arecaceae) and dicots (Bombacoideae, Strasburgeriaceae, Proteaceae, Olacaceae and Araceae). The continental interior was noticeably cooler (MAT 6–12 °C, MST 13–21 °C, MWT 3–7 °C, MAP >1000 mm yr^{-1}) supporting temperate rainforests characterised by *Nothofagus, Araucaria,* Proteaceae and podocarps. As climates cooled into the mid-Eocene (MAT dropped by ~2 °C) the coastal temperate biome expanded at the expense of the paratropical rainforests [Pross, J., Contreras, L., Bijl, P. K., *et al.* (2012). Persistent near-tropical warmth on the Antarctic continent during the early Eocene epoch. *Nature*, **488**, 73–77].

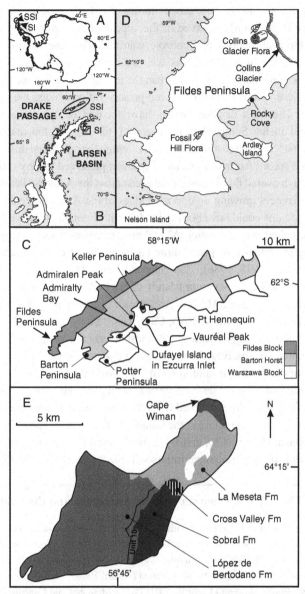

Figure 8.3 Maps indicating the location of South Shetland Island (SSI) and Seymour Island (SI) in relation to **A**, Antarctica and **B**, the Antarctic Peninsula. **C**, King George Island (in the South Shetland Islands) with the delimitations of the three tectonic blocks and place names mentioned in the text. **D**, enlargement of Fildes Peninsula (King George Island) with place names and fossil floras indicated in the text. **E**, Seymour Island in the Larsen Basin with the delimitations of the formations and place names mentioned in the text.

hiatus separates the sediments laid down during the Early Eocene Climatic Optimum (53–51 Ma), there is no record preserved of the vegetation that grew in these southern high latitudes during the peak periods of Cenozoic warmth[5]. Thus the maximum extent of this vegetation type into the high latitudes as well as its composition still remains uncertain. Nevertheless ideas concerning large-scale changes in high latitude vegetation during these greenhouse conditions can be gained from well-preserved floras along the northern Antarctic Peninsula region. The most detailed studies have focused on the floras of the terrestrial sedimentary record in the South Shetland Islands (~62° S) and the marine sediments on Seymour Island (~64° S) to the east (Figure 8.3). Although the latitudinal difference is relatively small, these islands record facets of dynamic vegetation preserved in very different geological settings. Even though overall floral similarity existed across the Antarctic Peninsula at this time, compositional differences growing on opposing sides of the Antarctic Peninsula indicate that the prevailing conditions could have been quite different. Enough is known about the vegetation for it to be described as a 'Palaeoflora Mixta' since thermophilous (neotropical) plants have been found alongside remains of plants more characteristic of cool temperate (subantarctic) biomes today (Romero, 1978, 1986). This Paleocene–early Eocene vegetation type extended northwards from at least 64° S (Seymour Island), through the connection with South American into Patagonia (and up to about 41° S – Argentina's Nahuel Huapi National Park) (Askin, 1988a; Birkenmajer and Zastawniak, 1986; Case, 1988; Poole, Mennega and Cantrill, 2003; Truswell, 1982), where the vegetation was at its most diverse floristically (Iglesias, *et al.*, 2007).

With time, the floras themselves show increasing compositional similarity and responsive dynamics to that of the Valdivian rainforests of southern Chile (see Case Study). This extant forest is now considered to be the last vestige of a vegetation that once extended across much of the southern latitudes at a time of global warmth. Here we will consider the floras preserved across the Antarctic Peninsula region, in the South Shetland Islands and on Seymour Island in the James Ross Basin (Figure 8.3), and try to understand the ecology across the Antarctic Peninsula during this time of global warmth (see Table 8.1 and Table 8.2 for summaries).

Case Study: The relictual Cenozoic flora of Antarctica and the Valdivian Model

Although some similarities can be found with the broad-leafed temperate forests of Tasmania and New Zealand (see Hill and Scriven, 1995 and references therein), the vegetation that once existed across Antarctica during the Paleogene at a time of global warmth is compositionally most similar to the Valdivian rainforests of southern South America (Askin, 1988a; Birkenmajer and Zastawniak, 1986; Case, 1988; Poole, Hunt and Cantrill, 2001; Poole, Mennega and Cantrill, 2003; Truswell, 1982) where neotropical and subantarctic elements coexist (Hueck, 1966). These Antarctic forests were characterised by podocarp and araucarian conifers alongside elements similar to *Dacrydium franklinii* (now growing in the cool temperate western Tasmanian rainforests), and southern beeches with an apparent scarcity of ferns and other gymnosperms. Similarity is not just confined to the floral assemblage but also to the tectonic setting: oceanic crust being subducted beneath a convergent continent margin resulting in mountain building, crustal melting at depth leading to volcanism and the development of large stratovolcanoes, and trade winds eastwards from across the Pacific bringing strong orographic rainfall. The stratovolcanic activity is a major cause of disturbance in both settings – along with associated events such as landslides, earthquakes, lahas, and damming and flooding of lake systems – such that plant community structure and composition of both the vegetation of Antarctica during the geological past and the Valdivian rainforests today are governed by disturbance, largely as a result of volcanic activity, high rainfall and altitude (Poole, Hunt and Cantrill, 2001; Veblen and Ashton, 1978).

The Valdivian rainforest is a cool temperate broad-leafed and mixed forest restricted to less than 440 000 ha in the central-northern part of Chile and extending into the western edge of Argentina. It is restricted to a narrow coastal strip between the Pacific Ocean to the west and the Andes Mountains in the east and extends from 37° 45' N to 43° 20' S (Veblen *et al.*, 1996). South of 45° this biome can be found on the chain of offshore islands which includes Chiloé Island. Above the tree line, about 2400 m in central Chile descending to 1000 m in the south, the forests are replaced by high Andean vegetation. To the north the Valdivian forests give way to a Mediterranean type vegetation, whereas to the south lies the Magellanic submicrothermal ecosystem. The most pronounced variations in average temperature occur latitudinally due to the elevation of the Andes. The minimum average annual temperatures vary between 4 and 7 °C, whereas the maximum annual average temperature can reach as high as 21 °C (CONAMA, 1999). Annual precipitation can vary between 1000 mm in the north to 6000 mm in the south, but decreases significantly on the eastern slopes of the Andes where rainfall levels drop to around 200 mm (Huber, 1979; Pérez, Hedin and Armesto, 1998).The Valdivian forests have evolved in isolation following their separation from other continental forest biomes since the mid-Cenozoic when global cooling led to the contraction of tropical vegetational belts in the Southern Hemisphere (Villagrán and Hinojosa, 1997). This isolation is responsible for the high levels of endemism (90% at the species level and 34% at the genus level for woody species), coupled with the high proportion of taxonomically isolated genera – several of them belonging to monogeneric families (of the 32 genera of trees exclusive to the forests, nearly 81% are monotypic such as Aextoxicaceae, Gomortegaceae, Desfontaineaceae, Eucryphiaceae, and Misodendraceae) – this suggests geological antiquity, long isolation and high rates of floral extinction during the Pleistocene (Arroyo *et al.*, 1996; Villagrán and Hinojosa, 1997). Phytogeographic studies of the Cenozoic fossil record in southern South America shows that 60% of the genera with tropical affinity disappeared from this region by the end of the Paleogene. Geographical barriers then prevented tropical species migrating back into this area during interglacials in the proceeding Pleistocene. Some areas in the Coastal Range remained ice-free during the Paleogene and served as refugia for the rainforest vegetation, but thermophilous taxa were displaced to other refugia to the north of 41° S (Smith-Ramírez, 2004). These forests are today probably more representative of the vegetation that grew across Antarctica during the Eocene than they were during the earlier Cenozoic when a higher number of tropical taxa would have been present.

The Valdivian rainforests are temperate broad-leaf and mixed forests with 700–800 species of vascular plants representing more than 200 genera. The forests are dominated by angiosperms (87–88%) with fewer ferns (10–12%) and scarce conifers (1%) (Smith-Ramírez, 2004). At least one third of the woody plants are of gondwanan origin with their closest relatives being found today in Australia, New Zealand, New Caledonia and Tasmania. Of the five main types of forest ecosystem in the Valdivian ecoregion, greatest similarity with the Antarctic vegetation lies with the forests occupying the middle region, namely (i) the laurel-leafed forests characterised by taxa such as *Laureliopsis philippiana, Aextoxicon punctatum, Eucryphia cordifolia, Caldcluvia paniculata* and *Weinmannia trichosperma*, with an understorey including *Myrceugenia* and *Luma*; (ii) the Andean forests distributed at high elevations dominated by conifers including *Araucaria araucana* and *Fitzroya cupressoides*, but giving way to scrublands of deciduous *Nothofagus* closer to the tree line – *Fitzroya cupressoides* dominates this forest in stature as well as age[6]; and (iii) the northern Patagonian forests that dominate the southern half of the region, with evergreen species such as podocarps (e.g. *Podocarpus nubigenus*), *Nothofagus dombeyi* and *Drimys winteri*. This vegetation has one of the highest incidences of pollination and dissemination by animals recorded in any temperate biome (Aizen and Ezcurra, 1998; Armesto and Rozzi, 1989). The vast majority of the woody plant genera (~85%) are pollinated by animals (Riveros, 1991; Smith-Ramírez, 1993; Smith-Ramírez and Armesto, 1998). About 20% of the woody plant genera produce red tubular flowers that are visited by a species of hummingbird unique to these forests, and many (50–70% of the woody plant species) produce fleshy fruit, presumably to attract frugivores highlighting the very strong dependence of plants on animals (Aizen and Ezcurra, 1998; Armesto and Rozzi, 1989; Armesto *et al.* 1987, 1996). Tragically, during the last decade 50% of these forests have been lost due to human activity with most of the timber harvested used for firewood and wood chips (Lara *et al.*, 2009).

The climax vegetation of the Valdivian rainforests occurs most extensively along the Coastal Range to the western flanks of the Andes. The higher, more xeric, slopes above the broad-leafed tree zone in the northern (~38° S) part of the Valdivian rainforest support Araucariaceae stands that give way to a more Podocarpaceae with cool-adapted Nothofagaceae dominated vegetation at slightly lower altitudes. Here, *Nothofagus* forms the emergent tree stratum and the podocarps, which are more diverse in height, contribute to both the dominant and subdominant stratum (Veblen *et al.*, 1996). In the old-growth, relatively stable forests (Veblen *et al.*, 1981) where catastrophic disturbance is absent, shade tolerant and slow regenerating tree species predominate rather than the typical pioneer *Nothofagus* species. On poorly drained sites, for example along ridge flanks at all elevations, and xeric sites some members of the Cupressaceae (e.g. *Fitzroya*, *Pilgerodendron* and *Austrocedrus*) contribute to the *Nothofagus* emergent tree stratum. On the lower slopes, *Nothofagus* is joined by emergent *Eucryphia* with other Cunoniaceae, Monimiaceae, Podocarpaceae (e.g. *Podocarpus nubigenus*, *Saxegothaea*) and some Cupressaceae (e.g. *Pilgerodendron*) forming a mixture of dominant and subdominant trees (Veblen, Schlegel and Oltremari, 1983; Veblen *et al.*, 1995). *Luma* and *Eucryphia* occur as canopy vegetation in low to mid-altitude positions (e.g. <700 m in the case of *Luma*). The understorey of these forests contains shade-tolerant tree species such as members of the Araliaceae, Lauraceae, Monimiaceae, Proteaceae and Myrtaceae along with tree ferns, other ferns and members of the Hydrangeaceae, which contribute to the scrambling, climbing element of the vegetation. The bamboo, *Chusquea*, tends to dominate the modern understorey vegetation in South America (Veblen, Schlegel and Oltremari, 1983), but is absent from similar forests in New Zealand and Tasmania. Cyatheaceae, Dicksoniaceae and some Osmundaceae are also found in these lower altitudinal regions where the humidity is relatively high, such as near watercourses. Here, in these damp habitats, *Equisetum* also grows. In undisturbed stands young or small trees of shade-intolerant species such as *Nothofagus*, *Eucryphia* and some Cunoniaceae are scarce but appear when light penetrates the canopy as a result of fine-scale gap dynamics. Such gap dynamics open up the canopy and allow shade-tolerant understorey species to gradually replace

---

[6] These conifers, although slow growing, are the tallest and oldest (3600+ years) species in South America, attaining heights of 60 m and diameters of 5 m, although individuals have been documented as reaching over 12 m in diameter before they fell victim to logging in the nineteenth and twentieth centuries.

scenescing emergent trees (such as *Nothofagus*) and which may, in turn, ultimately become dominant in the absence of any further disturbance (Veblen *et al.*, 1996). The fern component occurs in a number of different ecological niches. For example, in well-illuminated sites such as forest margins, Gleicheniaceae, Schizaeaceae and some Aspleniaceae predominate, whereas Dennstaedtiaceae dominates in similar localities at higher altitude. Blechnaceae, Hymenophyllaceae and other Aspleniaceae favour places with relatively low light regimes.

Situated at a boundary between two major crustal plates, the southern Andean region is an area of volcanic and tectonic activity. Disturbance in the Valdivian environment is largely driven by volcanic activity, with eruptions decreasing in severity and frequency further away from any volcanic edifice. Therefore, we would expect to see vegetation closer to the vent having greater abundance of taxa associated with early stages of secondary succession, while those further away would show a greater percentage of climax species. After widescale disturbances, the ensuing plant succession is governed by complex interlinked factors. These include the degree of damage sustained by the forest, initial taxonomic composition of the affected areas, time of year, presence or absence of a nearby regenerative flora, the stability of newly formed slopes, the degree of exposure and, in the case of volcanic disturbance, distance from the vent (Spicer, 1989; Veblen *et al.*, 1977). Vegetation close to the vent becomes buried or flattened (Spicer, 1989), whilst at decreasing altitude vegetation is partially buried or succumbs to forest fire resulting in complete to partial dieback of stand vegetation. Ash and lava often defoliates and/or buries plants and causes widespread tree fall. Fires and poisoning by noxious gases also result in plant death. In more distal locations the effects of the volcanism may be less severe, with limited fire and defoliation rather than death of the forest trees. Secondary effects of the eruptions can be equally disruptive with inundation and flooding of lakes leading to temporary damming of drainage systems and drowning valley-bottom floras.

Exposed sites in areas subject to relatively high precipitation enable *Gunnera*, with its nitrogen fixing algae, to initially dominate along with *Equisetum* (Spicer, 1989; Veblen *et al.*, 1996), whilst ferns of light-demanding genera, such as members of the Cyatheaceae and Gleicheniaceae, are good colonisers of ash. Substrates formed by recent glacial deposits such as moraines, supraglacial landslide debris and in-transit moraines, enable patches of temperate *Nothofagus*-dominated rainforest to establish (Veblen *et al.*, 1989). Often species of *Nothofagus* are the initial colonisers of these new deposits along with, or succeeding, *Gunnera* and some Ericaceae (*Pernettya*) on well-drained sites (Heusser, 1964). Sites in which the pre-existing vegetation has been removed and bare soils are exposed lends itself to the regeneration of most *Nothofagus* species. *Nothofagus antarctica*, for example, survives by sprouting adventitious roots on branches or from the base, whilst *N. pumilio* has the capacity to develop a new lateral root system closer to the new soil surface (Veblen *et al.*, 1977). Fires also promote vigorous resprouting of certain *Nothofagus* species thus enabling them to capitalise on the available light, resulting in even-aged cohorts of shade-intolerant *Nothofagus* (Burns, 1993; Veblen and Lorenz, 1987). *Drimys* also has the ability to become the dominant species in secondary forests after natural destructive events (Donoso, 1993; Navarro, Donoso and Sandoval, 1999; Willson and Armesto, 1996), although in older stands this species gradually looses its relative importance. Members of the Myrtaceae, *Laureliopsis philippiana* and *Eucryphia cordifolia*, on well-drained sites, are also pioneer species (Aravena *et al.*, 2002; Armesto and Figueroa, 1987; Armesto *et al.*, 1995; Donoso, 1993; Donoso, Escobar and Urrutia, 1985, Donoso *et al.*, 1984; Gutiérrez, Armesto and Aravena, 2004; Holdgate, 1961).

As succession continues, the pioneer taxa give way to shade-intolerant species of *Nothofagus* in those areas where conditions of constant disturbance favour pure stand *Nothofagus* growth. Elements of the flora further from the volcano, suffering from mass or partial defoliation, would releaf and continue to grow. *Nothofagus* ultimately becomes dominant in the successional forests, although substrate instability helps prevent canopy closure. This *Nothofagus* colonisation is consistent with the ability of certain *Nothofagus* species (e.g. *Nothofagus dombeyi*) to establish on weathered volcanic ash surfaces exposed by landslides in the mid-elevations of the Chilean Andes (Veblen *et al.*, 1989).

South Shetland Islands

The South Shetland Islands, lying to the northwest of the Antarctic Peninsula (Figure 8.3), have yielded the most complete Cenozoic terrestrial plant record. The main fossiliferous localities can be found within a ~3 km-thick sequence of volcanic and volcaniclastic deposits on King George Island. This is the largest of the South Shetlands Islands and is situated at 62° S, approximately coincident with its Cenozoic position (Lawver, Gahagan and Coffin, 1992). The Antarctic Peninsula Arc developed as a result of eastwards sub-duction of the Phoenix Plate along the western margin of Antarctica. This oblique sub-duction of the Phoenix Plate beneath the Antarctic Plate resulted in progressive ridge trench collision and transition from an active, to an inactive, margin (Eagles, 2003; Larter and Barker, 1991). In the South Shetland region the capture of the final ridge segment of the Phoenix Plate by the Antarctic Plate took place at 6.5 Ma and led to cessation of spreading on the Phoenix ridge system at 3.3 Ma. Following the cessation of subduction, rifting resulted in the formation of Bransfield Strait and the separation of the South Shetland Islands from the Antarctic Peninsula. Thus, although now geographically separated from the Antarctic Peninsula, the rocks on the South Shetland Islands record a change from intra-arc conditions in the Cretaceous to arc conditions in the Cenozoic.

Five distinct phases of magmatic activity have been recorded on King George Island with the earliest dating from the mid-Late Cretaceous (~108–107 Ma), followed by two in the early Eocene (~52 Ma and 51–45 Ma) during the Early Eocene Climatic Optimum, one in the middle–late Eocene (44–36 Ma) and the final one during the Oligocene (31–29 Ma) (Willan and Kelley, 1999). Although there are a number of fossiliferous localities on King George Island, many have either not been studied in detail, found to yield only fragmentary evidence, or the dating is uncertain. Therefore this discussion, focusing on temporal vegetational changes on the west of the Antarctic Peninsula, will concentrate on the well-studied Paleocene to Eocene sequences and the King George Island floras. (For a summary of published information regarding other known floras of uncertain age and/or compositions, see Appendix).

During much of the Mesozoic–Cenozoic, and spanning the Paleocene and Eocene, the environment on King George Island was dynamic and controlled predominantly by volcanic activity. Ongoing magmatic thermal events continued through the Paleogene (~61 to ~35.5 Ma) (Birkenmajer *et al.*, 1983; Park, 1989; Yeo *et al.*, 2004), such that today the island consists largely, with the exception of its uppermost part, of stratified calc-alkaline volcanic rocks with some nunataks representing the ancient volcanic centres (Barton, 1965; Birkenmajer *et al.*, 1986; Tokarski, Danowski and Zastawniak, 1987; Yeo *et al.*, 2004).

Towards the end of the Paleocene (Thanetian), the environment on King George Island was characterised by widely uniform environments resulting from a short interval lacking intense volcanic activity. Coupled with the ameliorating climate, a more diversified flora, in terms of diversity-indices and numbers of taxa, began to thrive – a similar scenario to that which was occurring across the world in both the Northern (e.g. Askin, 1990; Wise *et al.*, 1991; Wolfe, 1990) and Southern hemispheres (e.g. Greenwood, 1994; Hill, 1990). On King George Island, taxa number about 50 (including representatives of the Nothofagaceae, Proteaceae, Winteraceae, Lauraceae, Sapindaceae, Myrtaceae, Sterculiaceae and Smilacaceae, along with Cunoniaceae, Aquifoliaceae, Saxifragaceae, Anacardiaceae, Caesalpiniaceae, Melastomataceae, Araliaceae, Rhamnaceae, Rosidae, Monimiaceae and monocotyledons; Alves, Guerra-Sommer and Dutra, 2005), with the numbers of angio-sperms continuing to increase (Figure 8.4). Conifers were still represented by diverse podocarps as well as other taxa including Araucariaceae and Cupressaceae. This vegetation was more diverse than either the preceding Late Cretaceous or the superseding middle to late Eocene and may be a reflection of the relative climatic equability which peaked in the middle Eocene (Alves, Guerra-Sommer and Dutra, 2005). The floristic changes that took place during the late Paleocene–early Eocene were enhanced by a phase of uplift of the magmatic arc leading to a depositional break in the Eocene record. This break can be followed along the northern Antarctic Peninsula. The changing altitude resulting from this uplift gave rise to increasingly stratified forests. Change was also introduced through periodic phases of volcanic activity which resulted in distinct environmental changes that affected biotic diversity. All of these factors served to change the face of the ecological setting across the Antarctic Peninsula (Alves, Guerra-Sommer and Dutra, 2005).

The lithostratigraphic division of King George Island is complex and no single strati-graphic scheme exists. Birkenmajer (1981, 1989, 1990) and Birkenmajer *et al.* (1986) erected many local formations in comparison to a simpler scheme created by Smellie *et al.* (1984). Smellie *et al.* (1984) suggest two formations (the lower Fildes Formation in the western part of the island and the upper Hennequin Formation to the east) based on Barton's (1965) earlier work, whereas Birkenmajer (1982a, 1982b, 1982c, 1989; Birkenmajer, Soliani and Kawashita, 1990) suggest three tectonic blocks divided by large-scale strike-slip faults: the Barton Horst running along the central axis and forming the backbone of the island, the Fildes Block to the northwest, and the Warszawa Block to the southeast, with different lithologies for each block (Figures 8.3C and 8.5). Other more recent stratigraphic frameworks have been erected for specific areas of King George Island (e.g. Fildes Peninsula stratigraphy erected by Shen, 1999). Fossiliferous lithologies have been found on all three blocks, namely at Fildes Peninsula, Barton–Potter Peninsula (see Table 8.1), Ezcurra Inlet region and Point Hennequin (Figure 8.3C, D; see also Appendix).

Vegetation of the warm late Paleocene–early Eocene

The Barton Horst consists mainly of Paleogene volcanic-sedimentary rocks. The oldest Paleogene plant-bearing deposits can be found within the red tuffaceous layers outcropping on Barton Peninsula (Figure 8.3C) to the southwest end of the horst towards the middle of a

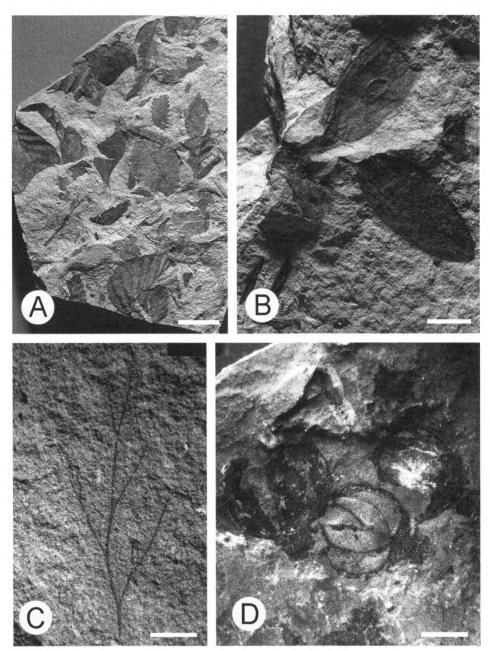

Figure 8.4 Plant material from King George Island. **A**, leaf mat from Dragon Glacier with a variety of different angiospermous leaf morphotaxa, BAS P. 3001.42. **B**, broad imbricate leaves of *Araucaria* (type 4 of Hunt, 2001) from Point Hennequin, BAS P. 2810.17.1. **C**, fronds of the fern ?*Lophosoria*, BAS P. 3029.19 from Rocky Cove, Fildes Peninsula. **D**, nothofagaceous cupules, BAS P. 3032.95 from Fossil Hill, Fildes Peninsula. Scale bars A = 5 cm, B, D = 1 cm, C = 2.5 mm. Photos courtesy of R. Hunt.

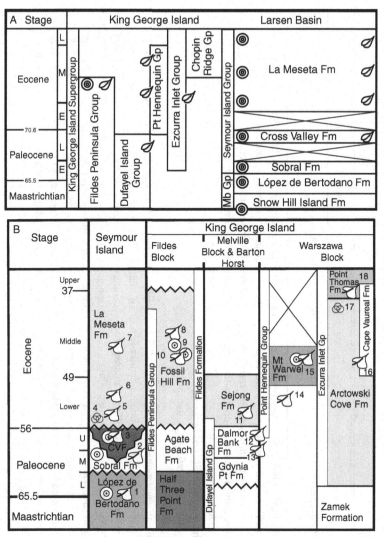

Figure 8.5 **A**, stratigraphic summary of key formations and groups for the Paleogene of the Antarctic Peninsula (East Antarctica not included). Units with fossil material are indicated. Adapted from Poole and Cantrill (2006). **B**, stratigraphic summary of A in more detail for Seymour Island and King George Island. Fossil floras discussed in the text are marked and include: 1 Lopez de Bertodano Flora Unit 10; 2 Sobral Flora; 3 Nordenskjöld Flora; 4 Valle de las Focas Allomember (pollen) Flora; 5 Acantilados Allomember (leaf, wood, pollen) Flora; 6 Campamento Allomember (wood and leaf) Flora; 7 Cucullaea I (pollen, leaf) Flora; 8 Fossil Hill Flora (leaves); 9 Collins Glacier Flora (*ex situ* wood); 10 Rocky Cove Flora (wood – youngest, leaves and pollen – older); 11 Sejong Formation (leaf) Flora; 12 Dufayel Island flora (leaf); 13 Barton Peninsula (leaf) Flora; 14 Potter Peninsula Flora (leaf); 15 Point Hennequin floras at Mount Wawel, Smok Hill, Dragon Glacier (*ex situ* leaf and wood); 16 Vauréal Peak Flora (leaf). Floras 17, Petrified Forest Creek Flora (pollen and spores), and 18 Cytadela plant beds (leaf), are included here for completeness but see Chapter 9 for discussion. See Appendix for floras not discussed in text. Adapted from Birkenmajer (2001); Birkenmajer and Zastawniak (1989a); Crame *et al.* (1991); Cunha, Dutra and Cardoso (2008); Dutra (2004); Elliot (1988); Hunt (2001); Kim, Sohn and Choe (2006); Riding and Crame (2002); and Shen (1994).

Table 8.1 *Summary of the vegetation and their modern vegetational and climatic analogues for King George Island (South Shetland Islands) as derived from nearest living analogues and multiproxy data. CMM = cold month mean, MAT = mean annual temperature, MAP = mean annual precipitation, MST = mean summer temperature, MWT = mean winter temperature, WMM = warm month mean. See text for further details and references.*

Stratigraphic age: Lithostratigraphy Location	Palaeovegetation	Modern analogue		Palaeoclimatic parameters
		Vegetation composition		
cool middle Eocene Middle Eocene Fildes Fm *Fildes Peninsula*	Floristically diverse *Nothofagus*-podocarp-fern association with other conifer taxa, angiospermous shrubs and diverse cycads, tree ferns, ferns and bryophytes forming the understorey.	Valdivian rainforests of southern Chile with the rare occurrence of conifers and ferns explained by taphonomical processes and/or local biases rather than a different vegetation type.		Warm to cool temperate MAT 10–15 °C MST 20–27 °C WMM ~20 °C CMM ~2.5 °C MAP ~875–~2100 mm
warm post-EECO early-middle Eocene Mount Wawel Formation Fildes Formation *Point Hennequin*	Changing diversity and in taxonomic composition reflecting a dynamic environment. Climax vegetation comprising a rich and diverse flora of the *Nothofagus*-conifer-fern assemblage with ~40 different angiosperm taxa, several conifer taxa (including podocarps) and diverse fern component becomes a taxonomically poor vegetation dominated by *Nothofagus* and Rosidae with few other angiosperms and rare conifers (including podocarps and *Araucaria*). Ferns, mosses and fungi form the groundcover.	Initially similar to the Valdivian rainforests of southern Chile; rare occurrence of conifers and ferns explained by taphonomical processes and/or local biases rather than a different vegetation type. As the climate cools so the vegetation assemblages begin to resemble those of western Patagonia where altitudinal stratification becomes more evident.		
Cape Vauréal Fm *Vauréal Peak*	*Nothofagus*-fern forest with conifers absent (this may be a local phenomenon reflecting environmental disturbance).			

Table 8.1 (cont.)

Stratigraphic age: Lithostratigraphy Location	Palaeovegetation	Modern analogue	
		Vegetation composition	Palaeoclimatic parameters
late Paleocene warm period late Paleocene– early Eocene Dalmor Bank Fm *Dufayel Island* Dufayel Island Group *Barton Peninsula*	Forests of angiosperms dominated by *Nothofagus*-Myrtaceae and laurophyllous angiosperms with other dicotyledonous angiosperms, rare ferns, monocotyledonous angiosperms and conifers.	Northern Valdivian rainforests of southern Chile (~37° 5′ to 43° S) and Mediterranean vegetation of Central Chile. Also the multi-specific broad-leafed forests of (warm) temperate regions.	Warm temperate High rainfall with distinct wet and dry seasons MAT 18–22 °C

45 m high cliff at the base of a well-exposed 30 m thick volcaniclastic sequence (Birkenmajer *et al.*, 1983; del Valle, Diaz and Romero, 1984; Tokarski, Danowski and Zastawniak, 1987). The age of this volcaniclastic sequence is still under debate. Potassium-argon and Ar-Ar dating have returned ages as old as ~61 Ma and as young as 41 Ma (Kim *et al.*, 2000; Lee *et al.*, 1996; Park, 1989; Smellie *et al.*, 1984), whilst K-Ar dating of the volcanics underlying the plant-bearing beds themselves return a late Paleocene (~58 Ma) age (Birkenmajer *et al.*, 1983; Tokarski, Danowski and Zastawniak, 1987). A second plant (leaf) assemblage preserved in fine-grained sandstones of the newly defined Sejong Formation on Barton Peninsula also suggests late Paleocene to Eocene deposition (Chun, Chang and Lee, 1994; Kim, Sohn and Choe, 2006). Even though the exact age of these rocks remains uncertain, there is general acceptance that the fossil material is preserved in strata laid down during the warming that culminated in the EECO. These floras are therefore considered to represent the vegetation that grew across the Antarctic Peninsula during the warming phase leading up to the Late Paleocene Thermal Maximum.

Floras of similar age and composition to those mentioned above are also found on Dufayel Island lying to the northeast of Barton Peninsula (Figure 8.3C). The floral assemblages are preserved in tuffaceous interbeds within the basaltic to andesitic lava sequences that cap the sequences of basal conglomerates and agglomerates of the Dalmor Bank Formation (Figure 8.5). The tuffaceous outcrops have been K-Ar dated to between ~51 and ~56 Ma (but has probably been affected by argon loss resulting in a younger bias), and 51.9 ± 1.5 Ma (Birkenmajer, 1989; Fisch and Dutra, work in progress) which, given the most recent dating, would suggest that they are more or less contemporaneous with either the PETM itself, or at least the period of early Eocene warming (Askin, 1992; Birkenmajer, 1980, 1985; Birkenmajer *et al.*, 1983; Birkenmajer and Zastawniak, 1986, 1989a).

The vegetation assemblages preserved on Barton Peninsula and Dufayel Island therefore represent the high latitude vegetation growing in the late Paleocene–early Eocene greenhouse world. The organs making up this flora became deposited in fluvial environments (del Valle, Diaz and Romero, 1984) and lacustrine settings (Birkenmajer, 1980) respectively, since the mineralised leaf impressions are small, fragmentary and lack cuticles. Both assemblages are dominated by dicotyledonous angiosperms and characterised in particular by the presence of *Nothofagus*. Compositional differences do, however, exist. On Barton Peninsula, the assemblage comprises entirely fragmentary dicotyledonous leaves. *Nothofagus* leaves have been identified as belonging to two species (*N. subferruginea* and *N.* cf. *densinervosa*) alongside leaves with affinities to the Lauraceae (*Ocotea, Nectandra*), Atherospermataceae (*Laurelia*), Sterculiaceae (*Sterculia*), Bixaceae and Cochlospermaceae (*Cochlospermum*) (del Valle, Diaz and Romero, 1984; T. Dutra personal communication, 2010; Orlando, 1963, 1964; Tokarski, Danowski and Zastawniak, 1987; Torres, 1990; Zastawniak, 1994). The only evidence of fern and conifer remains are a fertile frond (Hunt, 2001) and a few fragments of wood with araucarian affinity (Hee and Soon-Keun, 1991) and a podocarp leaf (Hunt, 2001). Conversely, on Dufayel Island, diversity is greater. Conifers are absent from the flora, and although ferns are present, they are rare. Monocots are also present although only as rare fragments (Birkenmajer and Zastawniak, 1986). As with the Barton Peninsula Flora, the vast majority of the vegetation on Dufayel Island was represented by diverse dicotyledonous angiosperm leaves from arborescent taxa. The remains are preserved as an allochthonous leaf assemblage of whole impressions. Some are complete with fine venation patterns, although many are represented simply by fragments with second-order venation patterns preserved. This dicot flora includes abundant *Nothofagus* again alongside taxa belonging to the Bixaceae (Cochlospermaceae), Dilleniaceae, Leguminosae, Myrtaceae, Sapindaceae, Sterculiaceae, Verbenaceae and leaf types of myrtaceous and laurophyllous affinity (i.e. similar to *Laurelia, Nectandra, Ocotea*) (Birkenmajer and Zastawniak, 1986, 1989a, 1989b; del Valle, Diaz and Romero, 1984; Tokarski, Danowski and Zastawniak, 1987). Fisch and Dutra (T. Dutra personal communication, 2010) undertook a revision of the flora based on new collections and also found that *Nothofagus* dominated the angiospermous leaf flora, represented by both evergreen and deciduous types, with a high proportion of Myrtaceae (including the first record of Leptospermoideae from the Antarctic Peninsula) and Lauraceae. They also found *Cupania* (Sapindaceae), *Myrica* (Myricaceae), *Sterculia* (Sterculiaceae), Monimiaceae, Celastraceae, Malvaceae and Anacardiaceae, but no Cochlospermaceae. The vegetation is thought to be reminiscent of the species-rich broad-leafed forests of mild temperate or warm temperate regions today, with a cool temperate element represented by *Nothofagus*. Similarities have been drawn with the extant Valdivian rainforests of southern Chile (Birkenmajer and Zastawniak, 1986; see Case Study). Fisch and Dutra agree with this assessment but are more specific and conclude (T. Dutra personal communication, 2010) that floristic similarity is shared with both the northern Valdivian rainforest and the drier Mediterranean vegetation of Central Chile.

From the evidence gained so far, the Barton Peninsula Flora and the flora of Dufayel Island both support the presence of warm temperate angiosperm taxa being able to grow in southern high latitudes during this globally warm period. The uniqueness of these two floras, which includes the rare occurrences of gymnosperms and ferns (both of which are known to have been an important element of the late Paleocene Antarctic vegetation) are not considered to be related to age or climate, but simply reflecting taphonomical processes (Hunt, 2001) and/or (local) compositional differences of the parent vegetation (Birkenmajer and Zastawniak, 1986).

Warm post-Early Eocene Climatic Optimum vegetation

Following the peak warmth at the EECO, conditions cooled. However, this took millions of years and evidence for vegetation that grew in the southern high latitudes during this time can be found recorded at various sites across King George Island. The newly discovered Vauréal Peak Flora located to the southeast of Point Hennequin (Figure 8.3C) on the eastern edge of the mouth of Admiralty Bay provides a glimpse of a unique vegetation that grew at this time (Hunt, 2001). Originally a late Oligocene date was proposed for the sediments exposed near Vauréal Peak, composed of local clasts overlying an incised volcanic basement (Birkenmajer, Soliani and Kawashita, 1989). However, this age is based on the flawed assumption that it is age equivalent with the basement (Troedson and Smellie, 2002; see Chapter 9). Hunt also dated the volcanics of the Cape Vauréal Formation that underlie the sedimentary beds where the plants are preserved. This returned a maximum radiometric age for the flora of ~49 Ma which dates the flora as latest early or earliest middle Eocene and thus just post-dating the peak of Early Eocene Climatic Optimum (Hunt, 2001; Zachos *et al.*, 2001). This flora is distinct in that it comprises leaf mats containing abundant ferns, such as *Gleichenia*, alongside angiosperm leaves belonging predominantly to a single *Nothofagus* species complete with evidence of herbivory (McDonald, 2009; McDonald *et al.*, 2007). No record has been recovered for an association with podocarps which would have suggested closer compositional similarity and, thus environmental similarity, to the more or less contemporaneous vegetation growing on Point Hennequin (Dragon Glacier; see below). This implies that the flora represents vegetation that grew at some distance from the source of the volcanic activity in a medial to distal setting (Hunt, 2001). It comprised a *Nothofagus* forest with fern understorey and represents a new facet of the overall vegetation, which in turn may reflect a different stage in the succession, possibly a pioneer vegetation type, recovering from ecological disturbance brought about by volcanic activity. *Nothofagus* may have quickly resprouted from multiple shoots in the crown as ferns recovered in the understorey. The absence of conifers might be explained by the longer time required by conifer taxa to recover from such events.

More detailed evidence of vegetational succession during this essentially warm, geologically active period between the EECO and the MECO can be derived from the floras preserved within the Mount Wawel Formation (Point Hennequin Group), located on the Warszawa Block. The macroflora deposits are collectively known as the Point Hennequin Flora after the locality (Figures 8.3C and 8.5), with individual localities named Dragon Glacier Moraine Flora (Figures 8.4A, B and 8.6), Smok (Hill) Flora and Mount Wawel Flora (Askin, 1992; Birkenmajer and Zastawniak, 1989a; Hunt, 2001; Hunt and Poole, 2003; Zastawniak *et al.*, 1985). These floras are either contemporaneous or slightly postdate the Vauréal Peak Flora. The vegetation was distinct from the forests that covered the land during the preceding warm period in that the composition had changed and diversity increased. This change in composition and diversity may be accounted for in part by the increase in specimens retrieved from these localities. Yet diversity changes are probably also due to shifts in prevailing palaeoenvironment as a result of ecological disturbance, changing climate, or a combination of driving forces. Ecological disturbance was probably the major driving force given the periodic volcanic activity that was taking place at this time. This dynamicism is illustrated by the two most diverse floras on King George Island, namely the Point Hennequin floras and the younger late middle Eocene Fossil Hill Flora on Fildes Peninsula (see below). These floras represent vegetation growing close to the source of volcanism, preserving a complex record of vegetational succession and environmental disturbance with elements of the pre-volcanic and syn-volcanic vegetation occurring together (Hunt, 2001).

The floras at Point Hennequin correspond to three successive plant-bearing sediment intercalations within the upper part of the Point Hennequin Group. The Point Hennequin Group consists mainly of andesite lavas which originally returned a K-Ar radiometric age of ~25 Ma (Oligocene–Miocene boundary) making them the youngest post-Polonez glaciation

Figure 8.6 Angiosperm plant material from Dragon Glacier Flora, King George Island. **A**, new species of *Dicotylophyllum* (i.e. *D. washburni* unpublished; Hunt, 2001), BAS P. 3001.5a. **B**, possible Proteaceae leaf, BAS P. 3001.67. **C**, angiosperm inflorescence, BAS P. 3001.61. **D**, Cunoniaceae leaf, BAS P. 3001.46. **E**, leaf of uncertain angiospermous affinity (Morphotype 1.5 of Hunt, 2001), BAS P. 3001.187. **F**, *Lomatia* (Proteaceae), BAS P. 3001.131. **G**, *Lauriphyllum nordenskjoeldii*, BAS P. 3013.7 with small elliptical feeding traces surrounded by a thin rim of reaction tissue. **H**, fossil leaf with greatest similarity to extant *Nothofagus antactica* of the subgenus *Nothofagus*, BAS P. 3001.100. **I**, leaflet probably belonging to Chloranthaceae, BAS P. 3001.83a. Scale bars A, B, D, F, G = 1 cm, C, E, = 2 cm, H = 5 cm, I = 5 mm. Photos courtesy of R. Hunt.

terrestrial flora in West Antarctica (Birkenmajer, 1997). As a consequence, they were interpreted as a low diversity post-glacial vegetation assemblage (Zastawniak, 1981; Zastawniak *et al.*, 1985). This vegetation was thought to have represented a ~30 Ma old vegetation type that had recolonised Point Hennequin from outlying refugia during a milder climate phase that succeeded peninsula-wide West Antarctic glaciation and extinction (Dingle and Lavelle, 1998a). However, K-metasomatism and Ar-loss is thought to have affected many K-Ar data for King George Island (Birkenmajer, 1990; Soliani and Bonhomme, 1994; Willan and Kelley, 1999), therefore a reassessment of the age data using ^{40}Ar/^{39}Ar was undertaken by Hunt (2001). These results indicate a much older pre-glacial middle Eocene age (~49–44 Ma) for the fossil floras from Point Hennequin (Hunt, 2001; Hunt and Poole, 2003). This corroborates another unpublished ^{40}Ar/^{39}Ar age of 47 ± 2 Ma, based on a sample originating from the Mount Wawel lavas (Dupre, 1982), and supports the argument put forward by Smellie *et al.* (1984) for an Eocene age for this strata. Moreover, the lack of any glacial sedimentation in the Dragon Glacier or Mount Wawel sediments also agrees with an older age. Analyses of the floras from Point Hennequin also reveal that the vegetation was much more diverse than previously realised and in turn supported an insect community (McDonald, 2009; McDonald *et al.*, 2007). Therefore, rather than being the species-poor post-glacial flora, the interpretation that they probably represent a relatively diverse vegetation growing under the milder conditions supports the revised older radiometric dates.

The Dragon Glacier Flora (Birkenmajer, 1981) (Figures 8.4A, B and 8.6) is concentrated in terminal moraines along the northern margin of Dragon Glacier, with the source beds lying 80 metres above sea level corresponding to Levels A and B of Fontes and Dutra (2010) (T. Dutra personal communication, 2010). The exotic plant-bearing blocks that contain the Dragon Glacier Flora can be subdivided into six facies. These are grouped into two facies associations, which are gradational in nature, suggesting that the exotic blocks sample through a complete fluvial-lacustrine section ranging from terrestrial fluvial/shoreline lacustrine sequences to deep water lacustrine facies (Hunt, 2001; Hunt and Poole, 2003). No evidence exists for the conditions being either brackish or marine given the abundance of palaeosols and plant material, and the lack of marine macro- or microfossils (Hunt and Poole, 2003). The sedimentological evidence derived from the fossiliferous blocks suggests that the volcanic source was medial to distal in proximity and the reworked volcaniclastic sediments suggest low levels of disturbance during the growth of this flora.

The second flora is located at Smok Hill where it is preserved in a separate sedimentary unit, more or less coeval with the Dragon Glacier Flora, that comprises a pale green, coarse air fall tuff. The beds from which these blocks originate lie 10–15 m above those that yield the Dragon Glacier Flora and correspond to Levels C and D of Fontes and Dutra (2010) (T. Dutra personal communication, 2010). The exotic blocks are sited close to the original beds and suggest a primary ash-fall deposit as a result of hydrovolcanic activity and a high level of disturbance (Hunt and Poole, 2003). As with the Dragon Glacier floral locality, the available sedimentological evidence derived from the fossiliferous exotic blocks at Smok Hill localities suggests a medial to distal volcanic source, but the primary ash-fall deposits possibly derived from hydrovolcanic activity imply high levels of disturbance. From comparisons with lithologies resulting from modern volcanic eruptions (see Burnham, 1994; Burnham and Spicer, 1986), the compositionally distinct sediments at these two sites could be interpreted as being laterally, rather than stratigraphically, distinct. However, based on their close relative geographical positions (all within ~1.5 km of the proposed vegetation source), coupled with the compositional differences of the preserved flora, they are interpreted as being stratigraphically distinct and thus representing different stages of the vegetation evolution in this area (Hunt, 2001).

In contrast the third flora, the Mount Wawel Flora, is preserved within *in situ* sediments that probably represent lacustrine sedimentation. Two facies can be distinguished: the lower coarse grey tuffaceous sandstone with coalified plant fragments including seeds and a cone scale, which may represent a leaf litter assemblage, and an upper fine grey-green siltstone with poorly preserved impressions, which may represent a lacustrine deposit (Hunt, 2001).

The volcanic centre that sourced these lavas and pyroclastic sediments on Point Hennequin, which resulted in the low-level volcanic disturbance, still remains unknown (Hunt and Poole, 2003). Birkenmajer (1981) and Zastawniak *et al.* (1985) originally proposed that the vent complex was at Point Hennequin itself. However, the new age for these fossiliferous lava sequences suggests that the lithologies originally inferred to be vent core strata by Birkenmajer (1981) and Zastawniak *et al.* (1985) may actually be younger intrusive bodies (Hunt and Poole, 2003). If this is the case, then it would imply a more distal source for the vent and volcaniclastic deposits at Point Hennequin than previously supposed (Hunt and Poole, 2003), with the source being possibly 5–6 km to the southwest, located between Point Hennequin and Point Thomas (J.L. Smellie personal communication in Hunt and Poole, 2003). However, with so many vents preserved on the island, more work is needed before any definite conclusions can be drawn.

Palaeoenvironmental reconstructions of the Point Hennequin strata suggest that intense pyroclastic volcanism led to localised damming of drainage networks. This gave rise to ephemeral lakes into which vegetative material and further sediments, derived from the erosion of loosely consolidated volcaniclastic deposits, were washed. Sedimentological characteristics found at the Dragon Glacier locality support the idea of pyroclastic volcanism, damming of local drainage networks and the formation of shallow, ephemeral lacustrine basins. Evidence for fluctuating water levels consistent with ephemeral lacustine basins comes from the cyclical development of palaeosols defined by rootlet beds, symmetrical wave-rippled horizons, sun-cracked surfaces and the absence of well-defined lacustrine fauna found at this locality (Barton, 1964a; Birkenmajer, 1981; Hunt, 2001). These conditions would be ideal for the preservation of organic material produced either through pyroclastic stripping of the surrounding vegetation or via normal taphonomic processes (cf. Burnham, 1994; Hunt, 2001; Spicer, 1989). Hydraulic activity by rivers or floods may have been responsible for the sorting of organic material. Some of this material (i.e. fruits, seeds and leaves) became trapped in the ephemeral lakes (Birkenmajer, 1980, 1981; Jardine, 1950), whilst other material (e.g. wood) may have been transported for some distance. This may account for the rare finds of petrified wood at Point Hennequin. Carbonised wood is, however, preserved at mid-altitude on Mount Wawel between the Smok Hill and Dragon Glacier beds and the Mount Wawel beds. The wood is very friable and has not been studied in any detail (T. Dutra personal communication, 2010). The amount of wood preserved is less than would be expected, since wood is often abundant in modern volcanic settings (Burnham, 1994), and occurs at many fossil localities on King George Island. Given that the volcanic source was possibly less than 6 km away, these three fossil assemblages provide insights into the vegetation recovery cycle following two periods of volcanic activity, with the best preserved material being located nearest to the volcanic source (cf. Burnham, 1994; Spicer, 1989).

A reconstruction of the palaeoecology of Point Hennequin was published by Hunt and Poole (2003) based on sedimentological and palaeobotanical interpretations of new fossil material collected from Point Hennequin by Hunt in 1998–1999. They describe a forest covering the slopes of an active stratovolcano. The two fossil localities are sourced from a single vegetational locality on the lower flanks of the volcano. The sedimentary data suggests that they occupied a fluvio-lacustrine environment. Since these three floras are geographically close, compositional variations are likely to reflect temporal rather than spatial variations in disturbance. In this respect the transitions from the high diversity flora (i.e. Dragon Glacier Flora) to the lower diversity Mount Wawel and Smok Hill floras could reflect different phases in the disturbance and successional cycle.

The cycle begins with the assemblage preserved in the lower Dragon Glacier plant beds, namely the remnants of a diverse climax vegetation characterising low environmental disturbance (Figure 8.7A). The sequence then grades through to the lower diversity vegetation characteristic of disturbed environments preserved at Smok Hill and Mount Wawel (Figure 8.7B, C). Each assemblage represents a snapshot in time of the vegetation as it recovered from the last eruption, and in turn may provide some indication of the frequency of volcanic activity.

The Dragon Glacier Flora originally described as a species-poor *Nothofagus*-podocarp-pteridophyte assemblage (Zastawniak, 1981) is now known to be a diverse dicotyledonous angiosperm vegetation complete with conifers, pteridophytes and sphenopsids (Cunha, Dutra and Cardoso, 2008; Dutra, 2004; Fontes and Dutra, 2010; Hunt 2001; Hunt and Poole, 2003; Zastawniak, 1981). Characteristically similar to the Valdivian rainforest today (see Case Study) it enjoyed a warm, wet climate with mean annual temperatures of 10–15 °C, mean annual precipitation of 885–1040 mm (possibly up to 2100 mm), warm month mean of ~20 °C and a cold month mean dropping to ~2.5 °C (Hunt, 2001; Poole, Cantrill and Utescher, 2005). The macro- and microfloral assemblages may represent the (near) climax vegetation of the time and includes nearly 40 different angiosperm morphotypes (including the first inflorescence found in Antarctica) (Cunha, Dutra and Cardoso, 2008; Duan and Cao, 1998; Dutra, 2004; Fontes and Dutra, 2010; Hunt, 2001; Hunt and Poole, 2003; Zastawniak, 1981). *Nothofagus* probably formed the emergent tree-stratum alongside conifers such as *Araucaria* and members of the Podocarpaceae (e.g. *Acmopyle*, *Dacrydium*, *Dacrycarpus*, *Microcachrys* and *Podocarpus*, including section *Stachycarpus*) (Figure 8.7A). The podocarps would also have contributed to the

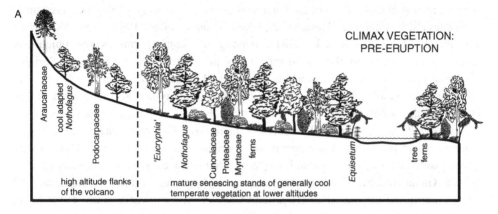

A

CLIMAX VEGETATION: PRE-ERUPTION

Araucariaceae

cool adapted *Nothofagus*

Podocarpaceae

'*Eucryphia*'

Nothofagus

Cunoniaceae
Proteaceae
Myrtaceae
ferns

Equisetum

tree ferns

high altitude flanks of the volcano

mature senescing stands of generally cool temperate vegetation at lower altitudes

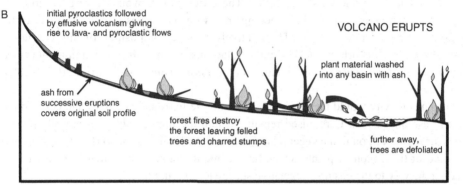

B

VOLCANO ERUPTS

initial pyroclastics followed by effusive volcanism giving rise to lava- and pyroclastic flows

plant material washed into any basin with ash

ash from successive eruptions covers original soil profile

forest fires destroy the forest leaving felled trees and charred stumps

further away, trees are defoliated

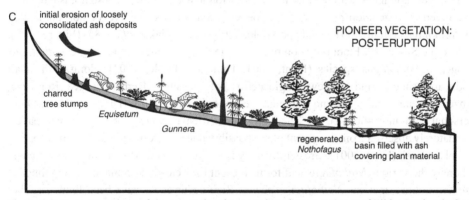

C

PIONEER VEGETATION: POST-ERUPTION

initial erosion of loosely consolidated ash deposits

charred tree stumps

Equisetum

Gunnera

regenerated *Nothofagus*

basin filled with ash covering plant material

Figure 8.7 Reconstructions of the vegetational succession along a transect of Fildes Peninsula from the high altitudinal (proximal) flanks of the volcanoes to the lower (more distal) altitudes derived from the palaeoflora and the Valdivian Model. **A**, climax vegetation composed mainly of cool temperate elements but some warmer temperate elements, such as Proteaceae, present. **B**, immediately post-disturbance a volcanic eruption with ash deposits, fires and tectonic movements which devastate the surrounding landscape. **C**, post-eruption pioneer vegetation of ferns, *Gunnera*, *Equisetum* and resprouting *Nothofagus*, which will, in time, return to the climax vegetation. Reproduced from Poole, Hunt and Cantrill 2001. See text for further details.

subdominant strata. Members of the Cupressaceae (*Libocedrus*) probably occupied more xeric sites (Birkenmajer and Zastawniak, 1989a; Duan and Cao, 1998; Hunt, 2001; Hunt and Poole, 2003; Zastawniak, 1981). Among the angiosperms is the ubiquitous *Dicotylophyllum* leaf morphotype, from a parent plant with unknown taxonomic origins (suggested affinities include extant Eucryphiaceae-Lauraceae-Monimiaceae-Myrtaceae-Sterculiaceae-Symplocaceae), represented by one morphotype that it also found in the Fossil Hill and Rocky Cove floras. There were also leaves similar to members of the Lauraceae (?*Lauriphyllum*), Proteaceae (?*Lomatia*), Eucryphiaceae, Sterculiaceae, Nothofagaceae and various Cunoniaceae all contributing to the understorey. Other understorey plants include *Equisetum* and tree ferns (Dicksoniaceae), with a diversity of ferns (such as Gleicheniaceae, Cyatheaceae, Adiantaceae, Lygodiaceae and Polypodiaceae) and fungi forming the ground cover (Duan and Cao, 1998; Zastawniak *et al.*, 1985; Hunt, 2001; Cunha, Dutra and Cardoso, 2008). These rich forests would have been home to abundant arthropods leaving their feeding traces on the leaves (Hunt, 2001; T. Dutra personal communication, 2010). The relatively diverse composition of this vegetation and inferred low disturbance levels suggest that this was a climax vegetation controlled only by small-scale, sucessional dynamics (e.g. Poole, Hunt and Cantrill, 2001; Veblen *et al.*, 1996).

Volcanic activity would then have devastated these forests, felled trees and covered the area in ash and debris. Carbonised remains within the flora suggest volcanically related forest fires swept through the vegetation preserving the local vegetation (Hunt, 2001). Over a period of time, pioneer plants would have come to colonise these newly exposed sites leading the way to the ensuing vegetative succession (Figure 8.7B).

The cycle continues with the Smok Hill and Mount Wawel floras. These floras indicate a low-diversity vegetation representing a post-eruption, successional vegetation dominated by *Nothofagus* (Hunt, 2001; Hunt and Poole, 2003; Zastawniak *et al.*, 1985) (Figure 8.7C). Indeed the Smok Hill Flora only comprises leaf remains belonging to a single *Nothofagus* subgenus *Nothofagus* species (Hunt, 2001; Hunt and Poole, 2003), along with small podocarp shoots and a few angiosperms of possible Berberidaceae, Rosaceae, Anacardiaceae, Proteaceae and Sapindaceae/Cunoniaceae affinity (Dutra, 2004; T. Dutra personal communication, 2010). These *Nothofagus* leaves were probably canopy leaves, prematurely abscised from the parent plant, possibly due to lack of light caused by a coating of volcanic ash (Hunt, 2001). Greater diversity is exhibited in the Mount Wawel assemblage whereby the various *Nothofagus* leaf forms present bear close resemblance to a number of modern South American *Nothofagus* species. Rare occurrences have been found of other angiospermous plants, including leaf fragments of rhamnaceous and cochlospermaceous affinity (Zastawniak *et al.*, 1985; Hunt, 2001), along with podocarps (including the more dominant aff. *Halocarpus* and *Dacrycarpus*; Fontes and Dutra, 2010) and rarer araucarian conifers (Hunt, 2001; Zastawniak *et al.*, 1985; T. Dutra personal communication, 2010). Bryophytes and the few fern types that were also present would have contributed to the ground cover (Hunt and Poole, 2003). This floral composition is characteristic of the low-diversity vegetation growing subsequent to periods of active volcanism, and thus provides evidence for the hypothesised raised level of disturbance in this area (Hunt and Poole, 2003; Poole, Hunt and Cantrill, 2001).

Further evidence for the composition of the volcanically disturbed successional vegetation growing in this dynamic environment comes from the Fildes Formation (Figure 8.5B), located on the Fildes Block to the west of Point Hennequin. The Rocky Cove Flora is preserved in the uppermost lacustrine tuffs of the Rocky Cove Submember (Middle Member of the Fildes Formation) based on Hunt's (2001) reconciliation of the Shen (1994) and Smellie *et al.* (1984) stratigraphies, which are indicative of continued volcanic activity ~45 Ma ago (Smellie *et al.*, 1984). Associated fluvial sandstones show evidence of braided river systems but no palaeosol features have yet been found. The assemblage comprises abundant fragmented and dispersed angiosperms, conifers and ferns (Figure 8.4C). The preservation of delicate ferns implies that they either formed part of the local vegetation, or low-energy depositional conditions prevailed. In contrast, given the fragmentary nature of most of the angiosperm remains, this suggests that either the angiosperm material does not represent the local vegetation but had been transported from outlying areas, or that the material was deposited in higher energy conditions. The assemblage suggests that a mixed conifer–angiosperm forest prevailed, comprising podocarps and *Araucaria* alongside *Nothofagus* and Proteaceae, with a fern understorey and copious amounts of fungi (Cao, 1992; Hunt, 2001; Shen, 1994) making it distinct from the Vauréal Peak Flora and more similar to the Dragon Glacier Flora, but further work needs to be undertaken to determine the degree of similarity.

Evidence for another vegetation type, this time growing at a greater distance from the centre of volcanic activity, and thus distal relative to the Point Hennequin and Vauréal Peak floras, is also preserved in the Fildes Formation. This flora is found in the tuffaceous and epiclastic sediments of the Middle Unit, situated on the southwestern flank of Collins Glacier. The Collins Glacier Flora (Figure 8.8A, D) has been dated as middle Eocene 48–43 Ma and is therefore contemporaneous with the Point Hennequin Flora and possibly slightly younger than Vauréal Peak Flora (Fensterseifer *et al.*, 1988; Smellie *et al.*, 1984; Soliani *et al.*, 1988; Shen, 1994, 1999[7]; J. L. Smellie personal communication in Poole, Hunt and Cantrill, 2001). The fluvio-lacustrine tuffaceous sandstone sediments, coupled with the fragmentary nature of the *in situ* material, suggest the presence of water bodies with moderate to high energy conditions (i.e. rivers) that may have transported the plant assemblage some distance to its site of deposition. This small flora is made up of silicified wood (~30 cm in diameter), fragmentary angiosperm leaves, ferns and conifer shoots (Hunt, 2001; Poole, Hunt and Cantrill, 2001). By contrast, the Collins Glacier Moraine Flora, comprising a rare excellently preserved entire angiosperm leaf and cone scales of *Araucaria* sect. *Eucata*, is located some 30 m away in a glacial outwash deposit and probably originates from an unidentified source bed beneath Collins Glacier (Hunt, 2001). The excellent preservation of the wood in particular has enabled identification and structural reconstructions of the arboreal vegetation in this distal locality.

---

[7] Shen (1999) preferred a Paleocene age for this flora based on lithological data, but a re-evaluation by Hunt (2001) agrees with the original interpretation by Shen (1994), and is supported by the petrographic studies of Smellie *et al.* (1984). No depositional environment has been suggested for these sediments.

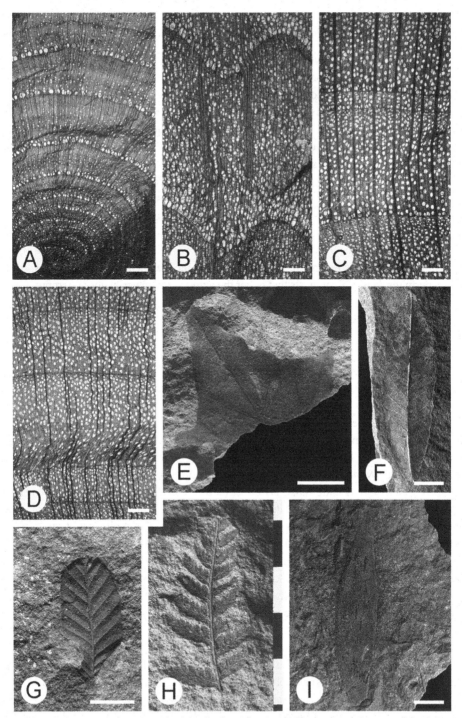

Figure 8.8 Paleocene and Eocene plant fossils from the South Shetland and Seymour islands. **A–D**, transverse sections of silicified wood. A, wood from a small branch of one of the *Nothofagus* morphotypes from the Collins Glacier Flora, King George Island, BAS P. 3023.19. B, a second morphotype of *Nothofagus* with characteristic undulating character to the growth rings from the La Meseta Formation, Seymour Island, BAS DJ. 1060.2. C, atherospermataceous wood from Cross Valley, Seymour Island, BAS D. 502.11. D, myrtaceous wood from the Collins Glacier Flora, King George Island, BAS P. 3023.4. **E–I**, leaves from the Paleocene Cross Valley of Seymour Island. E, sterculiaceous type leaf, BAS DJ. 1113.73. F, possible Cunoniaceae leaf, BAS DJ. 913.71. G, *Nothofagus* leaf with plicate vernation indicating its deciduousness, BAS DJ. 111.126. H, *Sphenopteris angustiloba* fern pinnae, BAS DJ. 1111.139. I, *Araucaria nathorstii* leaf, BAS DJ. 512–1a. Scale bars A–D = 1 mm, E–I = 1 cm.

Podocarpaceae and various *Nothofagus* types were abundant and, together with a minor contribution from *Araucaria* sect. *Eutacta* (Hunt, 2001), would have formed the upper canopy. The middle stratum probably included members of the Cupressaceae, Myrtaceae (*Myceugenelloxylon*) and Eucryphiaceae (*Eucryphiaceoxylon* – originally Cunoniaceae) (Poole, Hunt and Cantrill, 2001; Poole, Mennega and Cantrill, 2003; Zhang and Wang, 1994). *Equisetum* contributed to the lower shrub level although at relatively low abundance levels. Surprisingly, no proteaceous wood remains have yet been found. Considering their abundance in other middle Eocene leaf and pollen assemblages (see Alves, Guerra-Sommer and Dutra, 2005), this suggests that the Proteaceae were either shrubby in habit, and/or lived in more xeric habitats away from water courses – both characteristics serving to reduce their preservation potential. Evidence from the flora suggests that the mean annual temperature had dropped to 9 °C (from the 10–15 °C level in the EECO) by the middle Eocene, with an increase in precipitation. Since no immediate change from the semi-ring porous to ring porous condition is noticeable in the woods (Poole, Cantrill and Utescher, 2005), this would suggest that the changes were probably a result of environmental setting rather than overall climate cooling.

Vegetation of the cooler middle Eocene

The climate continued to cool from the relatively warm early middle Eocene through to the end of the middle Eocene at ~37 Ma. This cooling was only briefly interrupted by the transient warming at ~40 Ma marking the Middle Eocene Climatic Optimum (Zachos, Dickens and Zeebe, 2008). Younger floras dating from this cooler middle Eocene period can be found on Fildes Peninsula, to the west of Barton Peninsula[8], on the north of the Barton Horst, and at Admiralen Peak and Keller Peninsula to the west and north of Point Hennequin, respectively (Figure 8.3C) (see also Appendix). Whereas the assemblages at Admiralen Peak and Keller Peninsula are dominated by conifers with little/no evidence of angiosperms, the most diverse and complete fossiliferous assemblage is that from Fildes Peninsula and so forms the focus of this discussion.

On Fildes Peninsula, the Fossil Hill Flora (Figure 8.3D) is made up of abundant impressions and carbonised leaf fossils, complete with evidence of herbivory and (repro-ductive) organs (e.g. Figure 8.4D). This locality is an important one for Eocene Antarctic palaeoecological reconstructions during the relatively cool middle Eocene, prior to the MECO. The Fossil Hill sequence, in which the flora is preserved, comprises three members of the Fildes Formation. The Lower Member is characterised by basaltic–andesitic lavas interbedded with volcaniclastic rocks. The plant assemblage at Fossil Hill has most recently been placed within the Great Wall Bay Submember of the Lower Member of the Fildes Formation (more or less equivlalent to the Fossil Hill Formation; Figure 8.5) (Shen, 1999), but was initially regarded as being as young as Miocene (Orlando, 1964) or as old as

---

[8] Orlando (1963, 1964) described a diverse flora from Ardley Island similar in composition to those at Fossil Hill. However, the Ardley Island plant beds have never been relocated and other geological descriptions locate the flora at Fossil Hill (Schauer and Fourcade, 1964). Therefore, they are included within the Fossil Hill flora as suggested by T. Dutra (personal communication, 2010).

Paleocene–Eocene (Cao, 1992; Romero, 1978; Troncoso, 1986), based on comparative plant remains. Although numerous stratigraphies have been proposed for Fildes Peninsula (Barton, 1964a, 1965; Shen, 1999; Smellie *et al.*, 1984), we have adopted the broad divisions of Smellie *et al.* (1984) since it is supported by detailed petrographic studies, Ar-Ar geochronology and one of the most detailed geological maps published to date (Hunt, 2001; Pankhurst and Smellie, 1983), and the subdivisions of Xue, Shen and Zhou (1996) and Shen (1999), where these provide greater detail.

The plant-bearing sediments occur approximately in the middle of the volcanic complex, the lavas of which have been dated to 59 Ma (late Paleocene) in the south and 43 Ma (middle Eocene) in the north (Pankhurst and Smellie, 1983; Smellie *et al.*, 1984). Further radiometric dating of the lower lava beds indictate an age of 52–43 Ma (Li *et al.*, 1989). Although further dating of the sequence is needed to validate the age of the plant beds themselves, radiometric dating of the pyroclastic sediments of Unit 4 of the Great Wall Bay Submember (Xue, Shen and Zhou, 1996) have yielded a tentative maximum age of 40 ± 3 Ma for the plant assemblage (Hunt, 2001). Given that the flora was deposited in an active volcanic environment, this age could probably be applied to the entire section at Fossil Hill since the period of deposition is likely to have been smaller than the precision errors of the age date (Hunt, 2001). Much debate centres on the age of the fossiliferous beds and although we use the most recently derived date of 40 Ma for the plant beds in order to anchor it into its relative chronological position, we accept that, based on its compositional assemblage, it could indeed be late Paleocene–early Eocene. Nevertheless, if an age of ~40 Ma is accepted then this flora would represent a facet of the vegetation that grew during the slight amelioration in conditions (MECO) of the middle Eocene.

The Fossil Hill assemblage tells a story of ongoing volcanic activity. Preserved within the six laterally variable tuffaceous units of the Great Wall Bay Submember (Shen, 1999), it is the main plant locality on Fildes Peninsula and represents one of the youngest middle Eocene floras. The assemblage was discovered by Barton in 1964 and is variously described in Barton (1964a, 1964b), Cao (1992), Czajkowski and Rosler (1986), Hunt (2001), Li (1994), Orlando (1964), Rohn, Rösler and Czajkowski (1987), Torres and Méon (1990), Troncoso (1986) and Zhou and Li (1994a, 1994b). Since then it has been the focus of much palaeobotanical investigation.

At the east end of the Fossil Hill locality, the mudstones of Unit 2 provide evidence for a body of standing water that probably manifested itself as a tranquil lake system. The preservation of the footprints of wading birds (e.g. Covacevich and Lamperein, 1972; Covacevich and Rich, 1982), coupled with coarse sandstone deposits and the limited amount of plant material preserved here, suggest that the conditions were extremely shallow and flooding was seasonal. The lake was probably highly oxygenated and thus not conducive to organic preservation. Given the low energy conditions, any fossil plants preserved are likely to represent the lake margin vegetation with occasional representatives from more distant localities. Therefore, the flora preserved in the lake is considered to represent the local climax vegetation growing in a pre-volcanism environment. This vegetation was dominated by angiosperms such as the proteaceous *Lomatia*, with conifers and few ferns. This vegetation later succumbed to debris flow following a volcanic eruption.

Further lacustrine environments are evident higher up the sequence in Unit 3. These conditions yield well-preserved (carbonised) plant material (including leaves, fruits and seeds) and well-rounded wood fragments. The rounded nature of these fragments, coupled with the absence of any palaeosols, together provide evidence for the allochthonous nature of the assemblage probably having been brought here by rivers from more distant sources. Small-scale volcanic activity took place not far from this site and may have resulted in high levels of floristic change. Basal debris flows preserved carbonised wood fragments and suggest that pyroclastic stripping of the forest, similar to that observed following the eruption at Mount St Helens, along with the soil and possibly the bedrock also took place (Burnham and Spicer, 1986; Hunt, 2001; Spicer, 1989, 1991). This devastation would have led to a reorganisation of pre-existing drainage systems and the formation of new depositional basins. The plant material preserved in this unit is considered to represent an impoverished post-disturbance successional vegetation originating from different sources. This successional vegetation comprised mosses, post-disturbance coloniser ferns, such as gleicheniaceous ferns (Zhou and Li, 1994a) along with *Blechnum* (Hunt, 2001), conifers, such as Cupressaceae, and angiosperms including *Nothofagus*, Proteaceae and another disturbance coloniser, *Gunnera* (Cao, 1992; Torres and Méon, 1990). In Unit 4 there is a return to the pre-disturbance climax vegetation once again (Hunt, 2001).

Originally, the Fossil Hill Flora was considered to be *Nothofagus*-poor (Birkenmajer and Zastawniak, 1989a, 1989b), but more recent investigations have identified the presence of *Nothofagus* species throughout the entire sequence (Li, 1994; Li and Zhou, 2007; Shen, 1994). Interestingly, the *Nothofagus* leaves preserved here are much larger (up to 12 cm in length) than their modern counterparts (Dutra, 2000a, 2000b). This is probably a reflection of the local environmental conditions, although such a characteristic may be interpreted as an adaptation to warmer, more humid climate. Later, Cao (1992) described the vegetation as being a forest comprising a Podocarpaceae–Araucariaceae–*Nothofagus* assemblage, with an understorey growth of luxuriant hygrophilous and thermophilous ferns, and shrubs such as the Proteaceae. However, as more studies focused on these beds it became apparent that at its most diverse (i.e. 'climax' stage) the vegetation of the Fossil Hill Flora was a mix of Antarctic and neotropical taxa composed of a rich, mixed angiosperm vegetation. About 22 angiosperm morphotypes have been described belonging to the extant families Poaceae, ?Araliaceae, Proteaceae (including *Lomatia*, ?*Knightia*), Anacardiaceae (including *Schinopsis*, *Rhoophyllum*), Myrtaceae (*Myrtiphyllum*), Melastomataceae (*Pentaneurum*), ?Caesalpiniaceae (?*Cassia*), Dilleniaceae, Icacinaceae (including *Citronella*), Gunneraceae, Monimiaceae (?*Peumus*), Myricaceae (*Myrica*), Sapindaceae (*Cupania*, *Sapindus*), Sterculiaceae, Hydrangiaceae (*Hydrangeiphyllum*), *Brachychiton*, Melastomataceae (*Miconiiphyllum*, *Pentaneurum*), Cunoniaceae, Lauraceae, Rhamnaceae and Nothofagaceae (*Nothofagus fusca* and *N. brassii*), as well as morphotypes unassignable to extant taxa such as *Dicotylophyllum* (Birkenmajer and Zastawniak, 1989b; Cao, 1992; Czajkowski and Rösler, 1986; Dutra, 2000a, 2000b; Hunt, 2001; Li, 1992, 1994; Li and Song, 1988;

Lyra, 1986; Orlando, 1963, 1964; Rohn, Rösler and Czajkowski, 1987; Torres and Méon, 1990; Troncoso, 1986; Zhou and Li, 1994a, 1994b), and a single species of grass (Palma-Heldt, 1987). The dominant conifer component was the Podocarpaceae (including *Acmopyle, Phyllocladus, Dacrydium, Dacrycarpus s.l.* and *Podocarpus*), but leaves of *Araucaria* sect. *Eutacta* and cupressoid leaves with affinities to the *Austrocedrus-Libocedrus-Papuacedrus* group indicate the presence of Araucariaceae and Cupressaceae in the higher canopy (Cao, 1992; Hill and Brodribb, 1999; Fontes and Dutra, 2010; Li and Shen, 1989; Orlando, 1963, 1964; Torres and Méon, 1990; Troncoso, 1986; Zhou and Li, 1994b). Cycads belonging to the Zamiaceae, along with tree ferns of Cyatheaceae and Dicksoniaceae, and *Gunnera* contributed to the middle storey, whereas ferns of the Aspleniaceae, Blechnaceae, Gleicheniaceae, Lophosoriaceae, Thrysopteridaceae and Osmundaceae, and bryophytes made up the ground storey (Hunt, 2001; Li and Shen, 1989; Orlando, 1963, 1964; Torres and Méon, 1990; Zhou and Li, 1994a). This assemblage is thought to represent an evergreen broad-leafed forest with an admixture of conifers. The conifers may have inhabited drier edaphic conditions (Birkenmajer and Zastawniak, 1989a, 1989b), or were spatially separated with components such as *Halocarpus* occupying the higher slopes, whereas *Dacrycarpus* favoured the lower slopes nearer the sea during periods of volcanic quiescence (Fontes and Dutra, 2010). Li (1994) and later Hunt (2001) undertook various analyses of the floras to determine the palaeoclimate and concluded that this flora grew in a climate where the mean annual temperature fell between 8–14 °C, with a warm month mean 20–27 °C and mean annual precipitation was 1060–2100 mm – a climate very similar to that experienced by the Dragon Glacier Flora.

The final piece of evidence for the vegetation that grew across the Antarctica Peninsula during the Eocene comes from cores taken from seas around the South Orkney Islands to the northeast of the Peninsula. The ARA Islas Orcadas core 1578–59 was recovered from the southern flank of Bruce Bank (60° 33.6′ S, 40° 13.2′ W) near South Scotia Ridge. Bruce Bank, together with the South Orkney Microcontinent, represents the plate boundary between the Scotia and Antarctic plates. The South Scotia Ridge may represent continental fragments dispersed from the Antarctic–South American Isthmus during the opening of the Drake Passage (Toker, Barker and Wise, 1991). Palynological evidence from Bruce Bank has yielded information about the vegetation that grew 46–45 Ma ago during the Middle Eocene Climatic Optimum (Grube and Mohr, 2007; Mao and Mohr, 1995; Mohr, 2001).

The palynological assemblage shows similarity with the middle Eocene floras from King George Island. This microflora is represented by a very high diversity of spores and pollen yielding a high percentage of angiosperms and a very high abundance of ferns as well as gymnosperms and fungi. Varying percentages of fern and seed plant sporomorphs suggest some cyclicity in an otherwise homogeneous flora during the studied time interval. The flora was a *Nothofagus*-podocarp-fern forest with a diverse angiosperm component, but dominated by *Nothofagus* of all extant subgenera (comprising the *brassii, fusca*-a, *fusca*-b, and *menziesii* pollen types), and both evergreen as well as deciduous forms. The *Nothofagus brassii* pollen type is the most dominant. Other angiosperms include Gunneraceae,

Proteaceae, Asteraceae, Casuarinaceae, Myrtaceae, Apiaceae and Liliaceae. The conifer component is dominated by podocarps, high abundance and diversity, with *Podocarpus, Dacrycarpus, Dacrydium, Lagarostrobus, Phyllocladus* and *Microcachrys*. Additional gymnosperms include Araucariaceae (*Araucaria*) and Cupressaceae (?*Libocedrus*). Ferns are abundant and indicate a relatively humid environment (Mohr, 2001). They are represented by abundant Cyatheaceae (e.g. *Cnemidaria*) or Lophosoriaceae (e.g. *Lophosoria*), as well as Gleicheniaceae, Schizaeaceae (*Lygodium*), Dicksoniaceae (*Dicksonia*), Osmundaceae/Hymenophyllaceae. Various bryophyte spores (e.g. *Sphagnum*) along with evidence of *Lycopodium* are also present (Grube and Mohr, 2007; Mohr, 2001). The vegetation clearly indicates a warm temperate, humid, non-tropical climate at a latitude where sea surface temperatures reached 5–10 °C in winter and >14 °C in summer, which is in good agreement with other microfossil data (Kennett and Barker, 1990).

So as the climate cooled during the middle Eocene, the vegetation on the west side of the Peninsula began to show an increasingly striking resemblance to the Eocene vegetation of Tasmania (Carpenter, Hill and Jordan, 1994) and New Zealand (Pole, 1994), and in turn to the modern vegetation of Western Patagonia growing along the Andean mountain belt (Armesto *et al.*, 1995; Veblen, Schlegel and Oltremari, 1983; Villagrán, 1990), where specialisation to altitudinal stratification and cooler environments is evident. In the low-lying areas to the eastern side of the Antarctic Peninsula (see below), the vegetation adapted to the onset of the South Atlantic coastal environments (Askin, 1992; Baldoni and Barreda, 1986; Baldoni and Medina, 1989), and a good record of ecosystem dynamics is preserved on Seymour Island in the Larsen Basin.

Seymour Island: the 'Rosetta Stone' for Southern Hemisphere evolution

On the east side of the Antarctic Peninsula in the James Ross Basin (part of the larger Larsen Basin) lies Seymour (Marambio) Island (Figure 8.3), a mecca for scientists attempting to unravel the details of life at high latitudes during the early Cenozoic greenhouse phase. Its rich fossil record, of both plants and animals, alongside glimpses into the marine realm, has helped provide an intricate picture of a complete ecosystem that dominated this region during the Paleocene and early Eocene. As with the Shetland Islands, uplift and subsequent erosion have removed much of the magmatic arc that formed the spine of the Antarctic Peninsula (Elliot, 1988), but it had not been lost altogether. Erosion led to sediment being washed from the arc to form a 5–6 km-thick sequence of marine sediments in the Larsen and James Ross basins located to the east (Pirrie, Duane and Riding, 1992). Along with this eroded sedimentary material, the remains of the diverse terrestrial flora and fauna were also washed from the arc into the Weddell Sea to become waterlogged and finally buried along with other marine life in the sediments. Here they became permineralised with silica and calcite to form a record of the biota that once dominated the seas and land along the northern Antarctic Peninsula. Subsequent uplift has resulted in the re-exposure of these poorly consolidated marine sand, silt, mud and tuff sequences on Cockburn Island and nearby

Seymour Island to represent the only marine sequences of Paleocene to Eocene age to be found in Antarctica (e.g. Askin *et al.*, 1991; Elliot, 1988; Elliot and Trautman, 1982). The fossil localities associated with Seymour Island harbour the most complete fossiliferous southern high latitude succession known to date.

All too often, reconstruction of the sequences of terrestrial ecosystems through time is frustrated by the inadequacies of the fossil record, but on Seymour Island a plethora of Cenozoic plant and animal remains and some of the best known and most complete fossil assemblages in Antarctica found to date, have provided exciting insights into the high latitude ecosystem dynamics during the early Cenozoic. These, often exquisitely preserved assemblages, led Zinsmeister (1986) to refer to Seymour Island as a palaeontological 'Rosetta Stone' for understanding the evolution of Southern Hemisphere life. Recreating the vegetation that grew in this lost world began with the pioneering works of Dusén (1908) and Gothan (1908), who documented the diversity and quality of the preserved plant material available from the James Ross Basin sediments. With the sheer quantity and quality of fossil finds, focus has now switched to detailed reconstructions of biodiversity and ecosystem interactions and dynamics, thus providing the most complete ecological picture in the southern high latitudes at a time of polar warmth.

Geology and palaeoenvironment

Seymour Island is small (~20 km long) and seasonally ice-free, lying towards the north-eastern tip of the Antarctic Peninsula at ~63° S within the James Ross Basin (Hathway, 2000; Lawver, Gahagan and Coffin, 1992; Pirrie, Whitham and Ineson, 1991). During the Paleogene the Antarctic Peninsula crustal block comprised an inland region of highland to the west. Across this block, volcanic activity was sporadic and formed an extensive low mountain cordillera located at some distance from the eastern seaboard. Seymour Island occupied a coastal position of relatively low relief, supporting tidal channels and flats, and estuaries (Crame *et al.*, 1991; Marenssi, 1995; Marenssi, Santillana and Rinaldi, 1998a; Pirrie, Whitham and Ineson, 1991). The rocks exposed on Seymour Island range from late Campanian to late Eocene (Elliot, 1988) (Figure 8.5), with the Cretaceous and Paleocene sequences dipping southeast to crop out in the southern half of the island and at Cape Wiman (Elliot, Hoffman and Rieske, 1992; Olivero, 1995) (Figure 8.3E). The Cretaceous–Paleogene boundary is located in the upper part of the López de Bertodano Formation (between Units 9 and 10; Askin, 1988a, 1988b; Harwood, 1988; Huber, 1988; Macellari, 1988) (Figures 8.3 and 8.5), coinciding with a widespread glauconitic level corresponding to a peak in transgression (Brizuela *et al.*, 2007; Elliot *et al.*, 1994).

In the early Paleocene, Seymour Island represented a quiet, near shore, shallow marine environment that accumulated large amounts of fine to very fine sands, silt and clay-rich sediment (probably of volcanic origin), brought down by several major fluvial systems from a variety of terrestrial habitats in the mountain cordillera (Macellari, 1988). The habitats preserved include the coastal lowlands, moist fluvial/lacustrine/swamp areas, drier inter-fluvial areas and some upland regions. It is these deposits that form Unit 10 of the López de Bertodano Formation and the lower part of the 255 m-thick (Danian) Sobral Formation

(Macellari, 1988). The abundance of large fragments of wood devoid of encrusting bivalves in both Unit 10 of the López de Bertodano Formation and the overlying Sobral Formation, suggests that this region lay not far from the palaeocoastline. The contact between these formations is erosional, with evidence of basal channelling present in several sections (Harwood, 1988). The new depositional cycle that resulted in the Sobral Formation began as a basin filled by the progradation of the existing deltaic system. In time, the pro-delta facies were deposited, followed by clean sands from a coastal barrier and finally delta top facies including distributary channels and interdistributary marshes (Macellari, 1988). Delta progradation was not consistent and fluctuations produced an alternation of non-marine and very shallow marine facies (Macellari, 1988).

The Sobral Formation (Figure 8.5) comprises a coarsening upwards sequence that has been divided into five informal subunits (Sadler, 1988). According to Elliot, Hoffman and Rieske (1992), the lowest unit (Unit 1) consists of well-bedded silts, whereas Unit 2 is characterised by bioturbated fine sands that locally contain many thin clay-rich tuffaceous beds. These units were probably deposited in relatively quiet conditions. Unit 3 is marked by tuffaceous beds in the lower part, with the presence of glauconite and a coarsening to medium sands higher up, deposited above wave-base in a relatively high energy environment such as the foreshore of a beach close to the delta front, similar to the lower portion of the coastal barrier sand of the Niger Delta (Macellari, 1988). Final progradation of a deltaic sequence is recorded in Units 4 and 5 as the sequence changes from marine to non-marine facies, with coarsening upwards continuing through Unit 4 – the base of which is marked by relatively resistant tabular, cross-bedded glauconitic sands. A thin layer of coarse sands and tuffaceous beds constitutes the uppermost Unit (5) of the Sobral Formation.

By the early late Paleocene a river had cut a steep-walled channel down into the Sobral and upper López de Bertodano formations (Sadler, 1988). Over time this channel became infilled with both relatively coarse-grained volcaniclastic material deposited under high energy conditions, and fine deltaic sands (Askin, 1988a; Macellari, 1988). These channel fill deposits now represent the late Paleocene Cross Valley Formation that straddles the neck of the island (Wrenn and Hart, 1988) (Figure 8.3E).

The Cross Valley Formation represents the culmination of the shallowing trend initiated in the Sobral Formation (Macellari, 1988; Sadler, 1988). At the base of the Cross Valley Formation unconformities represent a regressive phase, which may have been short-lived and were eustatic and/or tectonic in origin (Marenssi, 1995; Marenssi, Santillana and Rinaldi, 1988b; Sadler, 1988). Within the estuarine, or enclosed bay conditions, the falling sea level would have led to a decrease in salinity, which in turn would have resulted in dinoflagellate blooms responding to the stressed anoxic conditions (Askin, 1988b, 1989; Huber, 1988). To the northeast at Cape Wiman (Figure 8.3E) the Paleocene 'Wiman formation' of Elliot and Hoffman (1989) is considered to be either the uppermost unit of the Sobral Formation (Elliot and Rieske, 1987) (Figure 8.3E), or the lateral equivalent of the Cross Valley Formation (Pirrie, Duane and Riding, 1992).

Following this regressive phase, during a time of tectonic quiescence and lull in volcanism, a river flowing in a southwesterly direction some 60 km from the foot of the mountains

Table 8.2 *Summary of the Paleogene vegetation and their modern vegetation and climate analogues for the James Ross Basin and South Orkney areas. CCM = cold month mean, MAT = mean annual temperature, MAP = mean annual precipitation, MST = mean summer temperature, MWT = mean winter temperature, min = minimum, max/mx = maximum. See text for further details and references.*

Stratigraphic age Lithostratigraphy Location	Palaeovegetation	Modern analogue	
		Vegetation composition	Palaeoclimatic parameters
cooling post-EECO Middle Eocene IO core 1578–1559 *Bruce Bank, near South Scotia Ridge*	Floristically diverse *Nothofagus*-podocarp-fern forest with abundant and diverse angiosperms including *Nothofagus* (all extant subgenera present) alongside other conifer taxa, angiospermous shrubs and diverse cycads, tree ferns, (diverse) ferns and bryophytes forming the understorey.	Valdivian rainforests of southern Chile with the rare occurrence of conifers and ferns explained by taphonomical processes and/or local biases rather than a different vegetation type. Also with vegetation of Tasmania and New Zealand.	MST 10–20 °C MWT 0–10 °C MAP 1000–2000 mm (from Tasmania and New Zealand analogue)
La Meseta Fm *Seymour Island*	Mixed mesophytic cool temperate rainforests with conifers (podocarps, araucarians and Cupressaceae) and angiosperms including increasing *Nothofagus*. Diversity had decreased (50%). Open heathland and disturbed environments also present.	Originally more similar to the cool temperate Valdivian rainforests grading to cool temperate *Nothofagus* forests and woodlands and Magellanic forests of southern South America.	Cool temperate, seasonal and wet climate MAP up to 1500+ mm MAT 7–13 °C WMM ~24 °C MinWT –3 to +2 °C
warm early Eocene Early Eocene La Meseta Fm *Seymour Island*	Grading from conifer (podocarp)-broad-leafed angiosperm rainforest to increasingly angiospermous (*Nothofagus*, especially *N. fusca* group)-dominated multistratal rainforests with Proteaceae contributing increasingly to the vegetation.	Cool temperate Valdivian rainforests.	Warm temperate aseasonal climate MAT 10–15 °C MaxST up to ~25 °C MinWT 7–8 °C MAP 1200–2000 mm (possibly up to ~3000 mm), initially seasonal becoming more aseasonal

Table 8.2 (cont.)

	Stratigraphic age Lithostratigraphy Location	Palaeovegetation	Modern analogue	
			Vegetation composition	Palaeoclimatic parameters
late Paleocene warm period	late Paleocene Cross Valley Fm *Seymour Island*	Paratropical multistratal rainforest with (podocarp) conifers and both thermophilous and cool temperate angiosperm taxa, including abundant *Nothofagus*, tree ferns in the mid-canopy and ferns to the groundstorey and fringing communities.	'Palaeoflora Mixta' Becoming more similar to the cool temperate Valdivian rainforest of southern Chile.	Warm, humid cool-warm temperate climate increasing in seasonality MAT 12–14 °C MaxST ~25 °C CMM ~3 °C MAP ~2100 mm
relatively cool early Paleocene	early Paleocene Sobral Fm Unit 10 López de Bertodano Fm *Seymour Island*	Mixed meso-megathermal community of multistratal conifer (podocarp) dominated lowland rainforest with abundant angiosperms including *Nothofagus*, Proteaceae, monocots, tree ferns, palms and abundant ferns forming the understorey; podocarps decline through this unit with a loss of palms and changing abundance within the tree fern community; filmy ferns increase.	'Palaeoflora Mixta' with no modern analogue.	MAP increasing Warm temperate, initially frost-free conditions but later possible frosts MAT 12–14 °C MaxST 20–24 °C MAP 1200 mm, possibly first decreasing then increasing

to the sea, cut a 7 km-wide valley into the emergent shelf (Stilwell and Zinsmeister, 1992). A subsequent period of eustatic sea level changes towards the end of the early Eocene led to the subsequent sedimentation of this incised valley. The seaward end of this 7 km-wide valley began to accumulate the deltaic, estuarine and shallow marine sediments composed of sandstones, mudstones and shell banks, which form the unconformity-bounded, 720 m-thick unit of the La Meseta Formation today (Marenssi, Net and Santillana, 2002; Marenssi, Santillana and Rinaldi, 1998b) (Figure 8.5). Provenance and palaeocurrent studies indicate that the sediments making up the La Meseta Formation originated from source rocks outcropping along the Peninsula to the westnorthwest. Along with these sediments, remains of plants and animals were also washed down from the terrestrial environments upstream (Marenssi, 1995; Marenssi, Santillana and Rinaldi 1998a; Marenssi *et al.*, 1999; Net and Marenssi, 1999). The plant material (e.g. Figure 8.8C, E–I) could therefore either have been transported downstream for quite a distance from the mountainous and hilly regions of the northern Antarctic Peninsula, and/or may represent relative local material growing in coastal environments near to the estuarine depositional basin (Doktor *et al.*, 1996; Gandolfo *et al.*, 1998; Porębski, 1995, 2000; Torres, Marenssi and Santillana, 1994a, 1994b). Although palaeogeographic interpretations indicate that these terrestrial facies had to be located nearby to the west, they are not yet known from either Seymour Island or neighbouring James Ross Island (Reguero, Marenssi and Santillana, 2002). Hence all terrestrial material found on Seymour Island had been transported into the marine setting of the back-arc basin to the east of the Peninsula. Evidence from the association of marine molluscs alongside leaves and wood, having been densely bored by the bivalve *Teredolites*, suggests that the plant material at least had been submerged for some time at the water–sediment interface prior to burial (Reguero, Marenssi and Santillana, 2002; Vizcaíno *et al.*, 1998). However, the presence of leaves, trunks and flowers suggests that, although they represent allochthonous rather than autochtonous assemblages, they originated from a relatively proximal locality on the northern Antarctic Peninsula region (Reguero, Marenssi and Santillana, 2002).

This composite fill of the La Meseta Formation represents a good example of a main transgression punctuated by short regressive periods (Marenssi, 2006) spanning a near complete record of the transition from the warm early Eocene through the transitional cooling of the middle Eocene to the greater cooling in the latest Eocene (and finally to the onset of the early Oligocene ice sheet development; Case, 2007; Ivany *et al.*, 2006). Since the deposition of the La Meseta Formation spanned much of the Eocene it has received considerable attention when trying to elucidate the effects of cooling on high latitude environments. Moreover, it has been subject to a number of geochronological analyses in an attempt to determine the time period covered by the formation, since its precise age has been the subject of ongoing debate (Dutton, Lohmann and Zinsmeister, 2002 and references therein; Ivany *et al.*, 2008). The formation has been subdivided into six lithological units, described as allomembers (Marenssi, 1995; Marenssi and Santillana, 1994; Marenssi, Santillana and Rinaldi, 1998a, 1998b) that approximate to the Telms (Tertiary Eocene La Meseta) of Sadler (1988). The base of each allomember represents a fluvial erosional surface reshaped during the following transgression (Marenssi, 1995; Marenssi, Santillana and Rinaldi, 1998a, 1998b) (Figure 8.7). The plants preserved in these units are detailed below.

Palaeovegetation

The palaeovegetational history of Seymour Island is recorded in plant-rich horizons of the Paleogene sedimentary sequences. The palaeoflora evidences forested vegetation having developed across the Antarctic Peninsula region at this time with the cordillera supporting a wide range of habitats from coastal to alpine. Yet the vegetation was dynamic and thus reflected its environment. The record preserved in the marine sediments on Seymour Island suggest environments changed drastically through the Paleocene and Eocene, from conifer dominated rainforest (early Paleocene), to paratropical rainforest growing in a warm, rainy climate (late Paleocene) and finally to cool temperate rainforest during the late early Eocene.

Vegetation of the early warm period of the Paleocene–early middle Eocene

The lower Danian sediments (Unit 10) of the López de Bertodano Formation (Figure 8.5) preserve abundant well-preserved palynomorphs from both marine and terrestrial origins. Although micro- and macrofossils within the López de Bertodano Formation are abundant when compared with the overlying Sobral Formation, overall diversity does not match that of the Late Cretaceous. The multistratal conifer-dominated lowland rainforests that prevailed immediately after the K-T boundary decrease up the sequence such that by the Danian the terrestrial pollen and spores reflect a variety of habitats from wet coastal lowlands, open heathland and disturbed environments, to higher altitude areas in the volcanic mountain chain further inland (Askin, 1990, 1994; Dettmann, 1994; Dettmann and Jarzen, 1988; Dettmann and Thomson, 1987; Dettmann et al., 1990; Greenhalgh, 2002).

Greenhalgh (2002) undertook a detailed analysis of the palynoflora from the López de Bertodano Formation and her results are summarised here. The vegetation was a mixed meso-megathermal community of multistratal conifer-dominated rainforest, with a minor microthermal community suggesting a mean annual temperature of 12–14 °C, perhaps reaching >20–24 °C in sheltered lowland environments, and an annual rainfall of 1200 mm. The rainforest was made up predominantly of podocarp conifers that formed the canopy, along with a secondary component of Proteaceae and an understorey of tree ferns and more shrubby proteaceous angiosperms. The ground cover comprised ferns and fungi. Although dominated by podocarp, other conifers (e.g. *Lagarostrobus*) and both dicotyledonous (Atherospermataceae, Chloranthaceae, Cunoniaceae, Casuarinaceae, Eucryphiaceae, Illiciaceae, Myrtaceae, Nothofagaceae, Proteaceae, and woods with anatomical similarity to members of the Juglandaceae) and monocotyledonous angiosperms (probable Liliaceae, or other monocotyledonous families such as Amaryllidaceae, Agavaceae, Bromeliaceae and Iridaceae), including palm pollen, were also present in varying abundances (Askin, 1994; Brizuela et al., 2007; Cantrill and Poole, 2005; Poole, Gottwald and Francis, 2000; Poole, 2002; Poole, Mennega and Cantrill, 2003). Being frost intolerant, some taxa, for example palms, Illiciaceae and Casuarinaceae, were probably at the limit of their climatic tolerance and gave a warm temperate aspect to the rainforest suggesting that frost-free conditions prevailed during the Danian, at least in more localised areas within sheltered aspects of the coastal lowlands (cf. e.g. Askin, 1992, 1994; Cranwell, 1969). Open freshwater bodies probably present on the coastal plains were dominated by *Azolla*, *Sphagnum* and other bryophytes (Greenhalgh, 2002).

Although the terrestrial palynofloral record within the Paleocene indicates a relatively stable spore and pollen assemblage (Greenhalgh, 2002), the pollen record throughout Unit 10 records an overall decline in conifers, due primarily to the decline in the abundance of the podocarps (particularly *Phyllocladidites mawsonii* pollen produced by the conifer *Lagarostrobus franklinii*), which would have opened up the canopy and allowed an increase in the understorey fern community (including tree ferns), or the expansion of fringe fern communities. However, the relatively local (wood) flora records a fairly stable conifer component, comprising abundant podocarps and Araucariaceae and relatively rare Taxodiaceae and Cupressaceae, having recovered from the decline in numbers seen at the end of the Cretaceous (Cantrill and Poole, 2005). The changes within the conifer pollen assemblage (in particular the distribution of *Lagarostrobus*, which today is linked to high levels of annual rainfall) suggest a decrease in the annual rainfall and a loss of balmy conditions. Moreover, the decrease in *Phyllocladidites mawsonii* in the upper López de Bertodano Formation also corresponds with the decline in Casuarinaceae and disappearance of the Palmae. The consistent and relatively abundant Cyathaceae-Dicksoniaceae (~10%) undergo a sudden sharp decline (to ~5%) before recovering to a peak abundance (23%) followed by a subsequent decline at the top of the formation. This shows an inverse relationship with the *P. mawsonii* pollen from *Lagarostrobus* conifers, implying that the tree ferns may be exploiting niches left exposed by the conifers as they succumbed to changing conditions (Greenhalgh, 2002). *Lagarostrobus* today requires an annual rainfall of 1200 mm and its decline in the López de Bertodano Formation might indicate a drying out of the rainforest. A decrease in moisture availability would also explain the changes among the ferns, with the opening up of the canopy allowing greater development of the fern under-storey (including tree ferns), and/or the expansion of a fringing fern community. Such changes may have occurred on a local or regional-scale. However, in the lower part of the Sobral Formation, wetter conditions returned but cooler conditions prevailed – an observation supported by the increase in *Microcachrys* spores, which indicate an expansion of the cold, high latitude flora (Greenhalgh, 2002).

Throughout the early Paleocene, from the uppermost López de Bertodano Formation and into the Sobral Formation, floral diversity remained relatively stable in terms of taxonomic composition, but more variable in terms of relative abundances. The multistratal conifer-dominated rainforest with podocarps (up to 39%) dominating and *Lagastrobus* contributing a significant (~15%), yet fluctuating, presence continued to thrive. Changes were, however, taking place in the understorey: *Nothofagus* was increasing although ferns continued to be the most abundant (up to 38–66%) component of the vegetation (with the filmy ferns, Hymenophyllaceae/Osmundaceae, contributing the most). The increase in abundance of ferns with multiple filicalean affinities to become the most dominant group suggests a more unsettled environment possibly relating to an increase in volcanism (Greenhalgh, 2002). In the upper canopy, *Podocarpus* remains dominant (with fluctuations from 2% to a peak of 39%), but *Lagarostrobus*, although still forming the secondary element, continued to increase in relative abundance (~15%), possibly at the expense of the tree ferns which were declining, along with *Microcachrys*, Phyllocladaceae and Araucariaceae. Angiosperms represent up to 35%, the majority of which is *Nothofagus* (at least three different species representing the *fusca* and *brassii* groups), with Proteaceae becoming increasingly abundant, possibly at the expense of the podocarps, and elements such as Eucryphiaceae, Illiciaceae, Atherospermataceae, Myrtaceae and other unknown angiosperms with primitive anatomy (Askin and Fleming, 1982; Cantrill and Poole, 2005; Dettman and Jarzen, 1991; Greenhalgh, 2002). Increasing annual rainfall in the lower part of the section to become high and constant throughout the remaining part of the Sobral Formation has been used to explain the increase following a decline in relative abundance of the Salviniaceae, coupled with an initial decline within the tree fern community, whilst the filmy ferns, ferns of multiple filicalean source groups, *Sphagnum* and bryophytes show a distinct increase(s) in abundance through the section. Therefore as annual rainfall increased the rainforest understorey and groundcover become an increasingly dominant component (Greenhalgh, 2002). It is possible that this coincides with an increase in volcanism and subsequent disturbance. The expansion in the altitudinal zones (as evidenced by the increase in *Microcachrys*) was caused by a fall in mean annual temperature or increased runoff from high altitudes, as a consequence of enhanced precipitation. The increase in bryophytes correlates with the interval of time during which incision of the basal channel unit occurred in the lower Sobral Formation, providing new niches for mosses to exploit and thus increasing the preservational potential of this group. Moreover, the disappearance of the Palmae and Casuarinaceae indicates the loss of frost-free conditions experienced in the earliest Paleocene, and the increase in *Microcachrys* suggests an expansion in the cold altitude flora (Greenhalgh, 2002).

The late Paleocene–early Eocene was the acme of Cenozoic warmth with tropical latitudinal belts extending 10–15° S and forests covering both poles. Coupled with the sea-level lowstand (identified at between 58.5 and 55.5 Ma; Haq, Hardenbol and Vail, 1987) this would have provided an ideal opportunity for animal and plant dispersal and radiations to have occurred between Patagonia and the Antarctic Peninsula (Reguero, Marenssi and Santillana, 2002). The ecology of the Antarctic Peninsula region at the peak of Cenozoic warmth is recorded in the late Paleocene Nordenskjöld Flora of the Cross Valley Formation on Seymour Island (Figures 8.5 and 8.8C, E–I). This leaf flora has been described (Dusén, 1908; Gothan, 1908) as a paratropical rainforest growing in warm, humid conditions (Gandolfo, Marenssi and Santillana, 1998; Gandolfo, Santillana and Marenssi, 1998). More recent investigations of the leaf, wood and pollen floras have found that it was dominated by angiosperms with thermophilous (subtropical) taxa, growing alongside plants more commonly associated with cool temperate climes today in a disturbed environment (Cantrill and Poole, 2005; Case, 1988; Francis *et al.*, 2008; Greenhalgh, 2002; Poole, unpublished data). The rainforest vegetation was still multistratal with podocarp conifers (e.g. *Podocarpus* and *Dacrydium*) dominating in abundance and statue. Other conifers would have included *Araucaria*, and *Lagarostrobus* (as evidenced by *Phyllocladidites mawsonii* pollen), and cupressoid types (Cantrill and Poole, 2005; Case, 1988; Francis *et al.*, 2008; Greenhalgh, 2002; Mohr, 2001), as evidenced in the wood and pollen flora. *Nothofagus* was the most dominant angiosperm alongside abundant lauraceous and proteaceous taxa (e.g. *Lomatia* and *Knightia*). Pollen and macrofossils belonging to Myrtaceae, Elaeocarpaceae, Moraceae, Cunoniaceae, Illiciaceae, Monimiaceae, Atherospermataceae, Lauraceae, Melastomataceae, Winteraceae (although the Onagraceae/Winteraceae type identified by Cranwell (1959) was not found in a reinvestigation of the flora by Mohr, 2001) and the tree fern, *Cyathea*, would have formed the mid-canopy (Cantrill and Poole, 2005; Case, 1988; Dusén, 1908; Francis *et al.*, 2008; Greenhalgh, 2002; Mohr, 2001; Poole, 2002; Poole and Cantrill, 2001; Poole and Francis, 1999; Poole and Gottwald, 2001; Poole, Richter and Francis, 2000; Poole *et al.*, 2000). Ferns were originally assigned by Dusén (1908) to *Polypodium*, *Asplenium*, *Taeniopteris*, *Alsophila* and *Dryopteris*, but later found to belong to a fewer number of types based on reassessment of the original material and new collections (Cantrill, Tosolini and Francis, 2011) and potentially include *Lophosoria antarctica* (Torres and Méon, 1993). Others tentatively related to ?Dennstaedtiaceae, Cyatheaceae, ?Osmundaceae and Blechnaceae (Mohr, 2001) contributed to the substantial understorey and fringing community. Water bodies may have been colonised by *Azolla* or *Salvinia* (Mohr, 2001). Lowland rainforest seems to be absent, with restricted communities of open heathland and disturbed communities. The mix of thermophilous and cool temperate taxa indicate that a paratropical/cool-warm temperate rainforest vegetation comprising both conifer and broad-leafed angiosperms prevailed under a warm temperate climate (Francis, Tosolini and Cantrill, 2003; Francis *et al.*, 2008; Poole and Cantrill, 2001; Poole and Francis, 1999; Poole and Gottwald, 2001; Poole, Gottwald and Francis, 2000; Poole *et al.*, 2000). Mean annual temperatures were probably in the region of 12–14 °C (with small local communities enjoying temperatures as high as 20–24 °C), with a warm month mean of ~25 °C, a cold month mean of ~3 °C, the probable absence of ice in winter and average precipitation levels of 2110 mm per year (Francis, Tosolini and Cantrill, 2003; Francis *et al.*, 2008; Gandolfo, Marenssi and Santillana, 1998; Greenhalgh, 2002; Poole, Cantrill and Utescher, 2005). The increasing incidence of well-defined growth rings in the wood flora during the Paleocene also suggests that one or more environmental factors, such as water availability or temperature for example, became limiting for short periods during the growing season, and thus indicate an increasingly seasonal environment relative to that in the late Cretaceous (Poole, Cantrill and Utescher, 2005).

The record of ecological evolution in the southern high latitudes continues with floras and faunas preserved within the La Meseta Formation (Figure 8.9). This formation has yielded material from most of its stratigraphic column with megafloral remains (including flowers, leaves and tree trunks) having been collected from all but the lowest 120 m (Brea, 1998; Gandolfo, Marenssi and Santillana, 1998; Gandolfo *et al.*, 1998; Torres, Marenssi and Santillana, 1994a, 1994b). Although the palynoflora documents a relatively stable

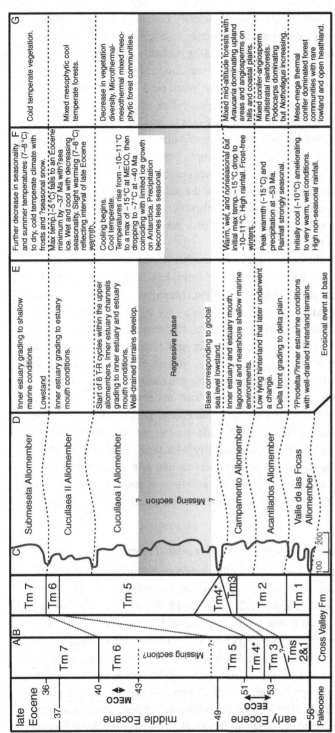

Figure 8.9 Comparison between the ecology and unconformity-bounded internal units of the La Meseta Formation (LMF) overlying the Cross Valley Formation showing proposed ages for the major erosive events at the base of each Telm (Tm) and eustatic control on sedimentation adapted from Marenssi (2006). The LMF extends from the beginning of the Eocene to 34.78–33.75 (Dutton, Lohmann and Zinsmeister, 2002; Ivany *et al.*, 2008). A, data in columns after Ivany *et al.* (2008) with ages based on ^{87}Sr/^{86}Sr dating of bivalves (age assignments in the lower part of the section are tenuous given the poor constraints on the global seawater curve and should be used with caution especially in the section below the presumed unconformity (missing section) since there were no fossils present in that interval and freshwater influx may have complicated the Sr ratios; L. Ivany personal communication, 2010, 2011); Telm 1 was not sampled by Ivany *et al.* (2008) and so is grouped here with Telm 2 for completeness. B, Sadler (1988) and Marenssi, Santillana and Rinaldi (1998b). C, the short-term eustatic sea level curve (in metres) of Haq, Hardenbol and Vail (1987). D, the six erosionally based allomembers of Marenssi and Santillana (1994), Marenssi (1995) and Marenssi, Santillana and Rinaldi (1998a, 1998b), with relationship to column B based on age estimates given by Marenssi (2006). E, data from Marenssi (2006) and references therein; T-R, transgressive-regressive cycle. F and G, summary of the palaeoclimate and vegetation respectively, see text for further details and references. MECO, Middle Eocene Climatic Optimum; EECO, Early Eocene Climatic Optimum. Note that the age estimates for Telm 3 overlap that of Telm 2 and are not time equivalent across Seymour Island. The transgressive lag phase sediments of Telm 4 provide a good marker bed separating Telms 3 and 5 and probably represent a transgressive lag at the base of Telm 5 (L. Ivany personal communication, 2010).

environment (Greenhalgh, 2002), during the course of the early Eocene there was a gradual change from the conifer (podocarp) dominated rainforests to angiosperm (*Nothofagus*) dominated forests (Reguero, Marenssi and Santillana, 2002), probably as a result of changing climate. Between 54 and 36 Ma sea surface temperatures dropped by 7 °C from ~18 to ~21 °C in the early Eocene to ~14 °C in the late Eocene based on $\delta^{18}O$ measurements of bivalve shells through the La Meseta Formation (Douglas, Affek and Ivany, 2010). This decrease in coastal temperatures is consistent with trends seen in benthic foraminifera $\delta^{18}O$ records and together they help confirm large-scale cooling of Antarctic coastal waters at this time (Douglas, Affek and Ivany, 2010). The several abrupt changes in floral composition recorded in the pollen and spore flora (e.g. fern abundance) probably relate to short-lived, localised changes in the conditions around the margins of the palaeoestuary, which encouraged the growth of these assemblies, or alternatively simply reflects the cyclical nature of the sedimentation in the palaeovalley (Greenhalgh, 2002). Either way this reflects the dynamic nature of the environment at the time.

Although the basal part of the Valle de las Focas Allomember (Figure 8.9) shows signs of possible (low-level) reworking following the initial (short-term) low sea levels at 56 Ma (an unconformity followed by a period of flooding) (Dingle, Marenssi and Lavelle, 1998; Greenhalgh, 2002), clay mineral data suggest that this was a very warm period with high, non-seasonal rainfall in well-drained hinterlands (Dingle, Marenssi and Lavelle, 1998; Roser and Korsch, 1986). By this time a mixed conifer and broad-leafed angiosperm rainforest had developed. Podocarps (*Podocarpus* and *Lagarostrobus*) still made an important contribution but *Nothofagus* (*fusca* group) had become the most dominant individual taxon (up to 50%) and *Nothofagus brassii* group had also undergone an increasingly significant presence. Of the fern community only the Cyatheaceae-Dicksoniaceae tree fern group maintains a consistent presence through the La Meseta Formation. The Proteaceae also forms a notable presence, which decreases erratically with time (Askin *et al.*, 1991; Greenhalgh, 2002). Casuarinaceae also has a low abundance and infrequent distribution. Lowland rainforest (that appears absent in the late Paleocene) was present, albeit rarely, during the early Eocene along with infrequently encountered open heathland and disturbed environments (Greenhalgh, 2002).

Vegetation of the warm to cooler middle middle Eocene

Within the Acantilados Allomember (Figure 8.9), further evidence for ameliorating climatic conditions are recorded in the quantitative pollen data (especially with respect to *Nothofagus* and *Lagarostrobus*; Askin, 1997, Greenhalgh, 2002; Pocknall, 1989) and increasing leaf size (Case, 1988). Initially the climate was warm, with mean annual temperatures having risen relative to the prevailing conditions of the underlying Cross Valley Formation to peak at ~15 °C during the Early Eocene Climatic Optimum, coupled with an increase in strongly seasonal precipitation (~1200 mm per year) in a low-lying hinterland that became seasonally waterlogged (Case, 1988; Dingle, Marenssi and Lavelle, 1998; Dutton, Lohmann and Zinsmeister, 2002; Greenhalgh, 2002; Ivany *et al.*, 2008). A slight cooling may have ensued, but these climatic conditions then remained relatively stable until the end of the earliest middle Eocene

(Dingle, Marenssi and Lavelle, 1998). The vegetation would have retained some similarity to that of the late Paleocene and earliest early Eocene, with a complex mixed conifer–broad-leafed angiosperm multistratal forest prevailing. Although podocarps formed the largest group (30–60%), the angiosperms were increasing in dominance, particulary *Nothofagus*. *Nothofagus* pollen assignable to *brassii*, *fusca* and *menziesii* types were all present with the *brassii* and *fusca* types being the most dominant and *menziesii* type being somewhat rare (Askin, 1997; Greenhalgh, 2002). Angiosperms would have ensured an increasing presence in the forest canopy with taxa belonging to the Proteaceae, Gunneraceae, Myrtaceae, Epacridaceae/Ericaceae, Casuarinaceae, Liliaceae, possible Arecaceae, Restoniaceae, Aquifoliaceae, Cunoniaceae/Elaeocarpaceae, Trimeniaceae, Droseraceae, Euphorbiaceae, Olacaceae and Sapindaceae (Cupanieae) (Askin, 1997). Podocarps (including *Podocarpus* and *Lagarostrobus franklinii*) and *Araucaria* contributed to the conifer component, whilst tree ferns (Cyatheaceae-Dicksoniaceae) and ferns comprised the most part of the understorey (Case, 1988; Greenhalgh, 2002). The coexistence, albeit spatially separated, of a cold, high altitude floral assemblage (e.g. with *Microcachys*) and a lowland tropical element (e.g. Sapindaceae) reflect both the initial warmth and subsequent cooler conditions.

Vegetation in the cool period of the late middle to early late Eocene

The change in climate and terrain conditions from those of the lower allomembers to that of the Campamento Allomember (51–49.5 Ma; Carpenter, 2008) occurred relatively rapidly. These changes could have resulted from progressive denudation or tectonic lowering of the hinterland, or alternatively, a eustatic sea level rise (Dingle, Marenssi and Lavelle, 1998) (Figure 8.9). Although the flora preserved in the Campamento Allomember records facets of the vegetation that grew following the Early Eocene Climatic Optimum, it was prior to any significant global cooling. It is dominated by organs (foliage, cone-scales and wood) from a single *Araucaria* conifer taxon, similar to *Araucaria araucana* (section *Columbea*) of South America and to the fossil species *Araucaria antarctica* and *A. nathorstii*, alongside relatively poorly preserved fragments of evergreen angiosperm taxa (Carpenter, 2008) (Figure 8.10). The angiosperm leaves are preserved as a thin layer of calcite and are mainly microphyllous and entire margined with only two morphotypes exhibiting toothed margins (Carpenter, 2008). The abundance of well-preserved (relatively undamaged) *Araucaria* leaves and other specimens contrasts with other floras from the La Meseta. This suggests that generally these conifers did not grow in or near an environment with good preservation potential (such as along the shoreline with preservation in shallow marine sediments). This flora is thought to represent a storm-torn assemblage composed of branches ripped from the *Araucaria* trees and leaves stripped from other taxa, which were then washed downstream to become channelled within an incised valley and flowed south-east towards the sea. From further studies of the palaeo-environment, Carpenter (2008) was able to recreate a more precise picture of the vegetation than that depicted by Reguero, Marenssi and Santillana (2002). Her study concluded that the flora was a mixed, ?predominantly evergreen mid-altitude forest, similar to evergreen coastal lowland or lower montane forests in warm to subtropical regions and those of central North Island New Zealand today. This vegetation comprised multiaged populations of an ancestral

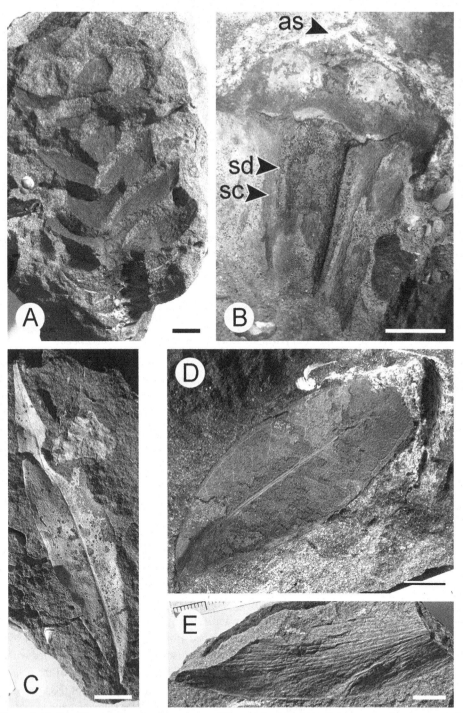

Figure 8.10 Fossil leaves from the Campamento Flora. **A**, leafy branch of *Araucaria* with characteristic flat, triangular leaves arranged in a helix BAS DJ. 1104.20a. **B**, fossil *Araucaria* cone scale clearly showing bract, scale (sc), seed (sd) and possibly the apical spine (as) BAS DJ. 1103.19a. **C**, angiosperm leaf of myricaceous-oleaceous affinity BAS DJ. 1106.9. **D**, cf. *Laurophyllum lanceolatum* BAS DJ. 1103.56. **E**, asteraceous leaf morphotype BAS DJ. 1106.12. Scale bars A, E = 10 mm, B, C, D = 5 mm. Photos courtesy of R. Carpenter.

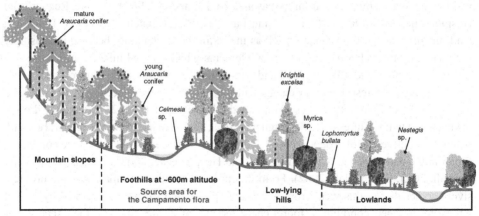

Figure 8.11 Reconstruction of the vegetation and environment of the northern Antarctic Peninsula during the latest early – middle Eocene, as recorded in the Campamento Flora (reproduced with permission from Carpenter, 2008).

Araucaria which dominated the eastern slopes and foothills of the Antarctic Peninsula mountain chain at ~600 m altitude, along with *Celmesia* (Asteraceae)-like herbaceous plants occupying sunny spots (Figure 8.11). Further downslope, tall trees of *Knightia* (Proteaceae) and *Nestegis* (Oleaceae) begin to appear as *Araucaria* decreased in abundance and eventually disappeared. At lower altitudes, the angiosperm component would have become more diverse, being made up of southern dicots that include Myrtaceae (cf. *Lophomyrtus*), Myricaceae (*Myrica*), and Lauraceae (cf. *Laurophyllum, Endiandra/Nectandra*), as well as possible Anacardicaeae, Leguminosae-Sapindaceae, Meliaceae, Violales and Asteraceae. The modern equivalent taxa are typically southern evergreen shrubs and trees with thick woody leaves and waxy cuticles (Carpenter, 2008). This high diversity vegetation lacks evidence for *Nothofagus* and ferns, both of which had been so abundant earlier in the La Meseta. This might suggest a taphonomical bias (i.e. the flora is allochthonous and may have undergone a prolonged and/or turbulent transport to the site of deposition), or that this unique vegetation assemblage occupied a different ecological niche (Carpenter, 2008). Sub-environments may have been present ranging from less fertile substrates to those prone to fire and others harbouring well-drained fertile soils.

This flora provides evidence for relatively mild terrestrial (warm temperate) conditions with warm summers (average temperatures of 17–19 °C) and frost-free winters (average temperatures of 7–8 °C) and a mean annual range in temperature of ~10 °C. Precipitation was probably seasonal and up to 2000 mm per year. Shallow marine temperatures ranged between ~5 and ~17 °C (Carpenter, 2008). Such conditions prevailed for several million years after the greenhouse conditions of the early Eocene, and high latitude cooling probably took place in a gradual manner (Carpenter, 2008).

By around 45 Ma, even though the water in adjacent shallow seas was relatively warm, the climate on land had begun to cool such that the very warm, wet, non-seasonal climate was being replaced by a wet, strongly seasonal, cool temperate climate (Cione, Reguero and Hospitaleche, 2007; Dingle, Marenssi and Lavelle, 1998; Kennett and Barker, 1990). On land, the diversity of the vegetation fell as the rainforest community became increasingly dominated by microthermal–mesothermal elements which formed mixed mesophytic forests (Case, 1988; Gandolfo, Marenssi and Santillana, 1998). Lowland rainforest and heathland became more restricted and changes in the vegetation were becoming obvious and apparent. The warm month mean still remained relatively high (~24–25 °C) and the drop in mean annual temperature ~7–13 °C (higher in local, sheltered habitats) was relatively small. Annual rainfall levels of around 1200–1500 mm per year were similar to those conditions found in the early Eocene (Cione, Reguero and Hospitaleche, 2007; Francis, Tosolini and Cantrill, 2003; Greenhalgh, 2002). From this point onwards the climate became progressively cooler during this wet, strongly seasonal period until it was interrupted by a brief return to warmer conditions, more characteristic of those seen in the Acantilados Allomember. This amelioration occurred between 42–41 Ma during the Middle Eocene Climatic Optimum, prior to the climate becoming cooler, wetter and less seasonal.

The subsequent cooling continued with temperatures dropping to an Eocene minimum of ~5 °C at ~37 Ma, consistent with the overall pattern of Eocene cooling (Dutton, Lohmann and Zinsmeister, 2002; Ivany, 2007; Ivany *et al.*, 2008). The change to a markedly seasonal climate, coupled with wet winters where temperatures hovered around freezing (−3 to +2 °C) during the Cucullaea II Allomember (Figure 8.11), impacted greatly on the environment, especially at higher altitudes where temperatures could have been colder still (Cione, Reguero and Hospitaleche, 2007; Gandolfo, Marenssi and Santillana, 1998). The 5 Ma of thermal decline towards the end of the Cucullaea II Allomember continued throughout the Submeseta Allomember, accompanied by levels of high rainfall (Dingle, Marenssi and Lavelle, 1998). The prevailing vegetation was mixed mesophytic cool temperate forests, with similar taxa growing today in the cool temperate Valdivian rainforests and Megallanic forests of southern South America. Elements such as podocarps (including *Lagarostrobus*), araucarian and cupressoid conifers, along with ferns and Nothofagaceae (e.g. *N. brassii*) were still present as were other angiosperm taxa such as Dilleniaceae (*Tetracera patagonica*), Hydrangiaceae (*Hydrangeiphyllum affine*), Betulaceae, Myrtaceae, Myricaceae, Lauraceae and Grossulariaceae, which are evident from the macro and microfloras (Brea, 1996, 1998; Case, 1988; Francis *et al.*, 2008; Gandolfo, Marenssi and Santillana, 1998; Gandolfo *et al.*, 1998; Gothan, 1908; Greenhalgh, 2002; Poole and Cantrill, 2006; Torres, Marenssi and Santillana, 1994a, 1994b). But the taxonomic composition was changing as diversity had dropped by nearly 50% relative to the end of the Paleocene. *Nothofagus* (especially *N. brassii*) and *Lagarostrobus*, however, were both increasing, which in turn supports the cooling trend leading up to the end of the Eocene (Dingle and Lavelle, 1998a, 1998b; Dingle, Marenssi and Lavelle, 1998; Dutton, Lohmann and Zinsmeister, 2002; Gaździcki *et al.*, 1992; Greenhalgh, 2002; Ivany *et al.*, 2008). A seasonal climate is recorded in the wood flora: narrow but regularly spaced, well-marked growth rings indicate that the trees grew slowly in a markedly seasonal climate (Torres, Marenssi and Santillana, 1994a, 1994b). As the similarities to the cool temperate Valdivian rainforests began to weaken through the late Eocene with the decrease in both floral diversity and the loss of subtropical and warm temperate taxa, similarities with the Magellanic forests of southern South America began to increase. The wood characteristics also indicate an increasingly cool temperate climate across Antarctica (Poole, Cantrill and Utescher, 2005) such that by the end of the middle Eocene rainfall had become seasonal (with 1000–3000 mm per year) falling during the spring and summer, the mean annual temperature rose no higher than 11–13 °C, whilst a freezing season developed that might have lasted several months as the cold month mean dropped to between −3 and +2 °C (Doktor *et al.*, 1996; Gandolfo, Marenssi and Santillana, 1998). The once significant forest cover was gradually eliminated, as evidenced by the lack of wood in sediments younger than ~43 Ma. Today similarities can be found in the relictual cool temperate conifer/broad-leafed Magellanic rainforests that comprise *Nothofagus betuloides*, together with evergreen taxa such as *Drimys* and *Pilgerodendron*, and a species-rich understorey in the shelter of established forest stands. Where conditions are more exposed, the landscape is more characteristic of moorland with pockets of deciduous *Nothofagus* (Veblen *et al.*, 1996).

What drove this change from warm temperate to cold temperate conditions during the early and middle–late Eocene is still under debate, but significant drops in atmospheric CO_2 levels, as well as an initial shallow breaching of the Patagonian–Antarctic Isthmus resulting in the opening of the Drake Passage, have been put forward as potential drivers (Case, 2007; DeConto and Pollard, 2003; Scher and Martin, 2006). Regardless of the cause, the ~10 °C drop in temperature towards the end of the Eocene could have imposed serious metabolic constraints upon lineages of organisms that had not experienced such climates for several tens of millions of years. This resulted in the greatest ecological and taxonomic implications for both the faunal and floral realm to have taken place during the Paleogene. Such consequences involved a substantial reorganisation of the ecosystem in the sea and on land, with a rapid decrease in vertebrate diversity, particularly with regard to the mammals and birds (Case, 2007), as life had to overcome the new physiological challenges of dealing with the ensuing decrease in temperatures towards the dry, cold climate, with possible frost and seasonal snow, that would characterise the end of the Eocene (Aronson *et al.*, 2009; Cione, Reguero and Hospitaleche, 2007; Ivany *et al.*, 2008).

The cooling trend that began during the Eocene not only changed the composition of the vegetation that had existed across the Peninsula region, it might also have added another barrier to the physical one that had been in existence since at least the early Eocene, and possibly even as early as the late Paleocene (T. Dutra and M. Reguero personal communication, 2010), that had been preventing animal migration. The isolation of the Antarctic biota that became physical with the development of the seaway between the Antarctic Peninsula and Patagonia began with climate deterioration (Reguero, Marenssi and Santillana, 2002). The faunal diversity of the La Meseta palaeoenvironment was diverse and included insects, worms, molluscs, echinoderms, corals, brachiopods, bryozoans, decapods and vertebrates. Soft-bodied organisms inhabited the sediment and were the food source for other animals and birds. (Schweitzer *et al.*, 2005). The land vertebrate faunal assemblage found on Seymour Island is the only Cenozoic land vertebrate fauna known from Antarctica – except for the bird tracks on King George Island (Covacevich and Lamperein, 1972; Covacevich and Rich, 1982; Hunt, 2001) – and represent the southernmost distribution of some South American land mammal lineages (Marenssi *et al.*, 1994; Reguero, Marenssi and Santillana, 2002; Vizcaíno *et al.*, 1997). It is assumed that there was no barrier to dispersal between Patagonia and the Antarctic Peninsula during the middle Eocene, such that those taxa able to adapt to cooler conditions and high altitudes could migrate into the southern high latitudes (although the presence of a high cordillera could have acted as a further geographical barrier). This would account for the close affinity between the La Meseta fauna and Paleogene faunas of Patagonia (Reguero, Marenssi and Santillana, 2002). As the vegetation that swathed the high, densely forested cordillera changed during the middle Eocene from one dominated by podocarps to principally *Nothofagus* dominated, so the habitats within them changed. These forests were populated by a diverse fauna – arboreal and browsers with different dietary requirements ranging from herbivory, omnivory, frugivory, folivory to insectovory, that had migrated to this area prior to the middle Eocene (Reguero, Marenssi and Santillana, 2002). The increase in arboreal habitats facilitated a radiation in insects and arboreal herbivorous/omnivorous mammals, an increase reflected in the proportion of small-sized mammals preserved in the Cucullaea I Allomember (Reguero, Marenssi and Santillana, 2002). Some of these animals were arboreal such as the small fruit- and insect-eating possums, and the large bodied sloths (weighing up to 10 kg), others were ground dwellers including small ('opposum-like') marsupials and rodent-like gondwanatherians (similar to the modern anteater), and the largest and most abundant terrestrial herbivore, the liptotern sparnotheriodontid, which fed on leaves and reached about 400 kg in weight (Vizcaíno *et al.*, 1998). Flightless, running ratites (one ostrich-like and another cursorial carnivorous bird) would have made their home amongst the trees, whilst small and large penguins congregated along the shores of the ancient Antarctic Peninsula as falcons dominated the skies above the forest canopy (Bond *et al.*, 2006; Case, 2006; Case, Woodburne and Chaney, 1987; Goin *et al.*, 2006; Tambussi *et al.*, 1994, 2006). This fauna comprises both large and small-sized mammals but lacks any evidence for medium-sized mammals. This suggests that only the small-sized mammals and the large mammals were able to adapt to the cold conditions (through torpor or hibernation and heat conservation via small surface area-to-volume ratio, respectively) and could thus migrate further into the southern high latitudes and survive the dark winter months (Case, 2006) through adopting a crepuscular and even extended nocturnal condition (Reguero, Marenssi and Santillana, 2002).

By the end of the Eocene, both fauna and flora had become relatively depauperate. Even though the northern Peninsula region was still well-vegetated, the first microphilic, deciduous *Nothofagus* (with plicate vernation) began to appear, pre-announcing the cold that was to characterise the Eocene–Oligocene transition (Alves, Guerra-Sommer and Dutra, 2005; Case, 1988; Ditchfield, Marshall and Pirrie, 1994; Gaździcki and Stolarski, 1992).

Comparisons between east and west Antarctic Peninsula

Even though the Antarctic Peninsula region is not very wide, especially at its northernmost extent, during the early Paleogene there were, as there are today, climatic differences between east (James Ross Basin) and west (South Shetland Islands). Although it is difficult to separate the regional signal from the local signal for both vegetation and climate, the plant record may provide certain clues. The Antarctic Peninsula was an active volcanic arc formed by uplift and metamorphism during the late Paleozoic and early Mesozoic, and is in essence an extension of the South American Andes. The mountains that divide east from west attained maximum heights of about 1000 m during the Mesozoic and have since been subject to erosion. With prevailing westerly winds, this mountain barrier would probably have been high enough to create a rain shadow effect to the east of the Antarctic Peninsula – a situation similar to that experienced on the east side of southern South America today. This would have influenced the climate and thus floras across an east–west transect. A possible second factor influencing vegetation, more difficult to determine from the fossil record, would have been the effect of altitude with associated trends in climate, also reflected in vegetational composition. Altitudinal effects can only be determined through inference without the preservation of a number of *in situ* fossil remains that have not been subject to later tectonic displacement.

When the east (e.g. Seymour Island) side of the Antarctic Peninsula is compared with the west (South Shetland Islands) side, general differences can be noted. On the east side the taxonomic composition of the vegetation appears to represent a drier climate (e.g. presence of *Casuarina* and Proteaceae) and a lower altitude flora, when compared with the floras of similar age on King George Island. However, the presence of *Sphagnum* in the Sobral Formation suggests damp environments, but this could be a local rather than a regional phenomenon. The leaves found in greatest abundance across the Antarctic Peninsula are those of *Nothofagus*, and on the west side they tend to be larger in size relative to those found in the east for the same time interval. This may suggest the climate on the east side was not just drier but also cooler. The most abundant podocarp on the east side is *Dacrydium* and, with its imbricate leaf form rather than needle-like leaves, may also indicate lower temperatures. With regard to the composition of the ferns and conifers, there is no marked difference between east and west (other than *Dacrydium*), which may suggest a more regional similarity. On the east side, palms are relatively rare whereas monocots are more abundant when compared with the west side. Estimates of temperature and precipitation also indicate a warmer, wetter climate on the western side of the Antarctic Peninsula relative to the east. Together these lines of evidence indicate that Seymour Island lying on the east side may well have occupied a location within the rain shadow cast by the uplifting mountains that ran along the Antarctic Peninsula. With further collection and investigation of the fossil floras from the northern Antarctic Peninsula region, any latitudinal trends will become more apparent.

Summary

- Vegetation records from the Paleocene–Eocene warm period of Antarctica come predominantly from the northern Antarctic Peninsula region in the South Shetland Islands and Seymour Island. Evidence for the vegetation that grew during peak warmth is missing, but fossil floras from the early Paleocene indicate mixed meso-megathermal, mutistratal rainforest communities prevailed with abundant angiosperms, including several species of *Nothofagus*, Proteaceae and palms, and an understorey of ferns. Since cool temperate elements were also present, this flora has been termed 'Palaeoflora Mixta' type (i.e. mixed neotropical taxa coexisting alongside those of more cool temperate–subarctic environments). These mixed forests seem to have survived for longer, or prevailed for longer, on the east side of the Antarctic Peninsula where evidence for this vegetation can still be found in the late Paleocene.

- By the late Paleocene, the relatively meso-megathermal elements were becoming more dominant and increasing similarities can be seen with the cool temperate Valdivian rainforests of southern South America today.

- During the early to middle Eocene, the angiosperms continued to increase and so too did the similarities with the Valdivian rainforests. These rainforests are considered to be the last vestige of the forests that once covered much of Antarctica during the Paleocene–Eocene. Variations within the composition of the flora can also be seen. These variations have been attributed to the environmental dynamics (e.g. volcanic disturbance) that were taking place at this time giving rise to successional gradations that are evident in the fossil record. This disturbance may have given the angiosperms an ecological advantage, as can be seen in similar environments in the Valdivian region of South America today.

- As climates cooled in the middle to late Eocene so diversity began to decrease. The forests became dominated by *Nothofagus*, represented by a number of different species, along with podocarps and ferns. Similarities with the Valdivian rainforests lessened and the vegetation became more similar to the cool temperate *Nothofagus* woods and forests of the Magellanic region of southern South America.

References

Agnini, C., Macri, P., Backman, J., *et al.* (2009). An early Eocene carbon cycle perturbation at similar to 52.5 Ma in the Southern Alps: chronology and biotic response. *Paleoceanography*, **24**, PA2209, doi:10.1029/2008PA001649.

Aizen, M. A. and Ezcurra, C. (1998). High incidence of plant-animal mutualisms in the woody flora of the temperate forest of southern South America: biogeographical origin and present ecological significance. *Ecología Austral*, **8**, 217–236.

Alves, L. S. D. R., Guerra-Sommer, M. and Dutra, T. (2005). Paleobotany and Paleoclimate. Part 1: Growth rings in fossil wood and paleoclimate. Part 2: Leaf assemblages (taphonomy, paleoclimatology and palaeogeography. In *Applied Stratigraphy*, ed. E. A. M. Koutsoukos. Netherlands: Springer, pp. 179–202.

Aravena, J. C., Carmona, M. R., Pérez, C. A. and Armesto, J. J. (2002). Changes in tree species richness stand structure and soil properties in a successional chronosequence in northern Chiloé Island, Chile. *Revista Chilena de Historia Natural*, **75**, 339–360.

Archangelsky, S. and Romero, E. J. (1974). Los registro más antiguos de pólen de *Nothofagus* (Fagáceas) de Patagonia (Argentina y Chile). *Boletin de la Sociedad Botanica de México*, **33**, 13–30.

Archangelsky, S. and Zamaloa, M. C. (1986). Nuevas descripciones palinológicas de las Formaciones Salamanca y Bororó, Paleoceno de Chubut, República Argentina. *Ameghiniana*, **23**, 35–46.

Armesto, J. J. and Figueroa, J. (1987). Stand structure and dynamics in the temperate rain forest of Chiloé Archipiélago, Chile. *Journal of Biogeography*, **14**, 367–376.

Armesto, J. J. and Rozzi, R. (1989). Seed dispersal syndromes in the rain forest of Chiloé: evidence for the importance of biotic dispersal in a temperate rain forest. *Journal of Biogeography*, **16**, 219–226.

Armesto, J. J., Rozzi, R., Miranda, P. and Sabag, C. (1987). Plant/frugivore interactions in South American temperate forest. *Revista Chilena de Historia Natural*, **60**, 321–336.

Armesto, J. J., Villagrán, C., Aravena, J. C., *et al.* (1995). Conifer forest of the Chilean coastal range. In *Ecology of the Southern Conifers*, eds N. J. Enright and R. S. Hill. Melbourne: Melbourne University Press, pp. 156–170.

Armesto, J. J., Aravena, J. C., Villagrán, C., Pérez, C. and Parker, G. (1996). Bosques templados de la cordillera de la costa. In *Ecología de los Bosques Nativos de Chile*, eds J. J. Armesto, C. Villagrán and M. Kalin. Santiago: Editorial Universitaria, pp. 199–213.

Aronson, R. B., Moody, R., Ivany, *et al.* (2009). Climate change and trophic response of the Antarctic bottom fauna. *PLoS One*, **4**, 1–6.

Arroyo, M. T. K., Cavieres, L., Peñaloza, A., Riveros, M., and Faggi, A. M. (1996). Relaciones fitogeográficas y patrones regionales de riqueza de especies en la flora del bosque lluvioso templado de Sudamérica. In *Ecología de los Bosques Nativos de Chile*, eds J. Armesto, C. Villagrán, and M. Kalin. Santiago: Editorial Universitaria, pp. 71–99.

Askin, R. A. (1988a). Campanian to Paleocene palynological succession of Seymour and adjacent islands, northeast Antarctic Peninsula. In *Geology and Paleontology of Seymour Island, Antarctic Peninsula, Geological Society of America Memoir*, **169**, eds R. M. Feldmann and M. O. Woodburne. Boulder: The Geological Society of America, pp. 131–154.

Askin, R. A. (1988b). The palynological record across the Cretaceous-Cenozoic transition on Seymour Island Antarctica. In *Geology and Paleontology of Seymour Island, Antarctic Peninsula, Geological Society of America Memoir*, **169**, eds R. M. Feldmann and M. O. Woodburne. Boulder: The Geological Society of America, pp. 155–162.

Askin, R. A. (1990). Cryptogam spores from the upper Campanian and Maastrichtian of Seymour Island. *Micropaleontology*, **36**, 141–156.

Askin, R. A. (1992). Late Cretaceous-early Tertiary Antarctic outcrop evidence for past vegetation and climate. In *The Antarctic Paleoenvironment: A Perspective on Global Change. Part 1, Antarctic Research Series*, **56**, eds J. P. Kennett and D. A. Warnke. Washington: American Geophyscial Union, pp. 61–73.

Askin, R. A. (1994). Monosulcate angiosperm pollen from the López de Bertodano Formation (upper Campanian-Maastrichtian-Danian) of Seymour Island, Antarctica. *Review of Palaeobotany and Palynology*, **81**, 151–164.

Askin, R. A. (1997). Eocene-?earliest Oligocene terrestrial palynology of Seymour Island Antarctica. In *The Antarctic Region: Geological Evolution and Processes*, ed. C. A. Ricci. Siena: Terra Antarctica Publication, pp. 993–996.

Askin, R. A. (1989). Endemism and heterochroneity in the Late Cretaceous (Campanian) to Paleocene palynofloras of Seymour Island, Antarctica: implications for origins, dispersal and palaeoclimates of southern floras. In *Origins and Evolution of the Antarctic Biota, Geological Society of London Special Publication*, **47**, ed. J. A. Crame. London: The Geological Society, pp. 107–119.

Askin, R. A. and Fleming R. F. (1982). Palynological investigations of Campanian to lower Oligocene sediments on Seymour Island, Antarctic Peninsula. *Antarctic Journal of the United States*, **17(5)**, 70–71.

Askin, R. A., Elliot, D. H., Stilwell, J. D. and Zinsmeister, W. J. (1991). Campanian and Eocene stratigraphy and paleontology on Cockburn Island Antarctic Peninsula. *Journal of South American Science*, **4**, 99–117.

Bains, S., Corfield, R. M. and Norris, R. D. (1999). Mechanisms of climate warming at the end of the Paleocene. *Science*, **285**, 724–727.

Bains, S., Norris, R. D., Corfield, R. M. and Faul, K. L. (2000). Termination of global warmth at the Palaeocene/Eocene boundary through productivity feedback. *Nature*, **407**, 171–174.

Baldoni, A. M. and Barreda, V. D. (1986). Estudio palinológico de las Formaciones López de Bertodano y Sobral, Isla Vicecomodoro Marambio, Antártida. *Boletin do Instituto Geológico-Universidade de Säo Paulo, Série Cientifica*, **17**, 89–98.

Baldoni, A. M. and Medina, F. (1989). Fauna y microflora del Cretácico, en Bahía Brandy, Isla James Ross, Antártida. *Serie Científica Instituto Antártico Chileno*, **39**, 43–58.

Barreda, V. D. and Palazzesi, L. (2007). Patagonian vegetation turnovers during the Paleogene-Early Neogene: origin of arid adapted floras. *The Botanical Review*, **73**, 31–50.

Barrera, E., Huber, B. T., Savin, S. M. and Webb, P. N. (1987). Antarctic marine temperatures: late Campanian through early Paleocene. *Paleoceanography*, **2**, 21–47.

Barton, C. M. (1964a). Significance of the Tertiary fossil floras of King George Island, South Shetland Islands. In *Antarctic Geology: Proceedings of the First International Symposium on Antarctic Geology, Cape Town, 16–21 September 1963*, ed. R. J. Adie. Amsterdam: North-Holland Publishing Company, pp. 603–608.

Barton, C. M. (1964b). The geology of King George Island, South Shetland Islands. Ph.D. thesis, University of Birmingham.

Barton, C. M. (1965). The geology of South Shetland Islands; III. The stratigraphy of King George Island. *British Antarctic Survey Scientific Report*, **44**, 1–33.

Berry, E. W. (1937). A Paleocene flora from Patagonia. *John Hopkins University Studies in Geology*, **12**, 33–50.

Bijl, P. K., Schouten, S., Sluijs, A., *et al.* (2009). Early Palaeogene temperature evolution of the southwest Pacific Ocean. *Nature*, **461**, 776–779.

Birkenmajer, K. (1980). Tertiary volcanic-sedimentary succession at Admiralty Bay, King George Island (South Shetland Islands, Antarctica). *Studia Geologica Polonica*, **64**, 7–65.

Birkenmajer, K. (1981). Lithostratigraphy of the Point Hennequin Group (Miocene volcanic and sediments) at King George Island (South Shetland Islands, Antarctica). *Studia Geologica Polonica*, **72**, 59–73.

Birkenmajer, K. (1982a). Mesozoic stratiform volcanic-sedimentary succession and Andean intrusions at Admirality Bay, King George Island (South Shetland Islands, Antarctica). *Studia Geologica Polonica*, **74**, 105–154.

Birkenmajer, K. (1982b). Report on geological investigations of King George Island and Nelson Island (South Shetland Islands, West Antarctica) in 1980/81. *Studia Geologica Polonica*, **74**, 175–197.

Birkenmajer, K. (1982c). Late Cenozoic phases of block-faulting on King George Island (South Shetland Islands, West Antarctica). *Bulletin de l'Academie Polonaise des Sciences, Series des Sciences de la Terre*, **30**, 21–32.

Birkenmajer, K. (1985). Onset of Tertiary continental glaciations in the Antarctic Peninsula sector (West Antarctica). *Acta Geologica Polonica*, **35**, 1–31.

Birkenmajer, K. (1989). A guide to Tertiary geochronology of King George Island, West Antarctica. *Polish Polar Research*, **10**, 555–579.

Birkenmajer, K. (1990). Geochronology and climatostratigraphy of Tertiary glacial and interglacial succession on King George Island, South Shetland Islands (West Antarctica). *Zentralblatt für Geologie und Paläontologie, Teil I*, **1990(1–2)**, 141–151.

Birkenmajer, K. (1997). Tertiary glacial/interglacial palaeoenvironments and sea level changes, King George Island, West Antarctica. An overview. *Bulletin of the Polish Academy of Sciences, Earth Sciences*, **44**, 157–181.

Birkenmajer, K. (2001). Mesozoic and Cenozoic stratigraphic units in parts of the South Shetland Islands and Northern Antarctic Peninsula (as used by the Polish Antarctic Programmes). *Studia Geologica Polonica*, **118**, 7–188.

Birkenmajer, K. and Zastawniak, E. (1986). Plant remains of the Dufayel Island Group (early Tertiary?), King George Island, South Shetland Islands (West Antarctic). *Acta Palaeobotanica*, **26**, 33–53.

Birkenmajer, K. and Zastawniak, E. (1989a). Late Cretaceous–early Tertiary floras of King George Island, West Antarctica: their stratigraphic distribution and palaeoclimatic significance. In *Origins and Evolution of the Antarctic Biota, Geological Society of London Special Publication*, **47**, ed. J. A. Crame. London: The Geological Society, pp. 227–240.

Birkenmajer, K. and Zastawniak, E. (1989b). Late Cretaceous–early Neogene vegetation history of the Antarctic Peninsula sector. Gondwana break-up and Tertiary glaciations. *Bulletin of the Polish Academy of Sciences, Earth Sciences*, **37**, 63–88.

Birkenmajer, K., Narębski, W., Nicoletti, M. and Petrucciani, C. (1983). Late Cretaceous through late Oligocene K-Ar ages of the King George Island Supergroup volcanics, South Shetland Islands (West Antarctica). *Bulletin de l'Académie Polonaise des Sciences, Série des Sciences de la Terre*, **30**, 133–143.

Birkenmajer, K., Delitala, M. C., Narębski, W., Nicoletti, M. and Petrucciani, C. (1986). Geochronology of Tertiary island-arc volcanics and glacigenic deposits, King George Island, South Shetland Islands (West Antarctica). *Bulletin of the Polish Academy of Sciences, Earth Sciences*, **34**, 257–272.

Birkenmajer, K., Soliani, E. and Kawashita, K. (1989). Geochronology of Tertiary glaciations on King George Island, West Antarctica. *Bulletin of the Polish Academy of Sciences, Earth Sciences*, **37**, 27–48.

Birkenmajer, K., Soliani, E. and Kawashita, K. (1990). Reliability of potassium-argon dating of Cretaceous-Tertiary island-arc volcanic suites of King George Island, South Shetland Islands (West Antarctica). *Zentralblatt für Geologie und Paläontologie, Teil I*, **1990(1–2)**, 127–140.

Birkenmajer, K., Gaździcki, A., Krajewski, K. P., *et al.* (2005). First Cenozoic glaciers in West Antarctica. *Polish Polar Research*, **26**, 3–12.

Bohaty, S. M. and Zachos, J. C. (2003) Significant Southern Ocean warming event in the late middle Eocene. *Geology*, **31**, 1017–1020.

Bohaty, S. M., Zachos, J. C., Florindo, F. and Delaney, M. L. (2009). Coupled greenhouse warming and deep-sea acidification in the middle Eocene. *Paleoceanography*, **24**, PA2207, doi:10.1029/2008PA001671.

Bond, M., Reguero, M. A., Vizcaíno, S. F. and Marenssi, S. A. (2006). A new 'South American ungulate' (Mammalia: Litopterna) from the Eocene of the Antarctic Peninsula. In *Cretaceous–Tertiary High-Latitude Palaeoenvironments, James Ross*

Basin, Antarctica, Geological Society of London Special Publication, **258**, eds J. E. Francis, D. Pirrie and J. A. Crame. London: The Geological Society, pp. 163–176.

Bowen, G. J., Koch, P. L., Gingerich, P. D., *et al.* (2001). Refined isotope stratigraphy across the continental Paleocene-Eocene boundary on Polecat Bench in the northern Bighorn Basin. In *Paleocene-Eocene Stratigraphy and Biotic Change in the Bighorn and Clarks Fork Basins, Wyoming, University of Michigan Papers on Paleontology*, **33**, ed. P. D. Gingerich. Ann Arbor: Museum of Paleontology, pp. 73–88.

Bowen, G. J., Clyde, W. C., Koch, P. L., *et al.* (2002). Mammalian dispersal at the Paleocene/Eocene boundary. *Science*, **295**, 2062–2065.

Bowen, G. J., Beerling, D. J., Koch, P. L., Zachos, J. C. and Quattlebaum, T. (2004). A humid climate state during the Paleocene/Eocene thermal maximum. *Nature*, **432**, 495–499.

Bowen, G. J., Bralower, T. J., Delaney, M. L., *et al.* (2006). The Paleocene–Eocene thermal maximum gives insight into greenhouse gas-induced environmental and biotic change. *Eos*, **87**, 165–169.

Bralower, T. J. (2002). Evidence of surface water oligotrophy during the Paleocene-Eocene thermal maximum: nannofossil assemblage data from Ocean Drilling Program Site 690, Maud Rise, Weddell Sea. *Paleoceanography*, **17**, 2, doi 10.1029/2001PA000662.

Bralower, T. J., Kelly, D. C. and Leckie, R. M. (2002). Biotic effects of abrupt Paleocene and Cretaceous climate events. Legacy of the Ocean Drilling Program. *JOIDES Journal: Joint Oceanographic Institutions for Deep Earth Sampling*, **28**, 29–34.

Brea, M. (1996). Análisis de los anillos de crecimiento de leños fósiles de coníferas de la Formación La Meseta, Isla Seymour, Antártida. *1st Congreso Paleógeno de América del Sur. Santa Rosa, Resúmenes*, p. 28.

Brea, M. (1998). Análisis de los anillos de crecimiento en leños fósiles de coníferas de la Formación La Meseta, Isla Seymour (Marambio), Antártida. In *Paleógeno de América del Sur y de la Península Antártica, Asociación Paleontológica Argentina Publicación Especial*, **5**, ed. S. Casadío. Buenos Aires: APA, pp. 163–175.

Brea, M., Bellosi, E. and Krause, M. (2009). *Taxaceoxylon katuatenkum* n. sp. en la Formación Koluel-Kaike (Eoceno inferior-medio), Chubut, Argentina: un componente de los bosques subtropicales paleógenos de Patagonia. *Ameghiniana*, **46**, 127–140.

Brizuela, R. R., Marenssi, S., Barreda, V. and Santillana, S. (2007). Palynofacies approach across the Cretaceous Paleogene boundary in Marambio (Seymour Island), Antarctic Peninsula. *Revista de la Asociación Geológica Argentina*, **62**, 236–241.

Brown, B., Gaina, C. and Müller, R. D. (2006). Circum-Antarctic palaeobathymetry: Illustrated examples from Cenozoic to recent times. *Palaeogeography, Palaeoclimatology, Palaeoecology*, **231**, 158–168.

Burnham, R. J. (1994). Plant deposition in modern volcanic environments. *Transactions of the Royal Society of Edinburgh: Earth Sciences*, **84**, 275–281.

Burnham, R. J. and Spicer, R. A. (1986). Forest litter preserved by volcanic activity in El Chichón, Mexico: a potentially accurate record of the pre-eruption vegetation. *Palaios*, **1**, 158–161.

Burns, B. R. (1993). Fire-induced dynamics of *Araucaria araucana-Nothofagus antarctica* forest in the southern Andes. *Journal of Biogeography*, **20**, 669–685.

Cao, L. (1992). Late Cretaceous and Eocene palynofloras from Fildes Peninsula, King George Island (South Shetland Islands), Antarctica. In *Recent Progress in Antarctic Earth Science*, eds Y. Yoshida, K. Kaminuma and K. Shiraishi. Tokyo: Terra Scientific Publishing Company, pp. 363–369.

Cantrill, D. J. and Poole, I. (2005). Taxonomic turnover and abundance in Cretaceous and Tertiary wood floras of Antarctica: implications for changes in forest ecology. *Palaeogeography, Palaeoclimatology, Palaeoecology*, **215**, 205–219.

Cantrill, D. J., Tosolini, A.-M. P. and Francis, J. E. (2011). Paleocene flora from Seymour Island, Antarctica: revision of Dusén's (1908) pteridophyte and conifer taxa. *Alcheringa*, **35**, 309–328.

Carpenter, R. J., Hill, R. S. and Jordan, G. J. (1994). Cenozoic vegetation in Tasmania: macrofossil evidence. In *History of the Australian Vegetation: Cretaceous to Recent*, ed. R. S. Hill. Cambridge: Cambridge University Press, pp. 276–298.

Carpenter, R. J., Truswell, E. M. and Harris, W. K. (2010). Lauraceae fossils from a volcanic Palaeocene oceanic island, Ninetyeast Ridge, Indian Ocean: ancient long-distance dispersal? *Journal of Biogeography*, **37**, 1202–1213.

Carpenter, R. S. (2008). Palaeoenvironmnetal and climatic significance of an *Araucaria*-dominated Eocene flora from Seymour Island, Antarctica. Ph.D. thesis, University of Leeds.

Case, J. A. (1988). Paleogene floras from Seymour Island, Antarctic Peninsula. In *Geology and Paleontology of Seymour Island, Antarctica Peninsula, Geological Society of America Memoir*, **169**, eds R. M. Feldmann and M. O. Woodburne. Boulder: The Geological Society of America, pp. 523–530.

Case, J. A. (2006). The late Middle Eocene terrestrial vertebrate fauna from Seymour Island: the tails of the Eocene Patagonian size distribution. In *Cretaceous–Tertiary High-Latitude Palaeoenvironments, James Ross Basin, Antarctica, Geological Society of London Special Publication*, **258**, eds J. E. Francis, D. Pirrie and J. A. Crame. London: The Geological Society, pp. 177–186.

Case, J. A. (2007). Opening of the Drake Passage: Does this event correlate to climate change and biotic events from the Eocene La Meseta Formation, Seymour Island, Antarctic Peninsula? In *Online Proceedings of the 10th International Symposium on Antarctic Earth Sciences*, eds A.K. Cooper, C.R. Raymond and the 10th ISAES Editorial team. U. S. Geological Survey Open-File Report 2007–1047, Version 2.0. Extended Abstract 117, available at http://pubs.usgs.gov/of/2007/1047/eq/0f2007–1047ea117.pdf. {Cited 2011}.

Case, J. A., Woodburne, M. O. and Chaney, D. S. (1987). A gigantic phororhacoid(?) bird from Antarctica. *Journal of Paleontology*, **6**, 1280–1284.

Chun, H. Y., Chang, S. K. and Lee, J. L. (1994). Biostratigraphic study on the plant fossils from the Barton Peninsula and adjacent area. *Journal of the Paleontological Society of Korea*, **10**, 69–84 (in Korean).

Cione, A. L., Reguero, M. A. and Hospitaleche, C. A. (2007). Did the continent and sea have different temperatures in the northern Antarctic Peninsula during the Middle Eocene? *Revista de la Asociación Geológica Argentina*, **62**, 586–596.

Clyde, W. C. and Gingerich, P. D. (1998). Mammalian community response to the latest Paleocene thermal maximum: An isotaphonomic study in the northern Bighorn Basin, Wyoming. *Geology*, **26**, 1011–1014.

CONAMA. (1999). *Estadísticas del Medio Ambiente 1994–1998*. Santiago: Instituto Nacional de Estadísticas.

Covacevich, V. and Lamperein, C. (1972). *Ichnites* from Fildes Peninsula, King George Island, South Shetland Islands. In *Antarctic Geology and Geophysics*, ed. R. J. Adie. Oslo: Universitetsforlaget, pp. 71–74.

Covacevich, V. and Rich, P. (1982). New bird ichnites from Fildes Peninsula, King George Island, West Antarctica. In *Antarctic Geoscience*, ed. C. Craddock. Madison: University of Wisconsin Press, pp. 245–251.

Crame, J. A., Pirrie, D., Riding, J. B. and Thomson, M. R. A. (1991). Campanian–Maastrichtian (Cretaceous) stratigraphy of the James Ross Island area, Antarctica. *Journal of the Geological Society of London*, **148**, 1125–1140.

Cramer, B. S. and Kent, D. V. (2005). Bolide summer: Paleocene/Eocene thermal maximum as a response to an extraterrestrial trigger. *Palaeogeography, Palaeoclimatology, Palaeoecology*, **224**, 144–166.

Cramer, B. S., Wright, J. D., Kent, D. V. and Aubry, M. P. (2003). Orbital climate forcing of $\delta^{13}C$ excursions in the late Paleocene-early Eocene (chrons C24n-C25n). *Paleoceanography*, **18**, 1097, doi 10.1029/2003PA000909.

Cranwell, L. M. (1959). Fossil pollen from Seymour Island, Antarctica. *Nature*, **184**, 1782–1785.

Cranwell, L. M. (1969). Palynological intimations of some pre-Oligocene Antarctic climates. In *Palaeoecology of Africa*, ed. E. M. Van Zinderen Bakker. Cape Town: Balkema, pp. 1–19.

Crouch, E. M. and Brinkhuis, H. (2005). Environmental change across the Paleocene–Eocene transition from eastern New Zealand: a marine palynological approach. *Marine Micropaleontology*, **56**, 138–160.

Crowley, T. J. and North, G. R. (1988). Abrupt climate change and extinction events in Earth history. *Science*, **240**, 996–1002.

Cunha, M. B., Dutra, T. L. and Cardoso, N. (2008). Uma Dicksoniaceae fértil no Eoceno da Ilha King George, Península Antártica. *GÆA Journal of Geoscience*, **4**, 1–13.

Czajkowski, S. and Rosler, O. (1986). Plantas fósseis da Península Fildes; Ilha Rei Jorge (Shetlands do Sul); Morfografia da impressões foliares. *Anais Academia Brasileira de Ciencias Supplemento*, **58**, 99–110.

DeConto, R. M. and Pollard, D. (2003). Rapid Cenozoic glaciation of Antarctica induced by declining atmospheric CO_2. *Nature*, **421**, 245–249.

del Valle, R. A., Diaz, M. T. and Romero, E. J. (1984). Preliminary report on the sedimentites of Barton Peninsula 25 de Mayo Island (King George Island), South Shetland Islands, Argentina Antártica. *Instituto Antártico Argentino Contribución*, **308**, 121–131.

Dettmann, M. E. (1994). Cretaceous vegetation: the microfossil record. In *History of the Australian Vegetation: Cretaceous to Recent*, ed. R. S. Hill. Cambridge: Cambridge University Press, pp. 143–170.

Dettmann, M. E. and Jarzen, D. M. (1988). Angiosperm pollen from the uppermost Cretaceous strata of southern Australia and the Antarctic Peninsula. *Association of Australasian Paleontologists Memoirs*, **5**, 217–237.

Dettmann, M. E. and Jarzen, D. M. (1991). Pollen evidence for Late Cretaceous differentiation of Proteaceae in southern polar forests. *Canadian Journal of Botany*, **69**, 901–906.

Dettmann, M. E. and Thomson, M. R. A. (1987). Cretaceous palynomorphs from the James Ross Island area, Antarctica—a pilot study. *British Antarctic Survey Bulletin*, **77**, 13–59.

Dettmann, M. E., Pocknall, D. T., Romero, E. J. and Zamaloa, M. C. (1990). *Nothofagidites* Erdtman ex Potonié, 1960; a catalogue of species with notes on the paleogeographic distribution of *Nothofagus* Bl. (Southern Beech). *New Zealand Geological Survey Palaeontological Bulletin*, **60**, 1–79.

Dickens, G. R. (2000). Methane oxidation during the Late Palaeocene Thermal Maximum. *Bulletin de la Société Géologique de France*, **171**, 37–49.

Dickens, G. R., O'Neil, J. R., Rea, D. K. and Owen, R. M. (1995). Dissociation of oceanic methane hydrate as a cause of the carbon isotope excursion at the end of the Paleocene. *Paleoceanography*, **10**, 965–971.

Diester-Haass, L., Billups, K., Gröcke, D. R., *et al.* (2009). Mid-Miocene paleoproductivity in the Atlantic Ocean and implications for the global carbon cycle. *Paleoceanography*, **24**, PA1209, doi:10.1029/2008PA001605.

Dingle, R. V. and Lavelle, M. (1998a). Late Cretaceous-Cenozoic climatic variations of the northern Antarctic Peninsula: new geochemical evidence and review. *Palaeogeography, Palaeoclimatology, Palaeoecology*, **141**, 215–232.

Dingle, R. V. and Lavelle, M. (1998b). Antarctic Peninsula cryosphere: Early Oligocene (*c.* 30 Ma) initiation and a revised glacial chronology. *Journal of the Geological Society of London*, **155**, 433–437.

Dingle, R. V. and Lavelle, M. (2000). Antarctic Peninsula Late Cretaceous-Early Cenozoic palaeoenvironments and Gondwana palaeogeographies. *Journal of African Earth Sciences*, **31**, 91–105.

Dingle, R. V., Marenssi, S. A. and Lavelle, M. (1998). High latitude Eocene climatic deterioration: Evidence from the northern Antarctic Peninsula. *Journal of South American Earth Sciences*, **11**, 571–579.

Ditchfield, P. W., Marshall, J. D. and Pirrie, D. (1994). High latitude temperature variation: new data from the Tithonian to Eocene of James Ross Island Antarctica. *Palaeogeography, Palaeoclimatology, Palaeoecology*, **107**, 79–101.

Doktor, M., Gaździcki, A., Jerezmanska, A., Porębski, J. and Zastawniak, E. (1996). A plant and fish assemblage from the Eocene La Meseta Formation of Seymour Island (Antarctic Peninsula) and its environmental implications. *Palaeontologica Polonica*, **55**, 127–146.

Donoso, C. (1993). *Bosques templados de Chile y Argentina. Variación, Estructura y Dinámica*. Santiago: Editorial Universitaria.

Donoso, C., Grez, R., Escobar, B. and Real, P. (1984). Estructura y dinámica de bosques del Tipo Forestal Siempreverde en un sector de Chiloé insular. *Bosque*, **5**, 82–104.

Donoso, C., Escobar, B. and Urrutia, J. (1985). Estructura y estrategias regenerativas de un bosque virgen de Ulmo (*Eucryphia cordifolia* Cav.) – Tepa (*Laureliopsis philippiana* Phil. Looser) en Chiloé, Chile. *Revista Chilena de Historia Natural*, **58**, 171–186.

Douglas, P., Affek, H., and Ivany, L. C. (2010). Eocene clumped isotope temperature estimates from Seymour Island, Antarctica. *Geochimica et Cosmochimica Acta*, **74**, A245.

Duan, W. and Cao, L. (1998). Late Paleogene palynofloras from Point Hennequin of the Admiralty Bay, King George Island, Antarctica with special reference to its stratigraphical significance. *Chinese Journal of Polar Science*, **9**, 125–132.

Dupre, D. D. (1982). Geochemistry and ^{40}Ar/^{39}Ar geochronology of some igneous rocks from the South Shetland Islands. Antarctica. M.Sc. thesis, Ohio State University.

Dusén, P. (1908). Über die Tertiare Flora der Seymour Insel. In *Wissenschaftliche Ergebnisse der Schwedischen Südpolar-Expedition 1901–1903, Volume 3(3), Geologie und Paläontologie*, ed. O. Nordenskjöld. Stockholm: P. A. Norstedt & Söner, pp. 1–27.

Dutra, T. L. (2000a). *Nothofgaus* no noroeste da Península Antártica. I. Cretáceo superior. *Revista Universidade Guarulhos Geociências*, **5**, 102–106.

Dutra, T. L. (2000b). *Nothofagus* no notre de Península Antártica (Ilha Rei George, Ilhas Shetland do Sul). II. Paleoceno superior-Eoceno inferior. *Revista Universidade Guarulhos Geociências*, **5**, 131–136.

Dutra, T. L. (2004). Paleofloras da Antártica e sua relação com os eventos tectônicos e paleoclimáticos nas altas latitudes do sul. *Revista Brasileira de Geociências*, **34**, 401–410.

Dutton, A. L., Lohmann, K. C. and Zinsmeister, W. J. (2002). Stable isotope and minor element proxies for Eocene climate of Seymour Island, Antarctica. *Paleoceanography*, **17**, 1–14.

Eagles, G. (2003). Plate tectonics of the Antarctic–Phoenix plate system since 15 Ma. *Earth and Planetary Sciences Letters*, **217**, 97–109.

Eagles, G., Livermore, R. and Morris, P. (2006). Small basins in the Scotia Sea: the Eocene Drake Passage gateway. *Earth and Planetary Science Letters*, **242**, 343–353.

Elliot, D. H. (1988). Tectonic setting and evolution of the James Ross Basin, northern Antarctic Peninsula In *Geology and Paleontology of Seymour Island, Antarctic Peninsula, Geological Society of America Memoir*, **169**, eds R. M. Feldmann and M. O., Woodburne. Boulder: The Geological Society of America, pp. 541–555.

Elliot, D. H. and Hoffman, S. (1989). Geologic studies of Seymour Island. *Antarctic Journal of the United States*, **24(5)**, 3–5.

Elliot, D. H. and Rieske, D. E. (1987). Field investigations of the Tertiary strata on Seymour and Cockburn Islands. *Antarctic Journal of the United States*, **22(5)**, 6–8.

Elliot, D. H. and Trautman, T. A. (1982). Lower Tertiary strata on Seymour Island, Antarctic Peninsula. In *Antarctic Geosciences*, ed. C. Craddock. Madison: University of Wisconsin Press, pp. 287–297.

Elliot, D. H., Hoffman, S. M. and Rieske, D. E. (1992). Provenance of Paleocene strata, Seymour Island. In *Recent Progress in Antarctic Earth Science*, eds Y. Yoshida, K. Kaminuma and K. Shiraishi. Tokyo: Terra Scientific Publishing Company, pp. 347–355.

Elliot, D. H., Askin, R. A., Kyte, F. T. and Zinsmeister, W. J. (1994). Iridium and dinocysts at the Cretaceous-Tertiary boundary on Seymour Island, Antarctica: implications for the K-T event. *Geology*, **22**, 675–678.

Farley, K. A. and Eltgroth, S. F. (2003). An alternative age model for the Paleocene–Eocene thermal maximum using extraterrestrial ^{3}He. *Earth and Planetary Science Letters*, **208**, 135–148.

Feary, D. A., Davies, P. J., Pigram, C. J. and Symonds, P. A. (1991). Climatic evolution and control on carbonate deposition in northeast Australia. *Palaeogeography, Palaeoclimatology, Palaeoecology*, **89**, 341–361.

Fensterseifer, H. C., Soliani, E., Hansen, M. A. F. and Troian, F. L. (1988). Geología e estratigrafía da associação de rochas so setor centro-norte da península Fildes, Ilha Rei George, Shetland do Sul, Antártica. *Serie Científica Instituto Antártico Chileno*, **38**, 29–43.

Fontes, D. and Dutra, T. L. (2010). Paleogene imbricate-leaved podocarps from King George Island (Antarctica): assessing the geological context and botanical affinitites. *Revista Brasileira de Paleontologia*, **13**, 189–204.

Francis, J. E., Tosolini, A-M. and Cantrill, D. J. (2003). Biodiversity and climatic change in Antarctic Paleogene floras. In *Antarctic Contributions to Global Earth Sciences, Proceedings of the 9th International Symposium on Antarctic Earth Sciences (ISAES IX), Potsdam, Germany*, eds D. K. Fütterer, D. Damaske, G. Kleinschmidt, H. Miller and F. Tessensohn. Berlin, Heidelberg: Springer, pp. 107.

Francis, J., Ashworth, A. C., Cantrill, D. J., *et al.* (2008). 100 million years of Antarctic climate evolution: evidence from fossil plants. In *Antarctica: A Keystone in a Changing World. Proceedings of the 10th International Symposium on Antarctic Earth Sciences*, eds A. K., Cooper, P. J., Barrett, H., Stagg, B. C., Storey, E., Stump, and W. Wise and the 10th ISAES editorial team. Washington: U. S. Geological Survey and The National Academies Press, pp. 309–368.

Freile, C. (1972). Estudio palinológico de la Formación Cerro Dorotea (Maastrichtiano-Paleoceno) de la Provincia de Santa Cruz. *Revista del Museo de La Plata Nueva Serie Seccion Paleontologie*, **6**, 39–63.

Frenguelli, J. and Parodi, L. R. (1941). Una *Chusquea* fósil de El Mirador (Chubut). *Notas del Museo de La Plata*, **6**, 235–238.

Gandolfo, M. A., Hoc, P., Santillana, S. and Marenssi, S. N. (1998). Una flora fósil morfológicamente afin a las Grossulariaceae (Orden Rosales) de la Formación La Meseta (Eoceno medio), Isla Marambio, Antártida. In *Paleógeno de América del Sur y de la Península Antártica, Asociación Paleontológica Argentina, Publicación Especial*, **5**, ed. S. Casadio. Buenos Aires: APA, pp. 147–153.

Gandolfo, M. A., Marenssi, S. A. and Santillana, S. N. (1998). Flora y paleoclima de la Formación La Meseta (Eoceno medio) Isla Marambio (Seymour) Antártida. In *Paleógeno de América del Sur y de la Península Antártica, Asociación Paleontológica Argentina Publicación Especial*, **5**, ed. S. Casadio. Buenos Aires: APA, pp. 155–162.

Gandolfo, M. A., Zamaloa, M. C., González, C. C., *et al.* (2004). Early history of Casuarinaceae in the Paleogene of Patagonia, Argentina. In *7th International Organisation of Paleobotany Conference, San Carlos de Bariloche, Abstracts book*. Trewlew: MEF, pp. 36–37.

Gandolfo, M. A., Hermsen, E. J., Zamola, M. C., *et al.* (2011). Oldest known *Eucalyptus* macrofossils are from South America. *PLoS One*, **6**, e21084, doi: 10.1371/journal. pone.0021084.

Gaździcki, A. and Stolarski, J. (1992). An Oligocene record of the coral *Flabellum* from Antarctica. *Polish Polar Research*, **13**, 265–272.

Gaździcki, A. J., Gruszczynski, M., Hofan, A., *et al.* (1992). Stable carbon and oxygen isotope record in the Paleogene La Meseta Formation, Seymour Island, Antarctica. *Antarctic Science*, **4**, 461–468.

Gingerich, P. D. (2006). Environment and evolution through the Paleocene–Eocene thermal maximum, *TRENDS in Ecology and Evolution*, **21**, 246–253.

Gingerich, P. D., Rose, P. D. and Krause, D. W. (1980). Early Cenozoic mammalian faunas of the Clark's Fork Basin-Polecat Bench area, northwestern Wyoming. *Papers on Paleontology, Museum of Paleontology, University of Michigan*, **24**, 51–68.

Goin, F. J., Pascual, R., von Koenigswald, W., *et al.* (2006). First gondwanatherian mammal from Antarctica. In *Cretaceous–Tertiary High-Latitude Paleoenvironments, James Ross Basin, Antarctica, Geological Society of London Special Publication*, **258**, eds J. E. Francis, D. Pirrie and J. A. Crame. London: The Geological Society, pp. 135–144.

Gonzalez, C. C., Gandolfo, M. A., Zamaloa, M. C., *et al.* (2007). Revision of the Proteaceae macrofossil record from Patagonia, Argentina. *The Botanical Review*, **73**, 235–266.

Gothan, W. (1908). Die fossilen Hölzer von der Seymour und Snow Hill-Insel. In *Wissenschaftliche Ergebnisse der Schwedischen Südpolar-Expedition 1901–1903, Volume 3(8), Geologie und Paläontologie*, ed. O. Nordenskjöld. Stockholm: P. A. Norstedt & Söner, pp. 1–33.

Greenhalgh, J. (2002). A palynological investigation into palaeoenvironmental changes in the early Cenozoic sediments of Seymour Island, Antarctica. Ph.D. thesis, University College of London.

Greenwood, D. R. (1994). Palaeobotanical evidence for Tertiary climates. In *History of the Australian Vegetation: Cretaceous to Recent*, ed. R. S. Hill. Cambridge: Cambridge University Press, pp. 44–59.

Grube, R. and Mohr, B. (2007). Deterioration and/or cyclicity? The development of vegetation and climate during the Eocene and Oligocene in Antarctica. In *Online Proceedings of the 10th International Symposium on Antarctic Earth Sciences*, eds A. K. Cooper, C. R. Raymond and the 10th ISAES Editorial team. U. S. Geological Survey Open-File Report 2007–1047, Version 2. Extended Abstract 075, available at http:/// pubs.usgs.gov/of/2007/1047/ea/of 2007–1047 ea 075. pdf. {Cited 2011}.

Grunow, A. M., Kent, D. V. and Dalziel, I. W. D. (1987). Mesozoic evolution of West Antarctica and the Weddell Sea Basin: new paleomagnetic constraints. *Earth and Planetary Science Letters*, **86**, 16–26.

Gutiérrez, A. G., Armesto, J. J. and Aravena, J. C. (2004). Disturbance and regeneration dynamics of an old-growth North Patagonian rain forest in Chiloé Island, Chile, *Journal of Ecology*, **92**, 598–608.

Haq, B. U., Hardenbol, J. and Vail P. R. (1987). Chronology of fluctuating sea levels since the Triassic. *Science*, **235**, 1156–1167.

Harwood, D. M. (1988). Upper Cretaceous and Lower Paleocene diatom and silicoflagellate biostratigraphy of Seymour Island eastern Antarctic Peninsula. In *Geology and Paleontology of Seymour Island, Antarctic Peninsula, Geological Society of America Memoir*, **169**, eds R. M. Feldmann and M. O. Woodburne. Boulder: The Geological Society of America, pp. 55–129.

Hathway, B. (2000). Continental rift to back-arc basin: Jurassic–Cretaceous stratigraphical and structural evolution of the Larsen Basin, Antarctic Peninsula. *Journal of the Geological Society of London*, **157**, 417–432.

Hays, J. D., Imbrie, J. and Shackleton, N. J. (1976). Variations in the Earth's orbit: pacemaker of ice ages. *Science*, **194**, 1121–1132.

Head, J. J., Jonathan, I., Bloch, J. I., *et al.* (2009). Giant boid snake from the Paleocene neotropics reveals hotter past equatorial temperatures. *Nature*, **457**, 715–718.

Hee, Y. C. and Soon-Keun, C. (1991). Study on the gymnospermous fossil woods from the King George Island. *Korean Journal of Polar Research Special Issue*, **2**, 179–185.

Heusser, C. J. (1964). Some pollen profiles from Laguna de San Rafael área, Chile. In *Ancient Pacific Floras*, ed. L. M. Cranwell. Honolulu: University of Hawaii Press, pp. 95–114.

Hickey, L. J., West, R. M., Dawson, M. R. and Choi, D. K., (1983). Arctic terrestrial biota: paleomagnetic evidence of age disparity with mid-northern latitudes during the late Cretaceous and early Tertiary, *Science*, **221**, 1153–1156.

Hill, R. S. (1989). Fossil leaf. In *Antarctic Cenozoic History from the CIROS-1 Drillhole, McMurdo Sound, New Zealand Departmental of Scientific and Industrial Research Bulletin*, **245**, ed. P. J. Barrett. Wellington: DSIR Publishing, pp. 143–144.

Hill, R. S. (1990). Evolution of the modern high latitude southern hemisphere flora: evidence from the Australian macrofossil record. In *Proceedings of the 3rd International Organization Palaeobotany Conference*, eds J. G. Douglas and D. C. Christophel. Melbourne: A-Z Printers, pp. 31–62.

Hill, R. S. (1998). Poor soils and a dry climate: the evolution of the Australian scleromorphic and xeromorphic vegetation. *Australian Biologist*, **11**, 26–29.

Hill, R. S. and Brodribb, T. J. (1999). Southern conifers in time and space. *Australian Journal of Botany*, **47**, 639–696.

Hill, R. S. and Merrifield, H. E. (1993). An Early Tertiary macroflora from West Dale, southwestern Australia. *Alcheringa*, **17**, 285–326.

Hill, R. S. and Scriven, L. J. (1995). The angiosperm-dominated woody vegetation of Antarctica: a review: *Review of Palaeobotany and Palynology*, **86**, 175–198.

Hill, R. S., Truswell, E. M., McLoughlin, S. and Dettmann, M. E. (1999). Evolution of the Australian flora: fossil evidence. In *Flora of Australia Volume 1, Introduction*, 2nd edition. Melbourne: CSIRO, pp. 251–320.

Holdgate, M. W. (1961). Vegetation and soils in the South Chilean Islands. *Journal of Ecology*, **49**, 559–580.

Hollis, C. J., Handley, L., Crouch, E. M., *et al.* (2009). Tropical sea temperatures in the high-latitude South Pacific during the Eocene. *Geology*, **37**, 99–102.

Huber A. (1979). Estimación empírica de las características hidrológicas de Chile. *Agro Sur*, 7, 57–65.

Huber, B. T. (1988). Upper Campanian–Paleocene foraminifera from the James Ross Island region (Antarctic Peninsula). In *Geology and Palaeontology of Seymour Island, Antarctic Peninsula, Geological Society of America Memoir*, 169, eds R. M. Feldmann and M. O. Woodburne. Boulder: The Geological Society of America, pp. 163–251.

Huber, B. T. and Webb, P. N. (1986). Distribution of *Frondicularia rakauroana* (Finlay) in the southern high latitudes. *Journal of Foraminiferal Research*, 16, 135–140.

Huber, B. T., Hodell, D. A. and Hamilton, C. P. (1995). Middle-Late Cretaceous climate of the southern high latitudes: stable isotopic evidence for minimal equator-to-pole thermal gradients, *Geological Society of America Bulletin*, 107, 1164–1191.

Huber, H. and Nof, D. (2006). The ocean circulation in the southern hemisphere and its climatic impacts in the Eocene. *Palaeogeography, Palaeoclimatology, Palaeoecology*, 231, 9–28.

Huber, M., Brinkhuis, H., Stickley, C. E., *et al.* (2004). Eocene circulation of the Southern Ocean: was Antarctica kept warm by subtropical waters? *Paleoceanography*, 19, PA4026, doi:10.1029/2004PA001014.

Hueck, K. (1966). *Die Wälder Südamerikas*. Stuttgart: Fischer.

Hünicken, M. (1955). Depósitos Neocretácicos y Terciarios del extremo SSW de Santa Cruz. *Revista del Museo Argentino de Científica Naturales, "Bernardino Rivadavia" Instituto Naccional Investigaciones Científica Naturales*, 4, 1–161.

Hunt, R. J. (2001). Biodiversity and Palaeoecological significance of Tertiary fossil floras from King George Island, West Antarctica. Ph.D. thesis, University of Leeds.

Hunt, R. J. and Poole, I. (2003). Revising Paleogene West Antarctic climate and vegetation history in light of new data from King George Island. In *Causes and Consequences of Globally Warm Climates in the Early Paleogene, Geological Society of America Special Paper*, 369, eds S. L. Wing, P. D. Gingerich, B. Schmitz and E. Thomas. Boulder: The Geological Society of America, pp. 395–412.

Iglesias, A., Wilf, P., Johnson, K. R., *et al.* (2007). A Paleocene lowland macroflora from Patagonia reveals significantly greater richness than North American analogs. *Geology*, 35, 947–950.

Ivany, L. C. (2007). Contributions to the Eocene climate record of the Antarctic Peninsula. In *Online Proceedings of the 10th International Symposium on Antarctic Earth Sciences*, eds A. K. Cooper, C. R. Raymond and the 10th ISAES Editorial team. U. S. Geological Survey Open-File Report 2007–1047, Version 2. Extended Abstract 068, available at http://pubs.usgs.gov/of/2007/1047/ea/of2007–1047ea068.pdf. {Cited 2011}.

Ivany, L. C., Van Simaeys, S., Domack E. W. and Samson, S. D. (2006). Evidence for an earliest Oligocene ice sheet on the Antarctic Peninsula. *Geology*, 34, 377–380.

Ivany, L. C., Lohmann, K. C., Hasiuk, F., *et al.* (2008). Eocene climate record of a high southern latitude continental shelf: Seymour Island, Antarctica. *Geological Society of America Bulletin*, 120, 659–678.

Jardine, D. J. (1950). Base G. Admiralty Bay Report (1949). *Geology Report (F.I.D.S. Bureau) No.*, 83/50, 1–31.

Jefferson, T. H. (1980). Angiosperm fossils in supposed Jurassic volcanogenic shales, Antarctica. *Nature*, 285, 157–158.

Jovane, L., Florindo, F., Coccioni, R., *et al.* (2007). The middle Eocene climatic optimum event in the Contessa Highway section, Umbrian Apennines, Italy. *Geological Society of America Bulletin*, 119, 413–427.

Katz, M. E., Pak, D. K., Dickens, G. R. and Miller, K. G. (1999). The source and fate of massive carbon input during the Latest Paleocene Thermal Maximum. *Science*, **286**, 1531–1533.

Kelly, D. C., Bralower, T. J., Zachos, J. C., Premoli Silva, I. and Thomas, E. (1996). Rapid diversification of planktonic foraminifera in the tropical Pacific (ODP Site 865) during the late Paleocene thermal maximum. *Geology*, **24**, 423–426.

Kennett, J. P. and Barker, P. F. (1990). Latest Cretaceous to Cenozoic climate and oceano-graphic developments in the Weddell Sea, Antarctica: an ocean drilling perspective. In *Proceedings of the Ocean Drilling Program, Volume 113 Scientific Results Weddell Sea, Antarctica Covering Leg 113 of the cruises of the Drilling Vessel* JOIDES Resolution, *Valparaiso, Chile, to East Cove, Falkland Islands, Sites 689–697, 25 December 1986–11 March 1987*, eds P. F. Barker, J. P. Kennett, S. O'Connell, *et al.* Online doi:10.2973/odp.proc.sr.113.195.1990, available at http://www-odp.tamu.edu/publications/113_SR/VOLUME/CHAPTERS/sr113_53.pdf. {Cited 2011}.

Kennett, J. P. and Stott, L. D. (1991). Abrupt deep-sea warming, palaeoceanographic changes and benthic extinctions at the end of the Palaeocene. *Nature*, **353**, 225–229.

Killops, S. D., Raine, J. I., Woolhouse, A. D. and Weston, R. J. (1995). Chemostratigraphic evidence of higher-plant evolution in the Taranaki Basin. *New Zealand Organic Geochemistry*, **23**, 429–445.

Killops, S. D., Hollis, C. J., Morgans, H. E. G., *et al.* (2000). Paleoceanographic significance of Late Paleocene dysaerobia at the shelf/slope break around New Zealand. *Palaeogeography, Palaeoclimatology, Palaeoecology*, **156**, 51–70.

Kim, H., Lee, J. I., Choe, M. Y., *et al.* (2000). Geochronologic evidence for Early Tertiary volcanic activity on Barton Peninsula, King George Island, Antarctica. *Polar Research*, **19**, 251–260.

Kim, S. B., Sohn, Y. K. and Choe, M. Y. (2006). The Eocene volcaniclastic Sejong Formation, Barton Peninsula, King George Island, Antarctica: evolving arc volcanism from precursory fire fountaining to vulcanian eruptions. In *Antarctica: Contributions to Global Earth Science*, eds D. K. Fütterer, D. Damaske, G. Kleinschmidt, H. Miller and F. Tessensohn. Berlin: Springer-Verlag, pp. 261–270.

Kurtz, A. C., Kump, L. R., Arthur, M. A., Zachos, J. C. and Paytan, A. (2003). Early Cenozoic decoupling of the global carbon and sulfur cycles. *Paleoceanography*, **18**, 1090, doi:10.1029/2003PA000908.

Lara, A., Little, C., Urrutia, R., *et al.* (2009). Assessment of ecosystem services as an opportunity for the conservation and management of native forests in Chile. *Forest Ecology and Management*, **258**, 415–424.

Larter, R. D. and Barker, P. (1991). Effects of ridge crest-trench interaction on Antarctic-Phoenix spreading: forces on a young subducting plate. *Journal of Geophysical Research*, **96**, 583–607.

Lawver, L. A. and Gahagan, L. M. (2003). Evolution of Cenozoic seaways in the circum-Antarctic region. *Palaeogeography, Palaeoclimatology, Palaeoecology*, **198**, 11–37.

Lawver, L. A., Gahagan, L. M. and Coffin, M. F. (1992). The development of paleoseaways around Antarctica. In *The Antarctic Paleoenvironment: A Perspective on Global Change. Part 1, Antarctic Research Series*, **56**, eds J. P. Kennett and D. A. Warnke. Washington: American Geophysical Union, pp. 7–30.

Leckie, D. A., Morgans, H., Wilson, G. J. and Edwards, A. R. (1995). Mid-Paleocene dropstones in the Whangai Formation, New Zealand evidence of mid-Paleocene cold climate? *Sedimentary Geology*, **97**, 119–129.

Lee, J. I., Hwang, J., Kim, H., *et al.* (1996), Subvolcanic zoned granitic pluton in the Barton and Weaver peninsulas, King George Island, Antarctica. *Proceedings National Institute of Polar Research Symposium on Antarctic Geoscience*, **9**, 76–90.

Li, H. (1992). Early Tertiary palaeoclimate of King George Island, Antarctica–Evidence from the Fossil Hill Flora. In *Recent Progress in Antarctic Earth Science*, eds Y. Yoshida, K. Kaminuma and K. Shiraishi. Tokyo: Terra Scientific Publishing Company, pp. 371–375.

Li, H. (1994). Early Tertiary Fossil Hill flora from Fildes Peninsula of King George Island, Antarctica. In *Stratigraphy and Palaeontology of Fildes Peninsula King George Island, Antarctica, State Antarctic Committee Monograph*, **3**, ed. Y. Shen. Beijing: China Science Press, pp. 133–171.

Li, H. and Shen, Y. (1989). A primary study of Fossil Hill flora from Fildes Peninsula of King George Island, Antarctica. *Acta Palaeontólogica Sinica*, **29**, 147–153.

Li, H. and Song, D. (1988). Fossil remains of some angiosperms from King George Island, Antarctica. *Acta Palaeontológica Sinica*, **27**, 399–403.

Li, H. and Zhou, Z. K. (2007). Fossil nothofagaceous leaves from the Eocene of western Antarctica and their bearing on the origin, dispersal and systematics of *Nothofagus*. *Science in China Series D: Earth Sciences*, **50**, 1525–1535.

Li, Z., Lui, X., Zheng, X., Jin, Q. and Li, G. (1989). Tertiary volcanism and formation of volcanic rocks in the Fildes Peninsula, King George Island, Antarctica. In *Proceedings of the International Symposium on Antarctic Research*, ed. K. Guo. Beijing: China Ocean Press, pp. 128–133.

Lourens, L. J., Sluijs, A., Kroon, D., *et al.* (2005). Astronomical placing of late Palaeocene to early Eocene global warming events. *Nature*, **435**, 1083–1087.

Lyra, C. S. (1986). Palinológica de sedimentos Terciários da Península Fildes, Ihla Rei George (Ilha Shetland do Sul Antártica) e algumas conciderações paleoambientais. *Anais de Academia Brasileria de Ciências (supplement)*, **58**, 137–147.

Macellari, C. E. (1988). Stratigraphy, sedimentology and palaeoecology of Upper Cretaceous/Paleocene shelf-deltaic sediments of Seymour Island. In *Geology and Paleontology of Seymour Island, Antarctic Peninsula, Geological Society of America, Memoir*, **169**, eds R. M. Feldmann and M. O. Woodburne. Boulder: The Geological Society of America, pp. 25–53.

Macphail, M. K., Alley, N., Truswell, E. M. and Sluiter, I. R. (1994). Early Tertiary vegetation: Evidence from pollen and spores. In *History of the Australian Vegetation: Cretaceous to Recent*, ed. R. S. Hill. Cambridge: Cambridge University Press, pp. 189–261.

Mao, S. and Mohr, B. A. R. (1995). Middle Eocene dinocysts from Bruce Bank (Scotia Sea, Antarctica) and their paleoenvironmental and paleogeographic implications. *Review of Palaeobotany and Palynology*, **86**, 235–263.

Marenssi, S. A. (1995). Sedimentología y paleoambientes de sedimentación de la Formación La Meseta, Isla Marambio, Antártida. Ph.D. thesis, Universidad de Buenos Aires.

Marenssi, S. A. (2006). Eustatically controlled sedimentation recorded by Eocene strata of the James Ross Basin, Antarctica. In *Cretaceous–Tertiary High-latitude Palaeoenvironments, James Ross Basin, Antarctica, Geological Society of London Special Publication*, **258**, eds J. E. Francis, D. Pirrie and J. A. Crame. London: The Geological Society, pp. 125–133.

Marenssi, S. A. and Santillana, S. N. (1994). Unconformity-bounded units within the La Meseta Formation, Seymour Island, Antarctica: a preliminary approach. In *XXI Polar Symposium, Warszawa Poland, Abstracts*. Warsaw: Polish Academy of Science, pp. 33–37.

Marenssi, S. A., Reguero, M. A., Santillana, S. N. and Vizcaíno, S. F. (1994). Eocene land mammals from Seymour Island, Antarctica: Palaeobiogeographical implications. *Antarctic Science*, **6**, 3–15.

Marenssi, S. A., Santillana, S. N. and Rinaldi, C. A. (1998a). Paleoambientes sedimentarios de la Aloformación La Meseta (Eoceno), Islas Marambio (Seymour), Antártica. *Instituto Antártico Argentino Contribución*, **464**, 1–51.

Marenssi, S. A., Santillana, S. N. and Rinaldi, C. A. (1998b). Stratigraphy of the La Meseta Formation (Eocene), Marambio (Seymour) Island, Antarctica. In *Paleógeno de América del Sur y de la Península Antártica, Asociación Paleontológica Argentina Publicación Especial*, **5**, ed. S. Casiado. Buenos Aires: APA, pp. 137–146.

Marenssi, S. A., Santillana, S. N., Net, L. I. and Rinaldi, C. A. (1999). Heavy mineral suites as provenance indicator: La Meseta Formation (Eocene), Antarctic Peninsula. *Revista de la Asociación Argentina de Sedimentología*, **5**, 9–19.

Marenssi, S. A., Net, L. I. and Santillana, S. N. (2002). Provenance, depositional and paleogeographic controls on sandstone composition in an incised valley system: the Eocene La Meseta Formation (Eocene), Seymour Island, Antarctica. *Sedimentary Geology*, **150**, 301–321.

McCarron, J. J. and Millar, I. L. (1997). The age and stratigraphy of fore-arc magmatism on Alexander Island, Antarctica. *Geological Magazine*, **134**, 507–522.

McDonald, C. M. (2009). Herbivory in the Antarctic fossil forests and comparisons with modern analogues in Chile. Ph.D. thesis, University of Leeds.

McDonald, C. M., Francis, J. E., Compton, S. G. A., *et al.* (2007). Herbivory in Antarctic fossil forests: evolutionary and palaeoclimatic significance. In *Online Proceedings of the 10th International Symposium on Antarctic Earth Sciences*, eds A. K. Cooper, C. R. Ragmond and the 10th ISAES Editorial team. U. S. Geological Survey Open-File Report 2007–1047, Version 2.0. extended abstract 059, available at http://pubs.usgs.gov/of/2007/1047/ea/of 2007–1047ea059.pdf. {Cited 2011}.

McKenna, M. C. (1980). Eocene paleolatitude, climate, and mammals of Ellesmere Island. *Palaeogeography, Palaeoclimatology, Palaeoecology*, **30**, 349–362.

Melchor, R., Genise, J. and Miquel, S. E. (2002). Ichnology, sedimentology and paleontology of Eocene calcareous paleosols from a palustrine sequence, Argentina. *Palaios*, **17**, 16–35.

Melendi, D. L., Scafati, L. H. and Volkheimer, W. (2003). Palynostratigraphy of the Paleogene Huitrera Formation in N-W Patagonia, Argentina. *Neues Jahrbuch für Geologie und Paläontologie*, **228**, 205–273.

Mildenhall, D. C. (1980). New Zealand late Cretaceous and Cenozoic plant biogeography: a contribution. *Palaeogeography, Palaeoclimatology, Palaeoecology*, **31**, 197–233.

Miller, K. G., Wright, J. D., Katz, M. E., *et al.* (2009). Climate threshold at the Eocene-Oligocene transition: Antarctic ice sheet influence on ocean circulation. In *A Late Eocene Earth: Hothouse, Icehouse and Impacts, Geological Society of America Special Papers*, **452**, eds C. Koeberl and A. Montanari. Boulder: The Geological Society of America, pp. 169–178.

Mohr, B. A. R. (2001). The development of Antarctic fern floras during the Tertiary, and palaeoclimatic and palaeobiogeographic implications. *Palaeontographica B*, **259**, 167–208.

Navarro, C., Donoso, C. and Sandoval, V. (1999). Los renovales de canelo. In *Silvicultura de los Bosques Nativos de Chile*, eds C. Donoso and A. Lara. Santiago: Editorial Universitaria, pp. 341–377.

Net, L. I. and Marenssi, S. A. (1999). Petrografía de las areniscas de la Formación La Meseta (Eoceno), Isla Marambio, Antártida. *Actas IV Jornadas sobre Investigaciones Antárticas*. Buenos Aires, pp. 343–347.

Nunes, F. and Norris, R. D. (2006). Abrupt reversal in ocean overturning during the Palaeocene/Eocene warm period. *Nature*, **439**, 60–63.

Olivero, E. B. (1995). Molluscan Cretaceous biostratigraphy of the James Ross Basin (Antarctica) and the Austral Basin of Terra del Fuego (Argentina). In *Inaugural Meeting of IGCP Project 381 (South Atlantic Mesozoic Correlations). Uberada, Brazil, 24–25 July 1995 Extended abstracts, SAMC (South Atlantic Mesozoic Correlations, IGCP Project No. 381) News*, **2**, 5.

Orlando, H. A. (1963). La flora fósil en las inmediaciones de la Península Ardley, Isla 25 de Mayo, Islas Shetlands del Sur. *Instituto Antártico Argentino Contribución*, **79**, 3–17.

Orlando, H. A. (1964). The fossil flora of the surroundings of Ardley Peninsula (Ardley Island), 25 de Mayo Island (King George Island), South Shetland Islands. In *Antarctic Geology: proceedings of the First International Symposium on Antarctic Geology, Cape Town, 16–21 September 1963*, ed. R. J. Adie. Amsterdam: North-Holland Publishing Company, pp. 629–636.

Pagani, M., Arthur, M. A. and Freeman, K. (1999). Miocene evolution of atmospheric carbon dioxide. *Paleoceanography*, **14**, 273–292.

Pagani, M., Zachos, J. C., Freeman, K. H., Tipple, B. and Bohaty, S. (2005). Marked decline in atmospheric carbon dioxide concentrations during the Paleogene. *Science*, **309**, 600–603.

Pagani, M., Calderia, K., Archer, D. and Zachos, J. C. (2006). An ancient carbon mystery. *Science*, **314**, 1556–1557.

Palazzesi, L. and Barreda, V. (2007). Major vegetation trends in the Tertiary of Patagonia (Argentina): A qualitative paleoclimatic approach based on palynological evidence. *Flora*, **202**, 328–337.

Palma-Heldt, S. (1987). Estudio palinológico en el Terciário de Islas Rey Jorge y Brabante, territorio insular Antártico. *Serie Científica Instituto Antártico Chileno*, **36**, 59–71.

Panchuk, K., Ridgewell, A. and Kump, L. R. (2008). Sedimentary response to Paleocene-Eocene Thermal Maximum carbon release: a model-data comparison. *Geology*, **36**, 315–318.

Pankhurst, R. J. and Smellie, J. L. (1983). K-Ar geochronology of the South Shetland Islands, Lesser Antarctica: apparent lateral migration of Jurassic to Quaternary island arc volcanism. *Earth and Planetary Science Letters*, **66**, 214–222.

Park, B. K. (1989). Potassium-argon radiometric ages of volcanic and plutonic rocks from the Barton Peninsula, King George Island, Antarctica. *Journal of the Geological Society of Korea*, **25**, 495–497.

Pearson, P. N. and Palmer, M. R. (2000). Atmospheric carbon dioxide concentration over the past 60 million years, *Nature*, **406**, 695–699.

Pekar, S. F., Hucks, A., Fuller, M. and Li, S. (2005). Glacioeustatic changes in the early and middle Eocene (51–42 Ma): Shallow-water stratigraphy from ODP Leg 189 Site 1171 (South Tasman Rise) and deep-sea $\delta^{18}O$ records. *Geological Society of America Bulletin*, **117**, 1081–1093.

Pérez, C. A., Hedin, L. O. and Armesto, J. J. (1998). Nitrogen mineralization in two unpolluted old-growth forests of contrasting biodiversity and dynamics. *Ecosystems*, **1**, 361–373.

Petriella, B. T. P. and Archangelsky, S. (1975). Vegetación y ambiente en el Paleoceno de Chubut. In *Actas Primer Congreso Argentino de Paleontóloga y Bioestratigrafa, Volume 2*. Tucamán: Asociacion Paleontologia Argentina, pp. 257–270.

Pirrie, D. and Marshall, J. D. (1990). High-paleolatitude Late Cretaceous paleotemperatures: New data from James Ross Island, Antarctica. *Geology*, **18**, 31–34.

Pirrie, D., Whitham, A. G. and Ineson, J. R. (1991). The role of tectonics and eustasy in the evolution of a marginal basin: Cretaceous-Tertiary Larsen basin, Antarctica. In *Sedimentation, Tectonics and Eustasy, Special Publications of the International Association of Sedimentologists*, **12**, ed. D. I. M. Macdonald. New York: Wiley, pp. 293–305.

Pirrie, D., Duane, A. M. and Riding, J. B. (1992). Jurassic-Tertiary stratigraphy and palynology of the James Ross Basin: a review and introduction. *Antarctic Science*, **4**, 259–266.

Pirrie, D., Marshall, J. D. and Crame, J. A. (1998). Marine high Mg calcite cements in *Teredolites*-bored fossil wood; evidence for cool paleoclimates in the Eocene La Meseta Formation, Seymour Island, Antarctica. *Palaios*, **13**, 276–86.

Pocknall, D. T. (1989). Late Eocene to Early Miocene vegetation and climate history of New Zealand. *Journal of the Royal Society of New Zealand*, **19**, 1–18.

Pole, M. S. (1994). The New Zealand flora–entirely long distance dispersal? *Journal of Biogeography*, **21**, 625–635.

Pole, M. S. (1998a). Paleocene gymnosperms from Mount Somers, New Zealand. *Journal of the Royal Society of New Zealand*, **28**, 375–403.

Pole, M. S. (1998b). The Proteaceae record in New Zealand. *Australian Systematic Botany*, **11**, 343–372.

Pole, M. S. and Macphail, M. K. (1996). Eocene *Nypa* from Regatta Point Tasmania, *Review of Palaeobotany and Palynology*, **92**, 55–67.

Poole, I. (2002). Systematics of Cretaceous and Tertiary *Nothofagoxylon*: implications for Southern Hemisphere biogeography and evolution of the Nothofagaceae. *Australian Systematic Botany* **15**, 247–276.

Poole, I. and Cantrill, D. J. (2001). Fossil woods from Williams Point Beds, Livingston Island, Antarctica: a Late Cretaceous southern high latitude flora. *Palaeontology*, **44**, 1081–1112.

Poole, I. and Cantrill, D. J. (2006). Cretaceous and Cenozoic vegetation of Antarctica integrating the fossil wood record. In *Cretaceous–Tertiary High-Latitude Palaeoenvironments, James Ross Basin, Antarctica, Geological Society of London Special Publication*, **258**, eds J. E. Francis, D. Pirrie and J. A. Crame. London: The Geological Society, pp. 63–81.

Poole, I. and Francis, J. E. (1999). Reconstruction of Antarctic palaeoclimates using angiosperm wood anatomy. *Acta Palaeobotanica Supplement*, **2**, 173–179.

Poole, I. and Gottwald, H. (2001). Monimiaceae *sensu lato*, an element of Gondwanan polar forests: evidence from the Late Cretaceous–early Tertiary wood flora of Antarctica. *Australian Systematic Botany*, **14**, 207–230.

Poole, I., Gottwald, H. and Francis, J. E. (2000). *Illicioxylon*, an element of Gondwanan polar forests? Late Cretaceous and Early Tertiary woods of Antarctica. *Annals of Botany*, **86**, 421–432.

Poole, I., Richter, H. and Francis, J. E. (2000). Evidence for Gondwanan origins for *Sassafras* (Lauraceae)? Late Cretaceous fossil wood of Antarctica. *International Association of Wood Anatomists Journal*, **21**, 463–475.

Poole, I., Cantrill, D. J., Hayes, P. and Francis, J. E. (2000). The fossil record of Cunoniaceae: new evidence from Late Cretaceous wood of Antarctica? *Review of Palaeobotany and Palynology*, **111**, 127–144.

Poole, I., Hunt, R. H. and Cantrill, D. J. (2001). A fossil wood flora from King George Island: ecological implications for Antarctic Eocene vegetation. *Annals of Botany*, **88**, 33–54.

Poole, I., Mennega, A. M. W. and Cantrill, D. J. (2003). Valdivian ecosystems in the late Cretaceous and early Tertiary of Antarctica further evidence from myrtaceous and eucryphiaceous fossil wood. *Review of Palaeobotany and Palynology*, **124**, 9–27.

Poole, I., Cantrill D. J. and Utescher, T. (2005). A multi-proxy approach to determine Antarctic terrestrial palaeoclimate during the Late Cretaceous and early Tertiary. *Palaeogeography, Palaeoclimatology, Palaeoecology*, **222**, 95–121.

Porębski, S. J. (1995). Facies architecture in a tectonically-controlled incised-valley estuary: La Meseta Formation (Eocene) of Seymour Island, Antarctic Peninsula. *Studia Geologica Polonica*, **107**, 7–97.

Porębski, S. J. (2000). Shelf-valley compound fill produced by fault subsidence and eustatic sea-level changes, Eocene La Meseta Formation, Seymour Island, Antarctica. *Geology*, **28**, 147–150.

Quillévéré, F., Norris, R. D., Kroon, D. and Wilson, P. A. (2008). Transient ocean warming and shifts in carbon reservoirs during the early Danian. *Earth and Planetary Science Letters*, **265**, 600–615.

Raigemborn, M., Brea, M., Zucol, A. and Matheos, S. (2009). Early Paleogene climate at mid latitude in South America: mineralogical and paleobotanical proxies from continental sequences in Golfo San Jorge Basin (Patagonia, Argentina). *Geologica Acta*, **7**, 125–145.

Reguero, M. A., Marenssi, S. A. and Santillana, S. N. (2002). Antarctic Peninsula and South America (Patagonia) Paleogene terrestrial faunas and environments: biogeographic relationships *Palaeogeography, Palaeoclimatology, Palaeoecology*, **179**, 189–210.

Riding, J. B. and Crame, J. A. (2002). Aptian to Coniacian (Early–Late Cretaceous) palynostratigraphy of the Gustav Group, James Ross Basin, Antarctica. *Cretaceous Research*, **23**, 739–760.

Riveros, G. M. (1991). Biología reproductiva en especies vegetales de dos comunidades de la zona templada del sur de Chile, 40° S. M.Sc. thesis, Universidad de Chile.

Robert, C. and Chamley, H. (1991). Development of early Eocene warm climates, as inferred from clay mineral variations in oceanic sediments. *Palaeogeography, Palaeoclimatology, Palaeoecology*, **89**, 315–333.

Robert, C. and Kennett, J. P. (1992). Paleocene and Eocene kaolinite distribution in the South Atlantic and Southern Ocean: Antarctic climate and palaeoceanographic implications. *Marine Geology*, **103**, 99–110.

Robert, C. and Kennett, J. P. (1994). Antarctic subtropical humid episode at the Paleocene–Eocene boundary: clay-mineral evidence. *Geology*, **22**, 211–214.

Röhl, U., Bralower, T. J., Norris, D. and Wefer, G. (2000). New chronology for the late Paleocene thermal maximum and its environmental implications. *Geology*, **28**, 927–930.

Röhl, U., Westerhold, T., Bralower, T. J. and Zachos, J. C. (2007). On the duration of the Paleocene-Eocene thermal maximum (PETM). *Geochemistry, Geophysics, Geosystems*, **8**, Q12002, doi:10.1029/2007GC001784.

Rohn, R., Rösler, O. and Czajkowski, S. (1987). *Fildesia pulchra* gen et sp. nov.–folha fósseis do Terciário inferior da península Fildes, Ilha Rei George, Antártica. *Boletin do Instituto Geológico-Universidade de Säo Paulo, Série Cientifica*, **18**, 13–16.

Romero, E. J. (1968). *Palmoxylon patagonicum* n. sp. del Terciário inferior de la Provincia de Chubut, Argentina. *Ameghiniana*, **5**, 417–431.

Romero, E. J. (1973). Pólen fósil de *Nothofagus* (*Nothofagidites*) del Cretácico y Paleoceno de Patagonia. *Revista del Museo de La Plata Nueva Serie Seccion Paleontologia*, **7**, 205–24.

Romero, E. J. (1978). Paleoecología y paleofitogeografía de las tafofloras del Cenofítico de Argentina y áreas vecinas. *Ameghiniana*, **15**, 209–227.

Romero, E. J. (1986). Paleogene phytogeography and climatology of South America. *Annals of the Missouri Botanical Garden*, **73**, 449–461.

Roser, B. P. and Korsch, R. J. (1986). Determination of tectonic setting of sandstone-mudstone suites using SiO_2 content and K_2O/Na_2O ratio. *Journal of Geology*, **94**, 635–50.

Royer, D. L. (2006). CO_2-forced climate thresholds during the Phanerozoic. *Geochimica et Cosmochimica Acta*, **70**, 5665–5675.

Sadler, P. M. (1988). Geometry and stratification of uppermost Cretaceous and Paleogene units on Seymour Island, northern Antarctic Peninsula. In *Geology and Paleontology of Seymour Island, Antarctic Peninsula, Geological Society of America Memoir*, **169**, eds R. M. Feldmann and M. O. Woodburne. Boulder: The Geological Society of America, pp. 303–320.

Sarmiento, J. L. and Gruber, N. (2002). Sinks for anthropogenic carbon. *Physics Today*, **55**, 30–36.

Schauer, O. C. and Fourcade, N. H. (1964). Geological-petrographical study of the western end of 25 de Mayo Island, South Shetland Islands. In *Antarctic Geology: Proceedings of the First International Symposium on Antarctic Geology, Cape Town, 16–21 September 1963*, ed. R. S. Adie. Amsterdam: North-Holland Publishing Company, pp. 487–491.

Scher, H. D. and Martin, E. E. (2006). Timing and climatic consequences of the opening of Drake Passage. *Science*, **312**, 428–430.

Schweitzer, C. E., Feldmann, R. M., Marenssi, S. and Waugh, D. A. (2005). Remarkably preserved annelid worms from the La Meseta Formation (Eocene), Seymour Island, Antarctica. *Palaeontology*, **48**, 1–13.

Shackleton, N. J. and Kennett, J. P. (1975). Paleotemperature history of the Cenozoic and the initiation of Antarctic glaciation: Oxygen and carbon isotope analyses in DSDP Sites 277, 279, and 281. In *Initial Reports of the Deep Sea Drilling Project, Volume 29, covering Leg 29 of the cruises of the Drilling Vessel* Glomar Challenger *Lyttleton, New Zealand to Wellington, New Zealand March-April 1973*, eds J. P. Kennett, R. E. Houtz, P. B. Andrews, *et al.*, pp. 743–55, doi:10.2973/dsdp. proc.29.117.1975. {Cited 2011}.

Shellito, C. J., Sloan, L. C. and Huber, M. (2003). Climate model sensitivity to atmospheric CO_2 levels in the Early-Middle Paleogene. *Palaeogeography, Palaeoclimatology, Palaeoecology*, **193**, 113–23.

Shen, Y. (1994). Subdivision and correlation of Cretaceous to Palaeogene volcano-sedimentary sequence from Fildes Peninsula, King George Island, Antarctica. In *Stratigraphy and Palaeontology of Fildes Peninsula, King George Island, State Antarctic Committee Monograph*, **3**, ed. Y. Shen. Beijing: China Science Press, pp. 1–36.

Shen, Y. (1998). A paleoisthmus linking southern South America with the Antarctic Peninsula during Late Cretaceous and early Tertiary times. *Science in China (Series D)*, **41**, 225–229.

Shen, Y. (1999). Subdivision and correlation of the Eocene Fossil Hill Formation from King George Island, West Antarctica. *Korean Journal of Polar Research*, **10**, 91–95.

Sluijs, A. (2006). Global change during the Paleocene-Eocene thermal maximum. *LPP Contributions Series No. 21*. Utrecht: LPP Foundation.

Sluijs, A. and Brinkhuis, H. (2008). Rapid carbon injection and transient global warming during the Paleocene-Eocene thermal maximum. *Netherlands Journal of Geosciences*, **87**, 201–206.

Sluijs, A., Schouten, S., Pagani, M., *et al.* (2006). Subtropical Arctic Ocean temperatures during the Palaeocene/Eocene thermal maximum. *Nature*, **441**, 610–613.

Sluijs, A., Bowen, G. J., Brinkhuis, H., Lourens, L. J. and Thomas, E. (2007a). The Paleocene-Eocene thermal maximum super greenhouse: biotic and geochemical signatures, age models and mechanisms of global change. In *Deep Time Perspectives on Climate Change: Marrying the Signal from Computer Models and Biological Proxies, The Micropalaeontological Society Special Publication*, eds M. Williams, A. Haywood, J. Gregory and D. Schmidt. London: The Geological Society, pp. 323–349.

Sluijs, A., Brinkhuis, H., Schouten, S., *et al.* (2007b). Environmental precursors to rapid light carbon injection at the Palaeocene/Eocene boundary. *Nature*, **450**, 1218–1221.

Sluijs, A., Röhl, U., Schouten, S., *et al.* (2009). Arctic late Paleocene–early Eocene paleoenvironments with special emphasis on the Paleocene-Eocene thermal maximum (Lomonosov Ridge, Integrated Ocean Drilling Program Expedition 302). *Paleoceanography*, **23**, PA1S11, doi:10.1029/2007PA001495. {Cited 2011}.

Smellie, J. L., Pankhurst, R. J., Thomson, M. R. A. and Davies, R. E. S. (1984). The geology of the South Shetland Islands: VI. Stratigraphy, Geochemistry and Evolution. *British Antarctic Survey Scientific Reports*, **87**, 1–85.

Smith-Ramírez, C. (1993). Los picaflores y su recurso floral en el bosque templado de la isla de Chiloé, Chile. *Revista Chilena de Historia Natural*, **12**, 65–73.

Smith-Ramírez, C. (2004). The Chilean coastal range: a vanishing center of biodiversity and endemism in South American temperate rainforests. *Biodiversity and Conservation*, **13**, 373–393.

Smith-Ramírez, C. and Armesto, J. J. (1998). Nectarivoría y polinización por aves en *Embothrium coccineum* (Proteaceae) en el bosque templado del sur de Chile. *Revista Chilena de Historia Natural*, **71**, 51–63.

Soliani, E. and Bonhomme, M. G. (1994). New evidence for the Cenozic resetting of K-Ar ages in volcanic rocks of the northern portion of Admiralty Bay, King George Island. Antarctica. *Journal of South American Earth Sciences*, **7**, 85–94.

Soliani, E., Kawashita, K., Fensterseifer, H., Hansen, M. A. and Troian, F. (1988). K-Ar ages of the Winkel Point Formation (Fildes Peninsula Group) and associated intrusions, King George Island, Antarctica. *Serie Científica Instituto Antártico Chileno*, **38**, 133–139.

Spalletti, L. A. and Mazzoni, M. M. (1978). Sedimentología del Grupo Sarmiento en el perfil ubicado al sudeste del lago Colhué-Huapi, Provincia Chubut. *Obra Centenario Museum La Plata*, **4**, 261–283.

Spicer, R. A. (1989). The formation and interpretation of plant fossil assemblages. *Advances in Botanical Research*, **16**, 96–191.

Spicer, R. A. (1991). Plant taphonomic processes. In *Taphonomy: Releasing the Data Locked in the Fossil Record*, eds P. A. Allison and D. E. G. Briggs. *Topics in Geobiology Volume 9*. New York: Plenum Press, pp. 71–113.

Stap, L., Sluijs, A., Thomas, E. and Lourens, L. (2009). Patterns and magnitude of deep sea carbonate dissolution during Eocene Thermal Maximum 2 and H2, Walvis Ridge, southeastern Atlantic Ocean. *Paleoceanography*, **24**, PA1211, doi:10.1029/2008PA001655.

Stickley, C. E., Brinkhuis, H., Schellenberg, S.A., *et al.* (2004). Timing and nature of the deepening of the Tasmanian Gateway. *Paleoceanography*, **19**, PA4027, doi:10.1029/2004PA001022.

Stilwell, J. D. and Zinsmeister, W. J. (1992). Molluscan systematics and biostratigraphy. Lower Tertiary La Meseta Formation, Antarctic Peninsula. *Antarctic Research Series*, **55**, 1–202.

Storey, M., Duncan, R. A. and Swisher, C. C. (2007). Paleocene-Eocene Thermal Maximum and the opening of the northeast Atlantic. *Science*, **27**, 587–589.

Suarez, M., de la Cruz, R. and Troncoso, A. (2000). Tropical/subtropical upper Paleocene-lower Eocene fluvial deposits in eastern central Patagonia, Chile (46°45′ S). *Journal of South American Earth Sciences*, **13**, 527–536.

Tambussi, C. P., Noriega, J. I., Gazdzicki, A., *et al.* (1994). Ratite bird from the Paleogene La Meseta Formation, Seymour Island, Antarctica. *Polish Polar Research*, **15**, 15–20.

Tambussi, C. P., Acosta Hospitaleche, C. I., Reguero, M. A. and Marenssi, S. A. (2006). Late Eocene penguins from West Antarctica: systematic and biostratigraphy. In *Cretaceous–Tertiary High-Latitude Palaeoenvironments, James Ross Basin, Antarctica, Geological Society of London Special Publication*, **258**, eds J. E. Francis, D. Pirrie and J. A. Crame. London: The Geological Society, pp. 145–161.

Taylor, G., Truswell, E. M., McQueen, K. G., and Brown, M. C. (1990). Early Tertiary palaeogeography, landform evolution and palaeoclimates of the Southern Monaro NSW, Australia. *Palaeogeography, Palaeoclimatology, Palaeoecology*, **78**, 109–134.

Thomas, E. and Shackleton, N. J. (1996). The Palaeocene–Eocene benthic foraminiferal extinction and stable isotope anomalies. In *Correlation of the Early Paleogene in Northwest Europe, Geological Society of London Special Publication*, **101**, eds R. W. O'B. Knox and R. M. Corfield. London: The Geological Society, pp. 401–441.

Thomas, E. and Zachos, J. C. (2000). Was the late Paleocene thermal maximum a unique event? *GFF*, **122**, 169–170.

Thomas, E., Zachos, J. C. and Bralower, T. J. (2000). Deep sea environments on a warm earth: latest Paleocene-early Eocene. In *Warm Climates in Earth History*, eds B. T. Huber, K. G. MacLeod and S. L. Wing. Cambridge: Cambridge University Press, pp. 132–160.

Thomas, D. J., Zachos, J. C., Bralower, T. J., Thomas, E. and Bohaty, S. (2002). Warming the fuel for the fire: Evidence for the thermal dissociation of methane hydrate during the Paleocene-Eocene thermal maximum. *Geology*, **30**, 1067–1070.

Thomson, M. R. A. and Burn, R. W. (1977). Angiosperm fossils from latitude 70° S. *Nature*, **269**, 139–141.

Tokarski, A. K., Danowski, W. and Zastawniak, E. (1987). On the age of the fossil flora from Barton Peninsula, King George Island, West Antarctica. *Polish Polar Research*, **8**, 293–302.

Toker, V., Barker, P. F. and Wise, S. W. (1991). Middle Eocene carbonate-bearing sediments from Bruce Bank off the northern Antarctic Peninsula. In *Geological Evolution in Antarctica*, eds M. R. A. Thomson, J. W. Thomson and J. A. Crame. Cambridge: Cambridge University Press, pp. 639–644.

Torres, T. (1990). Tertiary paleobotanical study in the King George and Seymour Islands, Antarctica. Ph.D. thesis, University of Lyon (in French).

Torres, T. and Méon, H. (1990). Estudio palinológico de Cerro Fósil, península Fildes, Isla Rey Jorge, Antártica. *Serie Científica Instituto Antártico Chileno*, **40**, 21–39.

Torres, T. and Méon, H. (1993). *Lophosoria* del Terciário de Isla Rey Jorge y Chile Central: origen y dispersión en el hemisferio sur. *Serie Científica Instituto Antártico Chileno*, **43**, 17–30.

Torres, T., Marenssi, S. and Santillana, S. (1994a). Maderas fósiles de la Isla Seymour, Formación La Meseta, Antártica. *Serie Científica Instituto Antártico Chileno*, **39**, 43–58.

Torres, T., Marenssi, S. and Santillana, S. (1994b). Fossil wood of La Meseta Formation, isla Seymour, Antártica. *Serie Científica Instituto Antártico Chileno*, **44**, 17–38.

Tripati, A. K. and Elderfield, H. (2005). Deep sea temperature and circulation changes at the Paleocene-Eocene thermal maximum. *Science*, **308**, 1894–1898.

Tripati, A., Backman, J., Elderfield, H. and Ferretti, P. (2005). Eocene bipolar glaciation associated with global carbon cycle changes, *Nature*, **436**, 341–345.

Troedson, A. L. and Smellie, J. L. (2002). The Polonez Cove Formation of King George Island, Antarctica: stratigraphy, facies and implications for mid-Cenozoic cryosphere development. *Sedimentology*, **49**, 277–301.

Troncoso, A. (1986). Nuevas órgano-especies en la tafoflora Terciária Inferior de península Fildes, Isla Rey Jorge, Antártica. *Serie Científica Instituto Antártico Chileno*, **34**, 23–46.

Truswell, E. M. (1982). Palynology of seafloor samples collected by the 1911–14 Australasian Antarctic Expedition: implications for the geology of coastal East Antarctica. *Journal of the Geological Society of Australia*, **29**, 343–356.

Truswell, E. M. (1990). Cretaceous and Tertiary vegetation of Antarctica: A palynological Perspective. In *Antarctic Paleobiology: its Role in the Reconstruction of Gondwana*, eds T. N. Taylor and E. L. Taylor. Berlin: Springer-Verlag, pp. 71–88.

Truswell, E. M. (1993). Vegetation changes in the Australian Tertiary in response to climatic and phytogeographic forcing factors. *Australian Systematic Botany*, **6**, 533–557.

Veblen, T. T. and Ashton, D. H. (1978). Catastrophic influences on the vegetation of the Valdivian Andes, Chile. *Vegetatio*, **36**, 149–167.

Veblen, T. T. and Lorenz, D. C. (1987). Post-fire stand development of *Austrocedrus-Nothofagus* forests in Patagonia. *Vegetatio*, **73**, 113–126.

Veblen, T. T., Ashton, D. H., Schlegel, F. M. and Veblen, A. T. (1977). Plant succession in a timberline depressed by volcanism in south-central Chile. *Journal of Biogeography*, **4**, 275–294.

Veblen, T. T., Donoso, C., Schlegel, F. and Escobar, B. (1981). Forest dynamic in south central Chile. *Journal of Biogeography*, **8**, 211–247.

Veblen, T. T., Schlegel, F. and Oltremari, J. (1983). Temperate broad-leaved evergreen forest of South America. In *Temperate Broad-Leaved Evergreen Forest*, ed. J. D. Ovington. Amsterdam: Elseiver, pp. 5–31.

Veblen, T. T., Ashton, D. H., Rubulis, S., Lorenz, D. C. and Cortes, M. (1989). *Nothofagus* stand development on in-transit moraines, Casa Pangue Glacier, Chile. *Arctic and Alpine Research*, **21**, 144–155.

Veblen, T. T., Burns, B. R., Kitzberger, T., Lara A. and Villalba, R. (1995). The ecology of the conifers of southern South America. In *Ecology of the Southern Conifers*, eds N. J. Enright and R. S. Hill. Washington: Smithsonian Institution Press, pp.120–155.

Veblen, T. T., Donoso, C., Kitzberger, T. and Rebertus, A. J. (1996). Ecology of Southern Chilean Argentinean *Nothofagus* forests. In *The Ecology and Biogeography of Nothofagus forests*, eds T. T. Veblen, R. S. Hill and J. Read. New Haven and London: Yale University Press, pp. 293–353.

Veevers, J. J., Powell, C. McA. and Roots, S. R. (1991). Review of sea floor spreading around Australia. 1. Synthesis of patterns of spreading. *Australian Journal of Earth Sciences*, **38**, 373–389.

Veizer, J., Godderis, Y. and François, L. M. (2000). Evidence for decoupling of atmospheric CO_2 and global climate during the Phanerozoic eon. *Nature*, **408**, 698–701.

Villagrán, C. (1990). Glacial climates and their effects on the history of the vegetation of Chile: a synthesis based on palynological evience from Isla de Chiloé. *Review of Palaeobotany and Palynology*, **65**, 17–24.

Villagrán, C. and Hinojosa, F. (1997). Historia de los bosques de Sudamérica II. Fitogeografía. *Revista Chilena de Historia Natural*, **70**, 241–267.

Vizcaíno, S. F., Bond, M., Reguero, M. A. and Pascual, R. (1997). The youngest record of fossil land mammals from Antarctica, its signicance on the evolution of the terrestrial environment of the Antarctic Peninsula during the late Eocene. *Journal of Paleontology*, **71**, 348–350.

Vizcaíno, S. F., Reguero, M. A., Goin, F. J., Tambussi, C. P. and Noriega, J. I. (1998). Community structure of Eocene terrestrial vertebrates from Antarctic Peninsula. In *Paléogeno de America del Sur of de la Península Antártica, Asociación Paleontológica Argentina Publicación Especial*, **5**, ed. S. Casadio. Buenos Aires: APA, pp. 177–183.

Weijers. J. W. H., Schouten, S., Sluijs, A., Brinkhuis, H. and Damste, J. S. S. (2007). Warm Arctic continents during the Palaeocene–Eocene thermal maximum. *Earth and Planetary Science Letters*, **261**, 230–238.

Westerhold, T. and Röhl, U. (2009). High resolution cyclostratigraphy of the early Eocene–new insights into the origin of the Cenozoic cooling trend. *Climates of the Past*, **5**, 309–327.

Westerhold, T., Röhl, U., Laskar, J., *et al.* (2007). On the duration of magnetochrons C24r and C25n and the timing of early Eocene global warming events: Implications from the Ocean Drilling Program Leg 208 Walvis Ridge depth transect. *Paleoceanography*, **22**, PA2201 doi: 10.1029/2006PA001322.

Wilf, P., Cúneo N. R., Johnson K. R., *et al.* (2003). High plant diversity in Eocene South America: evidence from Patagonia. *Science*, **300**, 122–125.

Wilf, P., Johnson, K. R., Cúneo, N. R., *et al.* (2005). Eocene plant diversity at Laguna del Hunco and Río Pichileufú, Patagonia, Argentina. *American Naturalist*, **165**, 634–650.

Willan, R. C. R. and Kelley, S. P. (1999). Mafic dike swarms in the South Shetland Islands volcanic arc: unravelling multiepisodic magnetism related to subduction and continental rifting. *Journal of Geophysical Research*, **104**, 23051–23068.

Willson, M. and Armesto, J. J. (1996). The natural history of Chiloé: on Darwin's trail. *Revista Chilena de Historia Natural*, **69**, 149–161.

Wing, S. L., Harrington, G. J., Smith, F. A., *et al.* (2005). Transient floral change and rapid global warming at the Paleocene-Eocene boundary. *Science*, **310**, 993–996.

Wise, S. W., Breza, J. R., Harwood, D. M. and Wei, W. (1991). Paleogene glacial history of Antarctica. In *Controversies in Modern Geology: Evolution of Geological Theories in Sedimentology, Earth History and Tectonics*, eds D. W. Müller, J. A. McKenzie and H. J. Weissert. London: Academic Press, pp. 133–171.

Wolfe, J. A. (1980). Tertiary climates and floristic relationships at high latitudes in the northern hemisphere. *Palaeogeography, Palaeoclimatology, Palaeoecology*, **30**, 313–323.

Wolfe, J. A. (1990). Palaeobotanical evidence for the marked temperature increase following the Cretaceous/Tertiary boundary. *Nature*, **343**, 153–156.

Woodburne, M. O. and Zinsmeister, W. J. (1984). The first land mammal from Antarctica and its biogeographic implications. *Journal of Paleontology*, **58**, 913–948.

Wrenn, J. H. and Hart, G. F. (1988). Paleogene dinoflagellate cyst biostratigraphy of Seymour Island, Antarctica. In *Geology and Paleontology of Seymour Island, Antarctic Peninsula, Geological Society of America Memoir*, **169**, eds R. M. Feldmann and M. O. Woodburne. Boulder: The Geological Society of America, pp. 321–447.

Xue, Y. S., Shen, Y. B. and Zhou, E. J. (1996). Petrological characteristics of the sedimentary volcanic rocks of the Fossil Hill Formation (Eocene) in King George Island, West Antarctica. *Antarctic Research*, **7**, 99–117.

Yeo, J. P., Lee, J. I., Hur, S. D. and Choi, B.-G. (2004). Geochemistry of volcanic rocks in Barton and Weaver peninsulas, King George Island, Antarctica: implications for arc maturity and correlation with fossilized volcanic centres. *Geosciences Journal*, **8**, 11–25.

Zachos, J. C. and Kump, L. R. (2005). Carbon cycle feedbacks and the initiation of Antarctic glaciation in the earliest Oligocene. *Global and Planetary Change*, **47**, 51–66.

Zachos, J. C., Lohmann, K. C., Walker, J. C. G. and Wise, S. W. (1993). Abrupt climate changes and transient climate changes during the Paleogene: a marine perspective. *The Journal of Geology*, **101**, 191–213.

Zachos, J. C., Quinn, T. M. and Salamy, K. A. (1996). High-resolution (10^4 years) deep sea foraminiferal stable isotope records of the Eocene-Oligocene climate transition. *Paleoceanography*, **21**, 251–266.

Zachos, J., Pagani, M., Sloan, L., Thomas E. and Billups, K. (2001). Trends, rhythms and aberrations in global climate 65 Ma to present. *Science*, **292**, 686–693.

Zachos, J. C., Dickens, G. R. and Zeebe, R. E. (2008). An early Cenozoic perspective on greenhouse warming and carbon-cycle dynamics. *Nature*, **451**, 279–283.

Zachos, J. C., McCarren, H., Murphy, B., Röhl, U. and Westerhold, T. (2010). Tempo and scale of late Paleocene and early Eocene carbon isotope cycles: Implications for the origin of hyperthermals. *Earth and Planetary Science Letters*, **299**, 242–249.

Zamaloa, M. C. and Andreis, R. R. (1995). Asociación palinológica del Paleoceno temprano (Formación Salamanca) en Ea. Laguna Manantiales, Santa Cruz, Argentina. *Actas 6e Congresso Argentino Paleontologia y Bioestratigrafia*, **1**, 301–305.

Zastawniak, E. (1981). Tertiary leaf flora from the Point Hennequin Group of King George Island (South Shetland Islands, Antarctica). Preliminary report. *Studia Geologica Polonica*, **72**, 97–108.

Zastawniak, E. (1994). Upper Cretaceous fossil leaf flora from the Błaszyk moraine (Zamek Formation), King George Island, South Shetland Islands, West Antarctica. *Acta Palaeobotanica*, **34**, 119–163.

Zastawniak, E., Wrona, R., Gaździcki, A. and Birkenmajer, K. (1985). Plant remains from the top part of the Point Hennequin Group (Upper Oligocene), King George Island (South Shetlands, Antarctica). *Studia Geologica Polonica*, **81**, 143–170.

Zeebe, R. E., Zachos, J. C. and Dickens, G. R. (2009). Carbon dioxide forcing alone insufficient to explain Palaeocene-Eocene Thermal Maximum warming. *Nature Geoscience*, **2**, 576–80.

Zhang, S. and Wang, Q. (1994). Palaeocene petrified wood on the west side of Collins Glacier in the King George Island, Antarctica. In *Stratigraphy and Palaeontology of Fildes Peninsula King George Island, Antarctica, State Antarctic Committee Monograph*, **3**, ed. Y. Shen. Beijing: China Science Press, pp. 231–238.

Zhou, Z. and Li, H. M. (1994a). Early Tertiary ferns from Fildes Peninsula, King George Island, Antarctica. In *Stratigraphy and Palaeontology of Fildes Peninsula King George Island, Antarctica, State Antarctic Committee Monograph*, **3**, ed. Y. Shen. Beijing: China Science Press, pp. 181–207.

Zhou, Z. and Li, H. (1994b). Early Tertiary gymnosperms from Fildes Peninsula, King George Island, Antarctica. In *Stratigraphy and Palaeontology of Fildes Peninsula King George Island, Antarctica, State Antarctic Committee Monograph*, **3**, ed. Y. Shen. Beijing: China Science Press, pp. 208–230.

Zinsmeister, W. J. (1986). Fossil windfall at Antarctica's edge. *Natural History*, **95**, 60–67.

9

After the heat: late Eocene to Pliocene climatic cooling and modification of the Antarctic vegetation

Introduction

The Cenozoic progression from Eocene to Oligocene at ~34 Ma marked profound global climatic and oceanographic changes. This was accompanied by a rapid transition from greenhouse to icehouse world that was reflected in the vegetation of the Antarctic continent through this time. The greenhouse mode that had characterised the early Cenozoic was replaced by an icehouse mode that was to typify the later Cenozoic. This was marked by strong pole to equator temperature gradients that resulted in greater planetary wind speeds and polar ice caps that drove oceanic circulation through thermohaline circulation (Kidder and Worsley, 2010). The well-oxygenated bottom water currents that drove the oceanic circulation were created by the generation of cold, dense, salty water as the oceans froze and the bases of ice sheets melted. Continental collision has been cited as a possible major forcing factor, responsible for shifting the state of the Earth from the normal greenhouse to the icehouse condition (e.g. Raymo and Ruddiman, 1992) as this draws down atmospheric CO_2 through increased chemical weathering and subsequent sequestering of silicate weathering products in the ocean[1] (e.g. Kidder and Worsley, 2010; Raymo, 1991; Ruddiman, 1997).

The strong oceanographic and atmospheric isolation of Antarctica over the South Pole today characterises the modern icehouse world. The major and most dominant oceanographic feature of the Southern Ocean is the suite of complex currents that together make up the Antarctic Circumpolar Current – the largest current on Earth. The Antarctic Circumpolar Current is key to maintaining the present day ice sheets as it reduces heat flow from the tropical regions into the high southern latitudes. As the Antarctic Circumpolar Current became established during the Cenozoic, oceanic isolation increased resulting in the inevitable cooling into icehouse conditions. As a consequence Antarctica and the surrounding Southern Ocean began to exert increasing influence on global climate through regulation of global sea level, the strengthening of both the thermohaline circulation and carbon cycle dynamics, as well as

---

[1] This hypothesis is supported by recent weathering estimates that suggest 50% of CO_2 drawdown today results from the weathering taking place in active mountain belts (Hilley and Porder, 2008), thus providing an important link between climate and mountain building, and in turn reinforces the role of the Himalayas in Cenozoic climate cooling (e.g. Raymo and Ruddiman, 1992). However, some researchers have questioned whether the changes in topographic relief that resulted in modifications to erosion or weathering flux globally are indeed the cause of major glaciations in Earth history (e.g. Gibbs *et al.*, 1999; Willenbring and von Blanckenburg, 2010). Consequently, this raises the possibility that other factors, such as deep water formation, may be just as important.

driving the southward flux of heat and hence atmospheric circulation (Shevenell and Kennett, 2007).

The development of the conditions that came to define the Antarctic cryosphere was one of the fundamental reorganisations of the global climate system. With this reorganisation came a radical change to the ecosystems across the Antarctic continent. The richly vegetated landmass that had enjoyed the warm temperate conditions of the earlier Eocene adapted to the cooling climate to become more characteristic of tundra ecosystems today, before entering deeper into the Oligocene icehouse. Therefore, critical to understanding the transition from warm Eocene to Oligocene icehouse is understanding the development of the Antarctic Circumpolar Current. Yet in order to create the possibility for an Antarctic Circumpolar Current, the gateways between South America and the Antarctic Peninsula, and Australia and Antarctica had to be open. For that reason timing of the opening of these two gateways is crucial. The opening of these two gateways has been suggested to be the most important factor for the initiation of glaciation but this is hotly debated. Recent modelling studies (DeConto and Pollard, 2003a) have suggested an alternative main driver – a reduction in atmospheric CO_2 brought about by a significant increase in continental weathering (Kidder and Worsley, 2010). This in turn raises the question as to whether in fact global weathering rates changed prior to the onset of glaciation, and whether this mechanism can provide a plausible mechanism for CO_2 drawdown and glaciation. In this chapter we discuss the possible drivers for the change in climate state and the effect that this downturn in climate had on the prevailing vegetation.

Palaeogeography of the southern high latitudes

Today the Antarctic Circumpolar Current moves unhindered around the isolated Antarctic continent, unheeded by any continental barrier, through deep water channels and eroding the sea floor around the globe. It courses through Drake Passage and the Tasman Gateway off East Antarctica (see below), enabling the free transfer of water masses between the Atlantic, Pacific and Indian ocean basins, keeping Antarctica thermally isolated from the rest of the world (Pfuhl and McCave, 2005). Changes to bathymetry driven by uplift due to tectonic processes impact on the amount of water, and thus heat, transported by the Antarctic Circumpolar Current such that the implications are global in proportion.

The development of the Antarctic Circumpolar Current was, however, only possible because of the separation of South America from the Antarctic Peninsula, and Australia from East Antarctica, with the formation of deep waters in the resultant Drake Passage and Tasman Gateway respectively (Figure 9.1). With increased flow speed and homogeneity of the Southern Ocean water masses, the Antarctic Circumpolar Current was born and with it came massive changes in global oceanic and atmospheric circulation patterns (e.g. Barker and Thomas, 2004). Understanding both the development and timing of these two gateways underpins our understanding of global oceanic circulation, climate and Antarctic cryosphere growth during the Cenozoic (Lagabrielle *et al.*, 2009; Stickely *et al.*, 2004).

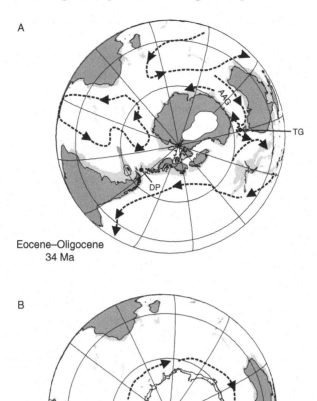

Figure 9.1 Illustration of increasing isolation of Antarctica during the Cenozoic. **A**, Eocene–Oligocene **B**, Miocene. Arrowed, dotted lines indicate direction of the palaeocurrents in existence at the time. Latitude intervals are 30°. Reconstructions provided by R. A. Livermore, British Antarctic Survey. AAG = Australo-Antarctic Gulf; DP = Drake Passage; TG = Tasman Gateway.

Formation of the South America–Antarctic Peninsula gateway (Drake Passage)

Since about 50 Ma, and thus predating the Australia–Antarctic opening, crustal stretching had been separating West Antarctica from South America. Multiple passageways opened in this region, including one shallow passageway between the Pacific and Atlantic, and a second between East and West Antarctica (Livermore *et al.*, 2007; Lyle *et al.*, 2007). This trans-Antarctic seaway would have persisted through the Oligocene (assuming there was no

West Antarctic ice sheet, see below) until the East and West Antarctic ice sheets had evolved enough to become grounded in the Ronne and Ross embayments during the middle Miocene (Figure 9.2). The seaway may, however, have closed prior to this if the general Cenozoic drop in sea level caused by ice-sheet extension had been great enough (Haq *et al.*, 1987; Lawver and Gahagan, 2003; Nelson and Cooke, 2001). Regardless of its duration, water exchange between the Ross and Weddell seas through the trans-Antarctic seaway would have been minimal or non-existent (Nelson and Cooke, 2001).

Major changes in South America-Antarctica plate motion during the Eocene and Oligocene resulted in the opening of two small basins in the southern Scotia Sea (Eagles, Livermore and Morris, 2006; Lagabrielle *et al.*, 2009; Livermore *et al.*, 2005, 2007). Subsequent subsidence in these basins produced a deepening rift that eventually led to seafloor spreading and ultimately gave rise to Drake Passage (Eagles *et al.*, 2005). The oldest oceanic crust in the Drake Passage region ranges from 34 to 29 Ma, implying that onset of spreading began at or near the Eocene-Oligocene (E-O) boundary thus opening the northern Drake Passage region to proto-Antarctic Circumpolar Current circulation (Lagabrielle *et al.*, 2009). Since the exact motions of certain microcontinental

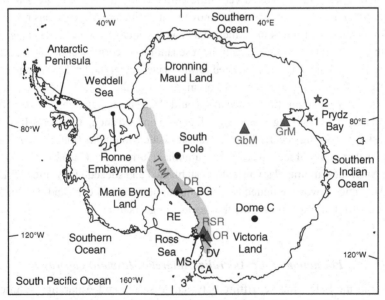

Figure 9.2 Map with names and localities mentioned in this chapter. Mountains include Transantarctic Mountains (TAM) and ranges (grey triangles) DR = Dominion Range, GbM = Gamburtsev Mountains, GrM = Grove Mountains, OR = Olympus Range, RSR = Royal Society Range; RE = Ross Embayment with the Ross Ice Shelf and including the McMurdo Ice Shelf that delimits the southern boundary of McMurdo Sound (MS) which in turn opens to the Ross Sea to the north. Fossil localities BG = Beardmore Glacier, DV = Dry Valleys (see Figure 9.4 for details) and geological drilling programs (grey stars) 1 = ODP Leg 188 site 1166, 2 = CRP-1 Site 1165 in Prydz Bay, and 3 = DSDP Site 274 off Cape Adare (CA).

fragments are not known, the exact timing of the opening of a deep seaway prior to 30 Ma remains somewhat circumstantial (Lagabrielle *et al.*, 2009). Isotopic and micropalaeonto-logical evidence support the hypothesis that Drake Passage opening predated that of the Tasman Gateway, implying a possible late Eocene inception of a circum-Antarctic pathway (Eagles, Livermore and Morris, 2006; Scher and Martin, 2006), and thus contra-dicting hypotheses for a relatively warm Antarctic continent at the beginning of the Oligocene (Stickley *et al.*, 2004). According to Lagabrielle *et al.* (2009) during the 26–14 Ma period major tectonic events affected the physiography of the northern region of the Drake Passage.

By 22 Ma, the reduction in through-flow was such that it has been suggested that the Antarctic Circumpolar Current may have undergone a partial shutdown altogether (Lagabrielle *et al.*, 2009). The seas along the newly formed North Scotia Ridge shallowed, the Fuegian seaway closed and uplift of the Patagonian Cordillera could all have con-tributed to a reduction in through-flow such that the resultant narrowing would have impacted significantly on the Antarctic Circumpolar Current by restricting water flow and contributing to global climate warming.

Although such tectonic processes had been at work since the Eocene, the southern tip of South America has actually moved very little latitudinally (i.e. a few degrees to the north) (Cunningham *et al.*, 1995). Longitudinal movement during the Miocene has resulted in the separation of Antarctica and South America that we see today (Lawver and Gahagan, 2003). Around the Oligocene–Miocene transition these landmass reorganisations at the Antarctic–Patagonia connection led to the suggested further widening of the Drake Passage during the middle Miocene coeval with the major cooling event of the middle Miocene (Lagabrielle *et al.*, 2009). Reopening this gateway would lead to reactivation of the Antarctic Circumpolar Current resulting in strong cooling both on land and in the ocean. It is estimated that at least 8 °C cooling occurred in the Transantarctic Mountains (Lewis *et al.*, 2008) and in surface waters south of the Antarctic Circumpolar Current area (Lagabrielle *et al.*, 2009). With the Antarctic Circumpolar Current re-established and subsequently intensifying as the circum-Antarctic seaway continued to widen throughout the Miocene and into the Pliocene (from 20 to 5 Ma), it grew to become the world's largest current.

Formation of the Australia–Antarctic (Tasman) gateway

Beginning in the Late Jurassic, rifting between Australia and Antarctica resulted in the development of a limited seaway between southwestern Australia and Antarctica. This rift system propagated eastward during the Late Cretaceous and Cenozoic, slowly widening the Australo–Antarctic Gulf between the two continents (Figure 9.1). Yet the Indian and Pacific Oceans were effectively blocked at the eastern end of the Australo-Antarctic Gulf by an almost continuous 'land bridge', the Tasman Rise, until ~35.5 Ma (Exon, Kennett and Malone, 2004; Stickley *et al.*, 2004). Sediments from ODP Leg 189 (Exon, Kennett and Malone, 2004) show that the Tasmanian margin consisted of predominantly terrestrially derived silicoclastic sediments in the middle Eocene, that were gradually replaced by more marine conditions through the Oligocene (Brown, Gaina and Müller, 2006; Exon, Kennet and Malone, 2004; Lawver and Gahagan, 2003; Stickley *et al.*, 2004). As the Tasman Rise slid away from Antarctica, a significant increase in deepening and associated bottom water current activity took place in the Tasman Gateway, just prior (~35.5 Ma) to the Eocene-Oligocene boundary and thus preceded significant Antarctic glaciations by ~2 Ma (Stickley *et al.*, 2004). This deepening of the Tasman Gateway, which was most significant at the Eocene–Oligocene transition, was associated with an important increase in bottom water activity and its associated erosive powers became increasingly energetic (Stickley *et al.*, 2004). This gateway was now able to allow proto-Antarctic Circumpolar Current circulation to exploit the seaway between East Antarctica and Australia (Ghiglione *et al.*, 2008; Lagabrielle *et al.*, 2009; Lawver and Gahagan, 2003). The development of the Antarctic Circumpolar Current resulted in the Northern Hemisphere warming by up to 3 °C (Lagabrielle *et al.*, 2009), whilst in the southern high latitudes widescale cooling led to severe glaciations across Antarctica during the Oligocene, and enrichment in ^{18}O of the surrounding oceans (Ghiglione *et al.*, 2008; Huber and Nof, 2006; Lagabrielle *et al.*, 2009; Sijp and England, 2004; Toggweiler and Bjornsson, 2000; Zachos *et al.*, 2001a). This enrichment, just above the Eocene-Oligocene boundary at 34 Ma, as evidenced by a pronounced positive oxygen isotope excursion reflects a brief (400 kyr), yet extreme, glacial event (the Oi-1; Figure 9.3). The appearance of large continental ice sheets on Antarctica and changes to ocean productivity patterns were both probably the result of the reorganisation of the climate-ocean system (Huber *et al.*, 2004; Zachos *et al.*, 2001a). This reorganisation allowed the eastward flow of relatively warm surface waters (the 'proto-Leeuwin' current) from the Australo-Antarctic Gulf into the southwestern Pacific Ocean (Figure 9.1A) accompanied by a significant warming of sea surface temperatures in the area around south and east Tasmania in the earliest Oligocene. At the same time, a weak cold westward current developed near Antarctica (Huber *et al.*, 2004; Lear, Elderfield and Wilson, 2000; Lyle *et al.*, 2007; Stickley *et al.*, 2004; Figure 9.1A). The seaway had by now opened substantially and was accompanied by slow deposition of deep water pelagic carbonates (Stickley *et al.*, 2004). Through the Oligocene, water depths did not continue to increase significantly because subsidence rates had fallen (Exon, Kennett and Malone, 2004). It remains debatable whether these ocean movements helped with the inception of a proto-Antarctic Circumpolar Current at this time. If the circumpolar current had not come into existence by the Oligocene, heat carried by the oceans and atmosphere would not have been blocked as it was transferred towards the pole and East Antarctica[2]. Although physically isolated from its once gondwanan neighbours, Antarctica was not thermally isolated at the beginning of the Oligocene (Stickley *et al.*, 2004).

---

[2] High latitudes receive little of their heat directly from the Sun – the greater part is received indirectly by atmospheric and oceanic transport from lower latitudes.

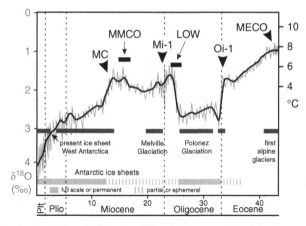

Figure 9.3 Climate change during the last 65 million years as expressed by the oxygen isotope curve based on benthic $\delta^{18}O$ and showing the computed ocean equivalent sea surface temperature (°C) for high latitudes based on an ice-free ocean and glaciations in Antarctica (grey lines). MECO = Middle Eocene Climatic Optimum (included for ease of cross reference to Figure 8.1); Oi-1 = Oligocene Oi-1 glaciation event marked by an oxygen isotope excursion that indicates the beginning of ice sheet coverage on Antarctic; Mi-1 = Mi-1 glaciation event marked by an oxygen isotope excursion at the Oligocene–Miocene boundary; MMCO = Middle Miocene Climatic Optimum. Evidence of glacial events on West Antarctica indicated by dark grey horizontal lines but breaks in the line do not necessarily mean no glacial events occurred (see text for details and references). Adapted from Zachos *et al.* (2001a) with permission from AAAS.

Inception of the Antarctic Ice Sheet

Although today the Antarctic Circumpolar Current is believed to exert considerable control on Antarctic Ice Sheet stability through the effective isolation of the southern polar region, this may not have always been the case. The exact link between global climate, gateway evolution and the onset of the Antarctic Circumpolar Current is still under discussion. Moreover their impact on the global ocean circulation, carbon cycling and hence climate has been strongly suspected and is supported by modelling data (Shevenell, Kennett and Lea, 2004; Sijp and England, 2005). The thermal isolation of Antarctica was initially thought to have been caused by the onset of the Antarctic Circumpolar Current, since proximity to the pole alone was not sufficient for glacial development. The production of cool waters sinking around Antarctica was hypothesised (Kennett, 1977) to form the Southern Origin Bottom Water (cf. Barker and Thomas, 2004) that became connected to the thermohaline circulation. However, the actual climatic impact derived from changes in the geometry of the Southern Hemisphere seaways during the entire Cenozoic is still under debate. Problems arise with such a simple scenario. If climate were driven solely by apparent continuous widening of both oceanic gateways after 30 Ma, global cooling would have been long term and regular, with no significant climatic aberrations in the late Paleogene and Neogene (Lagabrielle *et al.*, 2009). Yet this period of global cooling was indeed interrupted (Figure 9.3). A warming period began in the late Oligocene, manifesting itself as the late Oligocene Warming (~26 Ma), it lasted 12 Ma and ended progressively after a second peak warming in the mid-Miocene (~14 Ma) (Lagabrielle *et al.*, 2009; Shevenell, Kennett and Lea, 2004). Furthermore, the onset of the Antarctic Circumpolar Current does not seem to explain, by itself, the rapid rise of the Antarctic ice cap at the Eocene-Oligocene boundary, because glaciation may have predated the event by at least 5 Ma (Barker and Thomas, 2004; Eagles, Livermore and Morris, 2006; Tripati *et al.*, 2005). Therefore factors other than gateway opening alone must have played a role in climate evolution and polar ice formation during the Oligocene and ensuing Neogene.

Earth left its typical Eocene greenhouse state and entered its present glacial state, abruptly and essentially irreversibly, at the Eocene-Oligocene boundary. Evidence for cooling and the sudden growth of an East Antarctic ice sheet comes from marine records (Lear, Elderfield and Wilson, 2000; Zachos, Quinn and Salamy, 1996; Zachos *et al.*, 2001a), drilling on the East Antarctic margin (Wilson *et al.*, 1998), circum-Antarctic ice-rafted debris (Zachos, Breza and Wise, 1992) and shifts in the clay composition of sediments (Ehrmann and Mackensen, 1992), and Antarctic vegetation. Yet the precise history of Antarctic Ice Sheet dynamics during the Cenozoic remains uncertain, and is not confined to Antarctica. The nature and timing of glaciations globally is under debate since glaciations may have been bipolar (Tripati *et al.*, 2005), or they may have appeared at a different time in the Northern Hemisphere relative to those in the Southern Hemisphere (e.g. Edgar *et al.*, 2007; Eldrett *et al.*, 2007; Moran *et al.*, 2006; St John, 2008). Inferences from oxygen isotopic composition of carbonates suggest the presence of ice sheets on Antarctica 34 Ma (e.g. Kennett and Shackleton, 1976; Zachos, Quinn and Salamy, 1996; Figure 9.3), and much later (between 10 and 6 Ma) in the Northern Hemisphere (Driscoll and Haug, 1998). But more recently the geochemistry of sediments and foraminifera in the tropical Pacific and South Atlantic oceans suggest several small glaciations and one major, yet transient, glaciation occurred in both the Northern and Southern hemispheres during the middle to late Eocene, some 42 million years ago (Tripati *et al.*, 2005). Even though the suggestion of large Northern Hemisphere ice sheets in the Eocene is highly controversial, precursor glaciations foreshadowing the major transition to an icehouse state are theoretically expected of a dynamic system gradually moving from one stable state to another (Kump, 2005).

Antarctic cryosphere inception is no longer simply regarded as being the response to Antarctic Circumpolar Current inception, since problems occur over the timing of gateway opening and the inception of the Antarctic Ice Sheet. Most tectonic reconstructions place the opening of the Tasman Gateway close to the Eocene-Oligocene boundary, but Drake Passage may not have provided a significant deep water passage until several million years later (Nong *et al.*, 2000; Toggweiler and Bjornsson, 2000; see previous section). Recent studies have begun to show that the system was indeed more complex. DeConto and Pollard (2003a, 2003b) suggested that a combination of atmospheric CO_2, orbital forcing and ice-climate feedbacks was the primary cause of the Eocene–Oligocene climate transition. The authors emphasise that the opening of Southern Ocean gateways and the formation of the Antarctic Circumpolar Current undoubtedly cooled the high southern latitudes, but Drake Passage opening can only be regarded as a potential trigger (rather than a driver) for glacial inception when atmospheric CO_2 is in a relatively narrow range (i.e. 2.5–3 times pre-industrial levels). Tripati *et al.* (2005) also used declining pCO_2 levels to explain the stability of the Oligocene ice sheets relative to those in the earlier Eocene (see following section). Miller *et al.* (2009) proposed that growth of the continent scale Antarctic ice sheet was the primary cause of a dramatic reorganisation of ocean circulation and chemistry. Regardless of the seeming increasing complexity of the system, glacial history of Antarctica continues to improve with studies focusing largely on offshore shelf basins around Antarctica. These include both seismic surveys (e.g. Bart *et al.*, 2000; De Santis *et al.*, 1999) and geological drilling programs (e.g. Ocean Drilling Program (ODP), Cape Roberts Project (CRP), SHALDRIL, and ANDRILL; Naish *et al.*, 2008). Even though the number of sites drilled is small, and mainly confined to three areas, namely McMurdo Sound (e.g. Barrett, 1989, 2007; McKay *et al.*, 2009), Prydz Bay (e.g. Barron *et al.*, 1989; Shipboard Scientific Party, 2001), and the Antarctic Peninsula (e.g. Barker *et al.*, 1999), understanding of the timing and causes of fluctuations in the Antarctic Ice Sheet through the Cenozoic continues to advance.

Another factor that may have played an important influencing role in the timing of continental-scale ice sheet initiation and has not yet been mentioned is the type of vegetation cover growing across the continent at this time. The fossil evidence suggests that the vegetation evolved, in response to the changing climate conditions, from tall stature evergreen mixed broad-leaf and needle-leaf trees and shrubs dominated by *Nothofagus* in the late Paleocene (see Chapter 8), to a tundra-like vegetation of the early Miocene (this chapter). The Eocene-Oligocene boundary was a transition period between the two vegetation types. Thorn and DeConto (2006) explored this change from needle-leaf evergreen forests[3] to tundra, using global climate models, and found that the replacement of evergreen vegetation by tundra would result in a significantly cooler climate (up to 2.5 °C during the interglacial summer months), which supports other studies that also highlight the importance of vegetation feedbacks during glacial transition (e.g. Bonan, Pollard and Thompson, 1992; DeConto and Pollard, 2003a; Dutton and Barron, 1996). This in turn would have initiated a tundra-induced cooling effect reflected in the higher albedo of the landmass. Conversely, the forests of the early Eocene would have added to the warming effect and with this vegetation type in existence the inception of continental-scale ice sheet initiation could have been delayed. Moreover as the glaciation threshold level of greenhouse gases was approached, so the vegetation-climate feedbacks become ever more significant. With levels of pCO_2 suggested to have been between 2.5 and 3 times pre-industrial levels at the Eocene-Oligocene boundary (see below), according to Thorn and DeConto (2006) the vegetation type would appear to have had a significant effect on Antarctic continental climate. For example, a land surface

---

[3] Based on the fossil evidence, which illustrates relatively extensive deciduous *Nothofagus*/blade-leafed podocarp associations (this chapter), the needle-leaf evergreen vegetation type selected by Thorn and DeConto (2006) underplays the role of deciduousness and overplays the occurrence of needle leaves in the late Eocene forests, which in turn might impact on their conclusions.

completely covered in tundra vegetation could allow ice sheet formation at relatively high levels of atmospheric greenhouse gas concentrations. The authors also note that the climatic influence of vegetation cover across the Antarctic interior is greater than the difference between open *versus* closed Southern Ocean gateways (which in turn might have important implications for other Cenozoic climatic events). This therefore highlights the importance of climate feedbacks relating to shifting terrestrial ecosystems, and in turn emphasises the absolute necessity to fully understand the palaeovegetation using stratigraphically and taxonomically identified fossil plant material.

The growth of ice in Antarctica is therefore linked to a number of potential biotic and abiotic factors, including those brought about by changing palaeogeography and tectonic processes, further complicated by the effects of high atmospheric pCO_2 (e.g. Eagles, Livermore and Morris, 2006; Pak and Miller, 1992; Sijp, England and Toggweiler, 2009). These issues are only introduced here as the details are beyond the scope of this book. Nevertheless, the evolution of the Antarctic cryosphere cannot be separated from changes in palaeoclimate and this interplay forms the focus of the following section.

Late Paleogene palaeoclimate

The non-gradual, stepwise cooling that had begun in the middle Eocene (see Chapter 8) dramatically culminated in the earliest Oligocene, marked by an increase in $\delta^{18}O$ of deep sea benthic foraminifera[4] worldwide (oxygen isotope event Oi-1 at 33.5 Ma; Figure 9.3), and signalled the beginning of the icehouse Earth (Miller *et al.*, 2009 and references therein). The Oi-1 $\delta^{18}O$ shift that took place over a few hundred thousand years corresponded with an extended period of global climatic disruption, widespread extinction, biogeographic reorganisation, and the first major expansion of Antarctic ice in the Cenozoic (Miller *et al.*, 2009; Pearson, Foster and Wade, 2009; Zachos, Breza and Wise, 1992; Zachos *et al.*, 2001a). This cooling across the Eocene–Oligocene transition (EOT), as evidenced by a shift (+1.5‰) in oxygen isotopes of deep water carbonates, identifies three major steps associated with a geologically rapid (<40 kyr) phase of ice growth (DeConto and Pollard, 2003a). Firstly, during the latest Eocene at ~33.8 Ma, associated with global cooling and a little ice growth, a 2 °C decrease in deep water temperatures is recorded, associated with a minor sea level fall of ~25 m (EOT-1), before a partial return to pre-event values. Secondly, at ~33.6 Ma a second minor drop in sea level and deep water cooling is recorded (EOT-2). Finally, at 33.5 Ma (Oi-1) a further 2 °C cooling occurred coupled with a major fall (~80 ± 25 m) in sea level (Coxall *et al.*, 2005; DeConto *et al.*, 2008; Kennett and Shackleton, 1976; Miller *et al.*, 2009; Zachos, Quinn and Salamy, 1996).

The slow temperature decline at the Eocene–Oligocene transition is thought to be a result of a combination of changes in both oceanic circulation (see previous section) and a concomitant

---

[4] $\delta^{18}O$ of deep sea benthic foraminifera reflect changes in both temperature and the $\delta^{18}O$ of the seawater, which in turn relates to the changes in ice volume. Although local fractionation resulting from evaporation/precipitation changes affects the surface ocean, it has a minimal effect on deep-sea $\delta^{18}O$ records, which reflect ice volume and deep sea temperature that in turn reflect high-latitude surface temperature (Miller *et al.*, 2009).

long-term (10^7 yr scale) drawdown of pCO$_2$ concentration[5] of the atmosphere (DeConto and Pollard, 2003b; Huber and Nof, 2006; Pagani *et al.*, 2005; Tripati *et al.*, 2005). In the late Eocene, carbon dioxide levels of 1000–1500 ppmv dropped and crossed a critical threshold of ~750 ppmv at the E-O boundary (Miller *et al.*, 2009; Pagani *et al.*, 2005; Pearson, Foster and Wade, 2009; Pekar and Christie-Blick, 2008). This drop to below the threshold level of 750 ppmv served to precondition the system for the explosive ice sheet growth at the Oi-1. Conditions slowly changed from cool temperate to polar, with periods of humidity alternating with cooler and drier intervals (Ehrmann, Setti and Marinoni, 2005).

Recovery occurred during the earliest Oligocene, as the global carbon cycle adjusted to the presence of a large ice cap and pCO$_2$ rose again to levels of ~1000 ppmv. Once established, the Antarctic Ice Sheet withstood the subsequent increase in pCO$_2$ and associated warming[6] albeit with some shrinkage (Pearson, Foster and Wade, 2009). During the later part of the early Oligocene, pCO$_2$ continued its erratic decline to reach near-modern pre-industrial levels by the latest Oligocene. During this time, as carbon was being locked away for long time periods[7], the falling CO$_2$ and temperatures resulted in a near-global cooling phenomenon of 1–4 °C. Concentrations of pCO$_2$ then remained at depressed levels for the rest of the Miocene (DeConto and Pollard, 2003b; Miller *et al.*, 2009; Pearson and Palmer, 2000; Pekar and Christie-Blick, 2008).

On East Antarctica, sedimentological evidence suggests that initial ice expansion occurred on the craton at the Eocene-Oligocene boundary (e.g. Barrett, 1989; Cape Roberts Science Team, 2001; Hambrey, Ehrmann and Larsen, 1991; Whitehead *et al.*, 2006). As the ice masses built up so the glacial conditions were intensifying (Bo *et al.*, 2009), resulting in settings similar to those across Svalbard, southern Chile and Alaska today (Ehrmann, Setti and Marinoni, 2005). Initially, small highly dynamic ice caps began to form at the highest altitudes of Dronning Maud Land, the Gamburtsev Mountains (the key nucleation point for the vast East Antarctic Ice Sheet; Bo *et al.*, 2009; Young *et al.*, 2011) and Transantarctic Mountains (Figure 9.2). The extent of these ice caps would have fluctuated, alternately coalescing and separating in response to orbital variations (Coxall *et al.*, 2005; DeConto and Pollard, 2003a; DeConto *et al.*, 2008; Pearson, Foster and Wade, 2009; Sijp, England and Toggweiler, 2009). Modelling experiments show that with continued falling pCO$_2$, the Gamburtsev ice cap expanded to coalesce with the cap covering the Transantarctic Mountains to form a much larger central ice sheet. During orbital variations associated with minimal summer isolation, this ice sheet made repeated but temporary contact with the Dronning Maud Land ice cap.

In West Antarctica, the timing of the conception of the ice sheet remains uncertain since there is no complete record of the Eocene-Oligocene boundary. Recent studies however provide direct evidence for the presence of glacier ice at sea level on the Antarctic Peninsula

---

[5] Other greenhouses gases probably played a role but to date there are no proxies for these gases.

[6] Bright surfaces of the ice cap reflect more sunlight and once the ice cap has formed, melting at its margin is compensated by flow from the cold, high altitude interior (Pearson, Foster and Wade, 2009).

[7] Prior to ~52 Ma, organic carbon was stored in coals, coal swamps and peatlands and could easily be remobilised (through wildfire for example) and thereby sustained high atmospheric CO$_2$ concentrations. Later, carbon burial switched to the oceans, at first in the euxinic seas for a short period and then in the deep ocean sediments, where it could be locked away from the atmosphere for longer timescales (Eagles, Livermore and Morris, 2006; Kurtz *et al.*, 2003).

during the earliest Oligocene (Ivany *et al.*, 2006), yet deposits requiring an extensive marine-based ice sheet are only reported to have been on the Antarctic Peninsula during the middle to late Oligocene (Troedson and Smellie, 2002). This demonstrates that even though East and West Antarctica may not have shared the same history of ice sheet growth, the expansion of earliest Oligocene ice was not restricted to East Antarctica as was previously thought (Barrett, 1999; Dingle and Lavelle, 1998; Ivany *et al.*, 2006).

Regardless of the exact timing of initiation of the semi-permanent ice sheet at, slightly proceeding, or even preceding the Eocene-Oligocene boundary (e.g. Ivany *et al.*, 2006; L. Ivany personal communication, 2009), glacial conditions strengthened through the middle and late Oligocene. Ice extended across Antarctica becoming more permanent during the Oligocene. This continent-scale ice sheet, the first to cover Antarctica since the Permian (~280 Ma), has been estimated (from indirect oceanic proxies) to have increased to perhaps 30% greater than the present day East Antarctic Ice Sheet by the mid-Oligocene during glacial maxima, and generally not less than 40% during interglacials for the remainder of the Oligocene and early Miocene (Naish *et al.*, 2008; Pekar and Christie-Blick, 2008; Pekar and DeConto, 2006; Pekar, DeConto and Harwood, 2006). The ice sheet reached the coastline for the first time at ~33.5 Ma, and continued to do so again, albeit intermittently, throughout the Oligocene and early Miocene, all the while exerting its effect on both ocean circulation and chemistry. This in turn affected global climate and resulted in increased latitudinal thermal gradients and glacioeustatic sea level changes (Pekar and Christie-Blick, 2008; Miller, Wright and Fairbanks, 1991; Miller *et al.*, 2009).

Global climate continued to be affected by tectonic activity during the upper Oligocene and on into the earliest Miocene. In the 29–22 Ma time window, the partial shut down and subsequent restart of the Antarctic Circumpolar Current would have had complex and far reaching climatic implications (Lagabrielle *et al.*, 2009). It would have involved widespread cooling in the Northern Hemisphere and decrease in silicate weathering (e.g. Zachos and Kump, 2005) until atmospheric CO_2 started to rise once more and warmed the climate. With a warmer climate, silicate weathering would have increased serving to stabilise CO_2 once levels of weathering matched CO_2 degassing of the Earth (Lagabrielle *et al.*, 2009).

Enhanced pCO_2 and warmer temperatures culminated in the Late Oligocene Warming (Figure 9.3). At around 26 Ma the continental-wide East Antarctic Ice Sheet and West Antarctic glaciations receded. This event only lasted about 2 Ma, yet in East Antarctica the intermittent ice sheets or isolated ice caps persisted until ~14 Ma (Lawver and Gahagan, 2003; Zachos *et al.*, 2001a). The ice sheet across East Antarctica was probably reduced to 50% or less of the present day ice sheet volume (Zachos *et al.*, 1993). However, it is difficult to assess the actual glacial extent during this time and whether the trans-Antarctic seaway became reinstated as few late Oligocene or early Miocene sediments have been cored (Lawver and Gahagan, 2003). The warmer climates of the Late Oligocene Warming also served to enhance silicate weathering[8] that began mopping up atmospheric CO_2 until it

---

[8] For every two moles of CO_2 needed to weather silicates one mole of CO_2 is produced, thus serving to reduce pCO_2 in the atmosphere.

matched CO_2 production through Earth degassing. This resulted in a lowering of global temperatures through the early Miocene.

At the Oligocene–early Miocene transition (23.8 Ma), a time coincident with low eccentricity and low amplitude variability in obliquity of the Earth's orbit, climate deteriorated once more to become cooler and drier (Passchier and Krissek, 2008). This sustained period of unusually low seasonality (cold summers) encouraged a period of further ice growth at the poles that culminated in the Mi-1 glaciation event (~24 Ma; Pälike *et al.*, 2006; Wilson *et al.*, 2009; Zachos *et al.*, 2001b; Figure 9.3). The Antarctic Ice Sheet increased in volume, growing to slightly larger than it is today and extending as far north as the South Shetland Islands on the Antarctic Peninsula and grounding on Prydz Bay and the southwestern Ross Sea in East Antarctica (Pekar and DeConto, 2006; Pekar, DeConto and Harwood, 2006; Wilson *et al.*, 2009 and references cited therein). Ice rafted as far as the Maud Rise (to the north of Dronning Maud Land, Figure 9.2) and the central Ross Sea, but did not appear to reach the Kerguelen Plateau in the southern Indian Ocean (Barker *et al.*, 1988; Barker, Kennett and Scientific Party, 1988; Leckie and Webb, 1983; Schlich *et al.*, 1992). Cooling continued in a stepwise fashion (Roberts *et al.*, 2003) to become subpolar by the early Miocene, with subpolar ice masses prograding onto the shelf area and calving into the sea (Ehrmann, Setti and Marinoni, 2005; Passchier and Krissek, 2008; Pekar and Christie-Blick, 2008). A subsequent increase in eccentricity and high amplitude variability in obliquity during the early Miocene led to the re-establishment of warm mean summer temperatures (albeit a few degrees cooler than the pre-Miocene mean summer temperatures) soon after the Oligocene-Miocene boundary, allowing the recovery of vegetation on the craton (Wilson *et al.*, 2009). It was probably towards the end of the Miocene that the Dronning Maud Land ice cap permanently coalesced with the rest of the ice mass to form one giant spreading ice sheet – the continent-scale East Antarctic Ice Sheet (DeConto and Pollard, 2003a).

Later in the Miocene the postulated restart of the Antarctic Circumpolar Current would also be associated with complex and far reaching consequences. In the Northern Hemisphere, temperatures rose by up to 3 °C (Lagabrielle *et al.*, 2009). Weathering also increased across >65% of continental surfaces (i.e. those landmasses located in the Northern Hemisphere) and as silicate rocks weathered, atmospheric pCO_2 began to rise (Huber and Nof, 2006; Sijp and England, 2004; Toggweiler and Bjornsson, 2000; Zachos and Kump, 2005).

These conditions lasted until the middle Miocene when, at about ~17–14 Ma, the last long-lived warming event of the Cenozoic took place. The Middle Miocene Climatic Optimum (MMCO; Figure 9.3) was accompanied by a brief yet major warming event, similar to the LPTM, at around 15.5 Ma (Warny *et al.*, 2009) with sea surface temperatures ranging from 0 to 11.5 °C and annual summer temperatures rising to a peak of ~10 °C or higher (as compared with −3 to −5 °C today) (Franke and Ehrmann, 2010). This warming affected the ice sheet further inland with glaciers undergoing extensive basal and surface melting (Lewis *et al.*, 2007). The cause(s) of the MMCO and the warm excursion is still under debate. Some studies attribute fluctuations during this time to a poleward shift of the Southern Hemisphere jet stream (Warny *et al.*, 2009), whilst others describe an increase in pCO_2 of between 140 to 700 ppmv (Tong *et al.*, 2009 and references therein), with the latest modelling data refining this increase to between 460 and 580 ppmv (You *et al.*, 2009). Yet according to climate modelling studies the necessary reduction in the temperature could not have been produced by CO_2 alone (Tong *et al.*, 2009). Proxy data infer a reduced temperature gradient resulted as the effect of increasing global CO_2 was felt more keenly in the high latitudes than the tropics. Other arguments point to tectonic and oceanographic processes as being the main driver, where reconfigurations of ocean gateways may have altered oceanic circulation enhancing poleward heat transport (cf. Tong *et al.*, 2009), with CO_2 playing a secondary role. Whatever the driver(s), with a doubling of pCO_2 the result would have been a rise in global annual mean surface temperatures of up to ~18 °C (about 2–3 °C higher than at present and equivalent to the warming predicted for the next century), and possibly up to 6 °C warmer in the mid-latitudes (Flower and Kennett, 1994) accompanied by higher precipitation rates at high latitudes (increased by ~100 mm per year) (Steppuhn *et al.*, 2007; Tong *et al.*, 2009; You *et al.*, 2009).

The Miocene warm period was terminated by climate cyclicity (Nývlt *et al.*, 2011) and a major cooling event, at ~14.2 Ma (Figure 9.3), signifying a time of significant ice sheet expansion on Antarctica during the following middle and late Miocene (e.g. Shevenell, Kennett and Lea, 2008; You *et al.*, 2009). This shift is recorded by planktonic foraminifera in the southern high latitudes and by excursions in the $\delta^{18}O$ record (at 14.8 Ma) (Langebroek, Paul and Schultz, 2009; Shackleton, Hall and Boersma, 1984; Shevenell, Kennett and Lea, 2008; Zachos *et al.*, 2001a, 2001b). Mean annual temperatures dropped by 25–30 °C (Lewis *et al.*, 2007) as glaciers changed from wet- to cold-based and the Antarctic cryosphere underwent a permanent reorganisation with the expansion of the East Antarctic Ice Sheet over the Transantarctic Mountains (Denton *et al.*, 1984; Lewis *et al.*, 2007). This climate transition forms one of the most prominent shifts in Cenozoic climate although its cause is still a matter for debate (Kuhnert, Bickert and Paulsen, 2009). The paucity of well-dated terrestrial archives has led to the transition being only poorly constrained. Moreover uncertainty surrounds the extent of continental cooling and the effect that the cooling had on continental glaciations (e.g. the $\delta^{18}O$ shift may represent the onset of significant glaciations in West Antarctica, which is supported by the first occurrence of ice-rafted debris in the Bellingshausen Sea – on the west coast of the Antarctic Peninsula – during the early to middle Miocene; Haywood *et al.*, 2009 and references cited therein). From a biological perspective, constraining this event is important for understanding whether the tundra ecosystem succumbed during the middle Miocene or whether extinction was delayed until the Pliocene (Lewis *et al.*, 2008).

Various possible drivers have been put forward to account for the end of this warm period and these include: (i) the closure of the eastern portal of the Tethyan Seaway, which then enhanced the poleward heat transport and strengthened the Southern Ocean circumpolar circulation (Kennett, Keller and Srinivasan, 1985); (ii) declining atmospheric CO_2 to about 200 ppmv brought about by the intensification of the Asian monsoon after 15 Ma, the increase in weathering and erosion of the Himalayan belt and the uplift of the Andean Cordillera (e.g. France-Lanord and Derry, 1997; Kuhnert, Bickert and Paulsen, 2009; Lagabrielle *et al.*, 2009; Ramstein *et al.*, 1997; Tripati, Roberts and Eagle, 2009); (iii) orbital configurations

(Holbourn *et al.*, 2005, 2007); (iv) enhanced chemical weathering and burial of organic matter (e.g. Raymo, 1994); and (v) the re-widening of the Drake Passage in response to landmass reorganisation between the Antarctic Peninsula and southern South America at ~15–14 Ma (Lagabrielle *et al.*, 2009).

Regardless of the specific driver(s), the transition to moist, polar conditions with increased seasonality and associated change from wet-based to dry-based glaciers in the Dry Valleys of Victoria Land, East Antarctica (Figure 9.2), set the scene for a second ice sheet expansion across Antarctica. Between 13.9 and 13.8 Ma, mean summer temperatures and sea surface temperatures dropped by about 8 °C and 6–7 °C respectively. Small dynamic ice sheets on West Antarctica fused to form one large ice sheet (Holbourn *et al.*, 2005) and the large ice sheet on East Antarctica was reinitiated as maximum ice sheet extent was reached at 14–13.6 Ma (Jamieson, Sugden and Hulton, 2010; Kuhnert, Bickert and Paulsen, 2009; Scher and Martin, 2008; Shevenell, Kennett and Lea, 2004; Tripati, Roberts and Eagle, 2009). Since 13.6 Ma when it reached its full glacial configuration, the ice sheet has remained similar in scale until the present day (Hambrey *et al.*, 2008; Jamieson, Sugden and Hulton, 2010).

Climate conditions warmed slightly at around 5 Ma on the Antarctic Peninsula when volcanic activity increased (especially during glacial events) and sea ice became much reduced, especially off the west coast (Hillenbrand and Ehrman, 2005; Smellie *et al.*, 2008). Throughout the late Miocene and Pliocene the ice sheets have remained dynamic, responding to long-term regional shifts in climate and, for the Antarctic Peninsula Ice Sheet, short-term warming events (e.g. Hepp, Moerz and Gruetzner, 2006; Smellie *et al.*, 2008). Even though the climate was still fluctuating, there is no evidence for prolonged deglaciation during the late Neogene (Hambrey *et al.*, 2008; Smellie *et al.*, 2008).

During the mid-Pliocene (~3 Ma) global temperatures rose again, up to 5 °C higher than those of today, as heat transfer into the Southern Ocean intensified and the equator to pole gradient flattened (Bohaty and Harwood, 1998; Salzmann *et al.*, 2011a). During this time West Antarctica experienced long-lived clusters of relatively short glacials, when grounded ice masses advanced, alternating with longer interglacials that rendered the Antarctic ice-poor (Johnson *et al.*, 2009; Smellie *et al.*, 2008, 2009) rather than ice-free (e.g. Dowsett *et al.*, 1999). During the interglacial periods the relatively unstable West Antarctic Ice Sheet repeatedly collapsed into the Ross Sea embayment[9]. These warm periods were characterised by high surface water productivity, minimal summer sea ice, air temperatures above freezing and summer sea surface temperatures of >5.6 °C at 3.7 Ma falling to 2.5 °C by 3.5 Ma (Naish *et al.*, 2009; Pollard and DeConto, 2009). During the glacials, the Antarctic Peninsula Ice Sheet expanded across the region to the west until only the active volcanic edifices of James Ross and Vega Island emerged, which in time developed their own ice caps (Hambrey *et al.*, 2008).

The East Antarctic Ice Sheet, however, was in general relatively stable with a maximum elevation of ~100 m higher than the present ice surface (Huang *et al.*, 2008). Considerable

---

[9] The instability stems from the fact that it rests on rock below sea level (i.e. it is a marine ice sheet) and is grounded for much of its margin on a reverse slope (i.e. the slope is downwards from the margin to the interior), therefore any decrease in ice thickness above can lead to flotation of the ice sheet (Bentley, 2010).

controversy exists over whether or not East Antarctica experienced extensive deglaciation during the subpolar climate of this warm period. Sea ice cover strongly decreased as the ice sheets became less stable (e.g. Haywood *et al.*, 2002) and the margin is thought to have retreated some 50–450 km inland at the Prydz Bay region (Escutia *et al.*, 2009; Liu *et al.*, 2010; McKay *et al.*, 2009; Theissen *et al.*, 2003), and to within 500 km of the South Pole (Escutia *et al.*, 2009; Francis *et al.*, 2007; Weigelt *et al.*, 2009).

During the late Pliocene, between ~3.1 and ~2.4 Ma, as a result of a fall in pCO_2 to ~150 ppmv and subsequent drop in global temperatures by 1.3 °C (Lunt *et al.*, 2008), conditions became polar once more resulting in the intensification of glacial conditions (Tripati, Roberts and Eagle, 2009). The Antarctic Peninsula region underwent a major cooling, which in turn caused the annual sea ice coverage in the Southern Ocean to expand (Hillenbrand and Ehrmann, 2005) and the Antarctic Peninsula Ice Sheet to expand still further (Hambrey *et al.*, 2008; Figure 9.2).

Even though global climates rapidly cooled towards the end of the Neogene, the climate around Antarctica remained dynamic. During the Pleistocene, Antarctica and the Southern Ocean remained unstable as polar ice sheets underwent a cycle of slow build up to full glacial conditions, followed by a rapid deglaciation and relatively short interglacial periods (Convey *et al.*, 2008; Villa *et al.*, 2008). A brief interruption to this cycle occurred at 1 Ma with a short-lived warming event in the Prydz Bay region, followed by a partial collapse in the West Antarctic Ice Sheet (Scherer *et al.*, 1998). At about 430 kyr before present, conditions became more stable with Antarctica having remained in the glacial phase for approximately 90% of the time (e.g. Jouzel *et al.*, 1993; Masson-Delmotte *et al.*, 2010; Sime *et al.*, 2009).

The amount of interest associated with determining the degree of stability of the ice sheet during such warm periods, when ice sheets are known to grow as well as diminish in size[10], stems from the fact that the ice sheet is the largest store of freshwater, ice-equivalent to a 70 m global sea level rise, and any information that can be gained regarding ice sheet responses will be invaluable to issues we will face as a result of current changing climate (Huang *et al.*, 2008).

Antarctic palaeoecology

The Cenozoic glaciation of Antarctica resulted in the forced extinction of all its woody plant life at some point during the middle to late Neogene (Truswell and Macphail, 2009). This unique extinction event was not particularly rapid because at least 20 shrubs or low trees appear to have survived the expansion of the East Antarctic Ice Sheet down to sea level during the Oligocene. Moreover, possibly seven or eight may have survived in the fellfield communities on the Transantarctic Mountains as recently as the Pliocene. Nevertheless, the deteriorating climate had a major impact on the terrestrial realm, but

---

[10] For example the 'snow-gun' effect resulting from increased oceanic evaporation and continental precipitation (Hillenbrand and Ehrmann, 2005; Huybrechts, 1993; Nelson *et al.*, 2009; Prentice and Matthews, 1991).

what is less certain is whether East Antarctica and West Antarctica were affected in the same manner at the same time. Difficulties in trying to resolve this issue come from the relative lack of well-preserved fossil material described to date from across the continent. Work has focused on a number of sites yielding Oligocene and Neogene material across both East and West Antarctica, and based on these surveys initial interpretations can be made.

The majority of the palaeobotanical records of Antarctic vegetation growing through the late Eocene to Pliocene are located within three distinct geographic regions: the Antarctic Peninsula, the Ross Sea region and one further locality at Prydz Bay on the east side of East Antarctica (Figure 9.2). These records provide evidence for a flora responding to the overall cooling of the climate in both stature and diversity. Along with the falling temperature, the canopy too became lower and more open. By the early Oligocene the vegetation was much reduced and more tundra-like. In spite of the general sparseness and wide distribution of the plant fossil record from this period across East and West Antarctica, it can be assumed that a mosaic of vegetation types occupied various niches in the landscape from sheltered, low altitude valleys, perhaps filled with remnant temperate rainforest, to the more exposed plains of higher altitudes covered with shrub land and tundra (Thorn and DeConto, 2006). Then as Antarctica's climate cooled still further into the Pleistocene, vascular plants were eventually lost from the continent altogether[11], leaving only the rich fossil plant record as a legacy to the vegetation that once exploited a continent dominated by past warmth.

Landscapes of East Antarctica

The ice that covers much of East Antarctica today buries the topography of the Antarctic continent as it was prior to glaciation up to some 34 Ma (Jamieson, Sugden and Hulton, 2010). Beneath the ice – that now lies thousands of metres deep – is not, as was once assumed, a relatively flat landscape interrupted only by the three known mountain ranges, but one covered in mountains, plains, valleys and undulating landscapes with incursions of the sea. Through the implementation of aerogeophysical and seismic surveys, particularly to the mountain regions (e.g. Drewry, 1982; Jamieson and Sugden, 2008), the topography of this hidden world as it was some 35 Ma ago is finally being revealed.

Up until the end of the Eocene when a cool temperate, humid climate prevailed over Antarctica, the interior of East Antarctica was characterised by large river basins forming a low-lying, subdued landscape. Beneath Dome C (Figure 9.2) of the East Antarctic Ice Sheet, the terrain of the Antarctic Plateau is smooth, with a clear network of elongated valleys running through low-lying hills (Rémy and Tabacco, 2000; Siegert, Taylor and Payne, 2005). These rivers would have eroded much of the arid landscape as they

---

[11] The only native vascular plants of Antarctica today, *Deschampsia antarctica* (Poaceae) and *Colobanthus quitensis* (Caryophyllaceae), are more likely to have migrated into Antarctica during the Holocene or late Pleistocene by avian-aided long distance dispersal, rather than being relicts from the Cenozoic (Mosyakin, Bezusko and Mosyakin, 2007).

transported their bedload to the wetter coast, where as much as 7 km of rock was removed as compared with 1 km from inland (Jamieson and Sugden, 2008).

Towards the middle of the East Antarctic continent the Gamburtsev Mountain Range (Figure 9.2) presided over a fluvial landscape. The mountains lie buried beneath the ~1650 to ~3150 m of ice that form Dome A, the summit (at ~4100 m) of the East Antarctic Ice Sheet, but during the earliest Oligocene this range was very similar to that of the European Alps today. The mountains were characterised by very sharp peaks, reaching altitudes of 1500–3500 m (1500 to ~2500 m above the present sea level) and deep valleys (Bo *et al.*, 2009; Young *et al.*, 2011). River action cut more than 1 km into flanking mountains resulting in a branching network of river valleys (Bo *et al.*, 2009; Young *et al.*, 2011).

Separating East from West Antarctica along the western margin of the Victoria Land Basin, the Transantarctic Mountains also provide a window onto this relict world. The pre-glacial Eocene landscape of the Dry Valleys region (Figure 9.2) would have resembled those in semi-arid areas today (e.g. Arizona). Under the seasonally wet and dry climate, slope wash created gently sloping plains and steeper rectilinear slopes with weathered soils (Denton *et al.*, 1993). Rivers carved out long, sinuous valleys that ran towards the forested coast. Across the ranges of the Transantarctic Mountains, rivers continually gouged deeper valleys that dissected the high peaks. Small volcanic cones were also present and ash still lies preserved *in situ* in surficial deposits testifying to the stability of the climate, minimal erosion and virtually no uplift over the past ~15 Ma (Marchant, Denton and Swisher, 1993; Marchant, Denton and Sugden, 1993; Sugden *et al.*, 1995; Wilch, Denton and McIntosh, 1993). From these peaks the rivers flowed towards the sea cutting through the fertile coastal regions, transporting sediment and organic matter from further inland. The rivers deposited their load in fan deltas along the coastal terrestrial and near shore, marine environments (Fitzgerald, 1992; Harwood and Levy, 2000). This coastline was rugged with sandy beaches dissected by estuaries, drowned valleys (rias), embayments and rivers (Webb, 1994). With ice restricted to montane areas, and relatively dry, cool climate prevailing at low altitudes, the coastline would have been home to molluscs, decapods and birds (e.g. the false-toothed bird, *Pseudodontorn*) hunting out the rich phytoplankton assemblages along the seashore (e.g. Holmes, 2000; Jones, 2000; Levy and Harwood, 2000).

As the cool temperate climate conditions deteriorated across the Eocene–Oligocene transition the Neogene climate became subpolar. Nearer the sea, conditions remained relatively mild with summer temperatures still reaching ~10 °C in the earliest Oligocene, but there is still evidence for the presence of temperate glaciers calving into the sea (Ehrmann, Setti and Marinoni, 2005; Passchier and Krissek, 2008). In the mountains, for example the Gamburtsev Mountains, maximum summer temperatures fell to around 3 °C at higher altitudes and snowline elevations would have been lowered. Circumpolar storms brought increased precipitation to the mountains and, coupled with the falling temperatures, snow fall became more frequent and soon alpine glaciation followed. Small mountain top glaciers appeared and occupied the corries causing the development of discrete cliffed basins. Glaciation began in the East Antarctica interior discharging mainly via the Lambert Graben to Prydz Bay (DeConto and Pollard, 2003a). Small highly dynamic ice caps slowly began to form at the highest altitudes of the Gamburtsev Mountains, Transantarctic Mountains and Dronning Maud Land. The ice, accumulating either in the mountains or on ice caps centred over the mountains, would have fed the glaciers occupying the larger valleys. As the glaciers flowed downhill bringing their load of physically weathered glacial debris, valley floors and sidewalls were eroded to their present U-shape, leaving series of hanging valleys. As valley glacier activity intensified over-deepening of valley floors to >400 m resulted in steep trough sides, hanging tributary valleys and corries with steep cliffs (Bo *et al.*, 2009; Jamieson, Sugden and Hulton, 2010; Young *et al.*, 2011). The flat floors of the tributary valleys would have resembled those in the European Alps today, but the valleys would have been wider (Ehrmann, Setti and Marinoni, 2005; Passchier and Krissek, 2008; Pekar and Christie-Blick, 2008). Ice flow moving from the interior towards the Ross Sea would have been restricted by the Transantarctic Mountains until the ice sheets became larger during the middle Oligocene. In these mountains the warm-based glaciers, and subsequent glacial activity, modified the fluvial landscape removing much of the land surface and slowly creating a subglacial landscape (Jamieson and Sugden, 2008; Jamieson, Sugden and Hulton, 2010). In the Dry Valleys local warm-based glaciers further deepened the existing valleys over the ensuing 15 Ma between 34 and 19(?)Ma (Sugden and Denton, 2004).

By the mid-Miocene (~14 Ma), with temperatures declining sharply by 20–25 °C, the Antarctic Ice Sheet at its maximum expansion reached the edge of the continental shelf (Jamieson and Sugden, 2008; Jamieson, Sugden and Hulton, 2010). Only the high peaks (at 4000 m) of the Royal Society Range (Figure 9.2) were left exposed as the environment changed to one of cold desert conditions. The ice sheet with warm-based glaciers that had flowed northeast across the mountain rim became cold-based and frozen to the bed, thus preserving pre-existing deposits at high altitudes (e.g. those of the Sirius Group). After about 13.5 Ma and the onset of true polar, superarid conditions, coupled with tectonic stability and the thinning of the ice sheet, the Cenozoic landscape became exposed to the polar desert (hyperarid) climate resulting in little change to this fragile landscape to this day (Jamieson and Sugden, 2008; Jamieson, Sugden and Hulton, 2010; Sugden and Denton, 2004).

At the coast, warm-based glaciers persisted for longer, so too did the erosive influence. Glacier-abraded U-shaped valleys, which were later drowned to form deep fjords and coastal troughs, were excavated by up to 2800 m (Jamieson, Sugden and Hulton, 2010). After about 13.5 Ma the ice sheet retreated to its present lair, at least in East Antarctica, leaving exposed coastal fjords to become infilled with shallow marine sediments. Since this time the ice sheet has remained essentially intact under the hyperarid polar climate. The stability of the full ice sheet is demonstrated by the remarkably low erosion rates in the Transantarctic Mountains.

Although exciting revelations can be gleaned for the Cenozoic landscape, the fossil record of Antarctica becomes increasingly sparse through the Oligocene as much of Antarctica's Cenozoic geological record is covered by the Antarctic Ice Sheet. As a consequence the Neogene palaeoenvironments of both East and West Antarctica remain enigmatic. Late Paleogene and Neogene sequences across Antarctica are represented by only a few sections including the Eocene La Meseta Formation on Seymour Island on the Antarctic Peninsula (Chapter 8), and those recovered from drillholes, mainly in the Ross Sea (see Bohaty and Harwood, 2000), rendering the latest Eocene–Miocene geological history of Antarctica poorly understood. Therefore great attention has been given to isolated finds and glacial erratics eroded and transported from subglacial basins at times when the ice sheet expanded. These erratics contain important palaeontological and sedimentological information, documenting the palaeoclimate and palaeoenvironment of poorly known periods of Antarctic geologic history (Harwood and Levy, 2000) as it slid deeper into its icehouse state.

Vegetation of East Antarctica

Since much of Antarctica's late Paleogene and Neogene geological record is covered by the Antarctic Ice Sheet, our glimpses into these past environments are by means of mainly palynological assemblages supplemented by isolated macrofossil finds in glacial erratics eroded and transported from subglacial basins at times when the ice sheet expanded. From around the coast, drilling projects such as the Ocean Drilling Project (ODP) and the Cape Roberts Project have yielded palynological assemblages and occasional plant macrofossils (e.g. Hill, 1989; Truswell, 1990) that have contributed greatly to our understanding of the vegetation of East Antarctica during the climatic downturn. The McMurdo Sound provides one of the few sites on the entire continent where the complete palaeontology (vertebrates, invertebrates and plants) of Eocene to Neogene age can be studied (Harwood and Levy, 2000). Although these assemblages are sparse, the fossils recovered have been studied in detail such that even with relatively limited data we are still able to reconstruct aspects of the vegetation through time (Table 9.1) that existed on the East Antarctic craton.

Late Eocene to Oligocene cool temperate vegetation

Through the latter part of the Eocene *Nothofagus*-Podocarpaceae-fern forests had dominated Antarctica, and this association continued to flourish up until the Eocene-Oligocene boundary. The earliest post-Eocene warm period flora and the most southerly (78°25′ S 165°50′ E) Eocene plant assemblage known from East Antarctica comes from moraine deposits in the southern McMurdo Sound region. The McMurdo Sound is bounded on its western shoreline by the Royal Society Range of the Transantarctic Mountains, rising some ~4000 m from the sea level, and Ross Island to the east. The sound comprises the area to the southwest of Ross Island including Minna Bluff, Mount Discovery, Brown Peninsula, White Island and Black Island (Figure 9.4), and extends some 55 km before opening up into the Ross Sea. This region is an important site where the palaeontology of Eocene to Miocene age can be studied (Harwood and Levy, 2000). The fossil material is preserved within erratic boulders that are distributed (non-randomly) throughout the

Table 9.1 *Summary of the Cenozoic vegetation and their modern vegetational and climatic analogues as derived from nearest living analogues (unless clarified by other proxies) described from East Antarctica with data (in addition to those referenced in the text) from [1]Greene (1964) and [2]Puigdefábregas et al. (1999). MAP = mean annual precipitation, MAT = mean annual temperature, MaxAT = maximum annual temperature, MaxST = maximum summer temperature, MinWT = minimum winter temperature, MST = mean summer temperature, MWT = mean winter temperature. See text for further details.*

Stratigraphic age: Lithostratigraphy *Location*	Palaeovegetation	Modern analogue	
		Vegetation composition	Palaeoclimatic parameters
mid-Pliocene: Core 9, DSDP site 274, *Offshore from Cape Adare*	?Tundra community with *Nothofagus* of at least 5 types	↑	↑
?Miocene ?Pliocene: Meyer Desert Formation, Sirius Group *McMurdo Dry Valleys, Transantarctic Mts* *Oliver Bluffs, Dominion Range*	Cushion-forming mosses and vascular plants, including grasses and sedges, cover exposed ground with moss carpets and/or microbial mats on wetter sites; dwarf and prostrate deciduous *Nothofagus* and rare podocarps	(Alpine) tundra with *krumholtz*-type vegetation typical of subarctic environments such as Magellanic submicrothermal moorland	MAT −8 to −12 °C MST <10 °C (~4–5 °C) MinWT −20 °C
Post middle Miocene: Lake sediments *McKelvey Valley, Dry Valley region of the, Olympus Range*	Tundra-like community with shrubby *Nothofagus* and few other dicotyledonous plants at lower elevations or sheltered habitats, otherwise patchy ground cover of mosses, liverworts and rare *Lycopodium* and moss-rich mires		
AND-2A core (above 284 mbsf) *Southern McMurdo Sound*	Mossy-tundra communities return to colonise the coastal plain with mosaic of tundra with shrub-like *Nothofagus* and podocarps on warmer sites	↓	↓

Table 9.1 (*cont.*)

Stratigraphic age: Lithostratigraphy *Location*	Palaeovegetation	Modern analogue	
		Vegetation composition	Palaeoclimatic parameters
middle Miocene: AND-2A core (312–284 mbsf) AND-2B (above 225 mbsf) *Southern McMurdo Sound*	Woodland with trees of increased statue dominated by abundant *Nothofagus* and podocarps with understorey of diverse angiosperms, mosses and liverworts Lakes with *Isoetes*, mats of moss and algae	Megallanic temperate-submicrothermal temperate rainforest vegetation of Tierra del Fuego	MaxST 0–10 °C[+], increased precipitation relative to early Miocene (i.e. MAP 1500–3000 mm) Cooler than modern austral polar-alpine tree limit
early to mid-Miocene: *Pearse Valley in Dry Valley region of the Olympus Range*	Deciduous *Nothofagus*-dominated (several species) tundra vegetation with podocarps and few other angiosperms, mosses and liverworts	↑ Similar to the *Nothofagus dombeyi* dominated forests of Chile (e.g. Puyehue National Park) ↓	↑ Cold temperate, humid, high precipitation with dry periods of <1 month MAP up to 1600 mm MST >14 °C MinWT 0–5 °C MWT 7 °C ↓
Miocene: CRP-1 Site 1165 (above 147 mbsf) *Prydz Bay*	Similar to the late Oligocene; reduced, low-growing sparse tundra with *Nothofagus*, podocarps and mosses vegetation, distinctively different from contemporaneous vegetation in Australia and New Zealand		
Oligocene: CIROS-1 core *McMurdo Sound, Ross Sea region*	Angiosperm-moss-liverwort assemblage forming herb-moss tundra locally with *Nothofagus* and podocarp trees of prostrate to shrubby stature growing in more favourable habitats	Herb-tundra of Magellanic submicrothermal scrubland and moorland vegetation (Subantarctic zone)	MaxAT <10 °C MaxST 0–10 °C MinWT<−15 °C

Table 9.1 (*cont.*)

Stratigraphic age: Lithostratigraphy *Location*	Palaeovegetation	Modern analogue	
		Vegetation composition	Palaeoclimatic parameters
late Oligocene: CRP-2/2A core (above ~300 mbsf) *Off Cape Roberts, McMurdo Sound, Ross Sea region*	Very similar to the early Oligocene low diversity scrubby vegetation with several species of *Nothofagus* and podocarps and possible slight increase in angiosperm diversity; mosaic tundra vegetation at higher altitudes and on more exposed sites; shrubland in more sheltered sites at lower altitude	Magellanic temperate-submicrothermal rainforest and scrubland vegetation of Tierra del Fuego including islands (e.g. Staten Island) and mainland but differ in that herbs dominate rather than woody species; tundra or similar to that occurring in fellfields. *Nothofagus*–podocarp forests and shrubland of Fiordland, South Island, New Zealand	Cool to cold Antarctic zone[1] south of 60° but includes South Sandwich Island and Bouvet Island with MST ~5–7 °C MST Patagonia and Tierra del Fuego 5–12 °C MST *Nothofagus* habitat 5 °C[2] MST Periglacial 7–10 °C MWT 1–2–7 °C MAT ~8 °C MAP >1500 mm
early Oligocene: CRP-2/2A core (below ~300 mbsf) and CRP-3 core (above 410 mbsf *Off Cape Roberts, McMurdo Sound, Ross Sea region*	Low diversity *Nothofagus*-podocarp dominated low shrub or closed forest/ woodland of intermediate stature at lower altitudes or in sheltered regions; understorey of other angiosperms and rare tree ferns; ground storey of sedges, lycopods, mosses and liverworts; higher altitude or cooler regions with herb-moss-tundra and deciduous dwarf *Nothofagus* and podocarps; understorey of ferns, lycopods and mosses in damp habitats		Cold temperate (MWT <−3 °C, MST >10 °C) to periglacial (MST at sea level ~10–12 °C) MST Patagonia 8–12 °C MST *Nothofagus* habitat 5 °C in growing season (short and warm)[2] MinWT −15 to −22 °C Humid, cool, oceanic MAT 4 –12 °C MAP 1000–2000 mm MinWT −20 °C

Table 9.1 (*cont.*)

Stratigraphic age: Lithostratigraphy *Location*	Palaeovegetation	Modern analogue	
		Vegetation composition	Palaeoclimatic parameters
?Oligocene ?Eocene: CIROS-1 core *McMurdo Sound,* *Ross Sea* *region*	Angiosperm-moss-liverwort assemblage forming herb-moss tundra locally with *Nothofagus* and podocarp trees of prostrate to shrubby stature growing in more favourable habitats with other shrubby angiosperms	Herb-tundra of Magellanic submicrothermal scrubland and moorland vegetation (Subantarctic zone)	MaxAT <10 °C MaxST 0–10 °C MinWT<−15 °C
late Eocene – early Oligocene ODP Leg 189 site 1168 *Tasman Sea*	Tall stature *Nothofagus-Podocarpus*-fern forest association with increased amounts of *N.* subgenus *Brassospora* and a lower percentage and diversity of ferns and tree ferns; other angiosperms present (e.g. *Casuarina*) Changes occur in the vegetation composition following the Eocene–Oligocene transition, perhaps climate-driven, with collapse of taller woody ecosystems	Deciduous *Nothofagus*-podocarp forests of New Zealand	Cool temperate MST ~10 °C near the coast becoming cooler inland and with time
late Eocene: ODP Leg 188 Site 1166 (142–240 mbsf) *Prydz Bay*	Floristically rich, temperate rainforest or rainforest scrub of *Nothofagus*-podocarp association with a diverse mix of other angiosperms, including Proteaceae, lycopods, very few ferns, and rare mosses; *N.* subgenus *Fuscospora* dominate	Mosaic of *krumholtz* rainforest trees, scleromorphic shrubland and wetlands of herbs similar in structure to the taiga in Scandinavia, Russia and northern North America and alpine heathlands of Tasmania	Cool-cold temperate MAT <12 °C and freezing winters High humidity MAP ~1200–1500 mm

Table 9.1 (*cont.*)

Stratigraphic age: Lithostratigraphy *Location*	Palaeovegetation	Modern analogue	
		Vegetation composition	Palaeoclimatic parameters
middle late Eocene: Glacial erratic *McMurdo Sound*	Floristically rich, temperate rainforest or rainforest scrub of (deciduous) *Nothofagus*-podocarp-fern vegetation type with diverse angiosperms, *Araucaria* and cryptogams; *N.* subgenus *Fuscospora* dominate	Forests of Araucaría region of Chile (~38° S) where extensive forests of *Nothofagus* occur alongside *Araucaria*	MAT <12 °C MaxST ~20 °C MinWT <4 °C with no long periods below freezing MAP ~1200–2500 mm occurring mainly in winter

coastal moraines along the shores of Mount Discovery, Brown Peninsula and Minna Bluff (south of Black Island), as well as in moraine on Black Island and along the floors of Salmon and Miers Valley (Levy and Harwood, 2000; Figure 9.4). Many of the source beds for the erratics were deposited in fan deltas that formed along the rugged coastline not far from the Transantarctic Mountains, whilst the erratics themselves were eroded and transported either from a broad, deep basin on the edge of Victoria Land to the western side of the Ross Embayment, or from the Transantarctic Mountains themselves.

Early studies assigned an 'Eocene' age to the palynomorph assemblages recovered from the erratics (McIntyre and Wilson, 1966) and these became the focus of a number of early works (Cranwell, 1959; Cranwell, Harrington and Speden, 1960; McIntyre and Wilson, 1966; Wilson, 1967). However, later studies have been able to refine the age to that which spans the middle to late Eocene to Miocene (Levy and Harwood, 2000). Building on these early works, Truswell (1983, 1990) and Askin (2000) were able to compile a more complete picture of the flora. They describe the vegetation as being diverse in both composition and stature. *Nothofagus* dominated along with other angiosperms such as Proteaceae, Casuarinaceae, Ericales (Ericaceae), Euphorbiaceae, Sapindaceae (Cupanieae), Sparagniaceae/Typhaceae, Restoniaceae plus occasional Liliaceae, Gunneraceae, Sterculiaceae, Myrtaceae, Loranthaceae, Pedaliaceae. Conifers were represented by podocarps (including *Podocarpus*, cf. *Libocedrus* and *Dacrydium*), *Araucaria* and other seed plants such as *Ginkgo*. The lower storey comprised cycads and tree ferns, with ferns, bryophytes, lycopods and fungi forming the ground cover. In the macrofossil record wood has also been recorded, testifying to the presence of trees of large stature belonging to *Nothofagus*, *Araucaria* and podocarps (*Phyllocladus/Dacrydium*) (Francis, 2000).

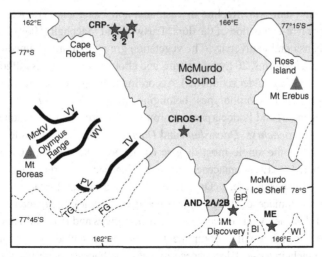

Figure 9.4 Map showing the localities and place names mentioned in this chapter. BP = Brown Peninsula, WI = White Island, BI = Black Island, Dry Valleys comprising three main valleys (black lines) Wright Valley (WV), Taylor Valley (TV) and Victoria Valley (VV) with associated glaciers Taylor Glacier (TG) and Ferrar Glacier (FG). Fossil localities are in McKelvey Valley (McKV), Pearse Valley (PV), and in the McMurdo Erratics (ME) and from the drill cores (stars) CIROS-1, CRP1, CRP2, CRP3, AND-2A and AND-2B.

Leaves of *Araucaria* and *Nothofagus* (two types) have also been described along with other (unidentifiable) dicot leaves and two types of *Nothofagus* fruits (Pole, Hill and Harwood, 2000). One of the fruits is thought to have been derived from an extant temperate *Nothofagus* subgenus, whilst one of the *Nothofagus* leaf types had plicate vernation implying that it was deciduous (Pole, Hill and Harwood, 2000). The presence of these identifiable taxa suggests that the climate was far from extreme. The vegetation was therefore floristically rich and probably resembled the forests that dominate the Araucaría region of Chile (~38° S), where *Araucaria* extends to the tree line alongside *Nothofagus* (Pole, Hill and Harwood, 2000).

The only reasonably well-dated microfloras published from the other side of the East Antarctic continent, which are contemporaneous with the McMurdo Eocene Erratics, come from the sector that faces the Indian Ocean. Sited here is the prominent re-entrant in the coastline of East Antarctica between 66° and 79° E, Prydz Bay (Figure 9.2). The bay is downstream from the Lambert Glacier–Amery Ice Shelf drainage basin which drains some 20% of East Antarctica, near the Gamburtsev Mountains. Sediment was probably brought into Prydz Bay from the inception of continental glaciations and subsequent phases of ice expansion. This offshore region is thus an ideal place to find evidence for the terrestrial vegetation growing within the drainage basin and has provided insights into the vegetation that grew near the ancient shoreline at ~70° S (Macphail and Truswell, 2004a, 2004b; Truswell and Macphail, 2009). It was originally described as floritically impoverished gymnosperm-*Nothofagus* rainforest scrub, analogous to the low-height *Nothofagus* communities found along the treeline in Tasmania and Patagonia (Truswell and Macphail, 2004). However on re-evaluation of the flora, Truswell and Macphail (2009) describe the flora as also one of considerable diversity. The vegetation grew on poorly drained acidic soils and comprised >80 angiosperms, 20 gymnosperms and about 25 cryptogams (although some may have been reworked from older sediments). According to the pollen record, the dominant taxa were *Nothofagus* (five morphotypes belonging to the *Lophozonia, Fuscospora* and *Nothofagus* subgenera) and Podocarpaceae (around six morphotypes including *Podocarpus, Lagarostrobus, Microcachrys, Dacrydium* and *Dacrycarpus*) although Proteaceae (~25 morphotypes) are by far the single most diverse (if not abundant) group. Other angiosperms represented include Escalloniaceae, Caryophyllaceae, Myrtaceae, Euphorbiaceae, Aquifoliaceae, Casuarinaceae, Droseraceae, Epacridaceae-Ericaceae, Sapindaceae, Callitrichaceae, Restoniaceae, along with sedges (Cyperaceae), Sparganiaceae and possible Arecaceae and Liliaceae. Araucariaceae (*Araucaria, Agathis* and *Wollemia)* and Cupressaceae also contributed to the conifer component. Ferns, represented by Hymenophyllaceae, Cyatheaceae, Gleicheniaceae, Blechnaceae and Polypodiaceae, were the least common in terms of relative abundance and diversity. Damp moorland habitats probably supported (surprisingly) rare mosses (*Sphagnum*), lycopods, and sundews (Droseraceae). As in McMurdo Sound a conifer *Nothofagus* (probably deciduous) temperate rainforest, or temperate rainforest scrub, flourished. Truswell and Macphail (2009) liken this diversity to the modern alpine heathland in Tasmania where a mosaic of dwarfed specimens of rainforest tree species, commonly found below the tree line, existed alongside scleromorphic shrubs and wetland herbs. In structure the vegetation was probably similar to the 'taiga' vegetation found in the transition zone between the boreal conifer forests and tundra biomes in Scandinavia, Russia and North America. Analogies can also be drawn with the 'teleocratic phase' of a late Quaternary interglacial–glacial cycle (cf. Birks and Birks, 2004; Truswell and Macphail, 2009). Although the dominant taxa in Antarctica at this time might not have been deciduous (see Chapter 6), the deciduous component would have been able to compete effectively in winter-dark, high-humidity environments. Two of the conifers, *Phyllocladus* and *Microcachrys*, found in this assemblage are today characteristic of freshwater swamps (Truswell and Macphail, 2009). Damp, open habitats are also evidenced by the presence of Droseraceae, which may have thrived on the nitrogen-poor soils along the banks of tidal channels bordering the bay (Truswell and Macphail, 2009). Today modern podocarps and *Nothofagus* are able to survive (and flower) as low shrubs under adverse conditions, and the same was almost certainly true in the Cenozoic. The temperature and rainfall at Prydz Bay would probably have been linked to sea surface temperatures and would have been subpolar (cool-cold temperatures <12 °C) with high rainfall regimes in the range 1200–2500 mm per year. Temperatures fell to below 0 °C during the dark seasons of winter and early spring, as evidenced by ice-rafted clasts found at the site (Macphail and Truswell, 2004a). These forests had probably collapsed as a viable ecosystem by the Eocene-Oligocene boundary (Truswell and Macphail, 2009).

At a familial level, the assemblages at McMurdo Sound and in Prydz Bay are basically similar with rare, poorly diversified crypotogams, common podocarp pollen, abundant and diverse *Nothofagus*, and diverse (if not abundant) Proteaceae. Although the palynological record indicates that ground ferns and possibly tree ferns were present at Prydz Bay, surprisingly they occurred in insufficient numbers to indicate that they were a prominent component of the vegetation at this time (Truswell and Macphail, 2009). Other taxa represented at both sites include Caryophyllaceae, Casuarinaceae (although less common in Prydz Bay), Epacridaceae, Euphorbiaceae, Myrtaceae, Restoniaceae, Sapindaceae (Cupanieae) (Askin, 2000; Truswell and Macphail, 2009). Differences are minor and include the presence of Liliaceae/Agavaceae, Anacardiaceae, Chenopodiaceae and Onagraceae in McMurdo Sound only, whereas Aquifoliaceae, Callitrichaceae, Escalloniaceae only occur at Prydz Bay (Truswell and Macphail, 2009), suggesting that a similar vegetation type may have existed around the palaeocoastline during the late Eocene[12].

Towards the Eocene–Oligocene transition changes, albeit minor, were taking place within the vegetation. Palynological assemblages in cores taken from the Tasman Sea (ODP Leg 189, Site 1168) further to the north show a clear response to Eocene–Oligocene climate change (Grube and Mohr, 2007). Although the vegetation composition remained very similar to that which had dominated Antarctica since the Paleogene, namely the typical *Nothofagus*-podocarp association with an understorey of ferns, minor differences serve to distinguish it from these earlier floras. In earlier floras *Nothofagus* subgenus *Fuscospora* tended to dominate, whereas on West Antarctica *Nothofagus* was represented by equal numbers of *brassii*, *fusca*-a, *fusca*-b and *menziesii*. Yet in the assemblage preserved in the Tasman Sea, *Nothofagus* was dominated by the evergreen *Nothofagus* subgenus *Brassospora* which had arisen before the Eocene-Oligocene boundary at about 34.5 Ma. Today this subgenus is restricted to the more tropical conditions of New Guinea and New Caledonia. Moreover, *Casuarina*, a dryness indicator, is also present along with a lower abundance and diversity of ferns, including the tree ferns Cyatheaceae, Dicksoniaceae, and Gleicheniaceae, Schizaceae and Osmundaceae (Grube and Mohr, 2007). Near to the Eocene-Oligocene boundary the angiosperm diversity, particularly *Nothofagus*, declined (from ~60% to 30%). The abundance of various other angiosperm groups such as Gunneraceae, Proteaceae and Myrtaceae also declined steeply along with the ferns. In addition the abundance of *Nothofagus* subgenus *Lophozonia* (*menziesii* pollen type), which includes evergreen and deciduous species, increased slightly. *Casuarina* also declined following the Eocene-Oligocene boundary. Araucariaceae and the ferns belonging to Dicksoniaceae and Cyatheaceae seem to benefit from this decrease in angiosperms. About 200 kyr after the Eocene-Oligocene transition the Osmundaceae were replaced by Schizaeaceae and Gleicheniaceae (Grube and Mohr, 2007). Although some changes to the vegetation have been recorded during the Eocene-Oligocene transition, these were not sustained. During the earliest Oligocene the vegetation more typical of the latest Eocene seems to have been restored as Casuarinaceae became less prominant, *Nothofagus* with the *fusca* and *menziesii* pollen type increased from about 33.8 Ma at the

[12] The assemblage at Prydz Bay may help to further our understanding of migration routes between Antarctica and Australia during the middle to late Cenozoic. At least two taxa including an extinct clade of Droseraceae, dispersed from Antarctica northwards into Tasmania during the Eocene–Oligocene transition (Truswell and Macphail, 2004; Macphail *et al.*, 1991). This migration was possibly facilitated by the rapid cooling event marking the Eocene-Oligocene boundary in the Southern Ocean. Dispersals from Australia into Antarctica include Chenopodiaceae-Amaranthaceae and Onagraceae (Truswell and Macphail, 2009).

expense of other angiosperm pollen, which declined slightly. However, by 32.9 Ma, it appears that the Araucariaceae increased in abundance, as it did about 1 Ma earlier at 33.7 Ma, along with ferns at the expense of the angiosperms, especially *Nothofagus*. This decline might be the result of fluctuating local conditions, or perhaps it reflects an episode of cold climate as the vegetation underwent its inevitable gradual response to long-term cooling that took place during the Oligocene (Grube and Mohr, 2007).

Oligocene microthermal vegetation

During the early Oligocene the Victoria Land coast is thought to have enjoyed a climate significantly warmer than that of today, but probably still cooler than southern Chile. On land, conditions were relatively wet and cool with an annual rainfall of 1000–2000 mm and a mean summer monthly temperature on the coast of 4–12 °C that dropped to a minimum annual temperature of –20 °C. Sea water temperatures ranged from 5 to 7 °C and winter sea ice may have been likely (Passchier and Krissek, 2008; Prebble *et al.*, 2006). Conditions were such that a diverse vegetation existed around the coastline. Palynological assemblages from the McMurdo Sound have been retrieved from a number of onshore and offshore drilling projects investigating the early history of the East Antarctic Ice Sheet[13].

From the Ross Sea some 10–15 miles north of Cape Roberts, the Oligocene sediment of the CIROS-1 (Figure 9.4) drillhole core was found to be dominated by a great variety of *Nothofagus* pollen, mainly *fusca* but also *brassii* and rare *menziesii* types, indicating that southern beech forests grew locally (as evidenced by the presence of pollen clumps representing whole anthers) along the coast (Mildenhall, 1989). *Nothofagus* subgenus *Fuscospora* is also represented by a leaf with venation patterns most similar to exant *Nothofagus gunnii*, which is today a coloniser of cold-disturbed habitats. *Nothofagus gunnii* is a small to medium-sized tree in alpine rainforest but becomes more stunted (<1 m high) in exposed alpine conditions (Hill, 1989). These forests also comprised minor components of podocarps, Proteaceae and other shrubby angiosperms. Herbs are also uncommon but represented by grasses, *Epilobium* (Onagraceae), Asteraceae, ?Convolvulaceae and Chenopodiaceae, possibly having occupied rocky substrates and gravelly soils. Mildenhall (1989) also concluded that a humid, oceanic, cool temperate climate probably prevailed during the Oligocene. However, a revision of the chronology has extended the age of the lower third of the core back into the early late Eocene (Wilson *et al.*, 1998), and now it is not certain whether the vegetation and climatic assemblage described by Mildenhall (1989) applies to the Oligocene or newly defined Eocene section of the core (Prebble *et al.*, 2006).

It appears that this vegetation type, similar to that of the Magellanic forests of Patagonia in southern South America or the *Nothofagus*-podocarp forest and shrubland of Fiordland New Zealand, continued to exist along the coast of East Antarctica throughout the early Oligocene. This essentially *Nothofagus* dominated vegetation coexisted with podocarps,

---

[13] For example, DSDP Site 270 (Kemp, 1975; Kemp and Barrett, 1975), MSSTS-1 (Truswell, 1986), CIROS-1 (Mildenhall, 1989), and the Cape Roberts Project cores CRP-1, 2 and 3 (Askin and Raine, 2000; Raine and Askin, 2001; Thorn, 2001).

including *Podocarpus, Dacrydium, Microcachrys, Phyllocladus* and *Lagarostrobus*. Angiospermous shrubs include Proteaceae and Myrtaceae along with rare Casuarinaceae, *Epilobium* (Onagraceae), Chenopodiaceae, possible Convolvulaceae and *Coprosma* (Rubiaceae). Rare cyatheaceous ferns, possible Cyperaceae, lycopods, mosses and liverworts formed the understorey (Askin, 2000; Askin and Raine, 2000; Hill, 1989; Kemp, 1975; Mildenhall, 1989; Prebble *et al.*, 2006; Raine and Askin, 2001; Truswell, 1983, 1990). Low-diversity *Nothofagus*-conifer forests dominated the coast terrain, supported by soils characterised by smectite (a clay mineral derived from chemical weathering associated with forest vegetation) (Ehrmann, Setti and Marinoni, 2005; Jamieson and Sugden, 2008). *Nothofagus* diversity is represented by a number of *Nothofagidites* species belonging to both the *Nothofagus fusca* type and *Nothofagus brassii* type (Prebble *et al.*, 2006; Raine and Askin, 2001). The long fossil record of *Nothofagus fusca* suggests that it was initially relatively common in the temperate Eocene, but became increasingly rare throughout the Oligocene and Neogene. Judging from the occurrences in Miocene–Pliocene deposits of the Transantarctic Mountains (see below), it may have been the only *Nothofagus* species to adapt, or perhaps able to survive, the deteriorating periglacial conditions of the Oligocene and ensuing Neogene (Askin, 2000; Hill and Truswell, 1993).

A similar vegetation type is also reported from the pollen and spore assemblage retrieved from the McMurdo Sound (CRP-2/2A and CRP-3; Figure 9.4). According to the Cape Roberts Science Team (1999, 2000, 2001) the diversity is low, but does include several species of *Nothofagus*, podocarps and a few other angiosperm taxa (Campanulaceae and Caryophyllaceae), and a few cryptogams (bryophytes, lycophytes and rare ferns in CRP-3). Many of the important taxa characteristic of the Eocene (such as the Proteaceae and certain cryptogams) are missing. Wetland vegetation was also present, as evidenced by the monocots Cyperaceae and *Phorium*. On exposed, more upland, sites the vegetation would have been stunted, whereas on more favourable sites the vegetation would have been intermediate in both stature and floristic abundance relative to the preceding Eocene and the late Oligocene–Miocene to come.

In summary, along the Ross Sea margin of Antarctica during the early Oligocene, *Nothofagus* forests of relatively tall stature were slowly giving way to a scrub forest supported by soils more characteristic of polar environments (Barrett, 2007; Ehrmann, Setti and Marinoni, 2005; Jamieson and Sugden, 2008; Raine and Askin, 2001). With cooling climates came a decrease in diversity. Analogies can be drawn with the vegetation and climate of the present day Magellanic region or the temperate rainforests in Tierra del Fuego, which grow next to coastal glaciers where low forests give way to *Nothofagus* scrubland above the altitudinal tree line (Macphail and Truswell, 2004a; Prebble *et al.*, 2006; Raine and Askin, 2001).

By the late Oligocene as the climate continued its deterioration towards submicrothermal conditions, the vegetation was originally thought to have become much less diverse and composed of all the elements found in the later (Pliocene) Meyer Desert Formation of the Sirius Group (see below). This led Askin and Raine (2000) to suggest that the vegetation in Victoria Land during this time had become restricted to sparse tundra growing in a periglacial climate. However, recent studies by Prebble *et al.* (2006) found a greater abundance of miospores and a greater species diversity in the upper Oligocene sediments than had been previously recorded, but with fewer forms indicative of the more temperate conditions of the early Oligocene (Prebble *et al.*, 2006). Therefore the contrast between the early and late Oligocene flora on the edge of Victoria Land was not as great as previously thought, and in general this distinctive flora remained largely unchanged throughout the Oligocene.

Although there seemed to be little temporal change in the flora, Prebble *et al.* (2006) suggested that any differences during this time interval probably reflect altitude. They conclude that two vegetation zones may have existed across the coastal flanks and up into the Transantarctic Mountains. Occupying the coast and low altitudes, as well as more sheltered regions of the mountains was the '*Nothofagidites flemingii* Vegetation Zone'. This would have comprised areas of bogs or mires with lycopods, mosses and ferns. Occupying drier areas, scrubby trees of the *Nothofagus*-podocarp assemblage (*Nothofagus* subgenus *Nothofagus* and/or *Fuscospora* — i.e. *fusca* type a) formed a closed canopy woodland of low stature. At higher altitudes or on cooler valley floors this zone would have given way to the low diversity 'Meyer Desert Vegetation Zone'. This upper zone would have been more akin to a herb-moss-tundra with deciduous dwarf trees. These trees would also have been characterised by *Nothofagus*, but this time with deciduous taxa from the *N.* subgenus *Fuscospora* (i.e. *Nothofagidites lachlaniae* pollen) along with podocarps, Caryophyllaceae and an understorey of ferns, lycopods and mosses on bogs and mires. Other rarer yet interesting components include sedges (Cyperaceae) and butter-cups (?Ranunculaceae). This vegetational zone remained essentially unchanged throughout the Oligocene with only a few monocots, including *Phormium* sp. (New Zealand flax), one other (unknown) dicot, and possible liverworts (Marchantiaceae) or lycopods (Selaginellaceae) having been introduced by the late Oligocene, and a possible Stylidiaceae (the trigger plant family with a herb or cushion-plant habit today) in the early Miocene (Askin and Raine, 2000).

Vegetational response to Miocene climatic fluctuations

The glacial expansion at the Oligocene-Miocene boundary, although significant in extent and volume, must have been relatively transient, and neither cold nor extensive enough to extinguish the *Nothofagus* vegetation completely (Askin and Raine, 2000; Roberts *et al.*, 2003; Wilson *et al.*, 2009). This flora persisted across the boundary responding to the slight drop in temperature (Passchier and Krissek, 2008) by increasing the mosaic tundra vegetation at higher altitudes, whilst in more exposed areas the low diversity shrubland became increasingly restricted to more sheltered sites (Prebble *et al.*, 2006). These tundra communities with the important component of herb and moss taxa was thus essentially a continental vegetation modified into a tundra formation (Truswell *et al.*, 2005), and differed from their modern analogues in being dominated by woody species and not herbs (Prebble *et al.*, 2006; Truswell and Macphail, 2009; Truswell *et al.*, 2005). Interestingly, if mineral soils had covered more than half the source area for the pollen, then a better analogue for this 'tundra' would be the modern fellfield *sensu* Polunin (1960) (Truswell and Macphail, 2009). Marine palynomorphs, however, indicate that coastal temperatures did not return to the warmth of the late Oligocene, as the freshwater melt input to coastal regions was much reduced (Barrett, 2007). So if, as Prebble *et al.* (2006) suggest, this vegetation type did survive through the Mi-1 glaciation event largely unchanged, then there seems no obvious reason for it not to have persisted for longer – possibly even until the middle Miocene when ice became significantly more extensive on the Antarctic craton (Flower and Kennett, 1994). Indeed the presence of *Nothofagus* in the lower Miocene microfossil and macrofossil record indicate that some elements of the vegetation, at least, were able to do so (Askin and Raine, 2000; Hill, 1989; Raine, 1998).

This vegetation type having survived the climatic downturn of the late Oligocene and the Mi-1 glaciation event, continued to survive on into the relatively warm early Miocene (Passchier and Krissek, 2008). Within the highlands adjacent to Taylor Glacier at ~1000 m above the floor of Pearse Valley (Figure 9.4) a local (?upland) vegetation with low diversity, more similar to the present day herb-tundra of the modern subantarctic zone (cf. Greene, 1964; Prebble *et al.*, 2006; Raine, 1998), has been preserved. These probable early to mid-Miocene lacustrine sediments, palaeosols and diamictites are interbedded with fossiliferous deposits exposed by wind action (Ashworth *et al.*, 2007). According to the work of Ashworth *et al.* (2007), the sediments record at least five wet-based glacial advances with soil having had time to develop between successive advances. The palynoflora records a vegetation dominated by several taxa of *Nothofagus* (including *Nothofagidites lachlaniae*, *Nothofagidites* cf. *flemingii* and several species of the undifferentiated *Fuscospora* type of *Nothofagidites*). Wood and leaves are also preserved and provide evidence for *Nothofagus* with *Nothofagus beardmorensis* leaves indicating a deciduous habit. Other angiosperms (belonging to *Tricolpites* sp. A of Prebble *et al.*, 2006) were also present. Spores of lycophytes, bryophytes and *Isoetes* complement the moss mats dominated by a semi-aquatic species of *Drepanocladus* (Amblystegiaceae). Remains of insects, including weevils, were also uncovered.

These relatively low diversity assemblages may reflect a contemporaneous, continental vegetation modified into tundra, rich in mosses and liverworts with the woody taxa particularly *Nothofagus* and podocarps, growing in favourable areas (Askin and Raine, 2000; Raine, 1998; Truswell *et al.*, 2005). Indeed the main elements of the flora are similar, with only minor differences, to those recovered from the Miocene strata offshore from Prydz Bay on the eastern side of East Antarctica, indicating that this vegetation was both temporally and geographically widespread. This vegetation type was distinctive in that many of the *Nothofagus* pollen types that occur here in East Antarctica differ from those of Cenozoic Australia and New Zealand (Truswell, 1990). The presence of the *Nothofagus fusca* morphotype, similar to the pollen of *N. dombeyi* today, suggests similarity to modern Chilean forests where *N. dombeyi* dominates, but the Antarctic forests would have been more *Nothofagus* rich in terms of species (Truswell, 1990).

At around 17–15 Ma, Antarctica warmed and entered the Middle Miocene Climatic Optimum, a period of relative warmth and humidity (Franke and Ehrmann, 2010). During this time an abrupt singular major warming event, similar to the Late Paleocene Thermal Maximum, is evidenced at 15.7–15.5 Ma (Warny *et al.*, 2009). This warming is reflected in the plant fossil record, where a massive increase in dinoflagellate cysts, freshwater algae and terrestrial pollen including a proliferation of woody plants, is recorded in the palynomorph assemblages recovered from the southern McMurdo Sound (AND-2A; Warny *et al.*, 2009). During this brief interlude as summer temperatures increased to a peak of ~10 °C or higher (Franke and Ehrmann, 2010), rainfall and meltwater also increased (Lewis *et al.*, 2007). In the relatively humid Dry Valleys region, glaciers warmed, increasing the volume of subglacial lakes and consequently meltwater discharged into the valleys that ran towards McMurdo Sound (Shevenell, Kennett and Lea, 2008; Warny *et al.*, 2009). These waters surged (at up to 15 m per sec) towards the sea, scouring the floors and walls and carving out channels. This resulted in floods reaching the coast and discharging large volumes of freshwater into the Ross Sea (Lewis *et al.*, 2006). Along the coastal areas adjacent to the Ross Sea the increased run-off resulted in ponds and lakes that provided a niche for increasing freshwater algae (Warny *et al.*, 2009). Woodlands were able to proliferate increasing in both number and stature from their prostrate to low shrubby habit to a more tree-like form. Podocarps increased by 16-fold and *Nothofagus* by >60-fold and was still represented by several taxa belonging to the *Fuscospora* type (Ashworth *et al.*, 2007). Moss and liverwort numbers also peaked around this time, along with ground covering plants such as Caryophyllaceae (*Colobanthus*), Stylidiaceae (trigger plants), Droseraceae (sundews), Campanulaceae, Ericales (heaths), Poaceae, Typhaceae and/or Sparganiaceae (bullrushes and burr reeds) (Ashworth *et al.*, 2007; Warny *et al.*, 2009). Shallow lakes or rivers were present, providing habitats for aquatic and semi-aquatic plants such as *Isoetes*, mats of moss species (e.g. Amblystegiaceae) and algae (Warny *et al.*, 2009). Insects such as listroderine weevils were also present (Ashworth *et al.*, 2007). The ecosystems would have established on soils that were able to develop during periods of glacial retreat, but succumbed each time to ensuing glacial advances (Ashworth *et al.*, 2007).

The Middle Miocene Climatic Optimum was immediately followed by an abrupt cooling in mid- to high latitudes (e.g. Shevenell, Kennett and Lea, 2008) heralding a permanent shift in climate. Atmospheric temperatures dropped by 20 °C, glaciers changed from wet- to cold-based as the East Antarctic Ice Sheet expanded (Lewis *et al.*, 2007). During this transition, the woody plants faded from the record and the open landscape returned to tundra characterised by the moss *Coptospora* with shrubby podocarps and *Nothofagus* (Warny *et al.*, 2009).

Middle Miocene lacustrine vegetation of the McMurdo Dry Valleys

A window onto the vegetation response to the post Middle Miocene Climatic Optimum cooling climate on continental Antarctica can be found within the Dry Valleys region of the Transantarctic Mountains. This location comprises *in situ* fossil-bearing strata that occur within several north facing valleys that open into the McKelvey Valley, one of the major ice-free troughs of the McMurdo Dry Valleys, nestled in the western Olympus Range of the Transantarctic Mountains (Lewis *et al.*, 2008) (Figures 9.2 and 9.4). Lacustrine deposits interdigitate with wet-based tills deposited just prior to the climate transition, providing evidence for meltwater forming a number of small lakes behind ridges of recessional moraines. Here well-preserved fossils and ash have been recovered. Radiometric dating of a 3 cm layer of *in situ* ash fall deposits within the deposit sequence of one of the lakes has returned an ^{40}Ar/^{39}Ar date of 14 Ma, which in turn provides an unambiguous age for both the last phase of wet-based alpine glaciations and the tundra ecosystem that colonised the locally deglaciated terrain (Lewis *et al.*, 2008).

The best preserved assemblage comes from within a small (14000 m^2) moraine-dammed basin at 1425 m near Mount Boreas (Figure 9.4) on locally deglaciated terrain. An exquisitely preserved *in situ* fossil lacustrine and terrestrial assemblage have been found in what was once a small alpine lake (Lewis *et al.*, 2008) named 'Palaeolake Boreas' by Williams *et al.* (2008). The fossils include diatoms, palynomorphs, plant material, ostracods and insects which represent one of the last vestiges of a tundra community that inhabited the mountains before the stepped cooling that brought a full polar climate to Antarctica (Lewis *et al.*, 2008; Williams *et al.*, 2008).

The centre of the basin has been infilled with glaciolacustrine silt and sand, followed by diatom- and moss-rich muds and subsequently fluvial sands and debris flows. Within this basin the lake had persisted for thousands of years, against a backdrop of volcanic activity, during which time water levels fell at least once and the floor of the basin became a bryophyte-rich mire (Ashworth *et al.*, 2007; Lewis *et al.*, 2007, 2008). With falling water levels obligate, freshwater organisms continued to thrive in the alpine lake (Lewis *et al.*, 2008). According to Lewis *et al.* (2008) many of the fossils are exquisitely preserved and represent aquatic taxa that inhabited the lake during an early shallow-water and later deep-water phase. The shallow-water assemblage is represented by layers of mosses, benthic diatoms, algae and ostracods (seed shrimps). The mosses are not so much fossilised but rather freeze dried, resembling herbarium specimens that can be rehydrated. The dominant moss can be confidently identified to the extant semi-aquatic species *Drepanocladus longifolius* (Amblystegiaceae), which can be found today on every continent in saturated (but not harsh) environments such as peat bogs and swamps. This is a different species from the one found in the early to mid-Miocene lacustrine sediments of Pearse Valley (see previous section) (Ashworth *et al.*, 2007). In Palaeolake Boreas *Drepanocladus longifolius* formed mats and provided a favourable habitat for the benthic crustaceans that can be found lodged between the stems and leaves. Carapaces of the ostracods, belonging to a single (unknown) species of the Cypridoidea, are preserved and are represented by both adults and larval instars, complete with the soft tissue of the appendages and mouth parts still preserved.

As this permanent, seasonally ice-free and productive lake shifted to deeper water (maximum depth of 8 m), it underwent progressive acidification over an inferred timescale of several millennia. This is reflected in changes in the diverse benthic and planktonic diatom community that evolved to become more similar to sequences found today in Arctic lakes, rather than the existing biota of the Dry Valleys, or elsewhere in the Antarctic region (Lewis *et al.*, 2008; cf. Moorhead and Priscu, 1998).

Pollen and spores and rare macrofossils provide an insight onto the terrestrial flora. No wood, leaves or seeds have been found, which perhaps indicates a sparse vegetation (Ashworth *et al.*, 2007). In the palynoflora *Nothofagidites lachlaniae*, the putative pollen of the deciduous *Nothofagus beardmorensis*, dominates the assemblage and with no evidence of gymnosperms suggests that this was now the only arboreal (*krumholtz* or otherwise) component of the vegetation. Other angiosperms include a species of Caryophyllaceae and two tricolpate pollen types whose natural affinity is unknown. The remainder of the flora comprised liverworts (Marchantiaceae) and species of the mosses including *Coptospora* (?Bartamiaceae), Dicranaceae/Bruchiaceae and the relatively rare club moss *Lycopodium* (Lewis *et al.*, 2008).

The vegetation covering the slopes of the valley is suggested to have been tundra-like, and mostly barren with patches of bryophytes (Lewis *et al.*, 2008). *Nothofagus* would have grown nearby in cold, dry soils, possibly at lower elevations. Midges would have flown over the lake and weevils lived around its edges. This lake ecosystem, coupled with the evidence for the existence of wet-based glaciers, all indicate that the climate of the western Olympus Range was warmer and wetter than that of today (Lewis *et al.*, 2007, 2008). Mean summer temperatures were probably in the region of 5 °C compared with −12 °C today. Interestingly, this flora appears most similar to the Sirius flora of the Meyer Desert Formation of equivocal Miocene or Pliocene age (see next section).

Between 14 and 13.8 Ma the climate cooled abruptly and mean summer temperatures dropped by about 8 °C. The transition to microthermal desert climate rendered the tundra plants and animals extinct at about 13.8 Ma (based on $^{40}Ar/^{39}Ar$ dating of ash falls within the sequence; Lewis *et al.*, 2007) and these conditions seem to have persisted uninterrupted in the Transantarctic Mountains, south of 77° S, until today (Lewis *et al.*, 2007, 2008). However, this scenario appears incompatible with both a marked reduction in the volume of the East Antarctic Ice Sheet inland of the McMurdo Dry Valleys region during the mid-Pliocene, and with the recolonisation of the land surface by tundra biota during the middle Pliocene (Lewis *et al.*, 2008). This once again serves to highlight the enigma of Antarctic biosphere dynamics.

Return of the tundra? The Micoene/Pliocene Sirius Flora from Oliver Bluffs

Considerable controversy exists over whether or not East Antarctica experienced extensive deglaciation during the mid-Pliocene (~3 Ma) when global temperatures are believed to have been a little warmer than today (e.g. Haywood *et al.*, 2002). In the east of East Antarctica the ice sheet is recorded to have retreated some 450 km inland (Liu *et al.*, 2010) and vegetation was able to re-establish itself on exposed surfaces. However, assemblages of late Miocene to Pliocene age are not confidently known from Antarctica as most post-glacial sediments are beset by problems of reworking of microfossils (Truswell *et al.*, 2005). One *in situ* Pliocene pollen and spore assemblage found in erratics inland from Prydz Bay in the Grove Mountains (Figure 9.2) records a vegetation that comprised taxa including ferns, conifers and angio-sperms, including podocarps, *Nothofagus* and Chenopodiaceae that supposedly grew in this area during polar–subpolar conditions (Fang *et al.*, 2005, 2009; Liu *et al.*, 2010). These spores and pollen grains are considered to originate mainly from local inland sources. However, given the overall warm temperate nature of this assemblage[14], and the absence of such taxa from older, coeval or younger assemblages in Antarctica, we suggest that this assemblage is highly contaminated and therefore will not be considered further.

---

[14] Fang *et al.* (2005, 2009) describe the presence of glacial erratics found in moraines derived from a suite of glaciogene strata hidden beneath the Antarctic Ice Sheet in the Lambert glacier drainage system of the Grove Mountains. The erratics contain a microfossil assemblage thought to be of Pliocene age, which are found to contain pollen and spores from taxa including: Lygodiaceae, ?Pteridaceae, Osmundaceae, Polypodiaceae, Parkeriaceae, *Deltoidospora*, Araucariaceae, Taxodiaceae, Podocarpaceae, Pinaceae, Chenopodiaceae, Asteraceae, Gramineae, Oleaceae, *Nothofagus*, Fagaceae, Juglandaceae, Hamamelidaceae, Ulmaceae, *Tilia*, Proteaceae and *Tricolpopollenites*.

Much of the mid-Pliocene deglaciation debate has centred on the strata found on the west side of the Antarctic continent. Reports of fossil wood, leaves and pollen from a site less than 500 km from the South Pole are one of the most important palaeobotanical discoveries to have been made in Antarctica. These discoveries provide the basis for palaeoclimatic interpretations that have become significant in an ongoing debate about the relative stability of the Antarctic ice sheets.

These fossil-bearing terrestrial glacial deposit sediments known as the Sirius Group (formerly the Sirius Formation) have been found in over 40 localities and are scattered along the inland flanks of the central and southern parts of the Transantarctic Mountains (McKelvey *et al.*, 1991). The localities fall within 1300 km reaching as far south as 86° S and include the Dominion Range and the McMurdo Dry Valleys (McKelvey *et al.*, 1991). According to Haywood *et al.* (2009) these deposits are exposed in two typical settings: (1) as thin erosional remnants at high elevation in palaeovalleys or on flat mountain summits, or (2) as thick (>100 m) sequences along the walls of broad trunk valleys, occupied by large outlet glaciers draining the modern East Antarctic Ice Sheet. Although sequences described as belonging to the Sirius Group have been found in a number of localities, age correlations between localities remain uncertain and thus have been variously ascribed to the Miocene, Pliocene or even early Pleistocene (see Ackert and Kurz, 2004; Ashworth and Cantrill, 2004; Marchant *et al.*, 1996 and references cited therein).

Within the upper part of the Sirius Group, in the Meyer Desert (about 130 km^2 of ice-free land situated at the north end of the Dominion Range) and Dominion Range regions, located on the south and north side of the Beardmore Glacier respectively, the topographically lowest and stratigraphically younger part of the Meyer Desert Formation has yielded exceptionally well-preserved assemblages of both plants (Figure 9.5) and animals (Ashworth and Cantrill, 2004 and references therein). The Meyer Desert Formation consists predominantly of non-marine glacial tillites with some thin mudstones, siltstones and sandstones, deposited in cool temperate to subpolar conditions. It crops out at numerous localities along the Beardmore Glacier, including Oliver Bluffs (latitude 85° 07′ S, longitude 166° 35′ E) in the upper valley of the Beardmore Glacier where a number of important discoveries have been made (Ashworth and Cantrill, 2004; Passchier, 2001, 2004; Webb *et al.*, 1996). The occurrence of glaciomarine deposits in the underlying marine Cloudmaker Formation, which outcrops further down the Beardmore Valley and forms the lower part of the Sirius Group in the Beardmore region, indicates local marine inundation and the existence at times of a fjord (the Beardmore Fjord) that penetrated inland as far as Oliver Bluffs (Ashworth and Cantrill, 2004). The deposits were formed in and around ice-proximal fluvial and lacustrine environments close to sea level (as evidenced by agglutinated foraminifera lying immediately below these fossiliferous beds) and represent periods of exposure, soil development and colonisation by plants and animals.

The presence of such an assemblage in the Sirius Group has resulted in two schools of thought. On one hand the stabilists argue that the Sirius Group deposits are relatively old and so the landscape in this region has remained stable at least since the middle Miocene

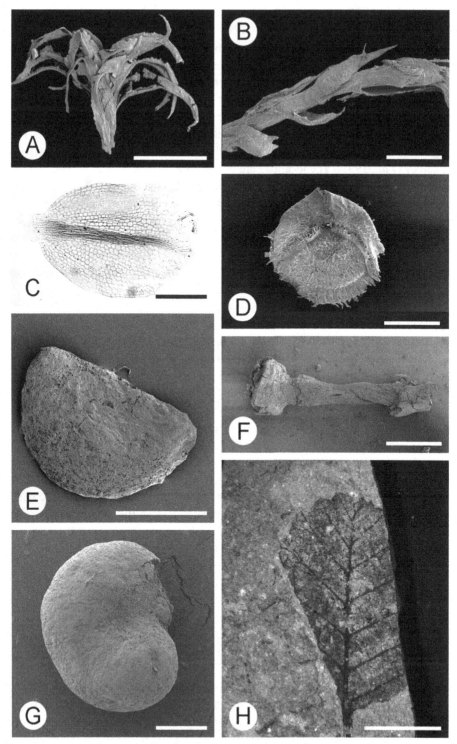

Figure 9.5 Meso- and macrofossils of angiosperms (A, C, E) and cryptogams (B and D) from the
Meyer Desert Formation of the Beardmore Glacier region, Transantarctic Mountains. **A**, overview of
habit of one type of moss with strongly recurved leaves. **B**, habit of axes of a second moss type with

(i.e. 14 Ma), with the subsequent mid-Pliocene warming event having had little impact on Antarctic ice sheets and local landscape. Whilst on the other hand, the dynamists propose that the ice sheet was unstable during the Neogene and did respond to the mid-Pliocene warming phase by undergoing partial deglaciation resulting in a major decrease of up to one-third of the present ice volume (e.g. Hambrey *et al.*, 2003; Haywood *et al.*, 2002; Huang *et al.*, 2008; Webb *et al.*, 1987). Both viewpoints are based on internally consistent evidence, but as yet the conflict concerning the timing of the switch from a dynamic temperate glacial regime to a cold stable one, remains unresolved (Hambrey *et al.*, 2003).

The age of the sediments comprising the Sirius Group remains controversial. The type locality is at Mount Sirius on the south side of the Beardmore Glacier and is the only location where robust evidence supports a Pliocene age (Webb *et al.*, 1996). Unfortunately no fossil material has been found here. At Oliver Bluffs, in the Meyer Desert region, a Pliocene age (estimated to be 3.8 Ma or younger) has been assigned to the Meyer Desert Formation on the basis of marine diatoms recovered from the glacial sediments (Webb *et al.*, 1996). Similar diatom taxa were also recovered from part of the CIROS-2 ice core in the Ross Sea region, interbedded with volcanic ash that provided a date of ~3 Ma (Barrett *et al.*, 1992). However, the age of the formation is intensely debated because of the questionable validity of the diatom evidence, since they may have been wind-deposited (Barrett *et al.*, 1997; Burckle and Potter, 1996), or originated from fallout from a meteorite impact (Gersonde *et al.*, 1997). However, geochemical and provenance data of the diatoms support the original hypothesis of Webb *et al.* (1996) that they were included in clasts that were derived from an inland source in East Antarctica (Passchier, 2004). Further support for a Pliocene age comes from palaeosols found in this formation, which have been assigned a maximum age of between 4.1 and 3.1 Ma (Retallack, Krull and Bockheim, 2001), and modelling studies (Francis *et al.*, 2007). However, surface exposure dating and geomorphology have further complicated the issue by suggesting that key Sirius Group fossiliferous sediments are much older than 3.8 Ma (Ackert and Kurz, 2004). The authors argue that a time period of longer than 3.8 Ma duration is needed to account for the uplift of the beds, which are thought to have been deposited near sea level at the head of the Beardmore Fjord, to their present day elevation of ~1800 m. Thus the beds were either deposited at a higher elevation or are significantly older than 3.8 Ma (i.e. at least 5 Ma), or both (Ackert and Kurz, 2004). Moreover, the available fossil plant data would also suggest an older middle Miocene age, given that only one species of *Nothofagus* pollen has been found here relative to the several taxa found in the early to mid-Miocene assemblage in Pearse Valley and the absence of *Nothofagus* from the lacustrine tundra assemblage dated

---

Caption for Figure 9.5 (cont.)
widely spaced appressed leaves. **C**, small ovate leaf of a third moss type with prominent central midrib and isodiametric cells. **D**, *Isoetes* megapore. **E**, fossil achene. **F**, overview of whole axis terminating in flower. **G**, reniform seed. **H**, *Nothofagus beardmorensis* leaf. Scale bars A, F = 1 mm, B = 200 μm, C, E = 0.5 mm, D, G = 250 μm, H = 1 cm.

as 14 Ma from Palaeolake Boreas in McKelvey Valley (see previous sections). Whatever the exact age of the ecosystem preserved in the Meyer Desert Formation, the sediments and fossil assemblage, especially the plants, indicate a period of relative warmth at some point during the Neogene where the ice sheet retreated to such an extent that the glacial margins were within 500 km of the South Pole.

Research on the Sirius material has focused on fossil assemblages found at localities where sediments have been assigned to the Sirius Group. At Oliver Bluffs on the eastern flank of the upper valley of the Beardmore Glacier, the Meyer Desert Formation has yielded abundant fossil material, including plants (Figure 9.5), insects, molluscs and even one vertebrate, some of these have already been described whilst many more are still under study. The ancestral Beardmore Glacier was probably already occupying the head of the valley, the Beardmore Fjord, at this time. The Beardmore Fjord was perhaps one of many along the coast that ran deep into the Antarctic continent allowing ancestors of existing bird species to extend their long-distance migrations far into the interior of Antarctica, carrying with them plants and invertebrates (Ashworth and Cantrill, 2004). As the glacier retreated, between the glacier margin at the head of the Beardmore Fjord and the edge of the ice sheet (that was probably already in existence), the landscape resembled that of western Greenland today. The surface would have been dominated by low-relief moraines through which meltwater streams would have run forming a wide, braided outwash plain (Ashworth and Cantrill, 2004). On the moraines, gravel bars and ridges between channels of braided streams, plants would have colonised, and in time formed thin peats and marls in abandoned channels along the margins of the plain. During periods of glacier stability, soils would have been able to develop that were very different from the type of soils expected to have formed under the warm conditions of the preceding middle Miocene (Retallack, Krull and Bockheim, 2001). Nor were they similar to either the soils found in this region today (Bockheim, Wilson and Leide, 1986) or to those associated with *Nothofagus* in Tierra del Fuego (Frederiksen, 1988). Instead they are more similar to the weakly developed structures found in landscapes characterised by a permafrost, such as those in Enderby Land, Antarctica (~66–70° S), and Devon Island (~75° N) in the Canadian Arctic (Retallack, Krull and Bockheim, 2001). Within the palaeosoil sequence, Retallack, Krull and Bockheim (2001) recognise two episodes of subdued climatic warming, one at about 5 Ma and the second at 3.5 Ma, even though Dominion Range remained flanked by large glaciers. Between these two intervals and then subsequently, the palaeoclimate dropped back to levels similar to those today. This is not at variance with indications of either relative stability in the Southern Ocean at this time (Burckle and Potter, 1996; Warnke, Marzo and Hodell, 1996), or with the polar dry climate and extensive glaciations experienced in other parts of Antarctica for the last 8 Ma (Bart, 2001; Retallack, Krull and Bockheim, 2001; Sugden *et al.*, 1995).

The palaeosols of the Meyer Desert Formation supported an ecosystem existing at 85° S. Here the landscape would have received no sunlight from the middle of March until the end of September. This low-level of annual solar radiation and absence of solar heating for 5.5 months, even though the ocean would have provided some additional heat, would have resulted in a low mean annual temperature and a short growing season, coupled with low levels of atmospheric moisture content, annual precipitation and nutrient availability (especially nitrogen and phosphorous), as well as slow rates of chemical weathering and a permafrost with restricted drainage and poor aeration. Moreover, strong adiabatic winds originating from the glacier, ice storms and frosts throughout the growing season, summer snow as well as winter snow cover, would have made growing conditions severe (Ashworth and Cantrill, 2004). Fossil assemblages of plants and animals indicate that mean annual temperatures were in the range of –8 to –12 °C, mean summer temperatures had to be at least 4–5 °C for two to three months of the year. This would have allowed a plentiful liquid water supply during the 6–12 week growing season whilst the rest of the year would have been characterised by freezing temperatures dropping to as low as –22 °C (Ashworth and Cantrill, 2004; Ashworth and Kuschel, 2003; Ashworth and Preece, 2003; Ashworth and Thompson, 2003; Francis and Hill, 1996). These values are similar to the estimates of palaeotemperature that resulted in wet-based glaciation and extensive surface melting ablation zones in the western Olympus Range (Lewis *et al.*, 2007). All evidence suggests that during the short period of amelioration, soils were able to develop and support, albeit for the short time whilst glaciers retreated briefly from the Oliver Bluffs region, a tundra-like ecosystem with a woody climax vegetation in an environment considerably warmer that the present glacial conditions.

Any plant life would have had to be adapted to surviving winter desiccation and a limited growing season, where maintaining an uptake of nutrients and a level rate of photosynthesis and respiration similar to those plants in the high Arctic today, was essential (Ashworth and Cantrill, 2004; Chapin and Shaver, 1985). The growing season would have been sandwiched between early summer snow melt and the harsh autumnal frosts. Above ground, vegetative growth would begin after snow melt, but root growth could only resume after the soils thawed and continued beyond the onset of autumnal frosts. For a few weeks each year the plants would have grown rapidly, taking advantage of the 24 hour photoperiod. Yet flowering, reproduction and seed setting probably only occurred in the most favourable summers.

The vegetation and ecology of the landscape at the head of the Beardmore Fjord is outlined in a detailed study by Ashworth and Cantrill (2004). The landscape would have comprised a mosaic of well-drained and poorly drained microsites in which nutrient availability would have been patchily distributed. Of these microsites only a few could have met the requirements that were critical for the survival of these last vestiges of plant life. One such criterion would have been the degree of warmth received from insolation. Sheltered, north-facing valley walls would probably have been favoured as they would have received the greatest amount of insolation and thus would have been slightly warmer. Indeed this is supported by the fossil record. Plant remains, including wood, in Sirius deposits of the Beardmore Glacier region, irrespective of age, are always found on the south side of the glaciers and so would have grown on the warmer northward facing valley walls.

The variety of growth forms found in the Sirius assemblage represents a typical tundra community in which there is selection in the harsh environment for phenological plasticity (Ashworth and Cantrill, 2004). The first colonisers of favourable niches in the newly exposed landscape during periods of glacier quiescence would have been the cushion-forming species of moss, morphologically adapted to the most exposed windswept locations. These mosses would have trapped organic matter and in turn contributed to soil formation (Lewis-Smith, 1994). Cushion-forming vascular plants were also present and these have been found preserved in their growth position in outwash deposits. Their woody tissue formed cushions of about 30 cm in diameter and 16 cm high and probably inhabited wind-exposed sites with well-drained soil in a similar way to the cushion plants in tundra habitats today (Ashworth and Cantrill, 2004; Bliss, 1997).

The fossil peat lenses that can be recognised from Oliver Bluffs are about 1–10 cm in thickness (Ashworth and Cantrill, 2004). Only occasionally are there recognisable moss stems preserved, suggesting that they are either strongly humified moss bank or mire deposits, or that they are microbial mats formed by cyanobacteria and algae. Weevils, similar to those living in Patagonian tundra moorlands today, and a puparium of a fly have also been found in these lenses, along with shell fragments and some achenes (*Ranunculus* type). Within the thin marl deposits, fragments of mosses along with rare achenes of the *Ranunculus* type, freshwater molluscs (shells of the bivalve *Pisidium* and a species of lymnaeid gastropod; Ashworth and Preece, 2003), and a single fish tooth have been identified. Calcareous lake deposits have yielded freshwater gastropods and clams (Ashworth and Kuschel, 2003; Ashworth and Preece, 2003; Ashworth and Thompson, 2003) testifying to the existence of a diverse ecosystem at some point during the Miocene–Pliocene.

Above ground fossilised remains preserved in a metre thick layer of sandstone, siltstone and mudstone can be found sandwiched between tillites. From here insects, ostracods, shell fragments and plant remains have been recovered. The growth forms of dwarf and prostrate trees, cushion and graminoid taxa would also be adapted to the different microtopographic sites that they inhabited (Ashworth and Cantrill, 2004). The plant diversity was low and included at least three types of liverworts, five species of moss, one species of podocarp conifer and at least seven species of vascular plants, including representatives of the grasses (Gramineae), sedges (Cyperaceae), buttercups (Ranunculaceae), mare's tail (Hippuridaceae), ?Caryophyllaceae, and ?Chenopodiaceae/?Myrtaceae (Ashworth and Cantrill, 2004; Askin and Raine, 2000; Hill and Truswell, 1993). This low-level floral diversity is to be expected for a tundra vegetation growing at 85° S. However the most impressive plants in terms of stature would have been the *Nothofagus* and podocarp conifers.

From the small diameter, gnarled and contorted appearance of the fossil *Nothofagus* wood (Carlquist, 1987; Francis and Hill, 1996), the trees appear to have adopted a habit similar to that of the prostrate *Nothofagus* species that grow in the Magellanic regions of southern South America today (Poole and Cantrill, 2006). With diameters of no greater than 1 cm, the asymmetrical nature of the dense growth rings indicate that these plants were slow growing and of great maturity (Francis and Hill, 1996). Axes of these dwarf trees preserved in their growth positions can be seen entwined around cobbles. Abrasion scars on the stem surface provide evidence of having experienced traumatic events, such as the abrasion by wind and outwash flood sediments. Together with the (probably) prostrate podocarps, the *Nothofagus* would have formed a low-lying, crooked and twisted *krumholtz*-type vegetation, typical of subarctic environments today (Poole and Cantrill, 2006), with their prostrate habit providing protection from the freezing winds. These *Nothofagus* trees were deciduous as evidenced by the strong plicate vernation of the thin fossil leaves (Hill, Harwood and Webb, 1996). The leaves occur in thin but dense mats representing a deciduous autumnal leaf fall that accumulated in a shallow pool of water, not long before they became incorporated into the sediment (Hill, Harwood and Webb, 1996; Hill and Truswell, 1993; Webb and Harwood, 1993). The leaf thickness is similar to that of extant deciduous *Nothofagus* species and therefore would not have been robust enough to remain intact had they been reworked from older sediments (Hill and Truswell, 1993). These leaves bear considerable similarity (although are not identical) to extant Tasmanian subalpine to alpine species *Nothofagus gunnii* (*Nothofagus* subgenus *Fuscospora*) and were assigned to the species *Nothofagus beardmorensis* (Hill, Harwood and Webb, 1996). The wood is most closely similar to that of *N. gunnii* (*Nothofagus* subgenus *Fuscospora)* and *N. betuloides* (*N.* subgenus *Nothofagus*) (Carlquist, 1987; Poole, 2002). Only one type of *Nothofagus* pollen, *Nothofagidites lachlaniae* of the *Nothofagus fusca*-type, is represented by the type which encompasses the extant subgenera *Fuscospora* and *Nothofagus* (Askin and Markgraf, 1986; Hill and Truswell, 1993).

Multiple phases of recolonisation of the landscape associated with glacial retreat are recorded in other palaeosoil horizons preserved in the Meyer Desert Formation higher in the Oliver Bluffs stratigraphic section (Retallack, Krull and Bockheim, 2001). Fossil wood of (presumably) *Nothofagus* and moss stems reported from the base of the Meyer Desert Formation about 90 km north of Oliver Bluffs suggest that the tundra vegetation had a much wider distribution than just the Beardmore region (Webb *et al.*, 1996). From reports of wood, woody tissue or moss stems from Sirius Group strata or their stratigraphic equivalents at localities separated by hundreds of kilometres across the Transantarctic Mountains, it is likely that the tundra may have also occurred in coastal locations in East Antarctica and extended inland up other fjords during warmer intervals, possibly until as recently as the late Pliocene (Ashworth and Cantrill, 2004; Hambrey *et al.*, 2003; Marchant *et al.*, 2002; Wilson *et al.*, 2002).

The last vestige of land vegetation (apart from lichens) from East (or West) Antarctica date from the mid-Pliocene (~3 Ma). In a core taken some 250 km offshore from Cape Adare (DSDP Site 274), on the furthest northeast peninsula of Victoria Land (Figure 9.2), a palynoflora dominated by *Nothofagus* has been recovered. The *Nothofagus* component is represented by five *Nothofagidites* morphotypes, but dominated by *Nothofagidites lachlaniae* (the main pollen species reported from the Sirius Group) that occurs as an order of magnitude higher than relative abundances for the other types of *Nothofagidites*. This pollen type is thought to originate from the same trees that produced the leaves *Nothofagus beardmorensis* and is assumed to be the sole remaining *Nothofagus* species of the tundra vegetation (Fleming and Barron, 1996). The occurrence of *Nothofagidites lachlaniae* in these offshore sediments suggests that the pollen may have been derived from trees of *Nothofagus* that still continued to grow in Antarctica during the mid-Pliocene and were then transported to an open ocean setting relatively free of sea ice (Fleming and Barron, 1996; Warny *et al.*, 2006). Since no *in situ* pollen and spores have been found from late Pliocene sediments (e.g. Warny *et al.*, 2006), this suggests that by this time the glacier had advanced along the fjords and destroyed all terrestrial communities from East Antarctica for the last time.

Landscapes and vegetation of West Antarctica

To the west of the Transantarctic Mountains today lie the largely marine-based West Antarctic Ice Sheet and the Antarctic Peninsula Ice Sheet. The Antarctic Peninsula Ice Sheet covers the mountainous Antarctic Peninsula region, a long, narrow, dissected plateau covered by small ice caps. Here valley, outlet and piedmont glaciers with tidewater glaciers and minor fringing ice shelves occur around much of the coast (Troedson and Riding, 2002). Ice cover on the Antarctic Peninsula is fairly insignificant when compared with the main West Antarctic Ice Sheet. These two ice sheets, the Antarctic Peninsula Ice Sheet in particular, both appear more dynamic and sensitive to external changes relative to the East Antarctic Ice Sheet (Bart and Anderson, 2000). Yet our knowledge about the Cenozoic history of the West Antarctic cryosphere from its inception is limited since few mid-Cenozoic records are available from this area and little is known about the inception and history of widespread West Antarctic ice.

Until fairly recently the scheme proposed by Birkenmajer, Soliani and Kawashita (1989) for local glacial stratigraphy for the northern Antarctic Peninsula had been widely referenced in reviews of Antarctic glacial history. This was based on exposures on King George Island and consisted of four mid-Cenozoic glacial episodes in the middle Eocene, middle Oligocene, latest Oligocene and earliest Miocene. But two of these episodes were later found to be untenable. The Eocene 'Krakow' event was based on an erroneous K/Ar age from lava atop Magda Nunatak, an outlier of the middle to late Oligocene Polonez Cove Formation (Birkenmajer *et al.*, 1985, 2005; Dingle and Lavelle, 1998; Troedson and Smellie, 2002). The latest Oligocene 'Legru' glaciation related to a tillite, exposed near Vauréal Peak, composed of local clasts overlying an incised volcanic basement. The proposed age of the tillite is based on the flawed assumption that it is age equivalent with the basement. Therefore, only two glacial episodes can be inferred from King George Island exposures – the middle to late Oligocene (~30–26 Ma) 'Polonez' event, based on the Polonez Cove Formation in the Lions Rump region to the west of Melville Peninsula, and the early Miocene 'Melville' event, identified from the Cape Melville Formation (Dingle and Lavelle, 1998; Troedson and Riding, 2002; Troedson and Smellie, 2002) (Figure 9.3).

A further source of evidence for Eocene glaciations in West Antarctica comes from King George Island in the north of the Peninsula region. A possible middle Eocene glacial event recorded at Ezcurra Inlet could predate, by about 15 Ma, the supposed rapid development of Antarctic Ice Sheets close to the Eocene-Oligocene boundary, as determined from the global deep sea oxygen isotope curve (Zachos *et al.*, 2001a). The Point Thomas Formation in Admiralty Bay preserves an Eocene to Oligocene sequence. At this time, the northern Antarctic Peninsula was still an active volcanic mountain range with lavas from two volcanic episodes having been dated as 45–41 Ma and 45–28 Ma. In between these two lava sequences is a tillite horizon that covers an uneven, scoured surface developed on the lower basalt sequence (Birkenmajer *et al.*, 2005). These valley-type tillites provide the first direct evidence for the appearance of mountain glaciers in the Antarctic Peninsula, suggesting a glacial period occurred some time between 45 and

41 Ma, i.e. during the middle Eocene (Birkenmajer *et al.*, 2005). If these dates are accurate then this would be the oldest record of alpine glaciers in West Antarctica (Barker, Diekmann and Escutia, 2007; Birkenmajer *et al.*, 2005). Data retrieved from Pacific Ocean cores may also indicate early ice sheet development at some point during the Eocene, on the coastal mountains of West Antarctica or the southern Antarctic Peninsula (as well as within East Antarctica), possibly as a result of abundant snowfall (Barker, Diekmann and Escutia, 2007). If this was the case then it is envisaged that against a backdrop of volcanic activity, under the seasonally cool and wet conditions of the middle Eocene, glaciers developed in mountains at altitude.

Evidence for the vegetation growing on the Antarctic Peninsula (see Table 9.2 for summary) during the late middle to late Eocene cooling comes from King George Island. In the late middle Eocene Petrified Forest Creek Member of the Arctowski Cove Formation, in Ezcurra Inlet (Figure 8.5), a pollen-spore assemblage reveals the presence of a once diverse *Nothofagus*-pteridophyte-conifer assemblage. At least nine types of *Nothofagus* (representing the three extant groups *fusca*, *brassii* and *menziesii*) comprised the upper canopy with rare podocarps and *Araucaria* conifers (Askin, 1992; Birkenmajer and Zastawniak, 1989a, 1989b; Cortemiglia, Gastaldo and Terranova, 1981; Torres and Lemoigne, 1988). The understorey was well-developed with a diversity of ferns and tree ferns (including Cyatheaceae, Dennstaedtiaceae, Polypodiaceae, Salviniaceae and Schizaeaceae), with rare Ephedraceae, Gramineae and Rhamnaceae. This vegetation may have resembled the species-rich fern bush communities of the southern oceanic islands (e.g. Auckland Island) today (Askin, 1992; Birkenmajer, 1997; Birkenmajer and Zastawniak, 1989a; Birkenmajer *et al.*, 2005; Mohr, 2001), where a remarkably high proportion (~50%) of the indigenous vascular plants comprise ferns (Manton and Vida, 1968). In the fern bush, evergreen shrubs or low-growing trees (<10 m in height), often with a prostrate habit, form the arboreal vegetation with ferns dominating the understorey (Wace, 1960). These communities, however, have little in common with the temperate *Nothofagus* rainforests of southern South America and New Zealand (Wace, 1960) and so can only be used as a structural, as opposed to a compositional, analogue for the fossil vegetation. The *Nothofagus-Araucaria* communities in the southern Andes may provide a better compositional analogue. Here both taxa are adapted to withstand natural wildfire (e.g. resulting from distal volcanic eruptions or lightning strikes), with *Nothofagus* quickly resprouting from multiple shoots in the canopy whilst *Araucaria* produce seedlings and root suckers (Burns, 1993). Once established, *Araucaria* grows through the sparse *Nothofagus* subcanopy (2–5 m) to form the upper canopy (10–20 m high) that eventually overtops the subcanopy. This suppression of *Nothofagus*, if uninterrupted for more than 150 years, may lead to pure stands of *Araucaria* (Burns, 1993). Since there is no evidence of ecosystems dominated by *Araucaria*, this suggests that either fire was a regular occurrence in the Antarctic ecosystem (since fire favours *Araucaria* expansion; Soares, 1979) and/or the (angiospermous) canopy was too dense to allow regeneration in the absence of disturbance (cf. Kershaw and Wagstaff, 2001). This palaeolandscape would have been inhabited by a rich array of

Table 9.2 *Summary of the Cenozoic vegetation and their modern vegetation and climate analogues described from West Antarctica. MAP = mean annual precipitation, MAT = mean annual temperature, MST = mean summer temperature, MWT = mean winter temperature, MaxST = maximum summer temperature, MaxAT = maximum annual temperature, MinWT = minimum winter temperature. See text for further details.*

Stratigraphic age: Lithostratigraphy *Location*	Palaeovegetation	Modern analogue	
		Vegetation composition	Palaeoclimatic parameters
Miocene: Cape Melville Formation *King George Island*	Impoverished *Nothofagus*-podocarp-fern vegetation with few other angiosperms (e.g. chenopods); *Nothofagus* diversity and abundance much reduced (locally)	Magellanic submicrothermal scrubland and moorland vegetation of Tierra del Fuego	Cold temperate rather than polar conditions MaxAT <10 °C MaxST 0–10 °C MinWT<−15 °C
MELVILLE GLACIATION EVENT			
late Oligocene: Destruction Bay Formation *King George Island*	*Nothofagus*-podocarp forests with reduced fern diversity and abundance	Magellanic temperate-submicrothermal rainforest and scrubland vegetation of Tierra del Fuego including islands (e.g. Staten Island) and mainland but differ in that herbs dominate rather than woody species. Tundra or similar to that occurring in fellfields; also similar to *Nothofagus*-podocarp forests of New Zealand	Ice cover regionally widespread and/or alpine glaciations
POLONEZ GLACIATION EVENT			
Eocene–?earliest Oligocene: La Meseta Fm, Seymour Island Group *Seymour and Cockburn islands*	Full stature *Nothofagus*-conifer-fern forests but differ from those of late Eocene by the decrease in *Nothofagus* diversity and increased conifer, fern and moss diversity; other angiosperms present	Forests of the Araucaría Region of Chile (~38° S) where extensive forests of *Nothofagus* occur alongside *Araucaria*; differ in that little/no *Araucaria* present possibly due to local ecosystem dynamics	MAT 5–8 °C MST <19 °C MWT <8 °C MAP 1000–3000 mm Seasonally cool and wet

Table 9.2 (*cont.*)

Stratigraphic age: Lithostratigraphy *Location*	Palaeovegetation	Modern analogue	
		Vegetation composition	Palaeoclimatic parameters
late Eocene: Point Thomas Formation, (Platt Cliffs and Cytadela floras) *King George Island*	Similarities with the middle late Eocene vegetation continues but forests here are comprised of Podocarpaceae-*Araucaria-Nothofagus* dominated with an understorey of hygrophilous and thermophilous ferns and other dicots; bogs and marsh also present	↑ Structurally similar to the forest and fern bush communities of southern ocean islands (e.g. Auckland Island); compositionally similar to	↑
middle late Eocene: Arctowski Cove Formation (Petrified Forest Creek Flora) *King George Island*	Diverse *Nothofagus*-conifer-fern forests with rare podocarps and *Araucaria*; other dicotyledonous angiosperms and tree ferns present; understory of diverse ferns and rare grasses	the forests of Araucaría Region of Chile (~38° S) where extensive forests of *Nothofagus* occur alongside *Araucaria* ↓	Mild and humid climate MAT 5–8 °C (possibly higher) MAP 1200–3225 mm ↓

vertebrate faunas comprising marsupials and ratite birds (e.g. Reguero, Marenssi and Santillana, 2002; Tambussi *et al.*, 1994).

   This flora survived into the late Eocene, as further evidenced by two floras, the Platt Cliffs and Cytadela floras. These floras originate from the same locality and have been described from tuffaceous intercalations of the Cytadela plant beds in the lower and upper members of the 500 m thick Point Thomas Formation of the Ezcurra Inlet Group (comprising the lower Arctowski Cove Formation and the upper Point Thomas Formation) (Birkenmajer, 2003) that outcrops along the southern coast of Ezcurra Inlet (Chapter 8, Figures 8.3 and 8.5) (Birkenmajer, 1980, 2003). These beds overlie, by some 150 m, basal basaltic lavas that return a radiometric K-Ar age of ~37 Ma (Birkenmajer, 1989; Birkenmajer and Zastawniak 1989a; Birkenmajer *et al.*, 1986). A recent study undertaken by Mozer (2012) focused on the plant remains preserved in shallow lake sediments of the Cytadela outcrop close to the Firlej Cove area. The fine grained, plant bearing sedimentary rocks found in the lower part of the Point Thomas Formation alternate with numerous lava flows (Mozer, 2012). Relatively abundant permineralised and coalified wood, and fragments of leaves evidence a Podocarpaceae-*Araucaria-Nothofagus* forest that grew across the lowlands and extended up the slopes of the stratovolcanoes overshadowing marshes, peat bogs and an understorey comprising hygrophilous and thermophilous ferns (including *Blechnum*) and pinnately-veined (but as yet unidentified) angiosperms (Birkenmajer and Zastawniak, 1989b; Hunt, 2001; Mozer, 2012). This vegetation was originally thought to have grown in a mild and moist prevailing climate with mean annual temperatures of 11–15 °C and precipitation between 1200 and 3225 mm (Birkenmajer and Zastawniak, 1989a; Cortemiglia, Gastaldo and Terranova, 1981; Stuchlik, 1981). However, considering the small leaf size and vegetational composition, estimated mean annual temperatures are more realistically in the region of 5–8 °C and thus slightly cooler than those on Seymour Island to the southeast (see below).

In this active volcanic environment, forests dominated by *Nothofagus* continued to thrive through to the Eocene–Oligocene boundary (~34 Ma) as the climate in the northern Antarctic Peninsula became colder, relatively dry, more prone to both frosts and possible seasonal snow by the early Oligocene (Dingle, Marenssi and Lavelle, 1998). By this time, the second phase of ice growth was underway across East Antarctica, yet for West Antarctica and the Antarctic Peninsula ice growth remained uncertain, with unequivocal direct evidence for glacial activity being sparse (DeConto and Pollard, 2003a, 2003b). Onshore exposures on Seymour Island indicate grounded ice to sea level in the early Oligocene on the Antarctic Peninsula (Barker, Diekmann and Escutia, 2007), whereas sediment from drill cores from the Central Ross Sea suggest onset of glaciations at sea level for West Antarctica during the late early Oligocene, shortly after 26 Ma (Barrett, 1974a, 1974b).

A small but intriguing window into the early stages of the icehouse development in West Antarctica can be found on the other side of the Peninsula in the James Ross Basin. On Seymour Island, at about 65° S, the shallow water sediments of the La Meseta Formation (see Chapter 8) extend up to the latest Eocene (Ivany *et al.*, 2008). Evidence for cooling has been recorded from both sediments and isotope ratio data, with a shift toward substantially more positive $\delta^{18}O$ values that began ~41 Ma (middle Eocene cooling) (Dingle, Marenssi and Lavelle, 1998; Dutton, Lohmann and Zinsmeister, 2002; Ivany *et al.*, 2008). This significant interval of cooling in the Seymour Island section is matched by a decreasing diversity in the flora – including reduced *Nothofagus* (Askin, 1992, 1997). Changes are also reflected in the increased conifer, fern and moss component as well as other aspects of the biota including terrestrial mammals (Reguero, Marenssi and Santillana, 2002; Stilwell and Zinsmeister, 1992). However, even though no late Eocene glacial sediments have been identified, Ivany *et al.* (2006) report an Eocene–Oligocene transition occurrence of glaciations at sea level from directly above the La Meseta Formation. These authors conclude that the ice responsible for the till and ice-rafted pebbles was not simply a localised valley glacier, but an ice sheet extending over much of the Antarctic Peninsula. With ice cover to this degree it is likely that the West Antarctic Ice Sheet was also present suggesting that initial expansion of ice in Antarctica encompassed the entire continent synchronously in the earliest Oligocene, even though the average temperature of the Seymour Island Shelf did not drop below zero until well into the Oligocene (Ivany *et al.*, 2006, 2008).

During the mid-Oligocene to mid-Miocene intermittent episodes of Antarctic ice expansion were superimposed on the long term warming trend (Miller, Wright and Fairbanks, 1991), with the 'Polonez' glacial episode possibly correlating with the Oi-2 glaciation (Birkenmajer, Soliani and Kawashita, 1989; Troedson and Smellie, 2002). The Polonez Cove Formation is the earliest known proven exposure indicating significant cryosphere development at this time (Troedson and Smellie, 2002). It provides evidence for a single episode of locally grounded ice in the shallow water, most likely nearshore, environment during the middle to late Oligocene (~30–26 Ma). Grounded ice may have been more widespread further south in the Weddell Sea region of West Antarctica, but the extent of such ice remains unclear (Troedson and Smellie, 2002). During the middle to late Oligocene, ice cover was widespread regionally and for a period also included significant marine-based ice. Further south in Marie Byrd Land pyroclastic deposits exposed at Mount Petras provide the only other terrestrial evidence for Oligocene ice. These deposits, dated at ~29–27 Ma, show characteristics of volcanic eruptions having taken place beneath the ice and provide evidence of local alpine glaciations rather than significant ice cover (Wilch and McIntosh, 2000). Non-glacial sediments overlying the Polonez Cove Formation suggest that a phase of climate warming followed, resulting in an interglacial period during the late Oligocene.

The latest Oligocene interglacial period (~26–24 Ma) is inferred from associated marine sequences of the Boy Point and Destruction Bay formations (Troedson and Riding, 2002) and provide indications regarding the palaeoenvironment. Volcanoes situated inland were being eroded and the sediment transported a short distance into the shallow marine coastal setting (Birkenmajer, 1984; Troedson and Riding, 2002). The palynological assemblage from the Destruction Bay Formation[15], provides evidence of vegetation dominated by *Nothofagus (fusca* group) alongside podocarps. Ferns (e.g. *Cyathidites* spp) were present but diversity and abundance had dropped relative to the floras of the late Eocene (Mohr, 2001). This typical Neogene vegetation type then succumbed to the ensuing regional ice advance across the continental shelf.

The second recorded glacial episode, the 'Melville' glacial event, took place in the earliest Miocene (23–22 Ma) (Birkenmajer *et al.*, 1985; Troedson and Riding, 2002). The West Antarctic cryosphere was originally thought to have been still confined to mountain glaciers and terrestrial ice caps at the Oligocene–Miocene transition (Anderson and Shipp, 1999), but further studies indicate that this Melville Glacial event may have been more extensive and may correlate with the major but short-lived Mi-1 glaciation (Troedson and Riding, 2002). Troedson and Riding (2002) suggest that the setting was temperate rather than polar, and differed from the modern subpolar regime of the South Shetland Islands in which vegetation is very limited. Ice cover was regional and accompanied by tidewater glaciers and/or fringing ice shelves. Erosive glaciers with rapidly calving ice margins fringed the south Weddell Sea, and ice rafted material from the Weddell Sea northwards before it melted around the South Shetland Islands. The low diversity miospore flora from the earliest Miocene Cape Melville Formation, the youngest flora known from King George Island, suggests an impoverished cold climate hinterland macroflora, comprised of ferns, podocarps, *Nothofagus* and a few other dicots (chenopods), continued to exist (Troedson and Riding, 2002). However, throughout this glacial event the once dominant *Nothofagus* had become increasingly rare and may have even disappeared locally during the early

---

[15] A piece of drift wood has also been recorded from this formation but its age and provenance is uncertain (Birkenmajer, 1984; Birkenmajer and Zastawniak 1989a).

Miocene. This would have been unusual since a diversity of *Nothofagus* species continued to dominate Victoria Land during the lower Miocene (Raine, 1998), and was able to persist until the mid-Pliocene in East Antarctica (Fleming and Barron, 1996). This may provide an indication of the severity of the early Miocene glacial episode. Marine evidence suggests that in West Antarctica significant ice sheet growth occurred during the middle Miocene around 14 Ma, such that by the late Miocene both ice sheets had reached their full glacial configuration (i.e. similar to that of today) (Hambrey *et al.*, 2008).

During the Miocene and into the Pliocene, a 10 Ma record of volcanic activity mainly in subglacial settings has been preserved in sediments of the James Ross Island Volcanic Group (Hambrey *et al.*, 2008; Marenssi, Casadio and Santillana, 2010). A single large stratovolcano, 1.5 km high and ~60 km diameter, together with small satellite centres dominated the landscape. About 50 individual eruptive episodes have been identified recording environmental information between 6.25 Ma and today, but given the length of time between successive eruptions the environmental record preserved is typically of coarse resolution (Hambrey *et al.*, 2008; Smellie *et al.*, 2008). No fossil assemblages (other than a reworked Late Cretaceous–Paleogene palynomorph assemblage; Salzmann *et al.*, 2011b) have been identified even from the warmer interglacial periods at ~6.5–6 Ma, ~5–4 Ma and <0.88 Ma (Smellie *et al.*, 2006).

From this time on, against a backdrop of increased volcanic activity, the ice sheets may have responded to long-term regional shift in climate in a similar fashion to the East Antarctic Ice Sheet, increasing in thickness from the latest Miocene onwards. Yet unlike the record from East Antarctica, the final vestiges of vegetation that once dominated West Antarctica have not been found. These preglacial terrestrial biotas did not necessarily succumb to the ice sheets. Some organisms may have survived repeated glaciations by being isolated in refugia, which were not overridden by successive ice sheets and provided favourable microhabitats suitable for supporting life (e.g. Convey *et al.*, 2008; Johnson *et al.*, 2009). However, by the late Pliocene the Antarctic Peninsula region underwent a major cooling coupled with a decline in biological productivity (due to reduced light availability) in the waters off the west coast of the Antarctic Peninsula and along the East Antarctic continental margin (e.g. Cowan *et al.*, 2008; Kennett and Barker, 1990; Murphy *et al.*, 2002; Warnke *et al.*, 1992). By the early Quaternary with conditions cooler than at present, all life from the continent at the bottom of the world had been long eradicated with ice covering all the evidence of a once green and flourishing land.

Comparisons between East and West Antarctica

During the late Eocene the vegetation across East and West Antarctica was compositionally and structurally similar. Tentative structural analogies include both the fern bush communities of southern ocean islands, and, along with compositional similarities, the forests of the Araucaría Region of Chile, where diverse *Nothofagus*-conifer-fern forests grow in relatively mild, moist climates. Although taxonomic changes are recorded at higher resolution in the

East Antarctic floras, similarities in both composition and decreasing level of diversity occur in both East and West Antarctica across the Eocene-Oligocene boundary and through the Oligocene. It appears, however, that the full stature forest vegetation existed for longer in the Peninsula region. This is not surprising considering its more maritime location and lower latitude situation. By the Miocene the vegetation in the northern Peninsula region consisted of an impoverished shrubby *Nothofagus*-podocarp-fern flora, whilst in East Antarctica the vegetation already comprised a herb-moss tundra with prostrate *Nothofagus* and podocarps. The vegetation responses to deteriorating climate were therefore similar across Antarctica, but decreasing in stature and diversity slightly earlier in East Antarctica relative to West Antarctica. When exactly the woody biota finally became extinct on Antarctica still remains uncertain.

Summary

- Little is known regarding the late Paleogene–Neogene vegetation of Antarctica as fossils have only been found from drill cores in the Ross Sea, an exposure in the Transantarctic Mountains, Prydz Bay and the Antarctic Peninsula region. As the climate cooled following the warmth of the mid-Eocene the vegetation was initially a floristically-rich temperate rainforest or deciduous scrub with diverse angiosperms, including Proteaceae and several species of *Nothofagus*, and often with an abundant fern understorey.

- Although slight variations exist across Antarctica, generally at the Eocene-Oligocene boundary, with the *Nothofagus*-podocarp dominated floras still prevailing, diversity overall had decreased relative to the late Eocene. Changes were probably climate driven as the Eocene–Oligocene transition marked the first appearance of some of the first (alpine) glaciers in Antarctica.

- By the late Oligocene *Nothofagus*-podocarp forests were still present, but now at lower altitudes, and abundance had declined further. The stature of the tree component had also reduced to become more shrubby and open even though more trees of an arboreal stature probably continued to thrive in sheltered valleys and coastal areas. At higher altitudes, in the Transantarctic Mountains, mosaic herb tundra replaced the shrubby-arboreal vegetation of the earlier Eocene. Trees here became lower in stature, even prostrate in exposed environments. The vegetation was of very low diversity and differed greatly from the diverse forested habitats of the preceding Paleogene.

- By the Miocene, tundra was present at lower altitudes as well as refugia in the Transantarctic Mountains. Diversity was much reduced but ecosystems thrived in lake systems during periods of glaciation. The tree component was represented by only one deciduous species of *Nothofagus*, and tundra, or possibly a fellfield type ecosystem, prevailed.

- Diversity and stature continued to decline through the Miocene and Pliocene until the time came when available moisture and temperature were significantly reduced such that not even the tundra vegetation could survive.

• When plant life became extinct from Antarctica requires further investigation. Long after the continent became isolated it was still able to support plant and animal life. Extinction may have followed the Middle Miocene Climatic Optimum or may have been as late as the mid-Pliocene. Nevertheless, the final biotic extinction marks a significant event and supports the argument that it was associated with a major reorganisation of the global climate system (Haywood *et al.*, 2009). The timing is in need of resolution as it continues to be one of the most outstanding problems in Neogene climate history.

References

Ackert Jr, P. R. and Kurz, M. D. (2004). Age and uplift rates of Sirius Group sediments in the Dominion Range, Antarctica, from surface exposure dating and geomorphology. *Global and Planetary Change*, **42**, 207–225.

Anderson, J. B. and Shipp, S. S. (1999). Evolution of the West Antarctic Ice-sheet. In *The West Antarctic Ice-sheet: Behavior and Environment, Antarctic Research Series*, **77**, eds R. B. Alley and R. A. Bindschadler, Washington: American Geophysical Union, pp. 45–57.

Ashworth, A. C. and Cantrill, D. J. (2004). Neogene vegetation of the Meyer Desert Formation (Sirius Group) Transantarctic Mountains, Antarctica. *Palaeogeography, Palaeoclimatology, Palaeoecology*, **213**, 65–82.

Ashworth, A. C. and Kuschel, G. (2003). Fossil weevils (Coleoptera: Curculionidae) from latitude 85° S Antarctica. *Palaeogeography, Palaeoclimatology, Palaeoecology*, **191**, 191–202.

Ashworth, A. C. and Preece, R. C. (2003). The first freshwater molluscs from Antarctica. *Journal of Molluscan Studies*, **69**, 89–92.

Ashworth, A. C. and Thompson, F. C. (2003). Palaeontology: a fly in the biogeographic ointment. *Nature*, **423**, 135–136.

Ashworth, A. C., Lewis, A. R., Marchant, D. R., *et al.* (2007). The Neogene biota of the Transantarctic Mountains. In *Online Proceedings of the 10th International Symposium on Antarctic Earth Sciences*, eds A. K. Cooper and C. R. Raymond and the 10th ISAES editorial team. U. S. Geological Survey Open-File Report 2007–1047, Version 2. Extended Abstract 071, available at http://pubs.usgs.gov/of/2007/1047/ea/of2007-1047ea071.pdf. {Cited 2011}.

Askin, R. A. (1992). Late Cretaceous–early Tertiary Antarctic outcrop evidence for past vegetation and climates. In *The Antarctic Paleoenvironment: A Perspective on Global Change. Part. 1, Antarctic Research Series*, **56**, eds J. P. Kennett and D. A. Warnke Washington: American Geophysical Union, pp. 61–73.

Askin, R. A. (1997). Eocene-?earliest Oligocene terrestrial palynology of Seymour Island, Antarctica. In *The Antarctic Region: Geological Evolution and Processes*, ed. C. A. Ricci. Siena: Terra Antarctica, pp. 993–996.

Askin, R. A. (2000). Spores and pollen from the McMurdo Sound erratics, Antarctica. In *Paleobiology and Paleoenvironments of Eocene Rocks, McMurdo Sound, East Antarctica, Antarctic Research Series*, **76**, eds J. D. Stilwell and R. M. Feldmann. Washington: American Geophysical Union, pp. 161–181.

Askin, R. A. and Markgraf, V. (1986). Palynomorphs from the Sirius Formation, Dominion Range, Antarctica. *Antarctic Journal of the United States*, **21(5)**, 34–35.

Askin, R. A. and Raine, J. I. (2000). Oligocene and Early Miocene terrestrial palynology of the Cape Roberts Drillhole CRP-2/2A, Victoria Land Basin, Antarctica. *Terra Antarctica*, **7**, 493–501.

Barker, P. F. and Thomas, E. (2004). Origin, signature and palaeoclimatic influence of the Antarctic Circumpolar Current. *Earth-Science Reviews*, **66**, 143–162.

Barker, P. F., Kennett, J. P., O'Connell, S., *et al.* (eds) (1988). *Proceedings of the Ocean Drilling Program, Initial Reports Volume 113, Weddell Sea, Antarctica covering Leg 113 of the cruises of the Drilling Vessel* JOIDES Resolution, *Valparaiso, Chile, to East Cove, Falkland Islands, Sites 689–697, 25 December 1986–11 March 1987*. Washington: U. S. Government Publishing Office, pp. 1–774.

Barker, P. F., Kennett, J. P. and Scientific Party (1988). Weddell Sea palaeoceanography: Preliminary results of ODP Leg 113. *Palaeogeography, Palaeoclimatology, Palaeoecology*, **67**, 75–102.

Barker, P. F. Camerlenghi, A., Acton, G. D., *et al.* (eds) (1999). *Proceedings of the Ocean Drilling Program Volume 178, Initial Reports, Antarctic Glacial History and Sea-Level Change Covering Leg 178 of the cruises of the Drilling Vessel* JOIDES Resolution, *Punta Arenas, Chile, to Cape Town, South Africa Sites 1095–11035 February–9 April 1998*, available at http://www-odp.tamu.edu/publications/178_IR/ 178TOC.HTM. {Cited 2011}.

Barker, P. F., Diekmann, B. and Escutia, C. (2007). Onset of Cenozoic Antarctic glaciations. *Deep-Sea Research II*, **54**, 2293–2307.

Barrett, P. J. (1974a). Textural characteristics of Cenozoic preglacial and glacial sediments at Site 270, Ross Sea, Antarctica. In *Initial Reports of the Deep Sea Drilling Project Volume 28 covering Leg 28 of the cruises of the Drilling vessel* Glomar Challenger *Fremantle, Australia to Christchurch, New Zealand December 1972–February 1973*, eds D. E. Hayes, L. A. Frakes, P. J. Barrett, *et al.* Washington: U. S. Governmental Printing Office, pp. 757–767.

Barrett, P. J. (1974b). Characteristics of pebbles from Cenozoic marine glacial sediments in the Ross Sea (Sites 270–274) and the southern Indian Ocean (Site 268). In *Initial Reports of the Deep Sea Drilling Project Volume 28 covering Leg 28 of the cruises of the Drilling vessel* Glomar Challenger *Fremantle, Australia to Christchurch, New Zealand December 1972–February 1973*, eds D. E. Hayes, L. A. Frakes, P. J. Barrett, *et al.* Washington: U. S. Governmental Printing Office, pp. 769–784.

Barrett, P. J. (ed.) (1989). Antarctic Cenozoic history from the CIROS-1 drillhole, McMurdo Sound, Antarctica. *New Zealand Department of Scientific and Industrial Research Bulletin*, **245**, 1–254.

Barrett, P. J. (1999). Antarctic climate history over the last 100 million years: *Terra Antarctica Reports*, **3**, 53–72.

Barrett, P. J. (2007). Cenozoic climate and sea level history from glacimarine strata off the Victoria Land coast, Cape Roberts Project, Antarctica. In *Glacial Sedimentary Processes and Products, International Association of Sedimentologists, Special Publication*, **39**, eds M. J. Hambrey, P. Christoffersen, N. Glasser and B. Hubbard. Oxford: Blackwell Publishing Ltd, pp. 259–287.

Barrett, P. J., Adams, C. J., McIntosh, W. C., Swisher, C. C. and Wilson, G. S. (1992). Geochronological evidence supports Antarctic deglaciation three million years ago. *Nature*, **359**, 816–818.

Barrett, P. J., Bleakley, N. L., Dickinson, W. W., Hannah, M. J. and Harper, M. A. (1997). Distribution of siliceous microfossils on Mount Feather, Antarctica, and the age of the Sirius Group. In *The Antarctic Region: Geological Evolution and Processes*, ed. C. A. Ricci. Siena: Terra Antartica, pp. 763–770.

Barron, J., Larsen, B., Baldauf, J., *et al.* (eds) (1989). *Proceedings of the Ocean Drilling Programme Volume 119 Initial Reports Kerguelen Plateau-Prydz Bay Covering Leg 119 of the cruises of the Drilling Vessel* JOIDES Resolution, *Port Louis, Mauritius, to Fremantle, Australia, Sites 736–746, 14 December 1987–20 February 1988.* Washington: U. S. Government Printing Office, pp. 1–942.

Bart, P. J., (2001). Did the Antarctic ice sheets expand during the early Pliocene? *Geology,* **29**, 67–70.

Bart, P. J. and Anderson, J. B. (2000). Relative temporal stability of the Antarctic ice sheets during the late Neogene based on the minimum frequency of outer shelf grounding events. *Earth and Planetary Science Letters,* **182**, 259–272.

Bart, P. J., Anderson, J. B., Trincardi, F. and Shipp, S. S. (2000). Seismic data from the Northern Basin, Ross Sea, record extreme expansions of the East Antarctic Ice Sheet during the Late Neogene. *Marine Geology,* **166**, 31–50.

Bentley, M. J. (2010). The Antarctic palaeo record and its role in improving predictions of future Antarctic Ice Sheet change. *Journal of Quaternary Science,* **25**, 5–18.

Birkenmajer, A. (1980). Tertiary volcanic–sedimentary succession at Admiralty Bay, King George Island (South Shetland Islands, Antarctica). *Studia Geologica Polonica,* **64**, 8–65.

Birkenmajer, K. (1984). Geology of the Cape Melville area of King George Island (South Shetland Islands, Antarctica): pre-Pliocene glaciomarine deposits and their substratum. *Studia Geologica Polonica,* **79**, 7–36.

Birkenmajer, K. (1989). A guide to Tertiary geochronology of King George Island, West Antarctica. *Polish Polar Research,* **10**, 555–579.

Birkenmajer, K. (1997). Tertiary glacial/interglacial palaeoenvironments and sea-level changes King George Island, West Antarctica: an overview. *Bulletin of the Polish Academy of Sciences, Earth Sciences,* **44**, 157–181.

Birkenmajer, K. (2003). Admiralty Bay, King George Island (South Shetland Islands, West Antarctica): a geological monograph. *Studia Geologica Polonica,* **120**, 5–73.

Birkenmajer K. and Zastawniak, E. (1989a). Late Cretaceous-early Tertiary floras of King George Island, West Antarctica: their stratigraphic distribution and palaeoclimatic significance. In *Origins and Evolution of Antarctic Biota, Geological Society of London Special Publication,* **47**, ed. J. A. Crame. London: The Geological Society, pp. 227–240.

Birkenmajer, K. and Zastawniak, E. (1989b). Late Cretaceous-early Neogene vegetation history of the Antarctic Peninsula sector. Gondwana break-up and Tertiary glaciations. *Bulletin of the Polish Academy of Sciences, Earth Sciences,* **37**, 63–88.

Birkenmajer, K., Gaździcki, A., Kreuzer, H. and Müller, P. (1985). K–Ar dating of the Melville Glaciation (Early Miocene) in West Antarctica. *Bulletin of the Polish Academy of Sciences, Earth Sciences,* **33**, 15–23.

Birkenmajer, K., Delitala, M. C., Narębski, W., Nicoletti, M. and Petrucciani, C. (1986). Geochronology of Tertiary island-arc volcanic and glacigenic deposits, King George Island, South Shetland Islands (West Antarctica). *Bulletin of the Polish Academy of Sciences, Earth Sciences,* **34**, 257–273.

Birkenmajer, K., Soliani, E. and Kawashita, K. (1989). Geochronology of Tertiary glaciations on King George Island, West Antarctica. *Bulletin of the Polish Academy of Sciences, Earth Sciences,* **37**, 27–48.

Birkenmajer, K., Gaździcki, A., Krajewski, K. P., *et al.* (2005). First Cenozoic glaciers in West Antarctica. *Polish Polar Research,* **26**, 3–12.

Birks, H. J. B. and Birks, H. H. (2004). The rise and fall of forests. *Science,* **305**, 484–485.

Bliss, L. C. (1997). Arctic ecosystems of North America. In *Polar and Alpine Tundra, Ecosystems of the World. Volume 3*, ed. F. E. Wielgolaski. Amsterdam: Elsevier, pp. 551–683.

Bo, S., Siegert, M. J., Mudd, S., *et al.* (2009). The Gamburtsev Mountains and the origin and early evolution of the Antarctic Ice Sheet. *Nature*, **459**, 690–93.

Bockheim, J. G., Wilson, S. C. and Leide, J. E. (1986). Soil development in the Beardmore Glacier region, Antarctica. *Antarctic Journal of the United States*, **21(5)**, 93–5.

Bohaty, S. M. and Harwood, D. M. (1998). Southern Ocean Pliocene paleotemperature variation from high-resolution silicoflagellate biostratigraphy. *Marine Micropaleontology*, **33**, 241–272.

Bohaty, S. M. and Harwood, D. M. (2000). Ebridian and silicoflagellate biostratigraphy from Eocene McMurdo erratic and the Southern Ocean. In *Paleobiology and Paleoenvironments of Eocene Rocks, McMurdo Sound, East Antarctica, Antarctic Research Series*, **72**, eds J. D. Stilwell and R. M. Feldmann. Washington: American Geophysical Union, pp. 99–159.

Bonan, G. B., Pollard, D. and Thompson, S. L. (1992). Effects of boreal forest vegetation on global climate. *Nature*, **359**, 716–718.

Brown, B., Gaina, C. and Müller. R. D. (2006). Circum-Antarctic palaeobathymetry: illustrated examples from Cenozoic to recent times. *Palaeogeography, Palaeoclimatology, Palaeoecology*, **231**, 158–168.

Burckle, L. H. and Potter Jr, N. (1996). Pliocene–Pleistocene diatoms in Paleozoic and Mesozoic sedimentary and igneous rocks from Antarctica: a Sirius problem solved. *Geology*, **24**, 235–238.

Burns, B. R. (1993). Fire-induced dynamics of *Araucaria araucana-Nothofagus antarctica* forest in the southern Andes. *Journal of Biogeography*, **20**, 669–685.

Cape Roberts Science Team (1999). Initial report on CRP-2/2A. 5 Palaeontology. *Terra Antarctica*, **6**, 107–144.

Cape Roberts Science Team (2000). Initial report on CRP-3. 5 Palynology. *Terra Antarctica*, **7**, 133–170.

Cape Roberts Science Team (2001). Studies from the Cape Roberts Project, Ross Sea, Antarctica: Scientific Report of CRP-3, Part II. *Terra Antarctica*, **8**, 311–620.

Carlquist, S. (1987). Pliocene *Nothofagus* wood from the Transantarctic Mountains. *Aliso* **11**, 571–583.

Chapin III, F. S. and Shaver, G. R. (1985). Arctic. In *Physiological Ecology of North American Plant Communities*, eds B. F. Chabot and H. A. Mooney. London: Chapman and Hall, pp. 16–40.

Convey, P., Gibson, J. A. E., Hillenbrand, C. D., *et al.* (2008). Antarctic terrestrial life – challenging the history of the frozen continent? *Biological Reviews*, **83**, 103–117.

Cortemiglia, G., Gastaldo, P. and Terranova, R. (1981). Studio di piante fossili trovate nella King George Island delle Isola Shetland del Sud (Antatide). *Atti Società Italiana Scienze Naturali e del Museo Civico di Storia Naturale di Milano*, **122**, 37–61.

Cowan, E. A., Hillenbrand, C. D., Hassler, L. E. and Ake, M. T. (2008). Coarse-grained terrigenous sediment deposition on continental rise drifts: a record of Plio-Pleistocene glaciation on the Antarctic Peninsula. *Palaeogeography, Palaeoclimatology, Palaeoecology*, **265**, 275–91.

Coxall, H. K., Wilson, P. A., Pälike, H., Lear, C. H. and Backman, J. (2005). Rapid stepwise onset of Antarctic glaciations and deeper calcite compensation in the Pacific Ocean. *Nature*, **433**, 53–57.

Cranwell, L. M. (1959). Fossil pollen from Seymour Island, Antarctica. *Nature*, **184**, 1782–1785.

Cranwell, L. M., Harrington, H. J. and Speden, I. G. (1960). Lower Tertiary microfossils from McMurdo Sound, Antarctica. *Nature*, **186**, 700–702.

Cunningham, W. D., Dalziel, I. W. D., Lee, T. Y. and Lawver, L. A. (1995). Southernmost South America-Antarctic Peninsula relative plate motions since 84 Ma: implications for the tectonic evolution of the Scotia Arc region. *Journal of Geophysical Research*, **100**, 8257–8266.

De Santis, L., Prato, S., Brancolini, G., Lovo, M. and Torelli, L. (1999). The eastern Ross Sea continental shelf during the Cenozoic: implications for the West Antarctic Ice Sheet development. In *Lithosphere Dynamics and Environmental Change of the Cenozoic West Antarctic Rift System, Global Planetary Change*, **23** (Special Issue), eds F. M. Van der Wateren, and S. A. P. L. Cloetingh. Amsterdam: Elsevier, pp. 173–196.

DeConto, R. M. and Pollard, D. (2003a). A coupled climate–ice sheet modelling approach to the early Cenozoic history of the Antarctic ice sheet. *Palaeogeography, Palaeoclimatology, Palaeoecology*, **198**, 39–52.

DeConto, R. M. and Pollard, D. (2003b). Rapid Cenozoic glaciation of Antarctica induced by declining atmospheric CO_2. *Nature*, **421**, 245–249.

DeConto, R. M., Pollard, D., Wilson, P. A., Pälike, H., Lear, C. H. and Pagani, M. (2008). Thresholds for Cenozoic bipolar glaciations. *Nature*, **455**, 625–656.

Denton, G. H., Prentice, M. L., Kellogg, D. E. and Kellogg, T. B. (1984). Late Tertiary history of the Antarctic ice sheet: evidence from the Dry Valleys. *Geology*, **12**, 263–267.

Denton, G. H., Sugden, D. E., Marchant, D. R., Hill, B. L. and Wilch, T. L. (1993). East Antarctic Ice Sheet sensitivity to Pliocene climatic change from a Dry Valley perspective. *Geografiska Annaler*, **75A**, 155–204.

Dingle, R. V. and Lavelle, M. (1998). Late Cretaceous-Cenozoic climatic variations of the Northern Antarctic Peninsula: new geochemical evidence and review. *Palaeogeography, Palaeoclimatology, Palaeoecology*, **141**, 215–232.

Dingle, R. V., Marenssi, S. A. and Lavelle, M. (1998). High latitude Eocene climatic deterioration: evidence from the northern Antarctic Peninsula. *Journal of South American Earth Sciences*, **11**, 571–579.

Dowsett, H. J., Barron, J. A., Poore, R. Z., *et al.* (1999). Middle Pliocene paleoenvironmental reconstruction: PRISM2. *U. S. Geological Survey Open File Report*, 99–535.

Drewry, D. J. (1982). Ice flow, bedrock and geothermal studies from radioecho sounding inland of McMurdo Sound, Antarctica. In *Antarctic Geoscience*, ed. C. Craddock. Madison: University of Wisconsin Press, pp. 977–983.

Driscoll, N. W. and Haug, G. H. (1998). A short circuit in thermohaline circulation: a cause for Northern Hemisphere glaciations? *Science*, **282**, 436–438.

Dutton, J. F. and Barron, E. J. (1996). GENESIS sensitivity to changes in past vegetation. *Paleoclimates*, **1**, 325–354.

Dutton, A. L., Lohmann, K. C. and Zinsmeister, W. J. (2002). Stable isotope and minor element proxies for Eocene climate of Seymour Island, Antarctica. *Paleoceanography*, **17**, 1–14.

Eagles, G., Livermore, R. A., Fairhead, J. D. and Morris, P. (2005). Tectonic evolution of the west Scotia Sea. *Journal of Geophysical Research*, **110** B02401, doi:10.1029/JB2004003154.

Eagles, G., Livermore, R. and Morris, P. (2006). Small basins in the Scotia Sea: The Eocene Drake Passage gateway. *Earth and Planetary Science Letters*, **242**, 343–353.

Edgar, K. M., Wilson, P. A., Sexton, P. F. and Suganuma, Y. (2007). No extreme bipolar glaciations during the main Eocene calcite compensation shift. *Nature*, **448**, 908–911.

Ehrmann, W. U. and Mackensen, A. (1992). Sedimentological evidence for the formation of an East Antarctic ice sheet in Eocene/Oligocene time. *Palaeogeography, Palaeoclimatology, Palaeoecology*, **93**, 85–112.

Ehrmann, W. U., Setti, M. and Marinoni, L. (2005). Clay minerals in Cenozoic sediments off Cape Roberts (McMurdo Sound, Antarctica) reveal palaeoclimatic history. *Palaeogeography, Palaeoclimatology, Palaeoecology*, **229**, 187–211.

Eldrett, J. S., Harding, I. C., Wilson, P. A., Butler, E. and Roberts, A. P. (2007). Continental ice in Greenland during the Eocene and Oligocene. *Nature*, **466**, 176–179.

Escutia, C., Bárcena, M. A., Lucchi, M. A., *et al.* (2009). Circum-Antarctic warming events between 4 and 3.5 Ma recorded in marine sediments from the Prydz Bay (ODP Leg 188) and the Antarctic Peninsula (ODP Leg 178) margins. *Global and Planetary Change*, **69**, 170–184.

Exon, N. F., Kennett, J. P. and Malone, (2004). Leg 189 synthesis: Cretaceous–Holocene history of the Tasmanian Gateway. In *Proceedings of the Ocean Drilling Program Volume 189 Scientific Results The Tasmanian Gateway: Cenozoic Climate and Oceanographic Development, Covering Leg 189 of the cruises of the Drilling Vessel JOIDES Resolution, Hobart, Tasmania, to Sydney, Australia Sites 1168–1172 11 March–6 May 2000*, eds N. F. Exon, J. P. Kennett, M. J. Malone, *et al.* Washington: U. S. Government Printing Offices, pp. 1–37.

Fang, A. M., Liu, X. H., Wang, W. M., *et al.* (2005). Preliminary study on the spore-pollen assemblage found in Cenozoic sedimentary rocks in Grove Mountains East Antarctica and its climatic implications. *Chinese Journal of Polar Research*, **16**, 23–32.

Fang, A. M., Liu, X. H., Wang, W. M., Huang, F. and Yu, L. (2009). Cenozoic terrestrial palynological assemblages in the glacial erratics from the Grove Mountains, east Antarctica. *Progress in Natural Science*, **19**, 851–859.

Fitzgerald, P. G. (1992). The Transantarctic Mountains of Southern Victoria Land: the application of apatite fission track analysis to the rift shoulder uplift. *Tectonics*, **11**, 634–662.

Fleming, R. F. and Barron, J. A. (1996). Evidence of Pliocene *Nothofagus* in Antarctica from Pliocene marine sedimentary deposits (Deep Sea Drilling Project site 274). *Marine Micropaleontology*, **27**, 227–236.

Flower, B. P. and Kennett, J. P. (1994). The Middle Miocene climatic transition: East Antarctic ice sheet development, deep ocean circulation and global carbon cycling. *Palaeogeography, Palaeoclimatology, Palaeoecology*, **108**, 537–555.

France-Lanord, C. and Derry, L. A. (1997). Organic carbon burial forcing of the carbon cycle from Himalayan erosion. *Nature*, **390**, 65–67.

Francis, J. E. (2000). Fossil wood from the Eocene high latitude forests, McMurdo Sound, Antarctica. In *Paleobiology and Paleoenvironments of Eocene Rocks, McMurdo Sound, East Antarctica, Antarctic Research Series*, **76**, eds J. D. Stilwell and R. M. Feldmann. Washington: American Geophysical Union, pp. 253–260.

Francis, J. E. and Hill, R. S. (1996). Fossil plants from the Pliocene Sirius Group, Transantarctic Mountains: evidence for climate from growth rings and fossil leaves. *Palaios*, **11**, 389–396.

Francis, J. E., Haywood, A. M., Ashworth A. C. and Valdes, P. J. (2007). Tundra environments in the Neogene Sirius Group, Antarctica: evidence from the geological record and coupled atmosphere-vegetation models. *Journal of the Geological Society of London*, **164**, 317–322.

Franke, D. and Ehrmann, W. (2010). Neogene clay mineral assemblages in the AND-2A drill core (McMurdo Sound, Antarctica) and their implications from environmental change. *Palaeogeography, Palaeoclimatology, Palaeoecology*, **286**, 55–65.

Frederiksen, P. (1988). *Soils of Tierra del Fuego: A Satellite-based Land Survey Approach*. Copenhagen: C.A. Reitzels.

Gersonde, R., Kyte, F. T., Bleil, V., *et al.* (1997). Geological record and reconstruction of the late Pliocene impact of the Eltanin asteroid in the Southern Ocean. *Nature*, **390**, 357–363.

Ghiglione, M. C., Yagupsky, D., Ghidella, M. and Ramos, V. A. (2008). Continental stretching preceding the opening of the Drake Passage: evidence from Tierra del Fuego. *Geology*, **36**, 643–646.

Gibbs, M. T., Bluth, G. J. S., Fawcett, P. J. and Kump, L. R. (1999). Global chemical erosion over the last 250 My: Variations due to changes in paleogeography, paleoclimate, and paleogeology. *American Journal of Science*, **299**, 611–651.

Greene, S. W. (1964). Plants of the land. In *Antarctic Research*, eds R. Priestley, R. J. Adie and G. de Q. Robin. London: Butterworths, pp. 240–253.

Grube, R. and Mohr, B. (2007). Deterioration and/or cyclicity? The development of vegetation and climate during the Eocene and Oligocene in Antarctica. In *Online Proceedings of the 10th International Symposium on Antarctic Earth Sciences*, eds A. K. Cooper, C. R. Raymond and the 10th ISAES Editorial team. U. S. Geological Survey Open-File Report 2007–1047, Version 2. Extended Abstract 075, available at http://pubs.usgs.gov/of/2007/1047/ea/of2007–1047ea075.pdf. {Cited 2011}.

Hambrey, M. J., Ehrmann, W. U. and Larsen, B. (1991). Cenozoic glacial record of the Prydz Bay continental shelf East Antarctica. In *Proceedings of the Ocean Drilling Program, Volume 119 Scientific Results Kerguelen Plateau-Prydz Bay covering Leg 119 of the cruises of the Drilling Vessel* JOIDES Resolution, *Port Louis, Mauritius to Fremantle, Australia, Sites 736–746 14 December 1987–20 February 1988*, eds J. Barron, B. Larsen, J. G. Baldauf, *et al.* Washington: U. S. Government Printing Office, pp. 77–132.

Hambrey, M. J., Webb, P. N., Harwood, D. M. and Krissek, L. A. (2003). Neogene glacial record from the Sirius Group of the Shackleton Glacier region, central Transantarctic Mountains, Antarctica. *Geological Society of America Bulletin*, **115**, 994–1015.

Hambrey, M. J., Smellie, J. L., Nelson, A. E. and Johnson, J. S. (2008). Late Cenozoic glacier-volcano interaction on James Ross Island and adjacent areas, Antarctic Peninsula region. *Geological Society of America Bulletin*, **120**, 709–731.

Haq, B. U., Hardenbol, J. and Vail, P. R. (1987). Chronology of fluctuating sea levels since the Triassic. *Science*, **235**, 1156–1166.

Harwood, D. M. and Levy, R. H. (2000). The McMurdo erratic: introduction and overview. In *Paleobiology and Paleoenvironments of Eocene Rocks, McMurdo Sound, East Antarctica. Antarctic Research Series, 76*, eds J. D. Stilwell and R. M. Feldmann. Washington: American Geophysical Union, pp. 1–18.

Haywood, A. M., Valdes, P. J., Sellwood, B. W. and Kaplan, J. O. (2002). Antarctic climate during the middle Pliocene: a model sensitive to ice sheet variation. *Palaeogeography, Palaeoclimatology, Palaeoecology*, **182**, 93–115.

Haywood, A. M., Smellie, J. L., Ashworth, A. C., *et al.* (2009). Middle Miocene to Pliocene History of Antarctica and the Southern Ocean. In *Developments in Earth and Environmental Sciences*, **8**, *Antarctic Climate Evolution*, eds F. Florindo and M. Siegert. Amsterdam: Elsevier, pp. 401–463.

Hepp, D. A., Moerz, T. and Gruetzner, J. (2006). Pliocene glacial cyclicity in a deep-sea sediment drift (Antarctic Peninsula Pacific Margin). *Palaeogeography, Palaeoclimatology, Palaeoecology*, **231**, 181–198.

Hill, R. S. (1989). Fossil leaf. In *Antarctic Cenozoic History from the CIROS-1 Drillhole, McMurdo Sound, New Zealand Department of Scientific and Industrial Research Bulletin*, **245**, ed. P. J. Barrett. Wellington: DSIR Publishing, pp. 143–144.

Hill, R. S. and Truswell, E. M. (1993). *Nothofagus* fossils in the Sirius Group, Transantarctic Mountains: Leaves and pollen and their climatic implications. In *The Antarctic Paleoenvironment: A Perspective on Global Change. Part 2. Antarctic Research Series*, **60**, eds J. P. Kennett and D. A. Warnke. Washington: American Geophysical Union, pp. 67–73.

Hill, R. S., Harwood, D. M. and Webb, P. N. (1996). *Nothofagus beardmorensis* (Nothofagaceae), a new species based on leaves from the Pliocene Sirius Group, Transantarctic Mountains, Antarctica. *Review of Palaeobotany and Palynology*, **94**, 11–24.

Hillenbrand, C. D. and Ehrman, W. (2005). Late Neogene to Quaternary environmental changes in the Antarctic Peninsula region: evidence from drift sediments. *Global and Planetary Change*, **45**, 165–191.

Hilley, G. E. and Porder, S. (2008). A framework for predicting global silicate weathering and CO_2 drawdown rates over geologic time-scales. *Proceedings of the National Academy of Sciences of the United States of America*, **105**, 16855–16859.

Holbourn, A., Kuhnt, W., Schultz, M. and Erlenkeuser, H. (2005). Impacts of orbital forcing and atmospheric carbon dioxide on Miocene ice sheet expansion. *Nature*, **438**, 483–487.

Holbourn, A., Kuhnt, W., Schultz, M., Flores, J. A. and Anderson, N. (2007). Orbitally-paced climatic evolution during the middle Miocene "Monterey" carbon-isotope excursion. *Earth and Planetary Science Letters*, **261**, 534–550.

Holmes, M. A. (2000). Clay mineral composition of glacial erratic McMurdo Sound. In *Paleobiology and Paleoenvironments of Eocene Rocks, McMurdo Sound East Antarctica, Antarctic Research Series*, **76**, eds J. D. Stilwell and R. M. Feldmann. Washington: American Geophysical Union, pp. 63–72.

Huang, F., Liu, X., Kong, P., *et al.* (2008). Fluctuation history of the interior East Antarctic Ice Sheet since mid-Pliocene. *Antarctic Science*, **20**, 197–203.

Huber, M. and Nof, D. (2006). The ocean circulation in the Southern Hemisphere and its climatic impacts in the Eocene. *Palaeogeography, Palaeoclimatology, Palaeoecology*, **231**, 9–28.

Huber, M., Brinkhuis, H., Stickley, C. E., *et al.* (2004). Eocene circulation of the Southern Ocean: Was Antarctica kept warm by subtropical waters? *Paleoceanography*, **19**, PA4026, doi:10.1029/2004PA001014.

Hunt R. H. (2001). Biodiversity and palaeoecological significance of Tertiary fossil floras from King George Island, West Antarctica. Ph.D. thesis, University of Leeds.

Huybrechts, P., (1993). Glaciological modelling of the Late Cenozoic East Antarctic Ice Sheet: stability or dynamism? *Geografiska Annaler*, **75A**, 221–238.

Ivany, L. C., Van Simaeys, S., Domack, E. W. and Samson, S. D. (2006). Evidence for an earliest Oligocene ice sheet on the Antarctic Peninsula. *Geology*, **34**, 377–380.

Ivany, L. C., Lohmann, K. C., Hasiuk, F., *et al.* (2008). Eocene climate record of a high southern latitudine shelf: Seymour Island, Antarctica. *Geological Society of America Bulletin*, **120**, 659–678.

Jamieson, S. S. R. and Sugden, D. E. (2008). Landscape evolution of Antarctica. In *Antarctica: A Keystone in a Changing World. Proceedings of the 10th International Symposium on Antarctic Earth Sciences*, eds A. K. Cooper, P. J. Barrett, H. Stagg, B. C.

Storey, E. Stump and W. Wise and the 10th ISAES editorial team. Washington: U. S. Geological Survey and The National Academies Press, pp. 39–54.

Jamieson, S. S. R., Sugden, D. E. and Hulton, N. R. J. (2010). The evolution of the subglacial landscape of Antarctica. *Earth and Planetary Science Letters*, **293**, 1–27.

Johnson, J. J., Smellie, J. L., Nelson, A. E. and Stuart, F. M. (2009). History of the Antarctic Peninsula Ice Sheet since the early Pliocene–Evidence from cosmogenic dating of Pliocene lavas on James Ross Island, Antarctica. *Global and Planetary Change*, **69**, 205–213.

Jones, C. M. (2000). First fossil bird from East Antarctica. In *Paleobiology and Paleoenvironments of Eocene Rocks, McMurdo Sound, East Antarctica, Antarctic Research Series*, **76**, eds J. D. Stilwell and R. M. Feldmann. Washington: American Geophysical Union, pp. 359–364.

Jouzel, J., Barkov, N. I., Barnola, J. M., *et al.* (1993). Extending the Vostok ice-core record of palaeoclimate to the penultimate glacial period. *Nature*, **364**, 407–412.

Kemp, E. M. (1975). Palynology of Leg 28 drill sites, Deep Sea Drilling Project. In *Initial Reports of the Deep Sea Drilling Project Volume 28, covering Leg 28 of the cruises of the Drilling Vessel* Glomar Challenger, *Fremantle, Australia, to Christchurch, New Zealand December 1972–February 1973*, eds D. E. Hayes, L. A. Frakes, P. J. Barrett, *et al.* Washington: U. S. Printing Offices, pp. 599–623.

Kemp, E. M. and Barrett, P. J. (1975). Antarctic glaciations and Early Tertiary vegetation. *Nature*, **258**, 507–508.

Kennett, J. P. (1977). Cenozoic evolution of Antarctic glaciations, the Circum-Antarctic Ocean, and their impact on global paleo-oceanography. *Journal of Geophysical Research*, **82**, 3843–3860.

Kennett, J. P. and Barker, P. F. (1990). Latest Cretaceous to Cenozoic climate and oceanographic developments in the Weddell Sea, Antarctica: an ocean drilling perspective. In *Proceedings of the Ocean Drilling Program, Volume 113 Scientific Results Weddell Sea, Antarctica Covering Leg 113 of the cruises of the Drilling Vessel* JOIDES Resolution, *Valparaiso, Chile, to East Cove, Falkland Islands, Sites 689–697, 25 December 1986–11 March 1987*, eds P. F. Barker, J. P. Kennett, S. O'Connell, *et al.* Online doi:10.2973/odp.proc.sr.113.195.1990, available at http://www-odp.tamu.edu/ publications/113_SR/VOLUME/CHAPTERS/sr113_53.pdf. {Cited 2011}.

Kennett, J. P. and Shackleton, N. J. (1976). Oxygen isotope evidence for the initiation of the cyrosphere 38 Myr ago. *Nature*, **260**, 513–515.

Kennett, J. P., Keller, G. and Srinivasan, M. S. (1985). Miocene planktonic foraminiferal biogeography and paleoceanographic development of the Indo–Pacific region. In *The Miocene Ocean: Paleoceanography and Biogeography, Geological Society of America Memoir*, **163**, ed. J. P. Kennett. Boulder: The Geological Society of America, pp. 197–236.

Kershaw, P. and Wagstaff, B. (2001). The southern conifer family Araucariaceae: History, status, and value for paleoenvironmental reconstruction. *Annual Review of Ecology and Systematics*, **32**, 397–414.

Kidder, D. L. and Worsley, T. R. (2010). Phanerozoic Large Igneous Provinces (LIPs), HEATT (Haline Euxinic Acidic Thermal Transgression) episodes, and mass extinctions. *Palaeogeography, Palaeoclimatology, Palaeoecology*, **295**, 162–191.

Kuhnert, H., Bickert, T. and Paulsen, H. (2009). Southern Ocean frontal system changes precede Antarctic ice sheet growth during the middle Miocene. *Earth and Planetary Science Letters*, **284**, 630–638.

Kump, L. R. (2005). Foreshadowing the glacial era. *Nature*, **436**, 333–334.

Kurtz, A. C., Kump, L. R., Arthur, M. A., Zachos, J. C. and Paytan, A. (2003). Early Cenozoic decoupling of the global carbon and sulfur cycles. *Paleoceanography*, **18**, 1090, doi:10.1029/2003PA000908.

Lagabrielle, Y., Goddéris, Y., Donnadieu, Y., Malavieille, J. and Suarez, M. (2009). The tectonic history of Drake Passage and its possible impacts on global climate. *Earth and Planetary Science Letters*, **279**, 197–211.

Langebroek, P. M., Paul, A. and Schultz, M. (2009). Antarctic ice-sheet response to atmospheric CO_2 and insolation in the Middle Miocene. *Climates of the Past*, **5**, 633–646.

Lawver, L. A. and Gahagan, L. M. (2003). Evolution of Cenozoic seaways in the circum-Antarctic region. *Palaeogeography, Palaeoclimatology, Palaeoecology*, **198**, 11–37.

Lear, C. H., Elderfield, H. and Wilson, P. A. (2000). Cenozoic deep-sea temperatures and global ice volumes from Mg/Ca in benthic foraminiferal calcite. *Science*, **287**, 269–272.

Leckie, R. M. and Webb, P. N. (1983). Late Oligocene–Early Miocene glacial record of the Ross Sea, Antarctica: evidence from DSDP Site 270. *Geology*, **11**, 578–582.

Lewis, A. R., Marchant, D. R., Kowalewski, D. E., Baldwin, S. L. and Webb, L. E. (2006). The age and origin of the Labyrinth, western Dry Valleys, Antarctica: Evidence for extensive middle Miocene subglacial floods and freshwater discharge to the Southern Ocean. *Geology*, **34**, 513–516.

Lewis, A. R., Marchant, D. R., Ashworth, A. C., Hemming, S. R. and Machlus, M. L. (2007). Major Middle Miocene climatic change: Evidence from East Antarctica and the Transantarctic Mountains. *Geological Society of America Bulletin*, **119**, 1449–1461.

Lewis, A. R., Marchant, D. R., Ashworth, A. C., *et al.* (2008). Mid-Miocene cooling and the extinction of tundra in continental Antarctica. *Proceedings of the National Academy of Science of the United States of America*, **105**, 10676–10680.

Lewis-Smith, R. I. (1994). Vascular plants as bioindicators of regional warming in Antarctica. *Oecologia*, **99**, 322–328.

Levy, R. H. and Harwood, D. M. (2000). Sedimentary lithofacies of the McMurdo Sound erratic. In *Paleobiology and Paleoenvironments of Eocene Rocks, McMurdo Sound, East Antarctica, Antarctic Research Series*, **76**, eds J. D. Stilwell and R. M. Feldmann. Washington: American Geophysical Union, pp. 39–61.

Livermore, R. A., Nankivell, A. P., Eagles, G. and Morris, P. (2005). Paleogene opening of Drake Passage. *Earth and Planetary Science Letters*, **236**, 459–470.

Livermore, R. A., Hillenbrand, C. D., Meredith, M. and Eagles, G. (2007). Drake Passage and Cenozoic climate: an open and shut case? *Geochemistry, Geophysics, Geosystems*, **8**, Q01005, doi:10.1029/2005GC001224.

Liu, X., Huang, F., Kong, P., *et al.* (2010). History of ice sheet elevation in East Antarctica: Paleoclimatic implications. *Earth and Planetary Science Letters*, **290**, 281–288.

Lunt, D. J., Foster, G. L, Haywood, A. M. and Stone, E. J. (2008). Late Pliocene Greenland glaciation controlled by a decline in atmospheric CO_2 levels. *Nature*, **454**, 1102–1106.

Lyle, M., Gibbs, S., Moore, T. C. and Rea, D. K. (2007). Late Oligocene initiation of the Antarctic Circumpolar Current: evidence from the South Pacific. *Geology*, **35**, 691–694.

Macphail, M. K. and Truswell, E. M. (2004a). Palynology of Site 1166, Prydz Bay, East Antarctica. *Proceedings of the Ocean Drilling Program, Scientific Results*, **188**, 1–29.

Macphail, M. K. and Truswell, E. M. (2004b). Palynology of Neogene slope and rise deposits from ODP Sites 1165 and 1167, East Antarctica. In *Proceedings of the Ocean Drilling Program, Volume 188 Scientific Results Prydz Bay-Cooperation Sea, Antarctica: Glacial History and Paleoceanography covering Leg 188 of the cruises of the Drilling Vessel* JOIDES Resolution, *Fremantle, Australia, to Hobart, Tasmania Sites 1165–1167 10 January–11 March 2000*, eds A. K. Cooper, P. E. O'Brien and

C. Richter. Online doi:10.2973/odp.proc.sr.188.012.2004, available at http://www-odp.tamu.edu/publications/188_SR/VOLUME/CHAPTERS/012.PDF. {Cited 2011}.

Macphail, M. K., Hill, R. S., Forsyth, S. M. and Wells, P. M. (1991). A late Oligocene–early Miocene cool climate flora in Tasmania. *Alcheringa*, **15**, 87–106.

Manton, I. and Vida, G. (1968). Cytology of the fern flora of Tristan da Cunha. *Proceedings of the Royal Society of London Series B*, **17**, 361–379.

Marchant, D. R., Denton, G. H. and Sugden, D. E. (1993). Miocene glacial stratigraphy and landscape evolution of the western Asgard Range: *Geografiska Annaler*, **75A**, 303–330.

Marchant, D. R., Denton, G. H. and Swisher, C. C. (1993). Miocene-Pliocene-Pleistocene glacial history of Arena Valley, Quartermain Mountains, Antarctica. *Geografiska Annaler*, **75A**, 269–302.

Marchant, D. R., Denton, G. H., Swisher, C. C. and Potter, N. (1996). Late Cenozoic Antarctic paleoclimate reconstructed from volcanic ashes in the Dry Valleys region of southern Victoria Land. *Geological Society of America Bulletin*, **108**, 181–194.

Marchant, D. R., Lewis, A. R., Phillips, W. M., *et al.* (2002). Formation of patterned ground and sublimation till over Miocene glacier ice in Beacon Valley, southern Victoria Land, Antarctica. *Geological Society of America Bulletin*, **114**, 718–730.

Marenssi, S. A., Casadio, S. and Santillana, S. N. (2010). Record of Late Miocene glacial deposits on Isla Marambio (Seymour Island), Antarctic Peninsula. *Antarctic Science*, **22**, 193–198.

Masson-Delmotte, V., Stenni, B., Pol, K., *et al.* (2010). Dome C record of glacial and interglacial intensities. *Quaternary Science Reviews*, **29**, 113–128.

McIntyre, D. J. and Wilson, G. J. (1966). Preliminary palynology of some Antarctic Tertiary erratics. *New Zealand Journal of Botany*, **4**, 315–321.

McKay, R., Browne, G., Carter, L., *et al.* (2009). The stratigraphic signature of the late Cenozoic Antarctic Ice Sheets in the Ross Embayment. *Geological Society of America Bulletin*, **121**, 1537–1561.

McKelvey, B. C., Webb, P. N., Harwood, D. M. and Mabin, M. C. G. (1991). The Dominion Range Sirius Group: a record of the late Pliocene-early Pleistocene Beardmore Glacier. In *Geological Evolution of Antarctica*, eds M. R. A. Thomson, J. A. Crame and J. W. Thomson. Cambridge: Cambridge University Press, pp. 675–682.

Mildenhall, D. C. (1989). Terrestrial palynology. In *Antarctic Cenozoic History from the CIROS-1 Drillhole, McMurdo Sound, New Zealand Department of Scientific and Industrial Research Bulletin*, **245**, ed. P. J. Barrett. Wellington: DSIR Publishing, pp. 119–127.

Miller, K. G., Wright, J. D. and Fairbanks, R. G. (1991). Unlocking the icehouse-Oligocene-Miocene oxygen isotope eustasy and margin erosion. *Journal of Geophysical Research Solid Earth and Planets*, **96**, 6829–6848.

Miller, K. G., Wright, J. D., Katz, M. E., *et al.* (2009). Climate threshold at the Eocene-Oligocene transition: Antarctic icesheet influence on ocean circulation. In *The Late Eocene Earth: Hothouse, Icehouse and Impacts, Geological Society of America Special Papers*, **452**, eds C. Koeberl and A. Montanari. Boulder: The Geological Society of America, pp. 169–178.

Mohr, B. A. R. (2001). The development of Antarctic fern floras during the Tertiary, and palaeoclimatic and palaeobiogeographic implications. *Palaeontographica B*, **259**, 167–208.

Moorhead, D. L. and Priscu, J. C. (1998). The McMurdo Dry Valley ecosystem: organization, controls, and linkages. In *Ecosystem Dynamics in a Polar Desert: The McMurdo*

Dry Valleys, Antarctica, Antarctic Research Series, **72**, ed. J. C. Priscu. Washington: American Geophysical Union, pp. 351–363.

Moran, K., Backman, J., Brinkhuis, H., *et al.* (2006). The Cenozoic palaeoenvironment of the Arctic Ocean. *Nature*, **441**, 601–605.

Mosyakin, S. L., Bezusko, L. G. and Mosyakin, A. S. (2007). Origins of native vascular plants of Antarctica: Comments from a historical phytogeography viewpoint. *Cytology and Genetics*, **41**, 308–316.

Mozer, A. (2012). Pre-glacial sedimentary facies of the Point Thomas Formation (Eocene) at Cytadela, Admiralty Bay, King George Island, West Antarctica. *Polish Polar Research*, **33**, 41–62.

Murphy, L., Warnke, D. A., Andersson, C., Channell, J. and Stoner, J. (2002). History of ice rafting at South Atlantic ODP Site 177–1092 during the Gauss and the late Gilbert Chron. *Palaeogeography, Palaeoclimatology, Palaeoecology*, **182**, 183–196.

Naish, T. R., Wilson, G. S., Dunbar, G. B. and Barrett, P. J. (2008). Constraining the amplitude of Late Oligocene bathymetric changes in western Ross Sea during orbitally-induced oscillations in the East Antarctic Ice Sheet: (2) Implications for global sea-level changes. *Palaeogeography, Palaeoclimatology, Palaeoecology*, **260**, 66–76.

Naish, T. R., Powell, R., Levy, R., *et al.* (2009). Obliquity-paced Pliocene West Antarctic ice sheet oscillations. *Nature*, **458**, 322–328.

Nelson, A. E., Smellie, J. L., Hambrey, M. J. *et al.* (2009). Neogene glacigenic debris flows on James Ross Island, Northern Antarctic Peninsula, and their implications for regional climate history. *Quaternary Science Reviews*, **28**, 3138–3160.

Nelson, C. S. and Cooke, P. J. (2001). History of oceanic front development in the New Zealand sector of the Southern Ocean during the Cenozoic — a synthesis. *New Zealand Journal of Geology and Geophysics*, **44**, 535–553.

Nong, G. T., Najjar, R. G., Seidov, D. and Peterson, W. (2000). Simulation of ocean temperature change due to the opening of Drake Passage. *Geophysical Research Letters*, **27**, 2689–2692.

Nývlt, D., Kosler, J., Mlcoch, B., *et al.* (2011). The Mendel Formation: Evidence for Late Miocene climatic cyclicity at the northern tip of the Antarctic Peninsula. *Palaeogeography, Palaeoclimatology, Palaeoecology*, **299**, 363–384.

Pagani, M., Zachos, J. C., Freeman, K. H., Tipple, B. and Bohaty, S. (2005). Marked decline in atmospheric carbon dioxide concentrations during the Paleogene. *Science*, **309**, 600–603.

Pak, D. K. and Miller, K. G. (1992). Paleocene to Eocene benthic foraminiferal isotopes and assemblages: implications for deepwater circulation, *Paleoceanography*, **7**, 405–422.

Pälike, H., Norris, R. D., Herrle, J. O., *et al.* (2006). The heartbeat of the Oligocene climate system. *Science*, **314**, 1894–1989.

Passchier, S. (2001). Provenance of the Sirius Group and related Upper Cenozoic glacial deposits from the Transantarctic Mountains, Antarctica: relation to landscape evolution and ice sheet drainage. *Sedimentary Geology*, **144**, 263–290.

Passchier, S. (2004). Variability in geochemical provenance and weathering history of Sirius Group strata, Transantarctic Mountains: Implications for Antarctic glacial history. *Journal of Sedimentary Research*, **74**, 607–619.

Passchier, S. and Krissek, L. A. (2008). Oligocene-Miocene continental weathering record and paleoclimatic implications, Cape Roberts Drilling Project, Ross Sea, Antarctica. *Palaeogeography, Palaeoclimatology, Palaeoecology*, **260**, 30–40.

Pearson, P. N. and Palmer, M. R. (2000). Atmospheric carbon dioxide concentrations over the past 60 million years. *Nature*, **430**, 695–699.

Pearson, P. N., Foster, G. L. and Wade, B. S. (2009). Atmospheric carbon dioxide through the Eocene-Oligocene climatic transition. *Nature*, **461**, 1110–1113.

Pekar, S. F. and Christie-Blick, N. (2008). Resolving apparent conflicts between oceanographic and Antarctic climate records and evidence for a decrease in pCO$_2$ during the Oligocene through early Miocene (34–16 Ma). *Palaeogeography, Palaeoclimatology, Palaeoecology*, **260**, 41–49.

Pekar, S. F. and DeConto, R. M. (2006). High resolution ice-volume estimates for the early Miocene: evidence for a dynamic ice sheet in Antarctica. *Palaeogeography, Palaeoclimatology, Palaeoecology*, **231**, 101–109.

Pekar, S. F., DeConto, R. M. and Harwood, D. M. (2006). Resolving a late Oligocene conundrum: deep sea warming versus Antarctic glaciations. *Palaeogeography, Palaeoclimatology, Palaeoecology*, **231**, 29–40.

Pfuhl, H. A. and McCave, I. N. (2005). Evidence for late Oligocene establishment of the Antarctic Circumpolar Current. *Earth and Planetary Science Letters*, **235**, 715–728.

Pole, M., Hill, R. S. and Harwood, D. M. (2000). Eocene plant macrofossils from erratics, McMurdo Sound, Antarctica. In *Paleobiology and Paleoenvironments of Eocene Rocks, McMurdo Sound, East Antarctica, Antarctic Research Series*, **76**, eds J. D. Stilwell and R. M. Feldmann. Washington: American Geophysical Union, pp. 243–251.

Pollard, D. and DeConto, R. M. (2009). Modelling West Antarctic ice sheet growth and collapse through the past five million years. *Nature*, **458**, 329–333.

Polunin, N. (1960). *Introduction to Plant Geography and some Related Sciences*. London: Longmans.

Poole, I. (2002). Systematics of Cretaceous and Tertiary *Nothofagoxylon*: implications for Southern Hemisphere biogeography and evolution of the Nothofagaceae. *Australian Systematic Botany*, **15**, 247–276.

Poole, I. and Cantrill, D. J. (2006). Cretaceous and Cenozoic vegetation of Antarctica implications from the fossil wood record. In *Cretaceous–Tertiary High-Latitude Palaeoenvironments, James Ross Basin, Antarctica, Geological Society of London Special Publication*, **258**, eds J. E. Francis, D. Pirrie and J. A. Crame. London: The Geological Society, pp. 63–81.

Prebble, J. G., Raine, J. I., Barrett, P. J. and Hannah, M. J. (2006). Vegetation and climate from two Oligocene glacioeustatic sedimentary cycles (31 and 24 Ma) cored by the Cape Roberts Project, Victoria Land Basin, Antarctica. *Palaeogeography, Palaeoclimatology, Palaeoecology*, **231**, 41–57.

Prentice, M. L. and Matthews, R. K. (1991). Tertiary ice sheet dynamics: the snow gun hypothesis. *Journal of Geophysical Research*, **96**, 6811–6827.

Puigdefábregas, J., Gallart, F., Biaciotto, O., Allogia, M. and del Barrio, G. (1999). Banded vegetation patterning in a subantarctic forest of Tierra del Fuego, as an outcome of the interaction between wind and tree growth. *Acta Oecologica*, **20**, 135–146.

Raine, J. I. (1998). Terrestrial palynomorphs from Cape Roberts Project Drillhole CRP-1, Ross Sea, Antarctica. *Terra Antarctica*, **5**, 539–548.

Raine, J. I. and Askin, R. A. (2001). Terrestrial palynology of Cape Roberts Project Drillhole CRP-3, Victoria Land Basin, Antarctica. *Terra Antarctica*, **8**, 389–400.

Ramstein, G., Fluteau, F., Besse, J. and Joussaume, S. (1997). Effect of orogeny, plate motion and land–sea distribution on Eurasian climate change over the past 30 million year. *Nature*, **386**, 788–795.

Raymo, M. E. (1991). Geochemical evidence supporting T. C. Chamberlin's theory of glaciations. *Geology*, **19**, 344–347.

Raymo, M. E. (1994). The Himalayas, organic carbon burial, and climate in the Miocene. *Paleoceanography*, **9**, 399–404.

Raymo, M. E. and Ruddiman, W. F. (1992). Tectonic forcing of the late Cenozoic climate. *Nature*, **359**, 117–122.

Reguero, M. A., Marenssi, S. A. and Santillana, S. N. (2002). Antarctic Peninsula and South America (Patagonia) Paleogene terrestrial faunas and environments: biogeographic relationships. *Palaeogeography, Palaeoclimatology, Palaeoecology*, **179**, 189–210.

Rémy, F. and Tabacco, I. E. (2000). Bedrock features and ice flow near the EPICA Ice Core Site (Dome C, Antarctica). *Geophysical Research Letters*, **27**, 405–408.

Retallack, G. J., Krull, E. S. and Bockheim, J. G. (2001). New grounds for reassessing palaeoclimate of the Sirius Group, Antarctica. *Journal of the Geological Society of London*, **158**, 925–935.

Roberts, A. P., Wilson, G. S., Harwood, D. M. and Verosub, K. L. (2003). Glaciation across the Oligocene-Miocene boundary in southern McMurdo Sound Antarctica: New chronology from the CIROS-1 Drill Hole. *Palaeogeography, Palaeoclimatology, Palaeoecology*, **198**, 113–130.

Ruddiman, W. F. (ed.) (1997). *Tectonic Uplift and Climate Change*. New York, Plenum Press.

Salzmann, U., Riding, J. B., Nelson, A. E. and Smellie, J. L. (2011b). How likely was a green Antarctic Peninsula during warm Pliocene interglacials? A critical reassessment based on new palynofloras from James Ross Island. *Palaeogeography, Palaeoclimatology, Palaeoecology*, **309**, 73–82.

Salzmann, U., Williams, M., Haywood, A. M., *et al*. (2011a). Climate and environment of a Pliocene warm world. *Palaeogeography, Palaeoclimatology, Palaeoecology*, **309**, 1–8.

Scher, H. D. and Martin, E. E. (2006). Timing and climatic consequences of the opening of Drake Passage, *Science*, **312**, 428–430.

Scher, H. D. and Martin, E. E. (2008). Oligocene deep water export from the North Atlantic and the development of the Antarctic Circumpolar Current examined with neodymium isotopes. *Paleoceanography*, **23**, PA1205, doi:10.1029/2006PA001400.

Scherer, R. P., Aldahan, A., Tulaczyk, S., *et al*. (1998). Pleistocene collapse of the West Antarctic Ice Sheet. *Science*, **281**, 82–85.

Schlich, R., Wise Jr, S. W., Palmer Julson, A. A., *et al*. (1992). The geologic and tectonic evolution of the Kerguelen Plateau: An introduction to the scientific results of Leg 120. In *Proceedings of the Ocean Drilling Program Volume 120 Scientific Results Part 1 Central Kerguelen Plateau, Covering Leg 120 of the cruises of the Drilling Vessel JOIDES Resolution, Fremantle Australia, to Fremantle, Australia, Sites 747–751, 20 February to 30 April 1988*, eds S. W. Wise Jr., R. Schlich, A. A. Palmer Julson, *et al*. Online doi:10.2973/odp.proc.sr.120.203.1992, available at http://www-odp.tamu.edu/publications/120_SR/VOLUME/CHAPTERS/sr120_01.pdf. {Cited 2011}.

Shackleton, N. J., Hall, M. A. and Boersma, A. (1984). Oxygen and carbon isotope data from Leg 74 foraminifers. In *Initial Reports of the Deep Sea Drilling Project Volume 74*, eds T. C. Moore Jr, P. D. Rabinowitz, P. Borella, A. Boersma, and N. J. Shackleton, Washington: U. S. Government Printing Office, pp. 599–612.

Shevenell, A. E. and Kennett, J. P. (2007). Cenozoic Antarctic cryosphere evolution: Tales from deep-sea sedimentary records. *Deep Sea Research II*, **54**, 2308–2324.

Shevenell, A. E., Kennett, J. P. and Lea, D. W. (2004). Middle Miocene Southern Ocean cooling and Antarctic cryosphere expansion. *Science*, **305**, 1766–1770.

Shevenell, A. E., Kennett, J. P. and Lea, D. W. (2008). Middle Miocene ice sheet dynamics, deep sea temperatures and carbon cycling: a Southern Ocean perspective. *Geochemistry, Geophysics, Geosystems*, **9**, QO2006. Doi:10.1029/2007GC001736

Shipboard Scientific Party (2001). Leg 188 summary: Prydz Bay–Cooperation Sea, Antarctica. In *Proceedings of the Ocean Drilling Program, Initial Reports, Volume 188, Prydz Bay–Cooperation Sea, Antarctica: Glacial History and Paleoceanography Sites 1165–1167 covering Leg 188 of the cruises of the Drilling Vessel* JOIDES Resolution, *Fremantle, Australia, to Hobart, Tasmania Sites 1165–1167 10 January–11 March 2000*, eds P. E. O'Brien, A. K. Cooper, C. Richter, *et al.* Online doi:10.2973/odp.proc.ir.188.101.2001 available at http://www-odp.tamu.edu/publications/188_IR/chap_01/chap_01.htm. {Cited 2011}.

Siegert, M., Taylor, J. and Payne, A. J. (2005). Spectral roughness of subglacial topography and implications for former ice-sheet dynamics in East Antarctica. *Global and Planetary Change*, **45**, 249–263.

Sijp, W. P. and England, M. H. (2004). Effect of the Drake Passage throughflow on global climate. *Journal of Physical Oceanography*, **34**, 1254–1266.

Sijp, W. P. and England, M. H. (2005). On the role of the Drake Passage in controlling the stability of the ocean's thermohaline circulation. *Journal of Climate*, **18**, 1957–1966.

Sijp, W. P., England, M. H. and Toggweiler, J. R. (2009). Effect of ocean gateway changes under greenhouse warmth. *Journal of Climate*, **22**, 6639–6652.

Sime, L. C., Wolff, E. W., Oliver, K. I. C., *et al.* (2009). Evidence for warmer interglacials in East Antarctic ice cores. *Nature*, **462**, 342–346.

Smellie, J. L., McArthur, J. M., McIntosh, W. C. and Esser, R. (2006). Late Neogene interglacial events in the James Ross Island region, Northern Antarctic Peninsula, dated by Ar/Ar and Sr-isotope stratigraphy. *Palaeogeography, Palaeoclimatology, Palaeoecology*, **242**, 169–187.

Smellie, J. L., Johnson, J. S., McIntosh, W. C., *et al.* (2008). Six million years of glacial history recorded in volcanic lithofacies of the James Ross Island Volcanic Group, Antarctic Peninsula. *Palaeogeography, Palaeoclimatology, Palaeoecology*, **260**, 122–148.

Smellie, J. L., Haywood, A. M., Hillenbrand, C.-D., Lunt, D. J. and Valdes, P. J. (2009). Nature of the Antarctic Peninsula Ice Sheet during the Pliocene: geological evidence and modelling results compared. *Earth-Science Reviews*, **94**, 79–94.

Soares, R. V. (1979). Considerações sobre a regeneração natural da *Araucaria angustifolia*. *Floresta*, **2**, 12–8.

St John, K. (2008). Cenozoic ice-rafting history of the central Arctic Ocean: terrigenous sands on the Lomonosov Ridge. *Paleoceanography*, **23**, PA1S05, doi:10.1029/2007PA001483.

Steppuhn, A., Micheels, A., Bruch, A. A., *et al.* (2007). The sensitivity of ECHAM4/ML to a double CO_2 scenario for the Late Miocene and the comparison to terrestrial proxy data. *Global and Planetary Change*, **57**, 189–212.

Stickley, C. E., Brinkhuis, H., Schellenberg, S. A., *et al.* (2004). Timing and nature of the deepening of the Tasmanian Gateway. *Paleoceanography*, **19**, PA4027, doi:10.1029/2004PA001022.

Stilwell, J. D. and Zinsmeister, W. J. (1992). Molluscan systematics and biostratigraphy. Lower Tertiary La Meseta Formation, Seymour Island, Antarctic Peninsula. *Antarctic Research Series*, **55**, 1–192.

Stuchlik, L. (1981). Tertiary pollen spectra from the Ezcurra Inlet Group of Admiralty Bay, King George Island (South Shetland Islands, Antarctica). *Studia Geologica Polonica*, **72**, 109–132.

Sugden, D. and Denton, G. (2004). Cenozoic landscape evolution of the Convoy Range to Mackay Glacier area, Transantarctic Mountains: Onshore to offshore synthesis. *Geological Society of America Bulletin*, **116**, 840–857.

Sugden, D., Marchant, D. R., Potter, N., *et al.* (1995). Preservation of Miocene glacier ice in East Antarctica. *Nature*, **376**, 412–414.

Tambussi, C. P., Noriega, J. I., Gaździcki, A., *et al.* (1994). Ratite bird from the Paleogene La Meseta Formation, Seymour Island, Antarctica. *Polish Polar Research*, **15**, 15–20.

Theissen, K. M., Dunbar, R. B., Cooper, A. K. and Mucciarone, D. A. (2003). The Pleistocene evolution of the East Antarctic Ice Sheet in the Prydz Bay region. *Global Planetary Change*, **39**, 227–256.

Thorn, V. C. (2001). Oligocene and Early Miocene Phytoliths from CRP-212A and CRP-3. Victoria Land Basin. Antarctica. *Terra Antarctica*, **8**, 407–422.

Thorn, V. C. and DeConto, R. (2006). Antarctic climate at the Eocene/Oligocene boundary–climate model sensitivity to high latitude vegetation type and comparisons with the palaeobotanical record. *Palaeogeography, Palaeoclimatology, Palaeoecology*, **231**, 134–157.

Toggweiler, J. R. and Bjornsson, H. (2000). Drake Passage and palaeoclimate. *Journal of Quaternary Science*, **15**, 319–328.

Tong, J. A., You, Y., Müller, R. D. and Seton, M. (2009). Climate model sensitivity to atmospheric CO_2 concentrations for the middle Miocene. *Global and Planetary Change*, **67**, 129–140.

Torres, T. and Lemoigne, Y. (1988). Maderas fósiles terciarias de la Formación Caleta Arctowski, Isla Rey Jorge, Antártica. *Serie Científica Instituto Antártico Chileno*, **37**, 69–107.

Tripati, A., Backman, J., Elderfield, H. and Ferretti, P. (2005). Eocene bipolar glaciation associated with global carbon cycle changes. *Nature*, **436**, 341–345.

Tripati, A. K., Roberts, C. D. and Eagle, R. A. (2009). Coupling of CO_2 and ice sheet stability over major climate transitions of the last 20 million years. *Science*, **326**, 1394–1397.

Troedson, A. L. and Riding, J. B. (2002). Upper Oligocene to lowermost Miocene strata of King George Island, South Shetland Islands, Antarctica: stratigraphy, facies analysis, and implications for the glacial history of the Antarctic Peninsula. *Journal of Sedimentary Research*, **72**, 510–523.

Troedson, A. L. and Smellie, J. L. (2002). Upper Oligocene to lowermost Miocene strata of King George Island, South Shetland Islands, Antarctica: Stratigraphy, facies analysis and implications for the glacial history of the Antarctic Peninsula. *Journal of Sedimentary Research*, **72**, 510–523.

Truswell, E. M. (1983). Recycled Cretaceous and Tertiary pollen and spores in Antarctic marine sediments: a catalogue. *Palaeontographica B*, **186**, 121–174.

Truswell, E. M. (1986). Palynology. In *Antarctic Cenozoic history from the MSSTS-1 drillhole. McMurdo Sound, New Zealand Department of Scientific and Industrial Research Bulletin*, **237**, ed. P. J. Barrett. Wellington: Science Information Publishing Centre, pp. 131–134.

Truswell, E. (1990). Cretaceous and Tertiary vegetation of Antarctica: A palynological Perspective. In *Antarctic Paleobiology: its Role in the Reconstruction of Gondwana*, eds T. N. Taylor and E. L. Taylor. Berlin: Springer-Verlag, pp. 71–88.

Truswell, E. M. and Macphail, M. K. (2004). Carnivorous plants at high latitudes: pollen evidence for Droseraceae growing in East Antarctica during the Late Eocene. *Association of Australasian Palaeontologists Memoir*, **29**, 85–97.

Truswell, E. M. and Macphail, M. (2009). Polar forests on the edge of extinction: what does the fossil spore and pollen evidence from East Antarctica say? *Australian Systematic Botany*, **22**, 57–106.

Truswell, E. M., Quilty, P. G., McMinn, A., Macphail, M. K. and Wheller, G. E. (2005). Late Miocene vegetation and palaeoenvironments of the Drygalski Formation, Heard Island, Indian Ocean: evidence from palynology. *Antarctic Science*, **17**, 427–442.

Villa, G., Lupi, C., Cobianchi, M., Florindo, F. and Pekar, S. F. (2008). A Pleistocene warming event at 1 Ma in Prydz Bay, East Antarctica: Evidence from ODP Site 1165. *Palaeogeography, Palaeoclimatology, Palaeoecology*, **260**, 230–244.

Wace, N. M. (1960). The botany of the southern oceanic islands. *Proceedings of the Royal Society of London Series B*, **152**, 475–490.

Warnke, D. A., Allen, C. P., Muller, D. W., Hodell, D. A. and Brunner. C. A. (1992). Miocene-Pliocene Antarctic glacial evolution: a synthesis of ice-rafted debris, stable isotope, and planktonic foraminiferal indicators, ODP Leg 114. In *The Antarctic Paleoenvironment: a Perspective on Global Change. Part 1, Antarctic Research Series*, **56**, eds J. P. Kennett and D. A. Warnke. Washington: American Geophysical Union, pp. 1–325.

Warnke, D. A., Marzo, B. and Hodell, D. A. (1996). Major deglaciation of East Antarctica during the early Late Pliocene? Not likely from a marine perpective. *Marine Micropalaeontology*, **27**, 237–251.

Warny, S., Wrenn, J. H., Bart, P. J. and Askin, R. A. (2006). Palynology of the NBP03–01A transect in the Northern Basin, western Ross Sea, Antarctica: a late Pliocene record. *Palynology*, **30**, 151–182.

Warny, S., Askin, R. A., Hannah, M. J., *et al.* (2009). Palynomorphs from a sediment core reveal a sudden remarkably warm Antarctica during the middle Miocene. *Geology*, **37**, 955–958.

Webb, P. N. (1994). Paleo-drainage systems of East Antarctica and sediment supply to West Antarctic Rift System basins. *Terra Antartica*, **1**, 457–461.

Webb, P. N. and Harwood, D. M. (1993). Pliocene *Nothofagus* (Southern Beech) from Antarctica: phytogeography, dispersal strategies, and survival in high latitude glacial-deglacial environments. In *Forest Development in Cold Climates NATO ASI Series A: Life Sciences*, **244**, eds J. Alden, J. C. Mastrantonio and S. Odum. New York: Plenum Press, pp. 135–165.

Webb, P. N., McKelvey, B. C., Harwood, D. M., Mabin, M. C. G. and Mercer, J. H. (1987). Sirius Formation of the Beardmore Glacier region. *Antarctic Journal of the United States*, **21(5)**, 8–13.

Webb, P. N., Harwood, D. M., Mabin, M. C. G. and McKelvey, B. C. (1996). A marine and terrestrial Sirius Group succession, middle Beardmore Glacier–Queen Alexandra Range, Transantarctic Mountains, Antarctica. *Marine Micropaleontology*, **27**, 273–297.

Weigelt, E., Gohl, K., Uenzelmann-Neben, G. and Larter R. D. (2009). Late Cenozoic ice sheet cyclicity in the western Amundsen Sea Embayment – Evidence from seismic records. *Global and Planetary Change*, **69**, 162–169.

Whitehead, J. M., Quilty, P. G., McKelvey, B. C. and O'Brien, P. E. (2006). A review of the Cenozoic stratigraphy and glacial history of the Lambert Graben–Prydz Bay region, East Antarctica. *Antarctic Science*, **18**, 83–99.

Wilch, T. I. and McIntosh, W. C. (2000). Eocene and Oligocene volcanism at Mount Petras, Marie Byrd Land: implications for middle Cenozoic ice sheet reconstructions in West Antarctica. *Antarctic Science*, **12**, 477–491.

Wilch, T. I., Denton, G. H. and McIntosh, W. C. (1993). Limited Pliocene glacier extent and surface uplift in Middle Taylor Valley Antarctica. *Geografiska Annaler*, **75A**, 331–351.

Willenbring, J. K. and von Blanckenburg, F. (2010). Long-term stability of global erosion rates and weathering during late-Cenozoic cooling. *Nature*, **465**, 211–214.

Williams, M., Siveter, D. J., Ashworth, A. C., *et al.* (2008). Exceptionally preserved lacustrine ostracods from the Middle Miocene of Antarctica: implications for high-latitude palaeoenvironment at 77° South. *Proceedings of the Royal Society of London Series B*, **275**, 2449–2454.

Wilson, G. J. (1967). Some new species of Lower Tertiary dinoflagellates from McMurdo Sound, Antarctica. *New Zealand Journal of Botany*, **5**, 57–83.

Wilson, G. S., Roberts, A. P., Verosub, K. L., Florindo, F. and Sagnotti, L. (1998). Magnetobiostratigraphic chronology of the Eocene–Oligocene transition in the CIROS-1 core, Victoria Land margin, Antarctica: Implications for Antarctic glacial history. *Geological Society of America Bulletin*, **110**, 35–47.

Wilson, G. S., Barron, J. A., Ashworth, A. C., *et al.* (2002). The Mount Feather Diamicton of the Sirius Group: an accumulation of indicators of Neogene Antarctic glacial and climatic history. *Palaeogeography, Palaeoclimatology, Palaeoecology*, **28**, 1–15.

Wilson, G. S., Pekar, S. F., Naish, T. R., Passchier, S. and DeConto, R. (2009). The Oligocene-Miocene boundary–Antarctic climate response to orbital forcing. In *Developments in Earth and Environmental Sciences 8, Antarctic Climate Evolution*, eds F. Florindo and M. Siegert. Amsterdam: Elsevier, pp. 371–400.

You, Y., Huber, M., Müller, R. D., Poulsen, C. J. and Ribbe, J. (2009). Simulation of the Middle Miocene Climate Optimum, *Geophysical Research Letters*, **36**, L04702, doi:10.1029/2008GL036571.

Young, D. A., Wright, A. P., Roberts, J. L., *et al.* (2011). A dynamic early East Antarctic Ice Sheet suggested by ice-covered fjord landscapes. *Nature*, **474**, 72–75.

Zachos, J. C. and Kump, L. R. (2005). Carbon cycle feedbacks and the initiation of Antarctic glaciation in the earliest Oligocene. *Global and Planetary Change*, **47**, 51–66.

Zachos, J. C., Breza, J. R. and Wise, S. W. (1992). Early Oligocene ice sheet expansion on Antarctica: stable isotope and sedimentological evidence from Kerguelen Plateau, southern Indian Ocean. *Geology*, **20**, 569–573.

Zachos, J. C., Lohmann, K. C., Walker, J. C. G. and Wise, S. W. (1993). Abrupt climate changes and transient climate changes during the Paleogene: a marine perspective. *The Journal of Geology*, **101**, 191–213.

Zachos, J. C., Quinn, T. M. and Salamy, K. (1996). High-resolution (10^4 years) deep sea foraminiferal stable isotope records of the Eocene-Oligocene climate transition. *Paleoceanography*, **11**, 251–266.

Zachos, J., Pagani, L., Sloan, E., Thomas, E. and Billups, K. (2001a). Trends rhythms and aberrations in global climate 65 Ma to present. *Science*, **292**, 686–693.

Zachos, J., Shackleton, N. J., Revenaugh, J., Pälike, H. and Flower, B. P. (2001b). Climate response to orbital forcing across the Oligocene-Miocene boundary. *Science*, **292**, 274–278.

Appendix

Alphabetical listing of the latest Cretaceous to Neogene fossil floras known from the Antarctic Peninsula region not discussed in the text due to poor age constraints, limited material and/or poor preservation, but included here for completeness. Further information is given in the key references provided.

Flora	Remarks	References
Admiralen Peak	Originally considered to be older than Late Cretaceous, now middle Eocene (43 ± 4.8 Ma). Located to the west of Point Hennequin. Vegetation dominated by gymnosperms (wood) including Araucariaceae (leafy shoots) with extremely rare angiosperms.	Barton, 1961, 1964a, 1964b; Birkenmajer and Zastawniak, 1986; Birkenmajer et al., 1986; Davies, 1982; Hunt, 2001
Ardley Island	Eocene. Diverse flora according to Orlando but plant beds have never been found. Now assumed to belong to the Fossil Hill flora (T. Dutra personal communication, 2010).	Orlando, 1963, 1964
Block Point	early Eocene. Located in Admirality Bay. Vegetation comprises Sapindaceae, Myricaceae, Rhamnaceae and several types of Nothofagus.	Dutra, 1989
Cape Shirreff	Late Cretaceous. Located on Livingston Island. Araucariaceae, Podocarpaceae, Proteaceae, Nothofagus, and epiphyllous fungi (Type B flora) typical of cool, humid temperate conditions.	Leppe et al., 2003; Palma-Heldt et al., 2004, 2005, 2007
	Early Cretaceous. Pteridophytes dominate mainly Cyatheaceae and Gleicheniaceae with podocarps and epiphyllous fungi (Type A flora) indicative of warm, humid conditions.	Palma-Heldt et al., 2007; Torres, 1993
Cardozo Cove	Paleocene. In Ezcurra Inlet. Petrified wood and leaves.	Barton, 1964a; Birkenmajer, 1989; Zastawniak, 1981
Ezcurra Inlet	Paleocene (younger than 66.7 ± 1.5 Ma). Wood fragments found in the conglomerates of the Hala Member of the lower Arctowski Cove Fm. See also Cardozo Cove.	Birkenmajer, 1985, 1989, 1990; Birkenmajer and Zastawniak, 1986, 1989

Half Three Point and Price Point	Late Cretaceous. Grey tuffaceous siltstones up to 5.5 m thick. Dominated by ferns (Cyatheaceae, Dicksoniaceae, Gleicheniaceae) with conifers (mainly podocarps) and angiosperms. Palynoflora comprises ferns (81.5% – Gleicheniaceae 28%, Cyatheaceae 20.5% and Adiantaceae 10.5%) angiosperms (3%), gymnosperms (2%), bryophytes (1%), arcritach cyst, (0.5%) and fungal spores (12%). The 38 types of fungal spores belong to 13 genera and 16 new species. Flora is considered to be indicative of warm humid climates and the vegetation is considered to be a fern-dominated lakeshore or possibly adjacent hillslope vegetation. Occasional *Baltisphaeridium* (algae) suggest a near coast location.	Cao, 1992; Dutra, 1997; Dutra and Batten, 2000; Liu, 1994; Song and Cao, 1994; Zhou and Li, 1994a, 1994b; Zichen and Liu, 1994
Hannah Point	Cretaceous (Coniacian–Santonian; 88–68Ma). Located on Livingston Island. Leaf imprints and fossil trunks. The assemblages are commonly dominated by ferns, mainly of the families Osmundaceae, Dicksoniaceae and Gleicheniaceae, and foliage of conifers (mostly referable to Podocarpaceae). The occurrence of angiosperm leaves is rare, but recently angiosperm pollen has been found in pollen and spore-bearing rocks.	Leppe *et al.*, 2007
Keller Peninsula	Originally considered older than Late Cretaceous, now middle Eocene (44–41 Ma). Located to the north of Point Hennequin. Wood flora. Angiosperm-free vegetation composed of *Araucaria* and podocarps.	Birkenmajer *et al.*, 1986; Lucas and Lacey, 1981; Pereira, Dutra and Almeida, 2003
Lions Rump	Late Cretaceous/Paleogene. Warszawa Block.	Birkenmajer and Zastawniak, 1986
Paradise Cove	Late Cretaceous (67.7±3.5 Ma minimum age). Located on the Warszawa Block. Lithologies included in the Creeping Slope Formation, Paradise Cove Group. Sedimentary unit of terrestrial origin consisting of red shale with a middle green tuff, flake-conglomerate horizon; the conglomerate contains large fragments of silicified wood with annual growth rings belonging to *Nothofagus* and podocarps. More wood fragments occur in conglomerates in the igneous rocks of the overlying Demay Point Formation.	Birkenmajer, 1980, 1985; Birkenmajer and Zastawniak, 1986, 1989; Jagmin, 1987
Potter Peninsula and Potter Cove	early (~52 Ma) or middle (48–42 Ma) Eocene. Located on the Warszawa Block to the east of Barton Peninsula. Two *in situ* plant beds in tuffs between andesite lavas at Stranger Point and close to Three Brothers Hill in tuffaceous sandstones; thought to be	Birkenmajer, 1982, 1985; Birkenmajer and Zastawniak, 1989; del Valle, Diaz and Romero, 1984;

stratigraphically equivalent to the Upper Member of
the Fildes Formation and more or less
contemporaneous with it, is located three ?early
Eocene floras:

1. The impression flora, preserved in purple tuffaceous
   sandstones, comprises large fragments of angiosperms (e.g. *Gunnera*) alongside plant debris and
   rare sterile fern leaves. Conifers absent and leaves
   highly degraded.
2. The compression flora, preserved in green tuffaceous sandstones comprises abundant (sterile and
   fertile) fern foliage and broad-leafed conifers
   (podocarps). Although collections of this poorly
   preserved flora have been undertaken and subsequently described as being rich in plant matter, no
   sedimentological or palaeobotanical data have yet
   been published.
3. The third flora is a moraine site north of the Argentine
   Teniente Jubany Station. Numerous deciduous
   angiosperm leaves and fern fronds occur in tuff fragments lithologically similar to those of the Zamek
   Formation.

Hunt, 2001;
Orlando, 1963,
1964; Smellie *et al.*,
1984; Tokarski,
Danowski and
Zastawniak, 1987;
Torres, 1985a, 1990;
Zastawniak, 1994

Precious Peaks	Cretaceous possibly older than Late Cretaceous	Birkenmajer and Zastawniak, 1986
Rip Point	Late Cretaceous. Located on Nelson Island. Leaf impressions and coalified wood.	Dutra *et al.*, 1996
Staszek Cove	Paleocene. *Ex situ* leaves.	Alves, Guerra-Sommer and Dutra, 2005
Stranger Point	Eocene. One of the Potter Peninsula floras located on Warszawa Block.	Birkenmajer and Zastawniak, 1986
Suffield Point, Profound Lake	Miocene now considered to be mid-Eocene. Equivalent to Rocky Cove Flora but originally deemed Oligocene–Miocene in age. Extremely fine-grained tuffaceous sediments with rare poorly preserved impressions of angiosperm, fern and conifer leaves along with a single cone, preserved by ash fall deposits. Charred plant material. Also wood of *Nothofagoxylon*, *Cupressinoxylon*, *Araucarioxylon* and *Podocarpoxylon*. Palynoflora comprising fungal spores, ferns (including *Lophosoria*), gymnosperms, angiosperms including a single monocot (Gramineae).The wood flora includes Monimiaceae, Cunoniaceae, Eucryphiaceae, Rubiaceae, Araucariaceae, Podocarpaceae, Cupressaceae.	Hunt, 2001; Palma-Heldt, 1987; Pereira, Dutra and Almeida, 2003; Shen, 1994; Smellie *et al.*, 1984; Torres, 1984a, 1984b, 1985a, 1985b, 1990; Torres and Méon, 1993; Torres *et al.*, 1984a, 1984b, 1984c

Three Sisters Point	Late Cretaceous. Leaf imprints occur in a tuff intercalation between olivine-basalt lavas corresponding to the Mazurek Point Formation.	Birkenmajer and Zastawniak, 1986; Tokarski, Danowski and Zastawniak, 1987
Wanda Glacier	Eocene to late Oligocene. Located on Point Hennequin. Volcanogenic sediments with plant rootlets, impressions of conifer branches and fragments of silicified and siderized wood.	Birkenmajer, 1980, 1981; Birkenmajer and Zastawniak, 1986
Wrona Butress	early Miocene. Destruction Bay.	Birkenmajer and Zastawniak, 1986
Zamek	Late Cretaceous to ?early Cenozoic (77–67 Ma). Located in the upper Zamek Formation, Warszawa Block, >40m thick, comprised of basaltic andesite lavas alternating with scoria and tuff. Plant horizons >1m thick and rich in plant remains (impressions). Angiosperms dominate (mainly large-leafed and deciduous *Nothofagus* with some laurophyllous taxa), conifer and fern fronds.	Birkenmajer, 1980, 1985, 1989; Birkenmajer, Soliani and Kawashita, 1990; Birkenmajer and Zastawniak, 1986; Birkenmajer *et al.*, 1982; Dutra 1989; Dutra and Batten, 2000; Jagmin, 1996; Zastawniak, 1994.

References

Alves, L. S. D. R., Guerra-Sommer, M. and Dutra, T. (2005). Paleobotany and Paleoclimate. Part 1: Growth rings in fossil wood and paleoclimate. Part 2: Leaf assemblages (taphonomy, paleoclimatology and palaeogeography. In *Applied Stratigraphy*, ed. E. A. M. Koutsoukos. Netherlands: Springer, pp. 179–202.

Barton, C. M. (1961). The geology of King George Island, South Shetland Islands. *Preliminary Reports of the Falkand Isands Dependencies Survey*, **12**, 1–18.

Barton, C. M. (1964a). Significance of the Tertiary fossil floras of King George Island, South Shetland Islands. In *Antarctic Geology: Proceedings of the First International Symposium on Antarctic Geology, Cape Town, 16–21 September 1963*, ed. R. J. Adie. Amsterdam: North-Holland Publishing Company, pp. 603–608.

Barton, C. M. (1964b). The geology of King George Island, South Shetland Islands. Ph.D. thesis, University of Birmingham.

Birkenmajer, K. (1980). Tertiary volcanic-sedimentary succession at Admiralty Bay, King George Island (South Shetland Islands, Antarctica) *Studia Geologica Polonica*, **64**, 7–65.

Birkenmajer, K. (1981). Lithostratigraphy of the Point Hennequin Group (Miocene volcanic and sediments) at King George Island (South Shetland Islands, Antarctica). *Studia Geologica Polonica*, **72**, 59–73.

Birkenmajer, K. (1982). Report on geological investigations of King George Island and Nelson Island (South Shetland Islands, West Antarctica) in 1980/81. *Studia Geologica Polonica*, **74**, 175–197.

Birkenmajer, K. (1985). Onset of Tertiary continental glaciations in the Antarctic Peninsula sector (West Antarctica). *Acta Geologica Polonica*, **35**, 1–31.

Birkenmajer. K. (1989). A guide to Tertiary geochronology of King George Island, West Antarctica. *Polish Polar Research*, **10**, 555–579.

Birkenmajer, K. (1990). Geochronology and climatostratigraphy of Tertiary glacial and interglacial succession on King George Island, South Shetland Islands (West Antarctica). *Zentralblatt für Geologie und Paläontologie, Teil I*, **1990(1–2)**, 141–151.

Birkenmajer, K. and Zastawniak, E. (1986). Plant remains of the Dufayel Island Group (early Tertiary?), King George Island, South Shetland Islands (West Antarctic). *Acta Palaeobotanica*, **26**, 33–53.

Birkenmajer, K. and Zastawniak, E. (1989). Late Cretaceous–early Tertiary floras of King George Island, West Antarctica: their stratigraphic distribution and palaeoclimatic significance. In *Origins and Evolution of the Antarctic Biota, Geological Society of London Special Publication* **47**, ed. J. A. Crame. London: The Geological Society, pp. 227–240.

Birkenmajer, K., Narębski, W., Nicoletti, M., Petrucciani, C. (1982). Late Cretaceous through late Oligocene K-Ar ages of the King George Island Supergroup volcanics, South Shetland Islands (West Antarctica). *Bulletin de l'Academie Polonaise des Sciences, Série des Sciences de la Terre*, **30**, 133–143.

Birkenmajer, K., Delitala, M. C., Narębski, W., Nicoletti, M. and Petrucciani, C. (1986). Geochronology of Tertiary island-arc volcanics and glacigenic deposits, King George Island, South Shetland Islands (West Antarctica). *Bulletin of the Polish Academy of Sciences, Earth Sciences*, **34**, 257–272.

Birkenmajer, K., Soliani, E. and Kawashita, K. (1990). Reliability of potassium-argon dating of Cretaceous–Tertiary island-arc volcanic suites of King George Island, South Shetland Islands (West Antarctica). *Zentralblatt für Geologie und Paläontologie, Teil 1*, **1990(1–2)**, 127–140.

Cao, L. (1992). Late Cretaceous and Eocene palynofloras from Fildes Peninsula, King George Island (South Shetland Islands), Antarctica. In *Recent Progress in Antarctic Earth Science*, eds Y. Yoshida, K. Kaminuma and K. Shiraishi. Tokyo: Terra Scientific Publishing Company, pp 363–369.

Cao, L. (1994). Late Cretaceous palynoflora in King George Island of Antarctica, with references to its palaeoclimatic significance. In *Stratigraphy and Palaeontology of Fildes Peninsula, King George Island, Antarctica, State Antarctic Committee Monograph, 3*, ed. Y. Shen. Beijing: China Science Press, pp. 76–90.

Davies, R. E. S. (1982). New geological interpretation of Admiralen Peak King George Island South Shetland Islands. *British Antarctic Survey Bulletin*, **51**, 294–296.

del Valle, R. A., Diaz, M. T. and Romero, E. J. (1984). Preliminary report on the sedimentites of Barton Peninsula 25 de Mayo Island (King George Island), South Shetland Islands, Argentine Antartica. *Instituto Antartico Argentino Contribución*, **308**, 121–131.

Dutra, T. L. (1989). Informações preliminares sobre a tafoflora do Monte Zamek (Baía do Almirantado, Ilha Rei George (Ilhas Shetland do Sul), Antártica. *Série Científica Instituto Antartico Chileno*, **39**, 31–42.

Dutra, T. L. (1997). Primitive leaves of *Nothofagus* (Nothofagaceae) in Antarctic Peninsula: an Upper Campanian record and a betulaceous more than fagaceous morphological character. In *Actas VIII Congresso Geológico Chileno Volume 1*. Antofogasta: Universidad Católica del Norte, pp. 24–29.

Dutra, T. L. and Batten, D. (2000). The Upper Cretaceous flora from King George Island, West Antarctica and their palaeoenvironmental and phytogeography implications. *Cretaceous Research*, **21**,181–209.

Dutra, T. L., Leipnitz, B., Faccini, U. F. and Lindenmayer, Z. (1996). A non-marine Upper Cretaceous interval in West Antarctica (King George Island, Northern Antarctic Peninsula). *SAMC (South Atlantic Mesozoic Correlations, IGCP Project No. 381) News*, **5**, 21–22.

Hunt, R. H. (2001). Biodiversity and palaeoecological significance of Tertiary fossil floras-from King George Island, West Antarctica. Ph.D. thesis, University of Leeds.

Jagmin, N. I. B. (1987). Estudo anatômico dos troncos fósseis de Admiralty Bay, King George Island (Península Antártica). *Acta Biológica Leopoldensia*, **9**, 81–98.

Jagmin, N. I. B. (1996). *Nothofagoxylon* sp. del monte Zamek, isla Rey Jorge, islas Shetland del Sur, Antártica. *Serie Científica Instituto Antártico Chileno*, **46**, 133–41.

Leppe, M., Fernandoy, F., Palma-Heldt, S. and Moisan, P. (2003). Flora Mesozoica en los afloramientos morrénicos de Cabo Shirreff, Isla Livingston, Shetland del Sur, Península Antártica. Simposio Evolución del Gondwana en el Margen Pacífico. In *Actas X Congreso Geológico Chileno*. Concepción: Universidad de Concepción pp. S-5. (CD).

Leppe, M., Michea, W., Muñoz, C., Palma-Heldt, S. and Fernandoy, F. (2007), Paleobotany of Livingston Island: The first report of a Cretaceous fossil flora from Hannah Point. In *Online Proceedings of the 10th International Symposium an Antarctic Earth Sciences*, eds A. K. Cooper, C. R. Raymond and the 10th ISAES editorial team. USGS Open-File Report 2007–1047, Short Research Paper 081, available at http://pubs.usgs.gov/of/2007/1047/srp/srp081/of2007–1047srp081.pdf. {Cited 2011}.

Lucas, R. C. and Lacey, W. S. (1981). A permineralized wood flora of probable Early Tertiary age from King George Island, South Shetland Islands. *British Antarctic Survey Bulletin*, **53**, 147–51.

Orlando, H. A. (1963). La flora fósil en las imediaciones de la Peninsula Ardley, Isla 25 de Mayo, Islas Shetlands del Sur. *Instituto Antarctico Argentino Contribución*, **79**, 3–17.

Orlando, H. A. (1964). The fossil flora of the surroundings of Ardley Peninsula (Ardley Island), 25 de Mayo Island (King George Island), South Shetland Islands. In *Antarctic Geology: proceedings of the First International Symposium on Antarctic Geology, Cape Town, 16–21 September 1963*, ed. R. J. Adie. Amsterdam: North-Holland Publishing Company, pp. 629–636.

Palma-Heldt, S. (1987). Estudio palinologico en el Terciario de isles Rey Jorge y Branbante, territorio insular Antartico. *Serie Científica Instituto Antartico Chileno*, **36**, 59–71.

Palma-Heldt, S., Fernandoy, F., Quezada, I. and Leppe, M. (2004). Registro palinológico de Cabo Shirreff, Isla Livingston, nueva localidad para el Mesozoico de las Shetland del Sur. In *V Simposio Argentino y I Latinoamericano sobre investigaciones Antárticas Ciudad Autónoma de Buenos Aires, Argentina 30 de agosto al 3 de septiembre de 2004 Resúmen Expandido* N° 104GP, 4pp. Buenos Aires, Argentina, available at http://www.dna.gov.ar/CIENCIA/SANTAR04/CD/PDF/104GP.PDF. {Cited 2011}.

Palma-Heldt, S., Fernandoy, F., Leppe, M., Rodríguez, M. and Salazar, C. (2005). Cretaceous palynoflora of Livingston Island (South Shetland Islands): relationships between moraine deposits of Cape Shirreff and Byers Peninsula. In *Gondwana 12 Conference. Geological and Biological Heritage of Gondwana*, eds R. Pankhurst and G. Veiga. Córdoba: Academie Nacional de Ciencias, pp. 1.

Palma-Heldt, S., Fernandoy, F., Henríquez, G. and Leppe, M. (2007). Palynoflora of Livingston Island, South Shetland Islands: Contribution to the understanding of the evolution of the southern Pacific Gondwana margin. In *Online Proceedings of the 10th International Symposium on Antactic Earth Sciences*, eds A. K. Cooper, C. R. Raymond and the 10th ISAES editorial team. U. S. Geological Survey Open-File Report 2007–1047, Extended Abstract 100, available at http://pubs.usgs.gov/of/2007/1047/ea/of2007–1047ea100.pdf. {Cited 2011}.

Pereira, F., Dutra, T. L. and Almeida, D. P. M. (2003). Ambientes vulcânicos associados à paleoflora no Cretáceo e Terciário da Ilha Rei George, Antártica. In *Caracterização e modelamento de depósitos minerais*, eds L. R. Ronchi and F. J. Althoff. São Leopoldo: Editora Unisinos, pp. 387–410.

Shen, Y. (1994). Subdivision and correlation of Cretaceous to Palaeogene volcano-sedimentary sequence from Fildes Peninsula, King George Island, Antarctica. In *Stratigraphy and Palaeontology of Fildes Peninsula, King George Island, State Antarctic Committee Monograph, 3*, ed. Y. Shen. Beijing: China Science Press, pp 1–36.

Smellie, J. L., Pankhurst, R. J., Thomson, M. R. A. and Davies, R. E. S. (1984). The geology of the South Shetland Islands: VI stratigraphy, geochemistry and evolution. *British Antarctic Survey Scientific Reports*, **87**, 1–85.

Song, Z. and Cao, L. (1994). Late Cretaceous fungal spores from King George Island, Antarctica. In *Stratigraphy and Palaeontology of Fildes Peninsula King George Island, Antarctica, State Antarctic Committee Monograph, 3*, ed. Y. Shen. Beijing: China Science Press, pp. 37–49.

Tokarski, A. K., Danowski, W. and Zastawniak, E. (1987). On the age of the fossil flora from Barton Peninsula, King George Island, West Antarctica. *Polish Polar Research*, **8**, 293–302.

Torres, T. (1984a). Identificación de madera fósil del Terciario de la Isla Rey Jorge, Islas Shetland del Sur Antártica. In *III Congreso Latinoamericano de Paleontología, México*, **2**, 555–565.

Torres, T. (1984b). *Nothofagoxylon antarcticus* n. sp., madera fósil del Terciario de la Isla Rey Jorge, Islas Shetland del Sur, Antártica. *Serie Científica del Instituto Antártico Chileno*, **31**, 39–52.

Torres, T. (1985a). Plantas fósiles en la Antártica. *Boletín Instituto Antártico Chileno*, **5**, 1–15.

Torres, T. (1985b). Coníferas y dicotiledóneas fósiles del Terciario de la Isla Rey Jorge, Islas Shetland del Sur, Antártica. *Paleobotánica Latinoamericana*, **7**, 10.

Torres, T. (1990). Tertiary paleobotanical study in the King George and Seymour Islands, Antarctica. Ph.D. thesis, University of Lyon.

Torres, T. (1993). Primer hallazgo de madera fósil en Cabo Shirreff, Isla Livingston, Antartica. *Serie Científica Instituto Antártico Chileno*, **43**, 31–39.

Torres, T. and Méon, H. (1993). *Lophosoria* del Terciario de isla Rey Jorge y Chile Central: origen y dispersion en el hemisfero sur. *Serie Científica Instiuto Antártico Chileno*, **43**, 17–30.

Torres, T., Hansen, M. A. F., Troian, F. L., Fensterseifer, H. C. and Linn, A. (1984a). Nota preliminar sobre plantas fósseis da Ilha Rei Jorge, Islas Shetland del Sur. In *Simpósio Nacional do Programa Antártico*. São Paulo: Instituto Oceanográfico da Universidade de São Paulo-USP, pp. 87–88.

Torres, T., Hansen, M. A. F., Troian, F. L., Fensterseifer, H. C. and Linn, A. (1984b). Flora fósil de alrededores de Punta Suffield, Isla Rey Jorge, Shetland del Sur. *Boletín Antártico Chileno*, **4**, 1–7.

Torres, T., Roman, A., Deza A. and Rivera, C. (1984c). Anatomía, mineralogía y termolu- miniscencia de madera fósil de la Isla Rey Jorge, Islas Shetland del Sur. In *III Congreso Latinoamericano de Paleontología, México, volume 2*, ed. M. de C. Perílliet. Mexico City: Universidad Nacional Autónoma de Mexico, pp. 566–574.

Zastawniak, E. (1981). Tertiary leaf flora from Point Hennequin Group of King George Island (South Shetland Islands, Antarctica). *Studia Geologica Polonica*, **81**, 143–64.

Zastawniak, E. (1994). Upper Cretaceous fossil leaf flora from the Błaszyk moraine (Zamek Formation), King George Island, South Shetland Islands, West Antarctica. *Acta Palaeobotanica*, **34**, 119–163.

Zhou, Z. and Li, H. M. (1994a). Early Tertiary ferns from Fildes Peninsula, King George Island, Antarctica. In *Stratigraphy and Palaeontology of Fildes Peninsula King George Island, Antarctica, State Antarctic Committee Monograph, 3*, ed. Y. Shen. Beijing: China Science Press, pp. 181–207.

Zhou, Z. and Li, H. (1994b). Early Tertiary gymnosperms from Fildes Peninsula, King George Island, Antarctica. In *Stratigraphy and Palaeontology of Fildes Peninsula King George Island, Antarctica, State Antarctic Committee Monograph, 3*, ed. Y. Shen. Beijing: China Science Press, pp. 208–30.

Zichen, S. and Liu, C. (1994). Late Cretaceous fungal spores from King George Island, Antarctica. In *Stratigraphy and Palaeontology of Fildes Peninsula, King George Island, Antarctica, State Antarctic Committee Monograph, 3*, ed. Y. Shen. Beijing: China Science Press, pp. 47–75.

Index

Printed in the United States
By Bookmasters